Periodensystem der Elemente mit Gmelin-Systemnummern

Each cell: atomic number (left), Gmelin system number (right), element symbol (below).

1	2	3	4	5	6	7	8	9	10	11	12	13	14	15	16	17	18
1 · 2 — H																1 · 2 — H	2 · 1 — He
3 · 20 — Li	4 · 26 — Be											5 · 13 — B	6 · 14 — C	7 · 4 — N	8 · 3 — O	9 · 5 — F	10 · 1 — Ne
11 · 21 — Na	12 · 27 — Mg											13 · 35 — Al	14 · 15 — Si	15 · 16 — P	16 · 9 — S	17 · 6 — Cl	18 · 1 — Ar
19 · 22 — K *	20 · 28 — Ca	21 · 39 — Sc	22 · 41 — Ti	23 · 48 — V	24 · 52 — Cr	25 · 56 — Mn	26 · 59 — Fe	27 · 58 — Co	28 · 57 — Ni	29 · 60 — Cu	30 · 32 — Zn	31 · 36 — Ga	32 · 45 — Ge	33 · 17 — As	34 · 10 — Se	35 · 7 — Br	36 · 1 — Kr
37 · 24 — Rb	38 · 29 — Sr	39 · 39 — Y	40 · 42 — Zr	41 · 49 — Nb	42 · 53 — Mo	43 · 69 — Tc	44 · 63 — Ru	45 · 64 — Rh	46 · 65 — Pd	47 · 61 — Ag	48 · 33 — Cd	49 · 37 — In	50 · 46 — Sn	51 · 18 — Sb	52 · 11 — Te	53 · 8 — J	54 · 1 — Xe
55 · 25 — Cs	56 · 30 — Ba	57** · 39 — La	72 · 43 — Hf	73 · 50 — Ta	74 · 54 — W	75 · 70 — Re	76 · 66 — Os	77 · 67 — Ir	78 · 68 — Pt	79 · 62 — Au	80 · 34 — Hg	81 · 38 — Tl	82 · 47 — Pb	83 · 19 — Bi	84 · 12 — Po	85 — At	86 · 1 — Rn
87 · 31 — Fr	88 · 31 — Ra	89*** · 40 — Ac	104 · 71	105 · 71													

Lanthanide 39

58 · 39 — Ce	59 · 39 — Pr	60 · 39 — Nd	61 · 39 — Pm	62 · 39 — Sm	63 · 39 — Eu	64 · 39 — Gd	65 · 39 — Tb	66 · 39 — Dy	67 · 39 — Ho	68 · 39 — Er	69 · 39 — Tm	70 · 39 — Yb	71 · 39 — Lu

***Actinide**

90 · 44 — Th	91 · 51 — Pa	92 · 55 — U	93 · 71 — Np	94 · 71 — Pu	95 · 71 — Am	96 · 71 — Cm	97 · 71 — Bk	98 · 71 — Cf	99 · 71 — Es	100 · 71 — Fm	101 · 71 — Md	102 · 71 — No(?)	103 · 71 — Lr

* NH$_4$ (23)

Reihenfolge der Gmelin-Systemnummern siehe Innenseite des hinteren Deckels

Gmelin Handbuch der Anorganischen Chemie

Achte völlig neu bearbeitete Auflage

Gmelin Handbuch der Anorganischen Chemie

BEGRÜNDET VON   Leopold Gmelin

Achte völlig neu bearbeitete Auflage

ACHTE AUFLAGE   begonnen im Auftrage der Deutschen Chemischen Gesellschaft
von R. J. Meyer
E. H. E. Pietsch und A. Kotowski

fortgeführt von
Margot Becke-Goehring und Karl-Christian Buschbeck

HERAUSGEGEBEN VOM   **Gmelin-Institut**
für Anorganische Chemie und Grenzgebiete der
Max-Planck-Gesellschaft zur Förderung der Wissenschaften

Springer-Verlag Berlin Heidelberg GmbH 1974

Gmelin Handbuch der Anorganischen Chemie

Achte völlig neu bearbeitete Auflage

SAUERSTOFF

Anhangband

Water Desalting
Wasser-Entsalzung

Mit 62 Figuren

VON

Anthony A. Delyannis und **Eurydike A. Delyannis**, Athen

REDAKTEUR DIESES BANDES

Walter Lippert, Gmelin-Institut (Frankfurt am Main)

System-Nummer 3

Springer-Verlag Berlin Heidelberg GmbH 1974

FACHLICHE BERATUNG DER REDAKTION:
KURT H. FISCHBECK, HEIDELBERG

DEUTSCHE FASSUNG DER STICHWÖRTER NEBEN DEM TEXT:
B. C. DRUDE, ERLANGEN

Die Literatur ist bis Ende 1973 ausgewertet,
in Einzelfällen darüber hinaus

Die vierte bis siebente Auflage dieses Werkes erschien im Verlag von
Carl Winter's Universitätsbuchhandlung in Heidelberg

Library of Congress Catalog Card Number: Agr 25-1383

ISBN 978-3-662-13338-5   ISBN 978-3-662-13336-1 (eBook)
DOI 10.1007/978-3-662-13336-1

Preface

Processes for recovering fresh water from the oceans — of which men have dreamed since antiquity — have changed markedly in the last 20 years. In fact, it has become possible so to increase the productivity of the technical steps involved that the cost of production of such water is almost three orders of magnitude smaller than for other large volume industrial products.

However, the monographs and comprehensive reviews which have appeared to date in this field have been prepared by specialists for specialists. In accordance with the tradition and objectives of the Gmelin Handbook, this bibliography has been prepared to provide access to all of the ways in which fresh water can be, and has been, obtained on an industrial scale from the ocean. Production of fresh water from sea and brackish waters amounts to almost two million cubic meters per day, and this is increasing by about 25% per year. This means that it will increase nearly tenfold in 10 years.

In the present volume the broadly scattered literature has been critically screened and arranged systematically. It was not possible to collect and publish all of the literature citations; a critical selection was much rather attempted. In this connection, about 5500 citations were evaluated for the years 1965 to 1969, and about 8300 in the 1969 to mid-1974 period, i.e., almost 14000 papers; over 3000 journals and other sources are cited. Specific reference was not made to the Gmelin Handbook, but it should be noted that many subject areas (for example, the properties of water, the CO_2–H_2O system, as well as salt systems mentioned in the present volume) have been exhaustively described in individual volumes of Gmelin. The patent literature was not considered.

In conclusion, I should like to note that, although the earth is by far the most water-rich of the planets, there is a serious danger that three or four billion people will so contaminate the water reserves of the earth (which are more than one billion cubic kilometers) that life itself could become very difficult or perhaps even impossible. This danger must be avoided, and the fresh water required for human support recovered from the oceans; irrigation of arid zones plays a very large role here.

Heidelberg, October 1974

Kurt H. Fischbeck

Vorwort

Die Verfahren zur Gewinnung von Süßwasser aus dem Meer, um die man sich seit dem frühen Altertum bemüht hat, haben in den vergangenen 20 Jahren eine Wandlung erfahren. Es ist nämlich möglich geworden, die Ergiebigkeit der dazu nötigen verfahrenstechnischen Schritte in einem solchen Ausmaße zu steigern, daß die Gestehungskosten des Wassers um fast drei Zehnerpotenzen kleiner sind als die bei anderen technischen Großprodukten.

Die bislang auf diesem Arbeitsgebiete erschienenen Monographien und zusammenfassenden Veröffentlichungen sind von Spezialisten für Spezialisten geschrieben. Hier ist der Tradition und dem Ziel des Gmelin Handbuchs entsprechend eine Bibliographie geschaffen worden, welche den Zugang zu allen Wegen eröffnet, auf denen im technischen Maße Süßwasser aus dem Meer gewonnen werden kann und auch bereits gewonnen wird. Die Produktion von Süßwasser aus Meer- und Brackwasser nähert sich einer Menge von täglich zwei Millionen Kubikmeter und nimmt pro Jahr um rund 25% zu. Das bedeutet, daß sie sich in zehn Jahren verzehnfachen wird.

In dem vorliegenden Band ist das weit verstreute Schrifttum kritisch gesichtet und systematisch geordnet worden. Es konnte nicht darum gehen, eine vollständige Aufstellung sämtlicher Schrifttumsstellen anzufertigen; es wurde vielmehr eine kritische Auswahl der Veröffentlichungen getroffen. Hierbei wurden aus den Jahren 1965 bis 1969 ca. 5500 und 1969 bis Mitte 1974 ca. 8300, das heißt fast 14000 Veröffentlichungen ausgewertet; in dem vorliegenden Band sind mehr als 3000 Publikationen zitiert. Nicht zitiert wurde Gmelin Handbuch; daher sei hier darauf hingewiesen, daß zum Beispiel die Eigenschaften des Wassers, das System CO_2–H_2O wie auch im vorliegenden Band erwähnte Salzsysteme in den einzelnen Bänden des Gmelin ausführlich beschrieben sind. Nicht berücksichtigt wurde die Patentliteratur.

Zum Schluß darf der Verfasser des Vorwortes sich die Bemerkung erlauben, daß, obwohl die Erde sicher der bei weitem wasserreichste Planet ist, die ernste Gefahr besteht, daß drei bis vier Milliarden Menschen die Wasservorräte der Erde, die mehr als eine Milliarde Kubikkilometer füllen, so weit denaturieren, daß das Leben erst zur Qual und dann unmöglich wird. Dieser Gefahr gilt es Einhalt zu gebieten, und es gilt das zur Ernährung der Menschen erforderliche Süßwasser aus dem Meer zu gewinnen; die Bewässerung arider Zonen spielt hierbei eine große Rolle.

Heidelberg, im Oktober 1974

Kurt H. Fischbeck

Table of Contents
(Inhaltsverzeichnis s. S. V)

II

Inhaltsverzeichnis

(Table of Contents see page I)

Panoramic views
of the Porto Torres, Sardinia, multi-stage
flash distillation plant.

Water Desalting

1 Introduction and General

1.1 The water problem

Water is the most important chemical compound on Earth. When men settled down to agriculture and farming, they built their houses near potable water resources, such as rivers and lakes. Increasing the size of settlements augmented the needs of fresh water supply. Conveying of water to the locations of urban settlements and agricultural operations was a usual practice in Egypt and Mesopotamia, in Crete and the Roman Empire. Although the annual precipitation on Earth's surface might be sufficient, the uneven distribution of rainfall does not meet the human needs in all regions of the world. In some arid areas, existing water resources are saline, exceeding the limits of potable water. Drought periods may also occur to make the situation worse.

The theoretical minimum water requirements, including agriculture, to sustain human life are about 1.1 m^3 per capita per day, assuming that man can live on bread alone. Introducing of 0.5 kg of animal fat and protein to the diet, the water requirement for subsistence increases to about 9.5 m^3 per capita per day [1]. Increasing demand of water, caused by the rising standard of living and by the increase in population, irresponsible wasting of water in many large cities and above all pollution of natural water reserves by industrial waste and sewage, have brought many regions in various countries close to the critical point, where existing resources can no longer satisfy the growing demand. The situation will become worse, if adequate drastic means are not adopted in time.

Forecasts have been made by Clodius for the situation, which might develop in the year 2000 [2]. The problems which might arise within the next quarter-century were reviewed by Burgess suggesting also solutions to these problems [3], and an estimation of the water needs in the United States within the next 100 years was given by Cywin. The population in the U.S. is expected to rise to 1000 million persons and tremendous water needs have to be met. It was suggested that water resources must be utilized through institutional arrangements that take all resources management into account [4].

Water demand and quality standards. In a modern urban agglomeration demand for water may be divided into domestic, municipal and industrial, as well as in agricultural water needs. There are no general standards adopted for the quality of water required by these various users.

Domestic requirements comprise all water consumed in housekeeping including gardening. The determination of taste thresholds and other data have led the United States Public Health Service to recommend a limit of 500 mg/l for total dissolved solids with a maximum of 250 mg/l for chloride and sulfate ions respectively. However, there is a large number of communities in various countries, which are still supplied with water containing over 1000 mg/l total dissolved solids and sometimes up to 3000 mg/l.

Historical perspective of water treatment standards and the history of U.S. drinking water standards were outlined by McDermott. Proposed 1973 standards were extensively examined [5]. The taste intensity of natural drinking water is directly related to the mineral content and to the water temperature [6]. A review of permissible chemical constituents in drinking water and limitations for safe public water supplies was given by Weigand [7]. The physiological changes which can be brought about by large intakes of the main as well as of the trace ions in water were analyzed by Berlyne and Yagil [8].

Municipal requirements, besides the supply of water for domestic use, include all water needed by offices, public and commercial establishments, fire-fighting and the irrigation systems of municipal parks. Although the standards set for the latter uses are not the same as for drinking water, in practice municipal water is equivalent to drinking water since it is generally supplied through the same distribution system.

Industrial requirements include all water that is used in industrial plants: A large variety of quality standards is involved, according to the specific use of water. Food processing needs high-quality drinking water. Process water for certain products imposes various limits of salt content. Boiling feed water and sometimes cooling water need special treatment. In some cases, seawater can also be used for cooling purposes, and this is the usual practice in plants located near the seashore.

Agricultural requirements are mainly accounted for by irrigation, but they also include water for animals. Irrigation water quality depends to a large extent on the nature of the soil, the crops and the climate. The quality of some crops can be affected, not only by the total amount of dissolved solids, but also by the presence of certain specific salts.

Importance of water quality. Most community water supplies in the United States meet Public Health Service standards. A survey of the water quality in communities of over 1000 population indicated that a total of 420 communities in 29 States were supplied with water exceeding 1000 mg/l total dissolved solids [24]. Webster, South Dakota, was one of these communities of about 2500 people. Since 1962 the municipal water supply in Webster has been primarily the product water from the electrodialysis demonstration plant (see section 3.7, p. 213). A study was made to compare the use of the water before and after the changeover from brackish water to a municipal supply of reduced mineral content. Reactions of users were generally favorable towards the product water from the conversion plant. Replacements of appliances, especially of heaters, were reduced sharply. There was not clear agreement on users preference when they compared water of different quality for drinking. However, the majority distinguished in favor of the water of reduced mineral content for laundry use and enterprises serving the travelling public were especially concerned with water quality and recognized it as one factor in the total economic growth of a region [25]. A survey was also made on candidate communities in the United States for saline water demineralization applications [26], and data were collected on water supply as well as of available regional water plants [27].

Investigating the economic value of water quality, Metcalf and Eddy collected data from a large number of water users [28]. The role of desalting in providing high quality water for industrial use was examined by Schmidt and Ross. Ten industrial categories were selected for study as to their need for demineralized water: conventional fuel power plants; nuclear fuel power plants; industrial boilers; electronics; primary metals; chemical process, including synthetic fibers; motor vehicles; drugs; photographic supplies; and miscellaneous others. The quality of demineralized water required for various purposes by each of the listed industrial categories was estimated and a summary of the determined total high quality water needs was given. Calculations were made to ascertain the economic feasibility of using desalting techniques to a greater extent for producing industrial high quality water. From the calculations, a series of curves were developed, which plotted cost versus raw water quality for various product water qualities and treatment equipment systems [29].

Water management. There are several possible solutions to the growing water problem, starting with the more efficient use of the available water resources. Steps might be taken to reduce per capita consumption in minimizing wastes, reduce losses in storage and transport, sometimes surprisingly high, develop more efficient industrial practices in the use of water, grow crops that need less water, develop water saving irrigation processes and grow crops that are more tolerant to brackish water.

One of the obvious means for augmenting the available water resources is the reclamation, purification and reuse of waste water, which at the same time is an effective means of pollution control. Possibilities and upper permissible limits of waste water purification regarding drinking water recovery were described by Märckli [9]. A research and development program of the U.S. Water Quality Office of the Environmental Protection Agency for industrial and water pollution control was outlined by Rey et al. Industrial waste water reuse is considered to be a tool for pollution control and abatement [10]. The reuse of industrial waste water is defined by Lücke as a positive method for reducing the consumption of water and indirectly increasing the available resources. Problems to be encountered with multiple use of water and fresh water conservation were examined using papermaking as an example [11].

The increased expense of waste water treatment coupled with the projected future shortage of quality water may make reusable water the most valuable product recoverable from industrial waste water. In order to determine the optimum quantity of water to recover and reuse, a generalized mathematical model was developed by Lyons and Eckenfelder. The model and the associated methodology was applied to the water management system of a medium size bleached kraft pulp and paper mill. The model allows evaluation of water reclamation and reuse, as it effects the economics of an industries total water management program [12].

The construction and operation of a pilot plant for advanced sewage purification were reviewed by Stander and van Vuuren. The plant was capable of producing potable water at reasonable cost [13]. Industry in Monterrey, Mexico, using the final effluent from the domestic waste water treatment plant, has solved the problem of lack of water and contamination [14]. The town of Windhoek, South West Africa, is supplied since 1969 with drinking water reclaimed from sewage. The plant capacity is 4540 m³/day (1.2 Mgd) and the cost was reported to be 19.8 cents/m³ (75 cents/kgal) [15].

Desalting is a valuable supplement for the reclamation of waste water. Desalting processes will be useful for the regeneration of used waters to make them appropriate for household and industrial supplies [16]. Economically beneficial, desalting of industrial wastes would also considerably alleviate pollution problems [17]. A review of desalting technology as applied to pollution control including an overall perspective of the general pollution problem and its technology, economic and social applications and the desalting methods, such as distillation, deionization by membranes, salt separation by freezing and chemical deionization methods was presented by Young [18]. Distillation of waste water, was considered by Patterson, might answer some water shortage problems. It appears easier to bring the public on using distilled water rather than reuse waste water [19]. The technical feasibility of evaporation of municipal sewage treatment plant effluent for the purpose of water reuse was investigated at the University of Florida. It was concluded that because of increased efficiencies, waste water evaporation should be more economical than seawater evaporation [20]. The possibilities of regenerating concentrated wastes to solve both water reuse and pollution problems were discussed ny Beaton and Dickson. The various processes, including electrodialysis, reverse osmosis, ultrafiltration, distillation and evaporation were briefly examined for this purpose and process economics given [21].

Probstein has estimated that it will shortly be necessary in industrialized countries to develop new supplies of fresh water by desalination and to recycle presently available supplies. About an order of magnitude reduction in present desalination and water purification costs will be required. The proper application of fluid mechanics research can help significantly to reduce the cost of purified and desalted water. A brief outline of the various desalting processes with emphasis to brackish water treatment, as well as a summary of some brackish water desalting experiences in the United States, were also given. Locations, quality and quantity of brackish water resources in the arid region of the American continent are outlined [22]. A study was carried out in 1967 by the Agency for International Development and the United States Public Health Service to evaluate the community water supply programs in underdeveloped countries. The conclusions reached and criteria set up were summarized by Harris [23].

The alternative of desalting. When all other possibilities to augment fresh water supply fail, desalting of seawater, of brackish water and/or of polluted water reserves might give the answer to local water problems. The cost of desalting has been drastically reduced over the past several years. However, desalted water is an industrial product and its cost would hardly compete with the cost of natural fresh water supplies, where available in the neighborhood of consuming agglomerations. Water desalting should therefore be considered as an alternate source of water supply, which might solve problems in such places, where natural and conventional water resources are exhausted. In arid regions desalting might be the only possibility of water supply.

Promising opportunities for desalting in the light of projected national and regional water supplies and in relation to alternatives were reviewed by O'Brien. Legal, political, institutional and environmental factors influencing the choice of desalting were examined and economic considerations in evaluating desalting as an alternative were explored [30].

A dynamic simulation model was developed by Rothermel, which would translate the relevant factors of water supply and demand into a forecast of desalting potential. The computerized model projects the needs for desalting in 20 major hydrologic regions in the United States on the basis of a comparison of projected demand to available supply in 100 subregions. A novel ingredient of this demand/supply balance is the consideration of the impact of pollution upon the total water requirements of a region. The results include the following national desalting potentials: about 1 250 000 m³/day (330 Mgd) of capacity in 1980, 8 517 000 m³/day (2250 Mgd) in 2000, and 29 337 000 m³/day (7750 Mgd) in 2020. Reasonable variations in desalting economics or other management parameters can affect these projections by a factor of four or five by 2000 to 2020 [31].

Regional studies on the application of desalting techniques in the United States include a study and analysis of the application of saline water conversion processes to acid mine waters [32], an engineering study of the potentialities and possibilities of desalting for Northern New Jersey and New York City [33], reports on the potential contribution of desalting to future water supply in Texas [34], and on the economics of desalting brackish waters for regional, municipal and industrial water supply in West Texas [35], on the economics of a regional municipal desalting system in the lower Rio Grande Valley in Texas [36], a review of California's regional water supply systems and possible applications of desalting [37], studies on improving municipal water supplies in Colorado by desalting [38], on the potential contribution of desalting to future water supply in New Mexico [39], and on the potentials for desalting in the Tularosa Basin, New Mexico [40], on the feasibility of desalting municipal water supplies in Montana [41], on municipal desalting for selected Kansas communities [42], a preliminary study to investgate feasibility of desalting groundwater in North Dakota [43], on the potential contribution of desalting systems to municipal water quality and supply in South Dakota [44], on improving municipal water supplies in Arizona by desalting [45], on the future role of desalting in Nevada [46], and on the potential of desalting for industrial water supplies in Northeastern Wyoming [47].

Literature to 1.1

[1] C. C. Bradley (Science [2] **138** [1962] 489/91). — [2] S. Clodius (Gas-Wasserfach **110** [1969] 1335/7; Wasser Luft Betrieb **14** [1970] 11/4). — [3] L. C. N. Burgess (J. Am. Water Works Assoc. **62** [1970] 211/4). — [4] A. Cywin (Mech. Eng. **93** No. 10 [1971] 7/10; Comments by D. E. Bluman, Mech. Eng. **93** No. 12 [1971] 56; Reply by Cywin, Mech. Eng. **94** No. 2 [1972] 54). — [5] J. H. McDermott (Proc. ASCE J. Environ. Eng. Div. **99** [1973] 469/78).

[6] R. M. Pangborn, L. L. Bertolero (J. Am. Water Works Assoc. **64** [1972] 511/5). — [7] R. W. Weigand (Military Eng. **64** No. 417 [1972] 25/7). — [8] G. M. Berlyne, R. Yagil (Desalination **13** [1973] 217/20). — [9] E. Märkli (Gas Wasser Abwasser **49** [1969] 13/21). — [10] G. Rey, W. J. Lacy, A. Cywin (Environ. Sci. Technol. **5** [1971] 760/5).

[11] F. Lücke (Zellstoff Papier [Leipzig] **19** [1970] 332/4). — [12] D. N. Lyons, W. W. Eckenfelder (Chem. Eng. Progr. Symp. Ser. **67** No. 107 [1971] 381/7). — [13] G. J. Stander, L. R. J. van Vuuren (Water Pollut. Contr. **68** [1969] 513/22). — [14] H. J. Gomez (Chem. Eng. Progr. Symp. Ser. **67** No. 107 [1971] 294/5). — [15] A. J. Clayton, P. J. Pybus (Civil Eng. **42** No. 9 [1972] 103/6).

[16] R. Colas (Proc. 3rd Intern. Symp. Fresh Water Sea, Dubrovnik 1970, Vol. 3, p. 471/4). — [17] E. D. Bovet (J. Am. Water Works Assoc. **62** [1970] 539/42). — [18] K. G. Young (J. Am. Water Works Assoc. **63** [1971] 21/4). — [19] G. P. Patterson (Water Wastes Eng. **8** No. 6 [1971] 44/5). — [20] University of Florida (U.S. Natl. Tech. Inform. Serv. PB 206145 [1971]).

[21] N. C. Beaton, D. A. Dickson (Process Technol. Intern. **18** [1973] 313/5). — [22] R. F. Probstein, J. M. Alvarez (Massachusetts Inst. Technol. Dept. Mech. Eng. Fluid Mech. Lab. Publ. No. 73-5 [1973]). — [23] R. R. Harris (J. Am. Water Works Assoc. **62** [1970] 561/2). — [24] W. L. Patterson, R. F. Banker (Off. Saline Water Res. Develop. Progr. Rept. No. 462 [1969]). — [25] A. J. Matson, N. G. Giebink, L. C. Rentschler (Off. Saline Water Res. Develop. Progr. Rept. No. 463 [1969]).

[26] Black and Veatch (Off. Saline Water Res. Develop. Progr. Rept. No. 162 [1966]). — [27] W. L. Patterson, H. J. Lobb (Off. Saline Water Res. Develop. Progr. Rept. No. 519 [1970]). — [28] Metcalf and Eddy (Off. Saline Water Res. Develop. Progr. Rept. No. 779 [1972]). — [29] C. J. Schmidt, D. Ross (Off. Saline Water Res. Develop. Progr. Rept. No. 819 [1972]). — [30] J. J. O'Brien (Chem. Eng. Progr. Symp. Ser. **67** No. 107 [1971] 196/201).

[31] T. W. Rothermel (Off. Saline Water Res. Develop. Progr. Rept. No. 784 [1972]), T. W. Rothermel, J. J. Strobel (Proc. 4th Intern. Symp. Fresh Water Sea, Heidelberg 1973, Vol. 2, p. 405/12). — [32] W. C. Schroeder, J. M. Marchelo (Off. Saline Water Res. Develop. Progr. Rept. No. 199 [1966]). — [33] Ralph M. Parsons Co. (Off. Saline Water Res. Develop. Progr. Rept. No. 207 [1966]). — [34] Southern Research Institute, Houston and Texas Water Development Board (Off. Saline Water Res. Develop. Progr. Rept. No. 250 [1966]). — [35] Ralph M. Parsons Co. (Off. Saline Water Res. Develop. Progr. Rept. No. 337 [1967]).

[36] Southern Research Institute, Houston and Texas Water Development Board (Off. Saline Water Res. Develop. Progr. Rept. No. 273 [1967]). — [37] W. E. Thompson (ORNL-NDIC-12

[1972]). — [38] F. J. Agardy, H. Daubert (Off. Saline Water Res. Develop. Progr. Rept. No. 702 [1971]). — [39] D. E. Morris, W. L. Prehn (Off. Saline Water Res. Develop. Progr. Rept. No. 767 [1971]). — [40] H. R. Stucky, W. C. Arnwine (Off. Saline Water Res. Develop. Progr. Rept. No. 776 [1971]).

[41] I. C. Watson, F. M. Heider (Off. Saline Water Res. Develop. Progr. Rept. No. 783 [1972]). — [42] R. E. Crawford, R. P. Selm, G. D. Starret, C. A. Roberts (Off. Saline Water Res. Develop. Progr. Rept. No. 869 [1973]). — [43] T. C. Owens, A. M. Cooley, G. O. Fossum, D. Schaaf (Off. Saline Water Res. Develop. Progr. Rept. No. 902 [1973]). — [44] South Dakota State Department of Health (Off. Saline Water Res. Develop. Progr. Rept. No. 918 [1973]). — [45] Arizona Water Commission (Off. Saline Water Res. Develop. Progr. Rept. No. 919 [1974]).

[46] W. L. Prehn (Off. Saline Water Res. Develop. Progr. Rept. No. 920 [1974]). — [47] R. L. Streeter (Off. Saline Water Res. Develop. Progr. Rept. No. 904 [1974]).

1.2 History of desalination

Geschichte der Entsalzung

The oldest reference of converting saline water into fresh water may be found in the Bible: " . . . so Moses brought the sons of Israel from the Red Sea und they went to the desert of Sur. And they marched three days in the wilderness and they found no water to drink. And then they arrived at Merra and they could not drink from the waters of Merra, because they were bitter. Therefore he gave to this place the name of Bitterness. And the people murmured against Moses, saying: 'What shall we drink?' And Moses cried onto the Lord. And the Lord shewed him a wood and he put it into the water and the water became sweet . . . " [1].

The importance of water as a matter of life was already known to the philosophers of the Antiquity. They distinguished the different nature of fresh water and saline water and tried to explain the transition of saline water into fresh. T h a l e s o f M i l e t u s considered water to be the origin of all beings and admitted that fresh water is in fact seawater filtered through the earth. The same theory was presented by D e m o c r i t u s, as stated in the works of Aristotle, Theophrastus and Plutarch. P l a t o, in the "Timaeus", referring to the vegetable juices, suggested that they originated from ground water filtered through small pores of plants. In the "Laws", he noticed that clay is impermeable to water. A r i s t o t l e dealt extensively in several of his works with the problem of water. He discussed the properties of fresh and saline water and gave a surprisingly accurate explanation of the water cycle in nature: "The sun, moving as it does, sets up processes of change and by its agency the finest and sweetest water is every day carried up and is dissolved into vapor and rises to the upper region, where it is condensed again by the cold and so water is formed, which falls down again to the earth . . . Saline water when evaporated becomes sweet and the vapor does not form salt when it condenses again . . . Saltness is concentrating in the remaining seawater, because saline is heavy . . . "

The most interesting statement of Aristotle is that referring to the possibility of seawater desalination: "Make a jar of wax and put it into the sea, having sealed its mouth in such a way as to prevent the sea getting in and let it down quite empty into the sea. In twenty-four hours it will be found to contain a quantity of water, which will be fresh and potable." This statement has been the object of many discussions, as attempts to obtain fresh water by filtration or osmosis through wax have been unsuccessful. The most probable explanation is that the Greek words "angeion keraminon" (pot of earth) have been miscopied from manuscripts as "angeion kerinon" (pot of wax).

P l i n y, in "Natural History", describes ways of meeting shortage on fresh water at sea. "Fleeces of wool, spread around a ship by night, become moist by absorption of evaporated seawater and deliver fresh water by squeezing."

P l u t a r c h in "Moralia" quotes several references to scientific problems and also to desalination phenomena. D i o s c o r i d e s describes the distillation process and the collection of condensed vapors. C l e o p a t r a gives full description of distillers with cooling tubes for condensing the vapors. A l e x a n d e r of Afrodisias presented in his commentary on Aristotle's "Meteorologica" the first description of distillation as a means of obtaining fresh water from the sea: "Pots containing seawater are placed on the fire and the escaping vapors are collected on appropriate covers. Fresh water is obtained by condensation of the vapors." St. Basil reported in his "Sermons" the method followed by seamen to obtain fresh water from the sea. They boiled seawater in a pot and suspended sponges over it. The vapors were condensed on the sponges. Fresh water was recovered by squeezing the sponges. The same method of obtaining fresh water is reported by Olympiodoros, a great commentator of Aristotle. Both, as well as several other ancient writers, are citing Aristotle's experiment with the pot of wax. Johannes Philoponus described the statement of Aristotle that fresh water might be obtained by filtration, but he clearly indicated the use of an earthen pot instead of a pot of wax.

The statements of the ancient writers on desalination were consequently repeated and enlarged by the Arab and Byzantine philosophers and other writers of the Middle Ages. Important citations can be found in the works of Al-Hirani, Al-Biruni, Nikephoros Blemmides, Gilbertus Angelicus etc. Leon Battista Alberti devotes a long chapter in "De re aedificatoria" to water purification and desalination. He cites again the Aristotle experiment of desalination by filtration through an earthen (not wax) pot.

Various citations on desalting may be as well found in works of Renaissance writers. The most important contributions are contained in the books on "Magiae naturalis" by Giovan Battista Della Porta, in which several means for obtaining potable water from seawater and from the humidity of the atmosphere are given, partly as citations from other authors and partly due to his own experiments. He also suggested the use of solar energy as a source of heat for distillation.

Until the time referred to, the world was concentrated in Europe and around the Mediterranean Sea. Geographical descoveries and long journeys made the existence of a fresh water supply on board the vessels a necessity. The ancient method of distilling seawater on board ship was used by Sir Richard Hawkins and by Pedro Fernandez de Quiros to survive during their journeys to the South Seas. The possibility of obtaining fresh water by distillation was also reported by Jan Huygen van Linschoten. The Dutch East India Company experienced a greater mortality on ships without seawater stills, than on those provided with such devices. The distillation method was developed for the Company by Aegidius Snoek. Cornelius Drebbel developed a portable still, which could produce fresh water from the sea.

Hauton suggested saving the condenser by passing the vapors in a pipe at the side of the ship, cooled by the sea. William Walcot in England was granted patent No. 184 for a distilling process to obtain potable water from seawater. A few years later Robert Fitzgerald obtained patent No. 226 for a similar device. In 1685 silver medals were coined, illustrating the art of making salt water potable on board ship and on land.

Samuel Reyer continued experiments by Thomas Bartholin and by Robert Boyle and reported that ice formed in saline water has a low salt content. Anton Maria Lorgna described experiments of consecutive freezing in his task to desalt seawater. However, any other attempt to desalt water, including filtration through earth and sand, except distillation, was unsuccessful in practice. Stephen Hales improved by 15 to 25% the distillate output by blowing air through salt water.

An excellent review on the history of water desalination with 212 references was given by Giorgio Nebbia and Gabriella Nebbia Menozzi [2]. The historical development of modern desalting processes is reviewed in the appropriate sections.

Literature to 1.2

[1] Exodus, chapter XV, 22-25. — [2] G. Nebbia, G. Nebbia Menozzi (Proc. 2nd Inchiesta Intern., Milano 1966, p. 129/72; Acqua Ind. Inquinamento 8 No. 41 [1966] 13/8, No. 42 [1966] 23/4).

Entsalzungs-
verfahren

1.3 Desalting processes

Many methods have been proposed for desalting saline solutions, but only a few of them have been developed to the point where they may be used as commercial processes. Desalting techniques or phenomena potentially capable of becoming useful separation processes may be classified into two general categories: processes that eliminate salts from solution and processes that isolate pure water from solution (Table 1/1). The applicability of each of these processes depends on the amount of salts contained in the available raw water or on the process economics.

Table 1/1

Classification of main desalination processes.

Processes that separate	
Water from solution	Salts from solution
1. Distillation	1. Ionic processes
Vertical tube evaporator	Ion exchange
Horizontal tube evaporator	Electrodialysis
Multi-stage flash evaporator	Transport depletion
Vapor compression	Osmionic
Solar evaporation	Piezodialysis
2. Reverse osmosis	Electrochemical
3 Crystallization	Biological systems
Freezing	2. Other processes
Hydrate formation	Liquid-liquid extraction

The most developed process of removing water from saline solution is distillation. It is applied up to very large capacities with various types of evaporators. Freezing is a process on the way to be developed, whereas hydrate formation is still in the experimental stage. Reverse osmosis, after a long period of experimentation mainly in membrane development, is entering the field of commercial operation.

The latent heat for changing phase mainly in distillation and lesser in freezing processes is an important factor in the overall process economics, but the degree of salinity of the raw water is of no importance. Both processes are therefore equally suitable for desalting seawater. The same does not apply for reverse osmosis, as the necessary counterpressure depends greatly upon the salt content of the raw water and imposes constraints on membrane life and performance.

The most developed process for eliminating salts from aqueous solutions is electrodialysis and to some extent ion exchange. Their economics depend closely on the salt content of the raw water, either as consumption of electric energy or of chemicals for the regeneration of the resins. Hence, they may preferably be applied for the purification of brackish waters.

Solar distillation is also a well developed process, but its possible application is restricted to plants of very small capacities. It is an excellent process for the supply of small, remote communities in areas with abundant solar radiation.

Minimum energy requirements. The thermodynamic minimum work done for desalting is theoretically the same, regardless of the method used. In an idealized process with zero velocity, in which the product water and the brine would leave the system at the same temperature as the raw saline water and the brine would have the same salinity as the saline water, the theoretical energy requirement for the production of $1 m^3$ of water is as low as 0.7 kWh (2.8 kWh/1000 gal). These idealized thermodynamic conditions do not apply in practice. The energy required varies in fact with the temperature of the feed water and the salinity increase in the residual brine. Pumping power is also necessary to perform the process. Under practical conditions the minimum energy requirement for desalting processes is estimated to be about 3.0 kWh/m^3 (11.4 kWh/1000 gal), not including the energy required to raise the feed water to the level of the desalting plant [1]. These figures are quite a long way far from those achieved under realistic operating conditions, which are significantly higher. Minimum work of desalination was calculated by Smirnov to 1.109 kWh/m^3 (4.2 kWh/1000 gal) for 25°C, 3.5% NaCl and 50% extraction [2].

International activities. According to the Desalting Plant Inventory Report No. 4, edited by the U.S. Office of Saline Water, there were 812 land based desalting plants of 95 m^3/day (25 000 gpd) capacity or larger in operation or under construction throughout the world as of 1 January 1972. These plants are capable of producing about 1 316 900 m^3/day (347.9 Mgd) of fresh water for cities and industries. Distillation is the most widely used desalting process [3].

Table 1/2, p. 8, shows the number and capacity of plants by process and by geographic location.

Table 1/2

Total number and capacity of plants exceeding 25000 gpd (95 m³/day). P = plants; a = data in 1000 m³/day; b = data in Mgd.

Geographic Location	Submerged tube evaporator			Flash evaporator			Vertical tube evaporator			Vapor compression evaporator			Multi-stage flash evaporator			Horizontal tube evaporator		
	P	a	b	P	a	b	P	a	b	P	a	b	P	a	b	P	a	b
United States	219	74.6	19.7	37	19.7	5.2	4	17.8	4.7	6	4.6	1.2	15	45.0	11.9			
U.S. Territories				1	0.8	0.2	1	3.8	1.0	1	0.1	0	13	43.1	11.4			
N. America except U.S.	6	1.1	0.3				2	1.1	0.3				5	30.3	8.0			
South America	7	1.5	0.4	1	0.8	0.2	2	1.1	0.3	2	0.3	0.1	12	14.8	3.9			
Caribbean Islands	2	14.4	3.8				1	0.4	0.1	6	1.1	0.3	28	83.3	22.0			
Europe Continental	11	4.9	1.3	2	1.1	0.3	46	22.3	5.9	3	0.8	0.2	39	145.4	38.4	1	0.4	0.1
England and Ireland	23	14.0	3.7	2	2.7	0.7	33	33.3	8.8				4	9.4	2.5	1	0.4	0.1
U.S.S.R.	3	1.9	0.5				9	109.8	29.0									
Asia	4	1.5	0.4	4	0.8	0.2	6	1.09	0.5	1	3.0	0.8	14	14.0	3.7			
Middle East and Persian Gulf	13	11.7	3.1	1	0.3	0.1	5	4.2	1.1	6	3.4	0.9	50	381.2	100.7	2	4.2	1.1
Africa	5	1.5	0.4	1	0.1		3	1.5	0.4	11	3.4	0.9	35	86.3	22.8			
Australia							2	1.5	0.4				3	3.0	0.8			
Total plants	293			49			114			36			218			4		
Total Mgd			33.6			6.9			52.5			4.4			226.1			1.3
Total 1000 m³/day		127.1			26.3			198.7			16.7			855.8			5.0	
Plants %	36.08			6.03			14.04			4.43			26.85			0.49		
Capacity %	9.66			1.98			15.09			1.26			65.00			0.37		

Geographic Location	Total distillation plants			Electrodialysis			Reverse osmosis			Freezing processes			Total desalination plant capacity			
	P	a	b	P	a	b	P	a	b	P	a	b	P	a	b	%
United States	281	161.7	42.7	15	21.6	5.7	24	9.1	2.4	1	0.3	0.1	321	192.7	50.9	14.63
U.S. Territories	16	47.8	12.6							1	0.3	0.1	17	48.1	12.7	3.65
N. America except U.S.	13	32.5	8.6				1	1.1	0.3				14	33.6	8.9	2.56
South America	24	18.5	4.9	2	1.9	0.5				1	0.4	0.1	27	20.8	5.5	1.58
Caribbean Islands	37	99.2	26.2	2	0.1	0.1	1	1.1	0.3				39	100.4	26.6	7.65
Europe Continental	102	174.9	46.3	12	16.3	4.3	2	0.2	0.1	1	0.4	0.1	117	191.8	50.8	14.60
England and Ireland	63	59.8	15.8	1	0.1	0							64	59.9	15.8	4.54
U.S.S.R.	12	111.7	29.5	1	0.1	0							13	111.8	29.5	8.48
Asia	29	21.2	5.5				4	3.4	0.9				33	24.6	6.4	1.84
Middle East and Persian Gulf	77	405.0	107.0	18	8.7	2.3	1	0.1	0				96	413.8	109.3	31.41
Africa	55	92.8	24.5	10	22.0	5.8							65	114.8	30.3	8.71
Australia	5	4.5	1.2	1	0.1	0							6	4.6	1.2	0.34
Total plants	714			61			33			4			812			
Total Mgd		1229.7	324.8		70.8	18.7		15.0	4.0		1.4	0.4		1316.9	347.9	
Total 1000 m³/day													100.0			
Plants %	87.93			7.51			4.06			0.49			100.0			
Capacity %			93.36			5.38			1.15			0.11		100.0	100.0	100.0

Table 1/3 shows the number and the capacity of plants exceeding 1 Mgd (3785 m³/day) for the same date [3].

Solar plants are not included as none of the existing plants exceeds the capacity of 25 000 gpd. Two regions, the islands in the Caribbean Sea and the Middle East, including countries around the Persian Gulf, represent large centres of desalting. A condensed review of desalting activities in various countries is given in the following.

In the United States, the Office of Saline Water was created in 1952 for the development of processes to economically convert saline water into fresh water. Research and development progress reports, covering in detailed form all activities of the Office, were published in a number of 924 up to the spring 1974. Nearly all of them are briefly reviewed in this book. In addition to these, extensive annual reports summarizing the Office activities were published up to 1970/1971. Only summary reports were published for the fiscal years 1971/1972 and 1972/1973.

Reports on the O.S.W. program and activities were also presented by McCoy on the research program of the Office of Saline Water [4], by Gillam and McCoy on desalination research at the O.S.W. [5], by Hunter on engineering developments in the O.S.W. program [6], by Hunter and Savage on the U.S. desalting program [7], by Wong on the activities of O.S.W. in research and operation of demonstration plants [8], by Strobel on cooperative studies sponsored by O.S.W. [9], and by O'Meara on current status of the desalting program in the United States [10].

Research and development work at the Oak Ridge National Laboratory published in various scientific and technical journals was compiled and reprinted in several Research and Development Progress Reports, edited by the Office of Saline Water [11]. Annual summaries of research work on various fields of desalination, conducted at the University of California, were published at the Campus of Los Angeles and of Berkeley, the latter covering also Riverside and La Jolla.

In two O.S.W. reports with the general title "Water Technology", a large number of reviews, comments and contributions on various fields of desalination are included [12].

The U.S. Bureau of Reclamation in cooperation with the Office of Saline Water has also published the Desalting Handbook for Planners, for use in preliminary planning as a source of up-to-date information on the state of the development, costs, economics and applicability of these desalting processes which are available for providing water supply. The Handbook has been prepared in looseleaf form to provide for incorporation on a periodic basis of data that reflect continued advances in desalting technology [41].

A Saline Water Conversion Engineering Data Book was compiled by the Kellogg Company, on behalf of the Office of Saline Water, and contains reference aids to engineers and designers engaged in the conversion of saline waters [42].

Herbert reported on research and development work on desalination in Australia [13].

French research and development in the field of seawater desalination was reported by Balligand et al. [14], Dutheil and Malissen presented a market survey of water desalination for small units [15], and Dutheil and Lambert outlined problems posed by the desalination of seawater in small coastal islands [16]. Michel and Martin have reported on the desalting test station at Toulon [17].

A report on research and development on seawater desalination was edited in Western Germany by an ad hoc Committee under the sponsorship of the Federal Ministry for Education and Science [18].

Desalting for Hong Kong was reported by Ford et al. [19]. Nohadani outlined water problems and desalination in Iran [20].

Desalting activities in Israel were reported by Vilentchuk giving a survey of water desalination in Israel [21], by Kantor on development problems of water supply [22], and by Wiener on the development of Israel's water resources [23].

Hashizume has reported on the utilization of desalinated water in Japan [24], and Ishizaka et al. on present and future of desalination in Japan [25].

Kuwait is one of the largest centres, where desalting is representing almost the unique water supply. Ali El-Saie reported on the water production experience of the city of Kuwait [26], and on the history, experience and economics of water production in Kuwait [27]. El-Shamy et al. reported on the potable water production [28].

Table 1/3
Number and capacity of plants exceeding 1 Mgd (3785 m³/day). P = plants; a = data in 1000 m³/day; b = data in Mgd.

Geographic location	Various distillation processes			Vertical tube evaporator			Multi-stage flash evaporator			Electro-dialysis			Total desalination plant capacity exceeding 1 Mgd			
	P	a	b	P	a	b	P	a	b	P	a	b	P	a	b	%
United States	1	3.9	1.0 VC	2	15.1	4.0	5	37.1	9.8	4	27.3	7.2	12	83.4	22.0	6.9
North America (except U.S.)							1	28.4	7.5				1	28.4	7.5	2.4
South America							1	5.4	1.4				1	5.4	1.4	0.4
Caribbean Islands	2	14.4	3.8 ST	3	20.8	5.5	17	111.3	29.4				22	146.5	38.7	12.1
Continental Europe				1	4.3	1.1	10	193.8	51.2	1	4.9	1.3	12	203.0	53.6	16.8
United Kingdom	1	4.9	1.3 ST				1	6.8	1.8				2	11.7	3.1	1.0
U.S.S.R.				8	108.6	28.7							8	108.6	28.7	9.0
Middle East and Persian Gulf	1	4.2	1.1 HTE				19	355.1	93.8				20	359.3	94.9	29.8
Asia							1	181.7	48.0				1	181.7	48.0	15.1
Africa							4	54.9	14.5	2	23.8	6.3	6	78.7	20.8	6.5
Total plants	5			14			59			7			85			
Total Mgd			7.2			39.3			257.4			14.8			318.7	
Total 1000 m³/day		27.4			148.8			974.5			56.0			1206.7		
Plants %	5.9			16.5			69.4			8.2			100			
Capacity %		2.3			12.3			80.8			4.6			100		100.0

VC = Vapor compression evaporator; ST = Submerged tube evaporator; HTE = Horizontal tube evaporator.

Desalination research as well as operation of the early desalting plants in Spain was reported by Suarez and Pliego [29]. The activities of the Junta de Energia Nuclear in various fields of desalination were reported by F. Pascual Martinez [30].

Some British achievements in desalination were presented by Smith [31], and Kronberger outlined the objectives of the United Kingdom research and development program for desalination [32].

The United Nations have published two Surveys on desalination plant operation, as a technical and economic analysis of the performance of desalination plants in operation [33]. A review on the first report was presented by Mawer [34]. Further to these, the United Nations have initiated a study on Water Desalination in Developing Countries, giving details on the water situation and possibilities for desalting in 43 countries [35]. Barnea has also published a survey on water costs in developing countries [36].

Desalination activities in Yugoslavia were reported by Jankovic [37], and by Arneri on the technical and economic aspects of water desalination on the Adriatic sea coast [38], by Ivekovic on the same subject [39] and on the necessity of obtaining drinking water from the sea on the Adriatic coast of Yugoslavia [40].

Further to specialized bibliographies, indicated in the appropriate sections, a general bibliography on saline water conversion literature up to 1965 was compiled by Schamus on behalf of the Office of Saline Water [43]. Delyannis and Piperoglou-Delyannis have also compiled a more or less complete bibliography on water desalting and related fields from Antiquity up to 1968 [44].

"Desalination", an international journal on the science and technology of desalting and water purification, is published bimonthly since 1966 by Elsevier, Amsterdam, Holland.

Two abstracting journals are also published. "Desalination Abstracts" appear four times per year since 1966 and is edited by the National Center of Scientific and Technological Information, Tel Aviv, Israel. "Water Desalting Abstracts" appeared bimonthly in 1970 and 1971 and monthly since 1972, are edited by A. Delyannis and E. Delyannis and contain abstracts of literature starting 1969. It is a continuation of the bibliography [44] edited by the same authors.

Literature to 1.3

[1] G. W. Murphy (Off. Saline Water Res. Develop. Progr. Rept. No. 9 [1956]). — [2] L. F. Smirnov (Vodosnabzh. Sanit. Tekhn. **1970** No. 2, p. 1/4). — [3] F. O'Shaughnessy (Desalting Plant Inventory Rept. No. 4 [1973]). — [4] W. H. McCoy (Proc. 1st Intern. Symp. Water Desalination, Washington, D.C., 1965 [1967], Vol. 1, p. 339/48). — [5] W. S. Gillam, W. H. McCoy (Desalination **2** [1967] 13/20).

[6] J. A. Hunter (Desalination **3** [1967] 384/91). — [7] J. A. Hunter, W. F. Savage (Desalination **6** [1969] 285/6). — [8] C. M. Wong (Proc. 3rd Intern. Symp. Fresh Water Sea, Dubrovnik 1970, Vol. 3, p. 133/40). — [9] J. J. Strobel (J. Am. Water Works Assoc. **63** [1971] 258/61). — [10] J. W. O'Meara (Proc. 4th Intern. Symp. Fresh Water Sea, Heidelberg 1973, Vol. 2, p. 381/5).

[11] K. A. Kraus, R. J. Raridon, J. S. Johnson, E. G. Bohlman, F. A. Posey (Off. Saline Water Res. Develop. Progr. Rept. No. 468 [1969]), J. S. Johnson, K. A. Kraus (Off. Saline Water Res. Develop. Progr. Rept. No. 508 [1970]), F. A. Posey, E. G. Bohlmann, S. S. Misry, D. V. Subrahmanyam, F. Nelson (Off. Saline Water Res. Develop. Progr. Rept. No. 852 [1973]), F. A. Posey, P. M. Lantz, R. E. Meyer, M. C. Banta, A. A. Palko (Off. Saline Water Res. Develop. Progr. Rept. No. 853 and No. 903 [1973]). — [12] W. E. Bell, D. R. Brenneman, R. M. Burd, R. E. Moore, J. K. Rice (Off. Saline Water Res. Develop. Progr. Rept. No. 536 [1970]), P. Goldstein, R. M. Burd, R. E. Moore, J. K. Rice (Off. Saline Water Res. Develop. Progr. Rept. No. 679 [1971]). — [13] L. S. Herbert (Austr. Chem. Proc. Eng. **22** No. 12 [1969] 16/7, **23** No. 1 [1970] 18/20). — [14] P. Balligand, J. J. Libert, A. Michel (Proc. 3rd Intern. Symp. Fresh Water Sea, Dubrovnik 1970, Vol. 3, p. 409/19). — [15] F. Dutheil, M. Malissen (Proc. Symp. Nucl. Desalination, Madrid 1968 [1969], p. 277/89).

[16] F. Dutheil, J. Lambert (Proc. 3rd Intern. Symp. Fresh Water Sea, Dubrovnik 1970, Vol. 3, p. 421/32). — [17] A. Michel, G. Martin (Bull. Inform. Sci. Tech. [Paris] No. 173 [1972] 21/34). — [18] E. Becherer, E. Bechinie, H. J. Hampel, B. Oberbacher, F. Reckefuss, G. Steinbach, H. Wolf (Forschung- und Entwicklungsförderung auf dem Gebiet der Meerwasserentsalzung, Bundesministerium Bildung Wissenschaft Rept. K-72-27 [1972]). — [19] S. E. H. Ford, M. C. D. LaTouche, F. A. Drake (Proc. 3rd Intern. Symp. Fresh Water Sea, Dubrovnik 1970, Vol. 3, p. 143/51). — [20] H. Nohadani (Proc. 3rd Intern. Symp. Fresh Water Sea, Dubrovnik 1970, Vol. 3, p. 265/71).

[21] I. Vilentchuk (Proc. 1st Intern. Symp. Water Desalination, Washington, D.C., 1965 [1967], Vol. 2, p. 87/101). — [22] S. Kantor (Gas Wasserfach **112** [1971] 531/40). — [23] A. Wiener (Am. Scientist **60** [1972] 466/74). — [24] M. Hashizume (Proc. 1st Intern. Symp. Water Desalination, Washington, D.C., 1965 [1967], Vol. 2, p. 475/91). — [25] S. Ishizaka, Y. Onaga, T. Kikuchi, T. Sakai, I. Suetsuna (Proc. Symp. Nucl. Desalination, Madrid, 1968 [1969], p. 63/73).

[26] M. H. Ali El-Saie (Proc. 1st Intern. Symp. Water Desalination, Washington, D.C., 1965 [1967], Vol. 3, p. 287/324). — [27] M. H. Ali El-Saie (Desalination **1** [1966] 77/95). — [28] H. K. El-Shamy, I. I. Mansi, O. A. Al-Fulaij (J. Am. Water Works Assoc. **63** [1971] 783/6). — [29] J. Suarez, J. M. Pliego (Proc. 1st Intern. Symp. Water Desalination, Washington, D.C., 1965 [1967], Vol. 2, p. 797/816). — [30] F. Pascual Martinez (Energia Nucl. [Madrid] **13** [1969] 206/31).

[31] A. C. Smith (Proc. 1st Intern. Symp. Water Desalination, Washington, D.C., 1965 [1967], Vol. 2, p. 407/13). — [32] H. Kronberger (Proc. 1st Intern. Symp. Water Desalination, Washington, D.C., 1965 [1967], Vol. 2, p. 493/7; Proc. Symp. Nucl. Desalination, Madrid 1968 [1969], p. 33/40). — [33] United Nations, First Desalination Plant Operation Survey, Rept. No. E 69.II.B.17 [1969]; Second Rept. No. E.73.II.A.10 [1973]). — [34] P. Mawer (Proc. Symp. Nucl. Desalination, Madrid 1968 [1969], p. 145/65). — [35] United Nations (Water Desalination in Developing Countries Publ. No. 64.II.B.5 [1964]).

[36] J. Barnea (Proc. 1st Intern. Symp. Water Desalination, Washington, D.C., 1965 [1967], Vol. 3, p. 737/52). — [37] S. Jankovic (Desalination **7** [1969] 122). — [38] G. Arneri (Krs Jugoslavije **6** [1969] 365/76). — [39] H. Ivekovic (Krs Jugoslavije **6** [1969] 357/63). — [40] H. Ivekovic (Proc. 4th Intern. Symp. Fresh Water Sea, Heidelberg 1973, Vol. 2, p. 349/56).

[41] U.S. Bureau of Reclamation and Office of Saline Water (Desalting Handbook for Planners, 1st Ed. May 1972, 2nd Ed. November 1972). — [42] M. W. Kellogg Company (Saline Water Conversion Engineering Data Book, Washington, D.C., 1965, Supplement 1, 1966). — [43] J. J. Schamus (Off. Saline Water Res. Develop. Progr. Rept. No. 146 [1965]). — [44] A. Delyannis, E. Piperoglou (Handbook of Saline Water Conversion Bibliography, Vol. 1: Antiquity-1940 [1967], Vol. 2:1941- 1950 [1967], Vol. 3:1951-1954 [1968], Vol. 4:1955-1956 [1968]), A. Delyannis, E. Delyannis (Handbook of Saline Water Conversion Bibliography, Vol. 5:1957-1958 [1968], Vol. 6:1959-1960 [1968], Vol. 7:1961-1962 [1969], Vol. 8:1963-1964 [1970], Vol. 9:1965-1966 [1972], Vol. 10:1967-1968 [to be published in 1975]).

1.4 Raw material seawater

Rohmaterial
Meerwasser

Seawater is an aqueous solution containing a variety of dissolved solids and gases. Some of the dissolved substances exist in very high concentrations, while others are present in minute quantities. The absolute concentration of the total dissolved solids varies largely with the location. In the open sea the chemical composition of seawater is more or less uniform, due to mixing by ocean streams. In certain confined areas the dissolved salt content is affected by the rate of evaporation and/or by the importance of incoming river flow. Characteristic mean values of total dissolved solids in various seas are given in table 1/4. Although the amount of dissolved solids varies largely, it is important to note that the ratios between the more abundant dissolved substances are practically constant. This fact permits the establishment of characteristic values for seawater, such as salinity, chlorinity and chlorosity (see section 1.4.3, p. 28).

Table 1/4

Total dissolved solids in various seas (g/kg).

Baltic Sea	7.0	Pacific Ocean	33.6
Caspian Sea	13.5	Atlantic Ocean	36.0
Black Sea	20.0	Mediterranean Sea	39.0
White Sea	28.0	Red Sea	43.0
Northern Adriatic	29.0	Kara Bogaz (Caspian)	164.0

1.4.1 Mineral content of seawater

Salzgehalt
des
Meerwassers

The constancy of composition of the dissolved salts affords a mean for estimating the concentrations of all the major constituents, if the concentration of any one of them is known. Furthermore, results of studies on the composition or the physical properties of seawater in any locality are generally applicable to the water in any other part of the oceans. Hence it was possible to the Hydrographic Laboratories of Copenhagen, Denmark, to prepare a standard composition of synthetic normal seawater.

Table 1/5 gives the ionic composition of normal seawater (Copenhagen) and ordinary seawater [1], arranged in the order of the abundance of elements. Only the major constituents of seawater are listed. If trace elements and dissolved gases are added, a total of over fifty elements are known to occur in seawater. Satisfactory data are available for the major elements, while data concerning trace elements vary to a greater or lesser extent.

Table 1/5

Ionic composition of seawater (g/kg).

Ions	Normal seawater		Seawater	
Chlorides	+ 01	1.93605	+ 01	1.89799
Sodium	+ 01	1.07678	+ 01	1.05561
Sulfates		2.7017		2.6486
Magnesium		1.2975		1.2720
Calcium	− 01	4.081	− 01	4.001
Potassium	− 01	3.876	− 01	3.800
Hydrogen carbonates	− 01	1.425	− 01	1.397
Bromides	− 02	6.59	− 02	6.46
H_3BO_3	− 02	2.65	− 02	2.60
Strontium	− 02	1.36	− 02	1.33
Fluorides	− 03	1.3	− 03	1.3
Iodides	− 05	5.0		
Silicon from	− 05	2.0		
to	− 03	4.0		
Others	− 03	1.3		
Total Solids	+ 01	3.51745	+ 01	3.44816
Water	+ 02	9.648255	+ 02	9.655184
Characteristics of seawater				
Salinity	+ 01	3.501	+ 01	3.4325
Chlorinity	+ 01	1.9381	+ 01	1.900
Chlorosity	+ 01	1.9862	+ 01	1.9462
Specific Gravity 20°C		1.0248		1.0243

The prefix refers to the powers of 10 of the decimal given. Accordingly − 01 means 10^{-1}, and + 03 means 10^3.

Further papers on the mineral content of seawater were presented by MacIntyre describing the geochemical cycles involved in conveying dissolved solids to the ocean [2], by Johnston reporting on the salinity and its estimation [3], by Sillen on the chemical composition of the oceans [4], by Chave on chemical reactions and the composition of seawater [5], by Ogata et al. on a chemical model of seawater [6], by Martin on the chemistry of the sea [7], by Pytkowicz and Kester on the physical chemistry of seawater [8], by Whitfield on progress towards a chemical model for seawater [9], by Martinova et al. on the calculation of the composition of saltwater [10], by Kitano on the origin of seawater [11].

A review of the most important constituents of the dissolved solids in seawater is given in the following in alphabetical order of elements:

Antimony. Ryabinin et al. reported on antimony content in Black Sea waters [12], and the same group of authors on the arsenic/antimony ratio in Atlantic Ocean and Mediterranean Sea waters [13].

Barium. Wolgemuth and Broecker reported on vertical profiles of barium content in the Atlantic and the Pacific Oceans [14], Bernat et al. on barium and strontium concentrations in Pacific and Mediterranean seawater profiles [15], Li et al. on barium in the Antarctic Ocean [16].

Boron is apparently present in seawater as undissociated boric acid. Ryabinin reported on the boron content in the tropical zone of the Atlantic Ocean [17], Barannik et al. on the content in waters of Caribbean Sea [18].

Bromine shows a very constant ratio to the chlorinity and is apparently all present as bromide ion. Foti reported on the concentration of radioactive bromide ions in seawater by isotopic exchange [19].

Calcium is present in small quantities and is permanently removed from the seawater by deposition of skeletal remains found in marine sediments. This removal does not necessarily imply a decrease of the calcium concentration, because a large supply is maintained by the river waters flowing into the sea. Detectable differences in the calcium/chlorinity ratio have been observed. The solubility of calcium carbonate in seawater is of great interest, as well as the factors that control its precipitation and solution. Knowledge of the calcium concentration is also important in an understanding of the carbon dioxide system in the sea and especially in preventing scale formation in desalination processes. Chave and Swess reported on calcium carbonate saturation in seawater [20], Lyankin on calcium and magnesium in the western tropical Atlantic [21], Duedall on partial molal volume of calcium carbonate in seawater [22], Tsunogai et al. on calcium content in the Pacific Ocean [23].

Carbon occurs in seawater partly in the form of carbonic acid and its salts but also in appreciable amounts as a constituent of organic material, either living or dead. The solubility of carbon dioxide depends upon the temperature and salinity of the water. An exchange of carbon dioxide with the atmosphere takes place at the surface. The quantities of carbon present in seawater as either free carbon dioxide, hydrogen carbonate or carbonate show a considerable range. Fairhall et al. presented data for ^{14}C content at several ocean locations throughout the world [24].

Cerium. Spitsyn et al. reported on the state of ultrasmall quantities of ^{144}Ce and ^{91}Y in seawater [25], Popov on the physicochemical state of ^{144}Ce in seawater [26].

Chlorine, present as chloride ion, is the most abundant ion and makes up about 55% by weight of the dissolved material. The chlorinity (s. p. 28) is of importance, because it is the basis of density computations and is the standard to which are referred substances present in major amounts.

Chromium. Elderfield discussed probable ion forms of chromium contained in seawater [27].

Cobalt. Robertson discussed the distribution of cobalt in oceanic waters [28], Preston and Dutton summarized the origins and nuclear properties of radionuclides of Co in marine environment [29].

Copper. Spencer and Brewer reported on the distribution of copper, zinc and nickel in seawater of the Gulf of Maine and the Sargasso Sea [30], Slowey and Hood on copper, manganese and zinc concentrations in Gulf of Mexico waters [31], Odier and Plichon on dissolved copper in seawater [32], Zirino and Yamamoto on a pH-dependent model for the chemical speciation of copper, zinc, cadmium and lead in seawater [33].

Fluorine is present as fluoride ion and bears a constant ratio to chlorinity. Warner discussed the normal fluoride content of seawater [34].

Gold. Jones reported on gold content including seawater [35], Sharma on the concentration of gold in seawater [36].

Indium. Baric and Branica discussed the behavior of indium in seawater [37], Chow and Snyder reported on the indium content of seawater [38], Matthews and Riley on the occurrence in seawater and in some marine sediments [39].

Iodine. Tsunogai and Henmi reported on the iodine in seawater of the Pacific [40], Paslawska and Ostrowski on the iodine content of the Baltic Sea [41].

Iron. Bernovskaya et al. discussed the effect of ^{55}Fe on the physicochemical state of radioactive elements in seawater [42], Head on the concentration of iron in seawater of Southampton [43].

Magnesium. Carpenter and Manella reported on magnesium to chlorinity ratios in seawater [44]. The ratio of magnesium to chlorinity is very uniform.

Manganese. Matsumoto reported on manganese content in the Suruga Bay and four rivers feeding the bay [45].

Mercury. Leatherland et al. reported on mercury in Atlantic Ocean water [46].

Molybdenum. Head and Burton reported on molybdenum content in some ocean and estuarine waters [47], Pilipchuk on the distribution in the Pacific Ocean waters [48], Volkov et al. on the concentration in the Atlantic Ocean, the Mediterranean and Black Seas [49].

Niobium and technetium. Spitsyn et al. discussed the state of trace amounts of [95]Nb and ^{99}Tc in seawater [50], Tikhomirov and Gromov on ^{99}Tc and [54]Mn in equatorial Pacific Ocean [51].

Nitrogen occurs in seawater both in compounds of various kinds, such as nitrate, nitrite, ammonia and organic compounds, as well as free dissolved nitrogen gas. Oxygen exists also in solution in seawater and its presence is of importance in biological processes, as well as in corrosion processes of metallic parts in desalination plants. Presence of argon, helium, neon, and probably hydrogen is also reported.

Potassium is present in amounts of only a few percent of that of sodium. The ratio of the potassium to chlorinity may also be modified by dilution with river water. Buyanov discussed the calculation of potassium content and potassium radioactivity of seawater from the measured salinity [52].

Radium and radioactivity. Moore reported on the measurement of [228]Ra and [228]Th in seawater [53], Kautsky on instruments for continuous monitoring the radioactivity of surface seawater [54], Broecker et al. on [226]Ra measurements in the North Pacific [55], Nakaya and Nakamura on the physical state of radionuclides in seawater [56].

Ruthenium. Guegueniat discussed the physicochemical behavior of ruthenium contamination in seawater [57].

Silicon in seawater might be present as soluble silicate but as well as in some colloidal compound. River water contains silicon both in solution and as colloidal particles. Colloidal silica in seawater may pass into true solution on ageing.

Sodium is the most abundant cation in seawater. The sodium to chlorinity ratio (≈ 0.5556) may be modified near river mouths. Kester and Pytkowicz reported on sodium, magnesium and calcium sulfate ion-pairs in seawater [58].

Strontium is jointly determined with calcium. Consequently the ratio of calcium to chlorinity reported for seawater represents the sum of Ca + Sr, which is expressed as calcium. Timoshchuk et al. reported on the distribution of ^{90}Sr in the surface layer of the Mediterranean Sea [59], Hildreth and Henderson on [87]Sr/[86]Sr in seawater [60], Nagaya et al. on strontium concentrations and strontium/chlorinity ratios in seawater of the North Pacific [61], Vdovenko et al. on concentrations of ^{90}Sr and ^{137}Cs in ocean waters [62], Chumichev on ^{90}Sr in the northwestern part of the Pacific Ocean [63], Eremeeva and Eremeev on strontium content in ocean waters [64].

Sulfur is present as sulfate ion. Under stagnant conditions, occurring in certain isolated basins, a part of the sulfate may be converted to sulfide ion. The sulfate/chlorinity ratio may also be modified by dilution with river water, which in general is relatively high in sulfate content. Clive and Richards discussed the kinetics of the reaction between sulfides and oxygen in seawater [65], Kester and Pytkowicz the effect of temperature and pressure on sulfate ion association in seawater [66], Spedding reported on the absorption of SO_2 by seawater [67].

Thallium. Matthews and Riley reported on the occurrence of thallium in seawater and marine sediments [68].

Thorium. Bhat et al. reported on [234]Th/[238]U ratios in the ocean [69], Kaufman on [232]Th concentration of surface ocean water [70], Kuznetsov on the forms of [230]Th and [232]Th in the ocean [71], Miyake et al. on thorium concentration and the activity ratios [230]Th/[232]Th and [228]Th/[232]Th in seawater in the western North Pacific [72], and on the disequilibrium between [228]Th and [232]Th in seawater [73].

Uranium. Heye reported on uranium, thorium and radium in ocean water [74], Miyake et al. on uranium content and the activity ratio [234]U/[238]U [75], Strogonov and Risik on the distribution of [238]U in the Mediterranean Sea water [76].

Zinc. Zirino and Healy reported on inorganic zinc complexes in seawater [77].

Other elements. Riley and Taylor reported on concentrations of cadmium, copper, iron, manganese, molybdenum, nickel, vanadium, and zinc in the tropical northeast Atlantic Ocean [78],

Mangel on the concentration of complex ions of 26 metals [79], Roskam presented a table of 31 potential biologically active elements polluting the sea [80].

Other minor constituents of the dissolved solids and gases in seawater include Ag, Al, La, Li, P, Rb, Sc, Se, Sn, oxygen and noble gases.

Synthetic seawater. Trace amounts present in seawater do not influence the basic properties of seawater. In preparing synthetic seawater, it is therefore not necessary to have present all ions occurring in natural seawater. On the other hand, impurities contained in chemicals to be used for the preparation of synthetic seawater, may be in excess of trace elements in natural seawater. As a rule only the more abundant salts are used in preparing synthetic seawater.

Table 1/6 contains the most probable compounds present in seawater, according to the ionic composition of normal seawater, given in Table 1/5, p. 14, as well as three suggestions for preparing synthetic seawater [1].

Table 1/6

Molecular composition of seawater.

Compound	Substitute ocean water (g/l)	Synthetic seawater (g/kg) McClendon (1917)	Synthetic seawater (g/kg) Brujewics, [Subow (1941)]	Synthetic seawater (g/kg) Lyman and Fleming (1940)
$NaCl$	+01 2.453	+01 2.6726	+01 2.6518	+01 2.3476
$MgCl_2$	5.20	2.260	2.447	4.981
Na_2SO_4	4.09			3.917
$MgSO_4$		3.248	3.305	
$CaCl_2$	1.16	1.153	1.141	1.102
KCl	−01 6.95	−01 7.21	−01 7.25	−01 6.64
$NaHCO_3$	−01 2.01	−01 1.98	−01 2.02	−01 1.92
KBr	−01 1.01			−02 9.6
$NaBr$		−02 5.8	−02 8.3	
H_3BO_3	−02 2.7	−02 5.8		−02 2.6
$SrCl_2$	−02 2.5			−02 2.4
Al_2Cl_6		−02 1.3		
NaF	−03 3.0			−03 3.0
Na_2SiO_3		−03 2.4		
$Na_2Si_4O_9$		−03 1.5		
NH_3		−03 2.0		
$LiNO_3$		−03 1.3		
H_3PO_4		−04 2.0		
$Ba(NO_3)_2$	−05 9.94			
$Mn(NO_3)_2$	−05 3.40			
$Cu(NO_3)_2$	−05 3.08			
$Zn(NO_3)_2$	−05 1.51			
$Pb(NO_3)_2$	−06 6.6			
$AgNO_3$	−07 4.9			
Total Solids	36.03	34.4424	34.421	34.481
Water	963.97	965.5576	965.579	965.519
Total	1000.00	1000.0000	1000.000	1000.000
Chlorinity	19.38	19.00	19.00	19.00

Literature to 1.4.1

[1] M. W. Kellog Co. (Saline Water Conversion Data Book, Washington, D.C., 1965). — [2] F. MacIntyre (Sci. Am. **223** No. 5 [1970] 104/15). — [2] R. Johnston (Oceanogr. Mar. Biol. **7** [1969] 31/48). — [4] L. G. Sillen (IVA [Ingenioersvetenskapsakad] Medd. No. 159 [1969] 9/18; C.A. **73** [1970] No. 28650). — [5] K. E. Chave (J. Chem. Educ. **48** [1971] 148/51).

[6] N. Ogata, N. Inoue, H. Kakihana (Nippon Kaisui Gakkai-Shi **24** No. 127 [1970] 19/25; C.A. **75** [1971] No. 25080). — [7] D. F. Martin (Quart. J. Florida Acad. Sci. **34** [1971] 175/86). — [8] R. M. Pytkowicz, D. R. Kester (Oceanogr. Mar. Biol. **9** [1971] 11/60). — [9] M. Whitfield

(Proc. Roy. Soc. Edinburgh B **72** [1972] 389/99). — [10] O. I. Martinova, I. G. Vasina, S. A. Pozdnyakova, E. S. Kolbasova (Tr. Mosk. Energ. Inst. No. 128 [1972] 121/9; C. A. **79** [1973] No. 9647). — [11] Y. Kitano (Kagaku No Ryoiki **27** [1973] 700/1). — [12] A. I. Ryabinin, A. S. Romanov, R. Khamidova (Dopovidi Akad. Nauk Ukr.RSR B **33** [1971] 834/6; C. A. **75** [1971] No. 143850). — [13] A. I. Ryabinin, A. S. Romanov, S. Katamov, R. Khamidova (Dopovidi Akad. Nauk Ukr.RSR B **34** [1972] 923/7; C. A. **78** [1973] No. 20043). — [14] K. Wolgemuth, W. S. Broecker (Earth Planet. Sci. Letters **8** [1970] 372/8). — [15] M. Bernat, T. Church, C. J. Allegre (Earth Planet. Sci. Letters **16** [1972] 75/80).

[16] Y. H. Li, T. L. Ku, G. G. Mathieu, K. Wolgemuth (Earth Planet. Sci. Letters **19** [1973] 352/8). — [17] A. I. Ryabinin (Geokhimiya **1972** 879/85; C. A. **77** [1972] No. 118047). — [18] V. P. Barannik, L. I. Man'kovskaya, A. I. Sheremet'eva (Morsk. Gidrofiz. Issled. **1972** No. 1, p. 139/47; C. A. **79** [1973] No. 34983). — [19] S. C. Foti (U.S. Natl. Tech. Inform. Serv. Rept. AD 734383 [1971]. — [20] K. E. Chave, E. Swess (Limnol. Oceanogr. **15** [1970] 633/7).

[21] Yu. I. Lyankin (Okeanologiya **11** [1971] 635/41). — [22] I. W. Duedall (Geochim. Cosmochim. Acta **36** [1972] 729/34). — [23] S. Tsunogai, H. Yamahata, S. Kudo, O. Sato (Deep Sea Res. **20** [1973] 717/26). — [24] A. W. Fairhall, R. W. Buddemeier, I. C. Yang, A. W. Young (RLO-2225-T-20-2 [1970]). — [25] V. I. Spitsyn, R. N. Bernovskaya, N. I. Popov (Dokl. Akad. Nauk SSSR **185** [1965] 111/4; C. A. **70** [1969] No. 109047).

[26] N. I. Popov (Tech. Rept. Ser. Intern. At. Energy Agency No. 118 [1979] 213/8). — [27] H. Elderfield (Earth Planet. Sci. Letters **9** [1970] 10/6). — [28] D. E. Robertson (Geochim. Cosmochim. Acta **34** [1970] 553/67). — [29] A. Preston, J. W. R. Dutton (Tech. Rept. Ser. Intern. At. Energy Agency No. 118 [1970] 223/41). — [30] D. W. Spencer, P. G. Brewer (Geochim. Cosmochim. Acta **33** [1969] 325/39).

[31] J. F. Slowey, D. W. Hood (Geochim. Cosmochim. Acta **35** [1971] 121/38). — [32] M. Odier, V. Plichon (Anal. Chim. Acta **55** [1971] 299/20). — [33] R. Zirino, S. Yamamoto (Limnol. Oceanogr. **17** [1972] 661/71). — [34] T. B. Warner (Deep Sea Res. **18** [1971] 1255/63). — [35] R. S. Jones (U.S. Geol. Surv. Circ. No. 625 [1970]).

[36] N. N. Sharma (Ind. J. Mar. Sci. **1** [1972] 151/2). — [37] A. Baric, M. Branica (Limnol. Oceanogr. **14** [1969] 796/8). — [38] T. J. Chow, C. B. Snyder (Earth Planet. Sci. Letters **7** [1969] 221/3). — [39] A. D. Matthews, J. P. Riley (Nature **225** [1970] 1242). — [40] S. Tsunogai, T. Henmi (Nippon Kaiyo Gakkai-Shi **27** No. 2 [1971] 67/72; C. A. **76** [1972] No. 27859).

[41] S. Paslawska, S. Ostrowski (Gdansk. Tow. Nauk. Rozpr. Wydz. No. 7 [1970] 77/84; C. A. **75** [1971] No. 143868). — [42] R. N. Bernovskaya, Yu. A. Bogdanov, V. V. Gromov, A. P. Lisitsyn, V. I. Spitsyn, V. N. Tikhomirov (Radiokhimiya **13** [1971] 157/9; C. A. **74** [1971] No. 146168). — [43] P. C. Head (J. Marine Biol Assoc. U.K. **51** [1971] 891/903). — [44] J. H. Carpenter, M. E. Manella (J. Geophys. Res. **78** [1973] 3621/6). — [45] K. Matsumoto (Chikyu Kagaku **26** No. 2 [1972] 58/63; C. A. **77** [1972] No. 156179).

[46] T. M. Leatherland, J. D. Burton, M. J. McCartney, F. Culknin (Nature **232** [1971] 112). — [47] P. C. Head, J. D. Burton (J. Marine Biol. Assoc. U.K. **50** [1970] 439/48). — [48] M. F. Pilipchuk (Geokhimiya **1971** 248/52; C. A. **74** [1971] No. 91012). — [49] I. I. Volkov, E. G. Sokolova, A. A. Tikhomirova, M. F. Pilipchuk (Geokhimiya **1973** 395/403; C. A. **79** [1973] No. 9621). — [50] V. I. Spitsyn, R. N. Bernovskaya, Yu. A. Bogdanov, V. V. Gromov, V. N. Tikhomirov (Radiokhimiya **11** [1969] 607/9; C. A. **72** [1970] No. 70508).

[51] V. N. Tikhomirov, V. V. Gromov (Radiokhimiya **13** [1971] 316/8; C. A. **75** [1971] No. 25121). — [52] N. I. Buyanov (Okeanologiya **10** [1970] 718/21; C. A. **74** [1971] No. 15609). — [53] W. S. Moore (J. Geophys. Res. **74** [1969] 694/704). — [54] H. Kautsky (Atompraxis **16** [1970] 316/20). — [55] W. S. Broecker, A. Kaufman, T. L. Ku, Y. C. Chang, H. Graig (J. Geophys. Res. **75** [1970] 7680/3).

[56] Y. Nakaya, K. Nakamura (J. Radiation Res. **13** [1972] 2/3). — [57] P. Guegueniat (Bull. Inf. Sci. Tech. [Paris] No. 151 [1970] 27/32). — [58] D. R. Kester, R. M. Pytkowicz (Limnol. Oceanogr. **14** [1969] 686/92). — [59] V. I. Timoshchuk, L. G. Kulebaskina, N. R. Filippov (Radioekologicheskie Issledovaniya Sredizemnogo Morya, Kiev 1970, p. 150/5; C. A. **75** [1971] No. 101138). — [60] R. A. Hildreth, T. W. Henderson (Geochim. Cosmochim. Acta **35** [1971] 235/8).

[61] Y. Nagaya, K. Nakamura, M. Saiki (Nippon Kaiyo Gakkai-Shi **27** No. 1 [1971] 20/6; C. A. **76** [1972] No. 37275). — [62] V. M. Vdonenko, A. G. Kolesnikov, V. I. Spitsyn, R. N. Bernovskaya, L. I. Gedeonov, V. V. Gromov, L. M. Ivanova, B. A. Nelepo, V. N. Tikhomirrov, A. G. Trusov (At. Energ. [USSR] **31** [1971] 409/22; Soviet J. At. Energy **31** [1972] 1164/76). — [63]

V. B. Chumichev (Tr. Inst. Eksp. Meteorol. **1970** No. 1, p. 45/50; C.A. **77** [1972] No. 168394). —
[64] L. V. Eremeeva, V. N. Eremeev (Morsk. Gidrofiz. Issled. **1972** No. 1, p. 185/90; C.A. **79** [1973]
No. 34981). — [65] J. D. Clive, F. A. Richards (Environ. Sci. Technol. **3** [1969] 838/43).

[66] D. R. Kester, R. M. Pytkowicz (Geochim. Cosmochim. Acta **34** [1970] 1039/51). — [67]
D. J. Spedding (Atmos. Environment **6** [1972] 583/6). — [68] A. D. Matthews, J. P. Riley (Chem.
Geol. **6** [1970] 149/52). — [69] S. G. Bhat, S. Krishnaswamy, D. Lal, Rama, W. S. Moore (Earth
Planet. Sci. Letters **5** [1968/69] 483/91). — [70] A. Kaufman (Geochim. Cosmochim. Acta **33**
[1969] 717/24).

[71] Yu. V. Kuznetsov (Geokhimiya **1969** 177/84; C.A. **70** [1969] No. 80738). — [72]
Y. Miyake, Y. Sugimura, T. Yasujima (Nippon Kaiyo Gakkai-Shi **61** No. 26 [1970] 130/6; C.A. **74**
[1971] No. 67465). — [73] Y. Miyake, K. Sarubashi, Y. Sugimura (J. Radiation Res. **13** [1972]
10/1). — [74] D. Heye (Earth Planet. Sci. Letters **6** [1969] 112/6). — [75] Y. Miyake, Y. Sugimura,
M. Mayeda (Nippon Kaiyo Gakkai-Shi **26** No. 127 [1970] 123/9; C.A. **74** [1971] No. 91017).

[76] A. A. Strogonov, N. S. Risik (Radioekologicheskie Issledovaniya Sredizemnogo Morya,
Kiev 1970, p. 187/92; C.A. **75** [1971] No. 101135). — [77] R. Zirino, M. L. Healy (Limnol. Oceanogr.
15 [1970] 956/8). — [78] J. P. Riley, D. Taylor (Deep Sea Res. **19** [1972] 307/17). — [79] M. S.
Mangel (Mar. Geol. **11** [1971] M24/M26). — [80] P. T. Roskam (Chem. Weekblad **66** No. 36
[1970] 56/61).

1.4.2 Thermodynamic and physical properties of seawater

Activity and osmotic coefficients. An experimental method has been developed by
Lindsay and Liu for determining osmotic coefficients of aqueous solutions by vapor-pressure lowering
measurements at 75 to 300°C. Activity coefficients have been obtained for NaCl and many important
thermodynamic functions have been calculated for water and salt for the complete concentration
range [1]. Measurements on NaCl solutions were carried out, thereby completing the thermodynamics of the NaCl–H_2O system. Precise solubility data of sodium chloride were obtained. The
systems $MgCl_2$–H_2O and $MgSO_4$–H_2O were studied. Boiling point elevations were compared with
previous estimates [2]. Osmotic coefficients of aqueous sodium chloride solutions were obtained
for the temperature range 125 to 300°C [3].

The importance of activity as opposed to concentration in the electrodialysis and reverse osmosis
processes was shown by Butler using the theory of irreversible thermodynamics. Activity coefficients
were measured in the aqueous multicomponent systems NaCl–Na_2SO_4, NaCl–LiCl, NaCl–$CaCl_2$,
NaCl–$MgCl_2$ and NaCl–$BaCl_2$ using amalgam electrodes [4]. Further measurements of the activity
coefficient of NaCl in electrolytes containing $NaHCO_3$ and Na_2CO_3 were reported by Butler and
Huston, as well as studies of acid-base equilibria of carbonate systems in the presence of NaCl, and
a theoretical analysis of the above data in terms of an ion-pairing model was given [5]. The use of
amalgam electrodes to measure activity coefficients in multicomponent salt solutions was reported [6].

A generalized correlation was presented by Bromley for activity coefficient, osmotic coefficient,
enthalpy and heat capacity of single and multicomponent strong aqueous solutions. Values presented
for the common ions at 25°C allow the estimation of activity and osmotic coefficients of many
unmeasured salt solutions [7].

Further papers were presented by Gardner on osmotic coefficients of some aqueous sodium
chloride solutions at high temperature [8], by Wu et al. on osmotic and activity coefficients for
binary mixtures of sodium chloride at 25°C [9], by Duclaux on osmotic pressure and density [10],
by Momicchioli et al. on the determination of activity coefficients from freezing point depressions
for alkali chlorides [11], by Tamamushi and Goto on the determination from the measurement of
membrane concentration potentials [12], by Mussini and Pagella on the activity coefficients and
transference numbers of aqueous NaCl and $CaCl_2$ solutions at various temperatures and concentrations
[13], by Raridon and Kraus on the activity coefficients of sodium chloride at saturation in aqueous
solutions [14], by Lietzke and Herdklotz on activity coefficient behavior in aqueous binary salt
mixtures [15], by Reilly and Wood on the prediction of osmotic and activity coefficients in mixed
electrolyte solutions [16], by Meissner et al. on the activity coefficients of strong electrolytes in
aqueous and in multicomponent aqueous solutions [17], as well as on the effect of temperature [18],
by Dejak et al. on the theoretical calculation of the osmotic coefficient [19], by Minutilli et al. on the
optimization of osmotic conditions for small desalinators [20], by Kopecky and Dymes on osmotic

*Thermo-
dynamische
und
physikalische
Eigen-
schaften
des
Meerwassers*

coefficients of some electrolytes at 40°C [21], by Lindenbaum et al. on osmotic and activity coefficients for mixtures of lithium chloride with barium chloride and caesium chloride with barium chloride [22], by Robinson and Wood on the calculation of the osmotic and activity coefficients of seawater at 25°C [23].

An equation for the electrostatic terms of activity and osmotic coefficients and of the Gibbs function was presented by Glueckauf [130]. The gradient of dielectric constant near the surface of an ion immersed in water is shown to interact with the electrical field generated by another ion in such a way that a repulsive force exists between every pair of ions. Activity coefficient data agree with the model over a very large concentration range [131].

Density. Densities of water solutions of NaCl, KCl, Na_2SO_4, $MgSO_4$ and their ternary combinations were measured by Fabuss in the 25 to 175°C temperature range. The data were correlated using apparent molal volumes and the experimental and calculated data were presented in tables [24]. Further density measurements on binary and ternary aqueous solutions of salts from the major components of seawater were reported by Fabuss and Korosi [25]. Tabulation of data obtained by Fabuss and Korosi on the density of pure water and of binary and ternary solutions of the above salts is also presented [26].

Kremling has described a new laboratory apparatus for determining the density of seawater electronically [27], Isdale and Morris have measured the density of seawater and its concentrates in the temperature range 70 to 180°C and gave an equation which fits the new measurements with selected values from the literature to provide density values over a wide range of temperature and salinity [28], and Aleksandrov and Trakhtengerts presented standardized tables of water density values of high precision [29]. Densities of seawater and its concentrates were also reported by Grunberg [30] and by Akhundov et al. on softened Caspian Sea water [31].

Electrical conductivity. The change in the total electrical conductivity produced by the addition of a small amount of a solute to a concentrated salt solution was analyzed by Mysels and Mysels in terms of the effects of the accompanying changes in volume, ionic strength and viscosity upon the conductivity of the salt as well as the contribution of the additive itself [32]. The variation of the specific conductances of electrolyte solutions has been investigated by Blyumenfel'd et al. A mechanism was proposed for this phenomenon [33]. Further work is reported by Castagnolo et al. on the conductance of sodium chloride in sulfolane-water mixtures [34], by Sacco et al. of potassium chloride in the same mixture [35], by Kulkarni on the conductivity of mixed aqueous electrolytes [36], by Quint and Viallard on the electrical conductivity of electrolyte mixtures [37], by Mazo and Teterina on the specific electrical resistance of desalinated water as an indicator of the removal of organic and inorganic electrolytes [38], by Akol'zin and Kuznetsova on the effects of the hydrodynamics of a stream of electrolytes on their electrical conductivity [39].

Heat capacity and enthalpy. Data for heat capacity at one atmosphere constant pressure for seawater and its dilutions and concentrates were presented by Bromley as a function of temperature in the form of data point plots, least squares linear and quadratic curves and equations. A nomogram has been constructed from which heat capacities can be readily estimated to a high degree of accuracy [40]. Heat capacity and enthalpies of seawater solutions to 200°C and up to 12% salinity are reported in tabulated form [41]. Further measurements of the heat capacity were presented by Bromley for NaCl, KCl, $MgSO_4$, Na_2SO_4 and $MgCl_2$ solutions [42] and by Bromley et al. for heat capacities and enthalpies of sea salt solutions from 80 to 200°C [43]. A nomogram was given by Singh and Upadhye to estimate the enthalpy of seawater at any salinity and temperature [44], based on data reported by Bromley et al. [43]. Heat capacities of aqueous NaCl, KCl, $MgCl_2$, $MgSO_4$ and Na_2SO_4 solutions were also reported by Likke and Bromley in the temperature range 80 to 200°C [45], and by Singh and Bromley on relative enthalpies of sea salt solutions at 0 to 75°C [46].

The enthalpies of sodium chloride solutions have also been calculated and tabulated by Chou and Rowe. Included are the enthalpy changes required for the phase transition of water in the solutions from liquid to vapor phase [47].

Further work is reported by Jamieson et al. on the heat capacity of seawater and its concentrates in the temperature range 80 to 180°C [48], by Grunberg on heat capacity for seawater and its concentrates [30], by Novoselov and Mishchenko on heat capacity of aqueous and nonaqueous solutions of electrolytes [49].

Heats of mixing. Heats of mixing were determined by Bromley at 50°C for solutions of salts contained in seawater [41], as well as at 73°C [42]. Obtained values were reasonably consistent with enthalpy values previously published [43]. Heats of dilution of aqueous NaCl solutions were determined by Ensor and Anderson [50].

Molal volumes. A high precision densitometer was devised and used to determine apparent and/or partial molal volumes of aqueous electrolyte solutions at various temperatures. An apparatus has also been constructed for measurement of the isothermal compressibility of water from 2 to 55°C [51]. Partial molal volumes of ions in seawater were examined by a simple model for ion-solvent interactions. An equation was given which represents the volume of transfer of an ion from water to seawater [52]. Ion-ion interactions are strongly related to the effect of temperature on the structure of the hydrated ions or the structure of water between the interacting ions [53]. A theoretical equation for the compressibility of seawater as a function of chlorinity has been developed in terms of the apparent equivalent compressibility of the major ionic components of seawater [54].

Solubilities. The solubilities of various modifications of calcium sulfate were determined by Power and Fabuss from 25 to 250°C in water and from 25 to 95°C in 0.25 and 1.0 molal sodium chloride solution [55].

The solubilities of some silicate minerals in saline waters were determined by Collins. Solubility of Si from clay minerals decreased with increasing concentrations at ambient temperature and pressure. Solubility of a serpentine mineral at elevated temperatures and pressures was determined in specially designed hydrothermal equipment, which proved to be of good design [56]. The solubility of amorphous silica in seawater was also measured by Jones and Pytkowicz as a function of pressure [57]. Silicate-seawater equilibria in the ocean system were discussed by Perry [58].

Most important in desalination applications is the solubility of some gases. The effects of gas pressure, temperature and electrolyte concentration on the solubility of O_2, CO_2, CO, H_2, Ar und N_2O in water was presented by Schröder, as well as a survey of research conducted up to the middle of 1972. It proved possible to describe the temperature dependence by a simple equation and to relate the solubility minimum with increasing temperature to the critical gas temperature in question [59].

Solubilities of gases in seawater were reported by Murray and Riley for oxygen in distilled water and seawater [60], by von Kuhlmann presenting a nomogram to measure oxygen saturation concentration in fresh, brackish, and salt water [61], by Culberson and Pytkowicz reporting on variations in dissolved oxygen and in total CO_2 in the eastern Pacific Ocean [62], by Weiss reporting on the solubility of nitrogen, oxygen and argon in water and seawater [63], by Murray et al. on the solubility of nitrogen [64], by Murray and Riley on the solubility of argon [65], by Weiss on the effect of salinity on the solubility of argon [66] and on the solubility of helium and neon in water and seawater [67].

A large amount of work has been conducted on the solubility of carbon dioxide in seawater and on exchange characteristics between carbon dioxide and seawater. Berger and Libby have reported on the equilibration of atmospheric carbon dioxide with seawater [68], Stewart and Munjal on the solubility of carbon dioxide in distilled water, synthetic seawater and synthetic seawater concentrates, presenting also a correlating equation [69], Akiyama on the carbon dioxide content in the atmosphere and in the adjacent seas of Japan [70], Miyake and Sugimura on carbon dioxide content in the surface water and the atmosphere in the Pacific, the Indian and the Antarctic Ocean areas [71], Murray and Riley on the solubility in distilled water and seawater [72], Miller et al. on a method for determination of reaction rates of carbon dioxide with water and hydroxyl ion in seawater [73], Li and Tsui on the solubility in water and seawater [74], Kelley and Hanson on carbon dioxide in the surface of the eastern North Ocean and the Bering Sea [75], Maksimova on partial pressure of carbon dioxide in Atlantic Ocean waters [76], Imai on the desorption of carbon dioxide in seawater with a bubble column [77], Plass on the relation between atmospheric carbon dioxide amount and properties of the sea [78], Pesret on the kinetics of carbonate-seawater interactions [79], Lyman on development of ideas concerning the carbon dioxide system in seawater [80], Edmond on the thermodynamic description of the carbon dioxide system in seawater [81], and Bradshaw on the effect of carbon dioxide on the specific volume of seawater [82].

Solute properties of water. Molecular complexes of water in organic solvents and in the vapor phase have been investigated by Christian and Affsprung, using a variety of classical and spectral techniques. Thermodynamic constants for the formation of complexes between water and methanol and other polar molecules in different media have been reported and correlated with solvent effects on the individual donor and acceptor molecules. A method was presented for predicting the influence of the medium on energies and equilibrium constants for formation of complexes. Progress toward the development of a general theory of the effect of solvents on molecular complex equilibria were described [83].

Structure of water. Studies related directly or indirectly to the structure of water, as well as to thermodynamic properties of electrolyte solutions, not elsewhere discussed, are briefly reviewed in the following. Greyson has reported on the influence of alkali halides on the structure of water [84], Greyson and Snell on a specific influence of dissolved ions on the structure of water [85], Greene et al. on an experimental study of the structure, thermodynamics and kinetic behavior of water [86], Bockris et al. on a study of ionic solvation [87], Henderson in two reports on the theory of liquids [88] and Henderson and Barker in one more on the same subject [89], Barker and Henderson on the structure and properties of water and aqueous solutions [90], Safford et al. on a neutron scattering study of water and ionic solutions [91], on neutron inelastic scattering studies of water and ionic solutions and on a neutron scattering study of the kinetics of diffusion and the relation to the structure of ionic solutions [92], Borgonovi et al. on thermal neutron scattering measurements in water [93].

Further studies include work by Schufle et al. on the behavior of water in various states [94], by Burnham and Barnes on the properties of water and aqueous solutions at high pressures and temperatures [95], by Prausnitz on intermolecular forces in systems containing water [96], by Malinowski on a study of aqueous solutions by nuclear magnetic resonance [97], by Litovitz on light scattering and ultrasonic investigations of relaxation in aqueous solutions [98], by Mueller on intermolecular forces and the liquid state [99], by De Bethune on heats and energies of transport of ions in aqueous saline systems [100], by Berg on interfacial properties of water solutions [101], by Brummer and Gancy on an investigation of ion mobility in aqueous solutions [102], by Gill et al. on dispersion and miscible displacement [103], by Arnett on salt and nonelectrolyte interactions in water [104], by Luck on structures of aqueous solutions of electrolytes [105].

Thermal conductivity. Thermal conductivity measurements were made by Korosi and Fabuss on water, NaCl and seawater solutions in the temperature range of 25 to 150°C. The molality of NaCl solutions was 0.7 and 3.5. The experimental measurements were correlated as functions of temperature [106]. Further work is reported by Emerson and Jamieson on distilled water and Ca-free artificial seawater in the range 0 to 75°C [107], by Drost-Hansen and Korson in their studies on viscosity and conductivity [108], by Jamieson and Tudhope reporting on measurements of seawater and its concentrates in the temperature range 0 to 175°C [109], by Castelli and Stanley reporting on the thermal conductivity of pure water as a function of pressure and temperature [110], and by Grunberg on the thermal conductivity of seawater and its concentrates [30].

Vapor pressure. Vapor pressure measurements on NaCl solutions up to 2.5 molal concentration were presented by Fabuss in the 25 to 150°C temperature range [24]. Vapor pressures of binary aqueous solutions of NaCl, KCl, Na_2SO_4 and $MgSO_4$ and ternary NaCl-KCl, NaCl-Na_2SO_4 and NaCl-$MgSO_4$ solutions, at concentrations corresponding to seawater and its concentrates up to five-fold concentration, have been determined by Fabuss and Korosi. The additivity rule for vapor pressure depressions was found valid over the entire temperature range of 75 to 150°C and concentration range investigated [111]. Tabulation of data on the vapor pressure of water and of binary and ternary solutions from above salts were also presented [26]. Data were also reported by Emerson and Jamieson on seawater and its concentrates in the temperature range 100 to 180°C [107], by Grunberg on seawater and its concentrates [30], by Wexler and Greenspan giving a vapor pressure equation for water in the range 0 to 100°C [112], by Pepela and Dunlop reporting on precise vapor pressure measurements of water and sodium chloride solutions at 25°C [113], by Gibbard and Scatchard presenting data on vapor pressures of synthetic seawater solutions from 25 to 100°C [114], by Meissner and Kusik presenting a relationship for determining the vapor pressure of water over aqueous solutions containing more than one strong electrolyte [115], and by Akhundov et al. on softened Caspian Sea water [31]. Bromley measured boiling points of sea salt solutions between 80 and 120°C to determine boiling point elevations [116]. The measured values appear to be accurate to ±0.001°C [117].

Viscosity. Viscosity measurements on binary and ternary solutions of NaCl, KCl, Na_2SO_4 and $MgSO_4$ were made by Fabuss and Korosi over the temperature range 25 to 60°C [25]. The temperature range was then extended up to 150°C and a new correlation was developed for calculating the dynamic viscosity of liquid water, which represents with good accuracy the measured data and the most reliable literature data [118].

Viscosity measurements on ternary solutions of several electrolytes present in seawater, on synthetic seawater and for seawater up to a concentration factor of 3 were reported by Korosi and Fabuss. The experimental results were correlated using Othmer's rule, which establishes a linear relationship between the logarithm of viscosity of an aqueous solution and that of water at the same temperature [119]. Tabulation of data obtained by Fabuss and Korosi on the viscosity of pure water and of solutions containing a single salt or two of the above mentioned salts were also presented [26].

The viscosity of standard seawater for the pressure range of 0 to 1400 kg/cm^2 and temperature range from 0 to 30°C was measured by Stanley and Batten. The relative viscosity of seawater is greater than that of pure water at comparable temperature and pressure [120].

Further work was reported by Drost-Hansen and Korson in their studies on viscosity and conductivity [108], by Matthäus on seawater viscosity as a function of temperature, salt content and pressure [121], by Yusufova et al. on the dynamic viscosity coefficient for seawater of various salinities [122], by Isdale et al. on the viscosity of seawater solutions in the temperature range 20 to 180°C for salinities up to 150 g/kg [123], by Chen et al. on the kinematic viscosity of seawater solutions and concentrates up to 11% salt by weight and from 0 to 200°C [124], by Bromley on the viscosity of seawater from 0 to 150°C and salinity up to 11% [41] with values believed to be accurate to ± 0.002 cSt [42], and by Grunberg on the viscosity of seawater and its concentrates [30].

Other physical properties. Tabulated values of various thermodynamic properties of water were presented by Burnham et al. They include tables of specific volume, Gibbs free energy, enthalpy, fugacity and fugacity coefficients [125]. A theory applicable to weak solutions and strong electrolytes was used by Tsybaneva in deriving empirical equations for the Gibbs function and the ratio of the differential change in the chemical potential to the differential change in salinity of seawater [126].

Other physical properties were reported by Hobson and Williams on infrared spectral reflectance of seawater [127], by Mehu and Johannin-Gilles on the variation of the specific refractivity of Copenhagen seawater with wavelength, temperature and chlorine content [128], by Ho and Hall on the dielectric properties of seawater and sodium chloride solutions [129].

Literature to 1.4.2

[1] W. T. Lindsay, C. T. Liu (Off. Saline Water Res. Develop. Progr. Rept. No. 347 [1968]). — [2] C. T. Liu, W. T. Lindsay (Off. Saline Water Res. Develop. Progr. Rept. No. 722 [1971]; J. Solution Chem. 1 [1972] 45/69). — [3] C. T. Liu, W. T. Lindsay (J. Phys. Chem. 74 [1970] 341/6). — [4] J. N. Butler (Off. Saline Water Res. Develop. Progr. Rept. No. 388 [1968]); J. N. Butler, P. T. Hsu, J. C. Synnott (J. Phys. Chem. 71 [1967] 910/4); J. N. Butler, R. Huston, P. T. Hsu (J. Phys. Chem. 71 [1967] 3294/300); J. N. Butler, R. Huston (J. Phys. Chem. 71 [1967] 4479); J. C. Synnott, J. N. Butler (J. Phys. Chem. 72 [1968] 2474/7). — [5] J. N. Butler, R. Huston (Off. Saline Water Res. Develop. Progr. Rept. No. 486 [1969]).

[6] J. N. Butler, J. C. Synnott, R. Huston (Off. Saline Water Res. Develop. Progr. Rept. No. 606 [1970]); J. N. Butler, R. Huston (Anal. Chem. 42 [1970] 676/9). — [7] L. A. Bromley (AIChE [Am. Inst. Chem. Engrs.] J. 19 [1973] 313/28). — [8] E. R. Gardner (Trans. Faraday Soc. 65 [1969] 91/7). — [9] Y. C. Wu, R. M. Rush, G. Scatchard (J. Phys. Chem. 72 [1968] 4048/52, 73 [1969] 2047/53); G. Scatchard, R. M. Rush, J. S. Johnson (J. Phys. Chem. 74 [1970] 3786/96). — [10] J. Duclaux (J. Chim. Phys. 67 [1970] 1025/9; Compt. Rend. C 270 [1970] 1257/60).

[11] F. Momicchioli, O. Devoto, G. Grandi, C. Cocco (Ber. Bunsenges. Physik. Chem. 74 [1970] 59/66). — [12] R. Tamamushi, S. Goto (Bull. Chem. Soc. Japan 43 [1970] 3420/4). — [13] T. Mussini, A. Pagella (Chim. Ind. [Milan] 52 [1970] 1187/91). — [14] R. I. Raridon, K. A. Kraus (J. Chem. Eng. Data 16 [1971] 241/3). — [15] M. H. Lietzke, R. J. Herdklotz (J. Tenn. Acad. Sci. 46 [1971] 133/5).

[16] P. J. Reilly, R. H. Wood (J. Phys. Chem. 75 [1971] 1305/15). — [17] H. P. Meissner, J. W. Tester (Ind. Eng. Chem. Process Design Develop. 11 [1972] 128/33); H. P. Meissner, C. L. Kusik (AIChE [Am. Inst. Chem. Engrs.] J. 18 [1972] 294/8). — [18] H. P. Meissner, C. L. Kusik, J. W. Tester (AIChE [Am. Inst. Chem. Engrs.] J. 18 [1972] 661/2). — [19] C. Dejak, O. Devoto, I. Mazzei Lalatta, G. Cocco (Desalination 10 [1972] 263/72). — [20] F. Minutilli, C. Dejak, I. Mazzei Lalatta, A. Loi (Chim. Ind. [Milan] 54 [1972] 514/20).

[21] F. Kopecky, A. Dymes (Chem. Zvesti **26** [1972] 327/32; C.A. **77** [1972] No. 118 668). — [22] S. Lindenbaum, R. M. Rush, R. A. Robinson (J. Chem. Thermodyn. **4** [1972] 381/9). — [23] R. A. Robinson, R. H. Wood (J. Solution Chem. **1** [1972] 481/8). — [24] B. M. Fabuss (Off. Saline Water Res. Develop. Progr. Rept. No. 136 [1965]). — [25] B. M. Fabuss, A. Korosi (Off. Saline Water Res. Develop. Progr. Rept. No. 189 [1966]; Desalination **2** [1967] 271/8; J. Chem. Eng. Data **11** [1966] 325/31).

[26] B. M. Fabuss, A. Korosi (Off. Saline Water Res. Develop. Progr. Rept. No. 384 [1968]). — [27] K. Kremling (Nature **229** [1971] 109/10; Comments by Crease, Nature **233** [1971] 329). — [28] J. D. Isdale, R. Morris (Desalination **10** [1972] 329/39). — [29] A. A. Aleksandrov, M. S. Trakhtengerts (Teploenergetika **17** No. 11 [1970] 86/90; Therm. Eng. [USSR] **17** No. 11 [1970] 122/30). — [30] L. Grunberg (Proc. 3rd Intern. Symp. Fresh Water Sea, Dubrovnik 1970, Vol. 1, p. 31/9).

[31] T. S. Akhundov, K. M. Abcullaev, F. I. Mamedov (Izv. Akad. Nauk Azer.SSR Ser. Fiz. Tekhn. i Mat. Nauk **1971** No. 3, p. 112/4; C.A. **76** [1972] No. 103 588). — [32] E. K. Mysels, K. J. Mysels (J. Colloid Interface Sci. **38** [1972] 388/94). — [33] L. A. Blyumenfel'd, G. N. Zatsepina, S. V. Tul'skii (Zh. Fiz. Khim. **46** [1972] 2073/7; Russ. J. Phys. Chem. **46** [1972] 1180/3). — [34] M. Castagnolo, L. Jannelli, G. Petrella, A. Sacco (Z. Naturforsch. **26 a** [1971] 755/9). — [35] A. Sacco, G. Petrella, M. Castagnolo (Z. Naturforsch. **26 a** [1971] 1306/8).

[36] A. G. Kulkarni (J. Indian Chem. Soc. **48** [1971] 3401/2). — [37] J. Quint, A. Viallard (J. Chim. Phys. **69** [1972] 1095/9, 1100/4). — [38] A. A. Mazo, E. G. Teterina (Elektron. Tekhn. Nauchn. Tekhn. Sb. Tekhnol. Organ. Proizvod. **1971** No. 2, p. 25/8; C.A. **77** [1972] No. 66 074). — [39] P. A. Akol'zin, S. R. Kuznetsova (Teploenergetika **19** [1972] No. 12, p. 78/80). — [49] L. A. Bromley (Off. Saline Water Res. Develop. Progr. Rept. No. 227 [1966]).

[41] L. A. Bromley (Off. Saline Water Res. Develop. Progr. Rept. No. 522 [1970]). — [42] L. A. Bromley (Off. Saline Water Res. Develop. Progr. Rept. No. 747 [1972]). — [43] L. A. Bromley, A. E. Diamond, E. Salami, D. G. Wilkins (J. Chem. Eng. Data **15** [1970] 246/53). — [44] D. Singh, R. S. Upadhye (Chem. Eng. **80** No 7 [1973] 136). — [45] S. Likke, L. A. Bromley (J. Chem. Eng. Data **18** [1973] 189/95).

[46] D. Singh, L. A. Bromley (J. Chem. Eng. Data **18** [1973] 174/81). — [47] J. C. S. Chou, A. M. Rowe (Desalination **6** [1969] 105/15). — [48] D. T. Jamieson, J. S. Tudhope, R. Morris, G. Cartwright (Desalination **7** [1969] 23/30). — [49] N. P. Novoselov, K. P. Mishchenko (Zh. Fiz. Khim. **45** [1971] 1254/7; Russ. J. Phys. Chem. **45** [1971] 709/11). — [50] D. D. Ensor, H. L. Anderson (J. Chem. Eng. Data **18** [1973] 205/12).

[51] W. Drost-Hansen, F. J. Millero (Off. Saline Water Res. Develop. Progr. Rept. No. 350 [1968]). — [52] F. J. Millero (Limnol. Oceanogr. **14** [1969] 376/85). — [52] F. J. Millero (J. Phys. Chem. **74** [1970] 356/62). — [54] F. K. Lepple, F. J. Millero (Deep Sea Res. **18** [1971] 1233/54). — [55] W. H. Power, B. F. Fabuss (Off. Saline Water Res. Develop. Progr. Rept. No. 104 [1964]).

[56] A. G. Collins (Off. Saline Water Res. Develop. Progr. Rept. No. 472 [1969]). — [57] M. M. Jones, R. M. Pytkowicz (Bull. Soc. Roy. Sci. Liege **42** [1973] 118/20). — [58] E. A. Perry (Deep Sea Res. **18** [1971] 921/4). — [59] W. Schröder (Chem. Ingr.-Tech. **45** [1973] 603/8). — [60] C. N. Murray, J. P. Riley (Deep Sea Res. **16** [1969] 311/20).

[61] D. H. H. von Kuhlmann (Acta Hydrophysica **16** [1971] 27/36). — [62] C. Culberson, R. M. Pytkowicz (Nippon Kaiyo Gakkai-Shi **26** No. 2 [1970] 95/100; C.A. **74** [1971] No. 79 361). — [63] R. F. Weiss (Deep Sea Res. **17** [1970] 721/35). — [64] C. N. Murray, J. P. Riley, T. S. Wilson (Deep Sea Res. **16** [1969] 297/310). — [65] C. N. Murray, J. P. Riley (Deep Sea Res. **17** [1970] 203/9).

[66] R. F. Weiss (Deep Sea Res. **18** [1971] 225/30). — [67] R. F. Weiss (J. Chem. Eng. Data **16** [1971] 235/41). — [68] R. Berger, W. F. Libby (Science [2] **164** [1969] 1395/7). — [69] P. B. Stewart, P. K. Munjal (Univ. Calif. Seawater Convers. Lab. Rept. No. 69-2 [1969]; J. Chem. Eng. Data **15** [1970] 67/71, **16** [1971] 170/2). — [70] T. Akiyama (Oceanogr. Mag. **21** No. 1 [1969] 53/9, 121/7, 129/35; C.A. **72** [1970] No. 124 936).

[71] Y. Miyake, Y. Sugimura (Records Oceanogr. Works Japan [2] **10** [1959] 23/8). — [72] C. N. Murray, J. P. Riley (Deep Sea Res. **18** [1971] 533/41). — [73] R. F. Miller, D. C. Berkshire, J. J. Kelley, D. W. Hood (Environ. Sci. Technol. **5** [1971] 127/33). — [74] Y. H. Li, T. F. Tsui (J. Geophys. Res. **76** [1971] 4203/7). — [75] J. J. Kelley, A. Hanson (U.S. Clearinghouse Fed. Sci. Tech. Inform. AD 712 545 [1970]).

[76] M. P. Maksimova (Khim. Resur. Morei Okeanov **1970** 56/66; C. A. **75** [1971] No. 25 083). — [77] T. Imai (Ishikawajima-Harima Giho **11** [1971] 282/7; C. A. **76** [1972] No. 6576). — [78] G. N. Plass (Environ. Sci. Technol. **6** [1972] 736/40). — [79] F. Pesret (U.S. Natl. Tech. Inform. Serv. AD 747 979 [1972]). — [80] J. Lyman (Proc. Roy. Soc. Edinburgh B **72** [1972] 381/7).

[81] J. M. Edmond (Proc. Roy. Soc. Edinburgh B **72** [1972] 371/80). — [82] A. Bradshaw (Limnol. Oceanogr. **18** [1973] 95/105). — [83] S. D. Christian, H. E. Affsprung (Off. Saline Water Res. Develop. Progr. Rept. No. 301 [1968]); S. D. Christian (Off. Saline Water Res. Develop. Progr. Rept. No. 706 [1971]). — [84] J. Greyson (Desalination **3** [1967] 60/5). — [85] J. Greyson, H. Snell (Off. Saline Water Res. Develop. Progr. Rept. No. 523 [1970]).

[86] F. T. Greene, T. A. Milne, A. E. Vandergrift, J. Beachey (Off. Saline Water Res. Develop. Progr. Rept. No. 493 [1969]); F. T. Greene, J. Beachey, T. R. Milne (Off. Saline Water Res. Develop. Progr. Rept. No. 772 [1972], No. 789 [1972]). — [87] J. O'M. Bockris, P. P. S. Saluja, G. Madan (Off. Saline Water Res. Develop. Progr. Rept. No. 569 [1970]). — [88] D. Henderson (Off. Saline Water Res. Develop. Progr. Rept. No. 210 [1966], No. 270 [1967]). — [89] D. Henderson, J. A. Barker (Off. Saline Water Res. Develop. Progr. Rept. No. 537 [1970]). — [90] J. A. Barker, D. Henderson (Off. Saline Water Res. Develop. Progr. Rept. No. 773 [1972]).

[91] G. J. Safford, A. W. Naumann, P. C. Schaffer (Off. Saline Water Res. Develop. Progr. Rept. No. 280 [1967]); G. J. Safford, P. S. Leung (Off. Saline Water Res. Develop. Progr. Rept. No. 485 [1969]). — [92] G. J. Safford, P. S. Leung, P. C. Schaffer (Off. Saline Water Res. Develop. Progr. Rept. No. 372 [1968]); G. J. Safford, P. S. Leung, S. M. Sandborn (Off. Saline Water Res. Develop. Progr. Rept. No. 590 [1970], No. 708 [1971]). — [93] G. M. Borgonovi, J. U. Koppel, J. A. Young (Off. Saline Water Res. Develop. Progr. Rept. No. 449 [1969]). — [94] J. A. Schufle, M. Venugopalan, R. Muller, N. T. Yu (Off. Saline Water Res. Develop. Progr. Rept. No. 262 [1967]). — [95] C. W. Burnham, H. L. Barnes (Off. Saline Water Res. Develop. Progr. Rept. No. 383 [1968]).

[96] J. M. Prausnitz (Off. Saline Water Res. Develop. Progr. Rept. No. 306 [1968]). — [97] E. R. Malinowski (Off. Saline Water Res. Develop. Progr. Rept. No. 335 [1968]). — [98] T. A. Litovitz (Off. Saline Water Res. Develop. Progr. Rept. No. 402 [1968]). — [99] C. R. Mueller (Off. Saline Water Res. Develop. Progr. Rept. No. 405 [1969]). — [100] A. J. De Bethune (Off. Saline Water Res. Develop. Progr. Rept. No. 412 [1969]).

[101] J. C. Berg (Off. Saline Water Res. Develop. Progr. Rept. No. 724 [1971]). — [102] S. B. Brummer, A. B. Gancy (Off. Saline Water Res. Develop. Progr. Rept. No. 643 [1971]). — [103] W. N. Gill, T. S. Lin, M. Posner, R. Sankarasubramanian (Off. Saline Water Res. Develop. Progr. Rept. No. 822 [1972]). — [104] E. M. Arnett (Off. Saline Water Res. Develop. Progr. Rept. No. 836 [1972]). — [105] W. A. P. Luck (Proc. 4th Intern. Symp. Fresh Water Sea, Heidelberg 1973, Vol. 4, p. 531/8).

[106] A. Korosi, B. M. Fabuss (Off. Saline Water Res. Develop. Progr. Rept. No. 363 [1968]). — [107] W. H. Emerson, D. T. Jamieson (Desalination **3** [1967] 213/4). — [108] W. Drost-Hansen, L. Korson (Off. Saline Water Res. Develop. Progr. Rept. No. 349 [1968]). — [109] D. T. Jamieson, J. S. Tudhope (Desalination **8** [1970] 393/401). — [110] V. J. Castelli, E. M. Stanley (U.S. Natl. Tech. Inform. Serv. AD 751 025 [1972]).

[111] B. M. Fabuss, A. Korosi (Desalination **1** [1966] 139/48, 149/55). — [112] A. Wexler, L. Greenspan (J. Res. Natl. Bur. Std. A **75** [1971] 213/30). — [113] C. N. Pepela, P. J. Dunlop (J. Chem. Thermodyn. **4** [1972] 255/8). — [114] H. F. Gibbard, G. Scatchard (J. Chem. Eng. Data **17** [1972] 498/501). — [115] H. P. Meissner, C. L. Kusik (Ind. Eng. Chem. Process. Design Develop. **12** [1973] 205/8).

[116] L. A. Bromley (Off. Saline Water Res. Develop. Progr. Rept. No. 522 [1970]). — [117] L. A. Bromley (Off. Saline Water Res. Develop. Progr. Rept. No. 747 [1972]). — [118] B. M. Fabuss, A. Korosi (Off. Saline Water Res. Develop. Progr. Rept. No. 249 [1967]). — [119] A. Korosi, B. M. Fabuss (Off. Saline Water Res. Develop. Progr. Rept. No. 363 [1968]); B. M. Fabuss, R. Korosi, D. F. Othmer (J. Chem. Eng. Data **14** [1969] 192/7). — [120] E. M. Stanley, R. C. Batten (J. Geophys. Res. **74** [1969] 3415/20).

[121] W. Matthäus (Monatsber. Deut. Akad. Wiss., Berlin **12** [1970] 850/5). — [122] V. D. Yusufova, R. I. Pepinov, N. V. Lobkova (Izv. Akad. Nauk Azerb.SSR Ser. Fiz. Tekhn. i Mat. Nauk **1971** No. 3, p. 69/72; C. A. **76** [1972] No. 103 587). — [123] J. D. Isdale, C. M. Spence, J. S. Tudhope (Desalination **10** [1972] 319/28). — [124] S. F. Chen, R. C. Chan, S. M. Read, L. A. Bromley (Desalination **13** [1973] 37/51). — [125] C. W. Burnham, J. R. Holloway, N. F. Davis (Off. Saline Water Res. Develop. Progr. Rept. No. 414 [1969]).

[126] T. B. Tsybaneva (Izv. Akad. Nauk SSSR Fiz. Atmos. Okeana **7** [1971] 814/7; C.A. **76** [1972] No. 17 694). — [127] D. E. Hobson, D. Williams (Appl. Optics **10** [1971] 2372/3). — [128] A. Mehu, A. Johannin-Gilles (Deep Sea Res. **16** [1969] 605/11). — [129] W. Ho, W. F. Hall (J. Geophys. Res. **78** [1973] 6301/15). — [130] E. Glueckauf (Proc. Roy. Soc. [London] A **310** [1969] 449/62).

[131] L. W. Bahe (J. Phys. Chem. **76** [1972] 1062/71).

Meerwasser-
Analyse

1.4.3 Analytical chemistry of seawater

The complex nature of the dissolved substances in seawater requires to determine the concentration of any constituent by specially developed techniques. The standard methods for the quantitative analysis of water and electrolyte solutions can not be applied to seawater analysis without checking for their accuracy. This refers in first case to elements present in extremely low concentration in seawater, as impurities contained in the reagents used may excced several times the amounts of elements existing in seawater. MacIntyre [1] presented a "thalossochemical" periodic table, which shows the forms in which the detectable elements appear in seawater. Robertson reports on the role of contamination in the trace element analysis of seawater [2] and Carpenter on the problems in the application of analytical chemistry to oceanography [3]. Sigalove and Pearlman developed a system, which pumps continuously seawater to analytical instruments [4].

Many authors reported on analytical procedures for seawater or brine analysis. Chemical methods, principles and procedures for seawater analysis were presented by Grasshoff [5]. Strickland and Parsons give a guide for practical analysis of seawater [6]. Extended analytical procedures and methods modified to eliminate or compensate for interferences normally present in seawater and analysis of scale component elements are reported by Newton and Atkins [7]. They evaluated also the use of differential thermal and thermogravimetric analysis as a method for analyzing scale components of seawater desalination equipment [8].

Moore reviewed the principles and limitations of the more important techniques in water and brine analysis [9]. Spencer and Brewer prepared a detailed review on inorganic analytical methods used in oceanography [10]. Andersen and Foyen reviewed methods used in the determination of some anions, salinity, temperature and density [11]. Detailed analytical methods in marine chemistry are described by Martin [12]. Information on new developments in instruments, which perform continuous and automatic analysis of seawater was presented by Walden et al. [13]. Trachtenberg and Baker reported on ion selective electrochemical sensors for determining and monitoring the concentration of electroactive species in aqueous solutions [14] and Trachtenberg presented developments of novel ion-selective sensors with specific respone to major components (Fe, Cu, Ca, Mg, Na, K, SO_4) in saline and brackish water [15]. Clerc et al. reviewed ion specific sensors, their theory and applications [16]. A review on the properties of ion-selective electrodes and their applications to seawater was also given by Warner [17]. Papers on ion selective electrodes presented in a Symposium were summarized by Pungor [18].

A Fortran 4 computer program for the analysis of water chemistry measurements was presented by Wigley. The program is suitable only for relatively dilute solutions (ionic strengths below 0.1) [19].

Another computer program, WATEQ, calculates the equilibrium distribution of inorganic species of major and important minor elements in natural waters using chemical analysis and in situ measurements of temperature, pH and redox potential [20].

Physical tests. Physical tests of seawater include determination of pH, specific conductance, specific gravity and turbidity, and temperature measurement.

The most common method for determining the pH of seawater is with a pH-meter and glass-calomel electrode system. The pH of seawater is approximately 8.1 [21]. The photoelectrocolorimetric method for pH determination gives error less than 0.01 to 0.02 [22]. In situ accurrate determination of pH was described by Ben-Yaakov [23]. Industrial semiautometic pH-meter permits also in situ determination of pH [24].

The specific conductance of seawater is determined by measurement with a self-contained conductance instrument. The specific conductance of seawater is usually $4.8 \times 10^{-2}/\text{ohm} \cdot \text{cm}$ at 20°C (68°F) and a salinity of 35 g/kg seawater. **Fig. 1–1** correlates the specific conductivity of seawater as function of salinity [25].

For the approximate determination of the specific gravity of seawater, a hydrometer in the proper range is satisfactory. Using resonance frequency changes caused by a seawater sample, the density can be determined rapidly to an accuracy of $\pm 2.5 \times 10^{-6}$ g/cm^3 [26]. The specific gravity of seawater in usually 1.024 kg/l at 20°C (68°F). **Fig. 1–2** shows the density of water, of normal seawater and its concentrates as a function of temperature [25]. The salinity of normal seawater is 34.483 g/kg.

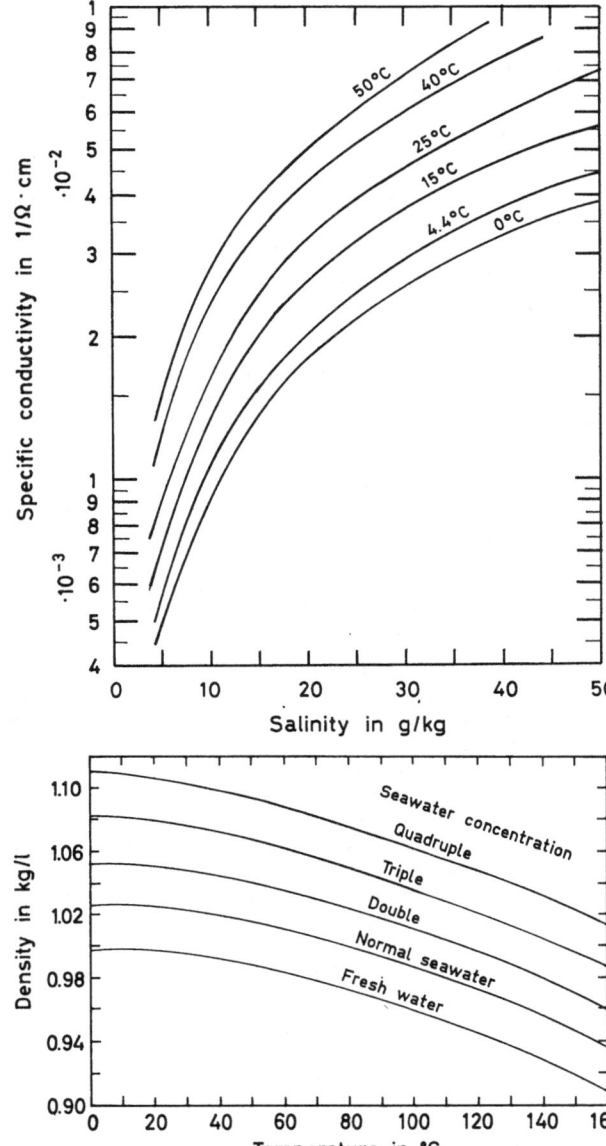

Fig. 1–1

Specific conductivity of seawater versus salinity.

Fig. 1–2

Density of seawater and its concentrates versus temperature.

Temperature measurements are made with a mercury filled thermometer, checked occasionally against a standard. The temperature is expressed in °C (or °F) and is reported to the nearest 0.1°. The turbidity may be determined by using a candle turbidimeter, photometer, nephelometer or similar instruments.

Chemical tests. Chemical tests include the analysis of all elements normally present in seawater, the determination of hardness and of salinity.

The salinity of seawater is defined as the total amount of solid material in g contained in one kg of seawater, when all the carbonate has been converted to oxide, the bromide and iodide replaced by chloride and all organic matter completely oxidized.

The chlorinity of seawater is defined as the total amount of chlorine ions in g/kg of seawater, after all bromide and iodide have been replaced by chloride.

The chlorosity of seawater is defined as the total amount of chlorine ions in gramms per liter of seawater at 20°C (68°F), after all the bromide and iodide have been replaced by chloride.

The concepts of salinity, chlorinity, chlorosity were discussed and the physical propertis of seawater were summarized by McIlhenny and Zeitoun [27].

Salinity and chlorinity are definitions that refer only to seawater, normal, diluted or concentrated and not to other saline waters. Salinity is one of the most commonly encountered determinations, important as an overall indicator of salt content in seawater and one of the necessary analytical procedures in the operation of certain types of desalination plants. It is arrived at by determining the chlorinity, converting to chlorosity using the following empirical equation [28]:

$$^o/_{oo} \text{ Salinity} = 1.8147 \times \text{chlorosity}$$

A new definition of salinity was included in the International Oceanographic Tables, published by UNESCO and the British Oceanographic Institute. A polynomial expression of salinity was computed [29]. Third order polynomial expressions of salinity, derived from measurements of the specific conductance by CTD sensors, were formulated and discussed by Knowles [30].

Chlorinity and chlorosity are determined by titration of the halide by silver or mercury(II) nitrate using any of the conventional methods of end point. Interferences in the first method are bromides and iodides, which are precipitated by silver nitrate and registered as equivalent chlorides; for the mercury(II) nitrate method sulfite, chromate, bromide, iodide and iron(III) ions, if present in sufficient quantities. Alternatively, salinity data are often obtained less directly by measurement of one or more associated parameters, such as electrical conductivity, density and index of refraction. The relationships of these parameters to actual salt content are not adequately established.

Methods for the determination of total salinity were reviewed by Andersen and Foyen [11] and total salinity and chlorinity by Spencer and Brewer [10]. Determination of salinity on the basis of relative electrical conductivity by prototype temperature compensated instruments gave an accuracy ± 0.02% [31]. Fedorov developed equations for computer calculations of salinities derived from in situ measured electrical conductivity of seawater [32]. Shaub and Denenok measured the salinity of seawater using a conductivity cell [33]. Proton activation analysis for the determination of halogens in seawater gives lower limits of detection, with accuracy ± 10%, in the range of 5×10^{-7} g [34]. A linear relation between Cl^- concentration and number of counts were found for the determination of chloride by neutron capture γ-ray [35]. Solid state electrode and a double junction type electrode were used by Ogata for measuring chloride concentration in seawater, brine and bitterns [36]. Rapid potentiometric determination using $AgNO_3$ solution gives standard deviation 0.33 to 0.34% [37]. Jägner and Åren used a semiautomatic determination of total halide concentration in seawater by potentiometric titration with $AgNO_3$ recorded directly on punched tape and tapewriter and evaluated by means of a computer program. Simultaneous titration of many samples is possible with a precision ± 0.02% [38]. D'Arrigo et al. used a transistorized system for automatic potentiometric titration of salinity and halogen content in seawater and brackish waters [39]. Meyer et al. discussed batch and continuous electroanalytical techniques and special instrumentation and cells, performed in flowing streams of seawater, for chloride analysis with an error ≤ 2% [49]. Grasshoff and Wenck determined chlorinity in natural seawater by titration with standardized silver nitrate solution using a motor driven buret and K_2CrO_4 as indicator [41]. Phenosafranine indicator gives a sharper end point than the chromate, for seawater samples containing more than 2% chlorides [42].

Alkalinity and carbon dioxide. Alkalinity is a measure of the hydrogen carbonate (bicarbonate), carbonate and hydroxide components of water. Chemical analysis of the carbonate system in seawater involves the determination of the total concentration of CO_3^{2-}, HCO_3^- and dissolved

CO_2. The alkalinity of seawater is determined by titration with a standard acid to a specified pH. For the analysis of the carbonate systems in seawater various methods have been investigated. A review of these methods is given by Park [43]. A semiautomatic apparatus built on the basis of an industrial pH-meter is suitable for the determination of carbon dioxide and total alkalinity [44]. Using EMF titration and Gran's graphical method for the evaluation of the results Dyrssen and Sillen determined total alkalinity and total carbonate concentration. They also suggested that dissociation constants of carbonic acid should be determined using the activity scale [45]. High precision titrations at sea give an accuracy of $\pm 0.17\%$ for alkalinity and $\pm 0.68\%$ for total carbon dioxide at the 95% confidence limit [46]. Carbonate/hydrogen carbonate ratios, total alkalinity and carbon dioxide was determined in seawater, treated by various ways for scale prevention, by an automatic technique [47]. Hansson has determined the concentration of CO_3^{2-}, H_2CO_3 and CO_2 by titration in synthetic seawater and by computation the dissociation constants [48]. The dissociation constants have been computed from experimental data by means of a computer for a salinity range of 20 to 40% and temperatures from 5 to 30°C [49]. The use of computer calculations was also applied to the potentiometric titration of the total alkalinity and carbonate content in seawater for the determination of the systematic errors associated with Gran plots [50]. In a differential potentiometric method for the determination of the alkaline reserve of seawater two polarized zirconium electrodes were used [51]. Simpson and Broecker determined total CO_3^{2-} content of saline and seawater by a method based on partial pressure measurements of CO_2. Good agreement has been observed to methods using dissociation constants of carbonic and boric acid [52]. Total carbon dioxide was determined by Wong with an accuracy $\pm 9.2\%$ by vacuum extraction followed by infrared spectrometric analysis [53].

Alkalinity is determined by titration with standard acid solution. The determination is free of interferences and the accuracy is ± 3 mg/l. Carbon dioxide is determined by liberation of the CO_2 gas and absorption in barium hydroxide, which is titrated by standard acid solution. There are no interferences and the method is reproducible to 0.2 mg/l seawater.

Hardness, calcium and magnesium. Seawater of 35°/$_{oo}$ salinity is approximately 0.05 M in magnesium, 0.01 M in calcium and 0.0001 M in strontium. Calcium and magnesium are usually determined as a sum by titration with EDTA or EGTA or by any other method, and separately calcium, the content of magnesium being calculated from the difference. Strontium is determined by one of the methods reported in p. 30. Precision end point titration is the aim of several studies.

Dyrssen et al. determined calcium in seawater by titration with EGTA using calcium ion specific membrane electrode and Gran extrapolation for the evaluation of the equivalence point. Deviations are as high as 1 to 2% [54]. Visual indicator methods for the determination of calcium, using empirical corrections for magnesium interference, have 0.1% accuracy [55]. A quick photometric method for titration of calcium with EGTA, in the presence of large amounts of magnesium, uses calcon as indicator. The deviation is about 2.0% [56]. Blake et al. have determined calcium by flame-photometer in solutions of high sodium content after separation of calcium in simulated seawater by means of a iminodiacetic acid resin in the sodium form. For 10 mg/l calcium there is an error of ± 0.2 mg/l [57]. Jägner used a computerized photometric titration method for the determination of calcium in the presence of magnesium with zinc-zincon as indicator [58]. Jägner also used a potentiometric method for the determination of magnesium as hydroxide. An accuracy of about 0.2% is obtained [59]. By titration with EDTA, Shah and Gagate determined magnesium in natural seawater as the difference of total alkaline earth and the sum of calcium and strontium content [60]. Direct gravimetric determination of magnesium using N-benzoyl-N-phenylhydroxylamine has a variation 0.55% for 5 mg of magnesium [61]. Riley and Skerrow reviewed methods for the determination of calcium and magnesium, including various EDTA titrations [62]. Culkin and Cox also used EGTA titrations for the determination of calcium and magnesium [63]. EDTA titrations with visual end points give an overall accuracy within $\pm 0.5\%$ for calcium and $\pm 0.3\%$ for magnesium [64]. Liquid ion-exchange electrodes have been used by Whitfield and Leyendekkers as end point detectors for the complexometric titration of calcium and magnesium [65]. They also titrated calcium using EGTA and the sum of calcium and magnesium by DCTA. Accuracy of 0.5% for both calcium and magnesium is reported. Errors due to strontium are small [66]. Equilibrium composition of the liquid ion-exchangers are reported by the same authors [67]. EDTA titration of calcium at pH 13 and magnesium at pH 10, using hydroxy naphthol blue indicator, is successful in ratios Ca:Mg 1:1 to 10:1 [68]. Titrations of total alkaline earth with EDTA using Eriochrome black T as indicator gives a precision in the range of 0.01% [69]. Huber et al. used a titration method based on the anodic response to EDTA and EGTA of the lead dioxide electrode [70]. A spectrophotometric method uses EGTA for calcium and EDTA or DCTA for magnesium and phthalein complexon as indicator with an error 0.5% in concentrations of 0.02 to 0.33 mg/l for magnesium and 0.07 to 0.71 mg/l for calcium [71]. For total alkaline earth determination Romanov and Eremeeva used spectral analysis with arc discharge in an aerosol atmosphere [72]. Viswanathan et al. used atomic absorption and fluorometric analysis for the determination of the calcium and magnesium in seawater [73]. Kurihara et al. determined calcium and magnesium and other elements in brines, using atomic absorption spectrophotometry at a limiting salt concentration of 3% [74]. Berge and Bruegmann determined calcium and magnesium by polarography [75].

Several types of voltammetric sensors have been used for the continuous sensing of complexones in a flowing stream. The determination of the calcium and magnesium has been based on the automation of two manual polarographic procedures [76].

Hardness or calcium and magnesium are titrated, in a diluted sample of seawater, with EDTA using a convenient indicator. Interferences present in seawater (Cu, Fe, Mn, Zn, Pb, Al and Sn) are removed by addition of thioacetamide solution. Results are reproducible to 20 mg/l.

Alkali metals. Sodium is a major element in seawater and as sodium chloride the main component. The sodium chloride concentration in seawater and its concentrates was determined by Vaidya by means of refractometer readings [77]. Maksyuta et al. used an ultrasonic concentration meter with an accuracy ± 3 g/l [78]. Sodium, potassium and lithium were determined by Anosova by flame photometry with error less than 1%. The method is applicable to high saline waters and brines [79]. A rapid determination of sodium is discussed by Kim and Kim, using a sodium-sensitive glass electrode [80]. A potentiometric method by measuring the activity coefficient is described by Afanas'eva and Oradovskii [81]. Kulev and Bakalov used an indirect polarographic determination [82]. Romanov and Eremeeva studied the accuracy of spectral methods for the determination of potassium in seawater and the variation coefficients are given for various experimental conditions [83]. Indirect polarographic determination in water with high salinity was used by Marczak and Ziaja for the determination of potassium [84]. For high precise determination of potassium, Anfält and Jägner developed a potassium selective valinomycin electrode [85]. Lithium in natural brines was determined by a fluorescence method by Markman and Strel'tsova [86]. Extraction of lithium with isoamyl alcohol and then determination by atomic absorption spectrophotometry is reported by Shigematsu et al. [87]. A flame spectrophotometric method used by Morozov gives a maximum sensitivity of 0.005 γ/ml for lithium determination and of 0.008 γ/ml for rubidium [88].

Alkaline earth metals. Strontium is usually determined simultaneously with magnesium and calcium in chelatometric titration of hardness as outlined in p. 29. A survey on analytical methods of stable strontium and ^{90}Sr is given by Bowen, also comparing various procedures and recommending the preferred procedures [89]. Gordon and Larson determined strontium in seawater by photon activation analysis [90]. Carr studied the salinity effect on the determination of strontium by atomic absorption spectrophotometry [91]. A radiometric device, with counting efficiency 50%, for the determination of ^{90}Sr is described by Vakulovskii and Silant'ev [92]. Bondarenko et al. determined strontium-90 after isolation, by ion exchange, of yttrium-90 [93]. Silant'ev et al. developed a method for determining the ^{90}Sr concentration in seawater according to the ^{90}Y daughter isotope [94]. A radiochemical method uses the magnesium and calcium of seawater as natural collectors for the determination of ^{90}Sr, ^{144}Ce and Cs [95]. Major et al. developed a method for shipboard analysis of ^{90}Sr using direct extraction of ^{90}Y [96]. Nagasawa and Miyahara describe a simple method for the determination of ^{90}Sr by scavenging other ions from the solution by means of coprecipitation [97]. Petrov presented a method for the simultaneous determination of Ca and Sr as well as for ^{90}Sr [98]. A spectral method for determining yttrium carriers during the radiochemical analysis of ^{90}Sr is described by Romanov and Eremeeva [99].

Aluminum. Trace amounts of aluminum in seawater were determined by Nishikawa et al. with a fluorometric method [100]. Shigematsu et al. used also fluorometric determination of trace amounts of aluminum by the lumogalion method [101]. The potential interferences on the determination of aluminum by atomic absorption spectrometry are given by Spencer and Sachs [102]. Gas liquid partition chromatography has been demonstrated by Lee and Burrel [103].

Arsenic and Antimony. Neutron activation analysis has been used by Ryabinin et al. for the determination of arsenic and antimony, after their concentration with ferric hydroxide [104]. Gohda used also activation analysis for the determination of arsenic, copper and antimony [105]. Direct determination of antimony and bismuth by anodic stripping voltammetry is reported by Gilbert and Hume [106].

B o r o n determination is made by spectrographic fluorimetric [197] and spectrophotometric [198] methods using various dyestuffs. Tube-spray emission spectroscopy gives good reproducibility for the boron content in the concentration range 4 to 5 mg/l B [199]. Nickolson described a spectro-photometric determination by complexing boron with Nile Blue A and extracting the complex with o-dichlorobenzene [209]. Basselt and Matthews described also a spectrophotometric method using ferroin and chloroform for the extraction of the complex [201].

Belyaev and Ovsyanyi determined boric acid as the mannitol complex by titration, after separation of the interfering ions [202]. A rapid potentiometric method suitable for application in laboratory and on board-ship was described by Sheremet'eva. For concentrations of 10^{-3} to $10^{-4}\%$ the observed error was $< 2\%$ [203].

Uppström used a manual method with curcumin [204] and the same method was adapted for automatic analysis. In the concentration range 0.1 to 6.0 mg/l B a standard deviation of 1.5% was obtained [205]. Column reversed-phase partition chromatography was used by Testa and Staccioli for isolation and determination of H_3BO_3 [206]. The acidity constant of boric acid for various salinities was determined by Hansson in synthetic seawater [207].

B r o m i d e a n d i o d i d e. Bromide and iodide ions are determined in seawater usually by titration with standard sodium thiosulfate solution. Total bromine in brines was determined by Dunton using X-ray fluorescence [107]. Wilkniss determined halides in a single seawater sample using proton activation analysis [108]. A rapid spectrophotometric method for the determination of bromine is reported by Saenger [109]. Berge and Bruegmann determined traces of bromide, in the presence of large amount of chlorides, using two polarographic techniques [110]. Photometric determination of bromides in concentration range 0.02 to 8.0 mg/l was developed by Morgen et al. [111]. Revel used an autoanalyser method based on the reduction of Ce^{4+} for the determination of total mineral iodide [112]. An improved procedure of Sugawata's method [113] for the determination of iodine minimizes the sources of errors of this method [114]. Tsunogai improved further and simplified Sugawata's method [115]. Total iodine and iodate were determined spectrophotometrically by Schnepfe [116]. Low concentrations of iodide in waters with salinities less than 8000 mg/l were determined polarographically by Khelashvili [117].

Iodates are always present in seawater with iodide ions and often in concentrations much higher than that of iodide. Iodates are easily reduced to iodide by biological activity [118]. It is therefore, necessary to determine or separate iodide from iodate immediately after seawater samples are collected.

Interferences in the standard determination methods are iron, manganese and organic matter, which are removed by addition of calcium oxide.

C a d m i u m. Cadmium can be determined in concentrations 0.05 to 10 ppb*) by atomic absorption spectroscopy, after concentration by chelating techniques [119]. Shiraishi et al. determined also cadmium and lead by atomic absorption spectroscopy [120]. Robinson et al. used atomic absorption [121]. An ion-exchange enrichment method for the determination of ppb level cadmium by atomic absorption spectrometry is reported by Itiiro et al. [122].

C o b a l t. A procedure was described by Kentner and Zeitlin for the determination of cobalt in seawater by solvent extraction followed by spectrophotometric determination of the chelate [123]. Motomizu determined cobalt spectrophotometrically after extraction and without preconcentration [124]. Minimal sample preparation is reported by Harvey and Dutton for analysing stable cobalt by atomic absorption spectrometry in a fuel-rich air-acetylene flame [125].

C o p p e r. Copper, a trace element in seawater, is determined mainly colorimetrically. Abraham et al. determined copper spectrophotometrically in the concentration range 2.4 to 3.1 ± 0.1 ppb*) [126]. A new spectrophotometric method, suitable for series analysis of seawater samples, was proposed by Balabanoff et al. [127]. To determine the chemical forms of copper in seawater as well as the total copper concentration, Odier and Plichon utilized the a. c. polarographic method [128]. Marvin and Proctor discussed the effects of filtration on the accuracy of the determination of copper

---

*) In this connection ppb means mg/m^3 solution.

[129]. Kamp recommends the use of ion selective electrodes or polarography for the determination of copper ions [130]. Spectrophotometric determination of copper in brines is presented by Valcarcel and Pino [131]. Dolmanova et al. reported on optimum conditions for the determination of copper by catalytic oxidation [132]. Flameless atomic absorption was used by Fairles and Bard [133]. Muzzarelli and Rocchetti utilized atomic absorption spectrometry with a graphite atomizer after elution from chitosan [134].

As little as 0.1 ppb*) Cu can be determined by using electron spin resonance [135]. Goncharov used spectrochemical determination of silver and copper by adsorption [136]. A rapid method was developed by Kuwata et al. concentrating cadmium and copper by chelation followed by atomic absorption spectroscopy [137]. Gohda determined copper, simultaneously with arsenic and antimony, by using activation analysis [138].

Anodic stripping a.c. voltammetry was used by Rojan for the determination of copper, as well as for lead, cadmium and zinc [139]. Trachtenberg investigated an ion-selective electrochemical sensor with response for 0.1 ppm total copper in seawater [140]. Solvent extraction-atomic absorption spectrophotometry was claimed to be satisfactory for ppb*) determination of Pb, Cr and copper in seawater [141]. Kerfoot and Vaccaro adopted adsorptive extraction by using carbon powder followed by determination of copper by atomic absorption [142]. APDC-MIBK extraction system for the determination of copper and iron by flameless atomic absorption spectrometry was used by Kremling and Petersen [143]. As little as 0.1 ppb*) copper can be determined by extraction of copper and measuring the electron spin resonance signal intensity of the organic phase containing the extracted copper [144]. Total soluble copper content was measured by Jasinski et al. with ion selective electrodes, calibrated to response directly in seawater [145].

Fluorine. Walker and Smith discussed several types of continuous fluoride analysers of either coulometric, potentiometric or amperometric principles [146]. The use of photon activation analysis is reported by Wilkniss and Linnenboom for the determination of fluorine in seawater [147], and by Wilkniss for the one sample determination of halogens [148]. Warner describes a quick method with lanthanum fluoride electrode for the determination of fluoride [149]. Kletsch and Richards report on a modified spectrophotometric method adapted for the determination of fluoride in seawater [150]. Anosova describes a colorimetric method for high saline-waters [151]. Anfaelt and Jägner determined total fluoride concentration with a lanthanum fluoride electrode. Equations for computer treatmant of titration data are given [152].

Iron. Alcoholic bathophenanthrolin solution was used by McMahon for the spectrophotometric determination of iron in seawater [153]. Total soluble iron was determined by Burton and Head using thioglycolic acid [154]. Kawamura et al. used inorganic ion-exchange chromatography in combination with a Ge[Li] detector for a rapid and simple determination of ^{59}Fe, as well as for ^{60}Co, ^{137}Cs and ^{95}Zr [155]. Solvent extraction-atomic absorption spectrometry was found to be suitable for the determination of ppb*) iron in seawater [156]. Lee and Burrell extracted iron and cobalt, indium and zinc from seawater by means of the trifluoroacetylacetone-toluene system [157]. Iron was determined by Dennen simultaneously with other cations by a double beam adsorption spectrograph developed and tested by the author [158].

Lead. UV spectrophotometry was used by Skurnik-Sarig et al. to determine traces of lead in seawater [159]. Shiraishi et al. determined lead and cadmium in seawater by atomic absorption [160]. Anodic stripping analysis permits determination of 0.1 ppb *) Pb [161].

Mercury. Radioactivation determination of mercury in seawater after radiochemical purification was reported by Weiss and Crozier [162]. Spectrophotometric atomic absorption after concentration was used by Topping and Pirie for the determination of inorganic mercury [163]. Carr et al. developed and tested a cold-vapor atomic absorption technique [164]. Determination by an improved dithizone colorimetric method is applicable for mercury content in the range of 1.4 to 26.0×10^{-3} mg/l [165]. A flameless atomic absorption procedure is reported by Olafsson, coupled to amalgamation on gold for the determination of nanogram quantities [166], and by Voyce and Zeitlin in combination with adsorption colloid flotation for the separation of mercury from other components [167].

Molybdenum. Kim and Zeitlin studied the role of iron(III) hydroxide [168] and thorium hydroxide [169] as collectors of molybdenum from seawater before its spectrophotometric determination. Kawabuchi and Kuroda used ion-exchange and a spectrophotometric method for the deter-

---

*) In this connection ppb means mg/m^3 solution.

mination of Mo and W in seawater [170]. A rapid adsorbing colloid flotation technique was reported by Kim and Zeitlin followed by spectrophotometric analysis [171]. Atomic absorption spectrometry after concentration was also reported by Muzzarelli and Rocchetti [172]. A new analytical method based on the catalytic activity of molybdenum was described by Kuroda and Tarui [173].

Nickel. The spectrophotometric determination of nickel with pyridine-2-aldehyde-2-quinolyl-hydrazone is reported by Afghan and Ryan [174]. Kentner et al. described a dimethylglyoxime method for the determination of nickel in seawater [175]. Rampon and Cuvelier introduced a modification of the extraction method of Kentner for the determination of nickel by atomic absorption spectro-photometry [176].

Nitrate, nitrite, and nitrogen compounds. Wagner compiled a review on methods for the determination of various forms of nitrogen in seawater [177]. A procedure for determining nitrates is described by Matsunaga and Nishimura [178]. Mertens and Massart reported on ultraviolet spectrophotometry for nitrate determination [179]. The optimum conditions for the spectrophoto-metric determination of nitrate and nitrite depend on pH values [180]. Matsunaga et al. used for seawater [181] a modified method investigated for fresh water [182]. Wada and Hattori described a spectrophotometric method of determination after concentration of nitrate [183]. For ammonia in seawater Grasshoff used a sensitive method of automatic or manual operation [184] and for ammonia nitrogen Manabe used indophenol for direct microanalysis [185]. Truesdale [186] used a modified method by Strickland and Parsons [187] for the determination of ammonia and amino acids. Ammonium concentrations in the range of 1 to 20 μg/ml were detected by Ellender et al. [188]. Matsunaga and Nishimura investigated a high sensitivity method using extraction by indophenol [189]. MacIsaac and Olund [190] and Head [191] adapted an automated extraction procedure for the determination of ammonium nitrogen. A novel method for the automatic determination of ammonia is given by Grasshoff and Johannson [192], and for ammonia and aminoacid nitrogen by Cioce et al. [193]. Srna et al. report on the use of ion specific electrodes for chemical monitoring of marine systems [194], and Gilbert and Clay on the use of an ammonia specific electrode [195]. Degobbis presented a method for storage of seawater samples for ammonia determination [196].

Organics. Analytical methods for monitoring the propane content of the potable water and effluent streams from the Koppers Hydrate Process were reported by Phillips [208]. Instrumental means for organic trace analysis applicable to the quality control of product water from desalination plants were investigated by Lysyj et al. [209]. A pyrographic technique for the analysis and identi-fication of organic matter and pollutants in aqueous media is reported by Lysyj et al. They also reported on analytical methods of total organic content of potable water resulting from desalination plants [210].

Oxygen and dissolved gases. The determination of the dissolved oxygen in seawater is of major importance, because the corrosion of desalination equipment depends strongly upon its content in the circulating streams. Precision methods, chemical and instrumental, have been de-veloped for this reason. A review on the methods for the determination of dissolved oxygen and other gases was given by Andersen and Foyen [11]. Comparative study of an electrochemical method and the Winkler method was made by Boutin et al. [211]. Apparatus and method for the automatic measurement of the concentration of dissolved oxygen up to 12 ml/l were developed by Khramov et al. [212]. They designed also diffusion sensors for the measurement of oxygen [213]. Dissolved oxygen in seawater concentrates was determined by Talreja et al. by a derivative polarographic method [214]. An assembly of an improved ppb*) analyser for seawater was evaluated by Behrens et al. [215]. Graefe used an oxygen sensor and a telemetering and digitizing system for the measurement of dissolved oxygen in great oceanic depths [216]. Kolesnikov and Khramov described also an apparatus for measuring the oxygen content in depths up to 70 m [217]. For the measurement in saline water Walker et al. used Miller's method [218]. Palermo investigated an amperometric method [219]. Mor and Beccaria described a modified Winkler method for seawater samples containing sulfides [220].

For measuring oxygen content in the range of 0 to 100 ppb*), Eddington and Roberts developed an oxygen calibrator [221]. A photometric modification of the Winkler method for low concentrations of oxygen was given by Smirnov [222]. Eddington described high, medium and low accuracy and precision procedures [223]. He also optimized Winklers method [224]. An electrochemical method for monitoring oxygen content in aqueous streams was described by Meyer et al. [225]. Reusmann [226] and Toknev [227] used gas chromatography for the determination of dissolved gases in seawater,

---

*) In this connection ppb means mg/m³ solution.

and Hahn used for nitrous oxide an improved gas chromatographic method [228]. Atkinson determined dissolved nitrogen in seawater by on-stream stripping gas-chromatography [229]. Finally, low concentration of dissolved hydrogen sulfide was determined by a photocolorimetric method by Smirnov [230]. Swinnerton et al. describe a sensitive gas-chromatographic method for the determination of carbon monoxide [231]. An apparatus is described by Baker for on-stream analysis of free chlorine, which combines features of amperometric and potentiometric analyzers. The analyzer can be applied to seawater and highly contaminated brackish and fresh water [232].

Interferences in the modified Winkler method are iron (II) and iron (III). In seawater their effects are negligible but in deareated seawater their effect becomes evident and a special technique for the titration is required. Precision and accuracy varies with interferences and technique. The polarographic method is free of interferences except if the seawater sample was treated with chlorine.

Phosphates. Small amounts, about 0.1 $\mu g/l$, of phosphates in seawater were determined by Isaeva by means of extraction [233]. Inorganic phosphate and total P in seawater were determined by Pakalns and MacAllister by an isobutyl acetate extraction procedure [234]. Suranova et al. carried out experiments on concentration of phosphates in chloride-form anion exchangers [235]. Cescon and Scarazzato determined also, by an isobutyl acetate extraction procedure, low phosphate concentration in seawater [236]. An improved method, by complexing with molybdenum, was used by Hosokawa and Ohshima [237]. A solvent extraction method and then determination in a low-background β-counter were used by Flynn and Meehan for ^{32}P content in seawater [238].

Radioactivity and nuclides. Methods for measuring the radioisotope content and application of these measurements for the study of biogeochemical processes in fresh and saline waters were reported by Perkins and Rancitelli [239]. An ideal system of quality control in radiochemical analysis was described by Harley and Volchok and data relating to the analysis of seawater were presented [240]. A collection of papers on methods and techniques for the determination of radioactive elements was published by the International Atomic Energy Agency [241]. Chesselet reviews therein the determination techniques of stable and radioactive $^{103}Ru-^{106}Ru$, ^{54}Mn, ^{51}Cr, ^{55}Fe, $^{95}Zr-^{95}Nb$ in marine environment [242], Hoegdahl an analytical procedure for the determination of radioactive and stable caesium [243], Preston and Dutton analytical methods for selected cobalt nuclides and for the separation and subsequent determination of stable cobalt [244]. Vinogradov and Vinogradova presented a theory of γ-activity measurement in seawater using an immersion detector [245]. Pelati and Mezzadri determined the total β-radioactivity by using radioactive tracers [246]. The use of a NaI (Tl) γ-counter for monitoring surface radioactivity in seawater was discussed by Kautsky [247]. Means for optimizing the design of a plastic scintillator for counting low-β-activities in ocean waters are reported by Kas'yanov et al. [248].

Inorganic ion-exchange chromatography was combined with a Ge (Li) detector for the determination of ^{59}Fe, ^{60}Co, ^{65}Zn, ^{137}Cs and ^{95}Zr [155]. A rapid method for collection and spectrometric determination with a Ge (Li) detector of radioactive nuclides in seawater was studied by Miyake and Mishijima. Collection efficiencies for ^{144}Ce, ^{103}Ru, ^{95}Zr, ^{95}Nb, ^{65}Zn and ^{60}Co with activated C were measured [249].

Determination of radioactive long-lived radionuclides in seawater released from nuclear installations is reported by Koda et al. [250]. Doshi et al. presented a sequential scheme to study the distribution of ^{144}Ce, ^{106}Ru, ^{95}Zr, ^{137}Cs and ^{90}Sr [251], and reported on the adsorption behavior of 14 radionuclides (Ce, Zr, Ru, Cs, Sr, Hf, Ag, Se, Zn, Co, As, Sb, Sn, and U) on MnO_2 for in situ activation analysis [252].

Ikeda et al. described a procedure for the radiochemical determination of $^{95}Zr-^{95}Nb$ [253]. ^{65}Zn and ^{54}Mn were separated and counted by Shah and Rao with a 512-channel analyzer with NaI (Tl) crystal [254]. ^{210}Po and ^{210}Pb in seawater were determined by Shannon et al. by a rapid method, using solvent extraction followed by α-counting or a spectrophotometric method [255] or by electrodeposition and an α-counting technique [256]. Nozaki and Tsunogai used a coprecipitation method for the separation of ^{210}Po and ^{210}Pb, which were determined simultaneously [257].

Shigematsu et al. reported on the fluorimetric determination of cerium in seawater [258], and the coprecipitation, extraction and final separation for the determination of ^{144}Ce was described by Yamato and Miyahara [259]. A two step simple method for the separation and determination of ^{137}Cs was described by Dutton [260]. Bauman presented an ammonium molybdate procedure for the separation of ^{137}Cs, which was determined with a multichannel analyzer using a NaI (Tl) crystal [261]. Folsom et al. presented a method for the determination of natural caesium in seawater [262], and Folsom and Sreekumaran discussed the above method [263]. Finally, Hodge et al. described methods for the determination of ^{137}Cs and [134]Cs and compared the results with methods used in 56 other institutions [264]. An isotopic exchange method for controlling ^{137}Cs concentrations in seawater by in situ measurements was reported by Nelepo and Domanov [265]. Ion exchange and α-spectrometric method is reported by Bojanowski et al. for the analysis of americium in marine environmental samples [266].

Wong determined plutonium by a radiochemical method [267], and Iwashima reported on a method for determination of [106]Ru [268]. Delos developed an apparatus for determining small quantities of [222]Rn [269]. A novel sampling and decomposition technique in conjunction with a modified mass spectrometer was used by Eremeev for the determination of [18]O [270].

Silica. Liss and Spencer discussed methods used for the determination of silicates in seawater [271], and Watanuki reported on automatic colorimetric methods for silicate Si and nitrate N [273].

Emmission spectroscopy was used by Latriant and Johannin-Gilles for the determination of silica, of boron and strontium [199]. Experimental use of new devices and methods for chemical analysis on shipboard was described by Oradovskii and Fedosov. An accuracy of ± 2% for silica was observed [273]. Oradovskii et al. determined silica either with spectrophotometer or photoelectric colorimeter by measuring the yellow Si-Mo complex [274]. A spectrophotometric investigation of the molybdosilicic acid reaction for application in the automatic analysis of seawater was reported by Lopez-Benito [275]. Fanning and Pilson studied the parameter that affects the precision and accuracy of the spectrophotometric determination of dissolved silica in seawater [276].

Silver. Carr determined silver, leached from Ag membrane filters, by atomic absorption spectrometry [277]. An aluminum hydroxide column was used by Goncharov for the concentration of traces of Ag and Cu, which were determined spectrochemically [136].

A neutron activation procedure has been developed for the determination of silver in seawater. The element is preconcentrated by an anion-exchange procedure. The concentrate is submitted to irradiation with thermal neutrons. Silver-110m is separated from other radionuclides by means of a conventional radiochemical separation scheme [278].

Sulfates. Sulfate content in seawater is of some importance because of scale formation. Normally sulfates are determined gravimetrically as the barium salt [7], but this method suffers from errors caused by co-precipitation. Macchi et al. determined sulfates by titration with $BaCl_2$ solution using thorin as indicator [279]. Jägner determined sulfates by means of photometric titration with hydrochloric acid in dimethylsulfoxide solution [280]. Berge and Bruegmann used an indirect polarographic determination with a relative deviation ± 0.5% [281]. Rasnick and Nakayama used nitrochromeazo titrimetric determination of sulfate in saline waters [282]. Titrimetric determination of sulfate by using the solid-state lead [-selective] electrode was also investigated by Mascini [283]. Jasinski and Trachtenberg used a sulfate-sensitive electrode for the determination of sulfates in seawater as well as in brackish waters and brines [284].

Uranium. Separation of uranium by adsorption on titanic acid followed by fluorimetric determination was reported by Ogata [285]. Muzzarelli et al. absorbed completely the uranium on a chitosan column [286]. Uranium content in seawater was determined by counting the fission tracks on muscovite immersed in a concentrated aqueous solution during the reactor neutron activation [287]. Activation-analytical determination of uranium with separation from [239]Np, by countercurrent ion migration, is reported by Bilal et al. [288]. Kim and Zeitling absorbed by a colloid flotation technique and analyzed uranium spectrophotometrically by using Rhodamine 8 [289]. Improvement of the adsorption technique in this method increased recovery of uranium from 82% to 90% [290]. Ryabinin and Lazareva used a two-stage carbonate leaching procedure for the separation of uranium and determined the uranium content by a photometric method [291]. Trace amounts of uranium were determined by Korkisch and Koch fluorimetrically or spectrophotometrically after concentration by an extraction and ion-exchange technique [292].

Vanadium. A standard addition method for the determination of ≤ 2,5 µg/l vanadium in brines, by atomic absorption spectroscopy, is reported by Crump-Wiesner et al. [293]. Combined ion-exchange spectrophotometric determination of vanadium in seawater was used by Kiriyama and Kuroda [294]. A simple method for direct determination of vanadium with sensitivity 0.025 µg/l is reported by Nishimura et al. [295]. Muzzarelli and Rocchetti determined the vanadium content in seawater by using hot graphite atomic absorption spectrometry on chitosan after separation from salt [296]. Vanadium and molybdenum were also determined by the use of a chelating ion exchange method [297].

3*

Zinc. A survey of the literature on the determination of stable Zn in seawater was given by Rozhanskaya and the sensitivities of various methods were compared with each other. A full description of spectrophotometric procedures using dithizone for Zn in the seawater was presented [298]. Various analytical methods for the determination of stable Zn were reviewed and compared by Branica. In particular, polarographic methods are described in detail. Procedures for determination of radio-active Zn are given [299].

Zirino [300] and Zirino and Healy [301] used differential anodic stripping voltammetry, pH controlled, to determine zinc in ocean waters. Sharipov et al. used an activation method for the determination of zinc, simultaneously with Fe and Se, in waters rich in mineral salts [302]. Tubular mercury graphite electrode has been used by Lieberman and Zirino for the determination of zinc in flowing solutions by anodic stripping voltammetry [303].

Multiple-elements analysis. Various techniques have been developed for the concentration of trace elements contained in seawater and their determination by instrumental analytical procedures. A report of adsorption of trace elements on various container surfaces in seawater was presented by Robertson [304]. Concentration by cocrystallization of certain trace elements followed by determination by atomic absorption or spectrophotometry was reported by Riley and Topping [305]. Muzzarelli et al. also separated trace elements in seawater and brines by chromatography on chitosan [306]. High yields are reported by Muzzarelli and Sipos for the collection of zinc, cadmium, lead and copper by using chitosan and anodic stripping voltammetry [307]. Solvent extraction was used by Armitage and Zeitling for trace element (uranium, copper, nickel, cobalt, iron and manganese) pre-concentration [308]. Extraction by a trifluoroacetylacetone-toluene system is rapid for iron, indium, cobalt and zinc determination [309]. 25 trace elements in seawater were separated by the use of cation and anion exchangers [310]. Eckhoff et al. reported on the theory of determination by neutron activation analysis [311], and Hoegdahl on the application in oceanography [312]. Piper and Goles reported on the determination of trace elements (Co, Cr, Cs, Fe, Rb, Sb, Sc, Sr and Zn) in seawater [313]. Mark described an apparatus for in situ determination of trace elements (Cu, Mn, Ag, Au) in simulated seawater [314]. Technology and applications of neutron activation analysis for the determination of potentially hazardous trace metals (Hg, Cd, Cu, Ag, As, Sb, Cr, Se und Zn) in seawater were discussed by Robertson and Carpenter [315]. Arsenic, copper and antimony were determined by activation analysis by Gohda [105].

The use of atomic absorption spectrophotometry in the field of marine research was reported by Burrel [316]. He also reported on atomic absorption spectrometric equipments for in situ analysis, especially for heavy metal pollutants [317]. Traces of Zn, Cu and Ni were determined by atomic absorption spectrophotometry by Brewer et al. [318]. Kurihara et al. described a method for the determination of trace elements (Ca, Mg, Fe, Cr, Mn, V and Mo) in brines by atomic absorption spectrometry [319], and by the same method Galle determined in seawater Mn, Fe, Co, Ni, Cu, Zn and Pb [320]. A simple and rapid method for the determination of both soluble and insoluble forms of Cu, Zn, Fe, Mg, K, Li and Rb is reported by Orren [321]. Tsalev et al. determined Mn, Fe, Co, Ni, Zn, Pb and Cu in seawater by extraction and atomic absorption analysis [322]. Kubo et al. reported on a solvent-extraction atomic-absorption spectrophotometric method for the determination of trace amounts of lead, chromium and copper [141]. Paus used flameless atomic absorption spectrophotometry with emphasis on the removal of interfering substances, mainly sodium salt [323]. Segar and Gonzales evaluated the use of atomic absorption with a heated graphite atomizer on the direct determination of trace transition metals in seawater [324].

Simultaneous determination of V, Cr, Mn, Fe, Co, Ni, Cu and Zn in seawater by X-ray fluorescence spectrometry was reported by Morris [325]. Atomic fluorescence spectrometry has been used by Jones et al., with preconcentration by an automated solvent extraction procedure, for the determination of Cu, Fe, Mn and Zn [326].

Whitnack and Sasseli reported on the application of anodic-stripping voltammetry in the determination of some trace elements in seawater [327]. The method was considered by Smith and Redmond as effective for the determination of trace metals in seawater [328]. Polished glassy C electrode was used by Florence for the determination by anodic stripping voltammetry of Bi, Sb, Cu, Pb, In, Cd, Tl and Zn [329]. Rojan determined Cu, Pb, Cd and Zn by using anodic-stripping alternating-current voltammetry [139]. Determination of Zn, Cd, Pb and Cu in the range of ppb*) is also reported by Zirino and Healy by pH-controlled differential voltammetry [330].

---

*) In this connection ppb means mg/m³ solution.

Analysis of seawater by difference chromatography is reported by Mangelsdorf and Wilson [331]. Polarography of the main components (Na, K, Mg and Ca) in seawater is reported by Berge and Bruegmann [75].

Literature to 1.4.3

[1] F. MacIntyre (Sci. Am. **223** [1970] 104/15). — [2] D. E. Robertson (Anal. Chem. **40** [1970] 1067/72). — [3] J. H. Carpenter (Natl. Bur. Std. [U.S.] Spec. Publ. No. 351 [1972] 393/430). — [4] J. J. Sigalove, M. D. Pearlman (Under Sea Technol. **13** No. 3 [1972] 24/8). — [5] K. Grasshoff (Method. Meeresbiol. Forsch. **1968** 13/31; C.A. **72** [1970] No. 47244).

[6] J. D. H. Strickland, T. R. Parsons (Fisheries Research Board Canada, Ottawa 1968). — [7] J. R. Newton, M. E. Atkins (Off. Saline Water Res. Develop. Progr. Rept. No. 450 [1969]). — [8] J. R. Newton, M. E. Atkins (Off. Saline Water Res. Develop. Progr. Rept. No. 514 [1969]). — [9] P. J. Moore (Inst. Mining Met. Trans. B **79** [1970] 107/15). — [10] D. W. Spencer, P. G. Brewer (Crit. Rev. Solid State Sci. **1** [1970] 409/78).

[11] A. T. Andersen, L. Foyen (Chem. Oceanogr. **1971** 111/49). — [12] D. F. Martin (Marine Chemistry, Vol. 1, 1972, Vol. 2, New York, N.Y., 1970). — [13] H. Walden, G. Weichart, H. Kautsky (Naturwissenschaften **59** [1972] 12/22). — [14] I. Trachtenberg, C. T. Baker (Off. Saline Water Res. Develop. Progr. Report No. 496 [1969], No. 619 [1970]). — [15] I. Trachtenberg (Off. Saline Water Res. Develop. Progr. Rept. No. 761 [1972], No. 844 [1973]).

[16] J. T. Clerc, G. Kahr, E. Fretsch, R. P. Scholer, H. R. Wuhrmann (Chimia [Aarau] **26** [1972] 287/98). — [17] T. B. Warner (J. Mar. Technol. Soc. **6** No. 2 [1972] 24/33). — [18] E. Pungor (Ion-selective Electrodes, Budapest 1973). — [19] T. M. L. Wigley (Dept. Mech. Eng. Univ. Waterloo Tech. Note No. 15 [1972]). — [20] A. H. Truesdell (U.S. Natl. Tech. Inform. Serv. PB 220464 [1973]).

[21] J. M. T. M. Gieskes (Limnol. Oceanogr. **14** [1969] 679/85). — [22] S. G. Oradovskii (Tr. Gos. Okeanogr. Inst. No. 113 [1972] 72/8; C.A. **79** [1973] No. 9670). — [23] S. Ben-Yaakov (Limnol. Oceanogr. **15** [1970] 326/8, **19** [1974] 144/51). — [24] S. V. Lyutsarev, S. S. Vladimirskii, V. F. Poluyaktov (Tr. Vses. Nauchn. Issled. Inst. Morsk. Ryb. Khoz. Okeanogr. **87** No. 7 [1971] 29/38; C.A. **78** [1973] No. 151451). — [25] M. W. Kellogg Company (Saline Water Conversion Engineering Data Book, Washington, D. C., 1965; C.A. **64** [1966] 17253).

[26] K. Kremling (Nature **229** [1971] 109/10). — [27] F. J. McIlhenny, M. A. Zeitoun (Chem. Eng. **76** No. 24 [1969] 81/6, No. 25 [1969] 251/6). — [28] H. U. Sverdrup, M. W. Johnson, R. H. Fleming (The Oceans. Their Physics, Chemistry and General Biology, 12th Printing, Englewood Cliffs, N.J., 1963). — [29] W. S. Wooster, A. J. Lee, G. Dietrich (Limnol. Oceanogr. **14** [1969] 437/8; Cah. Oceanogr. **21** [1969] 539/42). — [30] C. E. Knowles (NCSU-73-3 [1973]).

[31] Kozlovskaya, V. M. Ermakov (Ekspress-Inform. Morsk. Gidrofiz. Inst. Akad. Nauk Ukr. SSR No. 4 [1969] 113/8; C.A. **73** [1970] No. 28702). — [32] I. A. Fedorov (Okeanologiya **14** [1971] 739/43; C.A. **75** [1971] No. 132802). — [33] Yu. B. Shaub, V. I. Denenok (Geofiz. App. **1972** No. 48, p. 50/3; C.A. **78** [1973] No. 140324). — [34] P. E. Wilkniss (Radiochim. Acta **11** [1969] 138/42). — [35] P. F. Wiggins, K. J. Arthur (Naval Eng. J. **83** No. 5 [1971] 88/9).

[36] N. Ogata (Bunseki Kagaku **21** [1972] 780/7; C.A. **77** [1972] No. 92670). — [37] M. Ciabatti, F. Masini, E. Rabbi (G. Geol. **36** [1968] 599/616; C.A. **77** [1972] No. 9468). — [38] D. Jagner, K. Åren (Anal. Chim. Acta **52** [1970] 491/502). — [39] C. D'Arrigo, A. de Robertis, A. Casale (Proc. 5th Coll. Intern. Oceanogr. Med. **193** [1971]; Rass. Chim. **26** [1972] 393/6). — [40] R. E. Meyer, M. C. Banta, P. M. Lantz, F. A. Posey (Desalination **9** [1971] 333/50).

[41] K. Grasshoff, A. Wenck (J. Conseil Conseil Perm. Intern. Exploration Mer **34** [1972] 522/8). — [42] C. C. Cunningham, I. W. Duedall (J. Conseil Conseil Perm. Intern. Exploration Mer **33** [1970] 292/3). — [43] K. P. Park (Limnol. Oceanogr. **14** [1969] 179/86). — [44] S. V. Lyutsarev, S. S. Vladimirskii, V. F. Poluyaktov (Tr. Vses. Nauchn. Issled. Inst. Morsk. Ryb. Khoz. Okeanogr. **87** No. 7 [1971] 29/38; C.A. **78** [1973] No. 151451). — [45] D. Dyrssen, L. G. Sillen (Tellus **17** [1967] 113/21; C.A. **66** [1967] No. 118679).

[46] J. M. Edmond (Deep Sea Res. **17** [1970] 737/50). — [47] W. A. Stuart, A. R. Lister (Desalination **8** [1970] 69/72). — [48] I. Hansson (Chemistry of Seawater VIII, Univ. Göteborg, Dept. Anal. Chem. **1970**). — [49] I. Hansson (Acta Chem. Scand. **27** [1973] 931/44). — [50] I. Hansson, D. Jägner (Anal. Chim. Acta **65** [1973] 363/73).

[51] P. Deschamps, P. Deschamps (Trav. Centre Rech. Etud. Oceanogr. [2] **13** [1973] 7/11). — [52] H. J. Simpson, W. S. Broecker (Limnol. Oceanogr. **18** [1973] 426/40). — [53] C. S. Wong (Deep Sea Res. **17** [1970] 9/17). — [54] D. Dyrssen, D. Jägner, H. Johansson (Chemistry of Seawater V, Univ. Göteborg, Dept. Anal. Chem. **1968**). — [55] S. Tsunogai, M. Nishimura, S. Nayaka (Talanta **15** [1968] 385/90).

[56] L. H. Shaffer, R. A. Knight, D. A. Smith (Desalination **2** [1967] 125/9). — [57] W. E. Blake, M. W. R. Bryant, A. Waters (Analyst **94** [1969] 49/53). — [58] D. Jägner (Anal. Chim. Acta **68** [1974] 83/92). — [59] D. Jägner (Chemistry of Seawater III, Univ. Göteborg, Dept. Anal. Chem. **1967**). — [60] S. M. Shah, S. S. Gagate (Current Sci. [India] **38** [1969] 94/5).

[61] D. K. Das, M. Mazumbar, S. C. Shome (Anal. Chim. Acta **60** [1972] 439/41). — [62] J. P. Riley, G. Skerrow (Chemical Oceanography, Vol. 1, London 1965, p. 121/61). — [63] F. Gulkin, R.A. Cox (Deep Sea Res. **13** [1966] 789/804). — [64] B. J. Szabo (Bull. Mar. Sci. **17** [1967] 544/50). — [65] M. Whitfield, J. V. Leyendekkers (Anal. Chim. Acta **45** [1969] 383/98).

[66] M. Whitfield, J. V. Leyendekkers, J. D. Kerr (Anal. Chim. Acta **45** [1969] 399/410). — [67] M. Whitfield, J. V. Leyendekkers (Anal. Chim. Acta **45** [1969] 63/70). — [68] A. Ito, K. Ueda (Bunseki Kagaku **19** [1970] 393/7; C.A. **73** [1970] No. 10365). — [69] D. Jägner, K. Åren (Anal. Chim. Acta **56** [1971] 185/92). — [70] C. O. Huber, K. Dahnke, F. Hinz (Anal. Chem. **43** [1971] 152/4).

[71] H. Sato, K. Monoki (Anal. Chem. **44** [1972] 1778/80). — [72] V. I. Romanov, A. V. Eremeeva (Morsk. Gidrofiz. Issled. No. 4 [1969] 201/19; C.A. **74** [1971] No. 45435). — [73] R. Viswanathan, S. M. Shah, C. K. Unni (Bull. Natl. Inst. Sci. India **38** [1968] Pt. 1, p. 284/8). — [74] Y. Kurihara, K. Fukuda, Y. Sonehara, K. Fukuda (Soda To Enso **21** [1970] 321/8; C.A. **74** [1971] No. 130215). — [75] J. H. Berge, L. Bruegmann (Beitr. Meeresk. No. 26 [1969] 47/57).

[76] M. D. Booth, B. Fleet, S. Win, T. S. West (Anal. Chim. Acta **48** [1969] 329/37). — [77] V. H. Vaidya (Salt Res. Ind. **6** No. 1 [1969] 9/10; C.A. **71** [1969] No. 116402). — [78] V. I. Maksyuta, A. P. Movchan, Yu. V. Lozinskaya (Khim. Prom. Ukr. **1969** No. 5, p. 47/9; C.A. **72** [1970] No. 15658). — [79] I. M. Anosova (Tr. Nizhnevolzh. Nauchn. Issled. Inst. Geol. Geofiz. **1970** No. 8, p. 302/7; C.A. **75** [1971] No. 143839). — [80] J. S. Kim, S. W. Kim (Punsok Hwahak **8** No. 2 [1970] 31/3; C.A. **74** [1971] No. 115741).

[81] N. A. Afanas'eva, S. G. Oradovskii (Tr. Gos. Okeanogr. Inst. No. 113 [1972] 66/71; C.A. **79** [1973] No. 9668). — [82] I. I. Kulev, V. D. Bakalov (Dokl. Bolg. Akad. Nauk **26** [1973] 787/9; C.A. **80** [1974] No. 87354). — [83] V. I. Romanov, L. V. Eremeeva (Morsk. Gidrofiz. Issled. No. 3 [1970] 189/95; C.A. **76** [1972] No. 117354). — [84] M. Marczak, E. Ziaja (Chem. Anal. Poland **18** [1973] 99/104; C.A. **79** [1973] No 23446). — [85] T. Anfält, D. Jägner (Anal. Chim. Acta **66** [1973] 152/5).

[86] A. L. Markman, S. A. Strel'tsova (Tr. Tashkent. Politekhn. Inst. [2] No. 42 [1968] 50/8; C.A. **70** [1969] No. 31575). — [87] T. Shigematsu, T. Suzuki, M. Tabushi (Nippon Kaisui Gakkai-Shi **22** No. 5 [1969] 348/50; C.A **73** [1970] No. 101880). — [88] N. P. Morozov (Okeanologiya **9** [1969] 353/8; C.A. **71** [1969] No. 24651). — [89] V. T. Bowen (Tech. Rept. Ser. Intern. At. Energy Agency No. 118 [1970] 93/127) — [90] C. M. Gordon, R. E. Larson (Radiochem. Radioanal. Letters **5** [1970] 369/73).

[91] R. A. Carr (Limnol. Oceanogr. **15** [1970] 318/20). — [92] S. M. Vakulovskii, A. N. Silant'ev (Nucl. Instr. Methods **68** [1969] 42/4). — [93] G. N. Bondarenko, I. P. Bakheeva, L. N. Dubrova (Radiokhimiya **13** [1971] 314/6). — [94] A. N. Silant'ev, V. B. Chumichev, S. M. Vakulovskii (Tr. Inst. Eksp. Meteorol. Gl. Upr. Gidrometeorol. Sluzhby Sov. Min. SSSR No. 2 [1971] 15/8; C.A. **76** [1972] No. 37257). — [95] L. I. Gedeonov, L. M. Ivanova (Radiokhim. Anal. Ob'ektov Vnesh. Sredy **1972** 57/62; C.A. **79** [1973] No. 57486).

[96] W. J. Major, K. D. Lee, R. A. Wessman (Earth Planet. Sci. Letters **16** [1972] 138/40). — [97] K. Nagasawa, K. Miyahara (PNCT-831-72-01 [1972] 149/51; C.A. **78** [1973] No. 88440). — [98] Yu. M. Petrov (Issled. Teor. Prikl. Khim. Morya **1972** 89/91; C.A. **79** [1973] No. 45581). — [99] V. I. Romanov, L. V. Eremeeva (Ekspress-Inform. Morsk. Gidrofiz. Inst. Akad. Nauk. Ukr.SSR No. 13 [1969] 98/105; C.A. **73** [1970] No. 18398). — [100] Y. Nishikawa, K. Hizaki, K. Morishige, A. Tsuchiya, T. Shigematsu (Bunseki Kagaku **17** [1968] 1092/7; C.A. **70** [1969] No. 40565).

[101] T. Shigematsu, T. Suzuki, M. Tabushi (Nippon Kaisui Gakkai-Shi **22** No. 5 [1969] 348/50; C.A. **73** [1970] No. 101880). — [102] D. W. Spencer, P. L. Sachs (At. Absorption Newsletter **8** [1969] 65/8). — [103] M. L. Lee, D. C. Burrell (Anal. Chim. Acta **66** [1973] 245/50). — [104] A. I. Ryabinin, A. S. Romanov, Sh. Khatamov, A. A. Kist, R. Khamidova (Zh. Analit. Khim. **27** [1972]

94/9; J. Anal. Chem. USSR **27** [1972] 74/8; C. A. **77** [1972] No. 9469). — [105] S. Gohda (Bull. Chem. Soc. Japan **45** [1972] 1704/8).

[106] T. R. Gilbert, D. N. Hume (Anal. Chim. Acta **65** [1973] 451/9). — [107] P. J. Dunton (Appl. Spectry. **22** [1968] 99/100). — [108] P. E. Wilkniss (Radiochim. Acta **11** [1969] 138/42). — [109] P. Saenger (Helgoländer Wiss. Meeresunters. **23** No. 1 [1972] 32/7). — [110] H. Berge, L. Bruegmann (Beitr. Meeresk. No. 28 [1971] 19/32).

[111] E. A. Morgen, N. A. Vlasov, T. N. Mazko (Gidrokhim. Mater. **57** [1973] 164/7; C. A. **80** [1974] No. 52231). — [112] J. Revel (Cah. Oceanogr. **21** [1969] 273/81). — [113] K. Sugawata, T. Koyama, K. Terada (Bull. Chem. Soc. Japan **28** [1955] 494). — [114] A. D. Matthews, J. P. Riley (Anal. Chim. Acta **51** [1970] 295/301). — [115] S. Tsunogai (Anal. Chim. Acta **55** [1971] 444/7).

[116] M. M. Schnepfe (Anal. Chim. Acta **58** [1972] 83/9). — [117] K. V. Khelashvili (Soobshch. Akad. Nauk Gruz.SSR **65** [1972] 609/12; C. A. **76** [1972] No. 144690). — [118] S. Tsunogai, T. Sase (Deep Sea Res. **16** [1969] 489). — [119] K. Kuwata, K. Hisatomi, T. Hasegawa (Separ. Sci. **6** [1971] 505/13). — [120] N. Shiraishi, T. Hasegawa, T. Hisayuki, H. Takahashi (Bunseki Kagaku **21** [1972] 705/10; C. A. **77** [1972] No. 109092).

[121] J. W. Robinson, D. K. Wolcott, P. J. Slevin, G. D. Hindman (Anal. Chim. Acta **66** [1973] 13/21). — [122] K. Itiiro, A. Kawahara, T. Tanaka (Bunseki Kagaku **22** [1973] 1210/5; C. A. **80** [1974] No. 74191). — [123] E. Kentner, H. Zeitlin (Anal. Chim. Acta **49** [1970] 587/90). — [124] S. Motomizu (Anal. Chim. Acta **64** [1973] 217/24). — [125] B. R. Harvey, J. W. R. Dutton (Anal. Chim. Acta **67** [1973] 377/85).

[126] J. Abraham, M. Winpe, D. E. Ryan (Anal. Chim. Acta **48** [1969] 431/2). — [127] K. L. Balabanoff, R. H. Saelzer, V. J. Bartolome (Rev. Real. Acad. Cienc. Exact. Fis. Nat. Madrid **64** [1970] 621/9; C. A. **73** [1970] No. 123410). — [128] M. Odier, V. Plichon (Anal. Chim. Acta **55** [1971] 209/20). — [129] K. T. Marvin, R. R. Proctor (Limnol. Oceanogr. **15** [1970] 320/5). — [130] N. L. Kamp (Deep Sea Res. **19** [1972] 899/902).

[131] M. Valcarcel, F. Pino (Analyst **98** [1973] 246/50). — [132] I. F. Dolmanova, V. P. Poddubienko, V. M. Peshkova (Zh. Analit. Khim. **28** [1973] 592/5; J. Anal. Chem. USSR **28** [1973] 524/7). — [133] C. Fairles, A. J. Bard (Anal. Chem. **45** [1973] 2289/91). — [134] R. A. A. Muzzarelli, R. Rocchetti (Anal. Chim. Acta **69** [1974] 35/42). — [135] Y. P. Virmani, E. J. Zeller (Anal. Chem. **46** [1974] 324/5).

[136] A. F. Goncharov (Anal. Tekhnol. Blagorod. Metal. **1971** 220/2; C. A. **77** [1972] No. 147250). — [137] K. Kuwata, K. Hisatomi, T. Hasegawa (At. Absorption Newsletter **10** [1971] 111/5). — [138] S. Gohda (Bull. Chem. Soc. Japan **45** [1972] 1704/8). — [139] T. Rojan (Anal. Chim. Acta **62** [1972] 438/41). — [140] I. Trachtenberg (Off. Saline Water Res. Develop. Progr. Rept. No. 761 [1972], No. 844 [1973]).

[141] K. Kubo, N. Nakazawa, M. Sato (Sekiyu Gakkai-Shi **16** [1973] 588/91; C. A. **79** [1973] No. 108001). — [142] W. B. Kerfoot, R. F. Vaccaro (Limnol. Oceanogr. **18** [1973] 689/93). — [143] K. Kremling, H. Petersen (Anal. Chim. Acta **70** [1974] 35/9). — [144] Y. P. Virmani, E. J. Zeller (Anal. Chem. **46** [1974] 324/5). — [145] R. Jasinski, I. Trachtenberg, D. Andrychuk (Anal. Chem. **46** [1974] 364/9).

[146] R. J. Walker, R. B. Smith (J. Am. Water Works Assoc. **63** [1971] 246/50). — [147] P. E. Wilkniss, V. J. Linnenboom (EUR-3896 [1968] 147/60). — [148] P. E. Wilkniss (Radiochim. Acta **11** [1969] 138/42). — [149] T. B. Warner (Science [2] **165** [1969] 178/80). — [150] R. A. Kletsch, F. A. Richards (Anal. Chem. **42** [1970] 1435/6).

[151] I. M. Anosova (Tr. Nizhnevolzh. Nauchn. Issled. Inst. Geol. Gidrofiz. **1970** No. 8, p. 302/7; C. A. **75** [1971] No. 143839). — [152] T. Anfaelt, D. Jägner (Anal. Chim. Acta **53** [1971] 13/22). — [153] J. W. McMahon (Water Res. **3** [1969] 743/8). — [154] J. D. Burton, P. C. Head (Limnol. Oceanogr. **15** [1970] 164/7). — [155] S. Kawamura, S. Shibata, K. Kurotaki (Proc. Symp. Rapid Methods Meas. Radioactiv. Environ., Vienna 1971, p. 119/29; C. A. **77** [1972] Nr. 130451).

[156] K. Hiiro, T. Tanaka, T. Sawada (Bunseki Kagaku **21** [1972] 635/40; C. A. **77** [1972] No. 118052). — [157] M. L. Lee, D. C. Burrell (Anal. Chim. Acta **62** [1972] 153/61). — [158] W. H. Dennen (U.S. Natl. Tech. Inform. Serv. PB 220017 [1972]). — [159] S. Skurnik-Sarig, M. Zidon, I. Zak, Y. Cohen (Israel J. Chem. **8** [1970] 545/9). — [160] S. Shiraishi, T. Hasegawa, T. Hisayuki, H. Takahashi (Bunseki Kagaku **21** [1972] 705/10; C. A. **77** [1972] No. 109092).

[161] R. G. Clem (MPI [McKee-Pedersen Instr.] Appl. Notes **8** No. 1 [1973] 1/7; C. A. **80** [1974] No. 33442). — [162] H. V. Weiss, T. E. Crozier (Anal. Chim. Acta **58** [1972] 231/3). — [163]

G. Topping, J. M. Pirie (Anal. Chim. Acta **62** [1972] 200/3). — [164] R. A. Carr, J. B. Hoover, P. E. Wilkniss (Deep Sea Res. **10** [1972] 747/52). — [165] Y. Osajima, M. Tokubo, M. Nakasima, K. Matsumoto, F. Hashinaga, S. Furutani (Kyushu Daigaku Nogakubu Gakugei Zasshi **26** [1972] 525/8; C.A. **76** [1972] No. 158134).

[166] J. Olafsson (Anal. Chim. Acta **68** [1974] 207/11). — [167] D. Voyce, H. Zeitlin (Anal. Chim. Acta **68** [1974] 27/34). — [168] Y. S. Kim, H. Zeitlin (Anal. Chim. Acta **46** [1969] 1/8). — [169] Y. S. Kim, H. Zeitlin (Anal. Chim. Acta **51** [1970] 516/9). — [170] K. Kawabuchi, R. Kuroda (Anal. Chim. Acta **46** [1969] 23/30).

[171] Y. S. Kim, H. Zeitlin (Separ. Sci. **6** [1971] 505/13). — [172] R. A. A. Muzzarelli, R. Rocchetti (Anal. Chim. Acta **64** [1973] 371/9). — [173] R. Kuroda, T. Tarui (Z. Anal. Chem. **269** [1974] 22/6). — [174] B. K. Afghan, D. E. Ryan (Anal. Chim. Acta **41** [1968] 167/70). — [175] E. Kentner, D. B. Armitage, H. Zeitlin (Anal. Chim. Acta **45** [1969] 343/6).

[176] H. Rampon, R. Cuvelier (Anal. Chim. Acta **60** [1972] 226/8). — [177] R. Wagner (Vom Wasser **36** [1969] 263/318). — [178] K. Matsunaga, M. Nishimura (Anal. Chim. Acta **45** [1969] 350/3). — [179] J. Mertens, D. L. Massart (Bull. Soc. Chim. Belges **80** [1971] 151/8). — [180] M. I. V. Soares, P. G. S. Pereira, A. M. Antunes (Rev. Port. Quim. **13** [1971] 151/62; C.A. **77** [1972] No. 42822).

[181] K. Matsunaga, T. Oyama, M. Nishimura (Anal. Chim. Acta **58** [1972] 228/30). — [182] M. Nishimura, K. Matsunaga, K. Matsuda (Bunseki Kagaku **19** [1970] 1096/7; C.A. **73** [1970] No. 137082). — [183] E. Wada, A. Hattori (Anal. Chim. Acta **56** [1971] 233/40). — [184] K. Grasshoff (Z. Anal. Chem. **234** [1968] 13/22). — [185] T. Manabe (Nippon Suisan Gakkaishi **35** [1969] 895/906; C.A. **72** [1970] No. 47242).

[186] V. W. Truesdale (Analyst **96** [1971] 584/90). — [187] J. D. H. Strickland, T. R. Parsons (Fisheries Res. Board Can. Bull. Nc. 125 [1965]). — [188] R. D. Ellender, C. L. Armour, B. J. Camp (J. Fisheries Res. Board Can. **28** [1971] 788/9). — [189] K. Matsunaga, M. Nishimura (Bunseki Kagaku **20** [1971] 993/7; C.A. **76** [1972] No. 131258). — [190] J. J. MacIsaak, R. K. Olund (Invest. Pesquera **35** [1971] 221/32; C.A. **75** [1971] No. 121231).

[191] P. C. Head (Deep Sea Res. **18** [1971] 531/2). — [192] K. Grasshoff, H. Johannson (J. Conseil Conseil Perm. Intern. Exploration Mer **34** [1972] 516/21). — [193] F. Cioce, F. Dolci, G. Stocco, R. Toniolo (Arch. Oceanogr. Limnol. **17** [1972] 297/302). — [194] R. F. Srna, C. Epifanio, M. Hartman, G. Pruder, A. Stubbs (Delaware Univ. Coll. Mar. Studies Rept. DEL-SG-14-73 NOAA-730 91703 [1973]). — [195] T. R. Gilbert, A. M. Clay (Anal. Chem. **45** [1973] 1757/9).

[196] D. Degobbis (Limnol. Oceanogr. **18** [1973] 146/50). — [197] W. J. Barnes, C. A. Parker (Analyst **85** [1960] 828/38). — [198] R. Greenhalgh, J. P. Riley (Analyst **87** [1962] 970/6). — [199] J. Lotrian, A. Johannin-Gilles (Spectrochim. Acta B **24** [1969] 479/95). — [200] R. A. Nickolson (Anal. Chim. Acta **56** [1971] 147/9).

[201] J. Basselt, P. J. Matthews (Analyst **99** [1974] 1/11). — [202] L. I. Belyaev, E. I. Ovsyanyi (Okeanologiya **8** [1968] 920/6; C.A. **70** [1969] No. 60704). — [203] A. I. Sheremet'eva (Morsk. Gidrofiz. Issled. No. 4 [1970] 194/9; C.A. **72** [1970] No. 49719). — [204] L. Uppström (Anal. Chim. Acta **43** [1968] 475/86). — [205] P. Hulthe, L. Uppström, G. Ostling (Anal. Chim. Acta **51** [1970] 31/7).

[206] C. Testa, L. Staccioli (Analyst **97** [1972] 527/32). — [207] I. Hansson (Acta Chim. Scand. **27** [1973] 924/30). — [208] M. A. Phillips, W. R. Holden, E. W. Albaugh, W. E. McKinstry, H. A. Sweeny (Off. Saline Water Res. Develop. Progr. Rept. No. 115 [1964]). — [209] I. Lysyj, P. Newton, A. Counts (Off. Saline Water Res. Develop. Progr. Rept. No. 152 [1966]). — [210] I. Lysyj, K. N. Nelson, H. Snell (Off. Saline Water Res. Develop. Progr. Rept. No. 239 [1967], No. 327 [1968]); I. Lysyj (Symp. Org. Matter Natur. Waters, Alaska 1968 [1970], p. 333/49).

[211] C. Boutin, J. P. Grimaldi, J. P. Pencalet, J. Revel (Cah. Oceanogr. **21** [1969] 555/6). — [212] A. V. Khramov, N. B. Nechvalenko, L. V. Berkutova (Ekspress-Inform. Morsk. Gidrofiz. Inst. Akad. Nauk Ukr.SSR No. 14 [1969] 74/86; C.A. **73** [1970] No. 28709). — [213] A. V. Khramov, Yu. V. Pavlenko, I. B. Nechvalenko (Tr. Morsk. Gidrofiz. Inst. Akad. Nauk Ukr.SSR **41** [1969] 55/81, 81/9; C.A. **73** [1970] No. 28710 and No. 28696). — [214] S. T. Talreja, J. B. Bhalala, P. S. Rao (Salt Res. Ind. **6** [1969] 82/93; C.A. **74** [1971] No. 6305). — [215] H. C. Behrens, E. F. Sablatura, J. D. Theologos, B. P. Webb (Off. Saline Water Res. Develop. Progr. Rept. No. 572 [1970].

[216] V. Graefe (U.S. Clearinghouse Fed. Sci. Tech. Inform. PB 196145 [1970]). — [217] A. G. Kolesnikov, A. V. Khramov (Morsk. Gidrofiz. Inst. No. 4 [1970] 128/42; C.A. **76** [1972] No. 49733). — [218 K. F. Walker, W. D. Williams, U. T. Hammer (Limnol. Oceanogr. **15** [1970] 814/5).

— [219] P. J. Palermo (Diss. Northeastern Univ. 1971, p. 1/178; Diss. Abstr. Intern. B **32** [1971] No. 3224). — [220] E. Mor, A. M. Beccaria (Ann. Chim. [Rome] **61** [1971] 363/71).

[221] H. C. Eddington, R. M. Roberts (Off. Saline Water Res. Develop. Progr. Rept. No. 625 [1970]). — [222] F. Smirnov (Morsk. Gidrofiz. Issled. No. 2 [1970] 144/53; C.A. **76** [1972] No. 49728). — [223] H. C. Eddington (Off. Saline Water Res. Develop. Progr. Rept. No. 712 [1971]). — [224] H. C. Eddington (Off. Saline Water Res. Develop. Progr. Rept. No. 713 [1971]). — [225] R. E. Meyer, F. A. Posey, P. M. Lantz (Desalination **11** [1972] 329/40).

[226] G. Reusmann (Kiel. Meeresforsch. **24** [1968] 14/7). — [227] Yu. S. Tokuev (Tr. Gos. Okeanogr. Inst. No. 113 [1972] 30/41; C.A. **79** [1973] No. 9641). — [228] J. Hahn (Anal. Chem. **44** [1972] 1889/92). — [229] L. P. Atkinson (Anal. Chem. **44** [1972] 885/7). — [230] E. V. Smirnov (Morsk. Gidrofiz. Issled. No. 3 [1971] 195/200; C.A. **77** [1972] No. 38951).

[231] J. W. Swinnerton, V. J. Linnenbom, E. H. Check (Limnol. Oceanogr. **13** [1968] 193/5). — [232] R. Baker (Ind. Water Eng. **6** No. 1 [1969] 20/1). — [233] A. B. Isaeva (Zh. Analit. Khim. **24** [1969] 1854/8; J. Anal. Chem. USSR **24** [1969] 1505/8). — [234] P. Pakalns, B. R. MacAllister (J. Marine Res. **20** [1972] 305/11). — [235] Z. P. Suranova, O. Ya. Grabchuk, D. F. Osadchaya, V. M. Rutkovskii (Gidrokhim. Mater. **53** [1970] 144/6; C.A. **77** [1972] No. 66064).

[236] B. S. Cescon, P. G. Scarazzato (Limnol. Oceanogr. **18** [1973] 499/500). — [237] I. Hosokawa, F. Ohshima (Water Res. **7** [1973] 283/9). — [238] W. W. Flynn, W. R. Meehan (Anal. Chim. Acta **63** [1973] 483/8). — [239] R. W. Perkins, L. A. Rancitelli (BNWL-SA-3993 [1971]; C.A. **76** [1972] No. 144691). — [240] J. H. Harley, H. I. Volchok (At. Energy Comm. Health Saf. Lab., New York, Rept. No. CCC-3563-15 [1972]).

[241] International Atomic Energy Agency (Tech. Rept. Ser. Intern. At. Energy Agency No. 118 [1970]). — [242] R. Chesselet (Tech. Rept. Ser. Intern. At. Energy Agency No. 118 [1970] 275/84). — [243] O. T. Hoegdahl (Tech. Rept. Ser. Intern. At. Energy Agency No. 118 [1970] 187/211). — [244] A. Preston, J. W. R. Dutton (Tech. Rept. Ser. Intern. At. Energy Agency No. 118 [1970] 223/41). — [245] A. S. Vinogradov, K. V. Vinogradova (Tr. Morsk. Gidrofiz. Inst. Akad. Nauk Ukr.SSR **41** [1969] 122/31; C.A. **73** [1970] No. 69667).

[246] L. Tassi-Pelati, M. G. Mezzadri (Rend. Inst. Lombardo Sci. Lettere B **103** [1969] 281/6). — [247] H. Kautsky (Atompraxis **16** [1970] 316/20). — [248] A. V. Kas'yanov, A. D. Zemlyanoi, V. A. Klimenko (Morsk. Gidrofiz. Issled. No. 1 [1971] 204/7; C.A. **76** [1972] No. 49720). — [249] H. Miyake, M. Michijima (Kobe Shosen Diagaku Kiyo No. 20 [1973] 239/45; C.A. **80** [1974] No. 63708). — [250] Y. Koda, K. Iwashima, T. Koyanagi, K. Watari, M. Izawa (Radioisotopes [Tokyo] **21** [1972] 473/7).

[251] G. R. Doshi, T. N. Krishnasmoorthy, V. N. Sastry, T. P. Sarma (Arch. Oceanogr. Limnol. **17** [1972] 209/22). — [252] G. R. Doshi, T. N. Krishnasmoorthy, V. N. Sastry, T. P. Sarma (Indian J. Chem. **11** [1973] 158/61; C.A. **79** [1973] No. 9726). — [253] N. Ikeda, K. Kimura, K. Izawa, T. Yasni (Radioisotopes [Tokyo] **18** [1969] 8/11). — [254] S. M. Shah, S. R. Rao (Current Sci. [India] **41** [1972] 659/63). — [255] L. V. Shannon, M. J. Orren (Anal. Chim. Acta **52** [1970] 166/9).

[256] L. V. Shannon, R. D. Cherry, M. J. Orren (Geochim. Cosmochim. Acta **34** [1970] 701/11). — [257] Y. Nozaki, S. Tsunogai (Anal. Chim. Acta **64** [1973] 209/16). — [258] T. Shigematsu, K. Nishikawa, K. Hiraki, S. Goda, Y. Tsujimoto (Bunseki Kagaku **20** [1971] 575/81; C.A. **75** [1971] No. 80124). — [259] A. Yamato, K. Miyahara (PNCT-831-72-01 [1972] 155/58; C.A. **78** [1973] No. 88439). — [260] J. W. R. Dutton (FRL-6 [1970]).

[261] A. Bauman (Arh. Hig. Rada Toksikol. **21** [1970] 321/6; C.A. **76** [1972] No. 76322). — [262] T. R. Folsom, N. Hansen, D. E. Robertson (U.S. Natl. Tech. Inform. Serv. TID-25776 [1970]). — [263] T. R. Folsom, C. Sreekumaran (Tech. Rept. Ser. Intern. At. Energy Agency No. 118 [1970] 129/86). — [264] V. F. Hodge, T. R. Folsom, N. Hansen (U.S. Natl. Tech. Inform. Serv. TID-25777 [1971]). — [265] B. A. Nelepo, M. M. Domanov (Okeanologiya **13** [1973] 602/4; C.A. **80** [1974] No. 100084).

[266] R. Bojanowski, H. D. Livingston, D. L. Schneider, D. R. Mann (Woods Hole Oceanogr. Inst. COO-3563-8 [1973]). — [267] K. M. Wong (Anal. Chim. Acta **56** [1971] 355/64). — [268] K. Iwashima (J. Radiation Res. **13** [1972] 127/48). — [269] Y. Delos (NP-18404 [1970]). — [270] V. N. Eremeev (Morsk. Gidrofiz. Issled. No. 1 [1971] 199/203; C.A. **76** [1972] No. 76312).

[271] P. S. Liss, C. P. Spencer (J. Marine Biol. Assoc. U.K. **49** [1969] 589/601). — [272] K. Watanuki (Bunseki Kagaku **18** [1969] 1280/5; C.A. **72** [1970] No. 15657). — [273] S. G. Oradovskii, M. V. Fedosov (Khim. Resur. Morei i Oheanov 1970 52/5; C.A. **75** [1971] No. 25123). —

[274] S. G. Oradovskii, Yu. S. Tokuev, M. V. Fedosov (Issled. Teor. Prikl. Khim. Morya **1972** 83/8; C.A. **79** [1973] No. 9638). — [275] M. Lopez-Benito (Invest. Pesquera **34** [1970] 385/97).

[276] K. A. Fanning, M. E. Q. Pilson (Anal. Chem. **45** [1973] 136/40). — [277] R. A. Carr (At. Absorption Newsletter **8** [1969] 69). — [278] K. Kawabuchi, J. P. Riley (Anal. Chim. Acta **65** [1973] 271/7). — [279] G. Macchi, B. Cescon, D. Mameli-D'Errico (Arch. Oceanogr. Limnol. **16** [1969] 163/71). — [280] D. Jägner (Anal. Chim. Acta **52** [1970] 483/90).

[281] H. Berge, L. Bruegmann (Beitr. Meeresk. No. 27 [1970] 5/13). — [282] B. A. Rasnick, F. S. Nakayama (Comm. Soil Sci. Plant Anal. **4** [1973] 171/4). — [283] M. Mascini (Analyst **98** [1973] 325/8). — [284] R. Jasinski, I. Trachtenberg (Anal. Chem. **45** [1973] 1277/9). — [285] N. Ogata, N. Inoue (Nippon Kaisui Gakkai-Shi **23** No. 4 [1970] 148/53; C.A. **73** [1970] No. 38384).

[286] R. A. A. Muzzarelli, G. Raith, O. Tubertini (J. Chromatogr. **47** [1970] 414/20). — [287] T. Hashimoto (Anal. Chim. Acta **56** [1971] 347/54). — [288] B. A. Bilal, P. Braetter, B. Muehlig U. Roesic (Radiochim. Acta **16** [1971] 191/2). — [289] Y. S. Kim, H. Zeitlin (Anal. Chem. **43** [1971] 1390/3, **44** [1972] 181). — [290] G. Leung, Y. S. Kim, H. Zeitlin (Anal. Chim. Acta **60** [1972] 229/32).

[291] A. I. Ryabinin, E. A. Lazareva (Radiokhimiya **14** [1972] 919/21; C.A. **78** [1973] No. 88438). — [292] J. Korkisch, W. Koch (Microchim. Acta No. 1 [1973] 157/68). — [293] H. J. Crump-Wiesner, H. R. Felz, W. C. Purdy (Anal. Chim. Acta **55** [1971] 29/36). — [294] T. Kiriyama, R. Kuroda (Anal. Chim. Acta **62** [1972] 464/7). — [295] M. Nishimura, K. Matsunaga, T. Kudo, F. Obara (Anal. Chim. Acta **65** [1973] 466/8).

[296] R. A. A. Muzzarelli, R. Rocchetti (Anal. Chim. Acta **70** [1974] 283/9). — [297] J. P. Riley, D. Taylor (Anal. Chim. Acta **41** [1968] 175/8). — [298] L. I. Rozhanskaya (Tech. Rept. Ser. Intern. At. Energy Agency No. 118 [1970] 261/73). — [299] M. Branica (Tech. Rept. Ser. Intern. At. Energy Agency No. 118 [1970] 243/59). — [300] A. R. Zirino (Diss. Univ. of Washington 1970, p. 1/218; Diss. Abstr. Intern. B **31** [1970] 2155).

[301] A. R. Zirino, M. L. Healy (Limnol. Oceanogr. **16** [1971] 773/8). — [302] E. B. Sharipov, A. A. Abdullaev, L. I. Zhuk, A. S. Khuranov (Izv. Akad. Nauk Uz.SSR Ser. Fiz. Mat. Nauk **16** No. 5 [1972] 62; C.A. **78** [1973] No. 88441). — [303] S. A. Lieberman, A. Zirino (Anal. Chem. **46** [1974] 20/3). — [304] D. E. Robertsen (Anal. Chim. Acta **42** [1968] 533/6). — [305] J. P. Riley, G. Topping (Anal. Chim. Acta **44** [1969] 234/6).

[306] R. A. Muzzarelli, G. Raith, C. Tubertini (J. Chromatogr. **47** [1970] 414/20). — [307] R. A. Muzzarelli, L. Sipos (Talanta **18** [1971] 853/8). — [308] B. Armitage, H. Zeitling (Anal. Chim. Acta **53** [1971] 47/53). — [309] M. L. Lee, D. C. Burrell (Anal. Chim. Acta **62** [1972] 153/61). — [310] P. D. Novikov, M. U. Mizopol'skii, B. M. Talalaev (Okeanologiya **12** [1972] 161/7; C.A. **76** [1972] No. 131325).

[311] N. D. Eckhoff, T. R. Hill, W. R. Kimel (Trans. Kansas Acad. Sci. **71** [1968] 101/35). — [312] O. Hoegdahl (Activ. Anal. Geochem. Cosmochem. Proc. NATO Advan. Study Inst., 1970 [1971], p. 301/10; C.A. **76** [1972] No. 131250). — [313] D. Z. Piper, G. C. Goles (Anal. Chim. Acta **47** [1969] 560/3). — [314] H. B. Mark (J. Pharm. Belg. [2] **25** [1970] 367/99). — [315] D. E. Robertson, R. Carpenter (BNWL-SA-4455 [1972]).

[316] D. C. Burrell (At. Absorption Newsletter **7** [1968] 65/8). — [317] D. C. Burrell (Proc. 3rd Intern. Congr. At. Absorption At. Fluorex. Spectrom., 1971 [1973], Vol. 2, p. 409/28). — [318] P. G. Brewer, D. W. Spencer, C. L. Smith (Am. Soc. Testing Mater. Spec. Tech. Publ. No. 443 [1969] 70/7). — [319] Y. Kurihara, K. Fukuda, Y. Sonehara, K. Fukuda (Soda to Enso **21** [1970] 321/8; C.A. **74** [1971] No. 130215). — [320] O. K. Galle (Appl. Spectry. **25** [1971] 664/7).

[321] M. J. Orren (J. South African Chem. Inst. [2] **24** [1971] 96/102). — [322] D. L. Tsalev, I. P. Alimarin, S. I. Neiman (Zh. Analit. Khim. **27** [1972] 1223/4; J. Anal. Chem. USSR **27** [1972] 1100/1; C.A. **77** [1972] No. 105468). — [323] P. E. Paus (Z. Anal. Chem. **264** [1973] 118/22). — [324] D. A. Segar, J. G. Gonzales (Anal. Chim. Acta **58** [1972] 7/14). — [325] A. W. Morris (Anal. Chim. Acta **42** [1968] 397/406).

[326] M. Jones, G. F. Kirkbright, L. Ranson, T. S. West (Anal. Chim. Acta **63** [1973] 210/5). — [327] G. C. Whitnack, R. Sasseli (Anal. Chim. Acta **47** [1969] 267/74). — [328] J. D. Smith, J. D. Redmond (J. Electroanal. Chem. Interfacial Electrochem. **33** [1971] 169/75). — [329] T. M. Florence (J. Electroanal. Chem. Interfacial Electrochem. **35** [1972] 237/45). — [330] A. Zirino, M. L. Healy (Environ. Sci. Technol. **6** [1972] 243/9).

[331] P. C. Mangelsdorf, T. R. S. Wilson (NYO-3838-1 [1968]; J. Phys. Chem. **75** [1971] 1418/25).

2 Distillation Processes

Introduction. The term "evaporation", as used in this section, refers specifically to the vaporization of water from an aqueous saline solution, like brackish or seawater. It is a typical case applied to a solvent containing dissolved solids. In evaporating aqueous saline solutions, the solid constituents are practically nonvolatile, in the range of the working temperatures and pressures, and water alone is vaporized. Thus a concentration of the dissolved solids is obtained in the residual liquid, the brine.

In chemical industry, when an evaporation process is applied, the water vapors are usually discarded and emphasis is given to the recuperation of the dissolved solids. In desalting, the emphasis is given to the condensation of the vapors, except in some cases where the distillation process is applied for both, the recovery of fresh water and the salts contained in seawater.

The process is performed in evaporators, where at least the latent heat of evaporation is supplied to the solution. In any evaporation system, the input heat energy into the system must be rejected or discharged at other points of the system. In simple evaporation cycles the heat added to the salt water heater is discharged in both the product distilled water and the waste brine streams. In more complex distillation systems, with several effects or several stages, the input energy to the salt water heater is discharged, in addition to the standard brine and product distilled water streams, through heated last stage condenser water or heat rejection water.

Thermodynamic considerations lead to a common characteristic of all distillation processes, that the percentage of evaporated water in respect to the circulating seawater is as much larger, as higher is the difference between the maximum and minimum temperature of the saline solution. As the minimum temperature is defined by the temperature of the incoming seawater, enlarging of the temperature difference can only be obtained by increasing the initial maximum temperature of the salt water feed. Limitations due to the appearance of phenomena like scale formation and corrosion, which are becoming more important at higher temperatures, define an allowable maximum temperature for each distillation process. An appropriate pretreatment of the salt water is necessary to make an increase of the feed water temperature possible.

As a consequence, the economics of the process might be affected by the use of chemical additives for the feed water pretreatment, decrease of performance because of scale formation, major maintenance costs due to corrosion and/or increase of fixed charges, because of the possible use of more expensive materials of construction.

Definitions and terminology. The following formulae, in generally simplified form, and definitions of terms are the most frequently used in the evaluation and description of evaporation type equipment. The symbol "degr" refers to degree in centigrade scale.

1) The specific heat c_p is constant over the ranges of temperature, pressure, and concentration taken in consideration and has the values:

   for water   1.0   kcal/kg·degr
   for seawater and its concentrates   0.955   kcal/kg·degr

2) The specific total enthalpy for all changes in liquid temperature level, including those where minor increases in dissolved solids concentration occur, is defined as
$$H = c_p \cdot \Delta t \quad \text{in kcal/kg}$$

3) The vapor pressure of the dissolved salts is negligibly small and therefore the vapor above a seawater or seawater concentrate liquid phase is pure water.

4) The latent heat of vaporization of water from the liquid concentrate at temperature t_B has a value corresponding to the vapor temperature t_v, at which the vapor is in equilibrium with the concentrate, i.e. minus the boiling point elevation Θ:
$$t_v = t_B - \Theta$$

5) In the relation describing the overall heat transfer coefficient
$$k = \frac{1}{1/\alpha_1 + s/\lambda + 1/\alpha_2 + f} \quad \text{in kcal/m}^2 \cdot \text{h} \cdot \text{degr}$$
a fouling factor f is introduced for the resistance to heat transfer caused by the fouling or scale deposition on the heat transfer surface. This relationship requires the determination of the condens-

ing side heat transfer coefficient α_1 and the evaporating side heat transfer coefficient α_2. In addition the wall thickness s and the thermal conductivity λ of the heat transfer surface must be known.

6) The total heat transferred across a heat transfer surface equals:

$$Q = k \cdot A \cdot \Delta t_m \quad \text{in kcal/h}$$

where A is the heat transfer area and Δt_m the mean logarithmic temperature difference.

7) The total heat required for heating or cooling of a fluid is defined:

$$Q = M \cdot c_p \cdot \Delta t \quad \text{in kcal/h}$$

where M is the mass flow rate in kg/h.

8) The heat of evaporation or condensation transferred is defined as:

$$Q = M \cdot L_v \quad \text{in kcal/h}$$

where L_v is the latent heat of evaporation in kcal/kg.

9) The extraction ratio E represents the amount of water extracted from the feed solution and is determined by the initial salt concentration C_F in the feed and the final salt concentration C_B in the blowdown brine, both in wt.-%:

$$E = 1 - \frac{C_F}{C_B}$$

or by the amount of distillate produced D and the amount of seawater feed F into the system:

$$E = \frac{D}{F}$$

10) The performance ratio R is defined as the ratio of the mass of distillate produced to the mass of steam input S into the system

$$R = \frac{D}{S} = \frac{F \cdot E}{S}$$

The figure obtained is dimensionless (kg/kg or lb/lb). Some authors define the performance ratio as the mass of distillate produced per 1000 Btu of heat input to the brine heater in case of MSF (multi-stage flash) distillation or to the first effect in case of multiple-effect LTV (long tube vertical) evaporators. Although the latter definition is thermodynamically more accurate, as it refers to the enthalpy of the steam rather than to the mass, the resulting figures are not valid for the metric system as well. Care should, therefore, be taken when comparing figures, supplied by different authors, as to the definition given for the performance ratio.

11) The capacity of an evaporator is determined by the rate of water vapor produced, which depends on the available heat transfer area A, the effective temperature drop Δt and the overall heat transfer coefficient k:

$$D = \frac{k \cdot A \cdot \Delta t}{L_v} = \frac{Q}{L_v} \quad \text{in kg/h}$$

where L_v is the latent heat of vaporization and Q the rate of heat transferred to the system.

2.1 Common aspects of distillation processes

Gemeinsamkeiten der Destillationsverfahren

Some physical parameters are involved in and some technical features are associated with all distillation processes. Physical parameters include vapor pressure, boiling point elevation, viscosity changes, variation of heat exchange coefficient as concentration increases and variation of the temperature in the various distillation stages. A review of applied physics parameters in evaporation plants was recently given by Gull [1]. Technical features, as well as methods for the treatment of seawater and brine involved in the distillation process, corrosion problems and disposal of effluents will be discussed in the appropriate sections.

2.1.1 Enhancement of heat transfer

Erhöhung des Wärmedurchgangs

Uncertainties with the long tube vertical evaporator system may involve the mechanical reliability of the equipment, but also the inability to predict the scaling characteristics and the reliability of the methods used to design and predict heat transfer performance. Existing pilot plant data and theoretical coefficients for the evaporating and condensing side were combined to obtain theoretical overall coefficients. Optimum plant design requires tube diameters and lengths not be the same for all effects of a vertical-tube evaporator [18].

The removal of noncondensable gases is an important part in the operation of seawater evaporators to maintain the overall heat transfer coefficients. Noncondensable gases reduce the condensing heat transfer coefficient due to two effects acting simultaneously: lowering of the partial pressure of steam in the bulk stream, as well as of heat and mass transfer between the steam and the cooling surface. Gases are introduced into desalination plants as dissolved air in the feed, carbon dioxide released from decomposition of hydrogen carbonates and leakage into those regions operating under vacuum. Even when the feed seawater is carefully deaerated and decarbonated, there is a residual gas content in the seawater entering the high temperature regions, which accumulates in the evaporator. Noncondensable gases have, therefore, to be regularly evacuated from the distillation plant.

The amounts of steam-gas mixtures which should be removed from the evaporators vary with temperature and vapor velocity. A venting characteristic was derived by Hawes and Butcher, which relates the mean heat transfer coefficient to the steam/gas mass ratio in the vent [19]. Below a critical vent rate the condenser becomes blanketed with gas and ceases to function. Above the critical rate, the heat transfer coefficient increases rapidly at first and becomes asymptotic to the value which would be obtained in the absence of gas.

The effect of the presence of a small amount of noncondensable gas on a condensing vapor was investigated by Cunningham with various theoretical approaches to the problem [20]. Several equations, predicting the increase in gas concentration at the condensate surface were derived from the Reynolds flux theory.

Common equations were derived and uniform coefficients were defined for heat transfer by laminar free convection, film condensation and film evaporation. The coefficients are suitable for paying due attention to varying shapes of bodies. Simple approximative equations were also developed for heat transfer outside vertical and horizontal tubes with small diameters relative to the thickness of the boundary layer [21].

Reducing the temperature difference, which must be maintained between the condensing pure distilled water and the boiling seawater, and/or increasing the overall heat transfer coefficient are two main components for increasing the efficiency of seawater evaporators and ultimately for reducing the cost of product water. Several ways of enhancing heat transfer in the distillation of seawater have been investigated and tested.

2.1.2 Intake systems   *Einlaufwerke*

Brackish water is often supplied by wells, which provide essentially filtered water, free from fouling, marine life, trash and sand. However, the applicability of well intakes is limited and the most usual supply of feed water to a desalting plant is the open sea. There are three types of seawater intakes, depending on the shore formation [2].

The lagoon type intake may be applied to a location on a lagoon, canal, waterway, bay or any similar shore, free from direct influence of ocean or gulf waves and surf. The seawater is conveyed to the pump basin by a channel (**Fig. 2–1**, p. 46).

The pipe type intake withdraws water from the ocean a substantial distance from the shore, generally beyond the surf zone, where sand transport and bottom erosion are minimal. The seawater is conveyed to the pump basin by pipe (**Fig. 2–2**, p. 46), which is usually buried at a sufficient depth.

The shore type intake withdraws water from the surf zone or otherwise close to the shore. An offshore channel, formed by two parallel jetties or otherwise, makes an extension of the inland seawater conveying channel (**Fig. 2–3**, p. 46).

Before entering the pump basin, the seawater passes through a fixed bar trash rack and then through travelling spray washed screens, to remove solids carried with the seawater. Such solids might include floating objects, non-floating materials, marine animals and plants.

Fouling might be the most serious problem in intake systems. Fouling is the growth of marine plants and animals, mussels and barnacles, on the submerged surfaces, which reduce the effective inside diameter of pipes and increase friction losses. Fouling can also occur by bacteria and algae deposits on heat exchange tubes in forming a slime layer, which retards flow and reduces heat transfer efficiency. The usual method of marine fouling control is chlorination of the make-up and coolant water by injection of chlorine, as usual in water treatment plants.

Research was undertaken on the biology of fouling organisms and their response to technically and economically feasible control procedures has been determined. Specific control procedures examined include velocity, temperature, use of toxic surfaces and use of toxins including ozone and chlorine [22].

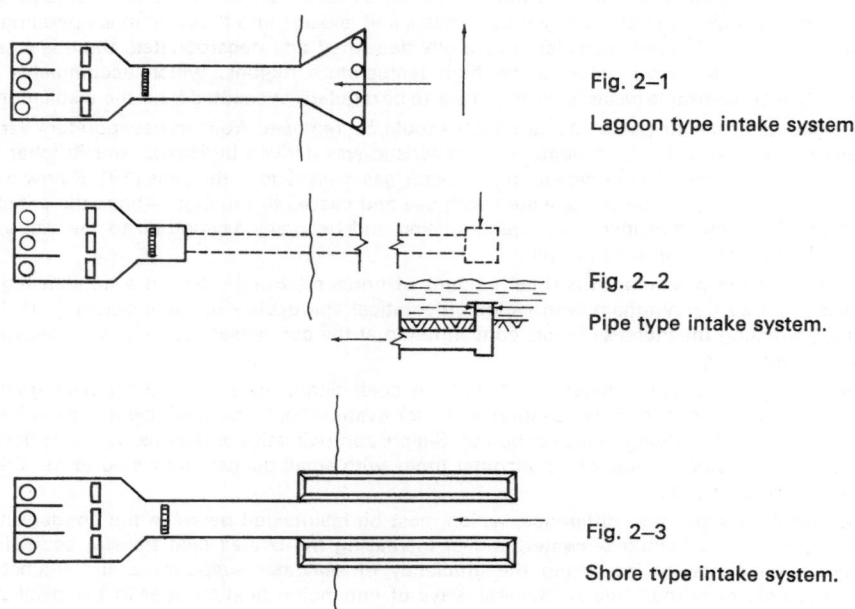

Fig. 2–1

Lagoon type intake system.

Fig. 2–2

Pipe type intake system.

Fig. 2–3

Shore type intake system.

*Siedepunkts-
erhöhung*

2.1.3 Boiling point elevation

Thermal properties of very dilute solutions might be considered as similar to those of water and any required data may be taken from steam tables, as there is no appreciable elevation of the boiling point. As the seawater becomes concentrated, the vapor pressure is lowered by the presence of dissolved solids and the boiling temperature, for a given pressure, is raised accordingly. The boiling point elevation is dependent upon the mole-fraction of the dissolved solids.

The greater the boiling point elevation, the more the thermodynamic characteristics of the solution, such as specific heat and latent heat of evaporation of the solvent, deviate from those of water. The vapor, coming from a solution with an elevated boiling point, is at the solution temperature and is therefore superheated by the amount of the boiling point elevation. However, the effective steam temperature in a following distillation effect is the saturation temperature of the vapor. Hence, in passing from a boiling solution in one effect to condensing steam in the next effect, the boiling point elevation is lost from the available temperature difference between effects. This phenomenon occurs in every effect of a multiple-effect evaporator and the resulting Δt loss is represented by the sum of the boiling point elevations in all effects.

The boiling point elevation for seawater is illustrated in **Fig. 2–4** for various concentrations and boiling points. The relationships are almost linear following the Dühring's rule.

Literature data on vapor pressures of seawater solutions and its concentrates were evaluated by Fabuss and Korosi, using the rule of additivity of molar vapor pressure depressions. A table of selected boiling point elevation data is presented in the 20° to 180°C range for seawater and its concentrates up to three-fold concentration [3]. The accuracy of the Dühring's rule for the determination of the boiling temperature of aqueous solutions was also investigated on the systems $H_2O–CaCl_2$, $H_2O–MgCl_2$, $H_2O–KOH$ and $H_2O–NaOH$ over a wide range of conditions [4]. The rule is also accurate at high pressures, temperatures and concentrations.

Fig. 2–4

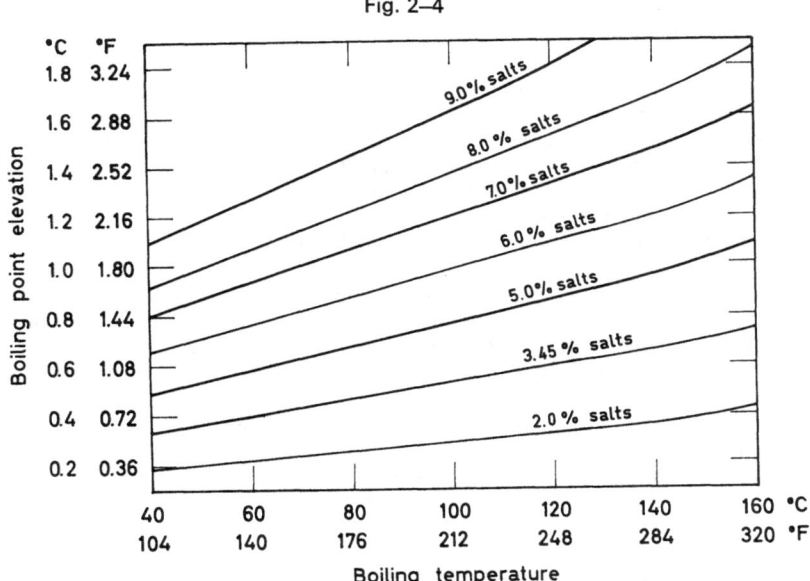

Boiling point elevation of seawater and its concentrates.

2.1.4 Stabilization of product water

Untreated product water from a seawater distillation plant will attack the materials of construction commonly used in storage and distribution systems. The distillate will dissolve lime in concrete and cause it to deteriorate. Similarly, it will remove the protective film from steel water lines, which were for years in service, and prevent protective film formation on new mains. Bare metal surfaces can corrode at high rates. Dissolved iron that is taken into solution by distillate will precipitate in contact with air and cause the so-called "red water".

To inhibit corrosion and reduce contamination by metal pick-up, treatment of the product water from distillation plants is required. Calcium carbonate stabilization is generally the most satisfactory corrosion inhibitor. The economics were considered of several methods of achieving a stable water containing 30 to 80 mg/l of calcium alkalinity expressed as $CaCO_3$ and a minimum of 5 mg/l of dissolved oxygen. If naturally hard water is unavailable for blending, the cheapest method is blending with a calcium carbonate solution prepared by absorption of carbon dioxide in a lime slurry. The estimated costs for calcium carbonate stabilization of product water are approximately 0.48 cents/m³ (1.8 cents/kgal) for a 37 850 m³/day (10 Mgd) plant and 0.37 cents/m³ (1.4 cents/kgal) for a plant producing 190 000 m³/day (50 Mgd) [5].

Two-year tests were reported by Egorov and Alekseev of a pilot plant, having two large marble filters, for the stabilization of desalted water by distillation [6]. Shtannikov suggests mixing with mineralized water or addition of $Ca(HCO_3)_2$ to improve the taste of the distillate. The $Ca(HCO_3)_2$ can be added by treating the distilled water with carbon dioxide and then filter through $CaCO_3$ [7]. Similar techniques are reported by Egorov and Alekseev [8] and by Alekseev [9].

The stabilizing treatment of distillate by the contact filtration method with marble and a magnesium compound decreased by a factor of 3 the corrosion of steel. At 3 m height of the column and 2.5 min contact, the calculated rate of filtration is 70 m/h and at 5 min contact with the ground marble the rate is 35 m/h [10].

Stabilisierung des erzeugten Wassers

2.1.5 Biomedical aspects

A maximum content of 500 mg/l total dissolved solids is recommended in standards for potable water quality. There is no similar recommendation for sanitary reasons of a minimum content of dissolved solids. It is reminded that in many islands in the Mediterranean Sea and in other arid areas, the inhabitants store rainwater during the rainy period and use this salt-free water throughout the year, very often as unique water supply.

However, there are reports that continuous use of distilled water is harmful to men and animals, as it affects the metabolism of water and salt in the organism [11]. El'piner et al. report on the hygienic assessment of distilled water by experiments on dogs and rats [12]. In another paper El'piner suggests, as the untreated distillate is not suitable as drinking water, to prepare potable water with a salinity of 15 to 30 mg/l by UV irradiation and addition of $NaHCO_3$, $NaHSO_4$, $MgSO_4$, $CaCl_2$ and NaF [13]. Pogosov et al. suggest the use of an anion exchange resin in the HCO_3^--form for the enrichment in hydrogen carbonate ions and of a mineralizer consisting of gypsum, MgO, $NaCl$ and KCl. Sanitary studies indicated that the resin and mineralizer are suitable for the stabilization of desalinated waters [14].

The taste of desalted water might be the most important parameter for its acceptance by consumers. In tests with 19 persons, Pangborn and Bertolero found that the taste intensity of drinking water with various contents of dissolved solids was directly related to the mineral content and to the water temperature [15]. O'Mahony reports that purity, as well as pre-adaptation, is an important factor in the distilled water taste [16].

On the other hand it appears certain that continuous drinking of brackish water may cause more or less serious metabolic effects. Berlyne and Morag report on findings on inhabitants of the Arava Rift Valley in Southern Israel, consuming a drinking water with a high sulfate and magnesium content [17].

Literature to 2.1

[1] H. C. Gull (Brit. Chem. Eng. **17** No. 1 [1972] 39/42, No. 2 [1972] 123/32). — [2] B. P. Shepherd, P. G. LeGros, J. C. Wiliams, D. C. Mangum, W. F. McIlhenny (Off. Saline Water Res. Develop. Progr. Rept. No. 678 [1971]). — [3] B. M. Fabuss, A. Korosi (J. Chem. Eng. Data **11** [1966] 606/9). — [4] Z. Rant, W. Moehle (Chem. Ingr.-Tech. **44** [1972] 261/5). — [5] C. D. Bopp, S. A. Read (Off. Saline Water Res. Develop. Progr. Rept. No. 709 [1971]).

[6] A. I. Egorov, L. S. Alekseev (Proc. 3rd Intern. Symp. Fresh Water Sea, Dubrovnik 1970, Vol. 4, p. 139/46). — [7] E. V. Shtannikov (Gigiena i Sanit. **35** No. 12 [1970] 10/3; C.A. **74** [1971] No. 79429). — [8] A. I. Egorov, L. S. Alekseev (Gigiena i Sanit. **36** No. 12 [1971] 19/22; C.A. **76** [1972] No. 63074). — [9] L. S. Alekseev (Tr. Vses. Nauchn. Issled. Inst. Vodosnabzh. Kanaliz. Gidrotekhn. Sooruzh. Inzh. Gidrogeol. **1971** No. 32, p. 107/11; C.A. **76** [1972] No. 131318). — [10] A. I. Egorov, L. S. Alekseev (Tr. Vses. Nauchn. Issled. Inst. Vodosnabzh. Kanaliz. Gidrotekhn. Sooruzh. Inzh. Gidrogeol. **1971** No. 29, p. 17/9; C.A. **78** [1973] No. 88479).

[11] A. I. Egorov, L. S. Alekseev (Proc. 4th Intern. Symp. Fresh Water Sea, Heidelberg 1973, Vol. 1, p. 237/44). — [12] L. I. El'piner, A. I. Bokina, Yu. B. Shafirov (Gigiena i Sanit. **34** No. 6 [1969] 22/6; C.A. **71** [1969] No. 53406). — [13] L. I. El'piner (Kosm. Biol. Med. **5** No. 6 [1971] 73/7; C.A. **76** [1972] No. 144699). — [14] D. P. Pogosov, N. I. Omel'yanets, L. V. Grigorieva, I. N. Medvedev, N. V. Mironets (Gigiena i Sanit. **50** No. 8 [1972] 19/22; C.A. **77** [1972] No. 130455). — [15] R. M. Pangborn, L. L. Bertolero (J. Am. Water Works Assoc. **64** [1972] 511/5).

[16] M. O'Mahony (Nature **240** [1972] 489). — [17] G. M. Berlyne, M. Morag (Desalination **10** [1972] 215/9). — [18] Prengle, Dukler and Crump Inc. (Off. Saline Water Res. Develop. Progr. Rept. No 74 [1963]). — [19] R. Hawes, A. A. Butcher (Proc. 4th Intern. Symp. Fresh Water Sea, Heidelberg 1973, Vol. 1, p. 293/301). — [20] J. Cunningham (Proc. 4th Intern. Symp. Fresh Water Sea, Heidelberg 1973, Vol. 1, p. 31/9).

[21] W. Roetzel (Chem. Ingr.-Tech. **43** [1971] 785/91). — [22] D. C. Mangum, B. P. Shepherd, W. F. McIlhenny (Off. Saline Water Res. Develop. Progr. Rept. No. 858 [1973]).

2.2 Single-effect distillation

The simplest conceptual design of distillation involves boiling of seawater in a vessel heated by condensing steam. Steam supplied by a boiler is condensed on one side of the heated metal surface to boil the seawater flowing on the other side of the heat exchanging surface, the evaporating surface. The evolving vapors pass to a second metal surface and are condensed by heat extraction to a supply of cooling seawater, which is heated. Part of the preheated seawater forms the feed to the evaporator and the remaining is used for the heat rejection.

In an once-through single-effect evaporator the thermal energy consumption is high, since all heat transferred through the evaporating surface is lost at the condensing surface, except the part which is recuperated in preheating the seawater feed.

Energy requirements for desalting in a single-effect distillation process are determined by the heat consumed in performing the transition from the liquid to the vapor phase. The heat input required to perform the evaporation process may be calculated from the prevailing heat balance. In **Fig. 2–5** the steam Q used as heat carrier has an enthalpy H_s at the condensation temperature t_s. The feed solution F has an enthalpy of H_F and an entrance temperature t_F. The evaporated solvent D has an enthalpy of H_D at the boiling point t. The heat input Q (in kg/h) to the system is then:

$$Q = \frac{D(H_D - t) + F(t - t_F)}{H_s - t_s}$$

Thermal losses are not included in this relation and have to be added in the heat input.

In the traditional distillation method the tubes were submerged in the boiling liquid. By adopting an embodiment of the system in which the boiling occurs in a thin falling film all the static heat problems associated with the traditional method vanish.

Fig. 2–6 shows the cross-section of a long tube vertical (LTV) evaporator, as used in the distillation of seawater. The ascending liquid column has been replaced by a falling liquid film. Hot seawater is fed into a chamber located at the top of the evaporator. At the bottom of this chamber a distributing device regulates the uniform distribution of the seawater in the vertical tubes. The seawater flows down filmwise inside the tubes, under partial evaporation, while steam is condensing outside the tubes. The pumping action of the bubbles formed increases considerably the velocity of the flowing

Fig. 2–5   Fig. 2–6

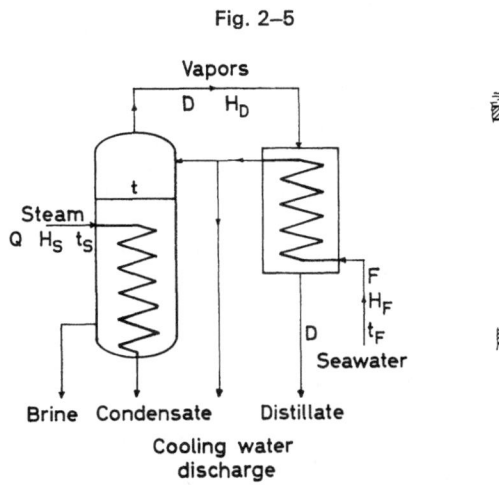

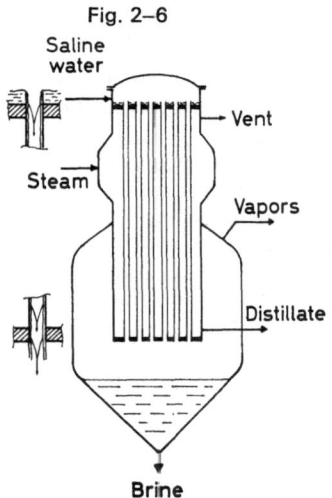

Flowsheet of single-
effect evaporation.

Cross section of falling
film LTV evaporator.

down liquid film. The staying time of the liquid inside the tubes is quite shorter than in the usual evaporators. Vapor and brine are collected and separated in a chamber at the lower part of the evaporator. Condensate is collected at the lower end outside the tube bundle.

In a single-effect distillation plant the steam condensate would be returned to the boiler and the distillate would be the product water.

Long tube vertical evaporators have the advantage that heat transfer is considerably higher per unit of surface, due to the characteristics of film flow, and that the rate of scale deposition is reduced, when compared with other older types of evaporators.

The exceptionally large amount of energy consumed in the single-effect distillation process makes the recycling of the heat of condensation and recuperation of the sensible heat of the distillate a necessity.

Mehrfach-
effekt-
Destillation

2.3 Multiple-effect distillation

Theoretically in single-effect distillation 1 kg of distillate will be produced for every kg of steam consumed and the performance ratio of the plant will be 1. In fact, despite preheating of the feed, a large part of the enthalpy of the vapors, evolved in the single-effect evaporator, is lost in the condenser. A better heat recuperation would be obtained if the heat, released by the condensing vapor, is not rejected in a condenser, but is used to heat the brine of a second evaporator and so on.

This leads to the concept of multiple-effect distillation, where the vapors from one effect are used as the heat source of the next effect, as long as the difference in temperature between the condensing vapor and the solution is high enough to act as the driving force in the evaporation process, each effect being at progressively lower temperature and pressure. Vapor condensing because of lower boiling temperature, in each effect. is producing fresh water as distillate, whereas the vapor from the final effect is condensed by a circulating seawater cooling stream.

Theoretically again, an additional kg of distillate would be obtained in each consecutive effect for the same kg of steam initially introduced into the first effect and the plant performance ratio would be equal to the number of effects in operation. However, this is not true in practice. Part of the condensation heat to be recovered is lost to the atmosphere, in design features and in the differences of temperature used as the plant's driving force.

Multiple-effect evaporators have been used in industry for more than 100 years and have been in use for seawater distillation for about 70 years. The latter were originally built for shipboard use, the main requirements being for compactness, simplicity in operation and reliability. In land based industrial evaporator plants the requirements are mainly directed to the cost of product water with emphasis on cheaper materials of construction, high boiling temperatures, efficient descaling methods and the use of the cheapest type of evaporator.

The usual method of feeding a multiple-effect evaporator is the forward feed. Seawater is introduced in the first effect and the concentration of the brine is increasing from the first effect to the last. Pumping work is reduced as the transfer of the brine from effect to effect is cascaded down by the decreasing pressure.

Pumping the feed backward from the last to the first evaporator needs a pump between each effect and has the additional disadvantage of giving the highest salt concentration at the highest temperature of the system. The brine is rejected at a higher temperature, thus requiring additional heat exchange devices.

The usual method of forwarding the feed water in the evaporator tubes is in a thin downflow film. The effect of downflow and upflow in a vertical-tube evaporation plant was evaluated on a 2.5 Mgd conceptual design plant [13]. The upflow plant was identical to the downflow plant, except where differences were necessary to carry out the upflow scheme. The principal difference between the two plants is that the upflow plant does not require intereffect pumps, but does require slightly more area. Capital cost for the 2.5 Mgd plant is about 6% lower for the upflow plant. This advantage improves as the size increases to 10 Mgd. Water cost is lowered by about 4.4% because of lower capital charges and lower electrical costs.

Another advantage of the multiple-effect evaporator design is the possibility to use extraction steam taken between the effects to increase thermal efficiency. Flash tanks arranged in the distillate stream between each effect may supply additional steam to the next evaporator effect, in using the available heat by flashing the distillate to the temperature and pressure of the succeeding effect. The downward cascading of brine also produces a fraction of vapor by flashing inside of each evaporator effect. The total amount of vapor flashed off from both sources is proportional to the temperature drop between effects. Nevertheless, flash of distillate does not represent additional product water.

The flow sheet of a typical multiple-effect vertical tube distillation plant is shown in Fig. 2–7, p. 52, describing the Freeport demonstration plant.

In general, the average efficiency K of each effect is usually between 0.85 and 0.95. From the relation

$$R = \frac{K\,(1 - K^N)}{1 - K}$$

it is found that the performance ratio R is slightly over 10, when K = 0.95 and the number N of effects is 15. Doubling the number of effects to 30, the performance ratio is increased only to 14.9 and it will attain a maximum of 19 for an infinite number of effects, when K = 0.95. The optimum number of effects has to be defined from case to case by an economic analysis.

Economic limitations in increasing the number of effects are as well valid. The investment costs and consequently the fixed charges are increasing almost linearly with the number of effects. The costs of steam and water fall off rapidly at first, but the savings are progressively diminishing. The total cost of operating an evaporator leads to an optimum number of effects, at the point where the sum of fixed charges and the cost of utilities shows a minimum. The most probable number of effects will be between 10 and 20.

A mathematical model and computer program for optimization of vertical-tube evaporator water desalting plants was developed by Houston Research Institute. The module approach was adopted in developing the complete system, which is divided into four major subsystems, namely: multi-stage evaporator process simulation, engineering design simulation, capital and operating cost calculation and finally optimum searching techniques [1].

It is reminded that in the vertical-tube evaporation process, the terms "effect" and "stage" are identical, each stage being an effect. The same can not be applied for the multi-stage flash distillation process (see section 2.6.2, p. 74).

The described process model basically consists of the following three components: Determination of the basic process variables of individual evaporator stages by solving the process material- and energy-balance matrix equations, design of an individual evaporator, i.e. heat transfer area and size to meet the process conditions determined above, and design of the auxiliary heat exchangers.

A generalized digital computer program, EVAPOCHALM, was also proposed by Alsholm [2]. The program is flexible and can be applied in operating plants, in design for new and enlargement of old plants.

A Fortran code, ORVEM, for the calculation of a desalination plant employing mixed feed in vertical effect evaporators was proposed by Ebel et al. All pertinent plant overall parameters such as temperatures, flows and concentrations are calculated along with a detailed cost analysis of the various plant components. The plant consists of a vertical-tube effect evaporator, enclosed in a steel and concrete shell (see section 2.11.9, p. 107). The program outputs all necessary plant operating parameters, including effect-by-effect values, as well as a detailed cost analysis [14].

A mathematical model, representing the multi-effect evaporation process and a computer program, MULT-EFFECT, for purposes of constructing seawater distillation plant flowsheets was presented by Guneratne. The more important equations are given in full, while the others are implied by a description of the factors to be considered [15].

Experimental results reported in the literature involving the evaporation of water and of brine in falling film evaporators have been compared to predictions in which the local heat transfer coefficients α were used, correlated as

$$\alpha = m\,(\Gamma/\eta)^n$$

where m and n are constants depending on the flow regime of the falling film, Γ is the mass flow rate per unit width of the wall and η is the absolute viscosity. The relative good agreement between theory and experiment supports the use of those coefficients for design [3].

The even distribution of brine between the several tubes in an effect is of primary importance. Various devices and weirs have been developed for this purpose [4]. A nozzle has been developed which controls the flow to each tube to within allowable limits and produces an even film around the circumference of each tube [5].

The first falling film vertical-tube evaporator was constructed by the Etablissements Kestner in France in 1905. Although at the beginning widely used only in chemical engineering, the vertical-tube evaporator is developing now to a leading contender for seawater distillation as well [6].

LTV evaporators with counter-current flow and plastic tube heat exchange surfaces were reported recently as having certain advantages over the classical design [7], mainly in the investment cost. Steam is condensed inside plastic vertical tubes and the brine flows down outside the tubes.

In concluding, it becomes apparent that there is a trend favoring vertical-tube evaporation over a multi-stage flash evaporation for reasons, which will be discussed in later section (2.9.1, p. 95).

The state-of-the art design, performance and economics of downflow thin film vertical-tube evaporator desalting plants were recently analyzed and reviewed by Blevitt and Curran. Both single- and dual-purpose plants ranging in size from 1 to 250 Mgd (3785 to 947000 m³/day) capacity were evaluated. The largest U.S. commercial plant is 1 Mgd with 11 effects at a performance ratio of 9.3 (Virgin Islands) and the largest experimental test bed is also 1 Mgd nominal capacity with 17 effects and a performance ratio of 13.2 (Freeport, Texas). Designs ranging up to 250 Mgd have been prepared. Recent innovations include integration of multistage flash preheaters with the VTE equipment (see section 2.11.4/5, p. 102/5, 2.11.11/14, p. 109/12), yielding apparent economies in materials, fabrication and operation. Current technology emphasizes the use of enhanced surfaces for improvement of evaporation/condensation heat transfer (see section 2.3.7, p. 58). Concrete lined steel appears to offer long term corrosion resistance to hot seawater brines. Current plants and designs yield from 0.75 to 0.85 lb of product water per 1000 Btu (initial heating steam) per effect at concentration ratios of 2.5 to 3.0 Aluminum-brass or 90-10 CuNi are the most attractive materials for evaporator tubes [16].

Demonstra-
tionsanlage
in Freeport,
Texas

2.3.1 Freeport, Texas, demonstration plant

In the original design (**Fig. 2–7**) the plant consisted of twelve long tube vertical (LTV) evaporators arranged for falling film operation. The twelve effects were operated in forward feed manner with seawater. The plant had been designed to produce 1 Mgd (Million gallons per day) potable water containing no more than 50 mg/l total dissolved solids. Low pressure steam is introduced in the first effect. Vapors generated from the seawater flowing down the heating element tubes in each effect are used in the next effect as the heating medium. Part of the vapors from the 11th effect are used to strip dissolved gases from seawater feed in the deaerator. Noncondensable gases are removed from the deaerator and the final condenser by means of a vacuum pump to maintain absolute low pressure conditions.

Fig. 2–7

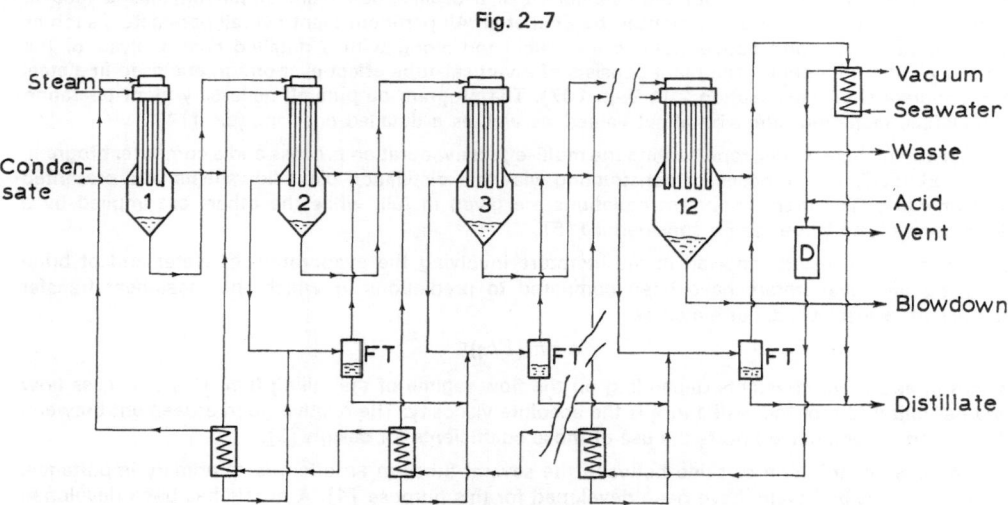

Flowsheet of a typical multiple-effect vertical
tube distillation plant. D = deaerator, FT = flash tank.

The raw seawater feed is preheated by heat exchangers, cooling the product water, and sprayed at the top of the deaerator. In the original design a slurry of $Mg(OH)_2$ was introduced in the lower portion of the deaerator and mixed with the deaerated seawater feed to an 1% by weight mixture. This liquor was pumped through a series of heat exchangers and preheated, by cooling the outgoing product water stream, up to a temperature slightly under the boiling point in the first effect. Final heating of the seawater feed to its boiling point was accomplished in the heating element of the first effect. The brine from each effect was pumped to the top waterbox of the next effect, in parallel with the vapors, until finally it was rejected from the 12th effect with a concentration factor of 4. The blowdown was clarified, according to the original design, in a thickener for the recuperation of the $Mg(OH)_2$, to be reused in the deaerator.

Product water was collected from the steam chest of each effect, as well as from the flash tanks. The condensate formed in the first effect from prime steam is at a temperature of 127.3°C. It is progressively cooled down to 31°C to be supplied to the consumer.

The Freeport plant was the first to use 12 effects in series, the first to use the falling-film-type evaporator for seawater and the first designed to operate at 121°C and above. The only important process design change was the abandonment of the $Mg(OH)_2$ sludge recycle method of scale prevention. This was replaced with the acid injection method of scale control, upstream of the deaerator, which now also acted as a decarbonator.

Table 2/1

Revised Freeport plant characteristics.

Plant	Ca removal in %	Brine temperature in °C	in °F	Concentra- tion factor	Gross capacity in m³/day	in Mgd*)	Performance ratio in kg/kg
Existing		111.1	232	2.87	3785	1.0	10.95
Revised	50	135.0	275	4.0	5424	1.433	11.1
Revised	75	148.9	300	5.0	6859	1.812	11.1

*) Million gallons per day

Table 2/2

Design comparison 12- and 17-effect plants.

Design variables		12-effect	17-effect
Nominal capacity	Mgd	1.0	1.0
Maximum brine temperature	°C	121.1	129.5
Heat rejection temperature	°C	46.1	37.8
Evaporation heat transfer surface	m²	5811	5342
Preheating heat transfer surface	m²	3508	3428
Heat rejection transfer surface	m²	389	232
Total heat transfer surface	m²	9708	9002
Number of heat exchangers		27	19
Number of installed pumps		32	25
Power demand	kW	360	450

A systematic study of the solubilities of calcium sulfate and a review of the scaling in the plant have led to the conclusion that the first effect temperature should not exceed 115.6°C (240°F) to prevent $CaSO_4$ formation, and that the last effect concentration factor should not exceed three times normal seawater to prevent $CaSO_4 \cdot 2H_2O$ formation [8].

Removal of 50% of the calcium in the feed water would allow operation at 135°C in the first effect and removal of 75% of the calcium would allow operation at 149°C. The calculated plant operation parameters as a function of the calcium removal from the feed water are shown in Table 2/1.

Table 2/3
Average operating achievements, 12- and 17-effect plants.

Plant parameter		Design value		Operation	
		12-eff.	17-eff.	12-eff.	17-eff.
Number of runs				12	4
Plant capacity	t/h	158.8	158.8	158.1	150.3
Seawater feed	t/h	222.3	231.3	215.6	214.1
Computed extraction ratio		0.710	0.680	0.673	0.661
Steam to effect 1	°C			127.8	137.2
Brine out	°C			50.0	38.3
Temperature difference	Δt			77.8	98.9
Steam to effect 1	kg/h	14.97	11.34	14.64	11.58
Performance ratio	kg/kg	11.5	14.8	10.97	13.2
Energy kWh/t seawater feed		1.62	1.95	1.80	2.45
Product water	kg/kWh	2.14	1.71	1.90	1.38
Operating time	%	90.5	90.5	92.83	86.54

The Freeport plant demonstrated the reliability of the vertical-tube falling-film multiple-effect evaporator desalting process. The plant was capable of producing in excess of the 1 Mgd rated capacity, utilizing only two-thirds of the installed evaporation heat transfer surface. The feasibility of further improvements in operation and cost was studied and it was decided to add five more effects to the existing twelve, bringing the total number of effects to 17. It became apparent that only minimal changes in the existing equipment and operating procedures were necessary to improve efficiency and reduce costs. A comparison of some plant design parameters [9] of the 12- and 17-effect plants are summarized in Table 2/2, p. 53. Average operating achievements during several development runs are given in Table 2/3.

Entsalzungs-anlagen in Shevchenko

2.3.2 Shevchenko desalting plants

A group of LTV (long tube vertical) desalting plants was erected at Shevchenko, on the Caspian Sea. A 4-effect advanced plant was constructed first, with a daily capacity of 4500 m³ and natural brine circulation. The heat transfer surface of each evaporator is 1000 m² and corrosion problems were avoided by using 18-8 stainless steel [10]. This plant was followed by one 5-effect LTV-plant with a nominal capacity of 13000 m³/day and two others with a capacity of 14000 m³/day each. All plants are using the seeding technique for preventing scale deposition. The cross-section of the Shevchenko evaporator is shown in **Fig. 2–8**.

A larger plant with 10 effects forced circulation fluted tubes evaporators in 3 units was put in operation in 1970. The plant is as well using the seeding technique for scale prevention. Each evaporator has a heat exchanging area of 1600 m². The average working capacity of each unit is 17000 m³/day. The quality of the product water is suitable not only for drinking water supply, but as well as feed water in high pressure boilers. The main characteristics of the plant operation are as follows [11]:

Steam consumption	78 to 82	t/h
Seawater feed	870 to 930	t/h
Boiling temperature in 1st effect	102	°C
Boiling temperature in 10th effect	37 to 40	°C
Heat transfer coefficient in 1st effect	5000	kcal/m² · h · degr
Heat transfer coefficient in 10th effect	3900	kcal/m² · h · degr
Performance ratio distillate/steam	8:1	kg/kg
Energy per t distillate	4.2	kWh/t
Seawater distillate ratio	5.1	t/t
Concentration factor	3.7	
Extraction factor distillate/seawater	0.73	t/t

Construction is now been completed to one more 5-effect LTV-evaporation plant with a capacity of 14000 m³/day to meet the requirements of the power plant, the oil refinery and the city of Krasnovodsk [12].

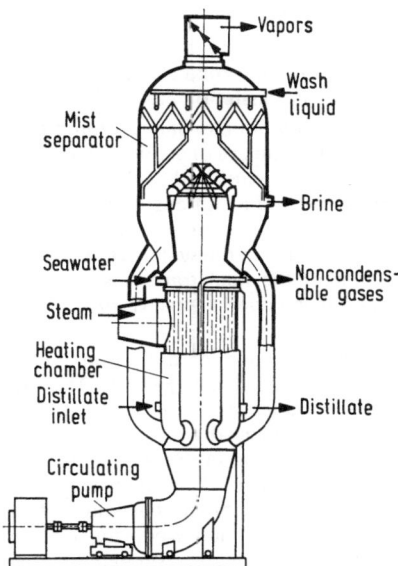

Fig. 2–8

Cross section of the
Shevchenko LTV evaporator.

Literature to 2.3 to 2.3.2

[1] A. E. Dukler, C. J. Huang, M. L. Lee (Off. Saline Water Res. Develop. Progr. Rept. No. 404 [1969]); C. J. Huang, H. M. Lee, A. E. Dukler (Desalination 6 [1969] 25/56). — [2] H. Alsholm (Chim. Ind. Genie Chim. 103 [1970] 1075/81). — [3] K. R. Chun, R. A. Seban (J. Heat Transfer 94 [1972] 432/6). — [4] D. Kays (Proc. 1st Intern. Symp. Water Desalination, Washington, D.C., 1965 [1967], Vol. 3, p. 23/56). — [5] R. I. Hawes (Proc. 3rd Intern. Symp. Fresh Water Sea, Dubrovnik 1970, Vol. 1, p. 275/89).

[6] R. Leleu, R. Emig, F. Lauro (Proc. 3rd Intern. Symp. Fresh Water Sea, Dubrovnik 1970, Vol. 1, p. 325/35). — [7] F. Lauro (Proc. 4th Intern. Symp. Fresh Water Sea, Heidelberg 1973, Vol. 1, p. 357/62). — [8] D. L. Preddy (Proc. 1st Intern. Symp. Water Desalination, Washington, D.C., 1965 [1967], Vol. 2, p. 337/66); Dow Chemical Co. (Off. Saline Water Res. Develop. Progr. Rept. No. 139 [1965]). — [9] G. H. Shroff, I. C. Watson, R. D. Cross (Off. Saline Water Res. Develop. Progr. Rept. No. 759 [1971]). For annual reports of plant performance see: Off. Saline Water Res. Develop. Rept. No. 71 [1963], 100 [1964], 123 [1965], 171 [1966], 253 [1967], 297, 298, and 299 [1968], 440 and 479 [1969], 559 and 739 [1971]. — [10] F. P. Zaostrovskii, E. P. Novikov, V. G. Shatsillo, S. I. Golub, V. B. Chernozubov, V. I. Tkach (Proc. 1st Intern. Symp. Water Desalination, Washington, D.C., 1965 [1967], Vol. 3, p. 97/106; Desalination 1 [1966] 165/77).

[11] E. A. Sobolev, S. I. Golub, V. B. Chernozubov, N. K. Tokmantsev, S. B. Ruchin, V. I. Tkach, V. L. Podbereznii, N. S. Konev, V. G. Shatsillo, E. P. Novikov, M. B. Viceblat, L. S. Mreyin (Proc. 3rd Intern. Symp. Fresh Water Sea, Dubrovnik 1970, Vol. 4, p. 103/10); A. P. Egorov, E. A. Sobolev, V. L. Podbereznii, S. I. Golub, V. I. Tkach (Proc. 4th Intern. Symp. Fresh Water Sea, Heidelberg 1973, Vol. 1, p. 245/52). — [12] N. M. Sinev, V. G. Shatsillo, I. P. Lazarev (Proc. 4th Intern. Symp. Fresh Water Sea, Heidelberg 1973, Vol. 1, p. 481/5). — [13] Fluor Corporation (Off. Saline Water Res. Develop. Progr. Rept. No. 698 [1971]). — [14] R. A. Ebel, R. O. Friedrich, I. R. Parsley (ORNL-TM-3314 [1971]). — [15] D. C. Guneratne (At. Energy Estab., Winfrith, Engl., AEEW R - 860 and 861 [1973]; Desalination 13 [1973] 343/58).

[16] R. Blevitt, H. M. Curran (Off. Saline Water Res. Develop. Progr. Rept. No. 896 [1973]).

2.3.3 Packing of evaporators

Verdampfer mit Füllkörpern

Packing the working space of evaporators with a spiral packing or Raschig rings increases the mixing of the liquid in the evaporator and as a result increases the heat transfer coefficient [5].

*Einfluß
akustischer
Schwin-
gungen*
2.3.4 Effect of acoustic vibrations

The feasibility of using acoustical energy to improve the economy of evaporator operation was investigated toward three objectives: a) increased water side film heat transfer coefficients, b) possible promotion of heat transfer on steam side and c) reduced scale formation.

A research program has demonstrated that improvement in water side heat transfer coefficients can be obtained by the use of acoustic vibrations. Pressure drop measurements indicated no significant difference for flows with and without vibration. Transverse vibration of the inner tube produced significantly increased heat transfer rates. Although percentage improvement was greater at low flow Reynolds numbers, the economic advantage increased with increasing Reynolds number. The effect of acoustic vibrations in the water was considerably less in improving heat transfer than that obtained with vibration of the heat transfer surface. Increases in steam condensation rates, as well as in boiling heat transfer coefficients, were also observed due to the effect of vibration of the heat transfer surface [6].

Mechanical vibration of a pipe might have two possible mechanisms for maintaining high heat transfer rates in a scaling liquor. The mechanical bending of the pipe might cause flaking of the scale and thus leave the pipe surface clean for efficient heat transfer. On the other hand the increased agitation caused by the vibrating pipe would reduce the stagnant water film next to the heat transfer surface and in turn reduce the tendency to scale. A calcium sulfate solution in distilled water was used as the scaling liquor in an experimental investigation. It was found that vibrations definitely cause flaking of the calcium sulfate scale. Better removal of scale was obtained under non-boiling conditions than when the water was boiling. Heat transfer coefficients under boiling conditions, however, were considerably higher with vibration than without vibration, indicating an improvement of 60%.

The influence of acoustic vibrations on heat transfer was found to be significant in the presence of acoustically induced cavitation, whereas in absence of such cavitation the effect was negligible. It was, therefore, concluded that the creation of acoustically induced cavitation is a more promising method of controlling heat transfer than a method using such vibrations in the absence of cavitation [7].

A further investigation was made in an "acoustic water tunnel" [8], which contained a saturated solution of calcium sulfate, on the influence of acoustic vibrations upon scale formation. The data obtained provided detailed information on the deposition of calcium sulfate scale on a heated cylinder in crossflow under conditions of constant heat flux and demonstrated that acoustically induced cavitation can remove deposits of calcium sulfate scale from heated surfaces.

The effect of pulsation in water flow on the overall heat transfer coefficient of a shell-and-tube heat exchanger with steam in the shell has been investigated by Keil and Baird [9].

*Tropfen-
kondensation*
2.3.5 Dropwise condensation

The dropwise condensation of steam is another means of enhancing heat transfer rates. It usually requires specially treated condensing surfaces. The bibliography on dropwise condensation has been compiled from 1930 to 1964 by Erb [10].

Dropwise condensation is promoted by forming a hydrophobic surface by chemisorbing an appropriate organic material on the metal condenser surface. Many of the materials used have short lifetimes and the treated surface trends to filmwise condensation within a few days. Some sulfur-bonded compounds have reported lifetimes up to several thousand hours under laboratory conditions. Much shorter lifetimes were found with the use of industrial steam. This disadvantage may be bypassed by periodic or continuous injection of organic promotor into the steam [11], but the injection technique does not appear to represent a practical solution to processes such as the distillation of seawater [12].

It was found in early work that sulfides formed on copper alloy and silver surfaces promoted dropwise condensation. However, their lifetimes usually limited to about 1000 hours, particularly with copper alloys. Theoretical considerations indicated that any metal surface substantially free of oxide film would exhibit hydrophobic behavior and on this basis the noble metals were found to promote dropwise condensation. Permanent dropwise condensing systems based on the use of thin coatings of noble metals, such as silver, palladium and rhodium, were developed and produced good dropwise condensation. Gold, above all, with a solid specimen and a plated specimen, has produced consistently good dropwise condensation through more than 18 months in continuing test in steam at 101°C [13]. The laboratory and pilot plant findings have been confirmed in a condensation test unit installed as part of the first effect in the Senator Clair Engle Test-Bed Plant in San Diego, California, as well as in the Wrightsville Beach Test Facility.

Discontinuous gold coating of tubes makes it possible to expect a 20% decrease of investment cost. Adding one or more pairs of fins fitted along the tube to drive off the liquid film is also a way to increase the condensation side heat transfer coefficient [14]. Addition of fluorinated organic promoters, especially a disulfide, maintained dropwise condensation for about 500 hours with a maximum promoter consumption of 40 mg/m^3 of condensate. Condensation heat transfer coefficient was drastically enhanced up to a factor of 12-fold [15].

Research work on dropwise condensation on metallic surfaces, with or without organic promoters, are continuing in various countries and on different tube materials, including copper, gold, admiralty metal, Cu-Ni 90-10, and Monel tubes [16]. Although the measured steamside dropwise heat transfer coefficients are very high, being in the range of 106 500 for Cu-Ni up to 194 000 kcal/m$^2 \cdot$ h $\cdot$ degr for copper, condensation promoted by paraffinic mercaptans or thiosilanes, dropwise condensation is still not reliably attained in evaporation equipment.

One of the reasons for the great increase in heat transfer and the condensation rate with dropwise condensation appears to be that the steam is cooled directly on the bare metal areas between the drops. In contrast to this a continuous film of liquid water, having a low thermal conductivity, covers the entire metal surface in the case of filmwise condensation.

Chiba investigated the surface temperature fluctuation during dropwise condensation. An abrupt temperature change was observed when a drop flowed over the measurement point. The heat flux in condensation showed that the bare surface, either created by sweeping drops or by coalescing drops, contributed more to the heat transfer [17].

Wenzel assumes that transfer of heat is taking place exclusively on the surface of the condensing droplets and not on the nonwetted cooling surface [18]. Graham concludes that at atmospheric pressure drop conduction was the limiting resistance, while at lower pressure interfacial heat transfer was as important as drop conduction. The most important drops for heat transfer were found to be those less than 10 μ in diameter [19].

Hurst and Olsen report that both experimental and analytical results pointed to the existence of an area of very high heat transfer around the droplet perimeter and to the importance of the condensing wall as a heat diffusing mechanism in dropwise condensation [20].

A physical explanation for dropwise condensation is given by Mikic, who is considering constriction phenomena in metallic surfaces, caused by a non-uniform heat flux distribution over the condensing surfaces. Surface thermal properties are one of the controlling factors in the dropwise condensation, especially the resistance occuring in the heat transfer plate [21]. The constriction-resistance theory was modified by Bromley to allow for the additional constriction-resistance when condensation occurs on a thin-walled tube [22]. If a promoter is used, the effectiveness of the dropwise condensation is dependent on the quality of the promoter and its ability to be strongly absorbed on the metal surface.

2.3.6 Parylene coatings for dropwise condensation

Another means of promoting dropwise condensation has been found in the use of Parylene vapor-deposited polymer films in extremely thin layers, from 1 μ down to 0.25 μ, on the metallic condensing surface with a chromium underlay. While the increases in heat transfer coefficient may not be as great as with the noble metal coatings, the added cost of the chromium/Parylene coating system is substantially less than that of the noble metal systems.

Parylene is a polymer formed from di-para-xylene dimer by a Union Carbide process:

*Parylen-
Überzüge
für die
Tropfen-
kondensation*

The effectiveness of Parylene coating was tested in a heat exchanger with Cu-Ni 90-10 tubes. Although this exchanger had shown a 13% increase in efficiency, maximum benefits derivable from the Parylene coating were restricted.

Other coating materials tested included Teflon TFE, Teflon FEP, and Ni phosphide [23].

*Erhöhung des
Wärme-
durchgangs
durch
Rillrohre*

2.3.7 Enhancement of heat transfer by fluted tubes

Vertical tube evaporators conventionally have been provided with smooth tubes. The overall heat transfer coefficient for these are 2440 to 2930 kcal/m² · h · degr (500 to 600 Btu/ft² · h · °F) at atmospheric pressure for both upflow and downflow, if the feed is not foaming.

Marked enhancement of film condensation on vertical surfaces was reported by Gregorig. By machining flutes (vertical grooves) in the outer surface of the tube, increases of 200 to 800% in the average film condensation coefficient were obtained [24].

Contouring the evaporating side surface of the tube was later shown to cause a marked increase in the heat transfer coefficient for thin film evaporation [25].

The primary objective of the fluting is to promote high heat transfer coefficients on the condensing side. The changing radius of curvature of the condensate surface on a fluted profile will, because of the high surface tension, cause a flow of condensate from the crests to the valleys of the flutes. The film of condensate on the crest, therefore, remains very thin and the condensing coefficient in this region is extremely high. **Fig.** 2–**9** shows the frontview and the cross-section of typical double fluted tubes and **Fig.** 2–**10** the condensate flow on the outer surface of the fluted tube.

Fig. 2–9

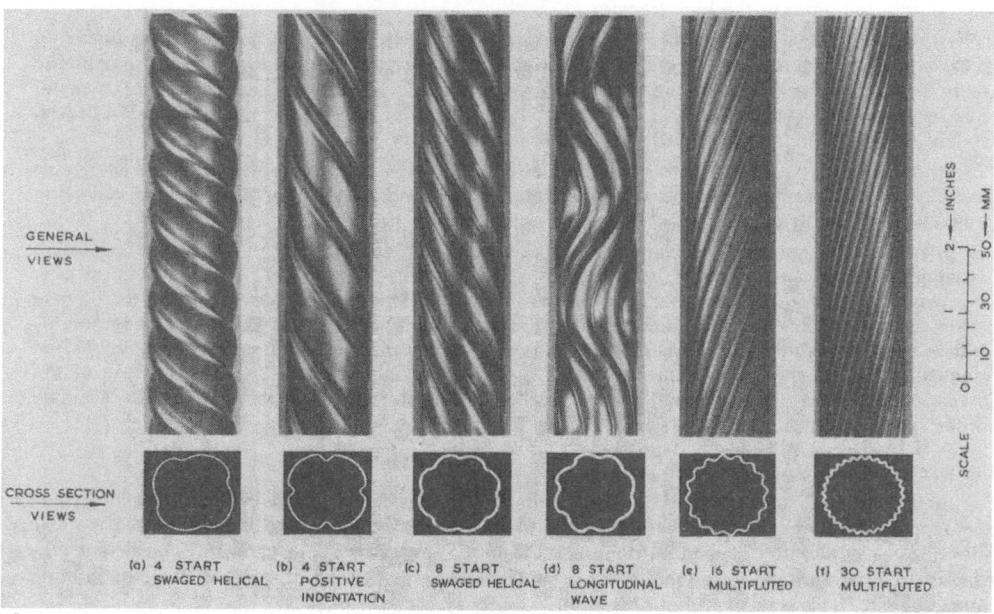

Cross section and general view of typical double fluted tubes.

A large part of the cost of a VTE (vertical-tube evaporator) plant (up to about 50%) is for its heat transfer surface. Fouling factors of 15 to 25% on heat transfer coefficients were used, because of inadequacies in design correlations. This excess can represent an important additional investment in large plants. To determine the optimum operating design and operating conditions, accurate experimental heat transfer, pressure drop, and entrainment data were obtained on a single smooth tube in a VTE process [26]. These experiments were continued on double fluted tubes in laboratory and pilot plant scale. The improved performance of a double fluted tube was considered to be a function of the pitch and depth of the grooves of the fluted surface. Profile shape of the surface was not decisive on heat transfer enhancement [27].

Extensive research was performed in a five-effect vertical-tube evaporator pilot plant at the Oak Ridge National Laboratory. The plant, which could evaporate a maximum of 20 000 gpd (gallons per day) of water under optimum conditions, has been tested in both the rising-film and the falling-film modes of operation using fresh water and a 3.5 wt.-% NaCl solution feeds. Improved heat transfer coefficients have been achieved using evaporator tubes having fluted or grooved surfaces [28]. The pilot plant was then moved to the O.S.W. Wrightsville Beach Test Facility for further test operations with seawater. The enhanced heat transfer coefficients, obtained at Oak Ridge, were confirmed. Enhanced tubes with surfaces modified in various ways to promote condensation and/or evaporation exhibited, under comparable conditions, overall coefficients several times larger than those of smooth tubes [29].

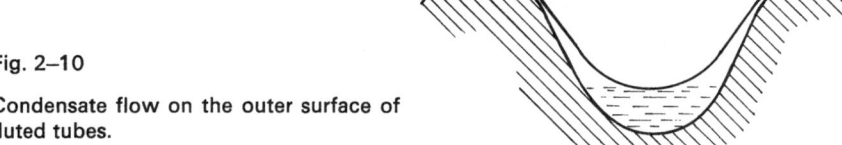

Fig. 2–10

Condensate flow on the outer surface of fluted tubes.

With fresh water feeds in downflow through double fluted and spirally corrugated tubes, the overall coefficients were two to three times larger than those of smooth tubes under the same conditions. With suitably foaming flow, coefficients five to six times larger were observed. Condensing-side coefficients were of the order of 48 900 kcal/m² · h · degr (10 000 Btu/ft² · h · °F) at atmospheric pressure. Evaporating-side coefficients were of the order of 10 750 kcal/m² · h · degr (2200 Btu/ft² · h · °F) for non-foaming flow and over 24 400 kcal/m² · h · degr (5000 Btu/ ft² · h · °F) for strongly foaming flow. Research work performed at Oak Ridge National Laboratory is reported in detail in three O.S.W. reports. The second contains, in tabular and graphical form, data obtained and is an addendum to the main report [30].

Further research on the performance of fluted tubes in heat transfer is reported by Hawes [31], including the development of suitable flute profiles. Increase of evaporative heat transfer coefficients of thin films of water flowing down the inside of vertical tubes was obtained by Thomas and Young by clamping longitudinal rectangular fins to the tube surface [32].

A general account is given by Mattern [33] on the performance of tubes with helically wound surface and variable cross-section in the axial direction.

The laboratory and pilot plant findings have been verified on industrial scale by including a bundle of 151 double fluted tubes in the 13th effect of the enlarged Freeport plant. An initial 250 % improvement in performance was observed, which was followed by a slight but gradual decrease. After over 6000 hours in evaporation service the bundle had a 2 : 1 performance ratio compared to the original smooth tube bundle [34].

Further research work reported at the Heidelberg Symposium by Beccari et al. [35] and by Newson and Hodgson [36] exhibit heat transfer coefficients on fluted tubes of the same order of magnitude.

The recently erected vertical-tube evaporator desalting plant in Gibraltar, with a capacity of 0.3 Imperial Mgd (1 Imperial gallon = 1.20095 U.S. gallons), has 13 effects all equiped with fluted tubes 3 inches in diameter. The length of the fluted tubes is 3.05 m (10 ft) for effects 1 to 12 and 3.96 m (13 ft) for the 13th effect [37].

In the commercially available fluted tubes, the usual surface configuration is an approximately 90° V-profile, the flutes being straight and parallel to the tube axis. The number of flutes varies according to the diameter of the tube. The tubes have integral plain ends and a minimum flute thickness of about 0.8 mm (0.030''). Fluted tubes are designed for use with their axis vertical, with steam condensing on the external surface and the brine feed evaporating internally.

Fluted tubes are also made with the flutes spirally-oriented instead of longitudinal. The spiral configuration has some advantages, e.g. a greater degree of rigidity than longitudinally double fluted tubes of equivalent thickness. A typical case of heat transfer enhancement of fluted tubes compared with smooth tubes is illustrated in **Fig. 2–11**.

A method was developed for determining local boiling heat transfer coefficients in falling film evaporators with corrugated surfaces, without the usual restriction of having to describe conditions on the condensing side as well. Corrugations increased heat transfer coefficients and channelled the flow while keeping the falling film distributed evenly across the surface under boiling conditions. Coefficients in the range 3400 to 22000 kcal/m² · h · degr (700 to 4500 Btu/ft² · h · °F) for deionized water and 14600 to 48800 kcal/m² · h · degr (3000 to 10000 Btu/ft² · h · °F) for saline water were obtained. Rougher surfaces increased the heat transfer coefficients by increasing nucleation [38].

In another report seven different enhanced tubes were evaluated. The Grob 54-flute flat-ridge aluminum-brass tube appears to exhibit the best heat transfer performance. It is closely followed by the Grob 54-flute sharp-ridge tube and the Yorkshire 50-flute tube. The brine feed rate shows a rather noticeable effect on the heat transfer performance. The effect of Δt on the heat transfer performance of a tube seems to depend upon the evaporating temperature level. For the Grob 54-flute tube, a factor of 1.55 should be applied to the average wall thickness for the correction of tube material from 90-10 CuNi to Al-brass and vice versa [39].

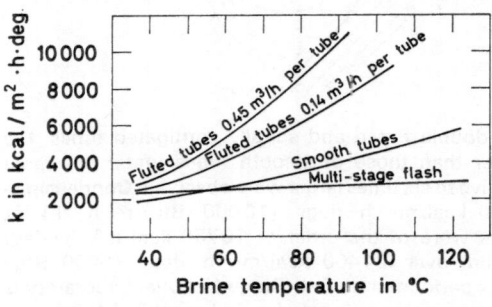

Fig. 2–11

Overall heat transfer coefficients with double fluted tubes.

The performance of horizontal tubes with transverse fins was investigated for two plain and six finned copper tubes for boiling heat transfer. Computation of the optimum transverse fin of constant thickness showed that in general they are short, thin fins as close together as possible without getting interference between fins [40].

The possibilities of cost savings in using tubes with specially formed surfaces in desalination plants were assessed by Hauck and Zemlin [41] and further fields of applications were examined.

The current state of the art of commercial enhanced surfaces for thin film evaporation and condensation on vertical tubes is outlined by Thomas [42]. Fluted tubes exhibited overall heat-transfer coefficients of 6350 to 9800 kcal/m² · h · degr (1300 to 2000 Btu/ft² · h · °F) at atmospheric pressure. Experimental surfaces have been tested which gave overall coefficients up to 24400 kcal/ m² · h · degr (5000 Btu/ft² · h · °F). Although the mechanism responsible for high performance on the condensing side of fluted or finned tubes seems relatively well understood and upper limits can be estimated, the mechanism on the evaporating side seems to be much more complicated. Consequently, an upper limit for heat transfer rates on the evaporating side was estimated assuming that the upper limit occurred when the water film thickness was just sufficient to prevent dry spot formation. Assuming upper limits of 88000 and 69400 kcal/m² · h · degr (18000 and 14200 Btu/ft² · h · °F) on the condensing and evaporating sides, respectively, gives an estimated upper limit on performance of enhanced surfaces of about 29300 kcal/m² · h · degr (6000 Btu/ft² · h · °F).

Vortex-
Fallfilmrohr

2.3.8 Vortex flash-tube

Two-phase flow control was applied to the distillation of saline water in a falling film vertical-tube evaporator with the purpose of enhancing heat flux through the wall of the vertical tube. Flow control was effected by imposing a continuous vortex upon the vapor phase flowing through the tube by means of a helically twisted metal ribbon of a width slightly less than the inside diameter of the distillation tube, inserted and mounted coaxially in the tube and leaving an annular gap adjacent to the tube wall. The brine annulus on the heat transfer surface is thus manipulated through shear effects at the liquid-vapor interface in vortex flow by the vapor phase core. Entrainment of liquid droplets in the vapor phase is reduced and a spiral flow pattern is induced in the liquid film, thus elongating its contact route on the heat transfer surface.

A test facility was erected at the University of California, Berkeley, to evaluate this process in a pair of copper tubes (50 mm diameter, 6.1 m length). In one of these, the vortex tube, helical baffles were tested, while the second plain tube provided reference data. Experimental results, under varying process variables, showed that heat flux gains between 30 and 90% were obtained under typical VTE (vertical-tube evaporator) process conditions by mounting a helical baffle of selected pitch and width in the vortex tube [43].

Further tests were made with tubes of 79.4 mm outer diameter at various operation conditions. Pressure drops were lower than those through the 50 mm tubes, but heat transfer gains were also lower. Significant gains were always associated with higher pressure drop in the vortex tube, compared to that through the reference tube and at relatively high Δt values.

Tests were also made with pairs of double fluted and spiral-corrugated tube types in downflow of the brines. Heat transfer gains were in general lower, compared to those in the smooth tube. Both types of enhanced tubes showed a rapid initial degradation in overall heat transfer coefficients down to stabilized conditions. Heat transfer with the double fluted tubes was more sensitive to a vapor pressure drop through these tubes than through the smooth tubes [44].

2.3.9 Interface enhancement

Another method of enhancing the brine-side heat and mass transfer is the addition of small quantities of selected surfactants to the feed brine. Bubble formation and bubble flow provide enhanced interfacial surface area and increase both heat and mass transfer in vertical-tube evaporators (VTE). This method was tested in a single effect, 5000 gpd (19 m³/day), 18-tube VTE pilot plant in upflow. Double fluted, spiral-corrugated and smooth distillation tubes were used.

Tests over the entire upflow VTE temperature range were performed under typical process conditions. The interface enhancement method was found to be effective with all kinds of tubes used, establishing that the method is independent of brine-side tube modifications. One of the major advantages is that an important increase in the brine evaporation rate is produced. About a 100% enhancement of the overall heat transfer coefficients was indicated for a multiple-effect upflow VTE utilizing enhanced distillation tubes, after the addition of about 5 mg/l of a selected surfactant to seawater feed.

In **Fig. 2–12**, p. 62, the heat transfer enhancement and the hydrodynamic pressure drop are shown for a double fluted aluminum-brass tube in response to the gradual addition of surfactant to seawater feed, while the process conditions were maintained constant at an evaporation temperature of 99°C. Addition of a few mg/l of surfactant causes a dramatic reduction in the tube side pressure drop Δp and a significant enhancement in the overall heat transfer coefficient k. Further addition of surfactant causes a further increase in heat transfer until a maximum effect is approached at about 50 mg/l of additive. The pressure drop usually goes through a minimum and shows a very slight rise with higher surfactant concentrations.

The use of surfactant additives in upflow VTE provides for substantial heat transfer enhancement and for good upflow stability with relatively low intereffect Δt values, as compared with previous work [45]. An increased number of effects can be accommodated in the interest of an increased performance ratio, when interface enhancement is utilized [1]. Neodol DMD-1 (Shell Chemical Co.) was the most effective and the least expensive surfactant of about ten commercial products examined.

Although best suited to upflow VTE, the interface enhancement method is also applicable to downflow VTE, horizontal-tube evaporation and flat-plate evaporation. It also provides advantages for multi-stage flash evaporation.

Similar findings were reported as well by other investigators. Photographs showed distinct differences in bubble size and dynamics in the nucleate boiling of dilute aqueous polymer solutions. Heat flux exceeded that of water [2].

Erhöhung des Wärme-übergangs an der Grenzfläche

Addition of valeric acid to a falling film system enhances the heat transfer coefficient by increasing the contact time [3].

Evaporative heat transfer coefficients for water with various concentrations of a surfactant were measured in vertical downward film flow over a range of flow rates and temperature differences typical of those employed in desaliration evaporators. The coefficients were found to be insensitive to flow rate, but strongly dependent upon temperature difference and surfactant concentration. The temperature dependence varied with the presence or absence of significant nucleate boiling. The dependence on surfactant concentration could not be correlated with surface tension, but was apparently due to foaming and could be explained in the light of a theory of foam stability [4].

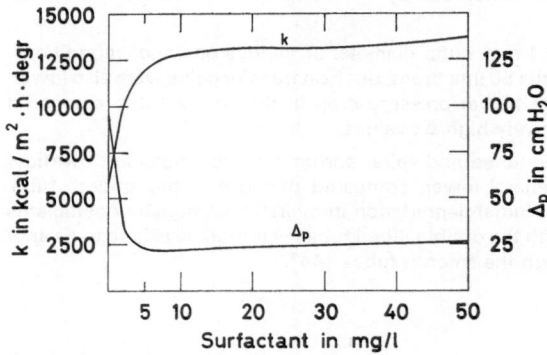

Fig. 2–12

Variation in overall heat transfer coefficient and pressure drop in response to surfactant addition.

Literature to 2.3.3 to 2.3.9

[1] H. H. Sephton (Off. Saline Water Res. Develop. Progr. Rept. No. 574 [1970]; Proc. 4th Intern. Symp. Fresh Water Sea, Heidelberg 1973, Vol. 1, p. 471/80). — [2] P. Kotchaphakdee, M. C. Williams (Intern. J. Heat Mass Transfer 13 [1970] 835/48). — [3] A. B. Ponter, S. Durepos, C. P. Haigh (AIChE [Am. Inst. Chem. Engrs.] J. 17 [1971] 1241/2). — [4] B. H. Shah, R. Dabry (Intern. J. Heat Mass Transfer 16 [1973] 1889/903). — [5] N. I. Gelperin, E. I. Korobkov, V. B. Kvasha, A. B. Bassel (Khim. Prom. 46 No. 2 [1970] 129/31; C.A. 73 [1970] No. 100387).

[6] Southern Research Institute (Off. Saline Water Res. Develop. Progr. Rept. No. 49 [1961], No. 65 [1962]). — [7] R. M. Fand (Off. Saline Water Res. Develop. Progr. Rept. No. 89 [1964]). — [8] R. M. Fand (J. Acoust. Soc. Am. 38 [1965] 561/9; Intern. J. Heat Mass Transfer 8 [1965] 995/1010; J. Heat Transfer 87 [1965] 309/10). — [9] R. H. Keil, M. H. I. Baird (Ind. Eng. Chem. Process Design Develop. 10 [1971] 473/8). — [10] R. A. Erb (Off. Saline Water Res. Develop. Progr. Rept. No. 119 [1964]).

[11] L. A. Minuhin, V. B. Chernozubov, E. P. Emelina, A. A. Bolotov, Z. M. Romanova (Desalination 2 [1967] 337/44). — [12] Yale University (Off. Saline Water Res. Develop. Progr. Rept. No. 8 [1955]). — [13] Franklin Institute (Off. Saline Water Res. Develop. Progr. Rept. No. 184 [1966]). — [14] G. Baudin, B. Lespinasse (Proc. 3rd Intern. Symp. Fresh Water Sea, Dubrovnik 1970, Vol. 1, p. 359/69). — [15] J. M. Niezborala, B. Claudel (Proc. 4th Intern. Symp. Fresh Water Sea, Heidelberg 1973, Vol. 1, p. 95/102).

[16] D. G. Wilkins, L. A. Bromley, S. M. Read (AIChE [Am. Inst. Chem. Engrs.] J. 19 [1973] 119/23). — [17] Y. Chiba, M. Ohwaki, S. Ohtani (Kagaku Kogaku 36 [1972] 412/8; C.A. 77 [1972] No. 22202). — [18] H. Wenzel (Wärme-Stoffübertrag. 2 No. 1 [1969] 6/18). — [19] C. Graham, P. Griffith (Intern. J. Heat Mass Transfer 16 [1973] 337/45). — [20] C. J. Hurst, D. R. Olson (J. Heat Transfer 95 [1973] 12/20).

[21] B. B. Mikic (Intern. J. Heat Mass Transfer 12 [1969] 1311/23). — [22] D. G. Wilkins, L. A. Bromley (AIChE [Am. Inst. Chem. Engrs.] J. 19 [1973] 839/45). — [23] R. A. Erb, T. I. Haigh, T. M. Dowing (Symp. Enhanced Tubes Desalination Plants, Washington, D.C., 1969 [1970], Papers, p. 179/201). — [24] R. Gregorig (Z. Angew. Math. Physik 5 [1954] 36/49). — [25] T. C. Carnavos (Proc. 1st Intern. Symp. Water Desalination, Washington, D.C., 1965 [1967], Vol. 2, p. 205/18).

[26] A. E. Dukler, L. C. Elliott (Off. Saline Water Res. Develop. Progr. Rept. No. 287 [1967]), A. E. Dukler, L. C. Elliott, C. V. Grana (Off. Saline Water Res. Develop. Progr. Rept. No. 487 [1969]). — [27] D. G. Thomas, L. G. Alexander (Desalination 8 [1970] 13/9). — [28] Oak Ridge National

Laboratory (Off. Saline Water Res. Develop. Progr. Rept. No. 367 [1968]). — [29] Stearns Roger Corporation (Off. Saline Water Res. Develop. Progr. Rept. No. 600 [1970], No. 618 [1970]). — [30] R. P. Hammond, L. G. Alexander, H. W. Hoffman (Off. Saline Water Res. Develop. Progr. Rept. No. 644 [1971], No. 699 [1971], No. 646 [1971]).

[31] R. I. Hawes (Proc. 3rd Intern. Symp. Fresh Water Sea, Dubrovnik 1970, Vol. 1, p. 275/89). — [32] D. G. Thomas, G. Young (Ind. Eng. Chem. Process Design Develop. 9 [1970] 317/23). — [33] K. Mattern (Dechema Monograph. 65 [1971] 189/200). — [34] K. S. Campbel, D. L. Williams (Off. Saline Water Res. Develop. Progr. Rept. No. 739 [1971]). — [35] M. Beccari, G. Boari, A. C. Di Pinto, R. Passino, M. Santori, L. Spinosa (Proc. 4th Intern. Symp. Fresh Water Sea, Heidelberg 1973, Vol. 1, p. 17/30).

[36] I. H. Newson, T. K. Hodgson (Proc. 4th Intern. Symp. Fresh Water Sea, Heidelberg 1973, Vol. 1, p. 69/94). — [37] D. C. Guneratne, R. I. Hawes, K. Mills (Proc. 4th Intern. Symp. Fresh Water Sea, Heidelberg 1973, Vol. 1, p. 267/80). — [38] G. Jansen, P. C. Owzarski (Off. Saline Water Res. Develop. Progr. Rept. No. 693 [1971]). — [39] W. S. Chia (Off. Saline Water Res. Develop. Progr. Rept. No. 733 [1971]). — [40] D. L. Bondurant, J. W. Westwater (Chem. Eng. Progr. Symp. Ser. 67 No. 113 [1971] 30/7).

[41] H. Hauck, H. Zemlin (VDI [Ver. Deut. Ingr.] Z. 115 [1973] 642/8). — [42] D. G. Thomas (Desalination 12 [1973] 189/215). — [43] H. H. Sephton (Off. Saline Water Res. Develop. Progr. Rept. No. 361 [1968]). — [44] Distillation Digest, Off. Saline Water Res. Develop. Progr. Rept. No. 731 [1971]. — [45] E. C. Hise, S. A. Thompson, R. van Winkle (ORNL-TM-2963 [1970]).

2.4 Vertical-tube evaporator process development

Entwicklung des Vertikal-Rohr-verdampfers

Since the first evaluation of the vertical-tube evaporation (VTE) process for its application in seawater desalination [1], the technology of the VTE-type desalting plant was well developed and allows upscaling to considerably larger units. The process has a relatively high average annual plant effectiveness. It appears that the VTE-process, in upflow or downflow, equiped with fluted tubes, will become one of the main future desalting processes, especially in combination with other distillation methods. Whereas the development was up to now limited to independent evaporator units, there is a trend to develop configurations having the effects placed in compact blocks either in horizontal or in vertical direction.

Variations of the conventional VTE-configuration have also been tested in pilot plant scale. They partly use flashing of brine. A review of these variations is given in the following. For combination of the multiple-effect vertical evaporator with other distillation processes see section 2.11.3/14., p. 102/12.

2.4.1 Wiped film evaporator

Verdampfer mit ver-wischtem Film

A fluted tube wiped-film seawater evaporator unit was designed as a compact distillation apparatus for marine use [2]. Seawater is distributed at various points on the inner smooth tube wall just in advance of each set of wiper blades so that it is spread by the wiper into a thin film over the entire inner surface of the cylinder. The vapor is drawn off at the top of the tube and is removed from the unit. Surplus feed water flows by gravity ahead of the wiper blades and is removed as a concentrated brine solution at the base of the unit. Approximately 43% of the feed evaporates to form vapor.

Heating steam is supplied on the fluted outer surface of the tube. The condensate formed on the fluted surface is removed from the annular chamber at the base of the unit.

A 37 000 gpd (gallons per day) two effect wiped film pilot plant was constructed and tested at the Wrightsville Beach Test Station of the Office of Saline Water [3]. During the test period one effect was modified to become a 10-foot long double fluted tube thin-film evaporator. It was concluded from the tests that the mechanical wiped film and double fluted thin-film distillation process is readily adaptable to large and intermediate size plants. The fluted tube design presented an advanced method for obtaining and maintaining heat transfer coefficients at values far in excess of those possible in conventional smooth or submerged tubes [4].

Another type of wiped film evaporator, using rotating flat discs as the heat transfer surface with stationary wipers, has been designed and tested at the University of California. The data obtained indicate high performance with scale-free operation at temperatures up to 82°C [5].

*Rotierender
Mehrfach-
effekt-
Verdampfer*
2.4.2 Multiple-effect rotating evaporator

A multiple-effect rotating evaporator has been developed at the University of California and operated at the La Jolla Sea Water Test Facility. The evaporator consists of a number of evaporator plates, up to 30, located directly above one another. The seawater, after treatment, enters a condenser coil above the top stage and is successively preheated in a heat exchanger located at the periphery of the evaporator. Preheated seawater is conducted to the center of each plate and flows outward across the top surface of the rotating plate. Up to 50% of the feed is evaporated by heat supplied by condensation on the underside of the plate. The evaporated water condenses on the plate above and flows outward, being held up by surface tension. The heat for the entire process is supplied by steam condensing on the bottom plate. Thus there is a pressure difference between stages set by the temperature of the liquids on the plates.

Performance on seawater has been found to depend on the operating parameters such as rotational speed, feed rate, number of plates, temperatures, etc. Up to 30 m^3/day (8000 gpd) of fresh water have been produced using steam of 121°C with a performance ratio of 16.5 [6].

*Mehrfach-
effekt-
Ent-
spannungs-
verdampfer
in Turmbau-
weise*
2.4.3 Multiple-effect vertical flash evaporator

A vertical evaporator, using partly brine flashing, has been also developed at the University of California. Short vertical brine evaporation tubes are used in effects located directly above each other. The effects are arranged to form a tall conical tower. Pretreated seawater (**Fig. 2–13**) enters feed tubes near the bottom of the evaporator and flows upward from effect to effect being successively heated to the highest temperature at the top of the evaporator, where it is further heated by indirect contact with steam. The hot feed flows downward into the top set of brine tubes, where boiling occurs. The vapor is condensed partly on the feed tubes and the rest on the brine tubes, which open into the effect below.

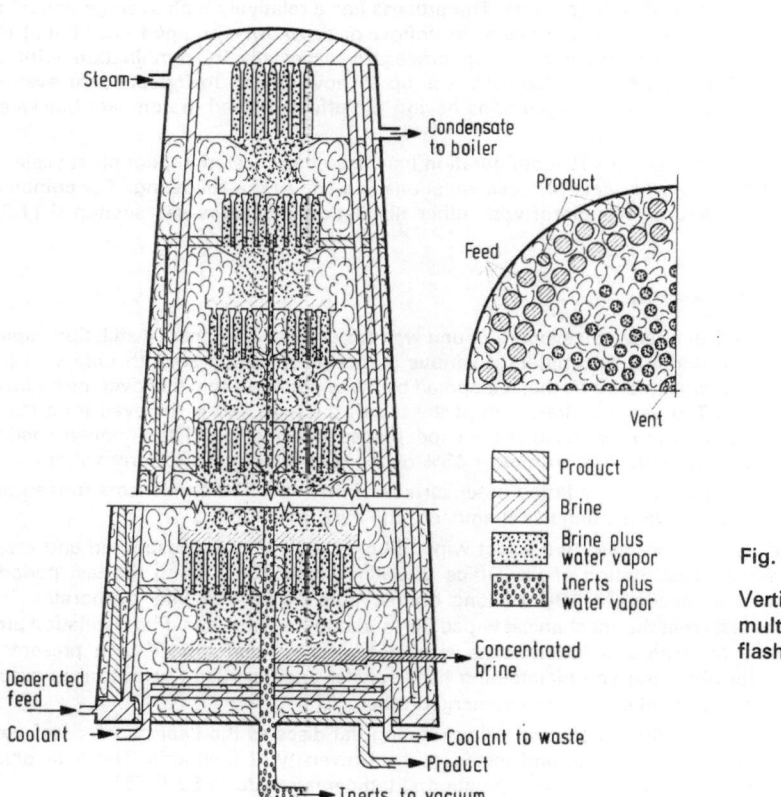

Fig. 2–13

Vertical
multiple-effect
flash evaporator.

The brine passes downward through plastic vortex level control and flow distributors into these brine tubes. After flashing, the brine boils and exits as a mixture of brine and vapor. The condensed vapor is the product, which is flashed from effect to effect for heat recovery. Concentrated brine and cooled product are removed from the lowest and coldest effect. No inter-effect pumps or controls are required.

Performance characteristics are claimed to be superior to other systems in showing reduced tube area requirements because of improved heat transfer. Dropwise condensation has been achieved. Economic projections for this process for seawater conversion appear very promising [7].

2.4.4 Vertical multiple-effect evaporator

A pilot plant with a daily capacity of 100 m³/day of the vertical multiple-effect falling film type evaporator with fluted tubes was constructed and tested in Japan. The effects are located directly one above the other. The pretreated and preheated seawater is introduced to the brine chamber of the top first effect and uniformly distributed to the evaporator double fluted tubes. Steam is condensed outside the tubes of the first effect as heat input medium. Brine flowing out of the tubes of the first effect is collected in the brine chamber of the next effect and fed inside the tubes of this and so on up to the last effect. Vapors evolved in each effect are condensed outside the tubes of the next effect. Condensate is conveyed to the next stage for flashing and from the last effect to the condenser. The process is an once-through system.

Mehrfach-effekt-Verdampfer in vertikalem Aufbau

Enhanced overall heat transfer coefficients were obtained with double fluted tubes ranging from 2 to 2.5 times that obtained in smooth tubes depending on the flow rate (Fig. 2–11, p. 60). It is concluded that the multiple-effect falling film evaporator needs a smaller heat transfer area as compared with the multi-stage flash evaporator. The cost of the heat transfer area with fluted tubes will be about 60% of that of the multi-stage flash evaporator of the same capacity [8].

2.4.5 Horizontal tube multiple-effect distillation

A considerable portion of the maintenance and initial cost of the LTV (long tube vertical) multiple-effect distillation plant is associated whith the pumps between effects. These pumps elevate the brine from the bottom of an effect to the top of the following effect. Therefore, designs were examined in which the tubes in the effects were arranged horizontally instead of vertically. In these arrangements, the vapor condenses inside the tubes while the boiling brine flows over the outside of the tubes in a more or less thick film. The horizontal effects are sufficiently low in height, that they can be stacked vertically, using a gravity brine flow from the hottest top effect down to the last coldest effect.

Mehrfach-effekt-Destillation mit horizontalen Rohren

Extensive tests in a single effect unit at Wrightsville Beach showed that horizontal tubes yield heat transfer coefficients approximately double of those of smooth LTV tubes. A mechanism for the heat transfer was proposed [9]. Coefficients for 2-, 2.5-, and 3-inch diameter horizontal aluminum-brass tubes were correlated as functions of tube diameter, tube spacing, feed rate, feed salinity, heat flux, boiling point, and vapor velocity. These parameters were utilized to design a 2.5 Mgd (million gallons per day) plant. The estimated plant parameters were [10]:

Plant capacity	2.5 Mgd	9460 m³/day
Capital cost total	1.625 Million $	
Capital cost (stream day)	65.0 cents/gal	17.2 cents/liter
Operating cost	53.7 cents/kgal	14.2 cents/m³
Electric power demand	390 kW	
Performance ratio	11.44	

A four effect demonstration plant with an output of 400 m³/day (100000 gpd) has been built in Ostende to produce make-up water for a power plant [11]. Another 3-effect test unit was erected at Wrightsville Beach [12]. After operating almost continuously for nine months, the testing of the pilot plant determined the heat transfer performance of tubes of various lengths, diameters, and materials. The performance, previously obtained from the single-effect unit was confirmed [13]. Further reports on the operation of the plant at Ostende are given by Upmalis [14] and Bongard, Muylle [15].

The process was also tested in a two effect prototype plant with a 0.5 Mgd capacity at the Dungeness 'A' power station of the British Electricity Generating Board [16]. A heat transfer coefficient of 6840 kcal/m² · h · degr was obtained when operating at 115.6°C, with a brine concentration of 1.5-times seawater. This coefficient was independent of the heat flux over the range of 10 850 to 118 000 kcal/m² and showed no appreciable change after 200 hours of operation. The operation showed that the plant was more representative of a 0.8 Mgd plant, as compared with the 0.5 Mgd design capacity.

The operation of the pilot plant has demonstrated that the horizontal tube falling film system is reliable and can be designed to produce fresh water efficiently and with high purity. Two more pilot plants were constructed and operated in Israel to develop and evaluate a multiple-effect horizontal aluminum tube evaporator. Salt water is sprayed over the bundles of tubes and heating steam is condensed inside the tubes.

The second pilot plant with ten effects was designed for a capacity of 150 m³/day. Heat transfer coefficients up to 3100 kcal/m² · h · degr could be obtained with a performance ratio of 7.5 to 6.5 kg/kg [17]. A demonstration plant with a capacity of 1 Mgd was also constructed at Eilat.

A mathematical model for the analysis of the transient behavior of this pilot plant was developed by Aschner and Schaal [18]. The model calculates time-dependent behavior of temperatures, pressures, liquid levels, brine concentrations, and product flows and can also be used for analysis of the performance at partial loads.

Literature to 2.4

[1] W. L. Badger and Associates (Off. Saline Water Res. Develop. Progr. Rept. No. 26 [1959]). — [2] General Electric Co. (Off. Saline Water Res. Develop. Progr. Rept. No. 54 [1961]). — [3] R. E. Anderson, C. W. Lotz (Off. Saline Water Res. Develop. Progr. Rept. No. 105 [1964]). — [4] General Electric Co. (Off. Saline Water Res. Develop. Progr. Rept. No. 181 [1966]). — [5] B. W. Tleimat (Univ. Calif. Seawater Convers. Lab. Rept. No. 69-3 [1969]).

[6] L. A. Bromley (Univ. Calif. Water Resour. Cent. Contrib. No. 100 [1965]; Desalination 1 [1966] 367/93). — [7] L. A. Bromley, S. M. Read (Desalination 7 [1970] 343/91). — [8] M. Ojima, M. Ayai, M. Tsujita, K. Tsukamoto, M. Numao (Proc. 4th Intern. Symp. Fresh Water Sea, Heidelberg 1973, Vol. 1, p. 429/37). — [9] R. B. Cox (Proc. 3rd Intern. Symp. Fresh Water Sea, Dubrovnik 1970, Vol. 1, p. 247/63). — [10] Universal Desalting Corp. (Off. Saline Water Res. Develop. Progr. Rept. No. 492 [1969]).

[11] W. Bongard, R. Muylle (Proc. 3rd Intern. Symp. Fresh Water Sea, Dubrovnik 1970, Vol. 1, p. 235/46). — [12] Universal Desalting Corp. (Off. Saline Water Res. Develop. Progr. Rept. No. 592 [1972]). — [13] G. A. Matta, J. Z. Karpf, A. S. Pascale, J. A. Cardello (Off. Saline Water Res. Develop. Progr. Rept. No. 740 [1971]). — [14] A. Upmalis (Wärme 78 [1972] 137/41; Brennstoff-Wärme-Kraft 24 [1972] 415/8; Entropie No. 49 [1973] 14/20). — [15] W. Bongard, R. Muylle (Proc. 4th Intern. Symp. Fresh Water Sea, Heidelberg 1973, Vol. 1, p. 197/206).

[16] D. C. Guneratne, R. I. Hawes, K. Mills (Proc. 4th Intern. Symp. Fresh Water Sea, Heidelberg 1973, Vol. 1, p. 267/80). — [17] M. Pachter, A. Barak, J. Weinberg (Proc. 4th Intern. Symp. Fresh Water Sea, Heidelberg 1973, Vol. 1, p. 439/50). — [18] F. S. Aschner, M. Schaal (Proc. 4th Intern. Symp. Fresh Water Sea, Heidelberg 1973, Vol. 1, p. 143/52).

Brüden-
kompression

2.5 Vapor compression

Vapor produced by evaporation from a salt solution is superheated, because of the boiling point elevation of the solution, and has a lower pressure than the saturation pressure of pure water. It will, therefore, losing the superheat, condense at a lower temperature than the boiling point of the solution. If this vapor is compressed to a higher pressure, the energy input results in a rise of temperature. With sufficient rise in pressure and temperature, the recompressed vapor might be used as a source of heat for evaporating the same salt solution.

Heat needs to be supplied to the system only at the start-up for elevating the temperature of the solution to the boiling point. Once boiling has started, it is really maintained by the external supply of power and no more by the addition of heat.

Start-up is also possible, without stand-by heat source, by using the temperature rise of the compressed air, which is sucked from the separator. The heated air is then introduced at the bottom of the vertical tubes and bubbles through the solution, which is progressively heated by direct contact. The cycle continues with vapor enriched air until the solution reaches the operating temperature. Then, the air is gradually removed through a special vent and the system is brought to a steady state [1].

As the energy input in some cases compares well with the energy requirements of a distillation plant with several effects, it might be concluded, that, under certain conditions, vapor compression plants combine advantages of single-effect and multiple-effect distillation plants.

Low thermodynamic efficiency is inherent to the process. Economic analysis shows that optimum conditions favor a high thermodynamic efficiency for each of the compressor, evaporator-condenser and main exchanger. Optimum tube diameter, tube length, and allowable pressure drops for the main exchanger may differ appreciably from conventional values. The optimum design depends upon factors, which vary from one locality to another [2].

The vapor compression process was first evaluated by Dodge and Eshaya for application in large scale distillation of saline water [3]. Dropwise condensation and forced circulation were studied to determine the optimum combination of the important variables. The erection of the demonstration plant at Roswell, New Mexico, with a capacity of 1 Mgd was the result of this study.

Although the vapor compression has several known and potential advantages, such as morphological simplicity, high performance ratio per unit of installed heat transfer surface, low pumping power, no cooling water requirements, and reduced capital cost, the process had a low penetration in the desalination plant market. Only 1.3% of the installed world capacity of land-based distillation plants, with an output of over 95 m³/day (25000 gpd), is represented by the vapor compression system of evaporation. This share is reduced to 1%, if the Roswell demonstration plant is excluded. The latter is in fact the largest vapor compression plant that exists. The majority of vapor compression desalting plants are relatively small units, not being included in the O.S.W. desalting plant inventory, with outputs below 200 m³/day and only one, except Roswell, above 600 m³/day [4].

In the early days of large scale desalination a compression still has been devised by Badger and Hickman, which contains a rapidly rotating evaporator-condenser of sheet metal. The chief attraction of this device has been its relatively high-heat transfer coefficient in the range of 10000 to 24000 kcal/m² · h · degr with fresh water feed, depending on operating temperature and speed of rotation [5].

Various sizes of Hickman stills up to a capacity of 64 m³/day have been extensively tested [6]. Although the expected performance characteristics have been demonstrated in smaller size stills, the increase of the rotor diameter and speed has been proven as not practical.

Distillation units with small capacities are usually operated in one effect, condensing the vapor at the appropriate temperature and supplying it as heating steam to the same evaporator. The evaporator, with natural or forced circulation, is of the tube bundle design. Manufacturers claim low capital cost, efficiency, and simplicity of operation and low maintenance cost. In modern designs a water to fuel ratio of about 365:1 is given [7].

Another design uses the horizontal falling film evaporator principle. Preheated seawater and/or recirculated brine is sprayed on the outside surfaces of a bank of horizontal tubes and is partly evaporated by steam condensing within the tubes. The vapor produced is separated from the brine and compressed in order to be delivered back to the condensing side of the tube bank.

Extended runs of a pilot plant at the O.S.W. Wrightsville Beach, N.C., Test Station demonstrated scale free operation for an extended period of time. Heat transfer coefficients from 5500 kcal/m² · h · degr (1125 Btu/ft² · h · °F) to 11700 kcal/m² · h · degr (2400 Btu/ft² · h · °F) were maintained successfully [8].

A similar process with high heat transfer performance, even at low temperature differentials, was reported later. In the described evaporator the presence of scale on the evaporating surface can be readily detected through inspection ports. The degree of brine entrainment is reduced, because of the large volume available for the separation of vapors on the outside of the tubes [9].

A transportable vapor compression distillation equipment, capable of producing 3600 gpd of potable water from seawater and suitable for military use is described by Schmitt and Hurley [10]. The unit uses a conventional vapor compression process, but is unique in its construction to attain light weight. It is self-contained and powered by a 20 HP Diesel engine. Scale control is effected by citric acid. Studies on the use of $CaSO_4$ seed crystals to prevent scaling are also reported.

5*

Transfer of heat across a liquid-liquid interface during vaporization and condensation, reported in another section (p. 98), is utilized in an alternative vapor-compression process [11]. Seawater, preheated by the outgoing product water and waste brine, is contacted in an atomized form with hot heat transfer medium. Fresh water is evaporated from the seawater droplets, producing superheated steam which stirs the heat transfer medium and holds the total mass of liquid in a turbulent state (**Fig. 2-14**). After demisting, the vapor passes to the compressor, which raises the pressure to that in the condenser. The brine-heat transfer medium mixture is drawn, at the temperature corresponding to the brine boiling point, from the evaporator into a separator. From there, the waste brine goes to the seawater preheater and the heat transfer medium is pumped to the condenser. The compressed steam is condensed by direct contact with the heat transfer medium, the latter being heated to a temperature several degrees above the boiling point of the waste brine in the low pressure evaporator. The liquid mixture passes from the top of the condenser to a decanter, where the potable water is separated from the heat transfer medium, which is recycled to the evaporator. The flowsheet of Fig. 2-14 refers to a heat transfer medium heavier than water. With appropriate minor changes the flowsheet is also valid for a heat transfer medium, which is lighter than water.

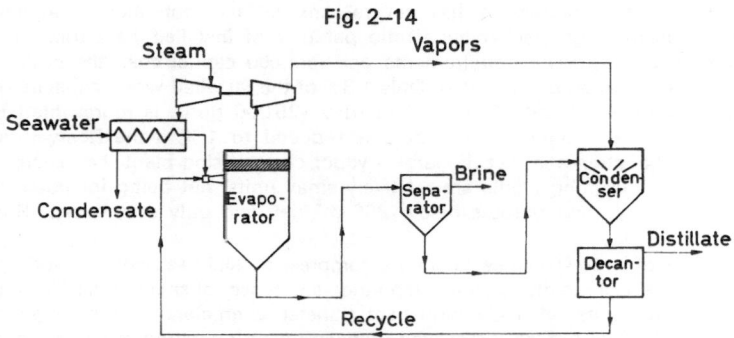

Fig. 2-14

Vapor compression with direct contact heat transfer.

Data on direct transfer of heat across a liquid-liquid interface, available from literature, have made it possible to estimate the cost of producing potable water by the described process. Capital and operating costs are reduced, compared to the costs of the standard vapor recompression process [12]. Added advantages could be obtained by combining the direct contact process with multiple-effect evaporation.

Demon-strations-anlage in Roswell, Neumexiko

2.5.1 Roswell, New Mexico, demonstration plant

The plant was designed to produce 1 Mgd of water. The feed water is produced from wells and contains about 1.5% total dissolved solids. The operation of the plant was a challenge to chemists and engineers, because of the feed water composition with excessively high calcium sulfate content. An extensive water treatment was necessary. Two different methods to prevent calcium sulfate scale deposits were installed: ion exchange pretreatment and calcium sulfate slurry seed recirculation. Only the ion exchange method has been demonstrated.

The plant incorporates two forced circulation rising-flow evaporators, designed to evaporate 1 kg of water for every 250 kg of brine recirculated. A stand-by boiler is used only for start-up. The feedwater is boiling in the first evaporator effect at 111.1°C and produces vapors of 106.4°C, which are conducted as heating steam to the second evaporator effect, operating at a lower pressure. The evolving vapors of the second effect at a temperature of 100.9°C are introduced to the compressor and recompressed to be used as heating steam in the first evaporator effect (**Fig. 2-15**).

The design rate of 1 Mgd at a concentration factor of 3 was demonstrated, but because of numerous equipment failures the plant never attained continuous operating times of greater than 28 days. Scale deposits, corrosion and mechanical failures were responsible for most of the plant shutdowns. Scale deposits were avoided except those scales containing silica [13]. The axial flow compressor proved to be a high maintenance item. Driven by an electric motor through a gear increaser, the compressor was plagued with many electrical and mechanical failures. Intermittent operation of the Test Facility compounded many of the problems.

An engineering analysis of the operation and the future aspects of vapor-compression distillation was also given by Geiringer [14].

Fig. 2–15

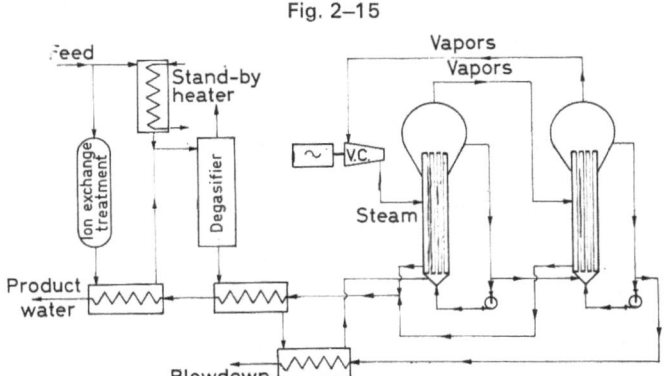

Flowsheet of a typical vapor compression distillation plant.

2.5.2 Vapor compression process development

Several studies have been initiated for the development of the vapor compression process. In an early attempt the thin film heat transfer concept was applied in a vapor compressor cycle. Basically the system consists of a low temperature difference, 1°C or 2°F, thin film vaporizer-condenser module constructed by accordion folding of a thin metal sheet, which is installed in an airtight box. Salt water is distributed over the top of the heat transfer surface by a spray system. The vapors are compressed and conducted the down side of the metal sheet, which is the condensing surface [15]. A significant improvement in heat transfer was obtained, but this was not sufficient to give further consideration to this process.

A vapor compression distillation system using thin-film evaporation and capillary fluted aluminum tubes was studied with the aim to approach the theoretical minimum work in obtaining pure water. It used a spiral flow-control insert in a tube to spread a uniform thin film over the inlet interior surface of a fluted tube [16].

The use of a secondary heat transfer medium for a vapor compression distillation plant gives lower cost water than the cycle compressing water vapor. The difference in cost between the two cycles is greater when the maximum temperature of operation is lowered [17].

A pilot plant with a capacity of 50000 gpd has been designed as a 10-effect forced circulation evaporator, operating with the water temperature of 74.1°C (165.4°F) at the hottest effect and 26.2°C (79.1°F) in the tenth effect. Heat to the first effect is supplied by condensing ammonia and exhaust steam from the compressor driver, both at 78.6°C (173.5°F). The ten effects are arranged in a vertical lay-out [18].

The most promising development is to be looked at the recompression of vapors in combined plants utilizing both vapor compression and other desalting processes. These studies are outlined in section 2.11.4/6, p. 102/5.

The multiple-phase ejector was studied as a vapor compressor in distillation desalination systems. The system performance is roughly comparable with that of a conventional vapor compression system operating at the same evaporation and condensation temperatures. An unusual aspect of multiple-phase ejector systems is their ability to operate as work input, heat input or combination work and heat input system [19].

An ideal pure work-input multiple-phase ejector desalination system was compared to the ideal minimum work requirement and a reversible adiabatic vapor compression system. The results indicate that on a work-input basis the multiple-phase ejector desalination system is more attractive than a vapor compression system [20].

Entwicklung des Brüden-kom-pressions-verfahrens

A high temperature water jet compressor was proposed as energy input in place of the expensive high grade shaft work, required in the conventional vapor compression process. Testing of the high temperature water jet compressor in a test stand with a capacity of 22.5 t/h of suction steam exceeded the predicted performance [21].

Further tests were made on a high temperature water jet vapor compressor at high discharge vapor pressures and at temperatures up to 138°C (280°F) to compare performance with that with lower discharge pressures, corresponding to 115.6°C (240°F). The mechanical efficiency was slightly lower than that achieved by the low pressure jet [22]. The high temperature water is expanded through a driving nozzle, where the fluid flashes and is accelerated to a high velocity. The kinetic energy of the driving fluid is used to compress the suction vapor. Because part of the water flashes to steam, more compressed steam, referred to as "excess" steam, leaves the jet than enters at the suction.

A special water jet compressor design of high temperature water jet was developed to compress 40.6 t/h (89500 lb/h) of steam at 101.1°C (214°F) to 105°C (221°F) for the Roswell plant. The hot water temperature can be varied from 284.4°C (544°F) to 301°C (574°F).

In the Roswell plant the excess steam is condensed in a feed preheat exchanger. The electro-mechanical vapor compressor was replaced with the high-temperature water jet compressor and the plant was re-sized from 1 Mgd to 300000 gpd (1135 m³/day). Further development program of the Roswell Test Facility included calcium carbonate slurry scale control and silica removal studies. Tests on the slurry method of scale control have shown that concentration ratios of 5:1 can be obtained and operation up to ratios of 8:1 are possible.

Verdampfung bei konstantem Gesamt-druck

2.5.3 Constant total pressure evaporation

A novel vapor reheat distillation process was developed, which increases the vapor temperature without increasing the pressure. An auxiliary system consisting of two organic substances is incorporated within the evaporating system so that the heat released in the condensation step can be utilized in the evaporating step under a constant total pressure condition. One of the organic substances is added to the boiling mass of the evaporator to lower its boiling temperature (boiling point depressor). The other is added on the condensing side to absorb the boiling point depressor vapor (absorbent) and raise the temperature of the condensing mass [23].

With the help of the auxiliary system, it is possible for the temperature of the condensing mass to become higher than that of the boiling mass even at the same pressure and heat reuse can be accomplished. Fluorocarbons have been found to be good boiling point depressors, but are rather expensive and have high molecular weights. Searching for low cost and low molecular weight boiling point depressors have not been successful. Liquid-liquid phase diagrams, the equilibrium vapor pressures, and heats of mixing have been determined for various binary systems.

Literature to 2.5

[1] S. N. Filippov, A. G. Khomenko, U. L. Zmeev (Desalination **4** [1968] 30/1). — [2] Y. M. El-Sayed, A. J. Aplenc (J. Eng. Power **92** No. 1 [1970] 17/26). — [3] D. F. Dodge, A. M. Eshaya (Off. Saline Water Res. Develop. Progr. Rept. No. 21 [1958]). — [4] F. O'Shaughnessy (Desalting Plant Inventory Rept. No. 4 [1973]). — [5] K. C. D. Hickman (Ind. Eng. Chem. **49** [1957] 786/800).
[6] Badger Manufacturing Co. (Off. Saline Water Res. Develop. Progr. Rept. No. 12 [1956], No. 15 [1957]), W. L. Buckel, W. D. Beck, J. R. Irwin, A. A. Putnam, J. A. Eibling (Off. Saline Water Res. Develop. Progr. Rept. No. 43 [1960]). — [7] C. H. Hughes, J. E. Pottharst (Proc. 4th Intern. Symp. Fresh Water Sea, Heidelberg 1973, Vol. 1, p. 341/6). — [8] Aqua-Chem. Inc. (Off. Saline Water Res. Develop. Progr. Rept. No. 209 [1966]). — [9] J. R. Howarth, F. C. Wood (Proc. 4th Intern. Symp. Fresh Water Sea, Heidelberg 1973, Vol. 1, p. 327/39). — [10] R. P. Schmitt, S. M. Hurley (Chem. Eng. Progr. Symp. Ser. **67** No. 107 [1971] 178/81).
[11] K. L. Pinder (Proc. 1st Intern. Symp. Water Desalination, Washington, D.C., 1965 [1967], Vol. 2, p. 181/92). — [12] K. L. Pinder (Desalination **4** [1968] 45/54). — [13] Phillips Scientific Corp. (Off. Saline Water Res. Develop. Progr. Rept. No. 717 [1971]). For annual reports see: Off. Saline Water Rept. No. 169 [1966], No. 170 [1966], No. 254 [1967], No. 362 [1968], No. 421

[1968], and No. 529 [1969]). — [14] P. L. Geiringer (Proc. 1st Intern. Symp. Water Desalination, Washington, D. C., 1965 [1967], Vol. 2, p. 659/94). — [15] A. Willenbrook, R. M. Chamberlin, D. R. Fraser, G. E. McGinnis, J. A. Cyphers, J. Wisnicki (Off. Saline Water Res. Develop. Progr. Rept. No. 85 [1964]).

[16] E. L. Kumm (Off. Saline Water Res. Develop. Progr. Rept. No. 103 [1964]). — [17] H. K. Ferguson Co. (Off. Saline Water Res. Develop. Progr. Rept. No. 56 [1961]). — [18] J. C. Chambers (Off. Saline Water Res. Develop. Progr. Rept. No. 122 [1966]). — [19] S. W. Gouse, G. F. Harper, C. A. Kemper, J. H. Leigh (Proc. 1st Intern. Symp. Water Desalination, Washington, D.C., 1965 [1967], Vol. 2, p. 733/49). — [20] C. A. Kemper, G. F. Harper, S. W. Gouse, J. H. Leigh (Off. Saline Water Res. Develop. Progr. Rept. No. 97 [1964]).

[21] P. L. Geiringer, L. T. Taylor (Off. Saline Water Res. Develop. Progr. Rept. No. 344 [1968]). — [22] L. Malfitani, A. Frenzel (Off. Saline Water Res. Devolop. Progr. Rept. No. 664 [1971]). — [23] Chen-yen Cheng, Hung-yean Sung, Kyung Chang Kwon (Off. Saline Water Res. Develop. Progr. Rept. No. 703 [1971]).

2.6 Flash evaporation

Ent-spannungs-verdampfung

When saline water is heated to a temperature slightly below its boiling point at a given pressure and then introduced into a chamber, where a sufficiently lower pressure exists, explosive boiling will occur. Bubbles are evolving from the whole mass of the liquid and part of the water will evaporate until equilibrium with its vapor at the prevailing pressure is reached. This evaporation lowers the temperature of the remaining brine. The liquid may then be passed into another chamber at an even lower pressure, where it flashes again to vapor. This phenomenon was used in the flash tanks placed between the effects of the multiple-effect LTV evaporators (Fig. 2–7, p. 52).

If a higher rate of saline water circulation is supplied, an increased proportion of flash will occur. The increased flow rate may be considered as a means of obtaining increased evaporative yield in a system, without increasing the evaporating surface. It is, therefore, equivalent to diminishing the evaporating surface. In the extreme case, where the evaporating heat transfer surface is entirely eliminated and all the product vapor is obtained only by flashing, the concept of the flash evaporator is reached [1].

Vapors flashed out in the evaporation chamber are channeled to condensing heat exchangers, in which the incoming salt water is preheated thereby condensing the vapor to fresh water. The heat of condensation is thus supplying a large part of the heat required to bring the salt water feed to its boiling point.

Fig. 2–16, p. 72, shows the cross section of a typical flash distillation chamber. The principle of the flash evaporation process in several stages is illustrated in Fig. 2–17, p. 72. Seawater is progressively preheated in the condensers C and finally brought to the required temperature in a steam operated heater under sufficient pressure to prevent boiling. It passes then to the evaporation chambers in a series of stages for flashing. Brine droplets entrained with the vapor are eliminated in a demister filter D. Vapors reaching the bundle of tubes are free of carry over brine and the distillate is collected at the bottom of the condensing department.

As the flash evaporation process is a continuous process involving several operating stages, an interacting flow of three separate streams through the whole plant, namely flow of energy, of concentrating brine, and of distillate, under permanent change of pressure and temperature conditions, has to be considered.

Flashing of vapor requires a finite residence time of the liquid in the evaporation chamber in order to achieve near equilibrium conditions. For a given flow rate the residence time is determined by the chamber length. Mass-transfer rates in the two-phase flow depend on the interfacial geometry of the two phases and on the degree of turbulence. They accordingly determine the residence time required and thus the size of the flashing chamber.

On the other hand, the length of the flashing chamber must be sufficient to achieve the required temperature rise of the incoming seawater, under the acceptable maximum velocity inside the condensing tubes. As there are limitations for both, brine and feed-water flow rates, the width of the flashing chamber becomes the important determinant for increasing the plant size.

Fig. 2–16

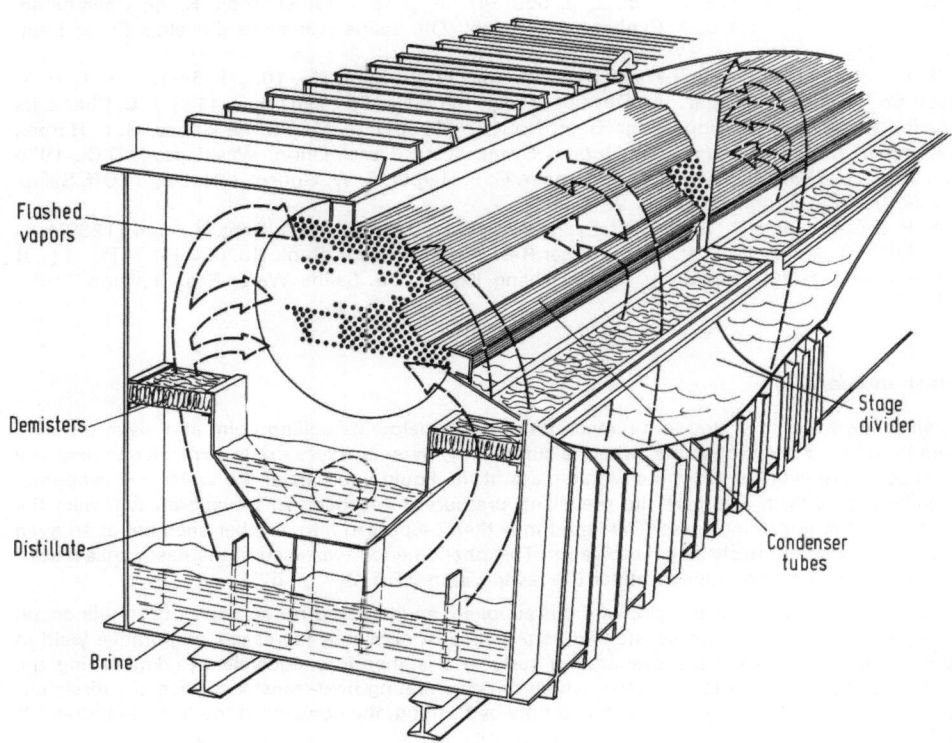

Cross section of a typical flash distillation chamber [29].

Fig. 2–17

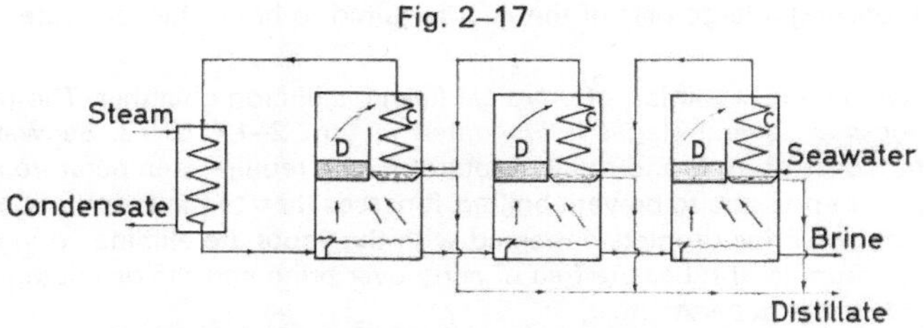

Flash evaporation process. C = condenser, D = demisters.

Strömung bei
Ent-
spannungs-
verdampfung

2.6.1 Flashing flow

The theory of flashing flow was investigated by several authors. During the flashing process the liquid is cooled by the evaporation of a fraction of it. The evaporation rate is governed by the temperature gradient in the liquid adjacent to the liquid-vapor interphase surfaces. The local rate of vapor formation determines the non-equilibrium character of flashing flow. Equations were developed by Dickson and Silver [1] to describe flashing flow and analytical solutions were presented for the simple frictionless flow case. Numerical solutions were given for the more general problem including friction effects.

A new theory of flashing flow was described by Hawes and Leslie [2], who assume that all bubble nuclei are present when the pressure is reduced and the liquid becomes superheated. The liquid and vapor are considered to be in thermodynamic equilibrium at the bubble surface and vapor formation occurs by conduction of heat from the bulk of the liquid to the surface of the nuclei, followed by evaporation at their surface. As the bubbles grow, the two phase mixture is accelerated along the duct in order to keep the mass flow constant. The resulting pressure drop leads to an increased superheat. When the superheat is small, nucleation of bubbles does not occur within the bulk of the fluid and vapor formation only occurs at existing interfaces. In a desalination plant, growth of bubbles swept from one stage to the next or flashing at the free surface is likely to be most important.

Porteous and Muncaster [3] assume that the developed models are probably quite valid at the high temperature end of a desalination plant at temperatures above roughly 65°C. Another model was developed, which predicts that flashing is an exponential decay process and correlates existing data for brine temperatures of 38 to 65°C.

The formation and release of vapor in the flash-distillation process was also studied by Dickson and Addlesee [4]. In an experimental investigation, substantial departures from thermodynamic equilibrium were observed and the addition of a wire mesh over the inlet orifice produced a marked increase in the rate of vaporization.

An experimental investigation into the mechanism of bubble nucleation in the flash boiling of seawater suggests that the naturally occuring nucleation sites give an effective site radius of the order of 10 to 20 μ [5]. Such sites with the help of dissolved gases would be effective only over the top quarter of the flash range in a typical plant. Diffusion of gas towards the sites is thought to be responsible for the time delays of the order of seconds between decompression and nucleation.

It is concluded in a further experimental investigation of the flash evaporation of superheated water, that the intensity of vaporization increases with the rise of both, the equilibrium temperature and the superheat temperature [6]. The seeding of the superheated water vapor bubbles intensifies the flashing process.

A model was proposed to predict the rate of phase transition and kinetic energy release in a non-equilibrium flashing mixture when the flow pattern comprises a metastable liquid core surrounded by an annulus of vapor [7]. Together with certain assumptions concerning the slip ratio, the critical flow rate may be predicted as a function of nozzle exit plane pressure and upstream stagnation conditions. It was shown that the model compares favorably with experimental results obtained with sets of nozzles of different lengths and entrance profiles and with varying initial subcooled water temperatures.

Flashing of brine was based in an experimental study essentially upon a vapor disengagement mechanism, which does not take into account bubbles as the source of vapor [8]. The proposed mechanism is connected with the flow structure as modulated by the flashing device. The effect of all parameters is shown and especially as most important some geometrical parameters. In the case of a flow without obstacles, the theory exhibited the importance of the liquid phase recirculation in the flash chamber, which diverts the hot liquid bottom layers from the bottom of the chamber to the free surface where evaporation occurs [9]. Bubble bursting at the interface initiates an intense turbulence in the liquid mass and explains the resulting high heat flux densities. It was shown that the amount of product vapor is not directly related to the number of bubbles, which reduces considerably the importance of bubbles, generally adopted in the various existing flashing models. The introduction of obstacles in the liquid flow modifies the flow structure and contributes to an important yield increase.

An analytical model of a flash stage was developed and an analysis of the physical phenomena taking place in an evaporator was made [10]. The proposed model, confirmed by comparison with the physical data, indicates that there are basically three different major regions of flow within the evaporation stage: a) The entrance jet region, wherein a submerged jet emanates from the orifice at the stage inlet. b) The central portion of the stage, wherein the velocity profile adjusts from a maximum velocity near the bottom of the channel at entrance to a maximum velocity near the top surface. In this region the major vertical transfer of heat appears to occur and the evaporation rate is highest. c) The exit region of the stage, wherein the fluid is gradually accelerated prior to exit from the stage for entrance into the orifice for the following stage.

The analytical model proposed requires that a turbulent thermal conductivity be experimentally determined since it is flow and temperature dependent. The partial wall jet at the entrance orifice, which is submerged by varying amounts, depending on the entrance temperature and flashdown, controls the magnitude of the turbulent thermal conductivity. A flow equation for the simple entrance orifice is proposed, which agrees reasonably well with physical data.

A simplified model of a flash stage, which assumes constant horizontal velocity, level free surface and various entrance temperature distributions was also proposed, using a constant value of turbulent thermal conductivity throughout the stage. The model appears to be useful in the design of seawater flash evaporators and provides approximate values for the significant design variables for properly specified input variables [11].

Gas bubbles released during the electrolysis of brine act as efficient nuclei for flashing [12]. The advantages of this technique of nucleation are the almost instantaneous removal of superheat and the lack of constriction of the flow. The nuclei can be produced either in the region of superheat or in a position, where they will be carried into this position by the flow.

2.6.2 Multi-stage flash (MSF) distillation

Mehrstufige Ent- spannungs- verdampfung

The flash-distillation process, as applied in large scale desalting of seawater, may be considered as consisting of three sections in handling heat: the heat input section, usually named brine heater, by condensing external steam; the heat recovery section, in which the heat of evaporation is recovered in the condensers at the various stages; and the heat rejection section, which maintains the thermodynamic process by reducing temperature and pressure and accounts for the last stages of the plant (Fig. 2–18).

Cold seawater is pumped to the inlet of the condensing tubes in the heat rejection section, where an increased stream of seawater is maintained for cooling purposes. The seawater is warmed up as it flows through the condenser tubes and a portion of it is returned to the sea, serving only as cooling water. The remainder is chemically treated to prevent a build-up of scale on the surfaces of the heat recovery section tubing. It flows then to the decarbonator to remove gaseous carbon dioxide and to the deaerator for the elimination of air from the make-up water. This is usually accomplished by allowing the flashed vapor from one of the heat rejection stages to strip the air from the seawater. The objective of deaeration is to prevent oxygen corrosion and to eliminate noncondensable blanketing of the condenser tubes.

The treated seawater flows through the tubes of the entire heat recovery section, receiving heat from the condensing product water and reflashing distillate. From the heat recovery section the seawater enters the heater, where it is further heated up to a temperature, established by requirements associated with the chemical treatment scheme used.

Leaving the heater the seawater passes through the flow control valve, which maintains the pressure required to avoid boiling. The hot seawater is then pumped to the flashing chamber of the first, the hottest, stage of the evaporator. A portion of the seawater flashes to vapor, which passes through demisters (Fig. 2–18) and condenses on the tubes of the heat recovery section. The amount of vapor evaporated and the brine saturation pressure in the stage are determined by the condensation rates.

Fig. 2–18

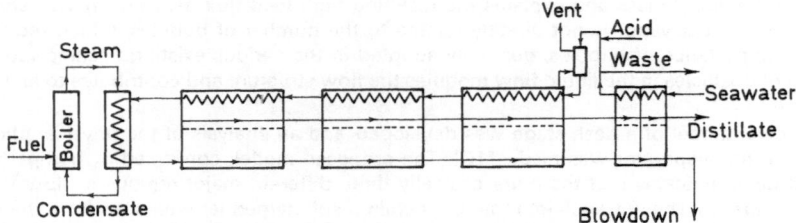

Once-through single-effect multi-stage flash distillation process.

As the brine flashes, it cools and then passes through flow control devices into the next lower pressure stage. The flashing process continues at progressively lower temperatures and pressures until the brine reaches the last, the coldest, stage of the heat rejection stage.

The quantity of water flashing into vapor at each stage is proportional to the enthalpy of the liquid or the difference in temperature between brine and vapor. An increase in the brine flow rate increases the evaporation rate. The distillate produced by the condensing vapor in the various stages is collected in trays and pumped to the desired location. In some plants the product water is sent to a cooler, where its temperature is reduced to the 24 to 26°C range.

The arrangement of the condensing tube bundles in the evaporator shell may be parallel (long tube design) or perpendicular (cross tube design) to the brine flow in the chambers. The tube length per stage has to be equated to the stage width for the latter and to the stage length for the long tube design. For evaporators with a capacity of 2.5 Mgd (9500 m³/day) the stage length and the stage width are approximately equal and either design would be compatible with tube length requirements. At plants with a capacity of over 5 Mgd the long tube design provides the more satisfactory solution.

Long tube design permits the use of tubes of the most economical length available and the placing of them over several stages. The number of tube-plates and water boxes is reduced and pumping costs may be reduced by as much as 25%. On the other hand the cross section design has the advantage, in cramped quarters, of minimum space for the tube withdrawal and replacement. In general both designs have their advantages and disadvantages. Long or cross tube design affects chamber geometry and consequently flashing characteristics too.

If the brine from the last stage of the heat rejection section is discarded to the sea in an open circuit, by the once-through method, or single-effect multi-stage process (SEMS), the circulation of large amounts of seawater is involved, which requires high pretreatment costs. This is one of the reasons why the once-through method is not preferred in large scale desalting of seawater. In a recent paper [13] the advantages of the once-through method are outlined and compared to processes using recirculation of brine. The method is said to be superior because of high initial temperature operation, required lower heat transfer surface and lower boiling point elevation, provided scale-free operation.

2.6.3 Recirculation of brine

Mehrfach-umwälzung der Sole

When the once-through system is applied, the distillation plant has usually a large number of stages, but only one effect, as rising vapor in the various stages is only once brought into contact with the incoming cold seawater. An improvement of the single-effect flash evaporator plants is the partial recirculation of brine. Upon leaving the heat rejection section, part of the concentrated brine is blown down as waste to maintain the proper salt content or brine concentration.

A portion of the seawater leaving the heat rejection section, after treatment and degassing, is mixed with the remaining brine and returned to the tube bundles in the heat recovery section. As the amount of seawater diverted to the evaporation process is significantly lower than in the once-through process, a large part of the pretreatment costs are saved. The brine flow rate is maintained in each recirculating stream by pumping (**Fig. 2–19**). To simplify the diagram pumps are not shown in the flow sheet.

Fig. 2–19

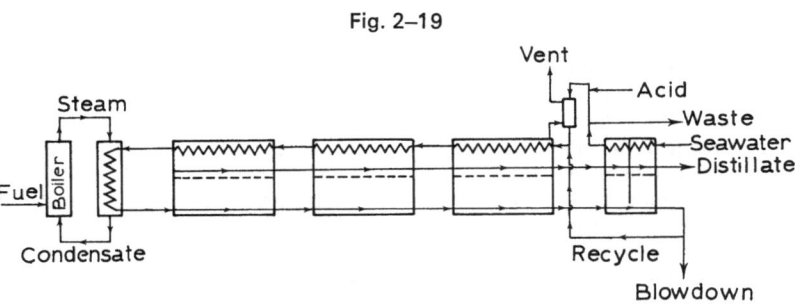

Multi-stage flash distillation with brine recirculation.

Recirculation can be applied so far as the concentration of the scale-forming compounds does not reach, after evaporation, the critical point. It is a disadvantage of this design that the brine concentration is at the hottest stages of the plant quite higher than the concentration of dissolved solids in the seawater. This fact limits the maximum brine temperature of the process.

Operating with this cycle arrangement, the maximum operating temperature with acid injection is limited to 121°C (250°F) with brine concentration 1.5-times. The number of stages is usually limited by 2°C flashdown per stage, because of the low pressure differential available at the deep vacuum conditions prevalent in the last stages [14]. On the other hand the advantage of the LTV-multiple-effect process in having the lowest salt concentration at the point of highest temperature is obvious. Under the same conditions the LTV-multiple-effect process can be operated at higher initial temperature.

This advantage can also be obtained in a multi-stage flash process by including more than one recirculation streams (**Fig. 2–20**). In this design each group of stages with its own recirculation stream is said to operate as one effect. The number of effects is defined by the number of recirculation streams, each effect consisting of a number of stages, the whole forming the multiple-effect multi-stage flash distillation process (MEMS).

This generally accepted definition might be somewhat misleading. In fact, in the LTV-distillation process the vapors produced in one stage are used as the heat source of the next stage and the heat of condensation is recuperated as many times as there are stages in the LTV-plant (see chapter 2.3, p. 51), Hence, the number of stages defines the times that recuperation of the heat of condensation is effected and every stage corresponds to an effect. However, in the MSF-distillation process, the heat of condensation is recuperated only once in the same stage, where the vapors are produced, and the concept of multiple-effect, as defined for the LTV-distillation process, is not existant. Instead of multiple-effects there are only multiple-recirculation streams in this variant of the MSF-distillation process. It appears, therefore advisable to define this process as a multi-cycle-multi-stage (MCMS) process, instead of the generally adopted term of multiple-effect-multi-stage (MEMS) distillation process, in order to avoid confusion.

Fig. 2–20

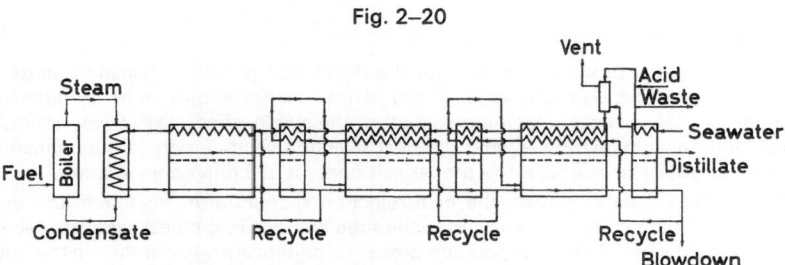

Multi-stage multiple-effect flash distillation process.

In this type of flash-evaporation plant the chemically treated and deaerated feed water is as well preheated in the series of condensers and mixed with the recirculation brine from the first effect, so that the brine with the lowest brine concentration is evaporated in the first stage at the highest temperature. The initial temperature can, therefore, be higher than that allowed for the single-effect flashing process. The top temperature in the first effect might be 135°C (275°F), a gain in operating temperature range of 14°C, while the concentration ratio is maintained at 1.15.

Another advantage of the multiple-effect flash distillation process is that the recirculation streams may be controlled separately. This makes it possible to choose the level of salt concentration in each effect and to fix the number of stages in such a way as to have more stages at the higher temperature ranges. By suitable design it is possible to operate at a higher concentration ratio in the low temperature effect, while operating at a lower concentration in the top temperature effect. This reduces considerably the amount of feed to be treated, the amount of acid required and the size of the deaerator, air ejector, feed pump, and acid pump.

A disadvantage of the system is the additional power for pumping, required to maintain the recirculation streams, as compared with the once-through system.

The initial heat input is in general so efficiently used, that the performance ratio is not less than 10- to 12-times the amount of steam supplied to the first stage. In some recent plants the performance ratio is even higher.

Performance data obtained during a test run of the Senator Clair Engle plant in San Diego, California, are summarized as follows to illustrate a typical operation of a multiple-effect multi-stage distillation plant:

Make-up

Seawater temperature	20.4°C	68.6°F
Rate	6776 l/min	1790 gpm
Brine heater concentration factor	1.19	
Maximum brine temperature	137.8°C	280°F

	1. Effect		2. Effect		3. Effect	
Recycle rate	14668 l/min	3875 gpm	12242 l/min	3234 gpm	9252 l/min	2444 gpm
Blowdown temperature	109.2°C	228.7°F	71.5°C	160.6°F	27.8°C	82°F
Concentration factor	1.26		1.56		1.96	
Product water					4656 m³/day	1.23 Mgd
Performance ratio lb/1000 Btu					20.7	

The design of a large multi-stage flash evaporation system requires a detailed analysis of heat and mass balances and plant cost factors. Computer programs are generally used to perform the large number of calculations, which are required. Optimization routines are used to evaluate the minimum water cost for the specified input conditions.

The number of stages might be increased as long as the addition of extra stages is economically justified. This is determined by the capital cost of providing additional stages in respect with the cost of heat thereby saved. In locations where the cost of fuel is very high, the incorporation of additional stages might be economically advantageous, but not in locations, where the cost of fuel is low.

2.6.4 Operation characteristics

Betriebs-charakteristik

The purity of product water depends mainly on the effectiveness of the entrained mist elimination. Apart from the pressure drop, the elimination efficiency is decisive for the valuation of a mist eliminator. Plant operation and feed-water conditions have a major effect on the quantity and quality of brine mist which is generated. Experimental measurements of dry pressure drop through demisters are reported by Dickson and Bradie and a correlation derived from a simple physical model is suggested [15]. This physical model is correlating liquid and vapor flow rates at the point where flooding of the demister occurs and re-entrainment takes place. Based on investigations into the mechanism of elimination and flow in connection with mist eliminators functioning according to the inertia principle, the reasons for the failure or the unsatisfactory efficiency of mist eliminators are discussed by Regher [16].

The content of dissolved solids is usually very low. For municipal water supplies the product water from flash distillation plants has a maximum of less than 50 mg/l, although such high water purities are not required. If a reasonable quality of brackish water is available, mixing up to as much as 500 mg/l with the evaporator distillate can increase the total potable water supply by up to 10 or 20% with corresponding reduction in cost. For special applications, such as for high pressure boiler feed make-up, plants can be designed to give distillate having 250 mg/m³ total dissolved solids with 10 mg/m³ copper and 30 mg/m³ iron content [17].

The amount of water that a desalting plant can produce each year is directly related to the availability of the plant. Availability defines the ratio of the actual annual production to the nominal annual plant production. This is in turn related to the reliability of each of the major plant components, such as brine heater, heat recovery and heat rejection tube bundles, pumps, evaporator enclosures, vacuum systems and instruments. The most important of these components was found to be the tube bundles. The tubing reliability was determined in a hypothesized model by extreme value statistics for tube failure rate of 0.1% failure in ten years. For this model and failure rate, the overall plant availability was found to be 95% [18]. In some of the commercially operating plants lower availability factors have been reported, but in some others the availability or on-stream factor was higher [19].

2.6.5 Process design parameters

Verfahrens-abhängige Parameter

The design of a multi-stage flash evaporation system requires a detailed analysis of heat and mass balances, as well as plant cost factors. The purpose of the heat and mass balances is to develop temperature, pressure, material and heat profiles through the individual stages of the MSF plant.

The heat transfer process is essentially identical to that encountered in a counterflow heat exchanger and it may be described by the following general relation:

$$Q = k \cdot A \cdot \Delta t_m$$

already given in p [44]. This relation can be used to determine the required heat transfer surface area A for a desired production rate, if the average overall heat transfer coefficient k and the logarithmic mean temperature difference Δt_m are known.

The heat transfer coefficient is determined by the sum of the thermal resistances between the condensing saturated vapor and the cooling brine stream. The individual resistances to be taken in consideration are: resistance to convection heat transfer between the flowing cooling brine and the inner tube wall; resistance due to the tube wall; resistance to convection heat transfer between the condensing vapors and the outer tube wall, and resistance due to fouling or scale formation on the brine and/or the vapor side of the tube wall, as well as to the presence of noncondensable gases near the outside surface of the condenser tubes.

Table 2/4 contains typical values of resistances determining the value of the overall heat transfer coefficient [21].

Table 2/4

Typical heat transfer resistance values in kcal/m$^2 \cdot$ h $\cdot$ degr.

Resistance values	Range*)			
	from		to	
Inside film	−04	1.023	−04	2.046
Tube wall	−05	1.023	−05	5.11
Outside film	−05	8.18	−04	1.227
Fouling	−05	8.18	−04	1.637
Total resistance	−04	2.761	−04	5.421

*) The prefix refers to the powers of 10 of the decimal given. Accordingly −04 means 10^{-4}.

The corresponding heat transfer coefficients (in kcal/m$^2 \cdot$ h $\cdot$ degr) are in the range from 3622 to 1845.

The logarithmic mean temperature difference, used for heat transfer calculations, is defined as:

$$\Delta t_m = \frac{t_c - t_b}{\ln \left[(t_v - t_b)/(t_v - t_c) \right]}$$

where t_v is the temperature of the condensing vapor, t_b the temperature of recycle brine entering the condenser and t_c the temperature of recycle brine leaving the condenser.

There is a finite difference between the flashing brine temperature and the condensing vapor temperature. This is due to several effects or resistances. No control is possible for some of these effects, as for the boiling point elevation. A reduction of the brine pressure entering the flash chamber, equivalent to the boiling point elevation, must be accepted.

Another resistance is due to the flashing phenomenon. Some work is required to compensate surface tension forces for bubble formation. This work manifests itself in the form of liquid superheat in the vicinity of a vapor bubble, which is at saturation temperature. To obtain liquid superheat the pressure is reduced by an increment equivalent to the required driving potential.

A third resistance to flashing occurs because of the hydrostatic pressure and lack of adequate convection in the brine pool. A pressure loss is also occuring in the demister and in the bundle of the condensing tubes.

The power required to maintain the MSF process might develop to an important cost factor. The pressure losses in the flash evaporation plant must be compensated by the pumping power input. Power must also be used for pumping seawater, product water, and chemical feed. In addition to these requirements, pumping provisions are sometimes made for blowdown and make-up pumps.

The most important power consumption, because of the tremendous amount of brine kept in recirculation, is for recycling the brine, which accounts for over two thirds of the total power required. This percentage is subject to significant variations with plant capacity. For very large capacity plants, the recycle pump power might increase to more than 80% of the total power. On the other hand efficiencies generally increase with pump capacity thereby decreasing the relative electrical energy requirements at higher pump outputs. **Fig.** 2–21 gives the electrical power requirements according to the plant capacity [22]. The approximate range of electrical energy consumption for almost any plant design is between 1.585 and 2.64 kWh/m³ of product water (6 to 10 kWh/kgal).

Fig. 2–21

Electrical power requirements in MSF plants. Plant capacity × 1000 m³/day.

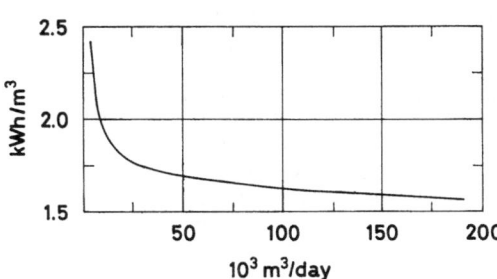

2.6.6 Effect of process parameters

Wirkung der verfahrens-abhängigen Parameter

The number of stages in a MSF distillation plant is related to the performance ratio. An increase in the number of heat recovery stages will generally result in a higher performance ratio, or, for a given product water output, in a decrease of the steam consumption at the brine heater.

Fig. 1–22 shows the temperature profile across a typical multi-stage flash plant. The stepped line shows the temperature fall at each stage of the heat recovery and heat rejection section. The effect of increasing the total number of stages, while maintaining the same initial maximum and final temperatures, t_{max} and t_f, will be to decrease the effective mean temperature difference available in each stage. A larger heat transfer surface area will be required to maintain the same performance ratio. An optimization must be made between increased tube costs and decreased stage construction costs.

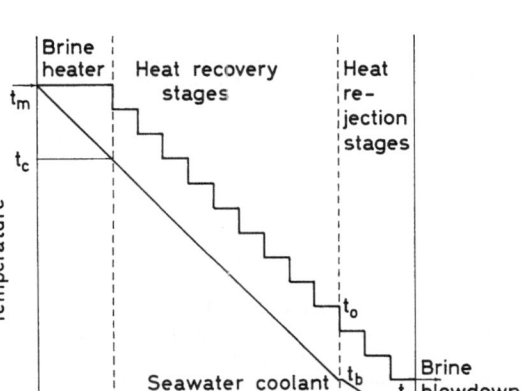

Fig. 2–22

Temperature profile across a typical multi-stage flash plant.

The performance ratio may be correlated to the number of stages by the following relation:

$$R = \frac{t_{max} - t_f}{t_o - t_f} = \frac{\text{No. of reject and recovery stages}}{\text{No. of reject stages}}$$

where t_o is the outlet temperature from the recovery stages to the rejection stages. Definitions of t_c and t_b see p. 78.

It might be concluded from this relation, that an increase in the available temperature difference within the heat recovery section, by raising the steam inlet temperature, will result in a decrease of the required heat transfer surface area. This may be accomplished by decreasing the stage length. Nevertheless, the designed stage length must be a balance between the length required to assume sufficient residence time of the flow ng brine stream in the flashing chamber to attain equilibrium and the length required for heating the brine to the desired temperature. The optimization will then be made between decreased tube and s:age costs and increased steam cost for the same water production.

Fig. 2–23 presents a plot of total number of stages versus performance ratio as a mean of data taken from various design studies [21].

The condenser tube diameter and the recycle brine velocity through the tubes influence the overall heat transfer coefficient and pumbing power. An increase in flow velocity can result in a significant increase of the average overall heat transfer. However, there is a drastic increase in required power with the flow velocity. The optimum conditions might be looked at the desired velocity which allows reasonable heat transfer without excessive pumping power requirements.

Velocities in the seawater stream inside the condenser tubes are generally specified to be between 1.5 and 2 m/s. The lower ve ocity limit is specified to minimize tube fouling. The upper velocity limit is specified to minimize tube material erosion effects.

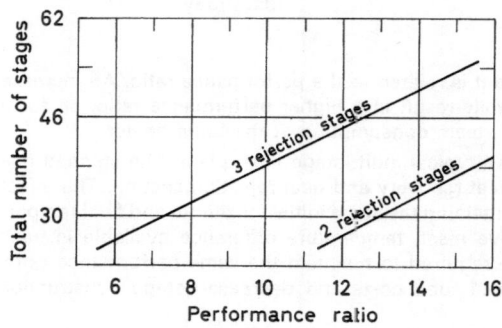

Fig. 2–23

Number of stages and performance ratio.

The brine temperatures are directly related to the amount of heat input required to produce distillate in the evaporator. The minimum heat input required is defined:

$$Q_{min} = W_{min}\left(\frac{t_{max}}{t_{max}-t_F}\right)$$

where W_{min} represents the minimum work required for the desalination process. t_F is the temperature of the incoming seawater.

It would appear that by increasing the brine heater outlet temperature t_{max}, better efficiencies would be obtained from the distillation plant. However, the above given function varies slowly at high temperatures and the advantages of brine temperatures above 200°C are small.

On the other hand scale formation is prohibiting operation at such high temperatures. With chemical treatment scale formation may be prevented as long as the evaporator does not exceed the temperature limits, at which supersaturation occurs. Commercial evaporator designs provide for maximum brine heater outlet temperatures around 145°C with sulfuric acid treatment and 89°C for polyphosphate treatment.

*Demon-
strations-
anlagen in
San Diego,
Kalifornien,
und Senator-
Clair-Engle-
Anlage*

2.6.7 San Diego, California, and Senator Clair Engle Test Bed plants

The San Diego plant has operated only 26 months, from January 1962 to February 1964, and was then dismantled to be transferred to the U.S. Naval Base, Guantanamo Bay, Cuba. The plant had a rated capacity of 1 Mgd (3785 m³/day).

The flow sheet of the plant corresponded more or less to Fig. 2–19, p. 75, which is drawn in simplified form. The plant had 34 heat recovery stages and two heat rejection stages. After passing the last two stages, part of the seawater was returned to the sea and the remainder was processed by sulfuric acid injection and mixed with brine from the last stage, which was recirculated to the condenser tubes of the heat recovery section. After final heating in the brine heater the brine mixture flashed down from the first to the last stage. The temperature drop was in the range of 93 to 121°C (200 to 250°F). The pressure drop from stage to stage varied from about 4 to 5 Torr in the first stages to about

1.5 Torr in the last stages. Product water was collected in a continuous trough and evacuated from the 36th stage at 26 to 30°C [23].

After the demonstration plant was dismantled and removed, a new plant of the same capacity was designed, using the multi-stage multiple-effect flash process [24], constructed and operated [25]. This plant is known as the Senator Clair Engle plant and it consists of three recirculation streams with a total of 68 stages. The first and second effects consist each of 20 heat recovery stages and 3 rejection stages. The third effect consists of 20 heat recovery stages and 2 heat rejection stages. The process flow sheet of the plant corresponds to Fig. 2–20, p. 76.

Fresh seawater is added to the first effect, at the highest temperature, as make-up. Each succeeding effect receives blowdown from the previous effect as its make-up. A steam heated brine heater supplies heat to the top temperature effect. Each succeeding effect makes use of the heat rejection stages of the previous effect as its brine heater.

The main advantages of the process may be summarized as: Ability to divide the flashing temperature range into several temperature intervals, each of which can be maintained at a different salt concentration; operation at higher maximum brine temperatures without scaling, and operation with reduced boiling point elevation resulting from lower salt concentrations at the higher temperatures.

The plant was designed to operate up to a maximum brine temperature of either 143°C (290°F) or 121°C (250°F). With the elevation of the maximum temperature from 121 to 143°C, the plant production rate was increased from 1 Mgd to 1.23 Mgd of desalted water, i.e. to 4656 instead of 3785 m³/day.

For the 121°C operation the feed seawater is treated with sulfuric acid. For operation at 143°C brine temperature, the make-up is further treated by the "lime-magnesium-carbonate" (LMC) process (see p. 158) to decrease the calcium concentration by 50 to 70% relative to that in the raw seawater. With subsequent design improvements a maximum brine temperature of 148°C (297°F) was realized. The feasibility of the LMC process was demonstrated.

A high temperature effect was constructed as an auxiliary effect to operate at higher operating maximum brine temperature (up to 176.7°C) than the main plant. This effect has 10 recovery stages, no rejection stages and is operated independently of the main plant. The plant was built to obtain data on thermodynamic and hydraulic behavior at high temperatures, to measure approaches to equilibrium and to determine scale thresholds using calcium-depleted seawater produced by the LMC plant. All operations were completed during 1972. Nonscaling operating was achieved at a temperature of about 39°C (70°F) higher than that used in present day commercial plants.

2.6.8 Porto Torres, Sardinia, plant

Entsalzungs-anlage in Porto Torres, Sardinien

Local water supply could not meet the increasing requirements of the petrochemical plant at Porto Torres in Sardinia, and the operating company, the Societa Italiana Resine, had to face the problem by desalting seawater, in a plant of their own.

The plant has two units with a maximum capacity of 700 and 1500 m³/h, respectively, both of the long tube multi-stage flash design with brine recirculation. The first module has been put into operation in September 1971 and the second, the large one, during spring of 1973. The latter is at the same time the largest desalting single unit in operation in the world (nearly 9.5 Mgd). The capacity of the two units was chosen to allow for a total fresh water production between 400 and 2200 m³/h, according to the requirements of the petrochemical plant [26].

The desalting plant design data are as follows:

	700 m³/h	1500 m³/h [27]
Maximum brine temperature	110°C	110°C
Heat recovery section stages	30	28
Heat rejection section stages	3	3
Tube bundles per stage	2	3
Performance ratio	8.5	8.2
Product-blowdown ratio	1	1
Scale control by injection of	H_2SO_4	H_2SO_4
Distillate conductivity μS/cm	2 to 4	1 to 3

The Porto Torres plant appears to be the first completely automated, controlled by an IBM 1800 on-line computer, that can act either in supervising or in direct digital control.

The computer operating features are: a) Routine on-line control in closed-loop mode and in emergency cases in open-loop mode, b) Periodic processing of the utility programs connected with the plant operation, and c) Aperiodic time-sharing processing of external programs either connected or not connected to the plant operation. The mathematical model for the multi-stage flash desalting plant control is described by Barba et al. [20]. This paper presents the basic criteria of the main models used in the plant control, with emphasis on programs for the calculation of fouling factors and evaporating efficiency. A simplified and a rigorous model for the evaporator simulation is introduced.

The computer input/output operations are carried out, beside standard devices such as teletypewriters and line printers, also by a process operator console that has been programmed at interrupt level and that allows instantaneous measurement of the parameter variations. In case of computer failure, part of the measurements and all control functions are carried out from the control room.

The plant automation ensures the right production for the needs of the petrochemical plant besides considerable savings in labor. The computer control guarantees not only optimized operating conditions but it also supplies important information on the plant performance.

Literature to 2.6

[1] A. N. Dickson, R. S. Silver (Desalination 2 [1967] 175/95). — [2] R. I. Hawes, D. C. Leslie (Desalination 2 [1967] 329/36). — [3] A. Porteous, R. Muncaster (Proc. 3rd Intern. Symp. Fresh Water Sea, Dubrovnik 1970, Vol. 1, p. 145/54), A. Porteous (Desalination 6 [1969] 337/47). — [4] A. N. Dickson, A. J. Addlesee (Proc. 3rd Intern. Symp. Fresh Water Sea, Dubrovnik 1970, Vol. 4, p. 31/41). — [5] W. T. Hanbury, W. McCartney (Proc. 4th Intern. Symp. Fresh Water Sea, Heidelberg 1973, Vol. 1, p. 281/91).

[6] N. K. Tokmantsev, V. B. Chernozubov (Proc. 4th Intern. Symp. Fresh Water Sea, Heidelberg 1973, Vol. 1, p. 497/505). — [7] R. Muncaster, G. M. Thomson (Proc. 4th Intern. Symp. Fresh Water Sea, Heidelberg 1973, Vol. 1, p. 391/401). — [8] J. C. Deronzier, F. Armand (Proc. 4th Intern. Symp. Fresh Water Sea, Heidelberg 1973, Vol. 1, p. 229/36). — [9] G. Coury, J. C. Deronzier, J. Huyghe (Desalination 12 [1973] 295/313). — [10] C. H. Coogan, D. A. Fisher, F. W. Gilbert, H. L. Ornstein (Off. Saline Water Res. Develop. Progr. Rept. No. 364 [1968]).

[11] C. H. Coogan, D. A. Fisher, P. S. Brewster, H. L. Ornstein (Off. Saline Water Res. Develop. Progr. Rept. No. 503 [1969]). — [12] J. C. Drake (Desalination 6 [1969] 335/6). — [13] W. Nishimoto, J. Sakuma (Proc. 4th Intern. Symp. Fresh Water Sea, Heidelberg 1973, Vol. 1, p. 419/27). — [14] W. R. Williamson, F. W. Gilbert, T. R. Scanlan (Proc. 1st Intern. Symp. Water Desalination, Washington, D.C., 1965 [1967], Vol. 2, p. 779/96). — [15] A. N. Dickson, J. K. Bradie (Proc. Symp. Nucl. Desalination, Madrid 1968 [1969], p. 859/77).

[16] U. Regher (Dechema Monograph. 65 [1971] 173/8). — [17] J. M. Stewart, W. R. Querns (Desalination 3 [1967] 139/46). — [18] N. Arad, S. M. Mulford, J. R. Wilson (Desalination 3 [1967] 378/83). — [19] R. H. Evans, S. M. Mulford (Proc. 3rd Intern. Symp. Fresh Water Sea, Dubrovnik 1970, Vol. 4, p. 43/51). — [20] D. Barba, G. Liuzzo, G. Tagliaferi (Proc. 4th Intern. Symp. Fresh Water Sea, Heidelberg 1973, Vol. 1, p. 153/68).

[21] Hittman Associates (Off. Saline Water Res. Develop. Progr. Rept. No. 490 [1969]). — [22] R. E. Childer, W. L. Prehm (Off. Saline Water Res. Develop. Progr. Rept. No. 257 [1966]). — [23] Burns and Roe Inc. (Off. Saline Water Res. Develop. Progr. Rept. No. 102 and No. 114 [1964]). — [24] Fluor Corporation (Off. Saline Water Res. Develop. Progr. Rept. No. 252 [1967]). — [25] Catalytic Construction Co. (Off. Saline Water Res. Develop. Progr. Rept. No. 596 [1970], No. 629 [1970], No. 666 [1971], and No. 668 [1971].

[26] D. Bogazzi, R. Cigna, S. De Zuccoli, A. Mastrodomenico, G. Spizzichino (Proc. 4th Intern. Symp. Fresh Water Sea, Heidelberg 1973, Vol. 1, p. 179/89). — [27] D. Barba (private communication).

Optimierung und Weiterentwicklung des vielstufigen Entspannungsverdampfungsverfahrens

2.7 Multi-stage flash distillation process optimization and development

As flash distillation is the most applied process in water desalination, there is a large number of studies made, as well as papers and reports have been published for the optimum plant design and process development.

2.7.1 Optimization   *Optimierung*

The trends of optimization aim at the increase of the scale of operations to very large size plants, the use of exhaust steam as low cost heat supply to the seawater conversion plant and the selection of such operating parameters as to obtain lower cost product water, including the use of less expensive materials of construction to reduce capital investment.

Five cases of possible plant designs with a capacity of 190 000 to 570 000 m³/day (50 to 150 Mgd) and a maximum brine temperature of 121 and 149°C were investigated in an early report. All cases were based in this study on an optimum balance between operating and capital costs to provide minimum cost product water [1]. The optimization involved a balance between the capital cost of increasing efficiency by using more condensing surface and stages and the savings from reduced energy requirements. Number of stages and terminal temperature difference were systematically varied until a configuration was reached which gave minimum cost of water.

At the time when studies for such large plants were initiated, the technology was not as much developed as to permit upscaling to the considered plant size. Means taken in the U.S.A. to face this problem are outlined in section 2.7.2 (p. 86).

A study made in the United Kingdom [2] has shown that it is technically feasible, on the basis of existing information, to build large desalination plants with units each having an output of 10 Imperial Mgd (12 U.S. Mgd; 45 460 m³/day). This capacity was about seven times greater than that of the largest unit in operation at the time of the study. The design parameters were optimized as follows:

Steam temperature	108.9°C		228°F
Top brine temperature	104.4°C		220°F
Reject brine temperature	37.8°C		100°F
Seawater temperature	10°C		50°F
Number of heat recovery stages		20	
Number of heat rejection stages		3	
Performance ratio kg/kg		8	
Brine concentration ratio		2	
Recirculation ratio		9.64	

The optimum brine heater outlet temperature was investigated for a single-purpose desalination plant with a capacity up to 190 000 m³/day [3]. With an efficient acid treatment the plant would operate at 143°C (290°F) maximum brine temperature. The optimum brine temperature of the plant would be 176.7°C (350°F), if appropriate scale control methods could be developed. Total water costs have been calculated for brine heater outlet temperatures from 121 to 204°C (250 to 400°F) for various fuel prices. Costs affordable for scale prevention at the different temperatures and fuel costs have been determined.

Parametric cost studies were made with computer optimization for the costs which affect the optimum plant design, including steam costs, tubing costs, shell costs, pump and power costs, as well as water treatment and water supply costs. An optimum plant design was computed and the effect of changes in component costs on the optimum plant design was determined [4].

Golubkov and Korneichev developed an analytical method for the determination of optimum parameters of thermal desalination plants, applicable for falling film evaporators, for flash distillation with and without heat transfer surfaces [5]. Analytical expressions were derived for the optimum number of stages, heat supply heating surface etc. The effect of various factors on the optimum value of the process parameters, as well as on the cost of product water was discussed.

Convex optimization was investigated as a method of solving for non-linear cost functions and was compared with the single-step technique in order to calculate the operating temperatures required to minimize the specific costs of multi-stage evaporators. The most favorable number of stages was determined in the second optimization step [6].

A simple cost model was developed by Wolberg for optimizing the design of flash evaporators [7]. The model was then used as a tool for studying the sensitivity of the optimum design in variations of the independent economic and physical parameters. For fixed water output the variable annual cost is minimized by varying the number of stages, the terminal temperature difference, and the condenser tubing diameter. The variable annual cost is assumed to consist of the variable portions of the amortization, steam, and pumping costs. Operating cost is assumed to be fixed. The optimum design is sensitive to the mode of optimization. By allowing the tube diameter to vary, in addition to the number of stages and the terminal temperature difference, a completely different optimum design is obtained.

To find an optimal solution for the design of a flash-distillation plant, the planning must be carried out in two stages [8]. In the first stage a series of optimum plant data is determined by means of a basic design procedure, an approximate estimation, and a minimum-seeking sub-program. In the second stage these data are used as initial values for the detailed design and cost estimation.

Mathematical relations are derived by Sterman et al. for the choice of the optimal parameters for a flash-desalination plant [9]. The optimum number of stages, the optimal amount of recirculation and a condenser surface, which corresponds to a minimum cost of distillate for given conditions, may be computed.

For optimum design of multi-stage flash evapcrators, the necessary heat and material balances are formulated and then an iterative dynamic programming procedure for the minimization of the annual operating cost is applied. A flow chart for the computer solution, a continuous formulation which represents the limiting design for an infinite number of stages and the optimum design of the heat rejection section are included in a paper by Itahara and Stiel [10].

Process analysis and optimization studies have been made for a multiple-effect multi-stage flash distillation system [11]. The heat supplied to the brine heater, the brine reject concentration, the recirculation rates within each effect and the allocation of brine temperature and composition among the effects are controlled to arrive at an optimal policy. A discrete form of the maximum principle in combination with other optimization techniques has been used for most of the optimization studies.

A mathematical model of a multiple-effect multi-stage flash distillation system, that can be used in process optimization studies, was then developed [12]. The model enables the heat input to the brine heater, the recycle flow rate in each effect, the concentration and temperature of the flashing brine leaving each cycle and the number of stages in each effect to be treated as variables in an optimization study. This model was used, together with an economic model of the system, to minimize the cost of water production. The heat input to the brine heater per unit of feed, the recycle flow rate in each effect and the concentration and temperature of the flashing brine leaving each effect are chosen so that the water production cost is a minimum.

A method for complex optimization of the parameters of single as well as dual-purpose desalination plants with adiabatic evaporation was developed by Korneichev et al. suitable for application with computers [13].

A new approach to the optimization of multi-stage flash evaporation plants was developed, which involves the use of relative values of economic parameters rather than their absolute values. The concept of minimum-possible energy required by the process is introduced [14].

A computer program was formulated, which might be used for economic optimization. Using measured parameters as initial data, the values or trends of certain quantities not directly accessible to experiment may be determined [15].

A mathematical model for a multistage-flash distillation plant was developed with a high degree of rigor and very few qualifying assumptions. References to the collection and reduction of thermophysical data and details of the differential equations and correlations describing the flash process and the associated algebraic relations were provided. The model takes into account variations in heat transfer with condenser tube geometry, tube surface fouling, the effects of non-condensable gases, boiling point elevation, revaporization of product, thermal inertia of the flash vessel walls and tubes, superheating of flashed vapor and others. The results are useful for dynamic and parametric studies as well as for the prediction of performance of a given plant under a wide range of possible conditions [16].

The most suitable value of design parameters for the multi-stage flash distillation process were determined from the calculation of optimization with authorized cost data. The controlling parameters for the water cost were also disclosed with the sensitivity analysis, by which the performance of the project can be evaluated [17].

A physical and economic model was built for a 570 000 m^3/day (150 Mgd) single-effect multi-stage flash desalination plant. The development of the model is reviewed and the limitations are discussed. The model is optimized using a combination of the discrete maximum principle and direct search procedures. Detailed optimization of design variables at each stage is made so that in the optimum plant over 100 variables are optimized. Previously typical optimization studies considered about five variables [18].

A mathematical and economic model of a single-effect multi-stage flash distillation system was developed by Coleman [19] and used to formulate a method for cost optimization. All process flow rates, heat transfer surface areas, flashing temperatures and the number of stages are allowed to vary in determining the minimum cost of producing water. The optimization technique allows for maximum design flexibility without reverting to an overly simplified model of the systems, so that the results provide a reliable basis for decision.

An once-through multi-stage flash evaporator process for desalination was simulated with a computer program by Satori et al. [20]. The rate of water production increased with increasing sea-water flow rate, decreasing seawater temperature, increasing brine-heater temperature, decreasing fouling factor, increasing heat-transfer area, and with increasing number of stages. A process simulation analysis was also made for the brine-recirculation-type flash evaporator. The relation of the system performance to characteristics of each stage were determined in the following effects: brine recirculation ratio, number of stages in the heat rejection section, ratio of the area of the heat recovery section to the area of the heat rejection section, the seawater feed and temperature, brine heater temperature, the fouling factor, and the number of stages.

The static characteristics of a multi-stage flash evaporator were analyzed by Hayakawa with the simulation method and the mathematical model of each stage of the evaporator was described. Calculations were made in regard to a 26-stage plant with a capacity of 5 Imperial Mgd (6 U.S. Mgd) and a performance ratio of 8. The operation curves assuring steady plant operation were obtained to-gether with useful information on the performance of the ejector, the accuracy of the controller and the fouling factor as the design value [21].

Barba and coworkers have developed in a series of publications a mathematical model, which could successfully be applied for the study of optimum sizing of a multi-stage flash distillation plant with brine recirculation [22]. The resulting computer program was applied in the design study of a desalting plant with a capacity of 4000 m³/day. In a further study the independent variables, which determine the flexibility problem of multi-stage flash distillation plant, were examined and the relations for the analysis of the behavior of the plant under varying load conditions other than rated were presented. The study resulted in the formulation of a computer program, based on a rapid converging iterative criterion. This mathematical model was then developed for on-line process computer control and applied at the Porto Torres desalting plant. The program is supplying information for the optimal plant operation to meet the varying water demands of the refinery, which is using the distillate as process water [23]. The control system is described in section 2.6.8, p. 81.

A digital computer code for simulating large multi-stage flash evaporator desalting plant dynamics was developed by Delene and Ball. This simulator can calculate the time response to various perturbations of the evaporator system, predicting liquid levels and temperatures etc. in each stage. The evaporator system, the equations used to describe the plant dynamics and the method of solving these equations are discussed. The simulator can calculate an one hour transient in a 50-stage multi-stage flash plant in a running time of 4.5 minutes [63].

A Fortran code for the rapid calculation or optimization of multi-stage flash plants was also presented by Fort. The code refers to a plant, consisting of a recovery section, reject section, brine heater and associated buildings and equipment. Operating costs, direct and indirect capital costs for plant, buildings, site and intakes are calculated. Design computations are based on the first and last stages of each section and a typical middle recovery. As a result, the program runs rapidly but does not give stage-by-stage parameters [64].

Simplified mathematical models were derived by Guneratne for representing various distillation processes, namely multiple-effect evaporation, vapor compression and multi-stage flash. Application of an ordering algorithm shows that the models can be solved by means of a suitable choice of design variables and by using each equation of the model in turn to evaluate another variable. Various computer programs were written, based on these models. A feature of these programs is the short time required for execution, about one second [65].

Enhancement of product water output rate by design modifications of existing evaporators may provide low cost alternative for incremental increase of water supply. The stretch capabilities of process parameters for an existing MSF (multi-stage flash) evaporator, consistent with techno-logical and process limitations, were analyzed by Arad et al. [24]. With the aid of a computer program, specifically developed for this purpose, product water output rates, as a function of brine recirculation rate, at varying brine temperature at heater outlet, were computed. These were compared with field test results, proving enhancement capability of 18% in one example and 16% in another. Incremental water cost was estimated to be 33% that of the average.

Weiter-
entwicklung
des MSF-
Verfahrens

2.7.2 MSF process development

One obvious definition of multi-stage flash distillation is simply a flash distillation plant with more than one stage. This definition leads to the concept that a plant having a performance ratio of 3.5 and 4 stages would be considered as a multi-stage flash distillation plant. Silver [25] restricted the de-finition to plants where the number of stages is greater than twice the performance ratio. For a given performance ratio the increase in the number of stages reduced the expensive heat transfer surface, at the cost only of relatively inexpensive steel divisions between stages. The repeated flash effect in the attempted designs conducted to increases in the size of plants and of their performance ratio to figures exceeding the number 10.

While the first multi-stage flash plants were designed for maximum brine temperatures of about 90°C only, the development of seawater treatment for scale prevention has permitted the raising of the top temperature to over 120°C This means an increase of about 50% in the temperature flash range and consequently an increase in the performance ratio and a reduction in pumping power requirements. A better control of calcium sulfate scale formation made it possible to use even higher initial temperatures in some plants.

An important development in MSF plant design is that the flash chambers could be substantially reduced in size for the same function. Plant capital costs were as well reduced. In **Fig.** 2–24 the capital cost factors,

$$\frac{\text{Total plant weight}}{\text{Capacity} \times \text{Performance ratio}}$$

are plotted for several installations built in the sixties [52]. It is seen that these capital cost factors have been gradually reduced to about half of the value in the first installations.

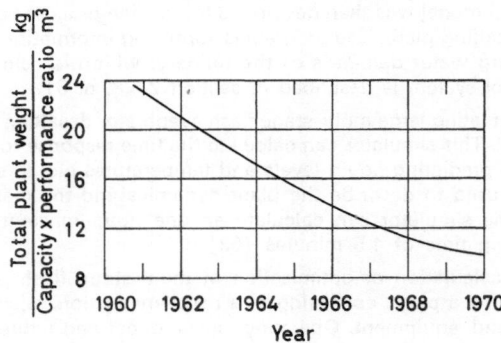

Fig. 2–24

Capital cost development in MSF plants.

The participation of the MSF process in the world capacity of distillation plants is shown in **Fig.** 2–25. Based on data given by the Office of Saline Water [26], the development of the various distillation processes is computed in m³/day capacity for two decades. The MSF process develops from 1960 an explosive growth capacity. At the same time the submerged tube boiling type of distillation plants, being previously the main distillation process, appears to be abandoned as far as desalting is concerned. Vertical tube evaporators have a growing participation in the late sixties. Vapor compression as independent process is insignificant. It has to be noted that all data used refer to plants with a daily capacity exceeding 25000 gallons (95 m³/day).

The large participation of the MSF process in the world desalting capacity is primarily due to the large size of most MSF plants. Out of 812 plants in total, operating or under construction on

January 1st 1972, with an overall capacity of 1 317 320 m³/day, 56 MSF plants had a total capacity of 713 018 m³/day.

The size of the first flash distillation demonstration plant at San Diego, California, ordered by the Office of Saline Water, U.S. Department of the Interior, early in 1961, was of a rated capacity of 1 Mgd (3785 m³/day). Many substantially larger plants have been built in the following decade, to reach a capacity of about 36 000 m³/day in the second unit of the Porto Torres plant, which is actually the largest single MSF unit. Table 2/5, p. 88, shows the increase in capacity of single units and the total plant capacity of the largest MSF distillation plants operating or ordered to be put in operation at the given time period. Within ten years the maximum size of a single unit increased more than six times. Whereas in the United Kingdom the trends are more conservative, in the United States conceptual designs are under consideration for desalting plants with capacities up to about 950 000 m³/day (250 Mgd) with several trains. Advanced studies bring the projected capacity up to 3.8 million m³/day (1000 Mgd).

Fig. 2–25

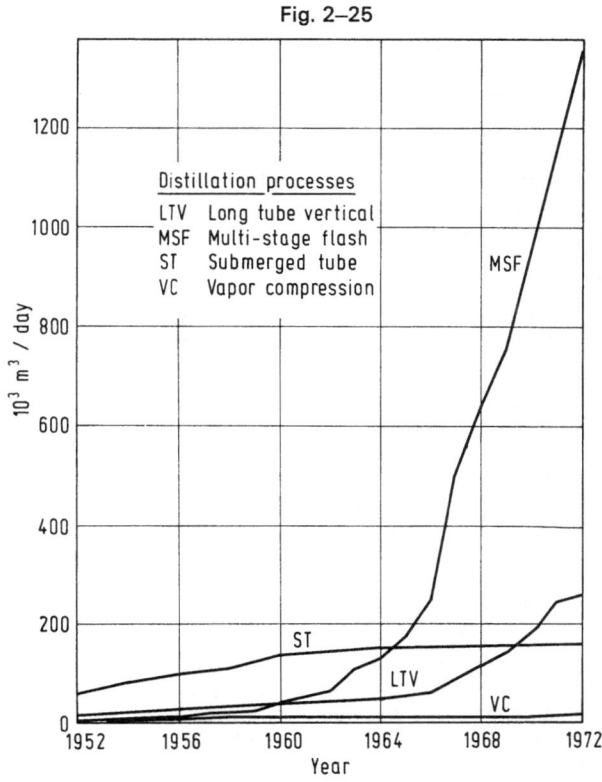

World capacity of distillation plants.

To assist in bridging the technology gap between the construction of units under operation and the very large capacity units projects, a test module of a 64 350 m³/day (17 Mgd) plant has been constructed and tested at the O.S.W. San Diego Test Facility [27]. The test module was designed to simulate full scale operation of a 190 000 m³/day (50 Mgd) desalting plant as described in a study by Fluor Corporation [28]. The module was able to simulate the high-temperature heat recovery section of the evaporator, as well as the heat rejection section. The facility served as a test bed for studying brine flows in full-scale flash chambers, heat transfer and venting of large condenser bundles, the hydraulics of a 290 m³/min (77 000 gal/min) recycle pump and a large number of equipment components for large plants [29].

Table 2/5
Increase in MSF unit size and plant capacity.

Time period	Location	Unit size in m³/day	Mgd	No. of units	Plant capacity in m³/day	Mgd
1961 to 1962	Cardon, Venezuela	5451	1.44	1	5451	1.44
1963 to 1964	Curacao, Netherlands Antilles	6511	1.72	2	13022	3.44
1965 to 1966	Shuwaikh, Kuwait	9085	2.40	2	18170	4.80
1967 to 1968	Terneuzen, Netherlands	14498	3.83	2	28996	7.66
1969 to 1970	Shuwaikh, Kuwait	18208	4.81	1	18208	4.81
1971 to 1972	Shuaiba, Kuwait	22500	6.00	5	112500	30.00
1973	Porto Torres, Italy	36000	9.50	2	52800	13.95
1974	Hong-Kong	30283	8.00	6	181700	48.00

Although the production capacity of the module is 2.5 Mgd (9500 m³/day), it seems more meaningful to describe it as being one-fourth the width and one-fifth the length of a complete 50 Mgd plant, but each major piece of equipment is full sized, as it would be for the complete plant. The module incorporates a six-stage heat recovery section and a three-stage heat rejection section in each of the three parallel trains with a common brine heater. The performance of the module is analyzed in three O.S.W. reports [30].

In a further investigation of multi-stage flash phenomena in a three-stage test system, equations were derived to express the empty stage non-equlibrium allowance for a typical flash stage simulating the 17 Mgd test module, the interstage brine flow for a typical flash stage, as well as for the discharge rate of submerged orifice [31].

A device for reducing the pressure in an evaporation chamber by varying the orifice area to reduce the non-equilibrium temperature difference in a flash evaporation plant was described by Sato [32]. The device was tested in a 5-stage flash distillation plant. The rate of evaporation was increased and the non-equilibrium temperature difference decreased by increasing the orifice area. An empirical equation for calculating the non-equilibrium temperature allowance is given.

As plant sizes increase, more emphasis must be paid to parasitic losses, because they may represent large capital costs, and affect the product water cost. The most common losses, which cause a flash evaporator to perform at a lower than design efficiency include non-condensable gas blanketing, scaling, fouling or surface oxidation of the stage condensers. These effects result in a depression of heat transfer rates and lower the cycle performance by decreasing the heat transfer efficiency. Interstage blow-through of vapors and non-equilibrium temperature allowance of flashing brine affect the temperature driving force required for heat transfer, by lowering the energy level of the condensing steam, thereby reducing the logarithmic mean temperature difference. Partial blockage of tubes with corrosion products or marine materials reduces the available heat transfer surface and lowers the plant performance by not making use of all the available tube surface for condensing.

In an early study, an investigation was made of the importance of various parameters on the length of a typical stage in a multi-stage seawater flash evaporator. An optimum chamber length was established, such that vaporization was complete. An empirical equation was derived from the basic test data to give an indication of the chamber length required for complete flash-off in a stage for a specified range of conditions [33]. In a further study the foam heights and entrainment heights for a range of the flashing conditions, normally met within evaporators, were investigated, as well as the pressure loss and de-entrainment characteristics of the separator mesh. The information obtained allows the separator height to be specified for a range of flashing conditions and indicates the maximum value of vapor loading to which the separator can be taken while still giving a reasonable vapor purity above the separator. Information is also given on the effects of brine circulation, interstage temperature difference, splash plate configuration, and brine level variations [34].

A computer study showed that the only type of perturbation, which seemed to require variable orifices, was turned down, if it was desired to minimize steam consumption in the turned-down condition. The main advantages were ability to optimize brine levels, to maximize output and to adjust efficiently to reductions in demand or to reductions in steam and recycle flow and easier initial adjustment. Better flow data for fixed orifices can adequately regulate large plants [35].

Further development of the flash distillation process was performed at the O.S.W. Test Station in Wrightsville Beach. In a 3-stage pilot plant tests covered the simple open interstage orifice under a wide range of operating conditions, as well as several of the more promising flash enhancers, including Dragon's teeth. Non-equilibrium values were related to the various operating parameters [36].

Results from controlled-trend tests to study non-equilibrium losses and interstage orifice brine flow behavior were analyzed and correlations were determined. The trend test method greatly enhanced the output rate of useful information. With special consideration given to measurement and data-acquisition problems, internally consistent and reproducible results were obtained [66]. Those portions of the tests in which the evaporator was in the blowthrough mode of operation were also discussed. Some observations were made on the operating characteristics in the blow-through mode and the transitions into and out of this mode [67].

A design variable which has a consistent and predominant effect on the magnitude of the non-equilibrium loss is brine level. A thermosyphoning device with a variable area riser was designed and tested at Wrightsville Beach. The hydraulic performance of this enhancer indicated a reduction in the non-equlibrium loss in the order of 50% [37].

A further study on flash enhancers recommended three types of such devices, which were performance tested at Wrightsville Beach. Each type of enhancer reduced the brine deviation losses in a different manner. The Dragon's Teeth design of flash enhancer was found as the most effective and appeared to have the greater water cost savings for large plants [38].

Head losses, incurred with the discharge of a given quantity of 5% salt-saturated brine at $47.7°C$ $(118°F)$ through the most critical of the interstage module piping, were determined for Reynolds numbers ranging from 170 000 to 1 200 000. The total head losses for the required flow, with and without a control valve, were 161.5 and 170.5 mm (0.53 and 0.56 ft) respectively. The differential head between modules must be increased to provide the desired flow [68].

The use of condenser tubing, enhanced by corrugations, in MSF evaporator plants was assessed by means of a computer program in an attempt to determine the economic consequences of using 0.75″ outer diameter bare tube, 0.75″ corrugated tube and 1″ outer diameter corrugated tube, in three 2.5 Mgd plants and a 50 Mgd plant. The computer runs were made with one set of economic conditions and the effects of changes in any of the conditions were considered. In general, large desalting plants would show an economic benefit for enhanced tubing [39].

Pressure drop of different combinations of orifices and weirs has been investigated in dependence of the distances between orifice and weir, the height of the weir, the free height of the orifice and the mass flow rate in the last cold stages of a MSF plant. The total pressure drop is caused by the acceleration on the orifice, the flow over the weir and the turbulence and vortices in the space between orifice and weir. The pressure drop between orifice and weir is mainly caused by carrying over of air and by two-phase turbulence. In reducing the distance between orifice and weir to a certain point, the pressure drop coefficient remains nearly constant. By passing this point, the air-water mixture disappears and the pressure drop coefficient decreases. By further reducing of the distance, the pressure drop is increasing again because of the resistance of the very small channel. For different free heights of the orifice the pressure drop coefficient rests rather constant. During the measurements hydrodynamic instabilities were observed. Oscillations were caused by water-air mixtures rolling between orifices and weir and behind the weir. The measurements showed that the range of instabilities is strongly dependent of the free height of the orifice [40].

A three-stage flash test system was used for a number of specially assigned tests and special plastic demisting material was evaluated in an attempt to enhance vaporization in a flash stage. The obtained results exhibit that brine injection enhances flash performance under certain conditions. Surfactants as additives to seawater feed showed no particular effect [41].

Preliminary screening tests on the influence of high molecular weight additives on the flashing process suggested that, although friction of the type experienced in pipes can be reduced with these additives, in flash systems the amount of drag reduction is not significant. The experiments did reveal an important secondary benefit of these additives. Additive causes a significantly more uniform bubble distribution downstream of the orifice. Such a change can be expected to cause improved mass transfer in the flash chamber, more rapid approach to equilibrium, and shorter metastable periods [42].

In a design study, a preliminary survey of vessels for MSF distillation units using concrete–polymer materials was made. Horizontal cylinders with hemispherical heads, inside diameter 9.15 m, length 27.5 m, resistant to full vacuum, were the design basis for the vessel study [43].

A method was developed for prediction of average plant-effectiveness for each year of plant life and for analysis of cost-effectiveness of variations in design concept and in maintenance procedures. The MSF-type desalting plant was found to possess a relatively high average annual plant-effectiveness even prior to the introduction of any of the variations in the design and maintenance concepts. Following the heat-rejection and the heat-recovery sections, the plant was made to operate continuously, as long as steam from the heat source is available.

The reduction in product water output rates, when operating the plant at failure states involving tubing subsystems, is rather small. Moreover, the plant may sustain a large number of tube failures with a relatively small effect on output. A loss of 10% of heat transfer surface in each of the tubing subsystems will reduce the plant output by 7.3%. The reduction in product water output rates at failure states involving pumping subsystems, being significant, justified the introduction of a redundant unit in each of the pumping subsystems. Degradation of the recycle pump has such an effect on the average annual plant-effectiveness, that a yearly overhaul for restoring pump performance is economically justified [44].

Further improvements in the MSF process might be in reducing the shell size, decreasing parasitic temperature losses and in developing improved heat transfer bundles. Other cost-reducing improvements are those that benefit all distillation processes, such as increasing plant life, on-stream time, improved feed pretreatment and lower cost materials.

A novel type of seawater multi-stage flash evaporator was described by Ricard [45]. In order to reduce the temperature difference between evaporating brine and cooling water, the brine is dispersed in each stage and surrounded by cooling surfaces. Experimental results on a prototype plant are discussed.

The multi-stage flash distillation process is not applied only with seawater as feed. It has been tested by U.S. Authorities on a pilot plant capable of producing 163 m³/day (43000 gpd) using the brackish and highly polluted waters of Newark Bay, New Jersey, as feed water for conversion to a water suitable for public consumption and industrial purposes. In general the operation of the pilot plant resulted in the production of low solids distillate without insurmountable operating problems. To achieve aesthetically acceptable product water some post-treatment of the distillate was necessary [46].

The flash distillation process is considered to be the best candidate for demineralizing mine drainage waters for potable use. A MSF demonstration plant with brine recirculation to process 5 Mgd (19000 m³/day) was under consideration in the United States [47].

Mine waters are desalted by flash evaporation in Poland to protect the water quality of the Oder river. Gypsum and common salt are recovered from the brine [48].

A system for the uninterrupted collection of distillate and for the accurate measurement of its production rate, developed for a flash evaporation research apparatus, is described by Lior et al. [49]. The system incorporates a non-contact level controller.

A review of the commercial applications of the multi-stage flash distillation process has been published by El-Ramly et al. [50]. Certain business, economic and technical aspects of the MSF technology were quantitatively analyzed. Data on many plants throughout the world, in operation or under construction as of January 1st 1970, are included [50].

Two typical MSF plants were described in sections 2.6.7, p. 80, and 2.6.8, p. 81. Design data for several others are given in Table 2/6.

Table 2/6

Design parameters of some large MSF plants.

Location	Guantanamo U.S. Naval Base		Ceuta Spanish Morocco	Jeddah Saudi Arabia	Al Khobar Saudi Arabia	Tijuana Mexico
Start-up	1964		1965	1968	1969	1969
Capacity (m³/day)	8520		4000	18930	28400	28400
(kgal/day)	2250		1055	5000	7500	7500
Number of units	2	1	2	2	3	2
Top brine temperature (°C)	90.6		86	121	121	113
(°F)	195		187	250	250	235
Recovery stages	26	12	20	39	30	40
Rejection stages	2	3	4	3	4	4
Feed treatment	H_2SO_4		Phosphate	H_2SO_4	H_2SO_4	H_2SO_4
Electricity	15 MW			50 MW		
Reference	[51]		[52]	[53]	[54]	[55]

Location	Terneuzen Netherlands	Las Palmas Canary Islands	Curacao Netherlands Antilles	Fuerteventura Canary Islands	Eilat 2 Israel
Start-up	1969	1970	1970/1972	1970	1970
Capacity (m³/day)	29000	20000	12000	2000	3875
(kgal/day)	7660	5280	3170	530	1000
Number of units	2	4	2	1	1
Top brine temperature (°C)	120	90	110	89	93
(°F)	248	194	230	192	199
Recovery stages	35	19	24	32	31
Rejection stages	3	3	3	3	3
Feed treatment	HCl	Phosphate	H_2SO_4	Phosphate	H_2SO_4
Electricity		2 × 12 MW			
Reference	[56]	[57]	[56]	[58]	[59]

Location	Shevchenko U.S.S.R.	Shuiaba Kuwait	Rotterdam Netherlands	Hong-Kong
Start-up	1972	1972	1973	1974
Capacity (m³/day)	15000	112500	32400	182000
(kgal/day)	3960	29720	8560	48000
Number of units	3	5	3	6
Top brine temperature (°C)	101	90	113	122
(°F)	214	194	235	252
Recovery stages	31	22	32	25
Rejection stages	3	3	3	3
Feed treatment	Seeding	Phosphate	H_2SO_4	H_2SO_4
Reference	[60]	[61]	[56]	[62]

Literature to 2.7

[1] Bechtel Corporation (Off. Saline Water Res. Develop. Progr. Rept. No. 116 [1964]). — [2] D. W. Clelland, J. M. Stewart (Desalination 1 [1966] 61/76). — [3] Bechtel Corporation (Off. Saline Water Res. Develop. Progr. Rept. No. 175 [1966]). — [4] F. De Winter, S. E. Sadek, J. M. Reynolds (Off. Saline Water Res. Develop. Progr. Rept. No. 251 [1967]). — [5] B. N. Golubkov, A. I. Korneichev (Proc. 1st Intern. Symp. Water Desalination, Washington, D.C., 1965 [1967], Vol. 3, p. 129/53).

[6] A. Hartkamph (Chem. Ingr.-Tech. **41** [1969] 238/44). — [7] J. R. Wolberg (Desalination **2** [1967] 299/307). — [8] B. Kunst (Desalination **3** [1967] 195/202). — [9] L. S. Sterman, S. I. Golub, V. B. Chernozubov, N. A. Mozarov, M. B. Viceblatt, V. M. Lavygin (Desalination **4** [1968] 94/102). — [10] S. Itahara, L. I. Stiel (Desalination **4** [1968] 248/57).

[11] L. T. Fan, C. Y. Cheng, L. E. Erickson, C. L. Hwang (Desalination **3** [1967] 225/36). — [12] L. T. Fan, C. Y. Cheng, C. L. Hwang, L. E. Erickson, K. D. Kiang (Desalination **4** [1968] 336/88; Off. Saline Water Res. Develop. Progr. Rept. No. 355 [1968]). — [13] A. I. Korneichev, A. V. Izvekov, A. A. Myagkov (Desalination **7** [1970] 179/86). — [14] M. A. Mandil, E. E. Abdel-Ghafour (Proc. Symp. Nucl. Desalination, Madrid 1968 [1969], p. 507/24; Chem. Eng. Sci. **25** [1970] 611/21). — [15] J. P. Agostini, R. Cattin-Vidal, R. Jacques (Proc. 3rd Intern. Symp. Fresh Water Sea, Dubrovnik 1970, Vol. 4, p. 21/9).

[16] A. R. Glueck, R. W. Bradshaw (Proc. 3rd Intern. Symp. Fresh Water Sea, Dubrovnik 1970, Vol.1, p. 95/108). — [17] S. Toyama (Proc. 3rd Intern. Symp. Fresh Water Sea, Dubrovnik 1970, Vol.1, p. 219/25). — [18] J. H. Beamer, D. J. Wilde (Desalination **9** [1971] 259/71). — [19] A. K. Coleman (Desalination **9** [1971] 315/31). — [20] H. Satori, K. Hayakawa, K. Konishi (Ishikawa-jima-Harima Giho **11** [1971] 353/9; C.A. **76** [1972] No. 37 281; Ishikawajima-Harima Giho **12** [1972] 127/36; C.A. **77** [1972] No. 66 075).

[21] K. Hayakawa, H. Satori, K. Konishi (Proc. 4th Intern. Symp. Fresh Water Sea, Heidelberg 1973, Vol. 1, p. 303/12). — [22] D. Barba, A. R. Giona, G. Tagliaferri (Quad. Ric. Sci. No. 49 [1968] 125/49, 150/66). — [23] D. Barba, G. Liuzzo, G. Tagliaferri (Petrolieri d'Italia **1970** November, p. 44/51), Anonymous (Oil Gas Intern. **10** No. 11 [1970] 87). — [24] N. Arad, P. Glueckstern, Y. Kantor (Proc. 4th Intern. Symp. Fresh Water Sea, Heidelberg 1973, Vol. 1, p. 135/42). — [25] R. S. Silver (Proc. 3rd Intern. Symp. Fresh Water Sea, Dubrovnik 1970, Vol. 1, p. 191/206).

[26] F. O'Shaughnessy (Desalting Plant Inventory Rept. No. 4 [1973]). — [27] W. A. Gardner (Proc. Symp. Nucl. Desalination, Madrid 1968 [1969], p. 127/43). — [28] Fluor Corporation (Off. Saline Water Res. Develop. Progr. Rept. No. 233 [1966]). — [29] E. N. Sieder, I. Spiewak (Proc. Symp. Nucl. Desalination, Madrid 1968 [1969], p. 829/46). — [30] Catalytic Construction Co. (Off. Saline Water Res. Develop. Progr. Rept. No. 615 [1970], No. 645 [1971], No. 705 [1971]).

[31] W. R. Williamson, F. W. Gilbert (Off. Saline Water Res. Develop. Progr. Rept. No. 525 [1970]). — [32] K. Sato, K. Kamiyama, K. Tahara (Nippon Kaisui Gakkai-Shi 25 No. 135 [1971] 198/206; C.A. **76** [1972] No. 142 762). — [33] Richardsons Westgarth and Co. (Off. Saline Water Res. Develop. Progr. Rept. No. 108 [1964]). — [34] Weir Westgarth Ltd. (Off. Saline Water Res. Develop. Progr. Rept. No. 271 [1967]). — [35] R. Van Winkle (ORNL-TM-2746 [1970]).

[36] R. A. Tidball, A. N. Rogers, R. N. Webb (Off. Saline Water Res. Develop. Progr. Rept. No. 562 [1970]). — [37] C. D. Hornburg (Off. Saline Water Res. Develop. Progr. Rept. No. 656 [1971]). — [38] A. Steinbruchel (Off. Saline Water Res. Develop. Progr. Rept. No. 749 [1971]). — [39] Off. Saline Water Res. Develop. Progr. Rept. No. 731 [1971]. — [40] F. Mayinger, H. Voos, J. Ulrich (Proc. 4th Intern. Symp. Fresh Water Sea, Heidelberg 1973, Vol. 1, p. 371/9).

[41] W. R. Williamson, J. R. Hefler (Off. Saline Water Res. Develop. Progr. Rept. No. 575 [1970]). — [42] A. E. Dukler, R. F. Castle (Off. Saline Water Res. Develop. Progr. Rept. No. 515 [1967]). — [43] P. Belzer, J. M. Hendrie, K. C. Hoffman, A. Oltman, M. Reich (Off. Saline Water Res. Develop. Progr. Rept. No. 687 [1971]). — [44] A. Brown (Off. Saline Water Res. Develop. Progr. Rept. No. 539 [1970]). — [45] J. Ricard (Rev. Gen. Therm. **9** [1970] 1255/73).

[46] U.S. State of New Jersey, New Jersey Depart. of Health, Aqua-Chem. Inc., S. T. Powell Associates and U.S. Geological Survey (Off. Saline Water Res. Develop. Progr. Rept. No. 728 [1971]). — [47] D. E. Maneval, S. Levezis (Trans. AIME **251** [1972] 42/5). — [48] I. Motyka (Technik in Polen **48** No. 2 [1972] 14/5). — [49] N. Lior, J. Leibovitz, A. D. K. Laird (Desalination **13** [1973] 91/4). — [50] N. A. El-Ramly, J. M. English, J. W. McCutchan (Univ. Calif. Rept. ENG-7079 [1970]).

[51] P. M. Rapier, W. H. Rowe, A. C. Ko, P. P. De Reinzo, H. Gitterman (Off. Saline Water Res. Develop. Progr. Rept. No. 769 [1972]). — [52] H. Laban (Chem. Ingr.-Tech. **41** [1969] 114/6). — [53] G. F. Leitner, R. W. Goeldner, A. Steinbruchel, A. A. Santilli (Proc. 3rd Intern. Symp. Fresh Water Sea, Dubrovnik 1970, Vol. 1, p. 115/26). — [54] F. Barluzzi, M. Galateri, E. Maraini (Proc. 4th Intern. Symp. Fresh Water Sea, Heidelberg 1973, Vol. 1, p. 169/78). — [55] H. M. Zuccolotto (Proc. 4th Intern. Symp. Fresh Water Sea, Heidelberg 1973, Vol. 2, p. 307/20).

[56] P. R. Bom (Proc. 4th Intern. Symp. Fresh Water Sea, Heidelberg 1973, Vol. 1, p. 191/6). — [57] J. Suarez, M. H. Ali-El-Saie (Proc. Symp. Nucl. Desalination, Madrid 1968 [1969], p. 239/60).

— [58] F. Cantera (Proc. 4th Intern. Symp. Fresh Water Sea, Heidelberg 1973, Vol. 1, p. 207/17). — [59] J. Lev-Er, M. Adler, Y. Kantor (Proc. Symp. Develop. Desalination Tech. Israel, Jerusalem 1971 [1972], p. 151/6). — [60] N. M. Sinev, V. G. Shatsillo, I. P. Lazarev (Proc. 4th Intern. Symp. Fresh Water Sea, Heidelberg 1973, Vol. 1, p. 481/5).

[61] R. Douvry (Proc. Study Group Nucl. Agro-Ind. Complexes, I.A.E.A. Rept. 139 [1971] 201/8). — [62] F. A. Drake, Y. N. Wu, J. C. Drake (Paper distributed at the 4th Intern. Symp. Fresh Water Sea, Heidelberg 1973). — [63] J. G. Delene, S. J. Ball (ORNL-TM-2933 [1971]). — [64] W. G. S. Fort (ORNL-TM-3535 [1972]). — [65] D. C. Guneratne (At. Energy Establishment, Winfrith 1973).

[66] S. J. Ball, N. E. Clapp, J. G. Delene (ORNL-TM-3393 [1970]). — [67] J. G. Delene, S. J. Ball, N. E. Clapp (ORNL-TM-3420 [1972]). — [68] G. L. Beighly (Off. Saline Water Res. Develop. Progr. Rept. No. 595 [1970]).

2.8 Other multi-stage distillation processes

Several other distillation processes have been proposed and either tested in pilot plant scale or reported in more or less advanced design studies. The principal processes of this group are summarized in the following pages.

2.8.1 Vertical direct contact flash evaporator

Analysis of the design of conventional flash distillation plant showed that the largest single component of capital cost and thermodynamic irreversibility is associated with transfer of heat from condensing steam through tubes into brine. A process was developed in the United Kingdom, known as the "Clementine" project, using the thermal enery available in the flashing brine to do work [1]. The energy released by flashing is partly extracted as work in making the foaming brine lift against gravity and steam is condensed directly on to a liquid heat-transfer medium. The brine circulates without aid of pumps. The deposition of insoluble solids is expected to occur at the liquid-liquid interface of the direct contact heat exchanger or on solid surfaces which are not called upon to transfer heat. Therefore higher top operating temperatures may be possible. The feasibility of the flashing lift has been demonstrated in laboratory experiments at atmospheric conditions. Energy recovery provided by the vertical arrangement reduces the pumping work required to about one half to one third that of a conventional horizontal plant operating over the same temperature interval.

Flashing brine flow in vertical ducts has been studied in support of the above vertical MSF plant concept. The experiments reported were conducted at atmospheric or near atmospheric conditions. Measurement of total brine throughput and pressure profiles along the duct have been compared with theoretical predictions based on a homogeneous flow model. Very good agreement between predicted and experimental brine throughput was obtained, when a nucleation device was used to initiate flashing at a precise level in the duct. Nucleation devices were found to stabilize flow conditions and provide increased throughput over unobstructed duct conditions [2].

In a further investigation flashing flow of seawater in a vertical parallel-sided duct of 76 mm bore has been studied under vacuum conditions, the physical lift for all experiments being 0.25 m. At the lowest temperature level, 60°C (140°F), measured throughput rates were up to 60% in excess of values predicted by a homogeneous flow theory. This discrepancy is due to slip between phases and slip velocities have been calculated. At the highest temperature, 93°C (200°F), measured rates were up to 40% less than predicted. The occurrence of sonic velocities, or choking flow, is tentatively suggested to explain this. Measurements of equilibration across the duct showed that there was a swifter approach to equilibrium at the lower temperature levels [3].

A reasearch program on vertical flash evaporators is also reported by Klaren [4]. The main objective was the determination of the mass flow of flashing water in vertical channels and through orifices, as a function of the channel or orifice geometry, the salt content of the water, the inlet temperature and the saturation temperature difference over the channel or orifice.

2.8.2 Vertical spiral arrangement MSF design

In a multi-stage flash distillation plant the condenser tube bundle is vertical arranged and the flash chambers are spirally accommodated around it. This configuration allows the design of a

Vertikaler Ent-
spannungs-
verdampfer
mit Direkt-
konden-
sation

plant, which corresponds best in dimensions to the economic optimization: appropriate length of the flash chambers and short tubes of small diameter for the condenser. The proposed use of thin plastic tubes for the heat exchange surfaces avoids corrosion risks and considerably decreases the investment cost. The necessary space to erect the plant is as well reduced. The development of spirally vertical type MSF plants is claimed to reduce notably the cost of product water [5].

Gesteuerte
Ent-
spannungs-
ver-
dampfung

2.8.3 Controlled flash evaporation

A system which would eliminate the losses caused in the usual turbulent MSF chambers has been described as controlled flash evaporation (CFE). Preheated seawater passes downwards through long and narrow vertical channels between a higher pressure stage and the next lower pressure stage. Rectangular orifice spacers at the top control the flow of the brine, which forms a flowing film on both sides of the channel (**Fig. 2–26**). As pressure lowers below the saturation value, water evaporates and cavitation forms a core stream of vapor, with films of liquid descending on each side wall of the channel. Vapor separates in the lower part of the channels, which has a considerable increase in cross section, passes at the two exits of the channels to condenser tubes and is recovered as product water. The brine flows from the skirt of the channels to the pool of the next lower stage. The evaporation from the liquid film being performed without bubbling and violence, there is no entrainment of seawater with the vapors and demisters are not required.

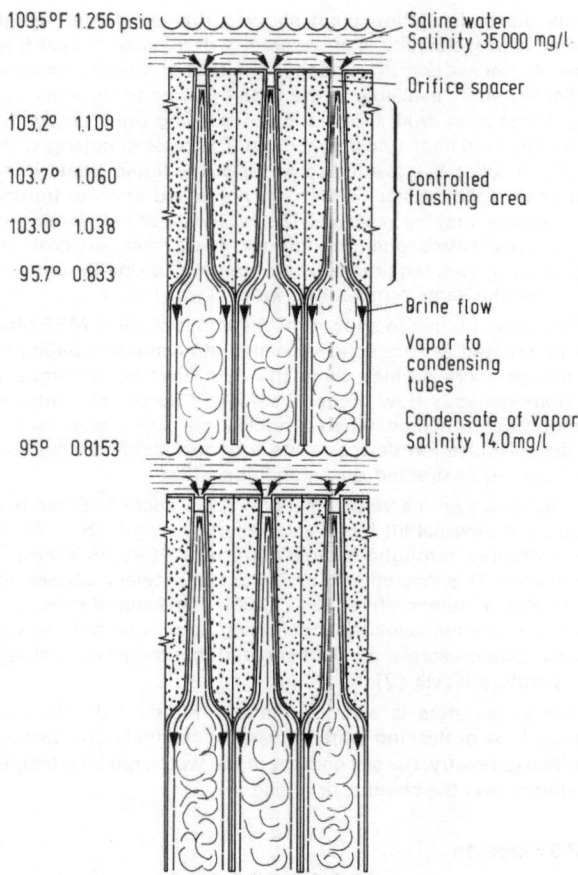

109.5°F 1.256 psia — Saline water
Salinity 35000 mg/l

— Orifice spacer

105.2° 1.109

103.7° 1.060   Controlled
                        flashing area
103.0° 1.038

95.7° 0.833

— Brine flow

Vapor to
condensing
tubes

Condensate of vapor
95° 0.8153   Salinity 14.0 mg/l

Fig. 2–26

Controlled flash evaporation.

Whereas the flash between stages in a conventional MSF evaporator is limited to about 2°C (3 to 5°F), the flash in the CFE can be very much higher. This means that the system can be built with fewer stages and therefore at lower capital cost [6].

Capital investment for the controlled flash evaporation plant is expected to be 80 to 85% of that of the corresponding conventional multi-stage flash distillation plant. A similar reduction is also expected in the cost of product water.

Literature to 2.8

[1] P. T. Walker, I. H. Newson, K. D. B. Johnson (Desalination **2** [1967] 196/206). — [2] I. H. Newson, M. H. Delve (Proc. 3rd Intern. Symp. Fresh Water Sea, Dubrovnik 1970, Vol. 1, p. 127/43). — [3] I. H. Newson (Proc. 4th Intern. Symp. Fresh Water Sea, Heidelberg 1973, Vol. 1, p. 403/17). — [4] D. G. Klaren (Ingenieur [The Hague] **83** No. 4 [1971] W1/W16). — [5] J. C. Deronzier, J. Huyghe, F. Lauro (Proc. 4th Intern. Symp. Fresh Water Sea, Heidelberg 1973, Vol. 1, p. 219/27). — [6] R. C. Roe, D. F. Othmer (Proc. 3rd Intern. Symp. Fresh Water Sea, Dubrovnik 1970, Vol. 1, p. 169/90; Chem. Eng. World **6** No. 9 [1972] 53/62; AIChE [Am. Inst. Chem. Engrs.] Symp. Ser. **68** No. 120 [1972] 130/9).

2.9 Comparison of the distillation processes

The comparison is based on the actual state of the art of the main distillation processes.

Vergleich der Destillationsverfahren

2.9.1 Long tube vertical evaporator versus multi-stage flash distillation

Vertical tube evaporation and multi-stage distillation have many similarities as evaporation processes. The feed liquid is passing up the temperature gradient through successive heaters and is gaining heat from the condensation of vapors. Brine and distillate streams are passing down the temperature and pressure gradient, the brine stream diminishing in quantity and the distillate stream increasing in quantity. Flashing of distillate and of brine occurs in both systems. The defining variable in the two systems is the rate of brine circulation relative to distillate product [1].

Vergleich des Vertikalrohrverdampfers mit dem mehrstufigen Entspannungsverdampfer

Despite the similarities, analytical studies showed that the LTV (long tube vertical) process requires less heat transfer area than MSF (multi-stage flash) distillation. The heat transfer area of a MSF distillation unit contributes up to 40% to the overall plant cost. Hence, the amount of heat transfer surface required is an important economic factor [2].

The MSF technology and component development have reached an advanced stage. Numerous plants ranging in size from small units up to as large plants as with a capacity of over 113 000 m³/day (30 Mgd) have been built and perform as designed. A plant with a daily capacity of over 180 000 m³/day is under construction. The technology of the VTE process is not developed to the same extent, as to permit direct projection to such very large size units.

Comparison of engineering characteristics and economics of three plant designs with a capacity of 2.5 Mgd was made to evaluate the VTE (vertical tube evaporator) and MSF process. Two of the designs were for VTE plants, one with conventional smooth tubes and one with enhanced surface tubes, and the third design was for a MSF plant. The performance ratio of each plant was approximately the same and the other plant characteristics were also very similar. With respect to maintenance, again the two types of plants appear to be about equal [3].

The estimated costs, based on 4.5% interest for capital and a 30-year plant life, are as follows:

	VTE plant smooth tubes	VTE plant enhanced tubes	MSF plant smooth tubes
Capital costs $ 1968	4 687 500	4 457 700	4 834 100
Escalated to $ 1970	4 900 000	4 680 000	5 060 000
Product-water costs:			
$ 1968 per 1000 gal	1.03	1.02	1.08
$ 1970 per 1000 gal	1.08	1.07	1.14
Cents 1968 per m³	27.2	26.9	28.5
Cents 1970 per m³	28.5	28.3	30.1

It might be concluded from the above study that capital costs and production costs for plants with a capacity of about 2.5 Mgd are very nearly the same. The enhanced tube VTE plant has the lowest cost, but the total spread in costs between the three plants evaluated is about 5%. This does not represent a significant economic advantage for either of the VTE plants over the comparable MSF plant. For larger plant sizes the plant economics favor the VTE plant, as it will be seen in the following.

A comparison of the economic data of the MSF distillation process and the VTE process was also made by Golub and his co-workers [4].

In a feasibility study of multiple-effect long tube vertical (LTV) evaporators for seawater desalination, at capacities over 40 000 m³/day, an optimization of temperature profiles and performance ratio was carried out for both smooth and double-fluted tubes. The calculations have confirmed the economic potential of multiple-effect LTV evaporators and indicate that, under certain economic conditions, these evaporators show an advantage over MSF evaporators. The use of double-fluted tubes in LTV evaporators has a potential for further reduction of cost of product water.

A comparison was made of the cost of water desalted in dual purpose plants using MSF and multiple-effect LTV evaporators. The results of this study are given in Table 2/7. It might be concluded that water costs from LTV evaporators compared with those from MSF evaporators are quite lower, especially with double-fluted tubes [5]. Prices in this table refer to 1967 U.S. dollars.

Table 2/7

Comparison of water cost breakdown of three types of evaporators in dual purpose plant.

		Multiple-effect smooth tubes	LTV fluted tubes	MSF
Heating steam temperature	°C	120	120	120
	°F	248	248	248
Maximum brine temperature	°C	116	116	113
	°F	241.3	241.3	235.4
Heating steam flow rate	ton/h	2140	2140	2140
	10^6lb/h	4.72	4.72	4.72
Annual fixed charge rate	%	7	7	7
Steam cost	cents/ton	6.39	6.39	6.39
	cents/10^3lb	2.90	2.90	2.90
Electricity to evaporator auxiliaries	MW$_e$	13	15	28
Optimal performance ratio	kg/kg	4.92	5.65	7.61
Steam charges	10^6\$/y	1.02	1.02	1.02
Fixed charges	10^6\$/y	2.04	2.24	4.28
Electricity charges	10^6\$/y	0.48	0.56	1.04
Operation and maintenance	10^6\$/y	1.77	1.98	2.91
Total charges	10^6\$/y	5.31	5.80	9.25
Water production	10^6m³/y	70.6	81.1	109.2
	Mgd	60.1	69.1	93.2
Water cost	cents/m³	7.52	7.15	8.50
	cents/kgal	28.4	27.0	32.0

Mehrfach-effekt-Destillation mit horizontalen im Vergleich mit vertikalen Verdampfer-rohren

2.9.2 Horizontal tube multiple-effect versus long tube vertical evaporator

The development of the horizontal tube multiple-effect (HTME) distillation process is a long way behind the development of the MSF and even the LTV process. HTME has been used in several small plants, in some of them combined with vapor compression, and is beginning to be developed for intermediate and large plants. A new double effect test unit with bundles equal in size to a 9500 m³/day (2.5 Mgd) plant will establish the feasibility of this equipment for intermediate and large plants.

The advantage sought in HTME evaporators is to obtain overall heat transfer coefficients comparable to the vertical enhanced tubes without the expense of fluting the tube to enhance heat transfer.

2.9.3 Vapor compression (VC) versus multi-stage flash (MSF) distillation

While MSF can profitably be scaled up to larger plants, it does not give economic advantages when scaled down. VC plants with small capabilities offer lower water costs than small MSF plants. In larger units the gap in the cost for the two processes is progressively vanishing. This is one of the reasons why vapor compression is usually preferred for small capabilities. On the other hand, the VC process might be optimized to require less energy input per unit of product than other processes, and the real economic advantage might be obtained if the shaft work is generated in-plant at low cost.

A comparison of the two processes was also made by Huyghe [6]. VC distillation plants for small capacities, about 500 m³/day, driven by Diesel engine, are usually more economic than MSF distillation plants. For larger plants, as well as for combined power and water plants, the economics depend on the cost of fuel and of fixed capital charges. Individual consideration for each case is necessary [7].

Vergleich der Brüden-kompression mit der vielstufigen Ent-spannungs-verdampfung

Literature to 2.9

[1] R. S. Silver (Desalination 9 [1971] 235/43). — [2] M. J. Burley (Desalination 2 [1967] 81/8). — [3] The Fluor Corporation (Off. Saline Water Res. Develop. Progr. Rept. No. 580 [1970]). — [4] S. I. Golub, I. L. Sergeenko, E. A. Sobolev, V. B. Chernozubov, V. G. Shatsillo, B. I. Tkach, N. K. Tokmantsev, N. N. Musikhin (Desalination 5 [1968] 29/33). — [5] M. Schaal, F. S. Aschner, D. Hasson (Desalination 10 [1972] 67/93).

[6] J. Huyghe (Chim. Ind. Genie Chim. 101 [1969] 575/85). — [7] A. Schweyer (Chem. Ingr.-Tech. 41 [1969] 100/4).

2.10 Direct contact heat transfer

The objective of the direct contact heat transfer is to eliminate metallic surfaces in the condensation step of the flash distillation processes. Direct contact is based upon a step-by-step countercurrent recovery of the heat from product vapors by direct condensation into a circulating fresh water stream. Heat is transferred with the best possible approach to equilibrium in various devices. The heat is removed from the hot fresh water stream and delivered to the salt water feed either by means of countercurrent liquid-liquid heat exchangers, using an immiscible fluid such as a hydrocarbon, or by conventional or plastic tubing. Scaling problems are reduced.

Wärme-transport mit Direkt-konden-sation

Make-up heat is supplied to the oil stream by conventional means. Conventional heat exchangers are also used to transfer most of the heat from the fresh water stream to the incoming salt water at the lower temperature levels, where no scaling problems appear.

2.10.1 Use of hydrophobic heat-carriers

Four commercial hydrocarbons were investigated in a two-column laboratory heat exchanger and operating parameters were determined [1]. The heat transfer coefficient depends upon the oil fluidity and the flow rate. Salt contamination of the fresh water is negligible. Su ,ended or scale formed matter collects at the oil-water interface and may be removed as an aqueous sludge. The process, called also vapor-reheat process, was further tested in a pilot plant, consisting of a two-column liquid-liquid heat exchanger system and a five-stage direct contact distillation system with a capacity of 5.5 m³/day, without solving the problem [2].

Verwendung hydrophober Wärmeträger

Process calculations have shown that a vapor-reheat distillation system using liquid-liquid heat exchangers exhibits, in its original design, no particular cost advantage compared with estimates of other distillation processes [3].

Liquid-liquid heat transfer for large-scale desalination was also investigated by Kern [4]. It was found that, from the various devices examined, heat transfer by direct contact in liquid spray columns exhibits considerable promise of feasibility and substantial economics over the conventional MSF process.

Heat transfer to mixtures of water and mineral oil passing through a double pipe heat exchanger in turbulent flow were analyzed by Somer [5] for the oil-in-water and water-in-oil systems. Operation of a single effect evaporation pilot plant for desalination is described. Preliminary design and cost estimation for a nine effect desalination plant reveals the possibility of producing fresh water at a cost of $ 0.80/1000 gal. Further data on the operation of the pilot plant were recently reported [24].

Experimental studies were performed on the hydrodynamic and thermal processes taking place in a heat exchanger column, using a hydrophobic heat carrier. The findings predict the applicability of this process for desalination purposes [6].

A 5-stage pilot plant with a capacity of 250 m³/day distillate was constructed at Shevchenko to test the application of an immiscible liquid heat carrier [7]. In seawater desalination studies, based on the use of a C_{16-20} liquid paraffin in a 3-liquid heat exchanger system heat transfer coefficients of 63 000 kcal/m² · h · degr were determined [8]. The flow sheet and the construction of the main apparatus of a pilot desalination plant using a hydrophobic heat carrier with the production of 3600 m³ of distillate per day was described by Podbereznii and technical and economical characteristics were presented [9].

Contact condensers exhibit high heat transfer rates, compactness, absence of scale, and can be operated even at very low temperature driving forces [10]. A technique to predict the condensation rate and height of a single bubble train is extended to predict the condenser height in counter- and co-current multi-bubble systems. This approach enables to distinguish the effect of the bubbles spatial density, i.e. effects of bubble frequency and horizontal spacing.

Processes for seawater distillation, which would substitute a stream of oil for the usual metallic heat transfer surface have generally required that the oil on the condensing part of the cycle shall precool a stream of product water which can then condense new steam. Inexpensive sensitizers were found, which when added to the oil promote direct condensation of steam with low intervals of Δp and Δt. The potential simplification of flow chart invites revival of interest in the "vapor reheat" process [11].

The Osdor process is a multi-stage direct contact flash distillation process, in which the stages are arranged vertically in a tower [12]. The process is applicable for the production water alone or water and marine salts. An inert liquid (oil) may be used as heat transfer medium. The vertical arrangement eliminates interstage pumps.

Direkt-
konden-
sation

2.10.2 Direct contact condensation

Another approach of the vapor reheat system was described by Othmer [13] and consists in a modification in vertical arrangement of the conventional multi-stage flash distillation process with complete elimination of condenser heat transfer surfaces. In the usual multiple flash evaporation, represented by **Fig. 2–27**, the seawater climbs the right side of a ladder of stages in each of which it is preheated. The heated feed-water flashes in stages at progressively lower temperatures and pressures on the left side of the ladder, giving off the vapors used in preheating. The condensate in each stage is flashed back in the next lower stage to recover its heat until it is discharged.

In the vapor reheat system, as shown in **Fig. 2–28**, a pump is used at each stage, to pass a stream of cooler fresh water up the ladder. The fresh water is heated by direct contact with the condensing vapors from the left half of each stage. The heated fresh water, from the top stage, passes through a heat exchanger, where it gives up its heat to preheat the entering seawater. Part of the fresh water, equivalent to the total condensate in all stages, is discharged as fresh water product. The balance is recycled starting again at the bottom of the column. On the other hand the incoming seawater, preheated by the fresh water stream, is further heated in a conventional steam heater and passes to the flashing side of the top stage. Vapor, in each stage, moves freely from left to right and each stage behaves as the plate of a distillation column.

An evaluation of the heat and mass transfer in flash distillation without metallic interfaces is reported by Kogan [14]. The mechanism of direct contact distillation was also studied using two separate streams of water simulating a single stage direct contact evaporation-condensation unit [15]. An increase of non-condensable gas concentration in the vicinity of the condensing surface by a factor of 20 was determined. The geometrical location of suction intakes for elimination of non-condensable gases is shown to have a marked influence upon mass transfer performance. These experimentally determined trends are explained theoretically by an analysis of diffusion in the vapor-non-condensable gas mixture.

A case study of a desalination process utilizing direct contact heat transfer in the evaporation and condensation stages was reported by Dajani [16]. Taking advantage of the abundance of low-cost fuel gas in oil producing areas, the process is designed to reduce to a minimum the problems resulting from scale deposition on heat transfer surfaces and eliminates the need for the pretreatment measures usually required in distillation processes. Evaporation is achieved by contacting the exhaust gases of a fuel gas burner firing downward through a tube with one end immersed in the water. The gas emanates in an annulus chamber surrounding the tube and vapors are condensed in a packed tower by a stream of recirculating fresh water.

A similar method for saline water conversion, using a carrier gas for the distillation process, was described by Lentz [17]. The gas is bubbled through the solution and carries the water vapor in a concurrent stream to the condenser. The use of hot gases from existing industrial processes is possible. The water produced is said to be of high purity.

The conceptual design and costs of a 2.5 Mgd direct contact condensation multi-stage flash distillation plant were determined by Marwede et al. [25]. The study indicated that the process is competitive with other processes such as VTE upflow or VTE downflow plants. Product water costs were estimated to $ 0.86 per kgal ($ 0.23 per m³).

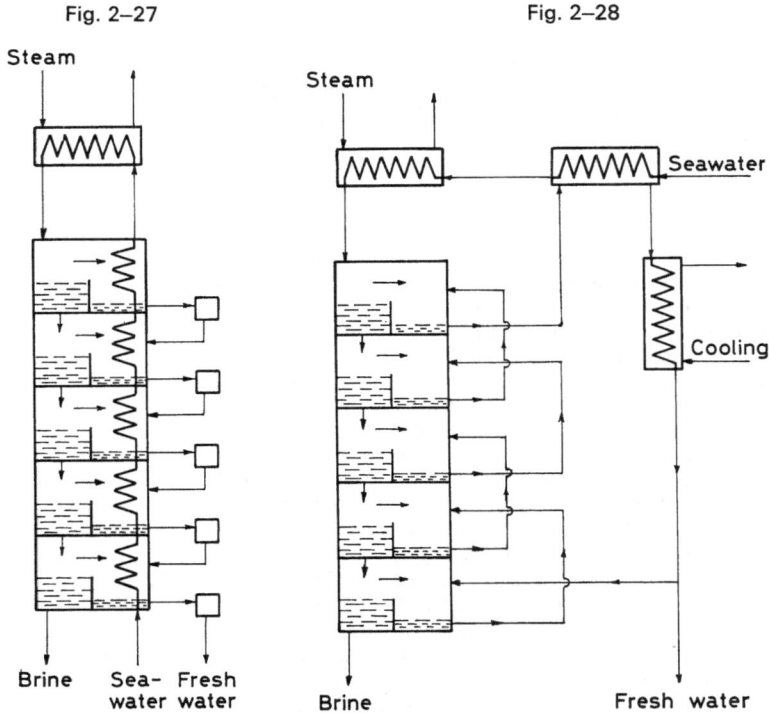

Fig. 2–27

Fig. 2–28

Multi-stage flash evaporator
in vertical arrangement.

Vapor reheat flash evaporation.

2.10.3 Use of nonmetallic heat exchanger surfaces

The use of plastics for heat transfer surfaces in heat exchanger systems of distillation plants was already suggested independently of direct contact condensation.

Plastic films can serve as a heat exchange surface within a limited field of conditions. A new design was presented by Beushausen [18] to produce heat exchanger elements from thin flexible plastic film. Lower total cost of heat exchanger may result, despite of lower heat conductivity and shorter lifetime of plastic film elements. A long-time test was run over 13 000 h.

Studies on a direct contact condensation module and a plastic film heat exchanger indicated the process to be feasible. Direct contact condensation of steam was accomplished with enhanced heat transfer coefficients by providing a high cross flow steam velocity and by the use of simple vane type turbulence promoters in the water system. Polyvinyl fluoride films were found to be adequate for

Verwendung nicht-metallischer Wärme-austausch-flächen

heat exchangers. No operational problems were encountered with the designed heat exchanger [19]. Tedlar tube heat exchangers were also successfully tested. Plastic films were evaluated as promising materials of construction in another report and conceptual designs were generated for a 100 Mgd plant. The process is considered as economically attractive [20]. A plastic film heat exchanger was successfully operated for a 6 month period under conditions simulating those at the high temperature end of a commercial plant. Hydraulic studies showed that a full scale model of a single interstage condensate passage would convey condensate from the trays of one stage to the trays of the next higher pressure stage without water back-up or other hydraulic problems. A pilot plant of a 63 m³/day (16700 gpd) capacity was designed. The design incorporates a 4-stage direct contact condensation MSF distillation unit and a plastic film heat exchanger [21].

The possibility of using small bore, thin-walled condensers made from selected types of glass and thermoplastics in MSF evaporators has been examined by Wood [22] from the economic standpoint and in experimental rigs. Tubular condensers capable of operation for extended life with overall heat transfer coefficients of 366 to 733 kcal/m² · h · degr (75 to 150 Btu/ft² · h · °F) have been tested. Their installed costs per unit of heat transmission in flash plants were estimated to be from 33 to 66% of those of conventional condensers.

The conceptual design of a 190000 m³/day (50 Mgd) multi-stage flash direct contact condensation desalting plant was presented by Kogan, including capital and operation costs [23]. Tedlar or crosslinked polyethylene is used in the heat exchangers.

Literature to 2.10

[1] W. S. Thompson, T. Woodward, W. A. Shrode, E. D. Baird, D. A. Oliver (Off. Saline Water Res. Develop. Progr. Rept. No. 63 [1962]). — [2] Food Machinery Corp. (Off. Saline Water Res. Develop. Progr. Rept. No. 78 [1963]); C. D. Watson (Proc. 1st Intern. Symp. Water Desalination, Washington, D.C., 1965 [1967], Vol. 2, p. 715/32). — [3] J. C. Orcutt, F. O. Mixon, D. R. Whitaker (Off. Saline Water Res. Develop. Progr. Rept. No. 198 [1966]). — [4] D. Q. Kern (Off. Saline Water Res. Develop. Progr. Rept. No. 261 [1967]). — [5] T. G. Somer, M. Bora, Ö. Kaymakcalan, Y. Arikan (Proc. 3rd Intern. Symp. Fresh Water Sea, Dubrovnik 1970, Vol. 1, p. 337/55).

[6] A. P. Yufin, E. D. Maltsev, L. S. Zhivotovskii, L. P. Ivanov, L. S. Skvortsov (Proc. 1st Intern. Symp. Water Desalination, Washington, D.C., 1965 [1967], Vol. 3, p. 107/22). — [7] L. L. Podbereznii, L. I. Trofimov, K. S. Siyanko, V. B. Chernozubov, A. M. Rozen, E. D. Maltsev, F. P. Zaostrovskii (Proc. 3rd Intern. Symp. Fresh Water Sea, Dubrovnik 1970, Vol. 4, p. 127/36). — [8] V. L. Podbereznii, A. M. Rozen, K. S. Siyanko (Vodosnabzh. Sanit. Tekhn. 1971 No. 6, p. 8/13; C.A. 75 [1971] No. 80137). — [9] V. L. Podbereznii, L. I. Trofimov, K. S. Siyanko, N. G. Narvatkina (Proc. 4th Intern. Symp. Fresh Water Sea, Heidelberg 1973, Vol. 1, p. 463/9). — [10] S. Sideman, D. Moalem (Proc. 4th Intern. Symp. Fresh Water Sea, Heidelberg 1973, Vol. 1, p. 103/12).

[11] J. R. Maa, K. Hickman (Desalination 10 [1972] 95/111). — [12] A. Osdor (Proc. Symp. Develop. Desalination Technol. Israel, Jerusalem 1971 [1972], p. 94/116). — [13] D. F. Othmer, R. F. Benenati, G. G. Goulandris (Dechema Monograph. 47 [1962] 73/98). — [14] A. Kogan (Proc. 1st Intern. Symp. Water Desalination, Washington, D.C., 1965 [1967], Vol. 2, p. 57/68). — [15] A. Kogan, M. Victor (Desalination 4 [1968] 80/93).

[16] M. T. Dajani (Proc. 3rd Intern. Symp. Fresh Water Sea, Dubrovnik 1970, Vol. 1, p. 455/60).— [17] H. Lentz (Desalination 7 [1970] 245/8). — [18] J. Beushausen (Proc. 3rd Intern. Symp. Fresh Water Sea, Dubrovnik 1970, Vol. 1, p. 369/81). — [19] A. L. Kohl, T. T. Shimazaki, J. R. Wetch, L. L. Bienvenue (Off. Saline Water Res. Develop. Progr. Rept. No. 475 [1969]). — [20] A. L. Kohl, T. T. Shimazaki, J. R. Wetch, W. B. Suratt (Off. Saline Water Res. Develop. Progr. Rept. No. 566 [1970]).

[21] A. L. Kohl, T. T. Shimazaki, W. B. Suratt, S. Westerman (Off. Saline Water Res. Develop. Progr. Rept. No. 647 [1971]). — [22] F. C. Wood (Proc. 3rd Intern. Symp. Fresh Water Sea, Dubrovnik 1970, Vol. 1, p. 383/402). — [23] A. Kogan, A. Lavie (Proc. Symp. Develop. Desalination Technol. Israel, Jerusalem 1971 [1972], p. 117/23). — [24] T. G. Somer, M. Bora, Ö. Kaymakçalan, S. Özmen, Y. Arikan (Desalination 13 [1973] 221/9). — [25] M. E. Marwede, P. J. Schroeder, A. Kohl, T. T. Shimazaki (Off. Saline Water Res. Develop. Progr. Rept. No. 821 [1973]).

Entsalzungs-
anlagen
großer
Kapazität

2.11 Large size desalting plants

For the development of desalting technology in commercial plants to yield low-cost fresh water a number of studies have been made in the United States and in other countries. The development program included conceptual designs of large plants, detailed feasibility and preliminary engineering studies and the construction and operation of modules. These are full size portions of large plants. The development programs for large plants are concerned with the multi-stage flash distillation process, the vertical tube distillation process, and combinations of both processes with vapor compression. Development of other processes are discussed in the appropriate sections of this book. The aim is to develop the MSF technology to capacities as high as 190000 to 4 million m³/day (50 to 1000 Mgd) and to make beneficial use of the heat transfer enhancement in vertical tube evaporator plants to yield a cost of product water acceptable even to such uses, as agriculture, for which the present cost of desalted water is still too expensive. The main large scale desalting studies and experience from the operation of modules and advanced technology plants will be summarized in the following sections.

2.11.1 Components of large desalting plants

Komponenten großer Entsalzungs-anlagen

A program was sponsored by Office of Saline Water (O.S.W.) at Oak Ridge National Laboratory (ORNL) to investigate the equipment components required for 50 Mgd and larger seawater distillation plants and to determine if the large components and suitable materials were available from industry. Included within the scope of the investigations were condenser tube bundles, condenser tubing, tubing fabrication methods, interstage tube seals, demisters, deaerators, deaerator jets and blowers, valves and vapor compressors.

It was concluded that suitable equipment for desalting plant applications was generally available without need of extensive development, except some cases indicated in the report [1]. Active development programs were initiated in most of the indicated areas.

The present state of the art for some of the components was examined and technical and economic problems were evaluated in several detailed reports and publications. These components are: deaerators [2], deaerators and dissolved gas analysis [3], deaerator ejectors and blowers [4], condenser tube bundle configurations [5], and high temperature water jet compressors [6]. In various publications several other plant components were evaluated: demisters by Golub [7], pumps by Lenhard [8] and Ibrahim [9], as well as water-jet and steam-jet ejectors by Golubkov and Stepanov [10]. A pilot plant was operated for demonstrating multiple-phase ejector topping units to augment the performance of distillation desalination systems [11]. Tests were also made on a high temperature water jet vapor compressor operating in the high temperature range (138°C) and at high discharge vapor pressures to compare performance with that with lower discharge pressures corresponding to temperatures of 115°C [12].

2.11.2 Universal desalting plant design

Allgemein einsatzfähige Entsalzungs-anlagen

The development of desalting by means of distillation wide over the world made it advisable to establish general rules for a "universal" plant, which is defined "as able to be installed at any appropriate spot in the world". The result of this concept of universality is that several plants were designed, one of which will be very close to the optimum desalting plant for the chosen site. Advantages may be taken of the desalting technology, which the prototype specifications and drawings represent. The Universal Desalting Plant Design was prepared by Burns and Roe Inc. on behalf of and published by the Office of Saline Water in 1969. The Manual consists of five volumes.

Volume 1 (Report on design of a 2.5 Mgd Universal desalting plant) explains the methods used to produce the technical specification and design studies required in producing the final series of optimum desalting plant designs. Volumes 2 and 3 (Users manual) were prepared to assist in the selection of either a dual-purpose or single-purpose desalting plant. They contain information and comparison of the cost of desalting with other alternatives, which may be available. Water costs of dual- and single-purpose plants are discussed and a self-explanatory step-by-step procedure for selection of the optimum plant is contained. Also included are procedures for correcting water costs for capital cost, operating, maintenance and chemical costs different from those used in the design. Volume 4 (Bidding specifications) establishes the conditions under which bidders may present formal proposals for the construction of a plant. Detailed data and material specifications for the evaporator modules and associated auxiliary equipment are included. Volume 5 (Technical specifications) deals with the engineering design of the desalting plant and covers all technical phases.

Hydraulic model studies of interstage module piping of the 2.5 Mgd Universal Desalination Desalting Plant are reported by Beichley [13]. Head losses, incurred with the discharge of a given quantity of salt-saturated brine through the most critical of the interstage module piping, were determined. It has been concluded that the differential head between modules must be increased to provide the desired flow.

The results of applying reliability and maintainability engineering analysis techniques to the 2.5 Mgd Universal Desalting Plant design for a multi-stage flash seawater distillation plant and evaluating the economics of applying these techniques in other Office of Saline Water programs were reported by Hittman Associates [35]. A 14% increase in predicted availability, from 77 to 88%, results from design modifications to critical component areas. The modifications result in a 9% decrease in product water cost, from 140 cents/1000 gallons to 128 cents/1000 gallons (37 to 34 cents/m³). An iterative analysis is defined for optimizing product water cost with respect to availability.

The first application of the Universal Desalting Plant design was at Jeddah, Saudi Arabia, for a 50 MW and 5 Mgd dual-purpose plant [14]. It was followed by a second plant in Saudi Arabia, at Al Khobar, with a capacity of 7.5 Mgd [15]. Design data of these plants are included in Table 2/6 (p. 91).

Weitere Unter- suchungen über Entsalzungs- anlagen großer Kapazität

2.11.3 Other large size desalting plant studies

Capital cost estimates of 1 to 10 Mgd multi-stage flash distillation plants were made by Oak Ridge National Laboratory [16]. In this report tables are given of cost estimates for various perform- ance ratios, tubing prices, flow velocities, number of vessels and stages in the plant, seawater feed temperatures and materials of construction for tubing. Breakdowns in plant components are as well tabulated.

A report is given of studies for the erection of a 34-stage flash distillation plant with a capacity of 15000 tons/day at the Caspian Sea. The choice of the scale preventing system, of the flowsheet and parameters of the brine recirculation, the number of recirculation loops, the number of flash stages, initial and final operating temperatures are outlined and the development of the principal con- struction of the flash apparatus is described [17].

The cost of steam supplied from coal fired boilers and the cost of desalted water was estimated for water-only and dual-purpose plants, for the latter by the power credit method (see section 2.12.3, p. 117). Water-only plants of 30000, 190000, and 950000 m³/day (8, 50, and 250 Mgd) capacities examined were VC-VTE-MSF design, in which the vapor compressors are driven by turbines. Dual- purpose plants were 25, 200, and 1000 MW$_e$ power plants, using back pressure turbines and combined with VTE-MSF desalting equipment with same capacity as mentioned. Thermal cycles and initial steam conditions were optimized for various fuel costs and fixed charge rates. Optimized coal fired electric generating units provided the base for establishing the power credit values. Water costs from dual-purpose plants are found to be consistently lower than those from single-purpose plants. Water costs from coal fired plants could be further reduced utilizing lower water/power ratios. For high water/power ratios the use of nuclear reactors or gas turbines should be considered [18].

In another early study the power and water requirements in Israel were investigated and an advance estimation to satisfy them is made [19]. The Israel study will be discussed in more detail in a later section on p. 124.

Destillations- anlage mit Diesel- antrieb

2.11.4 Diesel powered distillation plant

A study was made to determine wether the combination of a diesel engine with a vapor compressor (VC) and a vertical-tube evaporator (VTE) has good economic potential for desalting of seawater [38].

A diesel engine produces mechanical energy at 42% efficiency, which can be directly used to drive a vapor compressor. The latter, at 82% efficiency, supplies all the heat required for a vertical-tube evaporator with two effects. Over half of the waste heat of combustion of the fuel can also be recover- ed at temperatures high enough to serve as a heat source for a flash evaporator. A special heat ex- changer is used to recover heat from the exhaust gases and from the engine jacket coolant to generate sufficient steam, at 121°C, for the multi-stage flash (MSF) effect. Heat from the diesel lube oil can be used to preheat the incoming seawater. The recovered heat of combustion, plus the mechanical work of driving the vapor-compressor, permits a 4320 HP Diesel engine to serve as the complete power source for a 7700 m³/day (2.04 Mgd) desalting plant, using 72% of the energy content of the diesel fuel in the desalting plant operation. 3920 HP are required to power the vapor compressor and 400 HP are used to drive an auxiliary generator.

Fig. 2–29 shows the flowsheet of the described combined desalting process in simplified form. The prevailing temperatures at the main points of the process are market on the flowsheet in °C. It should be noted that the VTE is provided with double fluted tubes, operates as an independent VC cycle and vaporizes 18% of the feed in the first effect and 23% of the feed in the second effect. It should further be noted that the MSF plant is arranged in series with the VTE plant and vaporizes 11.3% of the brine entering the flash plant. The concentration ratio is 1.79 and the performance ratio is 27 lb of product water per 1000 Btu (48.65 kg per 1000 kcal) of fuel energy input.

Fig. 2–29

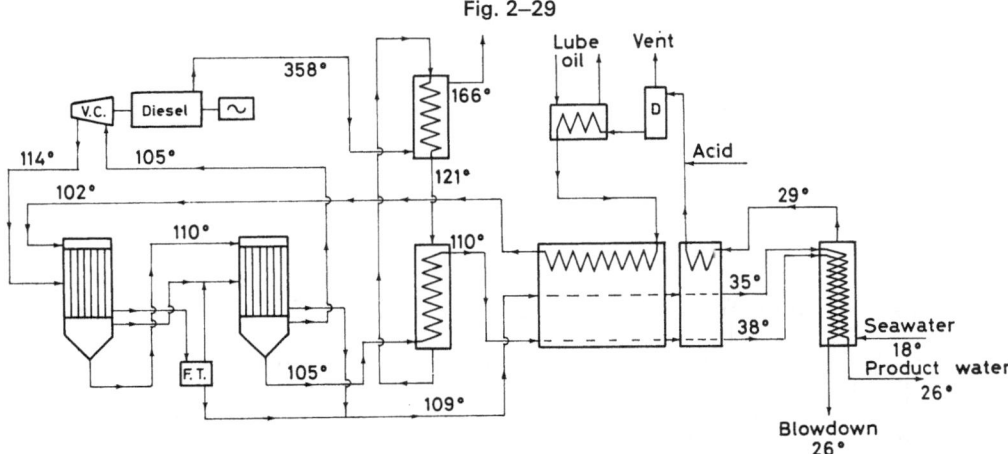

Diesel powered VTE-VC-MSF distillation cycle.

In another study MSF evaporator utilizing Diesel generator waste heat has been developed for seawater desalination. A unit was constructed to operate continuously from a variable heat supply and to produce between 9.5 and 23 m³/day (2500 and 6000 gpd) of fresh water. Interstage brine transfer is automatically regulated by level controllers in each stage. All-aluminum construction has reduced corrosion in relation to steel and the unit has performed satisfactorily during tests [39].

2.11.5 Gas turbine powered distillation plant

Destillations-anlage mit Gasturbinen-antrieb

A similar study was made for a larger desalting plant, with a capacity of 30000 m³/day (8 Mgd) powered by a gas turbine [40]. As seen in the flowsheet of **Fig. 2–30**, p. 104, a waste heat boiler recovers the heat from gas turbine exhaust. Steam at two energy levels is generated: one level is approximately 125°C saturated and the other is 343°C superheated. The latter is used to drive a back pressure steam turbine, which is coupled to an electric generator producing all the required auxiliary power for the facility. The exhaust steam at 125°C from the back pressure turbine is combined with the steam from the low pressure boiler. This steam is used as the heat input to the first effect of the four-effect vertical tube evaporator (VTE). A vapor compressor (VC) receives the vapors from the fourth VTE effect at 108°C and increases its temperature to 173°C superheated steam which is then desuperheated to 125°C. Temperatures at the most important points of the process are marked on the flowsheet in °C.

A multi-stage once through flash plant is used for feed heating the incoming seawater. The make-up feed is first preheated in several MSF stages to 32°C. At this temperature, sulfuric acid is injected for scale control and then the feed is decarbonated and deaerated. The make-up feed is then further pre-heated through the remaining stages of the MSF train to a temperature of 98°C and through four feed heaters, one between each VTE effect, by a portion of the vapors produced in each of the VTE effects, to its final temperature of 116.5°C. At this temperature the feed seawater is sprayed to the inside of the fluted tubes of the first VTE effect.

The steam from the low pressure boiler and the vapor compressor outlet is used to heat the incoming brine to the point where evaporation occurs. The reject brine from the first effect is used as the feed for the second effect and the vapors generated from the evaporating brine in the first effect are used as the heat input to the second effect. The vapor, produced in the first effect, thus becomes the product from the second effect. This process is repeated through all four vertical tube effects. The blowdown brine from the fourth effect then enters into the flash stages of the MSF train.

The product water from each of the vertical tube evaporators is flashed in flash tanks to the temperature of the next evaporator, giving some extra steam which is used in the next effect. Leaving the last VTE effect, the product water is further flashed down in the MSF train. Note that, besides the flashing down of the product water in order to recover its sensible heat, the four effects of the vertical tube evaporator and the multi-stage flash evaporator train are arranged in series.

Fig. 2–30

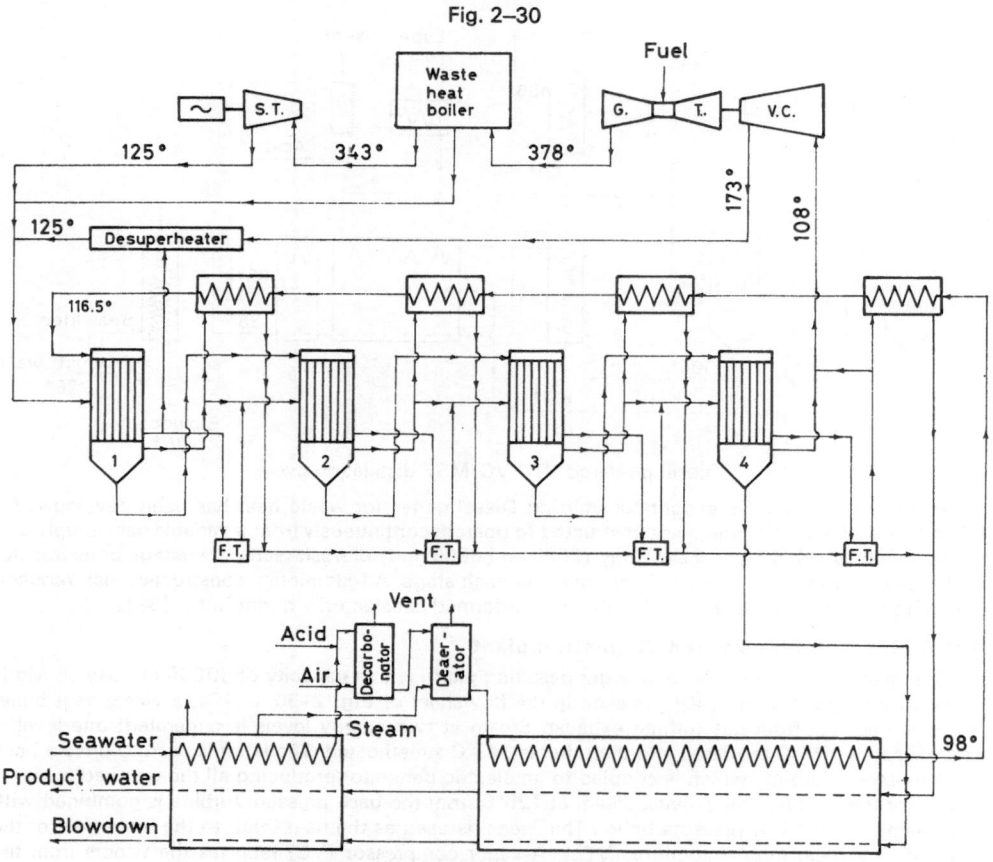

Gas turbine powered distillation cycle.

Chaffiotte [41] introduces an association parameter, α, which is defined as the ratio of the quantity of water produced to the electromechanical energy supplied by the turbine during the same time. The overall efficiency of such a water-producing plant is defined as the ratio of the sum of the thermal equivalents of the electricity and water that have been produced to the total heat introduced in the turbine cycle, including the reheat system if any.

Foster-Pegg [42] estimates that the addition of a gas turbine to dual-purpose power and water plant of approximately 10 Mgd capacity, can make reductions in total water production costs of approximately 15% compared with steam dual-purpose and 30% compared with single-purpose water plants. A gas turbine in the cycle provides additional power and subsidy, plus heat at a level usable at high efficiency for additional power generation and water production. Design and operating flexibility is achieved through supplementary firing in the heat recovery boiler.

The low efficiency of the gas turbine is a limiting factor in its further application. Part of the recent growth of industrial gas turbines is a result of the development of heat recovery boilers. The design and the characteristics of waste boiler are shown and the efficiency and performance of combined cycles of a steam- and a gas turbine installation are discussed in a paper by Van den Hoogen [43].

The advantages of a dual-purpose plant, in which the hot exhaust gases of a gas turbine are used in a desalination plant, are outlined by Taubman and Lebedev [44]. A combined desalination and gas turbine power plant is not economically inferior to a combined desalination and steam turbine plant. Local conditions may result in considerable advantages over steam turbines.

2.11.6 Brownsville, Texas, desalting project

Meerwasser-entsalzungs-projekt Brownsville, Texas

An engineering study of the design and the economics of a desalting plant, using a combination of MSF, VC, and VTE features, reported in section 2.11.5, p. 103, indicated that water could be produced for an order of magnitude of 10.6 to 13.2 cents/m³ (40 to 50 cents/kgal) depending on siting and fuel costs. Compared to the 1963 Point Loma cost of 31.7 cents/m³ (1.20 $/kgal), the 1967 Key West cost of 22.5 (0.85) and the 1969 reported Rosarito cost of 17.2 cents/m³ (0.65 $/kgal), the cost of producing water by this combination of equipment appeared very attractive [5].

In 1970, the Office of Saline Water, the State of Texas and the City of Brownsville agreed to study the feasibility and cost of building an 8 Mgd prototype seawater desalting plant embodying the combination design to serve the City of Brownsville [45].

The City of Brownsville, Texas, needs a source of water with low dissolved solids to improve the quality of its present water supplies. An 8 Mgd desalting plant fits in with the predicted future water requirements. The flowsheet used in the study is very similar to that of Fig. 2–30. In addition the feed brine to each effect is combined with a stream of brine recycling in the effect.

The process used is a single-purpose seawater distillation desalination with a nominal capacity of 8 Mgd. A 12500 HP gas turbine powered vapor compressor operates across four effects of a vertical tube evaporator. The compressor has a capacity to condense 270 t/h vapors from 108.7 to 124.7°C. A 24-stage flash evaporator is used as a low temperature brine heater. Brine recirculation is not used in the MSF portion of the plant. All process power requirements are furnished by a noncondensing steam turbo-generator. Steam is provided from a heat recovery boiler. Other design parameters of the plant are:

Performance ratio lb per 1000 Btu of fuel		19.05	
kg per 1000 kcal of fuel		34.33	
Maximum brine temperature	°C	121.1	
	°F	250	
Scale control H_2SO_4	kg/h	252	
Brine concentration factor from VTE maximum		2.0	
from MSF maximum		2.26	
Average heat transfer coefficient	kcal/m² · h · degr		Btu/ft² · h · °F
VTE effect 1	6938		1420
VTE effect 2	6840		1400
VTE effect 3	6742		1380
VTE effect 4	6645		1360

The total capital cost of the plant was estimated to $ 9730000, updated on purchasing power of 1970 dollars. The product water costs were estimated to 14.09 cents/m³ (53.35 cents/kgal), showing an increase of 2.90 cents/m³ (11 cents/kgal) taking in consideration inflation factors.

2.11.7 Conceptual design for a 50 Mgd MSF plant

Konzept eines vielstufigen Ent-spannungs-verdampfers für 190000 m³/d

The ground rules were provided by the O.S.W. to 15 contractors to prepare a conceptual design for a 50 Mgd (190000 m³/day) distillation plant. The designed plant should be capable of operating in conjunction with an electric generating plant and should use technology based on known concepts or data proven at the production or pilot-plant level.

A summary report on the submitted conceptual design evaluates the significant results obtained by these studies [20], which may be classified as follows: eleven studies propose single-effect multi-stage flash evaporator, with stages ranging from 28 to 98; one study proposes a multiple-effect multi-stage flash evaporator, with 4 effects and 61 stages; one study proposes a horizontal submerged tube multiple-effect evaporator, with 21 effects and 100 stages, and three studies propose a vertical falling film multiple-effect evaporator with multi-stage flash evaporator brine preheater, with 18 and 22 VTE-effects and with 21 and 51 MSF-stages or with integral preheaters.

The aim of each conceptual design is for a reliable plant to produce 50 Mgd of water at the lowest cost. It is of interest to note that in two of the studies the optimum conceptual design is looked at a combination of a VTE-plant and a MSF-plant serving as brine preheater. In one of the VTE-studies the performance is predicated on the attainment of exceptionally high heat transfer coefficients in the relatively short fluted evaporator tubes.

A detailed optimization procedure was made for a three effect multi-stage flash distillation plant with a capacity of 50 Mgd [21]. The plant would have 20 stages in each effect, 3 rejection stages in each of the first and second effect and 2 rejection stages in the third effect. The maximum brine temperature was chosen to 143°C (290°F). Carbon steel was proposed as shell material for the first and prestressed concrete for the two following effects.

Although all the conceptual designs proposed were generally feasible, there was a number of areas in which further detailed study and/or basic data were needed before it was possible to finalize design and to predict performance and cost. Uncertainties were involved primarily in extrapolating to the magnitude of process quantities and physical plant and in design requirements for the low temperature stages.

A final design study was made of the multi-stage flash distillation process, incorporating 36 recovery stages and 3 rejection stages and operating with a maximum brine temperature of 121°C (250°F). Concentration ratio of the brine circulated through the brine heater was 1.7, whereas the blowdown brine had a concentration ratio of 2.0. The design capacity of the plant was 50 Mgd (190000 m³/day). On the basis of this study a test module incorporating the important features of a 50 Mgd flash distillation plant was designed [22]. The module represents a significant fraction of the full-sized multi-stage flash plant. Operating experience with this module should insure that structural, material, and process selections in the large plants are sound. Some of the salient design features of both the 50 Mgd conceptual design and the test module are given in Table 2/8. Two modes of operation were possible with the test module: low temperature operation simulating the lower part of the flashing range and high temperature operation simulating the upper part of the flashing range.

Table 2/8

Design features of the 50 Mgd plant and the Flash Test Module.

	Plant			Test Module		
Stages: Recovery		36			6	
Rejection		3			3	
Total		39			9	
Evaporator width	38.43 m	126	ft	9.61 m	31.5	ft
Stage length: Recovery stages	3.46 m	11.33	ft	3.46 m	11.33	ft
Rejection stages	6.20 m	20.33	ft	6.20 m	20.33	ft
Bundle length: Recovery stages	20.74 m	68	ft	20.74 m	68	ft
Rejection stages	18.60 m	61	ft	18.60 m	61	ft
Condenser bundles per stage		12			3	
Tubes in bundle recovery stages		2950			2819	
Tubes in bundle rejection stages		2205			2122	
Brine circulation pumps		3			1	
Pump capacity, each	291.5 m³/min	77000 gpm		291.5 m³/min	77000 gpm	
Brine flashing range	121.1 to 26.7°C	250 to 80°F		121.1 to 96.1°C	250 to 205°F	
				52.8 to 31.7°C	127 to 89°F	
Production rate	189270 m³/day	50 Mgd		9842 to 12113 m³/day	2.6 to 3.2 Mgd	

A large number of tests were conducted to evaluate the performance of the plant and provide process data for the design of a prototype plant of 30 to 50 Mgd capacity. Operations included studies of flash enhancers and evaluation of stage performance over a wide range of operating conditions.

2.11.8 ORNL conceptual design of a 250 Mgd MSF plant

A multi-stage multilevel flash evaporator plant was designed by Oak Ridge National Laboratory to produce 250 Mgd (946 350 m³/day) of water. Incorporating technology expected to have been proven by 1975, the evaporator is coupled to a heavy-water-moderated organic-cooled reactor with a thermal output of 3300 MW and with a gross electrical output of 675 MW. The design provides for 40 heat recovery and 2 heat rejection stages with a brine concentration factor of 2.0. The plant consists of five desalting trains each having 50 Mgd capacity. Eight levels of trays are utilized to duct the brine flow throughout the heat recovery and heat rejection stages. The multilevel concept was thought to be an important factor in the efficiency of the flash chamber design, because it provided a low incremental cost of widening the brine stream. The condensers are situated on both sides of the brine evaporation trays. A maximum brine temperature of 121°C (250°F) was selected for the single-effect multi-stage reference design with acid treatment of the feed seawater. A temperature of 143.3°C (290°F) was chosen as the maximum brine temperature for the reference multiple-effect multi-stage plant [23].

In another study the 250 Mgd MSF plant is coupled with a back-pressure steam turbine operated from a power plant, which is driven by one 2500 MW$_{th}$ BWR reactor. The plant consists of four trains, each train having an output of 62.5 Mgd. The flash chamber design incorporates two flashing brine levels and a product water tray. The condenser tube bundles are located at the top of the evaporator structure and the product water condensate drains into the product tray immediately below the bundles. The product is flashed to downstream stages in the same fashion as the brine. The top brine temperature is designed to 121°C. The plant has 48 recovery and 2 heat rejection stages [24].

2.11.9 ORNL conceptual design of a 250 Mgd combined vertical tube evaporator and multi-stage flash distillation plant

Conceptual designs combining the VTE process with MSF distillation were already described in sections 2.11.4, p. 102, and 2.11.5, p. 103. In both these designs the MSF plant serves as brine preheating system, the VTE plant being the main desalted water producing component of the combined plant. In both designs as well the two plant components were arranged in series (Fig. 2–29, p. 103, and 2–30, p. 104).

A conceptual design was developed at the Oak Ridge National Laboratory [25] combining the two processes in the same functions, but the two plant components are arranged here in parallel. Tubes with fluted surfaces are used in the vertical tube evaporator plant to obtain enhanced heat transfer performance. Four parallel trains of 62.5 Mgd make the rated capacity of 250 Mgd. The reference plant process is a forward feed, 15-effect, falling-film vertical tube evaporator with double fluted vertical tubes and an integral flash evaporator for feed heating (**Fig. 2–31**, p. 108). About half the incoming seawater serves as coolant and is returned directly to the sea. The remainder is acidified with sulfuric acid and deaerated. The treated seawater is then pumped through the continuous condenser tubes of the 50 flash evaporator stages. In each evaporator stage, vapor from boiling seawater is flashed. In condensing, the vapors raise the temperature of the seawater from 18.3 to 119.2°C as it passes from stage 50 to stage 1. The seawater temperature is further raised with steam condensing in a brine heater to 126.7°C. Then the seawater flows through the flash evaporator stages on the floor at progressively lower pressures and temperatures in the stages from 1 through 50. As it enters each new stage, part of the hot seawater stream flashes into vapor, which condenses on the condenser tubes. The quantity of vapor, which flashes from the brine in stage 1 is 191.9 t/h for all four parallel trains, while in stage 50 the quantity is 90.7 t/h. A portion of the brine flowing through the flash evaporator is pumped from selected stages up into the brine chests, which feed the vertical tube bundles of the VTE evaporator. There are three or four flash stages available for each vertical tube effect. Brine is pumped up to a vertical tube effect only from a flash stage, which is at the same temperature as the vertical effect. A portion of this feed is evaporated isothermally as it flows downward through the vertical tubes. Brine emerging from the bottom of the vertical tubes flows back

ORNL Entwurf eines vielstufigen Entspannungs- verdampfers für 950 000 m³/d

ORNL Entwurf einer kombinierten Anlage für 950 000 m³/d mit Vertikalrohr und Ent- spannungs- verdampfer

into the flash stage from which it was pumped. Steam from the power plant is condensed on the outside of the vertical tubes of effect 1 and the condensate is returned to the power plant. Thereafter the vapor generated in each vertical-tube effect is condensed in the succeeding effect and the condensate is drained into the product stream in the flash evaporator. The vapor from the effect 15 is condensed in the final condenser and the heat is rejected to the sea.

The plant is housed in two rectilinear reinforced-concrete structures, each containing two parallel and independent trains. Each train has the multi-stage feed heater on the lower floor and the vertical-tube effects on the upper floor and is divided along the plant length into 50 flash evaporator stages and 15 vertical-tube effects. To accommodate the increasing vapor flow area required with decreasing pressure, each train is trapezoidal in plan view. The shell is designed to withstand full vacuum throughout. The vertical tubes concense 83% of the product water for the total plant output. The horizontal feed-heater condenser-tubes condense the other 17%. The overall plant performance ratio is 23.4 kg per 1000 kcal (13 lb/1000 Btu) of heat.

Fig. 2–31

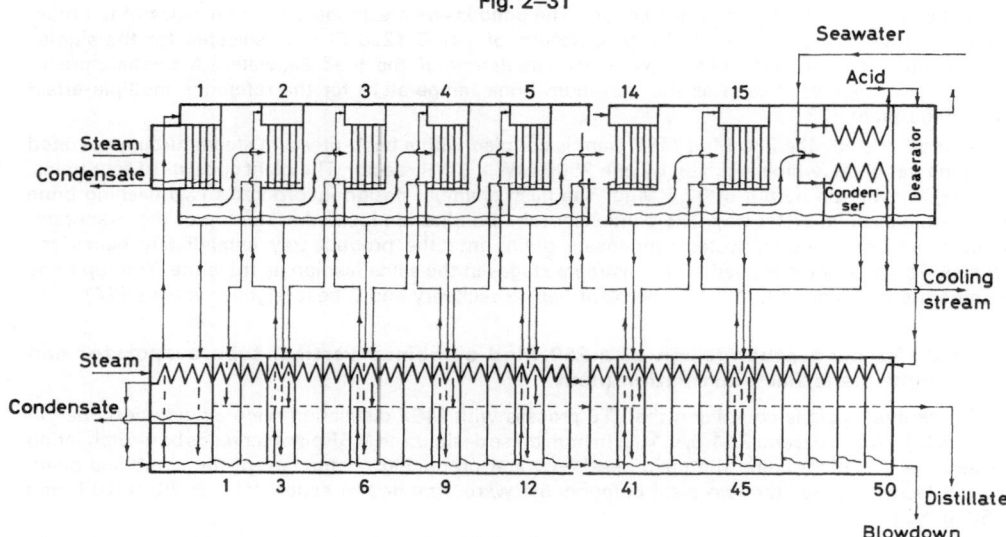

Combinec 62.5 Mgd VTE-MSF distillation plant.

It was concluded from the study that the combined vertical-tube multi-stage flash evaporator had a potential for 30% reduction in capital cost and 15% reduction in water cost relative to the multi-stage flash process for large plants [25].

Two Fortran codes have been written by Friedrich for the calculation of multi-stage flash desalination plant designs. The two plants consists of a brine heater, an evaporator heat recovery section, and an evaporator heat rejection section. The basic difference in the two plants is the shell design and material. ORSEF-2 has a steel shell with a single brine level, while ORSEF-3 has a concrete shell with two brine levels. The output from each program contains all pertinent plant operating parameters along with a detailed cost estimate [36].

Vergleich der Konzepte für 950000 m³/d

2.11.10 Evaluation of the 250 Mgd plant designs

The relative thermodynamic and engineering characteristics, as well as the economics of the two plant designs, described in the foregoing sections, were compared and evaluated, as far as the water plant portion is concerned [26].

Each plant consists of four trains with a capacity of 62.5 Mgd. The performance ratio of each plant is essentially the same, 5.85 kg (12.89 lb) and 5.82 kg (12.84 lb) per 252 kcal (1000 Btu) for the VTE and the MSF plants, respectively. Each train of the VTE plant combines a 50-stage flash feed-preheater with a 15-effect vertical tube evaporator. Each train of the MSF plant consists of 48 heat recovery stages and 2 heat-rejection stages.

An obvious difference between the VTE and the MSF plants is in the number of pumps needed for each. Each train of the VTE plant requires 20 major pumps, compared with only 5 for each MSF train.

Estimated costs, based on 4.5% interest of capital and a 30-year plant life, favor the VTE plant by a substantial margin:

Estimated costs		VTE plant	MSF plant
Capital cost	$	115 000 000	166 000 000
Product water cost:			
Cents/1000 gal		25.0	30.2
Cents/m³		6.60	7.98

Capital costs were adjusted to 1971-Dollars and product water costs were escalated to 1972-Dollars.

The VTE/MSF process advantages over MSF alone are more evident when plant size is sufficient to justify the additional complexity of piping, instrumentation, and equipment. This advantage is particulary valid where a high performance ratio design is necessary as a result of high costs on input steam. A multiple-effect process such as the VTE or VTE/MSF is capable of producing 15 and perhaps 20 kg of product water for each kg of input steam. Multi-stage flash plants experience considerable difficultly in operational stability at such high performance ratios and are usually not designed for more than 12 kg of product per kg of steam. The main advantage claimed for the combined VTE/MSF process over a plant using VTE alone is in the utilization of flash-down release between evaporator effects to preheat the seawater.

2.11.11 Vertical tube evaporator multi-stage flash test module at Freeport, Texas

The operations at the O.S.W. Freeport Test Bed plant were terminated in 1969 and it was decided to proceed with extensive modifications. The original twelve effects were removed, but the recently built five effects module was retained (see section 2.3.1, p. 52). A multi-stage flash preheating was combined with a new vertical tube evaporator module. The new design included a six-effect VTE module and eleven stages of MSF preheating. Capacity of the new plant was again in the order of 0.9 Mgd. Operating temperature range is 135°C (275°F) to 43°C (110°F) with a concentration ratio of 2.5. The modular plant utilizes advanced features of the multi-stage flash distillation process, developed at the O.S.W. San Diego Test Facility, including falling film evaporators with enhanced tube surfaces and concrete-lined vessels and multi-stage preheating arranged in parallel with the VTE plant, as shown in Fig. 2–29, p. 103. The existing five-effect VTE module was utilized for the cold end of the plant. The test results of the Freeport module and the conclusions reached regarding the benefits of combining the VTE and MSF concepts, the heat transfer performance of the enhanced surface VTE and MSF tubing, materials performance and overall plant operating stability are reported in a paper by Houle and Buhrig [27].

Heat transfer coefficients in enhanced surface VTE's operating between 77°C (170°F) and 124°C (255°F) decline quickly with time. This conclusion is based upon tests over a total of 11 134 operating hours, the longest period such a unit has been in service to date. Overall heat transfer coefficients were considerably below those predicted by other investigators for enhanced surface vertical tube evaporators. They are, however, in close agreement with data developed at San Diego on the two effect VTE-X experimental module (see next section) at temperatures from 66 to 93°C (150 to 200°F). Tubeside fouling consists mainly of copper oxides resulting from the oxidation or corrosion of the aluminum-brass tube surfaces. The fouling of the outside or steam side of the tubes is at least as large as that which occurs on the inside. Outside fouling is greatly responsible for the decline in overall heat transfer performance, since it is more difficult to clean these surfaces by conventional procedures. Chemical scale, in the form of $CaCO_3$, $Mg(OH)_2$, and calcium sulfate deposits, has not been a significant factor in the overall fouling. The rate of increase in fouling resistance is temperature dependent. Higher temperature evaporator tubes oxidize faster and more severely than do those at lower temperature. Cleaning techniques have been tested at Freeport which will allow at least partial restoration of heat transfer coefficients to those with new tubes. Coupling of MSF preheating with VTE results in a highly stable easy-to-operate plant. It is doubtful whether the benefits derived, in terms of heat utilization, are sufficient to justify this design in small plants.

Testmodul mit Vertikalrohr und Entspannungsverdampfer in Freeport, Texas

2.11.12 Vertical tube evaporator test vehicle at San Diego, California

A vertical tube bundle test vehicle was constructed at the O.S.W. San Diego Test Facility. Conceptual design studies for large distillation plants have indicated that the multiple-effect vertical tube process has a potential for significantly reducing evaporator capital costs and, in turn, product water costs. The test vehicle of the VTE process was used to investigate realistically some of the transients of large systems, its process hydraulics, controls, and operating characteristics.

The vertical tube bundle test vehicle, known also as vertical tube evaporator exchanger (VTE-X), consists of a vertical, three-compartment test cylinder with internal configuration that allows for housing two vertical tube elements with enhanced surface tubes. One of the two effects is equipped with 2-inch and the other with 3-inch tubes by 10 ft length. These are by far the largest tube bundles ever used for vertical tube falling film evaporation. Two modes of operation were included in the test procedure: a) high-temperature mode, in which pressures and temperatures throughout the assembly were high, and b) low-temperature mode, in which corresponding pressures throughout the assemblies were sub-atmospheric. Two major objectives of the tests were: a) to verify the heat transfer coefficients for double fluted tubes in large tube bundles and b) to determine the entrainment separation characteristics. In addition, the VTE-X was utilized to simulate the vapor-flow, pressure-drop characteristics of the large VTE/MSF module of Fountain Valley (see next section). As a direct result of the simulations, modifications were required in the conceptual design of the evaporator vessel and more definitive vapor-flow correlations were developed for the design of large plants.

A mathematical model has been developed by Aschner and Schaal in collaboration with Oak Ridge National Laboratory, which comprises expressions for fundamental process variables [28].

In late 1972, two large effects using the horizontal tube concept were coupled with the VTE-X. The horizontal tube experimental exchanger (HTME-X) can take exhaust vapor at reduced temperatures from the VTE-X effects or steam directly from the power plant.

Versuchs-
modul eines
Vertikalrohr-
Ent-
spannungs-
verdampfers
in Fountain
Valley,
Kalifornien

2.11.13 Vertical tube evaporator multi-stage flash Test Module at Fountain Valley, California

The second major program in the combined VTE/MSF process investigation was the development of conceptual designs for both a 760000 m³/day (200 Mgd) plant and a module or segment whereby data for such a plant can be developed.

Proposals were solicited from distillation desalting manufacturers for the conceptual design of a 200 Mgd desalting plant. The object of these studies was to establish the optimum distillation-type desalting process or combination of processes to achieve minimum water costs in a 200 Mgd seawater desalting plant. Two of the four firms selected the combined VTE/MSF process as the most economical. A third company chose the VTE process. The fourth company also selected a multiple-effect process, but with horizontal tubes [29]. Data from these studies are given in Table 2/9, From these conceptual designs the Office of Saline Water has selected Aerojet-General's design as the most appropriate for further development and the detailed design of a VTE/MSF was awarded to Aerojet.

Table 2/9

Conceptual designs of a 200 Mgd distillation plant.

Company	Aerojet-General	Westinghouse	Baldwin-Lima-Hamilton	Aqua-Chem
Process selected	VTE/MSF	VTE/MSF	VTE	HTME
Number of trains	8	4	2	4
Capacity per train m³/day	94650	189300	378540	189300
Mgd	25	50	100	50
Performance ratio lb/kBtu	13.6	10.1	10.8	12.4
Number of effects	16	11	13	15
Number of stages	30	35	11	15
Maximum brine temperature °C	131.2	122.2	126.7	115.6
°F	268	252	260	240

Company	Aerojet-General	Westinghouse	Baldwin-Lima-Hamilton	Aqua-Chem
Capital cost Million $	121	136	97	91
Water cost cents/kgal:				
Capital charges	15.1	16.7	13.7	11.2
Steam	7.7	11.3	10.7	8.2
Electricity	2.1	3.0	1.8	1.9
Labor	1.2	0.9	1.4	1.6
Materials and supplies	3.7	5.5	2.2	2.7
Replacements	—	0.5	0.3	—
Total cents/kgal	29.8	37.9	30.1	25.6
Total cents/m³	7.87	10.01	7.95	6.76

The construction and operations of the pilot plant at the O.S.W. Wrightsville Beach Test Facility, the 0.9 Mgd plant at the O.S.W. Freeport Test Facility and the experimental large bundle configuration of the falling film vertical tube evaporator exchanger (VTE-X) at the O.S.W. San Diego Test Facility were the forerunners of this module, which was erected at Fountain Valley, on a site offered by the Orange County Water District of Southern California. The term "module" rather than "proto-type" was applied to the unit, because it is a developmental plant simulating full size portions of a larger plant, whose prime purpose is to confirm critical design scale-up technology features required for subsequent prototype construction. The module represents the latest state of the art in the field of large scale seawater distillation, is capable of producing 11 350 m³/day (3 Mgd) and permits later expansion to an ultimate capacity of 47 500 to 56 500 m³/day (12.5 to 15 Mgd). The module is adapted to operation in various modes that simulate different sections or conditions of the expanded plant [30].

The major component of the plant is a four-effect vertical tube evaporator, with configurations to be utilized in the prototype plant and varying in tube diameter, total number of tubes and spacing. The tubes have enhanced surfaces of the double fluted type. The multi-stage flash plant has six stages. Five stages serve for flashing brine and product water, to provide heat to the incoming feed. The sixth stage utilizes boiler steam and functions as a trim heater for high temperature operations. The tubes are mechanically enhanced by a spirally indented surface. The combined process utilizes the vertical tube, forward feed, multiple-effect falling film in the evaporator. The multi-stage flash regenerative feed heating process is counter-current to the evaporator flow and an once-through multi-stage flash plant [31], similar to the flowsheet of Fig. 2–29, p. 103. Construction of the module was terminated in the summer of 1973 and there are no results of its operation published as yet.

Of interest is the planned use for the product water of the module. The Orange County Water District is constructing a 57 000 m³/day (15 Mgd) waste water treatment system to reclaim municipal sewage water, but the salinity of the reclaimed water will be in the range of 1350 mg/l. The near zero dissolved solids content product water of the module will be mixed with the reclaimed water providing a blended water of overall acceptable quality. Expansion of the module to the full size plant of 57 000 m³/day would provide sufficient blended water for injection into the ground to create a fresh water barrier to prevent seawater intrusion into the underground aquifer, which is a major source of fresh water in the area [32].

2.11.14 Multiple-phase ejector driven desalination plants

Pilot plant studies have indicated that in a combined multiple-phase ejector and vertical tube evaporator system the product water output increases by 44.1% and reduces the energy requirement by 23.4% per kg of product water in comparison with the VTE plant operating by itself [11]. The system has also shown various other operating advantages. In a further phase of this studies, large seawater distillation plants driven by multiple-phase ejectors (MPE) were evaluated and the character-istics and costs of the most promising configurations determined. A reference plant combining a multi-stage flash (MSF) feed heater with vertical tube evaporator (VTE) effects, several of which are driven by an ejector, was selected for mathematical modeling to facilitate parametric calculations by an overall system computer program. An optimum cost for an 8 Mgd MPE/VTE/MSF distillation plant was determined [33].

Entsalzungs-anlagen mit Zweiphasen-Ejektor

The optimum plant consists of 16 vertical tube effects with 42 multi-stage flash stages. A single ejector drives 4 of the top temperature VTE effects. Steam flow rate in these effects is more than double that in the remaining effects. Energy for the system is furnished by a fossil-fuel-fired boiler. Electrical power is produced by a steam turbine driven generator integral with the plant. The flowsheet of the process used is shown in **Fig. 2–32**. Water production from the four multiple-phase ejector driven VTE effects is 42.7% of the total plant production. The balance is produced by 43.9% in the remaining VTE effects and by 13.4% in the MSF stages. The overall plant performance ratio is 27.67 kg of product water per 1000 kcal (15.36 lb/1000 Btu). Capital cost of the plant is estimated to $ 8 153 800 and the product water cost to 11.80 cents/m³ or 44.68 cents per 1000 gallons.

Fig. 2–32

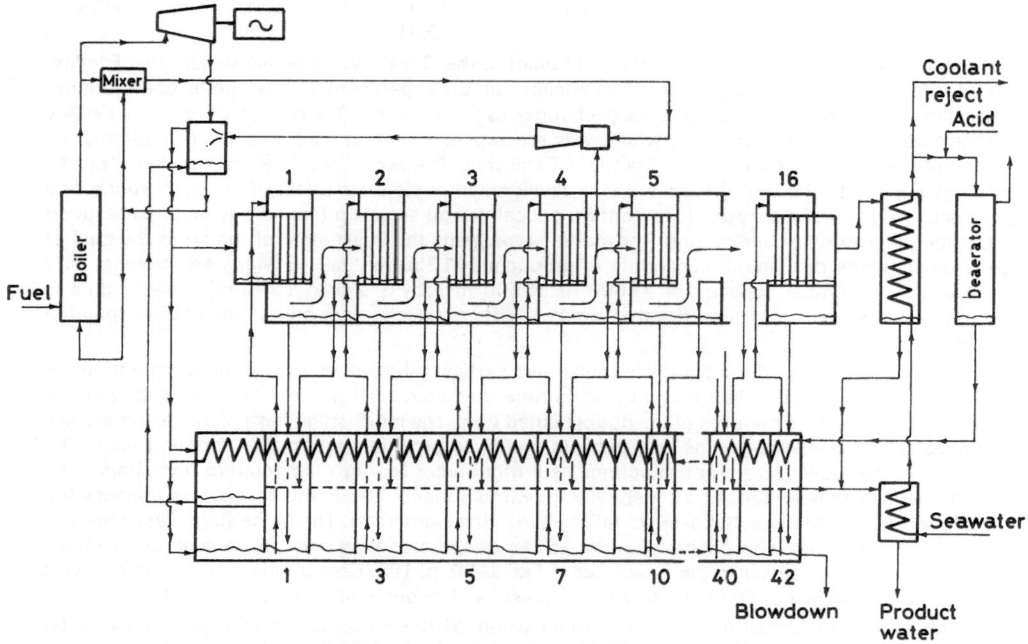

Multiple-phase ejector driven VTE-MSF distillation plant.

Hammond [34] has developed a new type of flash evaporator for seawater distillation that offers inherently stable operation, reduced parasitic losses and simpler construction compared to conventional flash plants. These changes should help especially to simplify the coupling, control and construction complexities of the combined process type of plant. The process uses a shallow, rapid brine flow that is self-regulating over a wide range of conditions. The number of stages can be increased. Pressure drop across each stage becomes small and the stage partition can be of inexpensive, flexible curtain instead of the usual rigid, reinforced pressure bulkhead. For matching to a large 15-effect VTE, the flash plant would have about 160 to 200 baffles, spaced a few inches apart at the high temperature end and a few feet apart at the other end.

Daten und Analyse industrieller Destillationsanlagen

2.11.15 Commercial distillation plant data and analysis

In a series of six reports, sponsored by the Office of Saline Water, operating data and experience are supplied for five commercial evaporation desalting plants. Volume 1 summarizes and compares the five plants giving excellent insight into actual operating experiences. The analysis includes background information covering water demand growth and plant history; design data with flowsheets and layouts; reports on operational tests, visual inspections and eddy current testing of tubes; performance data and analysis for three-year operation period; operating staff and their duties; documentation of material failures and corrosion of plant components; discussions of operating problems, including boilers, seawater supply, fouling and corrosion control problems; economic evaluations, including production cost analysis for a three year period; steam allocation methods for dual-purpose plants and recommendations on modifications, improvements and procedural changes. Chemical analysis of effluents are given and emphasis is placed on quantities of iron and copper being discharged. Volume 2 covers the 1.0 Mgd vertical tube evaporation plant St. Croix, Virgin Islands. Volume 3 covers the 1.0 Mgd multi-stage flash distillation plant St. Thomas, Virgin Islands. Volume 4 covers the 2.5 Mgd multi-stage flash distillation plant St.Thomas, Virgin Islands. Volume 5 covers the 2.62 Mgd multi-stage flash distillation plant Key West, Florida, and volume 5 covers the 2.4 Mgd multi-stage flash plant Nassau, Bahamas [37].

2.11.16 Conclusion

Schluß-bemerkungen

In the last three or four years the world-wide effort to advance desalting technology has moved ahead at an accelerated pace. A very large flash distillation plant producing 112500 m³/day in 5 units (22500 m³/unit) has been completed in Kuwait. Another plant with a capacity as large as 181 700 m³/day in 6 units (30 283 m³/unit) is under construction in Hong-Kong. In Sardinia a single unit capable to produce 36000 m³/day is already in operation. Except for plant sizes in the range of millions m³/day, the distillation process has reached the point, where private industry can satisfy the consumer needs without further large scale research and development work. The construction and operating experience obtained from the recently built large desalting plants will provide equipment manufacturers new insights to better design, fabrication, construction, and operating techniques. The leaning curve in distillation appears to have flattened out. Although there is still much to be learned, no major reduction in costs of distilled water due to improved technology can be expected. Future major economic improvements will be obtained through the availability of better plant operators, the selection of worthy materials, improved plant availability and the economy of scale-up [32].

As a consequence the Office of Saline Water has phased-down its distillation activities. In March 1973 the operation of the VTE/MSF test bed plant at Freeport, Texas, and all distillation work underway at the San Diego, California, test facility were terminated. The operation of the Fountain Valley module is the only continuing work of O.S.W. on distillation processes. Field development work is also continuing at the Wrightsville Beach, North Carolina, test station.

Literature to 2.11

[1] R. P. Hammond (Off. Saline Water Res. Develop. Progr. Rept. No. 283 [1967]). — [2] R. P. Hammond (Off. Saline Water Res. Develop. Progr. Rept. No. 314 [1967]). — [3] R. P. Hammond (Off. Saline Water Res. Develop. Progr. Rept. No. 310 [1967]). — [4] R. P. Hammond (Off. Saline Water Res. Develop. Progr. Rept. No. 311 [1967]). — [5] R. P. Hammond (Off. Saline Water Res. Develop. Progr. Rept. No. 315 [1967]).

[6] P. L. Geiringer, L. T. Taylor (Off. Saline Water Res. Develop. Progr. Rept. No. 344 [1968]). — [7] S. I. Golub, A. M. Rosen, A. N. Krasikov, I. F. Davidov, G. I. Gostinin (Proc. 3rd Intern. Symp. Fresh Water Sea, Dubrovnik 1970, Vol. 4, p. 53/66). — [8] M. Lenhard (VDI [Ver. Deut. Ingr.] Z. **113** [1971] 245/9). — [9] M. A. Ibrahim (Schweiz. Maschinenmarkt **73** No. 13 [1973] 48/55). — [10] B. N. Golubkov, R. E. Stepanov (Proc. 4th Intern. Symp. Fresh Water Sea, Heidelberg 1973, Vol. 1, p. 253/7).

[11] J. H. Leigh (Off. Saline Water Res. Develop. Progr. Rept. No. 578 [1970]). — [12] L. Malfitani, A. Frenzel (Off. Saline Water Res. Develop. Progr. Rept. No. 664 [1971]). — [13] G. L. Beichley (Off. Saline Water Res. Develop. Progr. Rept. No. 595 [1970]). — [14] G. F. Leitner, R. W. Goeldner, A. Steinbruchel, A. A. Santilli (Proc. 3rd Intern. Symp. Fresh Water Sea, Dubrovnik 1970, Vol. 1, p. 115/26). — [15] F. Barluzzi, M. Galateri, E. Maraini (Proc. 4th Intern. Symp. Fresh Water Sea, Heidelberg 1973, Vol. 1, p. 169/78).

[16] R. A. Greene, S. J. Senatore, R. A. Ebel (ORNL-TM-3083 [1970]). — [17] N. K. Tokmantsev, A. P. Egorov, L. A. Krasnyanskii, V. B. Chernozubov, E. A. Sobolev, V. L. Podbereznii, S. I. Golub, V. I. Tkach, V. G. Shatsillo, B. N. Borisov, M. B. Viceblatt, E. P. Novikov (Proc. 3rd Intern. Symp. Fresh Water Sea, Dubrovnik 1970, Vol. 4, p. 81/97). — [18] P. Sichel, G. Nagelberg (Off. Saline Water Res. Develop. Progr. Rept. No. 634 [1970]). — [19] N. Arad (Proc. 1st Intern. Symp. Water

Desalination, Washington, D.C., 1965 [1967], Vol. 3, p. 477/514). — [20] Technology Services Inc. (Off. Saline Water Res. Develop. Progr. Rept. No. 277 [1967]).

[21] P. A. Buckingham, S. G. Unitt (Proc. 1st Intern. Symp. Water Desalination, Washington, D.C., 1965 [1967], Vol. 3, p. 527/42). — [22] The Fluor Corporation (Off. Saline Water Res. Develop. Progr. Rept. No. 233 [1966]). — [23] R. P. Hammond (Off. Saline Water Res. Develop. Progr. Rept. No. 214 [1966]). — [24] R. P. Hammond (Off. Saline Water Res. Develop. Progr. Rept. No. 389 [1969]). — [25] R. P. Hammond (Off. Saline Water Res. Develop. Progr. Rept. No. 391 [1968]).

[26] The Fluor Corporation (Off. Saline Water Res. Develop. Progr. Rept. No. 580 [1970]).— [27] J. F. Houle, W. T. Buhrig (Proc. 4th Intern. Symp. Fresh Water Sea, Heidelberg 1973, Vol. 1, p. 313/25). — [28] F. S. Aschner, M. Schaal (Proc. 4th Intern. Symp. Fresh Water Sea, Heidelberg 1973, Vol. 1, p. 143/52). — [29] E. H. Sieveka, I. Spiewak (Proc. 4th U.N. Intern. Conf. Peaceful Uses At. Energy, Geneva 1971 [1972], Vol. 6, p. 229/42). — [30] F. W. Krebs, J. R. Cofer, E. H. Sieveka (J. Am. Water Works Assoc. 64 [1972] 749/60).

[31] C. Grua (Proc. 4th Intern. Symp. Fresh Water Sea, Heidelberg 1973, Vol. 1, p. 259/66). — [32] J. W. O'Meara (Proc. 4th Intern. Symp. Fresh Water Sea, Heidelberg 1973, Vol. 2, p. 381/5). — [33] G. Harper, J. Leigh (Off. Saline Water Res. Develop. Progr. Rept. No. 748 [1971]). — [34] R. P. Hammond (Chem. Technol. 1 [1971] 754/7). — [35] Hittman Associates Inc. (Off. Saline Water Res. Develop. Progr. Rept. No. 859 [1973]).

[36] R. O. Friedrich (ORNL-TM-3409 [1971]).— [37] C. D. Hornburg, O. J. Morin, R. E. Bailie, W. B. Suratt (Off. Saline Water Res. Develop. Progr., Vol. 1, Rept. No. 906 [1974], Vol. 2, Rept. No. 907 [1974], Vol. 3, Rept. No. 908 [1974], Vol. 4, Rept. No. 909 [1974], Vol. 5, Rept. No. 910 [1974], Vol. 6, Rept. No. 911 [1974]). — [38] R. P. Hammond (Off. Saline Water Res. Develop. Progr. Rept. No. 276 [1967]). — [39] J. S. Williams, A. S. Hodgson (Ind. Eng. Chem. Process Design Develop. 10 [1971] 460/6). — [40] Struthers Energy Systems Inc. (Off. Saline Water Res. Develop. Progr. Rept. No. 377 [1968]).

[41] P. Ph. Chaffiotte (Desalination 3 [1967] 46/59). — [42] R. W. Foster-Pegg (Proc. 3rd Intern. Symp. Fresh Water Sea, Dubrovnik 1970, Vol. 3, p. 253/67). — [43] B. Van den Hoogen (Ingenieur [The Hague] 84 No. 8 [1972] W13/W21). — [44] E. I. Taubman, Yu. N. Lebedev (Therm. Eng. [USSR] 17 [1970] 119/22). — [45] R. W. Newkirk, M. E. Marwede (Off. Saline Water Res. Develop. Progr. Rept. No. 691 [1970]).

*Doppel-
zweck-
Anlagen*

2.12 Dual-purpose plants

The main part of the steam enthalpy used in distillation is the heat of condensation. On the other hand turbines in a power plant use efficiently high temperature and pressure steam. The principle of a dual-purpose plant is to use high temperature steam for power production and low pressure steam, leaving the turbine, for the brine heater of the desalting plant. Thermodynamic considerations define the dual-purpose plant profitable when the ratio of water-to-power demand is between certain limits.

Electricity can not be stored and its production has to meet the actual power demand. Water can be stored, up to the existing facilities, and this gives to the dual-purpose plant a certain degree of flexibility in operation. The maximum advantage is obtained when both components of the dual-purpose plant, electricity generation and desalting, are operated as much as possible at rated capacities.

To obtain this requirement, the size and characteristics of each component of the dual-purpose plant must be selected in such a way, taking the power and water demand curves in consideration, as to arrive to the optimum cost. It should be noted that power and water demand may present daily and seasonal variations.

*Wasser/
Strom-
Verhältnis*

2.12.1 The water-to-power ratio

The ratio of the water output to the power output is an important parameter in the economic analysis of dual-purpose plants. As the two products are not necessarily consumed by a single market, it is important to consider what product ratio can be marketed and how the cost of the two products is affected by varying the product ratio.

The symbol ω is given to the product ratio and is defined, in anglo-saxon units, as

$$\omega = \frac{Mgd}{MW_e}$$

The value of ω may vary between 0 for power-only and ∞ for water-only plants, but normally will be between 0.1 and 1. In a combined plant, the power output is determined primarily by the inlet and outlet temperatures of the turbine and the water output is determined by the amount of heat transfer surface used and the supply and discharge temperatures of the heat. Thus the value of ω is normally limited to relatively low values and tends to be less than 0.5 [1]. There are various schemes to influence the water-to-power ratio.

All the high pressure steam is expanded in a back pressure turbine to a predetermined low pressure. In general the back pressure cycle favors plants with a high water to power ratio, which operate at base load for both utilities. Any variation in the power demand will result in a similar variation of the water output and this is the less flexible design. Additional steam, expanded and desuperheated, may be directed in a parallel stream to the brine heater for affecting the water to power ratio, if there is a larger demand of water. This flexibility is connected with a cost penalty.

In the extraction scheme the low pressure steam, delivered by the high-pressure turbine, is partly supplied to the brine heater and partly delivered to a low-pressure turbine followed by a condenser. A variation in power demand is met by appropriate modification of the steam flow in the low-pressure turbine. The steam flow through the brine heater is normally kept constant in order to operate the distillation plant at rated capacity. The scheme allows for good flexibility in the operation of the water plant, but the latter and the low-pressure turbine cannot be simultaneously operated at rated capacity. The water-to-power ratio can be varied from very low values to values nearly as high as those for the back pressure cycle.

Another possibility, when using two turbines, is to connect both at the high pressure steam cycle and to design one as back pressure and the other as condensing turbine. When the mechanical output of the steam cycle exceeds the need for power, the water-to-power ratio can be increased by using the excess work to drive the compressor of a vapor compression distillation plant.

2.12.2 Dual-purpose plant design and optimization

An analysis was made giving the methodology, which enables the selection of the optimal design parameters of thermal economy of power plants combined with large scale desalination [3]. A mathematical model for the optimization of dual-purpose plants was developed by Korneichev [4], in which the following parameters are optimized: number of evaporation stages, subheating of water in the condenser, brine velocity, flow rate of seawater, brine temperature in the last stage of the evaporator, pressure of the superheated steam and the type of power.

Konstruktion und Optimierung von Doppel- zweck- anlagen

The range of the applicability of dual-purpose plants is discussed by Taubman [5] in comparison with separate production of electrical power and fresh water. Minimum cost is determined for each version by means of solving a system of equations for steady-state operating conditions of the power and desalting units, taking into account a number of limitations resulting from their combined operation.

The main dual-purpose schemes were reviewed and their main features presented by D'Orival [6]. The limits of application and operating problems are given in the light of the thermodynamic and economic characteristics of the various plant types.

The optimum brine temperature [7] and the optimal steam operating temperature [8] were examined by Kunst for wide ranges of electrical performance and of desalination yield by means of a computer and limitations were established.

Technical and economic aspects of steam supply in combined power and desalination plants were investigated by Gat and Lavie [9]. Cost estimates are reported for various quantities and temperatures of steam from fossil-fueled power plants with 300, 400 and 500 MW net electrical output.

The technical and economic conditions, under which the electrical power output from a dual-purpose plant can be accommodated within the constraints of an existing electricity supply system, were examined by Goldsmith [10]. Steam cycles for varying heat-power ratios are reviewed.

The use of very large steam turbines in the United States has been accompanied by a trend toward designing for higher back pressures. These higher back pressures in the condenser allow economic production of desalted water by flashing the condenser outlet water through one or more stages of flash distillation [11]. Tables and graphs were prepared by Fuller [12] by means of the ORCENT computer code, for the rapid estimation of turbine cycle efficiency and heat rate, feedwater pumping power, throttle steam flow, exhaust steam flow and exhaust enthalpy.

Optimal selection among different schemes of power generation and water desalination systems was presented by Carasso and the selected schemes were optimized. Current methods of system design optimization and also of the general techniques of mathematical programming for the solution of large scale problems of special structure were given [22].

A dual-purpose plant in which the power plant is coupled with two water plants, a multi-stage flash plant and a reverse osmosis plant, is analyzed and optimized for the total cost of the system for producing given levels of power and water supply. The optimal design of several combinations of water and power demands were presented by Fan et al. [13]. A low cost of water, which may be used in agriculture water supply, is predicted in the optimal design.

The combination of a multi-stage flash (MSF) distillation plant with a vapor compression (VC) plant is a useful compromise, when no adequate demand of electric power exists. VC operating in conjunction with a MSF evaporator allows for a considerable flexibility in output for base load operation from a fixed heat source. The greatest advantage in terms of heat-transfer surface per unit of prime fuel is obtained for the hybrid process when the VC evaporator is coupled to the top of the MSF plant, either single of multiple-effect [14].

The feasibility and potential economic benefit resulting from non-base load designs applied to dual-purpose plants were analyzed by Glueckstern et al. [15]. Several system modifications to facilitate non-base load operation are evaluated parametrically at varying interest rates, peak power loads and ratio of peak load to total operation time. These modifications include flexible back-pressure turbines connected to MSF units coupled to a low pressure turbine for peak load generation, as well as constant back-pressure turbines connected to parallel oriented and to series oriented MSF units for base load and peak load operation, respectively, and coupled to a low pressure condensing turbine for peak load generation. The applicability of the dual-purpose concept to small plants was also demonstrated by Mandelzweig [16]. A cost analysis has been made of five different combinations of power and desalination units to supply remote areas with 2000 kW net power and 1250 m³/day of desalted water. In four cases power was supplied by Diesels and in the fifth by a gas turbine. Diesels are better suited for small plants. The desalting processes taken in consideration were multi-stage flash evaporation, multiple-effect distillation, and vapor compression.

The design possibilities and the cost-aspects for a dual-purpose plant to meet the requirements of a community of 60 to 70 thousand inhabitants were examined by Palm [17]. Three installation configurations covering a demand of 12000 m³/day and 50 MW are compared to a single-purpose plant. The effect of the cost items is investigated. It was concluded that the optimum water to steam ratio is lower for the dual-purpose models considered in comparison with single-purpose plants. For a specific plant concept with condensing turbines the fuel consumption will be unaffected by the top brine temperature in the evaporator. The plant concept with the lowest fuel costs depends mainly on the local quality and cost of fuels. A plant concept with unfired waste heat boiler will generally not meet the required flexibility in the production of electricity without also reducing the evaporator output.

Kosten-
vergleich
zwischen
Einzweck-
und
Doppel-
zweck-
anlage

2.12.3 Single-purpose water plant versus dual-purpose plant cost

The cost of water from a single-purpose plant is generally higher than that of a dual-purpose plant. The boiler required in a single-purpose plant will have a higher investment cost, per unit of thermal rating, and a lower combustion efficiency than the larger unit utilized in a dual-purpose plant. Single-purpose water plants must bear alone the expense of the boiler plant operation and maintenance labor. Dual-purpose plants share this cost with the power generating plant [2].

The pressure of steam to the brine heater commonly required is not more than 3 kg/cm². The usual outlet pressure from commercially available boilers is at least in the vicinity of 20 kg/cm². In a single-purpose plant the heating steam pressure must be reduced, thereby losing steam availability. The high pressure steam can also be utilized for pump turbine drives. The cost of steam to the evaporator of the dual-purpose plant is reduced.

Single-purpose water plants must either purchase power from an external grid or generate their own power. The cost to generate electricity is higher for the single-purpose plant than that for the dual-purpose plant.

The total cost of the site and site development is charged to the cost of water produced from a single-purpose plant. In dual-purpose plants, this cost is shared by both the water and power plants. In dual-purpose plants the investment cost of turbine condensers is reduced or eliminated, as well as that of the expensive low pressure stages of the turbines.

Cost accounting. In a single-purpose plant the fixed charges, as well as the maintenance and operation cost of the steam plant, power plant and off-site facilities are entirely charged to the water cost.

The steam supplied to the evaporator of a dual-purpose plant can not be considered as cost free. It has some value for the power section of the plant, because the steam has released in the turbine less energy than it normally does in a single-purpose power plant. A costing method had to be developed for allocating the appropriate amount of cost to either of the energy users, the power plant and the desalting plant. Water desalting is a power intensive industry and a considerable part of the power out of the combined power plant is consumed in the desalting plant.

The costing methods employed to evaluate the cost of product water and electrical power from a dual-purpose plant are, of necessity, fairly complex. Different costing methods have been developed. Some of them favor the water cost, others favor the power cost and others calculate both costs equitably and unbiased. The two major costing methods are the power credit and the available energy methods [18].

Power credit method. The total costs of a dual-purpose plant and the cost of a single-purpose electric plant having the same net electrical output are determined. The operating cost of the single-purpose power generating plant is considered as a power credit. The difference by which the cost of power produced by the dual-purpose plant exceeds the single-purpose electric plant is the energy cost charged to the water plant.

In the power credit method, no benefit is granted to the power because of the dual-purpose operation. It is assumed that the power plant would have the same cost, if operated as single plant.

The cost of water predicted by the power credit method is dependent upon the actual value of the power credit selected. Since much of the revenue of a dual-purpose plant comes from the sale of electricity, a slight increase in the electrical power price can significantly reduce the price, which must be charged for water.

Available energy method. The available energy method is more complicated and equitably allocates all costs to the electricity generation and water production plants on the basis of the available energy utilized by each.

All costs of the dual-purpose plant are segregated into four categories of cost: the steam plant, the turbine generator plant, the MSF plant, and the common site.

The cost of rejecting waste heat normally required when using power plant condenser, main circulating pump and associated equipment is allocated to the steam generator cost. The common facilities costs are allocated in proportion to the relative benefits received. The remaining portion of the common site cost is shared evenly between the turbine generator and the MSF plant.

A portion of the turbine generator plant cost is allocated to the steam generator and MSF plant cost. The amount allocated is dependent on the portion of the total gross power utilized by the pumps and auxiliaries in each of these systems.

The steam generator plant cost is allocated to the turbine generator power plant and the MSF water plant on the basis of available energy consumed by each. The separate power and water production accounts are adjusted to include indirect capital and operating costs.

In the available energy method of accounting, the fraction of steam cost attributed to power production is the ratio of the enthalpy drop across the turbine to the total enthalpy drop available for electrical power production in a single-purpose plant. The remaining fraction of the steam cost is attributed to water production. The annual cost of the steam generator plant is multiplied by the water and power fractions to obtain the cost attributable to each plant respectively. These fractions are defined as:

$$\text{Power fraction} = \frac{H_1 - H_2}{H_1 - H_3} \qquad \text{Water fraction} = \frac{H_2 - H_3}{H_1 - H_3}$$

where H_1 is the turbine inlet enthalpy, H_2 the enthalpy of the steam inlet to the brine heater and H_3 the turbine exhaust enthalpy.

The available energy method is based on thermodynamic premises and predicts water and power costs from a totally impartial standpoint. This method will typically predict water cost values higher than the power credit method for the same plant.

Other costing procedures. There are several other suggestions for an equitable allocation of the various items of cost between water and electricity. Barnea [19] suggested that the total annual costs of a dual-purpose plant should be allocated to water and power in proportion to the respective annual costs incurred to the cheapest possible alternative single-purpose plants producing the same quantities of marketable output as are obtained from the combined operation. The United Nations [20] published a detailed report on "Proposals for a costing procedure and related technical and economic considerations" incorporating the afore-mentioned suggestion.

The International Atomic Energy Agency published a report on "Costing methods for nuclear desalination" [21], which contains a review of the basic principles for costing desalination plants and of the various methods proposed for allocating costs in dual-purpose plants. These methods are: Prorating on the basis of the total costs of two single-purpose alternative plants; prorating on the basis of power generated; prorating on the basis of available energy; generation cost of a power-only station of the same net output; generation cost of a power-only station of larger net output; purchase price of the kWh by the electric utility; water credit as that corresponding to the optimum single-purpose desalination plant producing the same amount of water as the dual-purpose plant, and water credit as that corresponding to the least-cost alternative for water supply, which would be built in the absence of the dual-purpose plant. The possible limitations of each method are indicated. Numerical examples and further essential information are also provided in this report.

Literature to 2.12

[1] C. C. Burwell, R. A. Ebel, R. P. Hammond (Tech. Rept. Ser. Intern. At. Energy Agency No. 51 [1966] 74/88). — [3] Hittman Associates Inc. (Off. Saline Water Res. Develop. Progr. Rept. No. 490 [1969]). — [3] L. S. Sterman, V. V. Gubenko (Proc. 1st Intern. Symp. Water Desalination, Washington, D.C., 1965 [1967], Vol. 3, p. 363/72). — [4] A. I. Korneichev (Proc. 3rd Intern. Symp. Fresh Water Sea, Dubrovnik 1970, Vol. 3, p. 339/58). — [5] Y. I. Taubman, Y. N. Lebedev, A. R. Lyogky (Proc. 3rd Intern. Symp. Fresh Water Sea, Dubrovnik 1970, Vol. 3, p. 397/405).
[6] M. D'Orival (Desalination 4 [1968] 66/79). — [7] B. Kunst (Proc. Symp. Nucl. Desalination, Madrid 1968 [1969], p. 483/94). — [8] B. Kunst (Chem. Ingr.-Tech. 41 [1969] 104/9). — [9] Y. Gat, A. Lavie (Proc. Symp. Develop. Desalination Technol., Jerusalem 1971 [1972], p. 129/50). — [10] K. Goldsmith (Proc. Symp. Nucl. Desalination, Madrid 1968 [1969], p. 595/604).
[11] R. A. Tidball, J. G. Gaydos (Proc. 3rd Intern. Symp. Fresh Water Sea, Dubrovnik 1970, Vol. 1, p. 207/17). — [12] L. C. Fuller (ORNL-TM-2909 [1970]). — [13] L. T. Fan, C. L. Hwang, N.C. Pereira, L. E. Erickson, C. Y. Cheng (Proc. 3rd Intern. Symp. Fresh Water Sea, Dubrovnik 1970, Vol. 3, p. 231/51). — [14] F. C. Wood, R Herbert (Proc. Symp. Nucl. Desalination, Madrid 1968 [1969], p. 535/55). — [15] P. Glueckstern, N. Arad, I. E. Streifler-Shavit (Proc. 3rd Intern. Symp. Fresh Water Sea, Dubrovnik 1970, Vol. 3, p. 269/83).
[16] S. Mandelzweig (Proc. 4th Intern. Symp. Fresh Water Sea, Heidelberg 1973, Vol. 1, p.363/9). — [17] G. R. A. Palm (Proc. 4th Intern. Symp. Fresh Water Sea, Heidelberg 1973, Vol. 1, p. 451/62). — [18] C. C. Burwell, R. P. Hammond (ORNL-TM-1615 [1966]). — [19] J. Barnea (Water Resour. Res. 1 [1965] 143/5). — [20] United Nations (Water Desalination, Rept. No. 65.11.B.5 [1965]). — [21] International Atomic Energy Agency (Tech. Rept. Ser. Intern. At. Energy Agency No. 69 [1966]). — [22] M. Carasso (Univ. Calif. Seawater Convers. Lab. Rept. No. 71–2 [1971]).

Abwärme als Energie-quelle

2.13 Waste heat as energy source

Large quantities of heat, generated in various forms, are wasted and dissipated in the immediate environment. Such waste heat sources might include garbage and refuse incineration, cement plants, glass furnaces, metal smelters and especially exhaust from Diesels and gas turbines [1].

A feasibility study was made of a plant in San Diego, California, that would use as its heat source the solid wastes currently disposed by the city in land-fill operations. Since, in the foreseeable future, the city will be running out of economically feasible land-fill sites, it was determined that a solid waste incinerator would provide sufficient low-grade, low-cost steam to operate an 121 000 m^3/day (32 Mgd) plant with a performance ratio of 13. Assuming a credit of $ 6 per ton for refuse disposal, the cost of product water from this plant was estimated to be just less than 13 cents per m^3 or 50 cents per 1000 gallons [2].

Literature to 2.13

[1] H. Gitterman, S. Zwickler (Off. Saline Water Res. Develop. Progr. Rept. No. 235 [1966]). —
[2] J. W. O'Meara (Proc. 4th Intern. Symp. Fresh Water Sea, Heidelberg 1973, Vol. 2, p. 381/5).

2.14 Nuclear energy as heat source

Kernenergie als Wärmequelle

A nuclear desalination plant is practically a dual-purpose power and water plant, in which the turbine is driven by steam from a nuclear reactor. Coupling of nuclear reactors with a desalting plant involves some technical problems, but this coupling to be profitable, is largely affected by the cost of power and steam supplied to the water plant, as it is the case in any conventional dual-purpose plant.

Early studies referred to small capacity power reactors, in the range of about 400 MW_{th} [1] and 370 MW_{th} [2]. Interest in nuclear desalting became prominent when the increased size of nuclear plants and advancing nuclear technology were promising reduced energy costs. One of the main problems in using nuclear energy for desalting, as steam cost decreases considerably with increasing power level, is that only large size nuclear reactors were expected to be sufficiently economic for coupling with water desalination plants. Parallel advances in desalting technology gave hope for scale-up of the desalting plants to sizes suitable for reactor use. In a preliminary background study Hammond [3] expressed the view that very large reactors supplying heat to evaporators seem likely to be capable of producing fresh water from the sea, which would be cheaper than from any other method, especially if production of electric power is combined with production of water. An evaluation and preliminary design program was proposed for the application of large nuclear reactors to the desalination of seawater [4]. The matter was then taken up by the U.S. Office of Science and Technology and an appointed task group submitted a favorable report on "An assessment of large nuclear powered seawater distillation plants" (March 1964).

2.14.1 Nuclear versus conventional power

Vergleich Kernenergie– kon- ventionelle Energie

Nuclear power plants used to be more capital-intensive than conventional power plants. Accordingly fixed charges were higher for nuclear power plants, but fuel costs were considerably lower. Nuclear plant economics do not vary to any great extent with geographic location. This factor is of importance for those regions or countries, which have a low availability of fossil fuels. Thus, nuclear power offers an additional degree of freedom in securing local energy requirements.

Nuclear plants exhibit a more rapid decrease in unit costs with increased size than fossil fueled plants. Conventional power plants are tied to current fuel costs. Nuclear fuel costs are expected to decrease, as nuclear fuel technology is improved and larger fuelling industry is developed. These characteristics of nuclear energy is of particular advantage to developing countries, which lack of abundant natural energy resources.

Projections for the total electrical capacity and nuclear electrical capacity in industrialized and developing countries give the following percentages of nuclear electric capacity [5]:

	1970	1975	1980	1985
Industrialized countries	2.1	8	16	23
Developing countries	0.4	2	8	14
World, except mainland China	1.8	7	15	22

It is concluded from this table that the relative increase in nuclear power generation is expected to be higher in developing countries than in industrialized countries. The recent energy crisis may change this conclusion.

Other advantages of nuclear desalination may include that nuclear power units are operated at higher load factors than most fossil fuel power plants and provide a more compatible load pattern for combined water production.

The cost of water produced with conventional fuel was compared to that of a dual-purpose plant coupled with a nuclear reactor. As expected the cost was lower in the dual-purpose plant. An analysis was made to determine the break-even point, at which the cost of conventional fuel gives the same water cost with that of the nuclear dual-purpose plant [6].

Until the late 1960's the estimated costs of nuclear power plants declined steadily. It was believed at the time of growing interest for nuclear desalination, that a large light water reactor could be built for $ 125 per kW or less. Since 1968/1969 the estimated cost of nuclear power plants has risen sharply and this increase was followed by a similar increase in fossil fuel power plants.

From the numerous references concerning the coupling of nuclear reactors with desalting plants, only two are reviewed in the following. In the first paper the current and the near-term thermal converters and the long-range fast breeders for seawater desalination are analyzed, based on the fundamental principles, existing technology and future development of nuclear reactors. The trends of current and near-term thermal converters and long-range fast breeders are predicted. The unit plant cost and the total nuclear power cost versus the cumulative and the net plant capacity are respectively shown by curves. The comparison of water cost versus net plant capacity and plant load factor for the single-purpose and the multi-purpose nuclear desalting plants are also given. Although some of the figures might not be any longer valid, the correlation of data given is of interest [7].

In the second paper an analysis, which covers 1968 estimates for plants to be completed in the early 1970's on which adequate cost data could be compiled, shows that original cost estimates were about $ 150 per kW lower than experienced for those plants. Inflation, schedule stretchout, field labor, and additional safety measures are some of the components of the increase in cost. The cost of nuclear plants might be expected to reach $ 375 per kW in 1980. The comparable figures for fossil-fired plants using current technology are $ 325 per kW for a plant going into service in 1980 [8].

The cost of electricity, however, from large size nuclear reactors has already become competitive with conventional power generation plants. In a recent study the costs for a power plant of 1000 MW_e capacity, to go into operation in 1977 (see section 2.15.8, p. 133), have been estimated as follows:

	Fossil-fuel	Nuclear
Capital cost per kW_e net	$ 236	$ 323
Electricity generation in $-mills:		
Fixed charges	1.79	2.58
Fuel cost	2.34	1.38
Labor operation and maintenance	0.16	0.18
Materials and supplies	0.15	0.15
Sulfur removal from stack gases	0.30	—
Nuclear liability insurance	—	0.05
Total in $-mills/kWh net	4.74	4.34

The difference is expected to be larger at higher production capacities, favoring nuclear reactors especially when the technology of fast breeder reactors is adopted in the near future. On the other hand recent increases in the price of oil will certainly unfavorably affect the cost of conventionally produced power. Therefore the difference between the costs of fossil-fuel and nuclear power generation, as well as heating steam, might become higher in the years to come.

Optimierung der Entsalzung mittels Kernenergie

2.14.2 Optimization of nuclear desalting

Emphasis is given by the authors of papers, reviewed in the following, on nuclear desalination. Nevertheless, many of the conclusions stated therein are as well applicable to conventional dual-purpose plants. As a general rule, if a dual-purpose nuclear plant is designed to produce a steady water output, the power produced must also be kept steady or base-loaded. The base-loaded dual-purpose plant appears to give the lowest product cost. Means to form the power production more flexible or load-following, without upsetting water production, appear to be justified in any dual-purpose plant. Extra power output or peaking power can be added to a basic dual-purpose station at a cost which is attractive, compared with other forms of peaking power [9]. Plants using a variable back-pressure turbine can trade power production, over a 30 % range, for water production, over a 25 % range, with only a modest capital cost penalty. Flexible plant features can minimize the penalty associated with an insufficient market for all the plant electric power [10].

The optimization and range of application of nuclear reactors and flash evaporators for seawater desalination was investigated by Aschner et al. [11]. The research includes for comparison conventional boilers as heat sources for dual- and single-purpose plants. The distillation plants considered are of the multi-stage flash evaporator type. The dual-purpose plant field examined covers the range from 50 to 300 MW net power generation and the desalination of seawater quantities which can, under varying heating steam and evaporator design conditions, be obtained by heating steam exhausted from back-pressure turbines driving the electric generators. This corresponds to a range of water output between 37850 and 757000 m^3/day (10 and 200 Mgd). The single-purpose plant investigation covers the thermal ratings in the range from 200 to 1600 MW_{th} corresponding to a range of water output as for the dual-purpose plants. All plant types have been defined by mathematical models and optimized technically and economically in the study. Optimum water costs are computed for various reactor types, fixed charges and power-water plant outputs. For 10% fixed charges the cost of water for various parameters varies in the range of 12.9 to 17.6 cents/m^3 for dual-purpose plants against 20.2 cents/m^3 for single-purpose plant. The influence of design parameters on the water cost is also discussed.

The most desirable situation for the introduction of nuclear desalination is one, in which there is already an existing system of power and water supply into which the new dual-purpose plant can be integrated [12]. In the case of dual-purpose plants added to existing water and power systems new operating rules are required which take into account the effects on the load factors of all existing and future plants (both power and water), systems reliabilities and maintenance. Plans for future plant installation may be drastically modified by the method of cost allocation. For maximum economy it is essential that electricity generators in an associated system are no longer run to suit either the economics or the operating rules of the electricity system alone. Optimization of plant combinations and technical factors such as nuclear safety, relative plant availabilities and reliabilities, water/power ratios with various sources and brine disposal problems are to be taken in consideration in adding new units in an existing water and power system [13].

Some of the principal factors to be considered in the design of single- and dual-purpose nuclear power and desalting plants are the economic advantages in combining the two plants, the reliability and flexibility of operation, the effect of the load factor and the management, joint or diverse [14].

The state of the art of nuclear power operated desalination of seawater installations under nominal and partial load conditions and the principal development trends are reported by Kunst et al. with emphasis on the effects of reactor type and size. The main operating parameters of the nuclear power plant and desalination plant working under partial load are investigated with reference to an installation with a capacity of 400000 m^3 fresh water per day [15].

A method was developed by Glueckstern and Arad to modify the design of a given nuclear dual-purpose desalting plant in order to make it suitable for non base load applications. Three different systems were investigated in detail to demonstrate the feasibility and potential economic benefits resulting from such modifications [16].

A method for the analysis of optimum parameters and power of a nuclear power plant coupled with a high productivity desalination plant can be selected by a simultaneous consideration of the results of thermal economics, capital investment, and operation charges [17]. Computer calculations were also presented for the analysis of the thermo-economics and overall economics of a nuclear power plant combined with a large sea water distillation plant [18].

An analysis was given of recent estimates of energy and desalted water costs. The effect of parameters such as plant size, interest rate, cost escalation, and level of technology is outlined. The influence of cost escalation and of higher interest rates on desalination projects is compared with their influence on the cost of alternative water sources. Surface water developments are hurt most by rising money costs. Groundwater projects are least affected. Desalination projects are intermediate. The advantages and disadvantages of dual-purpose plants are discussed and a description is given of presently feasible routes to water-only plants [19].

A dual-purpose nuclear desalting plant was proposed, in which the evaporator is a single-effect vertical tube still, operating on very low temperature and placed between the turbine exhaust and the condenser of a conventional nuclear power plant. Water cost was estimated to be 38% less than for the best alternative dual-purpose plant [20].

The relevant design variables affecting the economics of a nuclear power and water system design are incorporated into a mathematical framework to show the system inter-relationship. The optimum product ratio is to be determined on the basis of maximizing the present worth of the revenue minus cost streams. The effects of budget and other economic constraints are considered. The merits of cost allocation policies are investigated and their effect upon investment are discussed. It is concluded that the proper method for allocating joint costs over the system economic life should be proportional to the revenue streams [21].

The methodology of costing computations for multi-purpose nuclear desalination plants involves both national and particular specific features. It is difficult to recommend an international method of making technical and economic evaluations. A methodology is outlined, which is based on the economic category of "evaluated expenditure", widely used in the USSR for carrying out alternative economic computations in the different branches of the national economy [22].

In an attempt to determine the optimal process design of a dual-purpose plant for producing power and water, a nuclear reactor and steam turbine power generator were coupled with two water plants, a multi-stage flash plant and a reverse osmosis plant. The total system cost for producing given levels of power and water is minimized. Optimal designs were presented for several combinations of water and external power demands ranging from 95 000 m³/day (25 Mgd) to 568 000 m³/day (150 Mgd) and in addition an analysis was carried out to determine the sensitivity of the system to selected process parameters of such a plant, designed and constructed to produce 150 Mgd of water and 50 MW of saleable power at the optimal conditions [24].

The problems of controlling large desalting plants were studied both theoretically and experimentally at the Oak Ridge National Laboratory. A general-purpose digital simulator was developed to study large multi-stage flash (MSF) plants coupled to a back-pressure turbine-generator plant, with a pressurized-water reactor (PWR) as a heat source. These studies of the dynamics of MSF and PWR systems indicate that the overall plant stability characteristics are highly dependent on the hydraulic design of the MSF plant and the means of coupling the MSF plant to the heat source [25].

The validity of conclusions are put under discussion on the basis of certain assumptions in a paper by Thiriet and Lievre. Unknown factors of different kinds, associated with technical, economic and financial data and also with other constraints are examined. Unknown factors, associated with the optimization method used and the criterion of choice selected, are discussed. It is shown that there are cases in which it is extremely difficult to reach a categorical conclusion as to whether or not nuclear energy is competitive for purposes of water desalination [26].

Sicherheit
und Umwelt

2.14.3 Safety and environment

An advantage of the nuclear plants is that they do not create a smoke or smog problem, but they have their own waste problems. Means to handle the waste products from nuclear plants have been developed and environmental effects are virtually eliminated [27].

The potential concentration of radioactive nuclides in the water plant brine, effluent and product water for nuclear dual-purpose plants, without steam leakage into the water plant, appears to be many orders of magnitude below maximum permissible concentrations. By maintaining the brine pressure above that of steam, radioactive nuclide leakage into the water plant might be prevented. However, means have also to be provided to prevent brine leakage into the nuclear steam supply system [28].

Additional safety is obtained if a loop be inserted between the heating steam and the brine to allow extended operation at full capacity with brine heater leakage. This involves additional capital cost and increase of the product water cost.

The effectiveness of safety features in reducing siting distance, the cost of site acquisitions and of delivering product water and electricity, as a function of distance, were studied in a report by Oak Ridge National Laboratory. Examples are given of economic trade-off between reactor engineered safety features and siting costs [29].

Regionale
und
ingenieur-
technische
Unter-
suchungen

2.14.4 Regional and engineering studies

It was a general belief for a period of time that nuclear desalting would be the most appropriate way of low cost desalination of water. Hence, a large number of studies were made either of regional interest or engineering background. Some of the more recent relevant publications are briefly reviewed in the following.

Egypt. The main features of power and water resources in Egypt are outlined in a paper by El-Mofty [30] and preparations and prospects for the future are given. Preparations for introducing nuclear energy applications were achieved. A dual-purpose 150 MW power station was chosen as the first nuclear station. In the foreseeable future, nuclear stations will be considered together with thermal stations, since the cheap hydro potential is almost exhausted. The capacity of the system now permits large stations. A need for desalted water exists to reclaim about three million acres of land along the northern coast. Also small desalination units are needed for mining industries and communities along the Red Sea coast and the Sinai peninsula [30]. The long range requirements and resources of the country are outlined in a paper by El Guebeily et al. [31]. Nile water is allocated to land reclamation projects and other demands. Desalination is envisaged as a promising future water source. A recent study of a prototype single-purpose desalination nuclear plant has been initiated. In the choice of reactor type and configurations, main consideration was given to the need for simple designs, simple materials and cheap construction within the capabilities of developing countries. The study led to a 50 MW_{th} natural uranium, heavy water moderated reactor with light water or CO_2 cooling. Reactor power and size of the desalination plant correspond to water needs of the agricultural experiment [31].

France. Design features of a 25000 m³/day fresh water unit were reported by Balligand et al. [32], the multiplication of which can form a several hundred thousand m³/day desalination plant in a dual-purpose plant supplying both water and power from a nuclear reactor. The chosen distillation cycle combines multiple-effect LTV (long tube vertical) evaporators in the high temperature part of the unit and multi-flash in the cold part. The choice of the multiple-effect process in the high temperature part allows the heating of seawater up to temperatures of 135°C without important scaling risk. The choice of the multi-flash process in the cold part of the unit allows the use of concrete flash chambers with long tube condensers. The connection between the two parts of the unit is realized through a condenser heater, in which the vapor coming from the last LTV effect condenses and heats the brine to be partly recycled into the multi-flash part of the unit. The combination of a vapor compression system with the highest temperature effects extends the operation to hours when the power demand is low [32].

Germany. The erection of seawater desalination plants in conjunction with heat supply from the various nuclear reactors built or planned in West Germany has been investigated. Market analysis for the estimated demand for electricity and fresh water in various regions shows the dual-purpose plants coupled with nuclear reactors to give the lowest costs. High temperature reactors as energy suppliers of dual-purpose plants appear to be particularly suitable as energy centres for large consumer complexes [33].

The economics of a dual-purpose electricity-desalination plant using a high-temperature reactor as the heat source, a gas turbine and multi-stage flash distillation were analyzed by the Nuclear Research Centre Jülich and compared with single-purpose plants [34].

Badische Anilin und Soda Fabrik (BASF) in Ludwigshafen is the first industrial company to order a nuclear station. Steam requirements of up to 2000 to 2500 tons/h and current requirements of 700 MW form the necessary potential for full load of a nuclear reactor of 2000 MW_{th} throughout the year. Such a nuclear power station must be erected inside the manufacturing plant. Problems related this tositing had to be solved, like: protection of the nuclear section of the power plant against outside effects; security measures against nuclear contamination products; security measures against the release of radioactivity in case of an accident; protection against sabotage [35].

Greece. In a study made with the assistance of the Office of Saline Water and the U.S. Atomic Energy Commission, a nuclear-fueled dual-purpose power and water plant appeared to be more economically attractive than fossil-fueled single-purpose or dual-purpose plants. A dual-purpose nuclear desalting plant of about 190000 m³/day (50 Mgd) and 300 MW_e capacity represented a potentially desirable adjunct to the Athens water and power systems. The power plant could operate under base load conditions and the power absorbed in the existing integrated power system [36]. The dual-purpose plant was not erected, but an aqueduct with a total length of 181 km, including about 51.5 km of tunnels and an artificial dam, to convey water from the river Mornos was given preference instead.

India. The present status of the electric power industry in India was reviewed in a paper by Thomas and the future electricity needs and implementation plans discussed with reference to the exploitable hydro, thermal, and nuclear energy resources of the country. Details of the prospects for nuclear power and desalination development were presented [37]. The concept of an agro-industrial complex with nuclear power stations as foci for simultaneous development of the industrial and agricultural sectors of the economy is discussed in section 2.15.9, p. 134.

Israel. The country is very poor in natural energy resources and most of the oil is imported. Israel is already using about 85% of all its natural water resources and for any further economic expansion will have to resort very soon to large-scale desalination. All power stations are inter-connected by a country-wide grid and what is less common is the additional integrated national water system which enables a large desalination plant to be base-loaded. All those facts lead to the impending introduction of nuclear power in Israel, either in single- or dual-purpose plants [38].

A joint United States–Israel technical team was nominated to review projected water and power needs of Israel and analyze the possibility of satisfying these needs with a large dual-purpose desalting and power generating plant. The preliminary findings being encouraging, a feasibility study has followed, showing that a dual-purpose plant with capacity of 200 MW$_e$ saleable power and 380 000 m^3/day (100 Mgd) of desalted water was technically feasible. A light water nuclear reactor was considered to supply heat to the desalting plant [39]. The study was later changed to a 300 MW$_e$ power plant, maintaining the same water output. Capital and annual costs estimates were developed, using a power credit of 5 $-mills/kWh, under varying nominal fixed charge rates, which are summarized as follows on a 1967 basis:

Fixed charge rate	5%	7%	10%
Capital cost in million dollars	212.7	224.1	238.4
Water costs in cents/m^3	6.2	10.6	17.6
Water costs in cents/kgal	23.6	40.2	66.6

Japan. The shortage of water in Japan is associated with the lack of sufficient supplies of common salt. A shortage of 1.5 billion m^3 of water is estimated in 1985. Means to face water shortage include rational utilization of river water, sea water desalination combined with power generation and erection of industrial complexes combined with nuclear desalination [40].

For the rational utilization of seawater and especially considering the situation of Japan, it is desirable to use the brine of a dual-purpose plant for the production of salt. The brine is concentrated by electrodialysis and thereafter evaporated. Scaling of the evaporator is greatly reduced, the distillation operation is made easier and savings in the cost of equipment are obtained. The products obtained by the process are electric power, fresh water, and industrial salt [41].

In another paper, conceptual designs of dual-purpose nuclear plants are shown. Creation of nuclear power complexes incorporating seawater desalination is of major significance to Japan in terms of effective utilization of energy and structural reform of industry and also from the viewpoint of environmental preservation [42].

Pakistan. The theoretical possibility is illustrated in a paper by Kamal of the use of an extraction cycle for the combination of a 1 Mgd multi-stage flash desalination plant with the 456 MW$_{th}$ heavy water nuclear power reactor, constructed near Karachi. The optimum performance ratio of the plant is shown to be 6.2 and the optimum number of stages is deduced as 25. Details of plant design including chamber geometry, heat transfer surface requirements, etc., are given. It is concluded that the desalination plant could be attached to the Kanupp power plant with a minor effect on the running cost of the plant [43].

The results of studies on the feasibility of nuclear desalination and power generation in Pakistan with particular reference to the Karachi-Sonmiani area are examined in another paper by Kamal [44]. Projections of water demand are worked out. Alternative sources of water supply from surface reservoirs are evaluated. The future power demand is considered in terms of integration with the northern grid of West Pakistan. It is concluded that a serious water shortage will develop in the metropolitan area in the 1980's and that there is adequate justification for a detailed feasibility study of a nuclear plant producing 400 to 500 MW$_e$ power and 380 000 to 568 000 m^3/day (100 to 150 Mgd) of fresh water to come into operation by 1980 [44].

Sweden. The need for additional water supply in Southern Sweden is estimated to about 20 million m³/year in 1980 and 180 million m³/year in 2020. In a project study concerning conveyance of lake water to the area over a distance of 120 km, water cost was estimated to 0.40 Skr/m³. The cost of water from a nuclear dual-purpose plant was estimated to 0.75 Skr/m³ [45].

In another study the generation of 500 MW electric power and 1000 MW heat for Stockholm residents was scheduled for 1975. The project was delayed because of urban siting problems.

United Kingdom. Britain faces not a water shortage problem, like many developing countries, but is seriously interested in nuclear desalting as exporter of both nuclear reactors and desalination plants.

Results of a design study of a large scale dual-purpose power and water plant for the production of 200 MW_e of saleable power and 377 000 m³/day (83 Imperial Mgd) of fresh water were presented by Clelland [46]. The plant comprises a steam generating heavy water reactor with associated power plant and eight multi-stage flash distillation units, each with a capacity of 47 250 m³/day (10.4 Imperial Mgd). The overall capital cost of the facility and the cost of water at a stated power price were estimated [46].

In another study the combination of steam generating heavy water reactor with a multi-stage flash destillation plant was investigated over the range of water outputs of 227 000 to 910 000 m³/day (60 to 240 Imperial Mgd) and of net electrical outputs from 100 to 400 MW. Optimization has been carried out and cost contours are drawn for plant combinations which could be built on current technology [47]. It was estimated in a further study that, in the range of 667 to 1500 MW_{th}, nuclear heat should be competitive with fossil fuel for base load. A hybrid VTE/MSF (vertical tube evaporation/ multi-stage flash) plant coupled to a steam generating heavy water reactor can supply up to 40 MW of peak power per 75700 m³/day (20 Mgd) of base load water output. The cost of peak power is about 2.3 times that from the same reactor designed for power production only [48].

Problems and economics of large-scale single-purpose multi-stage flash plants, whose thermal performance is boosted by vapor recompression, were considered in a paper by Starmer and Lowes [49]. In the cycle chosen there is no inter-effect heat-rejection or heat-input section between the two heat recovery sections used; instead a flash stage is used to provide the vapor to be compressed for condensation in part of the heat-input section. A comparison is given of the water cost from such a plant and from a large nuclear dual-purpose plant using either pass-out or back-pressure steam for the distillation plant [49].

U. S. A. Further early studies in the United States included the investigation of using radio-isotope power as heat source for saline water conversion [50] and the evaluation of desalting plants with a capacity of 15 to 150 Mgd (57 000 to 570 000 m³/day) and nuclear power plants of 200 to 1500 MW_{th} for combined water and power production [51]. The work on nuclear desalting was jointly sponsored by the Office of Saline Water and the Atomic Energy Commission.

Five small capacity desalting plants are operated by heat supplied from nuclear reactors: one in San Onofre, California, power plant with a capacity of 653 m³/day (172500 gpd), two at the Indian Point, New York, power plant with a capacity of 545 m³/day (144000 gpd) each, and two at the Surry, Virginia, power plant with a capacity of 655 m³/day (173000 gpd) each. The importance of these plants is local.

The most important feasibility study on nuclear desalting is that for the Metropolitan Water District of Southern California, also known as Bolsa Island nuclear power and desalting project. According to the design the plant was to be sited on a man-made island offshore from Bolsa Chica State Beach in Orange County, south of Los Angeles, California [25]. This unconventional siting was selected because the land cost in California was so high that several million dollars would be saved by this means. Approximately 1900 MW of gross electrical power from two reactors and 150 Mgd of desalted water would consist the outputs of the proposed dual-purpose plant. Three MSF trains would make the projected capacity with an initial installation of one train of 190000 m³/day (50 Mgd) and appropriate provisions for expansion of the desalting plant to the capacity of 570000 m³/day (150 Mgd) in four years. The cost of water at the plant site was estimated to 9.64 cents/m³ (36.5 cents/kgal). Although the expected cost of water was attractive the project was delayed for various reasons, including increase in costs since the feasibility studies [53].

It was concluded in another study on the application of dual-purpose nuclear power and desalting plants to provide supplemental water needs of the New York City metropolitan region, that large scale distillation plants, operating conjunctively with the existing surface water supply system, utilization off-peak power and steam, are the most economical alternative to accomplish this objective [54].

This problem was already examined in an engineering study of the potentialities and possibilities of desalting for Northern New Jersey and New York City [55], in a study of optimum operation of desalting plants as a supplemental source of time yield for New York City [56], and in a study to assess the long-term technical and economic feasibility of dual-purpose nuclear power and desalting plants to serve large metropolitan areas, with special relevance to existing and planned systems and sources of power and water supply, with emphasis on New York City. The study indicated that a 750 Mgd (2 840 000 m³/day) desalting plant operated in conjunction with the existing surface system would be required to meet the long range needs of the New York City region [57].

The most sophisticated multi-purpose nuclear plant being designed was also intended for erection in the New York area, at Riverhead, on Long Island's north shore. The project was named SURFSIDE, acronym for Small Unified Reactor Facility with Systems for Isotopes, Desalting and Electricity. When completed, it will produce 2500 kW net electricity, 1 Mgd (3785 m³/day) of fresh water and the equivalent of 400 000 Curie of cobalt-60 annually [58]. This project is as well delayed.

A large-scale prototype seawater desalting plant was proposed, to evaluate the technical and economic feasibility of a 40 Mgd (151 000 m³/day) desalting plant in two 20 Mgd MSF units. The project was planned to be located adjacent to nuclear power units now under construction at the site located along the central coast of California. Steam would be obtained from the nuclear reactors. The 40 Mgd desalter was estimated to cost about $ 66 million and the cost of desalting the seawater was estimated to be $ 0.73 per 1000 gallons (19.28 cents/m³) at the plant site [59]. To meet new environmental standards, such as thermal discharge, permissible copper in the effluent etc., it was necessary to increase the estimated capital investment for such a plant by 10 million dollars. The cost of the product water was substantially increased. In addition a 145 km pipeline was required to convey the water from the plant to Santa Barbara, the final consumer of the product water. The final result of this study revealed a cost of 92 cents/kgal (23 cents/m³) and the project was abandoned.

U.S.S.R. The principal engineering concepts are presented of a dual-purpose desalination plant for Southern Ukraine. As a heat source for the desalination plant the BB3P-400 reactor is used. The prototype of this reactor operates successfully at Novo-Voronezh nuclear power plant. Multi-stage flash evaporators are used for the desalination of water [60].

The Shevchenko desalination plant (see section 2.3.2, p. 54), originally operated with oil, was designed to be coupled with the local nuclear power plant.

Literature to 2.14

[1] D. B. Brice, M. R. Dusbabek, C. R. Townsend (Off. Saline Water Res. Develop. Progr. Rept. No. 19 [1958]). — [2] The Fluor Corporation (Off. Saline Water Res. Develop. Progr. Rept. No. 34 [1959]). — [3] R. P. Hammond (ORNL-TM-432 [1962]). — [4] G. Young, R. P. Hammond, I. Spiewak (ORNL-TM-465 [1963]). — [5] International Atomic Energy Agency (Power and Research Reactors in Member States, Vienna 1971).

[6] L. S. Sterman, W. M. Lavygin (Proc. 3rd Intern. Symp. Fresh Water Sea, Dubrovnik 1970, Vol. 3, p. 385/95). — [7] B. M. Ma (Nucl. Eng. Design 12 [1970] 9/17). — [8] J. McTague, G. J. Davidson, R. M. Bredin, A. A. Herman (Nucl. News 15 No. 2 [1972] 31/5). — [9] R. P. Hammond (Desalination 3 [1967] 243/51). — [10] I. Spiewak, J. K. Franzreb, J. C. Moyers (Storage and Transport of Water from Nuclear Desalting Plants, IAEA Rept. No. 141 [1972] 1/16).

[11] F. S. Aschner, S. Yiftah, P. Glueckstern (Desalination 3 [1967] 82/90), F. S. Aschner, G. Yiftah, P. Glueckstern, G. Frank, A. Lavie (Feasibility of Nuclear Reactors for Seawater Distillation 1964–1967, Technion-Israel Institute of Technology, Haifa 1967). — [12] S. Chambers (Storage and Transport of Water from Nuclear Desalting Plants, IAEA Rept. No. 141 [1972] 17/20). — [13] S. Chambers, F. A. Drake, E. F. O. Masters (Proc. 4th U.N. Intern. Conf. Peaceful Uses At. Energy, Geneva 1971 [1972], Vol. 6, p. 243/55). — [14] W. A. Homer, C. A. Scharpf (AIChE [Am. Inst. Chem. Engrs.] Symp. Ser. 67 No. 107 [1971] 184/99). — [15] B. Kunst, E. Gnam, K. Regnet, H. Laban, H. Schafstall, K. Baranowski (Brennstoff-Wärme-Kraft 23 [1971] 290/4).

[16] P. Glueckstern, N. Arad (Storage and Transport of Water from Nuclear Desalting Plants, IAEA Rept. No. 141 [1972] 21/37). — [17] L. S. Sterman, N. A. Mozharov, V. V. Gubenko (Desalination 5 [1968] 95/106). — [18] L. S. Sterman, N. A. Mozharov, W. M. Lavygin, V. V. Gubenko, A. S. Sedlov (Proc. 3rd Intern. Symp. Fresh Water Sea, Dubrovnik 1970, Vol. 3, p. 371/84). — [19] I. Spiewak, R. P. Hammond (Proc. Symp. Nucl. Energy Costs Econ. Develop., Istanbul 1969, Vienna 1970, p. 423/34). — [20] J. E. Jones, T. D. Anderson (Proc. Study Group Nucl. Agro-Ind. Complexes, IAEA Publ. No. 139 [1971] 133/40).

[21] N. A. El-Ramly, J. M. English (Proc. Symp. Nucl. Desalination, Madrid 1968 [1969], p. 585/94). — [22] Yu. I. Koryakin, A. A. Loginov, V. A. Chernyaev (Desalination 7 [1970] 323/42). — [23] L. T. Fan, C. L. Hwang, N. C. Pereira, L. E. Erickson, C. Y. Cheng (Desalination 11 [1972] 217/38). — [24] L. E. Stamets, L. T. Fan, C. L. Hwang (Desalination 11 [1972] 239/54). — [25] S. J. Ball, N. E. Clapp, J. G. Delene (Proc. 4th Intern. Symp. Fresh Water Sea, Heidelberg 1973, Vol. 2, p. 429/38).

[26] L. Thiriet, P. Lievre (Proc. Symp. Nucl. Desalination, Madrid 1968 [1969], p. 631/43). — [27] J. T. Ramey, J. A. Swartout, W. A. Williams (Proc. 1st Intern. Symp. Water Desalination, Washington, D.C., 1965 [1967], Vol. 1, p. 561/78). — [28] F. E. Crever (Off. Saline Water Res. Develop. Progr. Rept. No. 526 [1969]). — [29] C. W. Collins, J. O. Kolb (ORNL-TM-2380 [1970]). — [30] O. H. El-Mofty (Proc. Symp. Nucl. Energy Costs Econ. Develop., Istanbul 1969, Vienna 1970, p. 191/6).

[31] M. A. El Guebeily, K. E. Effat, M. F. El Fouly, A. E. El Kholy, Y. A. El Meshad, E. S. Gaddis, H. Y. Fouad, S. Marey, N. A. Haroun, A. A. Sayed (Proc. 4th U.N. Intern. Conf. Peaceful Uses At. Energy, Geneva 1971 [1972], Vol. 6, p. 115/30). — [32] P. Balligand, J. Huyghe, F. Lauro, P. Vignet (Proc. 4th U.N. Intern. Conf. Peaceful Uses At. Energy, Geneva 1971 [1972], Vol. 6, p. 213/27). — [33] W. Lenz (Meerestechnik 1 No. 2 [1970] 75/7). — [34] P. Schwegmann, H. Bonnenberg (Kernforschungsanlage Jülich Rept. JUL-656-PA [1971]). — [35] B. Frank (Chim. Ind. Genie Chim. 104 [1971] 1739/46).

[36] U.S. Department of the Interior and U.S. Atomic Energy Commission (Preliminary Study of Desalting for Athens, Washington, D. C., 1966). — [37] K. T. Thomas (Proc. Symp. Nucl. Energy Costs Econ. Develop., Istanbul 1969, Vienna 1970, p. 225/35). — [38] J. Adar (Proc. Symp. Nucl. Energy Costs Econ. Develop., Istanbul 1969, Vienna 1970, p. 215/24). — [39] Kaiser Engineers, Oakland, California, and Catalytic Construction Co., Philadelphia, Pennsylvania (Phase I Rept. No. 65-25-RE [1965]; Rept. No. 66-1-RE [1966]). — [40] I. Suetsuna, A. Ito (Proc. 2rd Intern. Symp. Fresh Water Sea, Dubrovnik 1970, Vol. 3, p. 193/202).

[41] H. Kakihana, Y. Tsunoda (Proc. Symp. Nucl. Desalination, Madrid 1968 [1969], p. 745/55). — [42] A. Ito (Chem. Econ. Eng. Rev. 4 No. 6 [1972] 7/13; C.A. 77 [1973] No. 92627). — [43] I. Kamal (Proc. 3rd Intern. Symp. Fresh Water Sea, Dubrovnik 1970, Vol. 3, p. 285/98; Nucleus [Lahore] 8 No. 3 [1971] 37/43). — [44] I. Kamal (Proc. 4th U.N. Intern. Conf. Peaceful Uses At. Energy, Geneva 1971 [1972], Vol. 6, p. 115/30). — [45] P. G. F. Hovsenius (Proc. 3rd Intern. Symp. Fresh Water Sea, Dubrovnik 1970, Vol. 3, p. 153/62).

[46] D. W. Clelland (Desalination 2 [1967] 215/9). — [47] S. Chambers (Proc. Symp. Nucl. Desalination, Madrid 1968 [1969], p. 331/40). — [48] S. Chambers, F. C. Wood (Proc. Symp. Small Medium Size Reactors, Oslo 1970 [1971]). — [49] R. Starmer, F. Lowes (Proc. Symp. Nucl. Desalination, Madrid 1968 [1969], p. 569/82). — [50] Chance Vought Corp. (Off. Saline Water Res. Develop. Progr. Rept. No. 68 [1962]), M. A. Welt (Dechema Monograph. 47 [1962] 373/414).

[51] Catalytic Construction Co. and Nuclear Utility Services Inc. (Off. Saline Water Res. Develop. Progr. Rept. No. 124 [1964]). — [52] W. K. Davis, L. Galstaun, C. Scharpf (Desalination 2 [1967] 227/33), H. J. Mills (Off. Saline Water Res. Develop. Progr. Rept. No. 570 [1970]). — [53] R. W. Durante (Proc. Symp. Nucl. Desalination, Madrid 1968 [1969], p. 169/87). — [54] S. Shiozawa (Storage and Transport of Water from Nuclear Desalting Plants, IAEA Rept. No. 141 [1972] 137/60). — [55] Ralph M. Parsons Co. (Off. Saline Water Res. Develop. Progr. Rept. No. 207 [1966]).

[56] Parsons-Jurden Corp. (Off. Saline Water Res. Develop. Progr. Rept. No. 553 [1970]). — [57] C. G. Clyde, W. H. Blood (Off. Saline Water Res. Develop. Progr. Rept. No. 528 [1970], Rept. No. 780 [1972]). — [58] R. L. Baer (Power 114 No. 3 [1970] 79/81). — [59] D. B. Brice, S. Shiozawa (Proc. 4th Intern. Symp. Fresh Water Sea, Heidelberg 1973, Vol. 2, p. 439/46). — [60] G. G. Alksnis, G. V. Krugilov (Proc. 3rd Intern. Symp. Fresh Water Sea, Dubrovnik 1970, Vol. 3, p. 205/15).

2.15 Nuclear powered agro-industrial complexes

Projections of the world population over the next 30 years show that about 6000 to 7400 million people have to be fed by the year 2000. Not only availability of crops and protein might create problems but even conservation of mineral resources, such as phosphorus, potassium and eventually fixed nitrogen, might become necessary. The irrigated agricultural area in the world will have to be considerably increased to meet the needs in food of the growing population.

2.15.1 The concept of an agro-industrial complex

The scarcity of land and water for irrigation might cause changes in the relative prices of industrial and agricultural commodities. Higher payments for irrigation water might result, which might allow wider use of desalted water. Nuclear powered industrial and agro-industrial complexes are a new concept that may make major contributions to industrial, agricultural and general economic advancement in developing countries. This kind of complex, in abbreviated form NUPLEX, is based on the assumption that both low cost energy and sufficiently low cost desalted water would be made available.

Fig. 2–33 illustrates the concept of a typical agro-industrial complex, combining the large scale production of electric energy and desalted water from a nuclear dual-purpose plant and the consumption of these commodities in industrial and agricultural operations [1]. The concept of the agro-industrial complexes was created at Oak Ridge National Laboratory.

Fig. 2–33

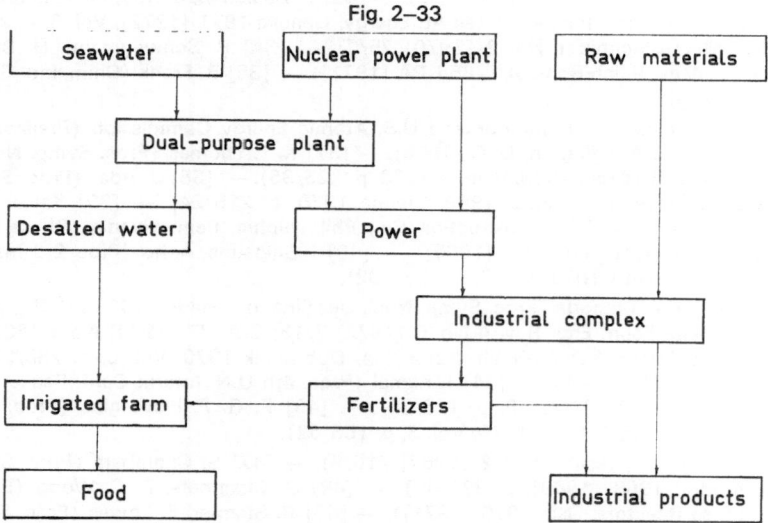

Concept of an agro-industrial complex.

A condition for the energy centre to become a success is that the selected location must be attractive for industries consuming large quantities of energy. Raw materials must be available and the site must offer an economic advantage in marketing the products. Moving low cost power to sources of raw materials rather than transporting raw materials to sources of energy is a novel concept. The combination of such energy centres with highly rationalized agriculture based on desalted water is as well a novel concept. A definite advantage of the industrial complex concept is that products from one plant can be used as raw materials for main or secondary processes in other plants represented in the complex. Byproducts, sometimes rejected as waste, may be made use of. The sharing of off-site facilities by the various industries in the complex will reduce investment costs significantly. Off-sites might include site preparation, roads, harbor, docks and dock facilities, offices, laboratories and housing, water and fire-protection system, communication system, mobile equipment such as cranes, bulldozers, trucks and so on. Offsite costs of an isolated plant may represent as much as 50% of the plant cost, but offsites could be reduced to less than 10% of the plant cost if this plant is integrated in a large complex.

2.15.2 Components of a complex

While the characteristics of an agro-industrial complex will vary with local conditions, a typical complex might include the following components:

Thermal energy centre. The power plant, sufficiently large to give low cost electricity, will always be the heart of the complex. Coal or oil, when available, could as well supply the necessary heat source, but nuclear energy was expected to be more economic in large scale operation as considered. Nuclear reactors of 1000 MW_e or greater are now commercially available. Such reactors, when used in dual-purpose power generation and desalting plant, correspond to a combined production of 1.3 million m^3/day of fresh water and 700 MW_e of electric power.

Large scale desalting plant. Desalting of seawater by distillation is both an energy intensive and a capital intensive process. Economies can be achieved in these cost factors by scaling up to large plant capacities. Taking into account the desalting plants already in operation, it is still quite far from the capacities of 1000 Mgd projected in the studies for large scale agro-industrial complexes.

Manufacture of fertilizers. If operation of large scale farming is to be made, the manufacture of nitrogenous and phosphatic fertilizers would contribute in increasing the yield of agricultural products. Ammonium nitrate, urea and ammonium phosphate would be the most probable products, by using the standard modern processes, except for phosphoric acid. In considering phosphoric acid production, preference might be given to the electric furnace process, to promote the consumption of electricity, available from the large scale power plant.

Energy intensive industries. As in the case of phosphoric acid, the projection of other energy intensive industries would justify the erection of large scale reactors. In this respect manufacture of electrolytic hydrogen was considered in some of the studies for the ammonia synthesis, and of caustic soda and chlorine by electrolysis. Smelting of aluminum is an energy intensive industry, which is favored in all studies on agro-industrial complexes. Industries, which have been considered in various studies, include soda ash, sulfuric acid and cement, oil refinery and petrochemicals, iron and steel, as well as plastics. There are several other industries too, energy intensive or not, which have been considered as constituents of agro-industrial complexes. The most relevant might be the manufacture of marine chemicals from the effluent of the desalting plant. The rejected brine has a concentration of 2 to 3 times that of seawater.

2.15.3 Agricultural use of desalted water

In the arid regions, where an agro-industrial complex is likely to be erected, irrigation is a condition without alternative. Irrigation with desalted water is a completely new and unpracticed technique in large scale operation. Novel farming and irrigation techniques have to be developed to make the use of the higher cost desalted water acceptable by increasing the efficiency of irrigation as well as the efficiency of water use. Great savings in water could be achieved by substituting the usual sprinkler irrigation by the modern method of subsoil or trickle irrigation, as most of the evaporation from the soil surface and deep percolation are both avoided. Further possibilities for improvements may be obtained in genetic development and production optimization. Environmental control offers additional possibilities for increasing the yield of crops. Greenhouse cultivation has shown that rice yield is almost doubled when the level of carbon dioxide is increased only from 320 to 2400 ppm. The yield of sorghum is increased by a factor of 3.5 under similar conditions. With several on-site abundant byproduct sources of carbon dioxide available in the agro-industrial complex, sizeable yield increases might also be achieved in open air cultivation.

Despite this optimistic description of the advantages that an agro-industrial complex would offer, there are many difficulties which should not be overlooked. A desalting plant is in principle a base load plant. The higher the load factor, the lower the cost of water in a given plant will be. On the other hand the water requirements of the crops would vary a great deal with the seasons of the year. Storage of water seems to be unavoidable. If natural storage possibilities do not exist, providing of man-made storage facilities would indirectly increase the cost of water. A part of the advantages of an agro-industrial operation, as already outlined, might be offset by the increased cost of water, if storage for large quantities of water must artificially be provided.

There are, however, some ways which might help to overcome these difficulties in coordinating the capacities of water production and water consumption. The seasonal peaks in water demand might be compensated to some extent by the appropriate choice of crop rotation. A temporary reduction of the irrigated land area at periods of high water demand might be another means to balance production and comsumption of water. From the engineering point of view, the variation of the water to power ratio, stand-by units etc. might be more economic than the erection of large capacity storage facilities.

Areas in the United States, which might benefit from salt water conversion were investigated by the Bureau of Reclamation [2]. Three irrigated areas in Arizona and California were studied to measure water quality benefits associated with desalted irrigation water at salt concentrations of 400, 900, and 1500 mg/l. Cost comparisons were made between electrodialysis and multi-stage flash distillation methods and between nuclear and non-nuclear sources of energy. Design capacity range was from 8300 to 325 000 m³/day (2.2 to 85.7 Mgd). High value crops were used in benefit studies. Crops were selected according to salinity and climatic tolerances. Crop enterprise budgets were used to develop direct, indirect and public benefits at each quality level. Benefits were found to exceed costs under several combinations studied [3].

The salinity tolerances of certain tropical soils and the relationships between sodium ion activities and soil physical properties were investigated by El-Swaify et al. [4]. Multi-stage flash distillation and electrodialysis were evaluated as potential processes for the preparation of small scale irrigation water. Investment costs per acre were found to be high. The most favorable benefit-cost ratio involved use of the electrodialysis process at a product water quality level of 400 mg/l [5].

There are no universally acceptable standard criteria for determining how much agriculture can pay for irrigation water. Social and political conditions often have as much or more influence on development cost decisions than do strictly economic analyses. Many studies indicate that 10 cents/1000 gal (2.64 cents/m³) is an upper limit of acceptable costs for irrigation water [6]. This of course depends to a large extent from the applied method of irrigation.

For the time being there seems to remain a very wide field of research work concerning the exact determination of water requirements under various conditions, the breeding and selection of salt-tolerant plants, the improvement and application of the most advanced irrigation techniques and the control of evaporation and seepage losses from surface-water storage [7].

In reviewing the possibilities of using desalted water in agriculture Barrada [8] suggested that first priority should be given to water conservation and the efficiency of irrigation. Where development of natural fresh-water resources is not feasible, desalination should receive close attention. Large-scale irrigation projects or agro-industrial complexes based on nuclear desalination are possible future solutions in arid coastal regions with high production potentials. Agro-industrial complexes might be realized when the cost of water would be less than 10 cents/1000 gal (2.64 cents/m³). The lack of the necessary capital investment, trained experts, skilled workers and educated farmers would make the development of such complexes in developing countries difficult but not an impossible task [8]. However, before proceeding to a large-scale agro-industrial complex operation, establishment of experimental farms irrigated with desalted water appears to be a necessity.

Another necessity is the improvement of water saving techniques for modern irrigation. Investigations have been made on subsurface and trickle irrigation and their possible uses. The need for further development coupled with the need for extensive testing over a number of agricultural cycles indicates that it will be several years before such systems are ready for widespread use. The capital cost of such systems appears to be high. It appears that the major application will be for high-value crops and for particular locations where water is scarce or expensive. The current use of similar irrigation in dry areas of Israel is good evidence that the high capital cost can be justified under the proper conditions [9].

Considering the quality of irrigation water, it might be mentioned that a method for reducing to nonphytotoxic levels the boron content of irrigation return waters, that have been renovated by reverse osmosis treatment, is based on a process modification, in which Amberlite XE-243 is used. A 3 kgal/day system has been designed for the field evaluation of the process [10].

2.15.4 Siting an agro-industrial complex

In selecting a site for an agro-industrial complex the agricultural requirements might prove to be the dominant parameters, as the constraints imposed by agricultural factors are more restrictive than those relevant to industrial processes [1]. The initial criteria for screening potential agricultural areas would include:

1) Proximity to and elevation above the sea, as large amounts of saline water are required for the desalting plant.

2) Frost-free growing season, favorable to the production of two or more crops per year. This condition limits possible sites to those situated between 35° north and 35° south.

3) Use of land not under intensive cultivation, such as arid and semi-arid regions with a rainfall less than 30 cm per year.

4) Physical suitability of the soil for agricultural production and topography of the soil avoiding expensive land levelling, and

5) Availability of land sufficient in quantity and quality to support a complex and to allow for possible future expansion.

A more detailed investigation should ensure the potential productivity of the soil by its chemical and physical properties, the eventual availability of brackish water for occasional mixing with desalted water to supplement irrigation, the possibility of storing water other than in expensive reservoirs etc.

As the economic use of desalted water for agricultural production has not yet been demonstrated on a large scale, an experimental farm that uses desalted water appears to be an unavoidable requirement in the planning of a complex. Data acquired on the experimental farm will be used to forecast the economics of large agricultural complexes accurately. An area of at least 100 ha is considered as necessary and the soil and the climate of the region should be representative of those predominant in the region under consideration for the agro-industrial complex.

2.15.5 General studies on agro-industrial complexes

*Allgemeine
Unter-
suchungen
über agro-
industrielle
Komplexe*

Most of the world's deserts are located near the oceans. Hence, a number of coastal desert regions were considered as potential areas for the location of an agro-industrial complex [1]. These areas are:

1) The Kutch Peninsula of India, 2) The Sharks Bay-Carnavon region in West Australia, 3) The Magdalena Plain of Baja California in Mexico, 4) The Sechura Desert of the northwest lowlands of Peru, and 5) Various Middle East desert regions, including Sinai, Mediterranean coastal areas in Egypt, as well as the Negev desert in Israel.

These preliminary studies were conducted to test the sensitivity of the assumptions made in relation to actual conditions in the world and to estimate the applicability of agro-industrial complexes. Other studies that followed have been made in varying degrees of detail or are underway and may be considered as more or less preliminary feasibility reports. These studies are:

1) Nuclear power and desalting plants for Southwest United States and Northwest Mexico, 2) Puerto Rico energy center study, 3) Nuclear powered agro-industrial complex in India, 4) Preliminary studies for the Middle East, and 5) Various studies in the U.S.A. for Louisiana, Kentucky and Texas.

2.15.6 Basic studies by Oak Ridge National Laboratory

*Grundlegende
Unter-
suchungen
des
Oak Ridge
National
Laboratory*

The study was initiated in 1967 and introduced the fundamental concept of combining large-scale industrial production in complexes with highly rationalized agriculture based on desalted water [11]. Supplementary information collected on several locales considered for the erection of agro-industrial complexes are given in another report [12]. This report is a supplement of chapter 8 of the main study, concerning siting of the potential complexes. Another supplement contains information to duplicate cost studies on the 17 chemical and metallurgical products considered in the main report. A number of worked examples illustrate the use of the data supplied [13].

The Oak Ridge studies have a very general approach to the subject and set out to prove that a nuclear-powered energy centre is a feasible undertaking. They try to provide sufficient information on the cost computation of the industrial processes considered, in order to make comparable cost estimates possible or to permit estimation of the economics of other product mixes.

The cost of producing desalted water can be substantially reduced by combining its production with that of ammonia. Between 7500 and 9500 m³/day (2 to 2.5 Mgd) of water can be recovered from low temperature heat [14].

The flow sheet of a thermodynamically balanced plant for the co-production of desalted water and anhydrous ammonia was also presented together with charts summarizing the preliminary cost estimates and unit costs for each primary product. The flow sheet could be adjusted in large scaled operations, to achieve an increased agricultural water-ammonia balance [15].

The possible role that electrolytic hydrogen production might play in a complex is examined in a paper by Mrochek et al. [16]. For a 1000 MW$_e$ nuclear power station and a 300 tons/day ammonia plant the credit for oxygen and ammonia enables the load-levelized plant to produce power for the same total annual costs as the power plant alone operating at a load factor of 79%. The adaptability of other electricity-intensive industrial processes such as chlorine-caustic soda, aluminum, magnesium metal, arc production of acelylene, electric iron and elemental phosphorus might also act as load-leveling devices [16].

Caustic soda and chlorine are the basic chemicals for a whole array of tertiary products, such as pulp and rayon from grain wastes of a nuclear-powered agro-industrial complex, as well as PVC plastics and chlorinated hydrocarbons when a source of petroleum is available. The economics of each process building block are discussed in a paper by Yee in the order of their sensitivity to the cost of electric power [17].

The general economics of combining a nuclear-electric generating station, an industrial complex and a large agricultural enterprise into an agro-industrial complex was evaluated at Oak Ridge for two levels of assumed technology for the nuclear desalination plants. The effects on the economic return of size, changes in product mix and sales price levels and of operation with bypass steam to reduce the generation of electricity were determined [18].

In a typical case the production of desalted water as a primary product is considered together with the production of elemental phosphorus, ammonia, and various chemicals from the electrical energy generated in a dual-purpose nuclear desalting plant:

In a complex involving the installation of a desalting plant with a capacity of 1000 Mgd (3785000 m³/day) and the production of 2000 MW electrical power, sufficient water would be available for the irrigation of 300000 or more acres under intensive agriculture. The fertilizers produced in the complex would cover the needs at optimum conditions of this newly cultivated area, as well as of an additional area of 2 million acres located outside the agro-industrial complex. The total food production of the considered agro-industrial complex, not including that obtained through the use of fertilizers outside the newly cultivated area, is sufficient to feed a population of 4 million people at a daily intake of 2500 calories. The Oak Ridge studies indicate that selected crops and industrial products can produce a return on investment estimated to be in the range of 10 to 20% per year. Total capital costs are estimated between 500 to 1500 million dollars for the agro-industrial complex considered [19]. Several of the improvements in intensive agricultural practices, such as subsoil irrigation, were not taken in consideration in these early studies.

It is estimated in another study that a complex covering 10 square miles (26 km²) could produce enough food for a million people, at prices comparable with current world market prices. It would require about 680000 m³/day (180 Mgd) of fresh water. The byproducts recovered from the seawater would provide the raw materials for the manufacture of fertilizers and a wide range of chemicals [20].

The application of dynamic programming algorithms to the optimal allocation of energy in large nuclear-powered complexes is discussed in a paper by Bouchey et al. The formulation and solution of the problem of optimal product mixes in both purely industrial and agro-industrial complexes is outlined. Simplified numerical examples of each case are presented to illustrate the solution method [21]. Two basic approaches, the mixed integer programming and the dynamic programming, are considered by Caplin. The first is a linear programming, which enables to determine capital investment strategies and operating programmes, which maximize present worth over the whole of the planning period; the second enables to establish not only optimal policies but as many near-optimal policies as desired [22].

The problems, which might arise in implementing an agro-industrial complex project are examined in an O.R.N.L. report [23]. Such items may be summarized as follows: The large amounts of capital required over a short time period may create financing problems particularly for developing countries in the area of foreign exchange. Recruting and/or training of a competent labor force, especially at the management level may prove to be a significant problem. Industrial integration and interchange of products offer advantages, but too much interdependence can cause difficulty during unscheduled shutdowns of individual plants. Added transport cost for delivery of products to markets may be incured, but may be offset by achieving lower delivered prices for raw materials. It is unlikely that many critical raw materials will be advantageously available at any one site so that the variety of industrial processes to be included in the complex might be limited. The industries considered produce basic products used in most cases as inputs in secondary or tertiary industries, which are typically more labor intensive. Meeting a goal of reducing unemployment will depend somewhat on the development of secondary industries, which should naturally occur within the environment of an energy center [24]. A poor estimate of the price and demand for the output of an industrial complex, that leads to an overly optimistic prediction of profitability reduces the prospect for a successful project. Once an industrial complex has been designed and built, there are few options remaining for management to exercise in order to compensate for a poor decision made in the beginning. In the case of the agricultural section, management has the option to reassess each year the crop mix, except for perennial crops, and select those crops that will yield the highest rate of return for the then existing conditions. The management of an agricultural complex can change the output each year while the management of an industrial complex is largely committed to the product mix that was designed into the plant initially [25].

A number of potential applications of nuclear desalting and energy centre projects have been the subject of other more specific studies. Those studies promote the agro-industrial concept from the field of pure theoretical consideration to the field of practical implementation. Some of them have an international character involving more than one country.

2.15.7 The U.S. – Mexico – I.A.E.A. study

A joint Mexico – U.S.A. – I.A.E.A. study was recently made of the applicability of nuclear dual-purpose desalting plants to the region comprising the southern portions of the states of Arizona and California in the United States and the northern portions of Baja California and Sonora in Mexico. Although this region is served by the Colorado river, an acute shortage of fresh water is expected in the years to come:

The proposed plant should have an initial fresh-water production capacity of 3 785 000 m³/day (1000 Mgd). Nuclear reactors would provide the energy for the water plant and, at the same time, would produce about 2000 MW of electricity. The study team concluded that the needed dual-purpose nuclear plants are technically feasible and that preliminary economic assessment showed sufficient incentive to justify more detailed investigations. The water deficit for the region is projected as increasing from about 5 680 000 m³/day (1500 Mgd) in 1980 to 26 500 000 m³/day (7000 Mgd) in 2000, assuming the projected municipal, industrial, and agricultural demand [26]. — A summary report on the U.S. – Mexico – I.A.E.A. study was also given by Hunter [27].

Die USA-Mexiko-I.A.E.A. Studie

2.15.8 The Puerto Rico study

The study was prepared for the Puerto Rico Water Resources Authority by two engineering firms in cooperation with the U.S. Atomic Energy Commission and the U.S. Department for the Interior. This study is oriented mainly to the industrial segment rather than to agricultural development. A multi-purpose reactor, producing 540 MW$_e$ and about 2500 tons per hour of process steam, is projected together with a 75 000 m³/day (20 Mgd) desalting plant, a petrochemical refinery and its derivative petrochemical plants, a 3000 tons/day salt recovery work from the waste brines of the desalting plant to feed a chlorine-caustic soda plant and an aluminum plant. At the time of the study, in 1970, nuclear power generation costs were estimated to 4.34 mills/kWh and were slightly lower than fossil-fuel power costs of 4.74 mills/kWh at the projected site as already mentioned in section 2.14.1, p. 120. The overall economic assessment indicated a rate of return of about 15%. The agricultural sector is represented by a 500 acre fully mechanized experimental farm to produce high value fruits and vegetables imported into Puerto Rico [28].

A considerable body of information was developed in this report relative to seawater pretreatment, desalting and the recovery of salt and other marine chemicals. Summary reports on the Puerto Rico study were also presented by Michel [29] and Hernandez Fragoso [30].

Der Entwurf für Portoriko

2.15.9 The Indian study

A very detailed study was made in India for two locations, involving industrial development and intensive agricultural practices. In one of these regions, the Kutch-Saurashtra project provides for desalted water for irrigation, whereas in the other region, in the Gangetic plain, the power produced by the energy centre will serve for pumping of existing underground fresh water resources. Both regions are equally oriented towards industrial and agricultural development. India is a typical example of a country, where availability of food is shorther than the needs of the population. Details of comparative economic evaluation of processes based on different feed stocks, including power for the production of nitrogenous and phosphatic fertilizers, are given and an assessment is made for aluminum production.

In the Kutch-Saurashtra project the power reactors will produce 1200 MW_e saleable power, apart from 100 MW_e for the nuclear island and 50 MW_e for the desalting plant with a capacity of 680000 m³/day (180 Mgd). In the industrial segment the manufacture of synthetic ammonia with a capacity of 496000 tons/year N and of phosphoric acid with a capacity of 272000 tons/year P_2O_5 for fertilizers, as well as 50000 tons/year aluminum, 50000 tons/year caustic soda and chlorine plants are projected together with the recuperation of 1 million tons of common salt and marine chemicals from the effluents of the desalting plant. Hydrochloric acid will be used to pretreat the raw seawater in a combined MSF-VTE distillation plant. In the agricultural sector a fully mechanized farm covering 16000 ha will be irrigated with desalted water.

In the Gangetic plain project the nuclear station will produce 1200 MW_e of saleable power to operate a 372000 tons/year N synthetic ammonia and a 206000 tons/year P_2O_5 phosphoric acid plant for fertilizer manufacture and a 50000 tons/year aluminum plant. A power distribution system is provided for the operation of about 2500 deep and shallow tube wells irrigating a total area of 7.7 million ha, out of which 1.5 million ha would be incorporated in the proposed agro-industrial complex.

Whereas the rule is that returns on the investment are higher in the industrial sector, in the Indian study the opposite appears to be the case. The overall return on the investment of the complex as a whole is estimated to be 17.85%. The part of the agricultural sector provides a return on investment of 99.5% for bullock farming and of 119.3% for fully mechanized farming. Another advantage of moderate and hot climates, like those prevailing in India, is the number of crops which might be harvested per year and the resulting increased agricultural production. The anticipated increase in grain production can meet the food requirements of about 30 million people if 3 crops are sown or of about 25 million people if only 2.25 crops are sown per year. Some of the difficulties, which developing countries may have to face in implementing such large projects, are also considered in the study [31].

In another paper the requirement of basic infrastructure, limitations of finance and indigenous technical knowhow, regional imbalances and other sociopolitical factors that have to be overcome are analyzed. The sequential decision-making steps that need to be taken for successful implementation of the nuclear-powered agro-industrial complexes in India are also presented [32]. Summaries of this study were also presented in some other international meetings [33] and Stout has written on the use of nuclear power for meeting critical power needs in modern agriculture, with special references to the Upper Indo-Gangetic plain [34].

2.15.10 Other studies on agro-industrial complexes

A study for the Middle East countries was initiated by the Oak Ridge National Laboratory, but only a preliminary draft report presenting data on the considered areas was published [35]. A dual-purpose plant producing 525 MW of electric power and 200 Mgd of distilled water would be the fundamental unit of the complex with emphasis placed on agriculture and industry relating to the needs of the region.

An anticipated gap of 10^9 m³ water per year at the end of the century between supply and demand will have to be filled in Israel by unconventional sources. The preliminary study made in 1967/68 (see section 2.14.4, p. 124) for a dual-purpose plant was reviewed and an energy center incorporating a power plant of 900 MW_e is considered. Eight chemical-metallurgical processes were selected for the study [36]. Adar [37], taking the economy of the country as a whole, is considering Israel as an example of applying the concept of an agro-industrial complex.

The most recent general study is probably that made at the Texas A & M University, called in abbreviated form the "Texas Nuplex". The study is exploring the potentialities for the State of Texas of producing electric power, desalted water, and chemicals. A new approach is introduced with this study that a Nuplex might prove to be more economic in an already industrialized region. The extensive reserves in oil and natural gas are exploited to a degree of saturation. Additional energy sources will be required and an offshore nuclear power plant combined with a desalting plant are under study. The climate is favoring intensive agriculture. Five locales along the coast are under consideration for the erection of the proposed agro-industrial complex, a large part of which is petroleum processing [38].

Other studies are underway for the Southeastern part of the United States, ranging from petrochemical operations in Louisiana to a nuclear center in Kentucky, using coal as the basic raw material.

The associated production of algae was suggested in the model of a algo-agro-industrial complex, which is balanced in its economy and includes the production of protein bio-mass as a substantial component of its structure. The cultivation of algae in the complex will produce 160 t protein per day on an area of 2000 ha, with a consumption of 200000 m^3 of water and 4000 MW$_{th}$ electrical energy [39].

Power production from deep seawater with associated desalting was suggested by Othmer. Deep seawater contains large quantities of nutrients. It was considered that these nutrients feed algae, which feed shellfish, ultimately shrimps and lobsters in shallow ponds. Wastes grow seaweed of value and combined revenues from desalination, power generation and marine culture will give substantial profit, it was concluded [40].

Balligand has proposed a variety of complex where the part of the industry is relatively large compared to the agricultural sector. The first use of desalted water would be urban and industrial. Sewage would be treated and devoted in a second step to irrigation. Some drawbacks in the utilization of water reclaimed from used water, such as the total dissolved matter, become for irrigation an advantage [41].

Literature to 2.15

[1] A. A. Delyannis (Nuclear Energy Centres and Agro-Industrial Complexes, IAEA Tech. Rept. Ser. No. 140 [1972]). — [2] U.S. Bureau of Reclamation (Off. Saline Water Res. Develop. Progr. Rept. No. 3 [1954]). — [3] U.S. Bureau of Reclamation (Off. Saline Water Res. Develop. Progr. Rept. No. 489 [1969]). — [4] S. A. El-Swaify, L. D. Swindale, G. Uehara (Off. Saline Water Res. Develop. Progr. Rept. No. 419 [1969]). — [5] H. L. Parkinson, J. T. Maletic, J. P. Wagner, M. S. Sachs (Water Resour. Res. 6 [1970] 1496/500).

[6] M. A. Hagood (Proc. Panel Value Agric. High-Quality Water, 1969, p. 145/50). — [7] H. G. F. Kreutzer (Proc. Panel Value Agric. High-Quality Water, 1969, p. 77/92). — [8] Y. Barrada (Proc. Symp. Nucl. Desalination, Madrid 1968 [1969], p. 659/80). — [9] T. E. Cole (ORNL-DNIC-9 [1969]). — [10] R. M. Roberts, L. E. Gressingh (Off. Saline Water Res. Develop. Progr. Rept. No. 579 [1970]).

[11] Oak Ridge National Laboratory (ORNL-4290-UC-80 [1968]). — [12] T. Tamura, W. J. Young, M. M. Yarosh (ORNL-4293-UC-80 [1971]). — [13] H. E. Goeller, I. E. Mrochek (ORNL-4296-UC-80 [1971]). — [14] S. A. Bresler (Off. Saline Water Res. Develop. Progr. Rept. No. 458 [1969]). — [15] S. A. Bresler, E. F. Miller (Proc. Symp. Nucl. Desalination, Madrid 1968 [1969], p. 683/95).

[16] J. E. Mrochek, J. M. Holmes, J. W. Michel (Proc. Symp. Nucl. Energy Costs Econ. Develop., Istanbul 1969, Vienna 1970, p. 639/52). — [17] W. C. Yee (Proc. Symp. Nucl. Energy Costs Econ. Develop., Istanbul 1969, Vienna 1970, p. 619/37). — [18] J. W. Michel (Proc. Symp. Nucl. Desalination, Madrid 1968 [1969], p. 713/44). — [19] R. P. Hammond (Proc. Panel Value Agric. High-Quality Water, 1969, p. 31/64). — [20] E. M. Walsh (Electronics Power 16 No. 2 [1970] 52/4).

[21] G. D. Bouchey, B. V. Koen, C. S. Beightler (Nucl. Sci. Eng. 41 [1970] 70/8). — [22] D. Caplin (Proc. Study Group Nucl. Agro-Ind. Complexes, IAEA Publ. No. 139 [1971] 127/31). — [23] J. A. Ritchey (ORNL-4295-UC-80 [1969]). — [24] J. Michel (Proc. Study Group Nucl. Agro-Ind. Complexes, IAEA Publ. No. 139 [1971] 1/17). — [25] J. O. Roberts (Proc. Study Group Nucl. Agro-Ind. Complexes, IAEA Publ. No. 139 [1971] 117/26).

[26] U.S. – Mexico – IAEA Study Team (Nuclear Power and Water Desalting Plants for Southwest United States and Northwest Mexico, U.S. Clearinghouse Fed. Sci. Tech. Inform. Rept. TID-24767

[1968]). — [27] J. Hunter (Proc. Symp. Nucl. Desalination, Madrid 1968 [1969], p. 189/206). — [28] Puerto Rico Energy Center Study, prepared and published by Burns and Roe Inc., Hempsted, N.Y., and the Dow Chemical Co., Midland, Mich., 1970; Off. Saline Water Res. Develop. Progr. Rept. No. 635 [1971]. — [29] J. W. Michel (Proc. Study Group Nucl. Agro-Ind. Complexes, IAEA Publ. No. 139 [1971] 57/74). — [30] J. Hernandez Fragoso, J. J. Byrnes (AIChE [Am. Inst. Chem. Engrs.] Symp. Ser. **67** No. 107 [1971] 217/23).

[31] K. T. Thomas, M. P. S. Ramani, V. B. Godse, P. L. Kapur, M. S. Kumza, M. G. Nayar, C. M. Shah, N. S. Sunder Rajan, R. K. Verma (Nuclear Powered Agro-Industrial Complex Bhabba Atomic Research Centre, Bombay 1970). — [32] K. R. Thomas, N. S. Sunder Rajan, M. P. S. Ramani (Proc. 4th U.N. Intern. Conf. Peaceful Uses At. Energy, Geneva 1971 [1972], Vol. 6, p. 131/53). — [33] K. T. Thomas (Proc. Symp. Nucl. Energy Costs Econ. Develop., Istanbul 1969, Vienna 1970, p. 459/82; Proc. Study Group Nucl. Agro-Ind. Complexes, IAEA Publ. No. 139 [1971] 35/55). — [34] P. R. Stout (ORNL-4292 [1968]). — [35] Oak Ridge National Laboratory, Middle East Study, Preliminary Draft 1968.

[36] J. Adar (Proc. 4th U.N. Intern. Conf. Peaceful Uses At. Energy, Geneva 1971 [1972], Vol. 6, p. 155/69). — [37] J. Adar (Proc. Study Group Nucl. Agro-Ind. Complexes, IAEA Publ. No. 139 [1971] 19/33). — [38] R. R. Davison (Proc. 4th Intern. Symp. Fresh Water Sea, Heidelberg 1973, Vol. 2, p. 479/89). — [39] I. Malek, J. Bartos, J. Simmer, B. Prokes (Proc. 4th U.N. Intern. Conf. Peaceful Uses At. Energy, Geneva 1971 [1972], Vol. 6, p. 171/83). — [40] D. F. Othmer, O. A. Roels (Proc. 4th Intern. Symp. Fresh Water Sea, Heidelberg 1973, Vol. 2, p. 497/506).

[41] P. Balligand, D. Alexandre, A. Fourcy (Proc. 4th Intern. Symp. Fresh Water Sea, Heidelberg 1973, Vol. 2, p. 465/77).

Geo-
thermische
Energie als
Wärmequelle

2.16 Geothermal energy as a heat source

Geothermal energy appears to be attractive as heat source for desalting, because it might be available at locations where fresh water is needed. However, in early estimations the cost of the hole represented about 75% of the total investment. Even if the hole would be available free of charge, the calculated water cost would still be between 1.32 and 2.64 dollars/m^3 (5 and 10 dollars/ kgal). This early estimation was made by Batelle Memorial Institute [1]. In an investigation of hot wells at Black Hills, South Dakota, a maximum well temperature of 75.6°C (168°F) was measured, which was considered as too low for practical purposes. Similar conclusions were drawn from an investigation of thermal springs in California [2].

Despite these early pessimistic studies, limited to predetermined regions of the United States, the interest in using geothermal energy remained vivid. Cheap desalinated water can be produced, if the thermal field, the source of raw water, and the distilled water market are situated close to one another. A single bore hole of characteristics typical of known geothermal fields would be capable of producing about 18000 m^3/day of fresh water [3]. A description of different applications using geothermal energy was given by Barnea with emphasis on exploring for new geothermal areas and developing of new ways to extract work from steam and hot water. Emphasis is given to the cost of exploitation of geothermal energy for various purposes [4].

Geothermal reservoirs from wet steam fields and geopressure reservoirs, both provide heat and brackish or saline water in great quantities at low costs. Some possibilities are described in a paper by Barnea and Wegelin on the basis of data on wells in El Tatio in northern Chile. Main problem areas will be found to be the prevention of deposits, interfering with proper heat transfer. The abundancy of geothermal and geopressure reservoirs is stressed and the possibility of obtaining distilled water at less than half the cost involved in current distillation operations is indicated [5].

Several new geothermal-well-based systems are either planned or are operating. Data from these systems will help to define the role geothermal resources may be able to play in meeting the world's future energy needs [6].

Brine at 260 to 370°C (500 to 700°F) has been obtained from depths of 1400 to 2450 m (4700 to 8000 ft) in exploratory wells, in the Imperial Valley. Fresh water recovery was estimated at 40 to 80% of the flow [9]. A proposed demonstration project would produce about 90 Mgd (340000 m^3/ day) from 72 production wells distributed among six groups at an estimated cost of 36 cents per kgal (9.5 cents/m^3). About 100 re-injection wells for the waste brine would be needed [10].

The geothermal resources in the Imperial Valley of the United State are extensive. Their successful development could have a profound influence on the region. Based on importance and timing, the priorities by subject were suggested in a report by Laird [7]. A pilot plant, the first to attempt production of pure water from geothermal brine, will be operated on a 1500 to 2100 m (5000 to 7000 ft) well in the Imperial Valley, California. The long term goal is a complex of wells and desalting plants to augment the flow of the Colorado river, enhancing supply to southwestern United States and parts of northern Mexico. The reject brine will be re-injected to the geothermal reservoir.

2.16.1 Use of low grade heat energy

Anwendung von Wärme niedriger Temperatur

Studies of various possible schemes were made for utilizing low-grade heat energy, including solar distillation and the low-temperature-difference plant concepts, involving temperature differences of the order 14 to 22°C (25 to 40°F). A single-stage flash distiller was designed and tested at the University of California. Design estimates for prototype plants were reported [8].

Literature to 2.16

[1] J. C. Bell, R. C. Crooks, W. T. Holser, G. F. Sachsel, J. R. Williams, R. B. Filbert, J. J. Stone (Off. Saline Water Res. Develop. Progr. Rept. No. 27 [1959]). — [2] J. R. MacDonald, E. R. Stensaas, P. M. Stafford (Off. Saline Water Res. Develop. Progr. Rept. No. 28 [1959]). — [3] H. C. H. Armstead, C. Rhodes (Proc. 3rd Intern. Symp. Fresh Water Sea, Dubrovnik 1970, Vol. 3, p. 451/9). — [4] J. Barnea (Sci. Am. **266** [1972] 70/7). — [5] J. Barnea, E. Wegelin (Proc. 4th Intern. Symp. Fresh Water Sea, Heidelberg 1973, Vol. 2, p. 449/61).
[6] G. Weismantel (Chem. Eng. **80** No. 6 [1973] 40/1). — [7] A. D. K. Laird (Off. Saline Water Res. Develop. Progr. Rept. No. 711 [1971]). — [8] E. D. Howe, A. D. K. Laird, B. W. Tleimat (Proc. 3rd Intern. Symp. Fresh Water Sea, Dubrovnik 1970, Vol. 1, p. 691/701). — [9] I. Spiewak, E. C. Hise, S. A. Reed, S. A. Thompson (ORNL-TM-3021 [1970]). — [10] J. J. O'Brien (J. Am. Water Works Assoc. **64** [1972] 694/700).

2.17 Solar energy as heat source

Sonnenenergie als Wärmequelle

Solar distillation exhibits a considerable economic advantage over other salt-water distillation processes because of its use of cost-free energy and its insignificant operating costs. Another advantage of the process is its simplicity. Since radiation per unit of surface is determined by nature, the output of a solar distillation unit can only be increased by increasing the surface or the efficiency. Capital costs and amortization rates are directly dependent on the intended output and the efficiency. They increase more or less proportionally with the increase in defined output for a given design, and most of the study in solar distillation is directed toward improving efficiencies and reducing construction costs through the design. Although the advantage of cost-free energy is partly offset by increased amortization expenses, distillation with solar energy remains the most favorable process for small-capacity water desalting at geographic locations, where there is considerable solar radiation. Most solar distillation plants are being (or will be) erected in less developed countries or in remote areas where there are limited maintenance facilities [1].

2.17.1. Solar radiation

Sonnenstrahlung

Not all the wave-lengths emitted by the sun are useful for solar distillation. Solar radiation is customarily measured in Langleys per minute, which is equivalent to 1 cal of radiation energy per cm^2 per minute. In technical units solar radiation is usually expressed in $kcal/m^2 \cdot h$ or the equivalent in $Btu/ft^2 \cdot h$.

The intensity of solar radiation reaching the earth varies from zero to about $1.5 \ cal/cm^2 \cdot min$ (or $900 \ kcal/m^2 \cdot h$ or about $332 \ Btu/ft^2 \cdot h$). In SI units this amount would be about $1047 \ W/m^2$. Part of this radiation may come directly from the sun, but sometimes as much as 10% of it comes as scattered light even when the sun is unobstructed by clouds. In cloudy weather the total radiation is greatly reduced and most of the light that gets through may be scattered light. Solar energy received on flat-plate collectors includes both direct and scattered or diffuse radiation.

The solar radiation striking a horizontal surface is greatest at noon, as the sun's rays pass through the atmosphere with a minimum length of passage through the air. In the morning and afternoon the rays are subject to increased absorption and scattering. Considering the latitude, maximum radiation is at the equator. Hence the radiation intensity depends on the hour of the day, the day of the year, and the clarity of the atmosphere for a given location, as well as of the latitude of the earth at the point of observation.

These limitations of the solar radiation render solar distillation obligatory a non steady state operation. Performance comparisons of horizontal stills can only be made with reference to radiation intensity. Tilting the solar radiation receiving surfaces from the horizontal position to be more nearly at right angles to the sun's rays is a considerable improvement.

Geschicht-
liches über
die
Destillation
mittels
Sonnen-
energie

2.17.2 History of solar distillation

The first large scale solar distillation plant was erected in Chile in 1872 and is described by Harding [2]. It produced almost 19 m³ per day of fresh water and was nearly 40 years in operation. The interest for solar distillation revived again during World War II, when a plastic film emergency apparatus, which was used in large amounts for life rafts of the U.S. Air und Marine Forces, was designed [3]. Research work continued at the Massachusetts Institute of Technology, at the Universities of New York and California in Berkeley. A commercial plant was erected in St. John on the Virgin Islands [4]. In the following, glass covered and plastic film covered solar stills and a variety of other designs were erected and extensively tested at the Daytona Beach Test Station of the Office of Saline Water, operated by the Battelle Memorial Institute, Cleveland, Ohio. The experience obtained is outlined in three reports [5].

The interest has in the meantime grown world-wide, expressed especially in the papers presented in two international meetings, the Conference in the use of solar energy, in Tucson, Arizona [6], and the World Symposium on Applied Solar Energy, in Phoenix, Arizona, both convened in 1955 [7]. Research work and interest was also reported in Cyprus, Algeria, Australia, Italy, Russia, and Greece. Further information on this research work will be given in the following sections.

Betriebs-
grundlagen
der
Destillation
mittels
Sonnen-
energie

2.17.3 Principles of solar still operation

Solar stills consist of an airtight space in which saltwater evaporation and condensation are performed simultaneously. Solar energy penetrates the airtight space through a tilted transparent cover and is partially absorbed by saltwater in a basin below the space. Thus the water is heated to a temperature higher than that of the transparent cover, but lower than the water's boiling point. Accordingly, the air-vapor mixture at the surface of the saltwater has a higher temperature and lower density than the air-vapor mixture immediately beneath the cover. Convection currents are formed between the cover and the water surface. Gliding over the saltwater surface, the air-vapor mixture is saturated with water, moves upwards, where it is cooled by contact with the cover, becomes saturated and its vapors partially condensed and then it moves back to the water surface again. The distillate is collected in gutters along the lower sides of the transparent cover (Fig. 2–34).

This process takes place in narrow layers between the two surfaces. The bulk of the air mass does not participate, because diffusion and heat conductance are low. Therefore, it is advantageous to keep the distance between the surface and the cover as well as the angle between the cover and the horizontal as small as possible. The larger the temperature difference between water surface and cover, the more intense becomes the circulation. Heat losses occur at several places: Direct and diffuse radiation is partly reflected from the outside and inside surfaces of the cover and from the water surface and bottom. Also, some radiation is absorbed by the cover. The remaining radiation is absorbed mainly by the saltwater and partly by the lining. Of this latter, most is transferred to the overlying water and only a small part is lost by conduction to the ground underlying the basin. In addition to the air-vapor conversion currents, heat exchange also takes place by radiation between the warm surface of the saltwater and the condensate film on the inside surface of the cover. The corresponding heat absorbed by this film is transmitted to the cover material and lost to the outside air by convection and radiation. Most of these heat losses are unavoidable. Furthermore, the heat lost by the cover to the outside air is necessary to keep the cover cool and thus maintain operation. Some sensible heat may also be supplied to the still in the feedwater and some sensible heat lost in the condensate and brine blowdown.

A typical distribution of energy in a solar still, based on an average daily solar radiation of 5400 kcal/m² (about 2000 Btu/ft²) and on the results of several investigators, is as follows (in %):

Evaporation	40	Internal convection	5
Radiation reflected by cover	10	Edge losses	5
absorbed by cover	5	Ground losses	5
reflected by liner	5	Vapor leakage	5
from brine	15	Unaccounted	5

Mass flow balances for solar stills involve the quantity of feedwater introduced into the still and the amounts of water leaving the still in form of distillate collected, blowdown, brine leakage, distillate leakage, and water vapor loss.

Fig. 2–34

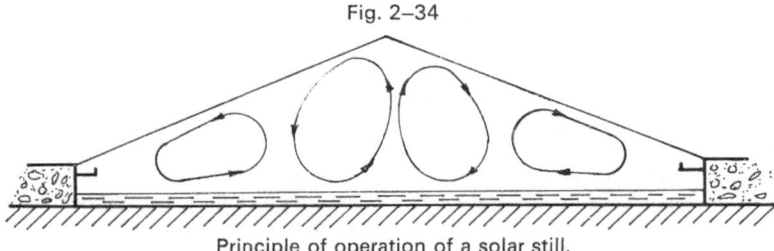

Principle of operation of a solar still.

2.17.4 Heat and mass balances

Wärme- und Stoffbilanz

The overall efficiency of a solar desalting unit is often determined by the ratio of product condensate to the theoretical amount of salt water vaporized, i. e. the ratio of the heat of evaporation of the condensate to the solar heat reaching apparatus. In terms of daily performance, the efficiency η can be expressed in either metric or anglosaxon units, as:

$$\eta = \frac{D \times 594 \times 100}{I} = \frac{D_1 \times 8913 \times 100}{I_1} \quad \text{in } \%$$

where D is the distillate production in kg/m² · day (D_1 the same in gal/ft² · day), 594 is the number of kcal required to preheat and evaporate 1 kg of distillate (8913 the Btu's required for 1 gal of distillate), and I is the amount of incident radiation on horizontal surface in kcal/m² · day (I_1 in Btu/ft² · day).

The operating efficiency of a solar still increases exponentially with increasing brine temperature, but heat losses increase also in the same way.

The productivity of a solar still expresses the amount of distillate produced per unit area of the water surface. It is a function of the brine-surface temperature and the Δt between the brine surface and the cover. Baum [8] developed the following equation relating hourly productivity to efficiency and solar radiation:

$$P = \frac{\eta \times I}{H_v - H_F} = \frac{\eta \times I}{600} \quad \text{in kg/m}^2 \cdot \text{h}$$

where η is the efficiency of the still in % and I the solar radiation on horizontal surface in kcal/m² · h. H_v is the enthalpy of the vapor in the distiller and H_F the enthalpy of the feedwater in the still, both in kcal/kg. The vapor-air mixture in the still has usually a temperature of 50 to 80°C and H_v = 618 to 630 kcal/kg and if the temperature of the feedwater is taken equal to 20 to 30°C, then $H_v - H_F$ might be taken in round figures as 600, which as an average corresponds to the previous given value of 594.

Several authors have developed formulae and relations for the computation of energy balances in a solar still, including Löf [9], Baum [10], and Morse [11]. These are partly very detailed equations, which refer individually to all kinds of losses. Most of these losses are quite small and their effect on the still performance is also small. A general equation describing the energy balance in a solar still is:

$$I (1 - \Sigma_r) = Q + D \cdot t_c + \Sigma_v$$

where D is the distillate outflow rate, t_c the cover temperature, Σ_r are the heat losses by reflection, Σ_v all other heat losses and Q is the total heat involved in the process, i. e., the sum of heat transfer by convection and radiation from the brine surface to the cover, the heat of evaporation and condensation of water, and the heat absorbed by the transparent cover.

These equations are valid under stationary conditions. When a continuous feed is used, the term $D \cdot t_c$ must be substituted by:

$$D(t_c - t_F) + B(t_B - t_F)$$

where B is the brine outflow rate and the subscripts $_F$ and $_B$ refer to the temperatures of saltwater supply and brine in the basin, respectively.

Morse and Read [11] have developed a nomogram, relating heat transfer by evaporation to the temperature of the cover as a function of several variables, such as brine temperature, wind speed and internal heat transfer between brine and cover by radiation and convection. This chart was made with data taken from the Australian solar stills and may be used to predict the productivity of a solar still similar to this design. The daily output can be obtained by using the graphical technique in a stepwise procedure by hourly intervals.

The mass balance is affected by the amount of liquid and vapor leakages and can be expressed in terms of a degree of tightness: Degree of tightness = (Amount of distillate + Blowdown)/Amount of feedwater. The productivity of the still may be seriously affected by a reduced degree of tightness.

A digital simulation method of analyzing solar still processes was presented by Cooper [12] and used to process a mathematical model describing the system. In a more thorough investigation of the proposed mathematical model, results from an experimental program were used to establish quantitative agreement between simulated and experimental operation. The experimental and simulated base heat flow characteristics indicated the importance of reducing the instantaneous base heat flow during periods of high still activity. Base heat flows were found to occur as expected largely at the physical boundaries of the still [13]. The simultaneous energy transfer modes within a solar still envelope have been further examined from a theoretical and experimental point of view over a wide range of operating conditions. Energy transfer rates up to three times that which can occur under normal conditions have been investigated. A relationship for evaporative mass transfer is deducted from dimensionless correlation of the experimental data together with that from another source. The distillation rate within a solar still envelope is shown to be primarily dependent on the energy input rate at steady conditions with negligible losses [14]. A theoretical evaluation of the heat and mass transfer interchange in an air inflated solar still has been studied by Lawand [15]. Experimental verification tests have been carried out and the results compared with theoretical predictions. Unaccounted heat losses on the overall balance were under 3%. A heat transfer model containing two parameters derived from experimental data has been used by Trayford to compare the calculated performance with the measured output of a solar still. Hourly calculations were made for extended periods [16]. A rigorous mathematical analysis of the solar distillation process was presented by Glueck. Attention is turned toward improving the relatively low (ca. 30%) efficiency of current units. The most significant increase comes from the realization that the condensing surface should be cooled, resulting in an efficiency of about 50% or more. Techniques for effecting the above and other improvements are discussed [17].

An analytical model was developed by Porteous, which portrays the heat and mass transfer process within the still. The influence of various still parameters on output is investigated and the results compared with data from several solar still installations. An economic analysis is performed, which places the fresh water costs from this method in perspective with those of conventional distillation process. Installations with capacities up to 38 m³/day can compete with other more complicated desalination processes [18]. Solar still outputs can be predicted to an accuracy of 6% from experimental data obtained in an electrically heated model. Daily outputs of fresh water of 2.0 to 29.76 kg/m² are compared to calculated values of 1.4 to 31.5 kg/m² obtained at average daily solar heating rates of 2520 to 44582 kcal/m² · day [20]. The performance of a small gap solar still was analyzed by Headley in terms of the convective heat transfer between the still and its surroundings and between the lower and upper parts of the still. Under steady state conditions, the rate of heat transfer from the lower to the upper part is equal to the difference in convective heat losses between the upper and lower parts [19].

Einfluß der
Entwurfs-
parameter
2.17.5 Effect of design parameters

Various design factors can affect the productivity of a solar still, but only few are important. Whatever the design of a solar still might be, it must be completely water- and airtight. Distillate leakage is a net loss. Brine leakage increases the thermal conductance of the soil beneath the still and thus increases heat losses to the soil. Vapor leakage, which might be increased by wind, causes loss of both heat and evaporated water. An important innovation in the Australian designs has been the use of silicon rubber sealant throughout. This has proved to be an excellent, if not the best, sealing material. Loss of condensate by dripping from the cover back into the basin is equivalent to distillate loss. The condensate will flow smoothly along the under surface of a glass cover, as long as a slope of about 10° is provided. In some designs even a lower slope is acceptable. Since many plastics are not wetted by water, drop condensation occurs on these unless they are treated to become hydrophilic.

High radiation absorption is aimed for by all designers, so as to reach high water temperatures and consequent high rates of distillate output. A shallow layer of brine in a basin-type still, because of the lower thermal capacity, reaches higher temperatures during the hours of peak sunshine and the output is accordingly higher at noon and the early afternoon hours. For the same reason the brine cooles up early and the nocturnal output is minimized. On the other hand, in a deep basin still the maximum temperature is lower during the noon hours, as well as the productivity, but the higher thermal capacity prolonges the operation for several hours after sunset and differences in productivity might become relatively small.

The bottom and eventually the sides of solar stills can be insulated to increase productivity by reducing heat losses. The advantage of insulating the bottom of a basin still is greater at shallower brine depths. A layer of dry earthy material beneath the basin liner is usually sufficient as insulating material.

The most common solar still cover material is glass, usually single strength (2.5 mm). If the inside glass surface is clean, the condensate will form a thin film, which will not reduce the transmission of sunlight. Glass is essentially opaque to the long wavelength radiation emitted by the brine. In Australia, cheap horticultural grade glass is used as cover material. This is available locally at unusually low cost, but in limited lengths. Because of this, the Australian still can hardly be wider than 90 cm (**Fig. 2–35**e, p. 142).

Some plastic covered stills have been built using a 0.1 mm thick polyvinyl fluoride film (Tedlar). The resistance to weather conditions was proved to be poor and most of these stills, if not all, were abandoned. In the early designs the plastic cover was supported by a small inside overpressure, which kept it inflated (Fig. 2–35b). Vapor leakage was magnified due to the inside higher pressure. V-shape (Fig. 2–35c) and stretched cover designs (Fig. 2–35d), which followed, indicated even shorter life-times. In the Greek installations, the V-covers were damaged by wind and rainfall and the stretched covers developed cracks after short time. Platic films must be treated for wettability to prevent dropwise condensation and distillate dripping back into the brine. Tedlar was made wettable by scratching the inside film surface, but this treatment increases the cost considerably. Nearly all other plastic films, tested under less severe conditions than in practical operation, were not found to exhibit a sufficient lifetime to be used in commercial large scale solar stills.

Whether of glass or plastic, the cover material affects the geometry of the still. With glass, the most common shapes of the cover are: symmetrical double-sloped cover (Fig. 2–35a), non-symmetrical double sloped cover (Fig. 2–35f), and single sloped cover (Fig. 2–35d). The slope of the glass cover varies with the design. The usual slope is about 10° to the horizontal. In the Australian still the slope was increased to 18° for improved structural stability, because no supports other than the glass itself are used. In some designs a slope of down up to 5° performs well, but this appears to be practicable only when the still width and consequently the length of the film under the cover are not too large. Otherwise dripping may occur. Glass covered stills with non symmetrical covers or single sloped covers should be oriented with their long axes East-West and the low sloped cover or single cover facing the equator. The orientation of stills with symmetrical covers does not influence the productivity as much, but it appears that orientation of the axis North-South is more advantageous.

Asphalt mats about 6.5 mm thick, joined by asphalt-sealed overlapping strips of the same material have given trouble free service in Florida and Spain. Asphaltic concrete and Portland cement concrete are not recommended, because they are not sufficiently water tight. Earthquakes have caused cracks in concrete-stills in Chile. Black polyethylene has been used in Australia, but later replaced by butyl rubber sheeting. Butyl rubber sheeting with a thickness of about 0.8 mm has been used in all stills designed by Delyannis and Piperoglou [21] or built according to their design. It is the most successful and trouble-free material commercially available. Vulcanized or adhesive-sealed, leak-proof joints are usually prefabricated to give the exact size required for a still unit. They can also be made on site. The material is not affected by solar radiation and withstands temperature associated with dry spots or the edges above the brine surface. The use of black fabric-type materials or Orlon mat as black bodies to increase radiation absorption and evaporation rates involves additional investment cost, which is not justified by the expected increase in productivity. Calcium sulfate crystals form on the Orlon mat and no significant gain in productivity was found [22].

Fig. 2–35 a to g

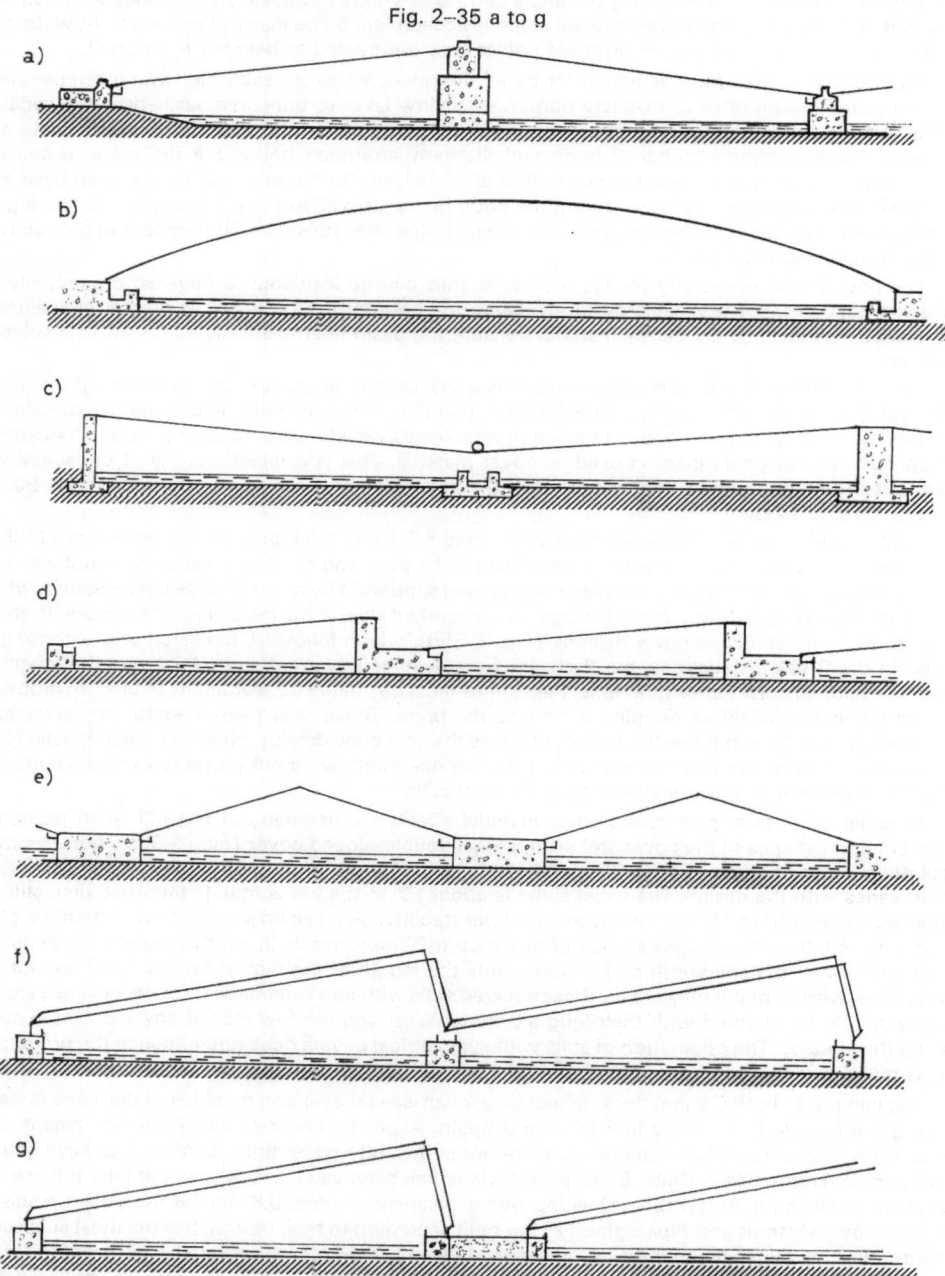

a) Battelle-Löf design. b) Inflated plastic cover design. c) V-shape plastic cover design. d) Streched plastic or inclined glass cover design. e) C.S.I.R.O. Australian design. f) Delyannis design. g) Delyannis modified design, unpublished.

2.17.6 Effect of atmospheric variables

The productivity of a solar still is usually plotted as a function of the amount of total solar radiation received on a horizontal surface. Comparisons of solar still productivities can be made only on the basis of a definite solar radiation value. Even then, accurate comparisons of performance in places with various weather conditions and angles of incident radiation are doubtful. An average output of 5 kg/m² · day (0.125 gal/ft² · day) in summer with an efficiency of 40 to 60% to 1 kg/m² · day (0.025 gal/ft² · day) with an efficiency of 30 to 50% in winter is given here only as a general indication. The actual yield might vary from 30% more than this during the hot summer months to about 30% less in cold winter months. Cooper has derived equations to compute the mean effective absorbance for any given combination of variables, such as day of the year, latitude, cover slope, orientation, percentage diffuse radiation, and insolation intermittency [23]. A method of quantitatively assessing the effect of reflecting layers of salt on the water surface and basin liner, taking into account the band absorption characteristics of water was also presented. A revised curve of the absorption of the solar spectrum in water was given [24].

Increased wind speed will remove more heat from the cover by convection. This will lower the cover temperature and increase the Δt between brine and cover. At higher brine temperatures an increase in productivity due to increased wind speed is to be expected. At low brine temperature increasing the wind speed might diminish the rate of evaporation, probably because of higher radiation from the brine to the cover [25]. When poorly sealed joints exist in the still cover, increased wind velocity will increase the vapor leakage considerably and thereby reduce the productivity.

There are several publications, in which an increased productivity is reported as the ambient temperature increases. This correlation does not appear to be relevant, as higher ambient temperature means higher incident radiation.

A solar still is not effective during heavily overcast or rainy days. However, the covers of basin-type solar stills are excellent catchment areas for rainfall collection with small additional investment. Collection of rain becomes interesting as an additional water supply in areas with sufficiently high rainfall. The importance of combining solar distillation with rain collection is discussed in section 2.17.11, p. 147.

2.17.7 Operating techniques

Various techniques have been used to preheat the saline water before supplying it to the still basin. The effect of preheating with external heat supply is small, as for mean still temperature of 55°C about 565 kcal/kg are required for evaporation and only 27 kcal/kg for preheating, which is less than 5% of the heat input. It is, therefore, cheaper to increase the length of the still by 5% than to provide additional heat exchangers. Internal condensers at the back side of the cover to combine preheating with additional condensation of vapors were also used [21]. Even then the increased investment cost for the condenser appeared not to be justified, although maximum performance of the still was reached. The repeatedly gained experience in operating solar stills suggests that the simplest design gives always the best still.

The production capacity of a solar still can be increased by extending its operation during the night. A waste heat source, such as cooling water from a power plant, can be used to preheat the feed. The nocturnal production seems to be influenced by several parameters. However, a simplified mathematical model suggests that the distillate depends only on the initial brine temperature, the drop in brine temperature and the brine depth. This was experimentally verified for different brine depths and for initial brine temperatures up to about 66°C (150°F) by Malik and Vi Tran [26]. In combining a solar still with a Diesel engine, the still replaces the cooling tower, ponds or radiators normally used to control the engine temperature. The advantages of using such a system compared with a conventional solar still are: the area occupied for the same output is much less, seasonal variation of the production is reduced, the efficiency of the still is considerably improved due to higher operating temperatures and the cost of water is reduced [27].

The salt content of the brine must be kept under control to avoid the build-up of a concentration to a level, where deposits would occur. This is usually the case when the concentration ratio exceeds

2. Two methods to replenish solar stills are used. The usual technique is the batch-type operation. Seawater feed is introduced at one end of the still and displaces the brine, which is overflowing a weir at the other end of the still. Flushing is performed either early in the morning or late at night, when the temperature of the brine is lowest. This avoids unnecessary heat losses. About twice as much seawater is required as was distilled during the period of operation. However, in shallow basin stills some stagnant areas may remain unchanged with local build-up of concentration and formation of deposits. An improved technique is to completely drain the stills from time to time and then to refill them with fresh seawater [28]. This procedure avoids also the build-up of algae. The other method of feedwater control uses a continuous supply of salt water and was mainly developed in Australia with the aim to avoid continual attention and to save labor. The brine is continuously overflowing at the other end of the still. As the feed rate is higher than the distillate output, some heat losses are unavoidable and the continuous operation involves a slight penalty in thermal efficiency. The formation of algae is also not avoided with the continuous operation technique.

Algae and bacteria growth may become a problem, particularly if contaminated feedwater is used. Sedimentation on the basin liner and/or growth of algae increase the solar reflection and reduce productivity. Filtering and eventually chemical treatment of the feedwater might be required in some cases. The described technique of completely draining the basin before refilling avoids the formation of algae [27], but does not eliminate sedimentation of suspended matter.

Periodic cleaning of the cover might be necessary if dust or other matter is deposited on its upper surface. This is especially true in the case of plastic films, which have an electrostatic property to attract dust. To further add to this disadvantage, distillate must be used to clean the covers because salt deposits would remain if seawater were used [22].

Decreases in productivity, reported by some investigators, are probably due to a gradual build-up of fine scale on the liner and also to progressive deterioration of sealing materials, which caused increased vapor leakage.

The concept of thermal inertia was introduced by Hirschmann and is defined as the time difference between the occurrence of distillate maximum output and maximum solar radiation [29]. Experimental results indicate a close relationship between heat capacity and thermal inertia on one hand and plant performance on the other. The analysis of the above relationships of a still simulation by computer shows what factors could contribute to an increase in production. There also exists an annual seasonal thermal inertia, which is defined as the time difference between the occurrence of the maximum average daily temperature and the maximum incident solar radiation. It exhibits strong displacements with geographical position. Computational analysis indicates a pronounced influence on the maximum distillate production of solar plants. Factors which were considered important for solar stills with low thermal inertia include: the significance of the thermal inertia of the air space between the water and the roof, the need for separate consideration of the water evaporation and condensation rates, the resulting time delay between their maxima during the daily cycle and the roof geometry and orientation in relation to the sun's altitude [30].

It should be borne in mind that continuous operation throughout the year is necessary for solar stills, independently if the product water is required or not. Abandonment of the plant during the winter period is not recommended.

Bauweisen ### 2.17.8 Principal basin type solar still designs

Various configurations of solar stills have been the object of intensive research and development, but practical considerations exclude all except the so-called greenhouse type. Moreover, the operation of any solar still of the greenhouse type is based on the same theoretical considerations, so that the actual designs differ only in geometry, in the means of cover support, in the art of feeding the saltwater and discharging the brine and in the materials of construction, principally the cover.

Table 2/10 shows the most important solar distillation plants in the world, as well as some of the principal plant characteristics, especially the cover material [34]. Figure number references in the table indicate similar geometry to the illustrations of Fig. 2–35, p. 142.

Table 2/10

The most important solar distillation plants.

Country	Location	Design Fig. 2–35	Year	Evapora- tion area in m²	Feed water	Cover material
Australia	Muresk I *)	e)	1963	372	Brackish	Glass
	Muresk II	e)	1966	372	Brackish	Glass
	Coober Pedy	e)	1966	3160	Brackish	Glass
	Caiguna	e)	1966	372	Brackish	Glass
	Hamelin Pool	e)	1966	557	Brackish	Glass
	Griffith	e)	1967	413	Brackish	Glass
Cape Verde Isl.	Santa Maria *)	c)	1965	743	Seawater	Plastic
Chile	Las Salinas *)	e)	1872	4460	Brackish	Glass
	Quillagua I *)	e)	1967	106	Brackish	Glass
	Quillagua II	e)	1968	104	Brackish	Glass
Greece	Symi I *)	b)	1964	2686	Seawater	Plastic
	Symi II *)	d)	1968	2600	Seawater	Str. plastic
	Aegina I *)	c)	1965	1490	Seawater	Plastic
	Aegina II *)	d)	1968	1486	Seawater	Str. plastic
	Salamis *)	c)	1965	388	Seawater	Plastic
	Patmos	f)	1967	8600	Seawater	Glass
	Kimolos	f)	1968	2508	Seawater	Glass
	Nisyros	f)	1969	2005	Seawater	Glass
	Fiskardo	f)	1971	2200	Seawater	Glass
	Kionion	f)	1971	2400	Seawater	Glass
	Megisti	f)	1973	2528	Seawater	Glass
India	Bhavnagar	e)	1965	377	Seawater	Glass
Indian Ocean	Aldabra	e)	1969	182	Seawater	Glass
Mexico	Natividad Isl.	d)	1969	95	Seawater	Glass
Pakistan	Gwadar I	f)	1969	306	Seawater	Glass
	Gwadar II	g)	1972	9072	Seawater	Glass
Spain	Las Marinas	a)	1966	868	Seawater	Glass
Tunisia	Chakmou	d)	1967	440	Brackish	Glass
	Mahdia	d)	1968	1300	Brackish	Glass
U.S.A.	Daytona Beach *)	a)	1959	228	Seawater	Glass
	Daytona Beach *)	a)	1961	246	Seawater	Glass
	Daytona Beach *)	b)	1961	216	Seawater	Plastic
	Daytona Beach *)	b)	1963	148	Seawater	Plastic
U.S.S.R.	Bakharden	e)	1969	600	Brackish	Glass
	Shafrikan	e)	1970	600	Brackish	Glass
West Indies	Petit St. Vincent	b)	1967	2100	Seawater	Plastic
	Haiti	d)	1969	223	Seawater	Glass

*) No more in operation

A large basin under glass panels (Fig. 2–35a) has been used in a design by Battelle Memorial Institute, Columbus, Ohio, for units built at the solar experimental station at Daytona Beach, Fla., [31], and with some modifications at Las Marinas, Spain, [32]. The dimensions of the single basin in this design are determined by the projected output. Asphalt matting was used as the basin lining material. The glass roof is peaked, with the sloping glass panels supported at their upper and lower sides on prefabricated concrete beams. Distillate is collected in gutters running along the lower edge of each panel. Saltwater is fed from time to time to give a depth of 10 to 30 cm (4 to 12 in.) and the concentrated brine is evacuated by overflow.

The Australian design provides for individual bays with a symmetrical glass roof [33]. The width of the basins is limited to about 90 cm, because of the available size of agricultural glass. In the latest design, two glass panels are supported along the ground on one side by concrete beams and attached by means of silicon rubber at the peak (Fig. 2–35e). A slope of 18° is necessary for structural reasons. The basin in this design has a slope of about 1:40 and the brine is subdivided by weirs into a series of cascades, so that saltwater is continuously fed and evacuated. The goal is unattended operation for long periods. Similar designs employing symmetrical glass roofs with varying geometry have been erected in Chile, India, and the U.S.S.R.

A design from the University cf California at Berkeley is similar to the Australian design in roof geometry, but the peaked glass roofs are oriented across the width of the basin and troughs below the roof edges run across the still to collection gutters along its length. Several family-size and larger units of this design have been installed in the Pacific islands [35].

Six large glass-covered stills, which have been erected by the Greek government on islands of the Aegean sea, are based on design (Fig. 2–35f) developed by A. Delyannis, E. Delyannis at the Technical University of Athens [34]. Non-symmetrical glass roofs with large low-sloping panes on one side and short, steep panes on the other, supported by aluminum alloy structures, ensure increased radiation absorption and longer plant life. The use of a supporting structure for the roof allows for larger glass panes to be used without sagging and permits the increase of the width of the basins. Several basins, each about 3 m wide by 20 to 25 m long, form a still. Butyl rubber sheeting is used as basin lining material.

One of these, the plant at Patmos, was the largest solar distillation unit in the world. An improved Patmos design was used in the solar still at Gwadar (Fig. 2–35g), A. Delyannis, E. Delyannis (unpublished). Roofs of similar geometry have also been adopted in Tunisia and Haiti, but those plants use glass in only the large panel, with brick or concrete forming the other side (Fig. 2–35d).

The first large scale solar still erected by Church World Service on the island of Symi, Greece, had an inflated Tedlar cover. Since this design (Fig. 2–35b) exhibited vapor leakage, due to the internal pressure, and especially poor weather resistance, a second still in Aegina was modified by laying steel pipes along the middle of the plastic spans stretching the plastic cover into an inverted peaked roof (Fig. 2–35c). This avoided the need for internal pressure and only one distillate gutter was necessary to collect condensate dripping from below the pipe. However, this design has also proved to have unsatisfactory resistance to weather. The stretched plastic cover design (Fig. 2–35d) has replaced both stills, in Symi and Aegina, but this design too could not withstand the required durability. All the plastic stills in Greece have been abandoned and dismantled.

Similar experience was gained at the University of California, Berkeley. Two small identical solar stills, one covered with glass and the other with Tedlar plastic film, were tested over a period of two years, as well as three more with Tedlar and one with glass of different design. Production of Tedlar covered stills was consistently less than glass covered [36].

A detailed description of all erected solar stills up to the time of its publication is contained in a Manual on Solar distillation of saline water prepared by the Battelle Memorial Institute [36]. This report was also published by the Office of Saline Water [37] and includes a review of factors that influence the technology and data on productivity of solar stills. A report was also published by the United Nations on solar distillation as a means of meeting small-scale water demands, defining the conditions under which solar distillation may provide an economic solution to the problems of fresh water shortage in small communities [38]. The U.S. National Academy of Sciences has also published a report of an ad hoc Advisory Panel charged to assess the state of the art in utilizing solar energy for developing countries and review current practical applications, to identify promising areas for research and development and to examine the desirability of establishing an international solar energy institute in North Africa [39].

Anlagen mit kleiner Leistung

2.17.9 Small size solar stills

There is a large variety in designs and sizes of small solar stills reported, either experimental or for residential use. Experimental stills aim at design improvements to increase performance, but very often the improved design loses one of the main solar still merits, which is simplicity. Small size solar stills are usually prefabricated in a design similar to the basin type still and vary in capacity to satisfy the needs of a wide range of consumers, from a small family to a resort hotel.

Family-size solar stills are described in a report of the Agency for International Development [40] and quite extensively in the Battelle Manual [37]. Therefore, only a few recent publications are briefly mentioned here. In a small-gap still, developed at the University of the West Indies, the brine is flowing in cascades on a corrugated aluminum bottom [41]. A solar still which produces 5 to 7 liters per day was developed in India [42] and a study for the development and application of prefabricated solar stills is reported from Canada [43]. A commercial firm in the United States has supplied approximately 150 solar stills for residential use in recent years [44]. Solar stills of inclined evaporating cloth were reported from Chile [45].

2.17.10 Other solar distillation techniques

Weitere Verfahren

Several designs of tilted or inclined solar stills have been proposed. Tilted evaporator surfaces intercept more energy per unit area and the covers reflect less radiation, because of a more direct angle of incidence. On the other hand, porous wicks or cascaded shallow steps provide stills with low thermal capacity and higher brine temperatures are achieved.

Several investigators have developed designs of improved tilted stills, but the advantage of claimed higher performance rates was offset by various disadvantages of operation, reduced durability to atmospheric conditions and higher investment cost. The most known are the tilted stills developed by Telkes [46] and the suspended envelope plastic solar stills developed by Bjorksten Laboratories [47]. Producing permanently hydrophylic as well as selectively infrared reflecting surfaces on plastic films for solar stills was also investigated [48].

A solar evaporator was constructed in which warm saline water flowed downward in the capillaries of hydrophilic gauze sheets. Countercurrent to the saline water flow, warm air flowed through spaces between the gauze sheets, thereby saturating it with water vapor in the aim to obtain increased distillate yields [54].

Another approach to increase productivity was that of blowing air through a solar still and condense the vapors in an external water-cooled condenser [49]. The increased investment and operating costs do more than offset the gain in yield. The same might be said for attempts to use the multiple-effect principle in the solar evaporation process.

The multiple-effect principle was also tested at the University of Arizona, Tucson, in pilot plant scale and at Puerto Peñasco, Mexico, in a larger pilot plant. The process was divided in three separate components: solar radiation collectors, brine evaporation, and condensation of vapors. The process is not a conventional solar distillation process, as the radiation is only used to preheat the brine. The evaporation and condensation steps are performed separately in packed humidification and dehumidification towers [50]. The separation of the process in the three basic processes of energy collection, evaporation, and condensation permits the recovery of the heat of condensation for return to the system. Evaporation of hot water from the energy collector occurs in a packed column by direct contact with a closed air cycle. The brine is further cooled and passed through a condenser, where it recovers the heat of condensation and returns it to the collector. Estimates indicate that eight effects are readily attainable [51]. Here again the increased investment cost and the pumping power required render questionable the economics of the process. A mathematical analysis was given by Barba [52] of a vertical multi-stage humidification and dehumidification evaporator process, operating at a temperature range of 30 to 80°C.

The heat pipe effect was also proposed for desalination of sea or brackish water by simultaneous use of solar energy. The method uses solar heated heat pipes and the process may be broken down in three groups. First is the energy absorption realized by a glass covered solar absorber. Then the energy transport section, applying the heat pipe effect of heat transportation at minimum temperature drop. Third the real desalination section, composed by an evaporation zone and condensing zone. A 20 m² pilot plant is under test [53].

2.17.11 Rainwater collection

Sammeln von Regen- wasser

Solar distillation plants are excellent rain catchment areas. Only a small additional investment is required to collect the rain falling on the cover. As rainfall coincides with the period of lowest distillate output, joint rain catchment and solar distillation is most profitable. The advantages of joint operation will be discussed with an illustration based on data collected at the Solar Experiment Station in Symi, operated by the Technical University of Athens [28].

Table 2/11 gives the measured distillate collection per day for the time from July 1965 to June 1967, as well as the amount of rain collected at the Symi prototype stills. The measurements refer to one year with lower and one year with higher rainfall. If these data are extrapolated to a solar still with 1000 m³ water evaporating surface, as shown in the lower part of Table 2/11, it is found that solar distillation alone would supply a community at a years average with 80 or 72 m³/month. An additional amount of about 55 or 115 m³/month could be obtained by rain catchment bringing the total to 135 or 187 m³/month. It is assumed in these calculations, that water demand is constant over the year's period, which is not the case in practice. Usually there is a larger water demand during the summer months. Taking in consideration the actual water demand pattern, sufficient storage capacity must be provided to store the excess water during the time of high productivity, so that the excess water might be used at the time of lower water availability. If it is assumed that during the peak summer season the water demand is by 50% higher than in winter, a minimum storage capacity, as shown in Table 2/11, will be required. If the community is supplied with rainwater catchment alone, the required minimum storage capacity (in m³ per m³ of annual output) is more than twice as large as required for solar distillation. In combining the two processes, a lower storage capacity as required for rain collection, will be sufficient for the supply of a larger amount of water.

Table 2/11

Operating data of the Symi prototype stills in kg/m² · day.

	Distillate		Rainfall		Total	
	1965–66	1966–67	1965–66	1966–67	1965–66	1966–67
July	5.87	4.30	—	0.20	5.87	4.50
August	4.60	4.12	—	—	4.60	4.12
September	3.34	3.05	—	1.60	3.34	4.65
October	2.01	1.65	0.13	1.45	2.14	3.10
November	0.98	0.78	0.46	3.97	1.44	4.75
December	0.54	0.54	4.80	20.46	5.34	21.00
January	0.51	0.54	11.21	8.96	11.72	9.50
February	1.20	1.10	2.91	1.70	4.11	2.80
March	1.65	1 95	1.28	2.80	2.93	4.75
April	2.63	2.80	0.30	3.10	2.93	5.90
May	3.85	3.15	0.10	0.55	3.95	3.70
June	4.27	4.30	0.14	0.15	4.41	4.45
Year	2.63	2.36	1.79	3.78	4.42	6.14
Total m³/year	962	862	655	1380	1617	2242
Average m³/month	80	72	55	115	135	187
Minimum storage m³	220	190	450	735	340	640

Wirtschaft-
liches

2.17.12 Economics

There are no sufficient data available, which might lead to comparable figures in estimating the investment cost of the various types of solar stills. For the time being, it can only be said that erection costs of large solar desalting plants fall between $ 10 to 15 per m² of evaporating area.

Since there are no substantial differences in unit investment costs between large and small plants, solar desalting is generally economical in small capacities compared with other distillation processes. And since the operating cost is of secondary importance, the most important factor to the cost of product water is the amortization rate. Although there are no records to estimate the life of solar desalting plants, the plant in Las Salinas, Chile, is said to have operated continuously for about 40 years, providing drinking water in a mine. The stability of the materials of construction (concrete, glass, asphalt, butyl rubber sheeting, aluminum alloys, sealants, etc.) is well known from other uses and there is no reason to believe that these materials will not perform equally well in solar distillation plants.

Literature to 2.17

[1] A. Delyannis, E. Delyannis (Chem. Eng. **77** No. 23 [1970] 136/40). — [2] J. Harding (Proc. Inst. Civil Engrs. [London] **73** [1883] 284/8). — [3] M. Telkes (U.S. Off. Tech. Serv. Rept. No. 5225 [1945]). — [4] G. O. G. Löf (Off. Saline Water Res. Develop. Progr. Rept. No. 5 [1955]). — [5] J. W.

Bloemer, R. A. Collins, J. A. Eibling (Off. Saline Water Res. Develop. Progr. Rept. No. 50 [1961]); Battelle Memorial Institute (Off. Saline Water Res. Develop. Progr. Rept. No. 147 [1965], Rept. No. 190 [1966]).

[6] Association for Applied Solar Energy (Trans. Conf. Use Sol. Energy, Tucson, Ariz., 1955 [1958], Vol. 1/5). — [7] Association for Applied Solar Energy (Proc. World Symp. Appl. Sol. Energy, Phoenix, Ariz., 1955 [1956]). — [8] V. A. Baum (Proc. U.N. Conf. New Sources Energy, Rome 1961 [1964], Vol. 6, p. 178/88). — [9] G. O. G. Löf, J. A. Eibling, J. W. Bloemer (AIChE [Am. Inst. Chem. Engrs.] J. 7 [1961] 641/9). — [10] V. A. Baum, R. Bairamov (Sol. Energy 8 No. 3 [1964] 78/82).

[11] R. N. Morse, W. R. W. Read (Sol. Energy 12 No. 1 [1968] 5/17). — [12] P. I. Cooper (Sol. Energy 12 No. 3 [1969] 313/31). — [13] P. I. Cooper (Sol. Energy 14 No. 4 [1973] 451/68). — [14] P. I. Cooper (Intern. Congr. Sun Service Mankind, Paris 1973, Paper E49). — [15] T. A. Lawand (J. Eng. Power 92 [1970] 95/102).

[16] R. S. Trayford, L. W. Welch (Intern. Sol. Energy Soc. Conf., Melbourne 1970, Paper No. 5/67). — [17] A. R. Glueck (Proc. 3rd Intern. Symp. Fresh Water Sea, Dubrovnik 1970, Vol. 1, p. 633/54). — [18] A. Porteous (Chem. Engineer No. 255 [1972] 406/11; Proc. Roy. Soc. Edinburgh B 73 [1972] 133/4). — [19] O. Headley, M. Sweeny (Proc. 4th Intern. Symp. Fresh Water Sea, Heidelberg 1973, Vol. 4, p. 493/8). — [20] K. Toilyev (Izv. Akad. Nauk Turkm.SSR Ser. Fiz. Tekhn. Khim. i Geol. Nauk 1971 No. 2, p. 109/11; C.A. 75 [1971] No. 40191).

[21] A. Delyannis, E. Piperoglou (Proc. 1st Intern. Symp. Water Desalination, Washington, D.C., 1965 [1967], Vol. 2, p. 627/40). — [22] A. Delyannis, E. Piperoglou (Sun Work 12 No. 1 [1967] 14/8). — [23] P. I. Cooper (Sol. Energy 12 No. 3 [1969] 333/46). — [24] P. I. Cooper (Sol. Energy 13 No. 4 [1972] 373/81). — [25] S. H. Soliman (Sol. Energy 13 No. 4 [1972] 403/15).

[26] M. A. S. Malik, Van Vi Tran (Sol. Energy 14 No. 4 [1973] 371/85). — [27] D. Proctor (Sol. Energy 14 No. 3 [1973] 433/49). — [28] A. Delyannis, E. Delyannis (Chem. Ingr.-Tech. 41 [1969] 90/5). — [29] J. Hirschmann, S. Roffler (Proc. 3rd Intern. Symp. Fresh Water Sea, Dubrovnik 1970, Vol. 1, p. 679/89). — [30] W. Szulmayer (Sol. Energy 14 No. 4 [1973] 415/21).

[31] J. W. Bloemer, J. R. Irwin, J. A. Eibling (Off. Saline Water Res. Develop. Progr. Rept. No. 112 [1964]). — [32] P. Blanco, C. Gomella, J. A. Barasoain (Proc. 1st Intern. Symp. Water Desalination, Washington, D.C., 1965 [1967], Vol. 2, p. 817/26); G. O. G. Löf (Off. Saline Water Res. Develop. Progr. Rept. No. 397 [1968]). — [33] R. N. Morse, W. R. W. Read, R. S. Trayford (Intern. Sol. Energy Soc. Conf., Melbourne 1970, Paper No. 5/34). — [34] A. Delyannis, E. Delyannis (Proc. 4th Intern. Symp. Fresh Water Sea, Heidelberg 1973, Vol. 4, p. 487/91). — [35] E. D. Howe, B. W. Tleimat (Desalination 2 [1967] 109/15).

[36] D. W. Tleimat, E. D. Howe (Sol. Energy 12 No. 3 [1969] 293/304). — [37] S. G. Talbert, J. A. Eibling, G. O. G. Löf (Manual on Solar Distillation of Saline Water, Battelle Memorial Institute, Columbus, Ohio, 1970; Off. Saline Water Res. Develop. Progr. Rept. No. 546 [1970]). — [38] United Nations (Solar Distillation as a Means of Meeting Small-Scale Water Demands, Department Economic Society Affairs, New York 1970). — [39] National Academy of Sciences (Solar Energy in Developing Countries, Perspectives and Prospects. Office of the Foreign Secretary, Washington, D.C., 1972). — [40] E. A. Cadwallader (Family Size Solar Stills. Office of Engineering, Agency for International Development, Washington, D.C., 1967).

[41] O. St. C. Headley, B. G. F. Springer (Proc. 3rd Intern. Symp. Fresh Water Sea, Dubrovnik 1970, Vol. 1, p. 669/77; J. Chem. Educ. 48 [1971] 49/51). — [42] S. D. Gomkale, R. L. Datta (Sol. Energy 14 No. 4 [1973] 387/92). — [43] T. A. Lawand, R. Alward (Proc. 4th Intern. Symp. Fresh Water Sea, Heidelberg 1973, Vol. 4, p. 511/21). — [44] H. McCracken (Intern. Congr. Sun Service Mankind, Paris 1973, Paper E6). — [45] G. Frick, J. v. Sommerfeld (Sol. Energy 14 No. 4 [1973] 427/31).

[46] M. Telkes (Off. Saline Water Res. Develop. Progr. Rept. No. 13 [1956], No. 31 [1959], No. 33 [1959]). — [47] Bjorksten Research Laboratories (Off. Saline Water Res. Develop. Progr. Rept. No. 24 [1959] ,No. 30 [1959]). — [48] R. A. Erb (Off. Saline Water Res. Develop. Progr. Rept. No. 29 [1959], No. 53 [1961]). — [49] W. N. Grune, R. A. Collins, R. B. Hughes, T. L. Thompson (Off. Saline Water Res. Develop. Progr. Rept. No. 60 [1962]). — [50] C. N. Hodges, T. L. Thompson, J. E. Groh, D. H. Frieling (Off. Saline Water Res. Develop. Progr. Rept. No. 194 [1966]).

[51] W. N. Grune (Proc. 3rd Intern. Symp. Fresh Water Sea, Dubrovnik 1970, Vol. 1, p. 655/68). — [52] D. Barba, D. Bogazzi, A. R. Giona, G. Tagliaferri (Quad. Ric. Sci. No. 58 [1969] 449/65). — [53] H. J. Preiss (Proc. 4th Intern. Symp. Fresh Water Sea, Heidelberg 1973, Vol. 4, p. 523/7). — [54] H. Ivekovic, S. Lendic, S. Lasic (Off. Saline Water Res. Develop. Progr. Rept. No. 841 [1973]).

2.18 Scale formation and its prevention

Formation of scale deposits on and fouling of heat transfer surfaces is one of the most serious problems of distillation equipment, operating with sea or brackish water. Extensive research work has been performed to understand the mechanism of scale formation and various methods were developed to prevent scaling.

A general review of the current status (1969) of scale control in saline water evaporators was given by McCutchan and Sieder [1].

2.18.1 Formation of alkaline scale

As the salt concentration increases during progressive evaporation, the critical point may be reached at which the solubility limit of scale-forming compounds, contained in the feed water, is exceeded and formation of scale occurs. The term of scale is particularly applied to describe hard, adherent, normally crystalline deposits on the heat transfer surfaces. Three simultaneous factors are required for the formation of scale:

a) Local supersaturation of the solution.

b) Nucleation, which when formed induces the rate of further scale deposition.

c) Sufficient contact time of the solution and the nucleus.

Under certain conditions a soft, amorphous material may be deposited or remain suspended in the brine, which is termed as sludge. If deposited, sludge is equally objectionable as scale, but is generally more easily removed than hard scale. Presence of solid particles, dust, seed crystals or the metal walls help nucleation by reducing the free energy barrier.

If the ions contained in seawater are combined in the form in which they usually deposit, the resulting compounds will be approximately:

$CaCO_3$	109 mg/l	$MgSO_4$	2 233 mg/l
$CaSO_4 \cdot 2 H_2O$	1 548 mg/l	$NaCl$	26 780 mg/l
$MgCl_2$	3 214 mg/l		

assuming that hydrogen carbonate decomposes to carbonate before precipitation occurs. Alkaline scale, $CaCO_3$ and $Mg(OH)_2$, results from the decomposition of the hydrogen carbonate ion. On heating seawater up to 82°C (180°F) the hydrogen carbonate ion decomposes and calcium carbonate is formed. At temperatures above 82°C the carbonate ion undergoes further reaction with water to produce CO_2 and OH^- ions. Magnesium hydroxide, therefore, precipitates around 82°C and above [2]. Both $CaCO_3$ and $Mg(OH)_2$ have negative solubilities.

Fig. 2–36 shows the effect of temperature on scale composition in a submerged tube boiler. Up to temperatures of 77°C deposition of $CaCO_3$ predominates. In the range of 77 to 85°C both scale constituents are present and at temperatures higher than that, deposition of $Mg(OH)_2$ will predominate.

Hasson et al. presented a general rate model, which takes into account physical and chemical steps involved in scale deposition. The dependence of scale growth rate from the pertinent basic parameters flow velocity, scale surface temperature and water composition is considered. Scale growth rate varies with Reynolds number and is only slightly dependent on surface temperature. $CaCO_3$ scale deposition is diffusion controlled within the range of surface temperatures of 67 to 85°C and Re = 13 000 to 42 000 [3].

Dooly and Glater have studied alkaline scale formation in boiling seawater brines by means of a laboratory method [4]. Experiments showed the amount and composition of alkaline scale to be a function of temperature, brine concentration, hydrogen carbonate ion concentration and flow conditions through the evaporator. The transition between calcium carbonate and magnesium hydroxide was shown to be influenced by factors other than temperature. A new mechanism for alkaline scale formation is proposed, in which the first step involves a unimolecular breakdown of hydrogen carbonate ion to form hydroxide ion.

Goto and Shirasaki have found that $CaCO_3$ scale deposited on the heated surface by thermal decomposition of HCO_3^-. The concentration of dissolved CO_2 was constant through the process of this reaction. These observations imply that the formation of $CaCO_3$ is the rate-determining step in this reaction. The deposition of alkaline scale in seawater near the boiling temperature showed that

$CaCO_3$ scale deposited initially and successively $Mg(OH)_2$ scale deposited by hydro-decomposition of $CaCO_3$ scale. The other salts in seawater may not affect directly the deposition of $Mg(OH)_2$ scale [5].

Fig. 2–36

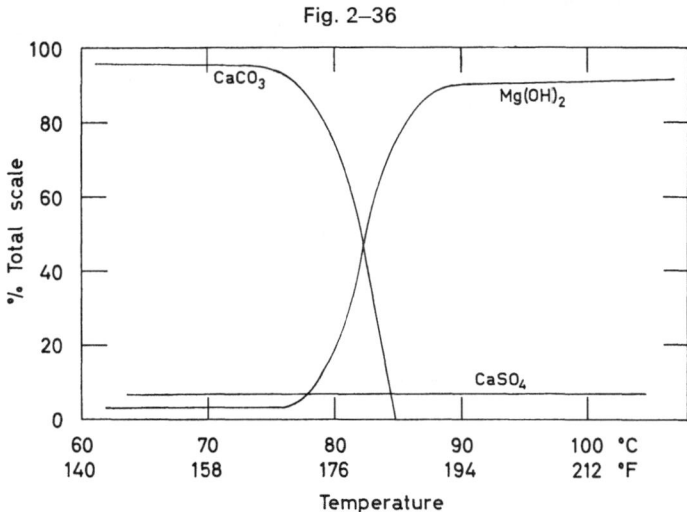

Alkaline scale composition as a function of temperature.

2.18.2 Formation of sulfate scale

Bildung sulfatischer Krusten

The second type of scale, called the acid scale, is due to three forms of calcium sulfate: the anhydrite $CaSO_4$, the hemihydrate $CaSO_4 \cdot 1/2 H_2O$, and the dihydrate $CaSO_4 \cdot 2H_2O$ or gypsum. While the precipitation of $CaCO_3$ and $Mg(OH)_2$ is mainly affected by CO_3^{2-} concentration, pH and temperature, the solubility of calcium sulfates is in addition affected by the concentration of other ions present. $CaSO_4$ has as well a decreasing solubility in the temperature ranges of interest. The solubility increases in chloride solutions, as the concentrate approaches 4 to 5% chloride concentration and then decreases to values comparable to those in chloride-free water, as the chloride concentration becomes 10 to 15%.

The solubility of calcium sulfate in NaCl-solution and in simulated seawater at temperatures up to 100°C was investigated by Furby et al. The results are in satisfactory agreement with those of similar experiments by other workers. A tabulation of the best molal solubility products of $CaSO_4$ (anhydrite) in sea salt solutions for the temperature range from 50 to 200°C is given [6]. The solubility of calcium sulfate in natural and artificial seawater, its concentrates, as well as in concentrated NaCl, $MgCl_2$ and $MgSO_4$ solutions in the temperature range of 50 to 170°C was also studied by Makinskii. A formula is given for the determination of calcium ion concentration in saturated solutions [7]. The solubility of $CaSO_4$-hemihydrate is greater than anhydrite, which is the stable form over 100°C. The rate of conversion of hemidrate to anhydrite was studied by Glater and Fung [8]. Thermodynamic data of the calcium sulfate solution process are given and the free energies, enthalpies, and entropies of the solution process were calculated by Gardner and Glueckauf [9].

A curve of the top temperatures of operating evaporators versus the corresponding concentration factor of the seawater, below which the safe operation of evaporators can be guaranteed, was presented by Simpson and Hutchinson and compared with the solubility limits of calcium sulfate [10]. The solubility of calcium sulfate in seawater was measured for the range of temperatures between 121 and 204°C (250 and 400°F) and seawater concentration factors of 2 to 8, extending the existing data to a higher temperature range.

The solubilities of the three forms of calcium sulfate in seawater are given in **Fig. 2–37**, p. 152, as a function of temperature and concentration factor of normal seawater.

Fabuss and Lu have made an extensive investigation to obtain basic information about the equilibria and rate processes controlling calcium sulfate precipitation. The investigation covered the calcium sulfate – distilled water system and the system in carbonate-free synthetic seawater. Based on the solubility data of the calcium sulfate modifications in sodium chloride solutions, magnesium chloride solutions, and in mixed solutions of these salts, a method was developed to calculate their solubilities in seawater concentrates. The precision of this method is estimated to be about ±2% [45].

The composition of solid and liquid phases in the system $CaSO_4$–Na_2SO_4–H_2O was investigated by Block and Waters also in the presence of sodium chloride in the temperature range of 25 and 100°C. An evaluation of a sodium sulfate recovery process, based on a 3:1 seawater brine concentrated further to 10:1 by solar evaporation, was also made [46].

Flint has presented experimental data, which indicate that uptake of calcium sulfate by seawater brines can be increased from about 0.8 to 3 or 6% by aluminum chloride or hydrochloric acid additions respectively. The advantage of such increased solubility might be found in the postponement of calcium sulfate scaling and of descaling of seawater evaporation plants [47].

Fig. 2–37

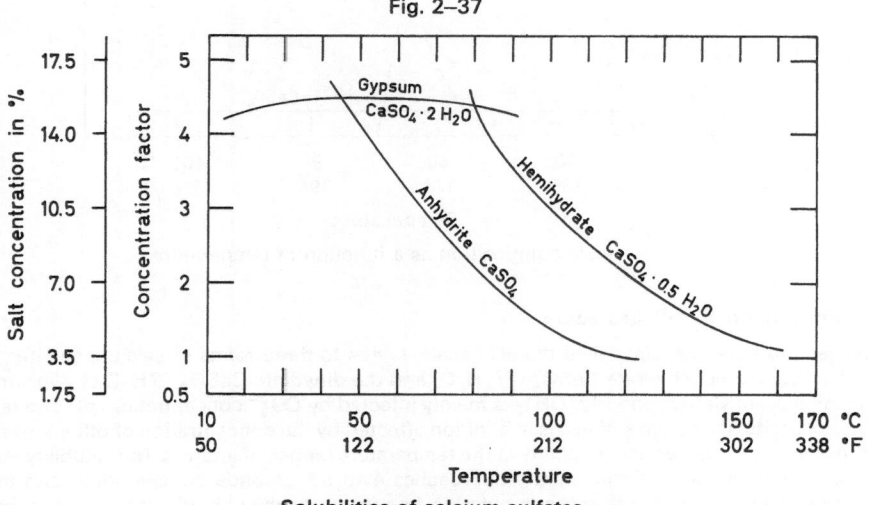

Solubilities of calcium sulfates.

A complex analysis was performed by Pribicevic of the behavior of calcium sulfate in sodium chloride solution in view of developing a technical process for the removal of the SO_4^{2-} ions from salt water by its sedimentation by $CaCl_2$ [48]. The possibility of using equimolar quantities of $CaCl_2$ for removal of the sulfate ions from brines was investigated. The optimum conditions were studied theoretically and experimentally investigated [49]. Scale formers and heavy metals must first be removed and are precipitated. The concentration of $CaSO_4$ can be reduced by several methods below saturation at any temperature [50].

A novel system of feed water treatment was discovered at the University of California. Calcium sulfate scaling thresholds in natural seawater could be markedly enhanced by enrichment with $MgCl_2$. The effect of magnesium ion on $CaSO_4$ solubility has been studied in natural seawater concentrates. Experiments were carried out at magnesium concentrations up to 3 times the ambient level, concentration factors between 1 and 4 and temperatures up to 171°C (340°F). Equilibrium solubilities of hemihydrate were measured in pressure vessels under non-boiling, non-evaporating conditions. The effect of salinity and magnesium ion concentration on hemihydrate – anhydrite phase transition was studied, at temperatures and concentrations normally encountered in high temperature evaporators [51].

Scaling threshold enhancement is based on complexing a portion of dissolved sulfate ion in the form of stable $MgSO_4$ ion pairs. Complexing may be accomplished by augmentation of feed water with magnesium ion recovered and recycled from distilling plant blowdown. Magnesium ion augmentation at levels up to three times ambient results in markedly improved resistance to hemihydrate scaling in natural seawater. Significantly higher levels of operating temperature and concentration factor may be achieved by this technology [52].

A research program to inhibit the formation of calcium sulfate anhydrite scale was carried out by Nelsen and Taylor. In the study ultra-thin layers of organic materials were deposited on heated metallic surfaces by means of controlled electrical potentials. The organics studied include polymers and low molecular weight organic compounds. Both types of coatings resulted in significant reduction in scale. The studies suggest that the deposited layers provided a passivation of the surface preventing scale nucleation [91].

Studies of calcium sulfate deposits in an annular test cell, simulating heat exchanger operation, were reported by Courvoisier et al. Photomicrographs of the tube taken in situ allowed the detection of the first needle-like crystals. The hydrodynamic conditions affect primarily the morphology of the deposit. The overall rate of hemihydrate precipitation was determined during unsteady-state operation. The growth rate of the deposit perpendicular to the surface was determined during steady-state operation. Iron in the solution results in the significant reduction in the rate of growth of the needles [11]. Investigation of $CaSO_4$ scale deposition in an annular heat exchanger shows that surface nucleation along a heated tube occurs at a nonuniform rate, highest at the downstream edge, and leads to the formation of a correspondingly nonuniform scale layer, thickest at the downstream edge. Rates of nucleate front propagation and scale-layer growth increase with surface temperature, decrease with increasing flow velocity, diminish with fine filtration, but are unaffected by the degree of recirculation. Transient nucleation is a predominant mechanism in $CaSO_4$ scale formation. A kinetic model based on this mechanism was presented by Hasson and Zahavi [12].

An experimental investigation to provide a better understanding of the kinetic mechanism during scaling of calcium sulfate on heating surfaces was also made by Lammers. In the important temperature range of 90 to 150°C in which mainly $CaSO_4 \cdot 1/2 H_2O$ develops, the parameters oversaturation, temperature of the heating wall, and conditions of flow were varied. The results of the measurements were analyzed and presented with the aid of equations for the crystallization and the convective mass transfer [13].

The formation of calcium sulphate scale in a vertical falling film evaporator has been investigated by Hodgson et al. It was found that the scaling threshold is dependent not only on the brine concentration and temperature, but also on the heat flux to the evaporating film. With normal seawater concentrations and a high heat flux, scale was observed at 115°C (240°F), whereas with a low heat flux scale-free operation was possible at 130°C (266°F). The effect of heat flux is attributed to the presence of local areas of high concentration in the falling film which form as a result of uneven evaporation from the film; a simple model has been developed to explain the results [14].

Surface potential was shown to be a pertinent variable in scale formation. Both the quantity of $CaSO_4$ deposited and the density of the deposits were reduced by cathodically polarizing the platinum electrodes [15].

A study of ultrasonic reflectivity at liquid-metal interfaces, where $CaSO_4$ scale is also present, has led to a technique for scale detection on the inside surfaces of heat transfer tubes. The technique appears to be capable of detecting deposits with thickness of 0.025 mm (1 mil) or more [16].

The effect of ultrasound on scale formation was also studied by Nevstrueva et al. during boiling of supersaturated $CaSO_4$ solution at atmospheric pressure. The critical thickness of the scale decreased with an increase in the amplitude of curved fluctuations. It depends on amplitude and on number of cycles and the rate of crystallization [85].

2.18.3 Prescaling deposition

Initialphase der Verkrustung

An experimental study was made with the purpose of investigating the chemistry and physics of the induction period usually observed in the formation of scale on heat-exchange surfaces. The induction period is that time at the beginning of particle nucleation, during which no bulk scale deposition is observed. It is usually followed by a period of high deposition activity. The specific objective of the study was to elucidate the mechanism and energetics of the surface influenced processes which are under way during the induction period.

Copper sheet was chosen as the first material for investigation and electropolishing was selected as the preferred technique for preparing surfaces with uniform characteristics. Sorption isotherms for calcium sulfate as a function of temperature and degree of saturation were determined. Adsorption of $CaSO_4$ increased with time and concentration and maximum values were found in the region of 75 to 90% saturation in the range from room temperature to 93°C (200°F). The sorption was lower at 65 and 80°C than at either 38 and 93°C and a hump was observed in all but the 24°C isotherm at about 5% of saturation. Adsorption increases with surface roughness [17].

The prescale deposition of calcium sulfate on 90—10 Cu-Ni alloy surfaces, using radioactive [45]Ca as a tracer was found to depend on pH, being greater at pH = 6.5 than at 4.5. Smooth surfaces were shown to inhibit scale formation. Surprisingly, adsorption of the [45]Ca species was enhanced by scale inhibiting additives, which act by complexing with the Ca ion. As expected, it was reduced by those additives, which act by forming protective films on the metal surface [18].

Verhütung
der Krusten-
bildung

2.18.4 Prevention of scale formation

As major components, calcium carbonate and magnesium hydroxide have been defined for alkaline scale and calcium sulfate for acid scale. In addition to these, sludge may include organic material and phosphates, derived from additives during treatment. Sodium chloride, silica, copper and iron are also present, but are considered as trapped materials. A scale deposit lowers the efficiency of heat transfer surfaces and increases pressure drop. Maximum brine temperature provided in the design, maximum allowable brine concentration and brine recirculation rate are also affected by the formation of scale. These operating variables and the plant availability are closely tied to the economics of the process, as the production rate is generally lowered. Periodic plant shut-downs for descaling would be required, either by an acid clean or, in extreme cases, by mechanical cleaning of the tubes. Incrustation allowances to reduce the frequency of shut-downs are made in designing evaporators, which are provided with a sufficiently larger heat exchange surface in order to maintain the design capacity. The term of fouling is often extended to this type of admissible scaling.

The formation of deposits on heat exchanger surfaces and their effect on heat transfer rates is reviewed by Taborek et al. and various mathematical models for the prediction of fouling in heat transfer equipment are evaluated. Considerable progress still needs to be made in formulating more reliable methods for fouling curve prediction [19].

The usual method of allowing for the heat transfer resistance of fouling is to employ some fixed value of the resistance. This may be mathematically unsound since it imposes a steady state situation on what is generally a transient condition. Models which have been proposed to overcome these objections and the difficulties in their application are discussed by Bott and Walker. It is suggested that some form of cooperative effort in industry could provide comprehensive data, which could be used in the development of realistic models for the assessment of fouling potential in heat transfer equipment under different conditions [20].

Efficient pretreatment of feed water is therefore the best method to prevent as much as possible deposition of scale. The main techniques used are briefly reviewed.

Säureinjektion

2.18.5 Acid injection

When seawater is treated by simple deaeration, the hydrogen carbonate ion is only partially decomposed and only part of the carbon dioxide is removed. Carbonate is formed, a large decrease in the solubility of calcium results and part of the calcium content in the seawater is precipitated.

Adding sulfuric acid, in proportion corresponding stoichiometrically to the hydrogen carbonate ion concentration, breaks down the calcium carbonate which is transformed to sulfate: $CaCO_3 + H_2SO_4 \rightarrow CaSO_4 + H_2O + CO_2$. Any other available acid, if less expensive, may also be used. The use of other acids, instead of sulfuric, is highly desirable, because it would avoid the increase of the sulfate ion concentration in the brine. Magnesium hydroxide does not precipitate as the pH is kept below 7.

After thorough mixing with the acid, the seawater feed is introduced into the decarbonator for countercurrent stripping with atmospheric air. The decarbonation step is followed by vacuum deaeration, seawater being stripped with steam to remove dissolved oxygen and noncondensable gases. In some designs excess acid, if present, is neutralized with caustic soda. With the acid treatment process, correctly designed and applied, no alkaline scale should be deposited and the scaling threshold is determined by the solubility of the calcium sulfate only.

An advantage of the flash distillation process is that each stage acts also as a deaerator. The major part of the brine is recirculated and the incoming feed is always diluted with a large volume of degasified brine. This tends to reduce also the corrosiveness of the make-up stream, if the incoming seawater is poorly degasified. The following tabulation gives maximum temperatures and concentration factors for scale free multi-stage flash operation [21]:

Brine temperature		Brine	Brine	Operation
in °C	in °F	pH	concentration	
148.9	300	6.0	1.00	once-through
138.9	282	6.5	1.00	Recycle
135.0	275	6.5	1.40	Recycle
128.4	263	6.8	1.65	Recycle
121.1	250	6.8	2.0	Recycle
112.8	235	6.8	2.65	Recycle

pH values are taken after decarbonation and should be based on 98% or better decarbonation. Brine concentration is based on 1.0 = 35000 mg/l.

In the vertical tube evaporator process, the feed water flows from the treatment system directly to the preheater tube bundles. Ineffective treatment would result in scaling and/or corrosion.

2.18.6 Carbon dioxide injection

Injection of carbon dioxide into the cold brine stream as a means of preventing alkaline scale has been proposed and tested in pilot plant. The kinetics of this process has been described by Langelier et al. [22]. If a sufficient partial pressure of carbon dioxide is maintained during the heating of seawater to the desired temperature, with a sufficient overpressure to solubilize the CO_2, the decomposition of the hydrogen carbonate ion will not occur. Calcium and magnesium are transformed to hydrogen carbonates [23]. The required carbon dioxide increases with temperature, e.g. 37 mg/l at 125°C and 64 mg/l at 140°C [24].

After heating and before introducing the feed water into the evaporator, the carbon dioxide is removed by steam stripping in a packed column. The alkaline scale precipitates on the column packing and can be periodically removed and discarded. Rapid precipitation of calcium carbonate was observed in pilot plant tests following release of the carbon dioxide overpressure at 135 to 143°C. The alkalinity of seawater could be reduced from about 115 mg/l equivalent $CaCO_3$ to about 32 mg/l by countercurrent stripping of the seawater at 131°C. By decreasing the seawater rate, alkalinities as low as 25 mg/l were obtained. The alkalinity was further reduced to 11 mg/l by passing the exit seawater from the stripping column through a bed of activated alumina. Similar tests performed at the University of California exhibited a drastic reduction in the rate of magnesium hydroxide growth. Alkaline scale was completely avoided up to 140.5°C [25]. It was confirmed in an other investigation, that carbon dioxide injected to a slight overpressure into the seawater feed will prevent alkaline scale formation in the brine heater by suppressing the decomposition of the hydrogen carbonate ion. By flashing off the CO_2 and precipitating solids in a separate vessel upstream of the first stage of the evaporator, the alkalinity entering the evaporator is low enough to avoid scale deposition. Results of preliminary scouting experiments are presented showing the effects of steam stripping rate and liquid loading rate on discharge alkalinity of a 19 liters per minute (5 gpm) pilot plant [26].

The decarboxylation temperatures of CO_2 dissolved in pure water and artificial seawater were determined by differential thermal analysis; they were 31.5 and 18.0°C respectively [27].

2.18.7 Sulfur dioxide injection

The process, involving treatment of seawater by sulfur dioxide or sodium hydrogen sulfite, was tested for several thousand hours in a pilot plant, where the main parameters, such as temperature, flow rate, residence time of seawater, and degassing were studied. Conditions have been found where acidic and reducing properties of sulfur dioxide or sodium hydrogen sulfite, respectively, avoid scale formation and minimize corrosion. SO_2 is required in a rate of 45 and 70 g/m³ of seawater respectively in direct and recirculating brine flow [28].

2.18.8 Treatment by polyphosphates

Nucleation of scale on the heat transfer surfaces is inhibited and precipitation of solids is dispersed in suspension by the addition of small amounts of certain commercial preparations. The sludge formed is then removed in the blowdown.

The first commercially available scale control compound was Hagevap, a mixture of sodium tripolyphosphate, lignin sulfonic acid derivatives, and various esters of polyalkylene glycols. Polyphosphates act as sequestrants for calcium and magnesium ions. Lignin sulfonic acid, starch, tannin etc. act as dispersants in coating surfaces such that scale adherence and crystal growth are inhibited. Polyalkylene glycols are surface active agents, which tend to retard foaming of the seawater.

The use of polyphosphate based additives is limited to temperatures below 88°C (190°F). Above this temperature polyphosphates undergo chemical change, which restrict their effectiveness as anti-scaling agents. It has also been reported that presence of a few mg/l of iron ions can poison polyphosphates and lead to very complicated sludge compounds. Although polyphosphates have in general performed well for retarding alkaline scale deposition, reports on the effectiveness of dosing vary. Generally, 2 to 4 mg/l are sufficient for their action. Overdosing appears to cause sludge deposition and the formation of magnesium slimes. Plant design criteria have been made responsible for unsuccessful use of polyphosphate treatment. In general, this type of treatment is attractive because the required facilities do not add considerably to the plant investment cost.

Flint has examined by X-ray powder diffraction analysis precipitates obtained by treating seawater with sodium tripolyphosphate. The form of polyphosphate most common to industrial scales is β-calcium orthophosphate or Whitlockite [29].

Various other anti-scaling compounds containing phosphates are marketed. Two of them were compared with Hagevap-LP, and in all cases tested a soft sludge was deposited [21]. The analysis of the scale deposits was as follows:

		SP-1230	Hagevap-LP	EES-VAP
Iron	as Fe_2O_3	20%	11%	14%
Copper	as CuO	5%	4%	3%
Phosphate	as P_2O_5	12%	12%	17%
Calcium	as CaO	9%	7%	6%
Magnesium	as MgO	36%	20%	30%
Chloride	as Cl	11%	28%	18%
Sodium	as Na	7%	18%	12%

SP-1230 is a corn starch containing 4% phosphorus as orthophosphate and EES-VAP is a sodium diphosphate compound.

Elliot had developed a laboratory technique to test surface coatings and seawater additives for the control of alkaline scale deposition in seawater evaporators. With a seawater feed at 93°C the untreated scale can be either mainly calcium carbonate or magnesium hydroxide depending upon the operating conditions used. Calcium carbonate was more effectively controlled than magnesium hydroxide and sodium tripolyphosphate was the most effective additive for calcium carbonate. Additives which are better than tripolyphosphate in controlling magnesium hydroxide have been identified [30].

2.18.9 Phosphate precipitation with recovery of byproducts

When phosphoric acid or sodium phosphates and ammonia are added to seawater, magnesium, calcium, and various trace elements such as iron, zinc, copper, manganese, and cobalt, all necessary plant nutrients, are precipitated. The amounts of phosphoric acid and ammonia utilized are stoichiometric to the magnesium and calcium concentrations in the seawater as needed for the precipitation of magnesium ammonium phosphate and dicalcium phosphate. Concentrations of magnesium and calcium in the seawater were reduced to 0.5 mg/l and 4 mg/l respectively. Due to the replacement of the calcium and magnesium ions with ammonium ions, the treated seawater contained about 2000 mg/l of ammonia as ammonium chloride and sulfate, as well as about 200 mg/l of free ammonia for pH adjustment. A residual concentration of 200 mg/l of P_2O_5 was also observed. The evaluation of the descaled seawater indicated that it may be evaporated up to very high concentration factors without evidence of scale formation. However, it was found that up to 200 mg/l of free ammonia were absorbed in the distillate product. The resulting slurry after precipitation is heated to dehydrate magnesium ammonium phosphate hexahydrate to the monohydrate. After filtration, granulation, and drying a fertilizer was obtained containing 7% N, 43% P_2O_5, 21% MgO, and 5% CaO. Agronomic evaluation of the fertilizer indicated that it might be a reliable source of nitrogen and phosphorus. Nearly complete recovery of potash can be obtained when seawater is treated with trisodium phosphate at pH 9.5 for descaling. The descaled seawater contained less than 2 mg/l calcium [31].

2.18.10 Treatment by polyacrylates

Small concentrations of low molecular weight polyacrylic acid or 10% ethyl acrylate-acrylic acid copolymer have been found to be effective in preventing scale deposition during seawater evaporation. An optimum polymer concentration of about 2 mg/l was observed to be most effective. Tests showed that polymers acted as deflocculating or dispersing agents. It appears that adherence and growth of scale particles is suppressed due to adsorption of polymer molecules at the metal surface, thereby substantially reducing the amount of scale deposited. This has a high content of magnesium poly- acrylate and the deposited scale is capable of self-stripping from the surface by the regenerating film mechanism [32].

The synthesis of a range of acrylic acid homopolymers and copolymers was investigated by Solomon and Rolfe and their capacity in preventing calcium sulfate precipitation was evaluated [33]. Conductance measurements revealed the marked effect of the polymer on the crystallization kinetics of calcium sulfate and are consistent with observations of the modification of crystal habit [34].

The rate of scale formation on copper heat transfer surfaces by the addition of about 3 mg/l of partly esterified polymers of acrylic acid was also investigated. In the range of molecular weight 1000 to 6100, the polymers of lower molecular weights are more effective. Self-cleaning scales, which peel from the heat transfer surface on exposure to air, are formed if about 0.2 mg/l of aluminum, manganese or zinc, as a soluble salt, is added with the polymer to the seawater [35]. These polymers have been as well evaluated as inhibitors for calcium sulfate crystallization. At 90°C a good correla- tion was found between effectiveness for scale prevention and for inhibiting crystallization. The poly- mers were more effective at the temperature of 30°C [36].

In screening tests by Flesher et al. of the effectiveness of homopolymers and copolymers for pre- venting precipitation of alkaline hardness salts and calcium sulfate, the influence of chemical structure, ionic functional groups, molecular weight, and copolymerisation has been studied. The effect of temperature, pressure, pH, and concentration of the additive were investigated with the most efficient polymers [37].

Laboratory tests have been used to screen chemical additives suitable for alkaline scale control at temperatures between 93 and 121°C (200 and 250°F). The tests have shown that the quantity of solids deposited and the heat transfer performance depends in turn upon the type of chemical used. Low molecular weight polyelectrolytes have been found the most suitable additives and a polymeric carboxylic acid has been found particularly effective. Field tests and initial results indicate a 36-day operating period between acid cleans [38].

Four general types of organic threshold inhibitors are reported by Ralston as especially useful in controlling water-formed deposits. Three of these compositions are phosphonates, diphosphonates, and phosphate esters, and the fourth type is a polyacrylate [39].

When a plant is run on threshold agents of the polyacrylic type, the resultant high brine pH can cause iron, if present, to precipitate on heat transfer surfaces as $Ca(Mg_{0.67}Fe_{0.33})(CO_3)_2$ (feroan dolomite) or as $Mg_6FeCO_3(OH)_{13} \cdot 4H_2O$ (brugnatellite). Both deposits severely inhibit heat trans- fer. The soluble iron contributing to these formations can be complexed by adding agents such as the phenol-substituted ethylenediamine diacetic acids. Above concentrations of 0.2 to 0.4 mg/l, copper will precipitate the polymers onto heat transfer surfaces. This too can be alleviated by chelating agents [40].

Additives, which retard the formation of alkaline scale from seawater, delay the precipitation of calcium carbonate but have no effect on the precipitation of magnesium hydroxide. Boiling aqueous solutions containing magnesium and hydrogen carbonate ions produce an initial precipitate of hydro- magnesite ($3MgCO_3 \cdot Mg(OH)_2 \cdot 3H_2O$), which converts on further heating to magnesium hydroxide. Crystals of calcite and hydromagnesite, produced in the presence of an additive which is effective in preventing alkaline scale formation, show marked distortion, tending to make the crystals larger and less regular in shape [41].

Among the polymers evaluated which had some efficacy are: polyacrylic acid, polymethacrylic acid, copolymers of maleic anhydride with such monomers as methyl-vinyl-ether, styrene and vinyl acetate, as well as hydrolized polyacrylamide. More recently polymaleic acid has been found as an extremely effective polyelectrolyte for scale control [42].

An attempt was made by Shaheen and Dixit to establish the optimum dosage of organic polymers, such as sodium polyacrylate and sodium polymethacrylate in saline water feed, by experiments with a single effect evaporator. Three different molecular weights of each polymer were investigated for their effectiveness. Sodium polymethacrylate with a molecular weight of 4500 was found to be most effective as a scale preventive additive. In the case of sodium polyacrylate, both a lower molecular weight of 84500 and a higher one of 190000 were found to be almost equally effective. However, a molecular weight of 125000 gave very poor results [86].

*Kalk-
Magnesium-
carbonat-
verfahren*
2.18.11 Lime-magnesium carbonate process

Distilling plants can be operated free of scale at temperatures up to 121 to 127°C (250 to 260°F) by the use of acid. Higher scale-free operation temperatures can only be safely attained if other means are used to prevent the precipitation of calcium sulfate hemihydrate and anhydrite.

The lime-magnesium carbonate (LMC) process for pretreating seawater was developed to circumvent this problem. The process is a modification of the lime-soda ash treatment to soften fresh water. The modification consists of replacing Na_2CO_3 by $MgCO_3$, which can be produced from constituents already present in seawater. If no material losses occur, the process would be self-sufficient and would require no added chemicals. A further advantage is offered by lowering the amount of sulfuric acid required to prevent alkaline scale formation since the dissolved carbon dioxide of the treated seawater is reduced. The process has been first tested in laboratory and then in a pilot plant with a capacity of 19 m³/day (5000 gpd) [43]. The process has been further tested in a plant with a capacity of 430 tons/h (950000 lb/h) soft water, which was annexed to the multi-stage multiple-effect Senator Clair Engle distillation plant (see section 2.6.7, p. 81).

The process has two main operating parts: the precipitation train and carbonation train. The first train performs the actual softening of the make-up seawater and the second train generates the active reactant for the first part of the process. **Fig. 2–38** shows a simplified flow-sheet of the LMC process.

Fig. 2–38

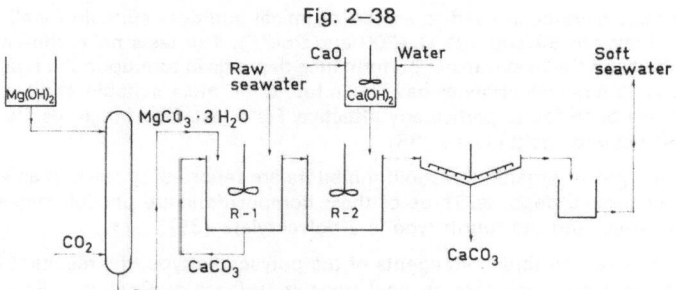

Lime-magnesium carbonate descaling process.

In the precipitation train, the make-up seawater emerges from stage 21 of the third recirculation steam of the Senator Clair Engle plant at below 38°C and is pumped to the LMC plant for pretreatment. Soluble $MgCO_3 \cdot 3H_2O$ and $CaCO_3$ seeding crystals are added in reactor R-1, which is vigorously agitated to keep all solids in suspension and provide favorable conditions for the carbonation of the dissolved calcium. The overflow from reactor R-1 flows by gravity into reactor R-2, where hydrogen carbonate ions react with lime to form more calcium carbonate. The overflow from reactor R-2 passes into the clarifier, where $CaCO_3$ settles out and the clarified liquor is pumped back to the

Senator Clair Engle plant from a surge tank. In the carbonation train, magnesium hydroxide slurry is diluted with raw seawater to a concentration of 2% and fed to the carbonation tower. The dilute magnesium hydroxide slurry encounters a countercurrent gas flow of 15% CO_2, magnesium carbonate trihydrate slurry forms and is pumped to the reactor R-1. The lime slurry is prepared by mixing quick-lime with domestic water.

In the overall process, both the hydrogen carbonate alkalinity and the total alkalinity are reduced and most of the calcium ions in solution is replaced by magnesium ion. The concentration of the two ions in inlet seawater and in the LMC product water is illustrated by the following figures, in mg/l, based on seawater treated in the Senator Clair Engle plant:

	Inlet seawater	LMC product water
Calcium ion	388	113
Magnesium ion	1233	1386

Approximately 70% of the dissolved calcium is removed. The pH of the LMC product water is between 9.0 and 9.5. Successful operation of the carbonation tower is probably the most critical factor in the performance of the LMC process. The equilibrium concentration of carbon dioxide in contact with magnesium hydroxide-carbonate mixtures in seawater at several temperatures and kinetic studies on the reaction were thoroughly investigated [44].

The model developed predicts that unless LMC product water is acid treated to remove alkalinity, precipitation will occur under the distillation plant operating conditions. Operation of the LMC and the Senator Clair Engle plants in series allowed a maximum brine temperature of 146.7°C (296°F). This limitation was imposed by the plant rather than by the process.

2.18.12 Alumina-lime-soda water treatment process

Aluminium-oxid-Kalk-Soda-Verfahren

A new pretreatment process for brackish water has been developed and evaluated in the laboratory. The alumina-lime-soda process involves treating raw saline water with sodium aluminate and lime, separating precipitated solids and neutralizing the separated effluent with carbon dioxide. The water recovered after the carbon dioxide neutralization may be used as feedwater to a desalting process. The sodium aluminate treatment precipitates a calcium sulfoaluminate. Both hardness cation and sulfate anion are removed simultaneously. Therefore, the feedwater is softened and reduced in total salinity. In addition to hardness and sulfate the sodium aluminate-lime treatment also removes boron, silica, iron and manganese. The process is more efficient in removing hardness than the conventional lime-soda treatment [87].

2.18.13 Scale prevention by seeding techniques

Krusten-verhütung durch Impf-methoden

A relatively low concentration of finely divided seed crystals in seawater, which is heated above the saturation temperature, will cause scale forming compounds to precipitate out on the particles rather than on the heat transfer surfaces, even though these surfaces are at a higher temperature. The heat transfer surfaces will remain clean under conditions for which they would otherwise scale up rapidly in the absence of the seeds.

The technique for scale control—the addition to the brine of fine solid seeds, as an alternative surface for scale deposition, and the elimination of part of the recycle seeds with the blowdown— was tested on a pilot vertical tube evaporator. Calcium carbonate has been found effective for retarding scale formation up to temperatures of 135°C, but not for long time. The use of calcium sulfate seeds to prevent scaling was less successful. Deposition of magnesium hydroxide silt caused as well difficulties. Seeds have accumulated in stagnant areas and on shut downs of the plant, posing important design problems. It appeared for some time that the seeding technique would be of reduced interest for desalting. However, it was demonstrated in more recent studies, that the injection of seeds prevents practically scale formation during an almost three-fold brine concentration. In evaporating seawater to a point where the concentration of sulfates and calcium exceeds the solubility of the anhydrite, the prevention of scale was achieved by injecting in each evaporator stage seeds of such crystalline structure as to correspond to the crystalline structure of the scale, which would form at the same evaporator stage under the prevailing temperature and degree of concentration [53].

A similar technique was successfully tested in a flash distillation pilot plant at a seawater temperature up to 95°C [54]. The process has been applied at the Shevchenko, Caspian Sea, desalting plant [55]. This process was also used successfully in evaporators for the production of common salt [56].

In investigating the calcite growth mechanism, stable supersaturated solutions of $CaCO_3$ have been prepared by Nancollas and Reddy by careful control of experimental conditions. The crystallization kinetics have been examined upon the addition of seed crystals of calcite. Photomicrographic evidence suggests that the initial surge in the growth curves results from additional nucleation at the surface of the added calcite crystals and in the bulk of the supersaturated solution [57].

Seeded crystal growth experiments near the dihydrate – hemihydrate phase transition temperature (103°C in salt free solutions at 4 atm) showed that phase transformation accompanies the seeded growth of the less stable modification [88]. The presence of additives, such as sodium benzoate, EDTA, cetyltrimethylammonium bromide, which inhibit the growth of other sparingly soluble salts, is almost without effect, whereas other additives markedly retard the rate of crystallization [89].

A dilatometric study of the crystallization of calcium sulfate dihydrate in the absence of seed crystals indicated that nucleation was heterogeneous. An extensive conductometric study of seeded crystallization allowed a comparison of various empirical expressions for crystallization kinetics which have been suggested for similar systems [58].

The tendency of the evaporator heat transfer surface to scale was compared by Simpson and Hutchinson with the solubility data and the value of the scale limitation curve was confirmed with a seed free brine. Results with seed recycle showed that the tolerable brine temperature can be greatly increased with calcium sulfate deposited on the seeds in the condenser tubes and flash chambers from supersaturated solutions. Seeding also inhibits scale build-up by erosion of the tube walls, but for seeding to be effective adequate residence time must be available [10].

Studies on the seeding technique conducted by Charuit et al. have been made in two steps. In the first one temperature was 100°C or below and synthetic seawater was used. The simultaneous prevention of calcium sulfate and calcium carbonate deposition has been confirmed. In a second step seeding was made with natural seawater, at 150°C, in industrial conditions. No deposit on heating tubes was observed [59].

It is important to know regularities of the scale deposition on heat transfer surfaces for the choice of desalting plant operating conditions using seed crystals for scale prevention. Laboratory experiments on the investigation of a carbonate scale deposition rate out of the calcium hydrogen carbonate solution at different temperatures, hydrodynamic conditions and other variable factors affecting the scale deposition have been investigated by Chernozubov et al. and the results of this investigation were recently reported [60].

The seeding technique was also investigated in the pilot and demonstration plants of the Office of Saline Water. It was found that the circulation of 1.5 to 2% of $CaCO_3$ sludge in a recycle brine will maintain acceptable heat transfer rates for long periods of time. The approach temperature from the heater did not vary more than 0.5°C (1°F) during a 200 h operating period. The condenser heat transfer rates remained constant. At a 920 h operation period a decline in heat transfer rates was observed. At 149°C (300°F) brine outlet temperature from the heater, calcium sulfate and magnesium hydroxide deposited together with the calcium carbonate in heater tubes. Little or no scale was found in the condenser tubes. A disadvantage of the method is the additional cost of design and maintenance caused by the handling of a slurry [21]. In a long tube vertical evaporator pilot plant, after acidification, decarbonation and deaeration of the feed water, calcium sulfate seed crystals were added to the hot feed and the slurry was passed through a delay column. The slurry then passed through the evaporator and a concentrator to concentrate the brine and recover the seed crystals. This mode of operation indicated the probability of a successful hot end calcium sulfate sludge system for multi-stage flash plants. pH control to prevent all saline scale formation combined with 1.5 to 2% calcium sulfate sludge did not prevent the deposition of heavy scale at the hot end [21].

The effects of such factors as temperature, composition, fluid dynamics, heat flux, size and weight of seeds on calcium sulfate precipitation were studied in a laboratory unit. Crystal growth was extremely slow and appeared to be the limiting process in scale formation. The total precipitation, scale plus suspension, increased with wall temperature of the heater and the amount of seawater circulated through the heater. The presence of seeds in the fluidized bed heater improved the heat transfer coefficient and, hence, reduced the temperature difference between the solution and the heater wall accordingly. The presence of fluidized seeds did not affect the total precipitation, but did reduce the amount of scale formed. With proper size and amount of seeds, scale-free operation could be achieved. The dissolved organic compounds in natural seawater had no significant effect on scale formation and total precipitation [61].

2.18.14 Scale prevention by ion exchange techniques

Krusten-verhütung durch Ionen-austausch-verfahren

An experimental program to partially soften seawater by cation exchange in fluidized flow has been conducted in pilot plant scale. The resin was separated in a hydraulic cyclone and recycled. The spent resin was regenerated downflow in a contact column by concentrated blowdown from the seawater evaporator. No $CaSO_4$ scale was found when soft seawater was evaporated with and without pH control. $Mg(OH)_2$ was the major scaling constituent when the alkalinity was not neutralized. A combination of a fluidized softening and a fixed bed regeneration was found to be both workable and desirable. Dowex 50W resin, 50 to 100 mesh, was quite stable, both chemically and physically [62].

Further work on continuous countercurrent ion exchange proved that a 65% calcium removal from seawater can be obtained using a three-fold concentrated blowdown as regenerant with a 12% cross-linked strong-acid type resin. The calcium content was reduced from 800 to 280 mg/l expressed as $CaCO_3$, which permits raising the evaporator temperature to 149°C (300°F). If 50% calcium is removed, the temperature in the evaporator can be raised to 135°C (275°F) [63].

A similar experimental work on partial softening of seawater by means of countercurrent ion exchange, using the brine concentrated in the evaporators as a regenerant, is reported by Kunz. The results of the operation of a pilot plant are given with particular reference to the exchange efficiency. The maximum operating temperatures can be as high as 160°C without the solubility limit of the calcium sulfate hemihydrate being exceeded [64]. Calcium concentration of evaporator feed seawater was also reduced from 400 to 120 mg/l in a pilot plant producing 4 m³/h soft water [65].

Optimization of seawater distillation is based on the use of a porous sulfonated polystyrene ion exchange resin to remove about 93% of the calcium present. Efficient utilization of the ion exchange resin was attained by using concentrated brines, formed in the process, for the regeneration of the resin in counterflow at high flow rates. A resin lifetime of 37 000 cycles is indicated for a plant with a desalination capacity of 190 000 m³/day (50 Mgd) [66].

Publications with similar content were presented by Boari and Passino [67], Mirkis et al. [68], Feiziev [69] and Mikhailov [70], who studied the removal of sulfates and hydrogen carbonates by means of anion exchange resins. A sulfate removal process and dealkalization on weak anion resins was also tested in a pilot plant by Aveni et al. The annexed evaporation plant was operated for 7500 h at temperatures up to 145°C and with concentration factors varying between 2.0 and 2.5 with no scale on the heat exchange surfaces being observed at any time [71].

Another approach to scale control was developed by the elimination of sulfates from seawater. The U.S. Bureau of Mines had conceived a process which removed the sulfates from Salt Lake brines as a preliminary step to mineral recovery. The process with some modifications to optimize operating conditions has been evaluated for seawater application and found to be technically feasible and economically desirable [72]. A pilot plant with a capacity of 284 m³/day (75 000 gpd) was constructed and tested.

The raw seawater is first decarbonated by acidification followed by degasification. The acidified seawater is conveyed to the ion exchange resin in barium form. Aided by agitation, the resin is contacted by the seawater. A slight excess of barium ion is maintained in solution, causing precipitation of barium sulfate to take place in the seawater. The resin is thus converted to the sodium form. The suspended resin is retained on a screen and the seawater containing the precipitated sulfate passes through. The resin is rinsed with process water in a second screen and is then conveyed to the barium loading operation. The exhausted ion exchange resin is loaded with approximately 58% sodium and 10% calcium with the balance consisting mainly of magnesium ions. The ion exchange resin, while descending countercurrent to a 2.5 N solution of barium chloride in the loading column, is regenerated into the barium form. The effluent from the top of the column is a 2.0 N solution containing primarily sodium, calcium and magnesium chlorides and a trace of barium chloride. All traces of barium are removed from the effluent prior to its ultimate disposal. The regenerated resin is rinsed and the rinsing liquor is utilized for making up the barium chloride regeneration solution, while the resin is available for desulfating. The production of byproducts such as sulfur or sulfuric acid, caustic soda, and soda ash is possible [73].

The composition (in mg/l) of the feed and the product seawater showed the following change:

	Feed seawater	Product seawater
Sulfates	2702	Trace
Hydrogen carbonates	142	Trace
Chlorides	19360	18817
Magnesium	1297	1064
Calcium	408	293
Sodium	10768	9794
Potassium	388	362

The operational program of the 75000 gpd continuous ion exchange desulfating pilot plant was also evaluated. Production of desulfated seawater for use as feedwater permitted multi-stage flash distillation plant operation at 177°C (350°F) and a concentration ratio of 3.5 for 336 h with no deposition of scale or sludge on the heat exchange surfaces. Capital and operational costs for desulfating plants furnishing feedwater to 1, 2, 5, and 10 Mgd distillation plants were displayed along with results of pilot plant test runs [74].

A direct contact desulfation process to remove over 95% of $CaSO_4$ and 32 to 90% of the sulfate ions from seawater was also developed. The softening is achieved by reacting hard seawater with a recirculated barium hydrogen carbonate. Two versions of the process are possible: 1) CO_2 controlled, wherein both alkaline and hard scale are removed, and 2) acid controlled, wherein acid decarbonation removes alkaline scale and the direct contact desulfation removes the hard scale, both operating in series. As much sulfate needs to be removed, as required to eliminate the desired amount of calcium [75].

Elimination of sulfates from brines was as well studied by Suzuki in adding $BaCO_3$ or $BaCl_2$. From an economical viewpoint, the addition of barium salts appears to be superior to that of $CaCl_2$ [76].

Weitere Krusten verhütungs- verfahren

2.18.15 Other scale prevention techniques

A novel approach is reported in the concept of utilizing fluidized beds of granular particles in heat exchangers as a means of preventing scale formation in the high temperature distillation of saline waters. The tube surfaces remained free of scale throughout extended tests with seawater in the temperature range of 165 to 171°C. Calcium sulfate remained in suspension in the brine stream under recirculation and did not accumulate in the bed particles. Heat transfer coefficients were high [77]. Further tests at the Brookhaven National Laboratory are reported in an O.S.W. report [78].

A possible method of controlling calcium sulfate scale is to operate the first effects of an evaporator not with conventional impervious heat-transfer tubes, but rather with tubes of controlled permeability, such as graphite tubes, through which a portion of the steam condensate flows at a specified rate. The scale was much thinner and smoother. Many areas were scale free, whereas in the impervious tubes the deposited scale consisted of a compact layer [79]. In a second phase, the test apparatus was modified to simulate the higher operating temperatures of the first VTE effect. Reduction of $CaSO_4$ scale was obtained by using a controlled permeability graphite tube. However, alkaline scale was not reduced [90].

By using a desalinating membrane, calcium and sulfate ions of the seawater are effectively rejected and distillation can be carried out therefore at higher temperatures without danger of scaling [80]. Donnan membrane softening is a technique to selectively remove divalent hardness ions from feed waters as a pretreatment to desalination processes. The feasibilities of continuous recirculation of the regenerant solution, of sulfate removal and seawater softening were examined in a 7.5 m³/day (2000 gpd) prototype exchange diffusion unit and a spiral wound module. The concept of a continuously and permanently recirculating regenerant stream was successfully demonstrated. The regenerant side of the membrane was periodically flushed to remove the $CaSO_4$ slurry. Sulfate removal was shown to be comparable to calcium removal efficiencies. Seawater softening appeared possible, but exchange rates were much lower than those attained with brackish feeds [81].

Removal of silica from brackish water feeds was investigated using materials with sorptive or chemisorptive properties for monomeric silica [82], and a review of silicate reactions in nonsaline and saline aqueous media was compiled [83].

Three types of scales containing aluminum and magnesium have been shown to form during the corrosion of aluminum in seawater in the pH range 6.3 to 8.5. The type most frequently occuring has a structure similar to that of the natural mineral hydrotalcite. The Al/Mg ratio in this compound increases with pH, resulting in a form of buffering action. In the presence of these scales anodic breakdown of the corrosion film takes place at a potential 100 to 200 mV more anodic than in their absence. The structures, mechanisms of formation, and effects of these scales in the operation of distillation plants with aluminum surfaces are discussed, as well as their effects on the operation and interpretation of laboratory corrosion tests [84].

Literature to 2.18

[1] J. W. McCutchan, E. N. Sieder (Off. Saline Water Res. Develop. Progr. Rept. No. 411 [1969]). — [2] S. F. Mulford (Off. Saline Water Res. Develop. Progr. Rept. No. 133 [1964]). — [3] D. Hasson, M. Avriel, W. Resnick, T. Rozenman, S. Windreich (Desalination 5 [1968] 107/19). — [4] R. Dooly, J. Glater (Desalination 11 [1972] 1/16). — [5] T. Goto, T. Shirasaki (Nippon Kagaku Kaishi 1972 2309/13; C.A. 78 [1973] No. 62030).

[6] E. Furby, E. Glueckauf, L. A. McDonald (Desalination 4 [1968] 264/76). — [7] I. Z. Makinskii, V. V. Shishchenko, I. A. Geivandov (Opyt. Expluat. Teplosilovogo Oborud. Sist. Azglavenergo 1970 59/69; C.A. 75 [1971] No. 112700). — [8] J. Glater, K. Fung (Environ. Sci. Technol. 3 [1969] 580/4). — [9] A. W. Gardner, E. Glueckauf (Trans. Faraday Soc. 66 [1970] 1081/6). — [10] H. C. Simpson, M. Hutchinson (Desalination 2 [1967] 308/24).

[11] P. Courvoisier, C. Duffau, R. Guillermin, E. Muratore, P. Vignet (Proc. 3rd Intern. Symp. Fresh Water Sea, Dubrovnik 1970, Vol. 1, p. 439/54). — [12] D. Hasson, J. Zahavi (Ind. Eng. Chem. Fundamentals 9 [1970] 1/14). — [13] J. Lammers (Verfahrenstechnik [Mainz] 7 No. 4 [1973] 114/8). — [14] T. D. Hodgson, M. N. Elliot, T. W. J. Jordan (Proc. 4th Intern. Symp. Fresh Water Sea, Heidelberg 1973, Vol. 2, p. 143/59). — [15] B. D. Marcus, H. P. Silverman, W. T. Tsukamoto (Off. Saline Water Res. Develop. Progr. Rept. No. 393 [1969]).

[16] F. R. Rollins (Off. Saline Water Res. Develop. Progr. Rept. No. 444 [1969]). — [17] H. L. Recht (Off. Saline Water Res. Develop. Progr. Rept. No. 521 [1970]). — [18] H. L. Recht (Off. Saline Water Res. Develop. Progr. Rept. No. 632 [1971]). — [19] J. Taborek, T. Aoki, R. B. Ritter, J. W. Palen, J. G. Kundsen (Chem. Eng. Progr. 68 No. 2 [1972] 59/67, No. 7 [1972] 69/78). — [20] T. R. Bott, R. A. Walker (Chem. Engineer No. 225 [1971] 391/5).

[21] Baldwin Lima Hamilton Corp. (Off. Saline Water Res. Develop. Progr. Rept. No. 186 [1966]). — [22] W. F. Langelier, D. H. Caldwell, W. B. Lawrence (Univ. of California, Berkeley, Inst. Eng. Res. Rept. Ser. 4, Issue No. 12 [1950]). — [23] E. A. Cadwallader (Ind. Eng. Chem. 59 No. 10 [1967] 42/7). — [24] R. D. Ellis, J. Glater, J. W. McCutchan (Environ Sci. Technol. 5 [1971] 350/6). — [25] R. D. Ellis (UCLA-69-61 [1969]).

[26] D. M. Eissenberg, C. C. Littlefield, R. P. Hammond, S. A. Reed, I. Spiewak (Proc. 3rd Intern. Symp. Fresh Water Sea, Dubrovnik 1970, Vol. 1, p. 479/91). — [27] T. Yamabe, M. Kubo, N. Takai, K. Umezawa (Nippon Kaisui Gakkai-Shi 24 No. 130 [1971] 163/5; C.A. 75 [1971] No. 80135). — [28] C. Duffau, J. L. Imbert, P. Vignet (Proc. 4th Intern. Symp. Fresh Water Sea, Heidelberg 1973, Vol. 2, p. 85/96). — [29] O. Flint (Desalination 6 [1969] 319/34). — [30] M. N. Elliot, T. W. J. Jordan, M. Hutchinson (Proc. 3rd Intern. Symp. Fresh Water Sea, Dubrovnik 1970, Vol. 1, p. 461/77).

[31] W. R. Grace and Co. (Off. Saline Water Res. Develop. Progr. Rept. No. 57 [1962], No. 66 [1962], No. 192 [1966]). — [32] L. S. Herbert, P. F. Rolfe, U. J. Sterns (Proc. 1st Intern. Symp. Water Desalination, Washington, D.C., 1965 [1967], Vol. 2, p. 39/56). — [33] D. H. Solomon, R. F. Rolfe (Desalination 1 [1966] 260/6). — [34] P. F. Rolfe (Desalination 1 [1966] 359/66). — [35] F. Sweett, B. R. Smith, P. Casamento, R. N. Taylor (Desalination 8 [1970] 167/75).

[36] B. R. Smith, F. Sweett (Desalination 9 [1971] 277/83). — [37] P. Flesher, E. L. Streatfield, A. S. Pearce, O. D. Hydes (Proc. 3rd Intern. Symp. Fresh Water Sea, Dubrovnik 1970, Vol. 1, p. 493/504). — [38] M. N. Elliot, T. D. Hodgson, A. Harris (Proc. 4th Intern. Symp. Fresh Water Sea, Heidelberg 1973, Vol. 2, p. 97/110). — [39] P. H. Ralston (Mater. Prot. Performance 11 No. 6 [1972] 39/44). — [40] J. Block, B. M. Watson, L. A. Burkholder, B. Albano (Proc. 4th Intern. Symp. Fresh Water Sea, Heidelberg 1973, Vol. 2, p. 33/45).

[41] A. Harris, M. A. Finan, M. N. Elliot (Proc. 4th Intern. Symp. Fresh Water Sea, Heidelberg 1973, Vol. 2, p. 131/42). — [42] D. R. Sexsmith, E. Q. Petrey (Desalination **13** [1973] 89/92). — [43] W. R. Grace and Co. (Off. Saline Water Res. Develop. Progr. Rept. No. 192 [1966]). — [44] R. W. Lawrence (Off. Saline Water Res. Develop. Progr. Rept. No. 754 [1971]). — [45] B. M. Fabuss, C. H. Lu (Off. Saline Water Res. Develop. Progr. Rept. No. 258 [1967]).

[46] J. Block, O. B. Waters (Off. Saline Water Res. Develop. Progr. Rept. No. 305 [1968]). — [47] O. Flint (Desalination **4** [1968] 328/35). — [48] S. Pribicevic, B. Stancic (Tehnika [Belgrade] **25** No. 11 [1970] 2175/80; C.A. **74** [1971] No. 67559). — [49] S. Pribicevic (Tehnika [Belgrade] **26** No. 2 [1971] 347/52, C.A. **75** [1971] No. 23277; **26** No. 4 [1971] 153/7; C.A. **75** [1971] No. 52652). — [50] S. Pribicevic, L. Milosavljevic (Tehnika [Belgrade] **26** No. 5 [1971] 973/7; C.A. **75** [1971] No. 132853).

[51] K. S. Murdia, J. Glater, J. W. McCutchan (UCLA-Eng-7232 [1972]), J. Glater, R. L. Dooly, J. W. McCutchan (Off. Saline Water Res. Develop. Progr. Rept. No. 888 [1973]). — [52] J. Glater, K. S. Murdia, R. Dooly (Proc. 4th Intern. Symp. Fresh Water Sea, Heidelberg 1973, Vol. 2, p. 119/30). — [53] V. B. Chernozubov, F. P. Zaostrovskii, V. G. Shatsillo, S. I. Golub, E. P. Novikov, V. I. Tkach (Desalination **1** [1966] 50/60). — [54] V. B. Chernozubov, E. P. Novikov, Y. V. Kartovskii, S. I. Golub, V. G. Shatsillo (Desalination **5** [1968] 34/9). — [55] F. P. Zaostrovskii, E. P. Novikov, V. G. Shatsillo, S. I. Golub, V. B. Chernozubov, V. I. Tkach (Desalination **1** [1966] 165/77), E. A. Sobolev, S. I. Golub, V. B. Chernozubov, N. K. Tokmantsev, S. B. Ruchin, V. I. Tkach, V. L. Podbereznii, N. S. Konev, V. G. Shatsillo, E. P. Novikov, M. B. Viceblat, L. S. Mrejin (Proc. 3rd Intern. Symp. Fresh Water Sea, Dubrovnik 1970, Vol. 4, p. 103/10).

[56] P. Charuit, R. Marchand, M. Viard, P. Courvoisier, P. Vignet (Proc. 3rd Intern. Symp. Fresh Water Sea, Dubrovnik 1970, Vol. 1, p. 417/38). — [57] G. H. Nancollas, M. M. Reddy (J. Colloid Interface Sci. **37** [1971] 824/30). — [58] B. R. Smith, F. Sweett (J. Colloid Interface Sci. **37** [1971] 612/8). — [59] P. Charuit, R. Marchand, M. Viard, C. Duffau, J. Ravoire (Proc. 4th Intern. Symp. Fresh Water Sea, Heidelberg 1973, Vol. 2, p. 47/56). — [60] V. B. Chernozubov, L. P. Karnauhov, A. P. Egorov (Proc. 4th Intern. Symp. Fresh Water Sea, Heidelberg 1973, Vol. 2, p. 57/67).

[61] C. H. Lu, B. M. Fabuss, A. Korosi, T. R. Middleton, J. P. DeMonico (Off. Saline Water Res. Develop. Progr. Rept. No. 461 [1969]). — [62] W. F. McIlhenny (Off. Saline Water Res. Develop. Progr. Rept. No. 62 [1962]). — [63] Chemical Separations Corp. (Off. Saline Water Res. Develop. Progr. Rept. No. 309 [1967]). — [64] G. K. Kunz (Desalination **3** [1967] 363/72). — [65] J. Arod, P. Charuit (Proc. 3rd Intern. Symp. Fresh Water Sea, Dubrovnik 1970, Vol. 2, p. 15/45).

[66] G. Klein, S. Cherney, E. L. Ruddick, T. Vermeulen (Desalination **4** [1968] 158/66), K. M. Makar, T. Vermeulen, G. Klein (Ion Exch. Process Ind. Papers Conf., London 1969 [1970], p. 174/81). — [67] G. Boari, R. Passino (Chem. Eng. Progr. Symp. Ser. **67** No. 107 [1971] 241/9). — [68] I. M. Mirkis, A. Yu. Dykhno, I. P. Mikhailov, G. A. Mitlitskii, Yu. Sh. Kegamyan (Vodosnabzh. Sanit. Tekhn. **1972** No. 10, p. 1/5; C.A. **78** [1973] No. 47615). — [69] G. K. Feiziev (Teploenergetika No. 3 [1972] 74/5; C.A. **76** [1972] No. 129474). — [70] I. P. Mikhailov (Tr. Vses. Nauchn. Issled. Inst. Vodosnabzh. Kanaliz. Gidrotekhn. Sooruzh. Inzh. Gidrogeol. **1971** No. 32, p. 52/9; C.A. **76** [1972] No. 131303).

[71] A. Aveni, G. Boari, L. Liberti, M. Santori, B. Monopoli (Proc. 4th Intern. Symp. Fresh Water Sea, Heidelberg 1973, Vol. 2, p. 13/31). — [72] C. E. Ennis, G. P. Gelblum, W. Pechenik (Off. Saline Water Res. Develop. Progr. Rept. No. 289 [1967]). — [73] C. E. Ennis, G. Gelblum (Off. Saline Water Res. Develop. Progr. Rept. No. 554 [1970]), C. E. Ennis, P. B. Pruett (Chem. Eng. Progr. Symp. Ser. **67** No. 107 [1971] 224/8). — [74] K. M. Garrison, J. E. Gugler (Off. Saline Water Res. Develop. Progr. Rept. No. 746 [1971]). — [75] G. P. Gelblum (Off. Saline Water Res. Develop. Progr. Rept. No. 735 [1971]).

[76] T. Suzuki (Soda To Enso **22** No. 3 [1971] 111/22; C.A. **75** [1971] No. 112767). — [77] L. P. Hatch, G. G. Weth, S. J. Wachtel (Desalination **1** [1966] 156/64). — [78] L. P. Hatch, G. G. Weth (Off. Saline Water Res. Develop. Progr. Rept. No. 571 [1970]). — [79] L. E. Vaaler, C. E. Hulswitt (Off. Saline Water Res. Develop. Progr. Rept. No. 272 [1967]). — [80] E. Glueckauf, L. A. McDonald, P. J. Russell (Desalination **3** [1967] 155/68).

[81] J. D. Smith (Off. Saline Water Res. Develop. Progr. Rept. No. 655 [1971]). — [82] G. R. Bell, J. P. Leineweber, J. L. Yang (Off. Saline Water Res. Develop. Progr. Rept. Pt. 1 No. 286 [1968]). — [83] A. G. Collins, L. R. Fisher (Off. Saline Water Res. Develop. Progr. Rept. No. 307 [1969]). — [84] C. E. Austing, A. M. Pritchard, N. J. M. Wilkins (Desalination **12** [1973] 251/72). — [85] E. I.

Nevstrueva, I. M. Romanovskii, K. Ya. Segreeva (Inzh. Fiz. Zh. **24** [1973] 120/5; C.A. **79** [1973] No. 7415).

[86] E. I. Shaheen, S. N. S. Dixit (Desalination **13** [1973] 187/206). — [87] J. W. Nebgen, E. P. Shea, S. Y. Chiu (Off. Saline Water Res. Develop. Progr. Rept. No. 820 [1973]). — [88] G. H. Nancollas, M. M. Reddy, F. Tsai (J. Cryst. Growth **20** [1973] 125/34). — [89] S. T. Liu, G. H. Nancollas (J. Colloid Interface Sci. **44** [1973] 422/9). — [90] L. E. Vaaler, S. V. Desai (Off. Saline Water Res. Develop. Progr. Rept. No. 396 [1968]).

[91] L. Nelsen, A. H. Taylor (Off. Saline Water Res. Develop. Progr. Rept. No. 835 [1972]).

2.19 Materials of construction. Corrosion

Werkstoffe.
Korrosion

The selection of suitable metals to construct desalination plants is of prime importance as the use of inadequate materials may lead to shutdowns, increased replacement and maintenance costs and affect the overall economics of the plant. An extensive research work on the corrosive mechanisms and the behavior of metals, alloys and various materials of construction has been made and is still under way. A short review of literature will be given in the following pages and only the more recent, out of the large number of publications, will be considered. The importance of the resistance to corrosion of the materials used in a desalination plant might be understood from the following illustration. Design studies for large multi-flash desalination plants showed that over 3000 km of condenser tubing would be required for a 950000 m³/day (250 Mgd) plant. The cost of smooth tubes for the condensers is a major item in the total capital cost of such plants. The achievement of reductions in the cost of tubing would reduce capital costs and therefore should lead to lower costs of fresh water distilled from the sea.

Various kinds and degrees of corrosion are encountered in a seawater desalination plant, depending on the nature of the metal and the conditions of its exposure in salt water. The choice of a metal or alloy for a given structure depends on its initial cost, divided by the expected life. The less expensive material of construction is steel. The use of more expensive metals or alloys is justified only if they have a longer life and they perform equally well or better.

The oxygen content ranks highest in importance among the many factors that affect the corrosiveness of seawater. Oxygen causes depolarization of cathodic areas, it may oxidize metal ions in solution to a higher valence state and it may also promote formation of a protective film on the metal surface. The seawater must be deaerated for improving the overall thermal efficiency, as well as for corrosion control. The rate of attack is expected to increase, as the temperature of deaerated seawater is increased. On the other hand the oxygen solubility is lowered with increasing temperature. There is a maximum temperature for each metal above which it is likely to be attacked in seawater at an excessive rate.

Seawater has usually a pH close to 8.0. Scale control of seawater by the acid method can lead to greater corrosion unless the pH is adjusted to near neutral after elimination of the carbon dioxide. The corrosion rate will increase when lower values of pH are maintained in the treated seawater.

The rate of corrosion is dependent on the velocity of the seawater flow or the degree of turbulence. Severe localized attack might be experienced at points of turbulence whether the seawater is heated or not.

Corrosion by concentrated salt solutions is usually less severe than by more diluted solutions, probably because the solubility of oxygen decreases as the salt concentration increases. In seawater brines, the corrosion behavior will depend on whether protective deposits form.

Iron(III) and copper(II) ions in seawater can be corrosive to other metals in the system. A higher build-up of metal ions may result in seawater recirculating streams. Copper ions promote pitting of aluminum. Copper(II) ions will attack metallic copper to form copper(I) ions. Titanium and austenitic stainless steels may show reduced rates of attack on the presence of copper(II) ions. Pitting of stainless steels might be promoted by the presence of iron(III) ions [1]. The corrosion mechanisms in saline water were investigated by Miller et al., using reversible and irreversible electrodes [2].

Bacterial corrosion. The corrosion of steel in seawater is subject to seasonal variations, the corrosion maxima being in spring and autumn. The activity of bacteria becomes obvious only at temperatures above 10°C. At these temperatures metabolism is activated and the products of metabolism may attack iron. During summer times bacteria action is inhibited by fouling of the iron surfaces. In autum, however, fouling organisms are killed by lower temperatures, so that there is another peak of bacteria activity [161]. Corrosion rates have been compared between specimens exposed to aerated seawater and Millipore filtered to remove living microbiota. No differences were found for carbon steel in the two environments. A limited study suggests that the corrosion of 5052 aluminum may be accelerated by the presence of bacteria. Microbiological aggregates including slimes were observed on aluminum and stainless steel specimens using light microscopy and dye stain techniques. Corrosion rates were estimated by the polarization resistance method [162]. Steel rusts very slowly in oxygen-free seawater, unless sulfate reducing bacteria are present. Bacterial nutrients might exist in the layer of slime on the surface of structures in seawater [163].

Korrosions-
unter-
suchungen
an ver-
schiedenen
Metallen

2.19.1 Corrosion studies on various metals

An early report on the corrosion of various metals in saline environment was given by Fink [3]. In a second report the corrosion of metals was studied in hot neutral deaerated seawater, in an experimental loop apparatus [4]. A corrosion test program study has revealed that a large quantity of corrosion data in seawater and information on materials perfomance has been developed from the operation of the various demonstration plants erected by the Office of Saline Water [5] and a materials evaluation program was developed by Bell and Bramer [6]. Corrosion probes that will detect and measure galvanic and crevice corrosion by resistance and polarization measurements were developed by Fitzpatrick to provide a valuable tool for in-line measurement of non-uniform corrosion. In pitting environment a galvanic couple will shorten the incubation period before pitting starts and increase the intensity of the non-uniform attack. Formation of a protective film at high temperatures significantly reduces galvanic effects [7].

The corrosion test program of the Office of Saline Water, in order to provide a sounder basis by which materials of construction may be selected, was implemented by constructing a small highly automated pilot plant with two loops which could simulate most of the water conditions encountered in a desalting plant [8]. The important variables of dissolved oxygen, temperature, velocity, pH, and refreshment rate were investigated [9]. Results on tested materials are presented in the appropriate sections in the following pages.

In non-destructive materials evaluation programs representative selected tubing materials have been tested in the Roswell, Freeport, Senator Clair Engle plants and the Module, operated by the Office of Saline Water [10]. A Materials Test Centre was erected at the Freeport Test Facility, incorporating a 750 l/min (200 gpm) seawater treating plant, a 375 l/min (100 gpm) seawater softening plant, for the purpose of corrosion testing in the environments encountered in distillation-type desalination of seawater. Corrosion test units in the form of miniplants have been constructed on the site by both the government and industry. The purpose of this program was to obtain baseline data during short-term coupon-stage corrosion tests. Effects of desalination plant operation variables include seawater temperature to 121°C (250°F) and dissolved oxygen levels from 5 mg/m³ to 1 mg/l (5 ppb to 1 ppm). Other factors examined were the effects of chlorination and of hydrogen sulfide pollution on waterside corrosion. The metals studied were mild steel and selected commercially available alloys of copper and aluminum [11].

A corrosion test facility, MEWAK I, erected 40 km east of Hamburg, West Germany, provides several possibilities in corrosion testing of desalination plant materials, e. g. condenser tubes, piping, valves and fittings and specimens of various shape and size. Main operation parameters are: brine concentration, threefold seawater salt concentration, temperature 115 to 125°C, brine velocity within condenser tubes 4 m/s, etc. [12].

A survey of materials behavior in multi-stage flash distillation plants was made by Newton and Birkett [13]. Corrosion rates of various structural metals in seawater were reported by Southwell [14], by Heitz [15], by Boisdé et al. for temperatures up to 150°C [16], by Yamashita et al. in high velocity seawater [17], by Compton with emphasis on dissolved oxygen for corrosion [18], by Fink et al., presenting results of closed seawater systems corrosion tests of aluminum base alloys coupled to Monel 400, brass, titanium, and stainless steel [19], by Sukach et al., with reference to water velocity and temperature [20], by Cigna, emphasizing the importance of deaeration and temperature [21], by Reinhart and Jenkins after exposure of various alloys for 12 and 18 months in surface seawater and in a depth of 1800 m (5900 ft) in the ocean [22], and by Coriou et al., examining the effect of various parameters and especially oxygen on the corrosion of several metals and alloys [23].

The reasons for desalination plant shut-down are considered in a paper by Bailie et al. It is shown that proper selection of materials can minimize down-time and lead to lower water costs. Particular attention is paid to the selection of materials for key components in water intake systems, heat rejection and heat recovery sections of the desalination plant [24].

The application of the proper material to the proper service as the best method of preventing corrosion was emphasized by Osborn et al. [25], by Desensy [26], by Bom [27], by Wangnick [28] and by Westphal [29], pointing out that materials used in desalination plant can be used as a guideline in the selection of materials for extreme applications in marine technology.

Table 2/12 shows the composition of alloys tested and evaluated in the hot seawater for potential use in desalination plant. A review on the choice of materials for desalination plant was also published recently in a brochure of the International Nickel Ltd. [30].

Table 2/12
Composition of alloys evaluated for corrosion in hot seawater. ("Bal" indicates balance).

	Fe	C	Cr	Ni	Mo	Cu	Si	Mn	S	P	Other	Lit.
Structural and low alloy steels:												
Steel 1010	Bal	0.17	0.027	0.009	—	0.018	0.014	0.44	0.026	0.009	—	[4]
Cor-ten	Bal	0.12	0.76	0.22	—	0.27	0.46	0.34	0.024	0.081	—	[4]
Wrought stainless steels:												
AISI 304	Bal	0.056	18.2	9.11	0.19	0.17	0.40	1.30	0.011	0.029	—	[9]
AISI 316	Bal	0.08	17	12	2.5	—	0.5	1.5	0.02	—	—	[9]
AISI 316L	Bal	0.022	16.9	13.23	2.73	—	0.55	1.73	0.009	0.023	—	[4]
AISI 409	Bal	0.07	11.08	0.24	0.14	0.09	0.4	0.5	0.005	0.016	0.8 Ti	[9]
AISI 430	Bal	0.07	16.1	0.32	0.046	0.075	0.38	0.50	0.007	0.022	—	[9]
Cast stainless steels:												
CF-8	Bal	0.068	19.75	8.40	0.09	0.07	1.35	0.68	0.027	0.020	—	[9]
CF-8M	Bal	0.064	19.65	10.05	2.25	0.09	1.05	0.77	0.029	0.016	—	[9]
CN-7M	Bal	0.055	19.25	28.29	2.10	3.10	0.76	0.72	0.009	0.011	—	[9]
CA-15	Bal	0.10	11.65	0.87	0.46	0.09	0.23	0.58	0.023	0.022	0.062 Al	[9]
Wrought high-nickel alloys:												
Incoloy 600	7.02	0.04	16.16	76.23	—	0.08	0.23	0.21	0.007	—	—	[4]
Incoloy 800	45.56	0.04	20.76	31.82	—	0.52	0.34	0.93	0.008	—	—	[4]
Incoloy 825	30.4	0.02	20.21	42.33	3.12	1.81	0.26	0.65	0.007	—	0.12 Al 1.05 Ti	[4]

	Cu	Ni	Fe	Mn	Sn	Al	Zn	As	Si	Other	Lit.
Wrought copper-nickel alloys:											
90-10 CuNi10Fe	Bal	10	1.25	0.5	—	—	—	—	—	—	[30]
90-30 CuNi30Fe	Bal	30	0.6	0.5	—	—	—	—	—	—	[30]
90-30 CuNi30Fe2Mn2	Bal	30	2	2	—	—	—	—	—	—	[30]
Wrought brasses:											
Naval brass	61	—	—	—	0.75	—	Bal	—	—	—	[30]
Al-brass CuZn22Al	76	—	—	—	—	2	Bal	0.05	—	—	[30]
Admiralty brass	71	—	—	—	1	—	Bal	0.05	—	—	[30]
Wrought nickel-copper alloys:											
MONEL 400NiCu30Fe	31	Bal	1.25	1	—	—	—	—	—	—	[30]
MONEL K500NiCu30Al	30	Bal	1	0.75	—	2.8	—	—	—	0.5 Ti	[30]
Cast nickel-copper alloys:											
MONEL 410	30	66	1	1	—	—	—	—	1.5	—	[30]
MONEL 411	33	62	1.5	1	—	—	—	—	1.6	1.3 Nb	[30]
MONEL 505	29	64	2	—	—	—	—	—	4	—	[30]

	Al	Si	Fe	Cu	Mn	Mg	Zn	Ti	Ga	Cr	Other	Lit.
Aluminum alloys:												
3003-H14	Bal	0.24	0.59	0.15	1.15	0.01	0.05	0.01	0.02	—	0.004 V	[9]
6061-T6	Bal	0.62	0.38	0.26	0.06	0.89	0.04	0.03	0.016	0.19	—	[9]
5052-H34	Bal	0.10	0.24	—	0.07	2.41	0.06	0.03	—	0.18	0.07 Ca	[9]
5454-H34	Bal	0.07	0.21	0.05	0.79	2.68	—	0.02	—	0.10	0.02 Ca	[9]
6063-T52	Bal	0.47	0.24	0.04	—	0.50	0.01	0.03	—	0.01	—	[9]
5005	Bal	0.3	0.4	0.025	<0.05	0.50	<0.01	—	—	<0.1	—	[9]
6061 clad	Bal	0.6	—	0.25	—	—	1.0	—	—	—	—	[9]

Case histories of corrosion in various desalination plants are also reported. The O.S.W. demonstration plants at Freeport and San Diego provided an excellent opportunity to observe the behavior of the materials used in the construction of the equipment [31]. Yamada and Kibe have examined specimens cut off from various parts of an evaporator after two years of operation and determined the importance of corrosion [32]. Condition of equipment after some years of operation of the Ceuta desalination plant in Spanish Morocco is reported by Hirschfeld et al. [33]. A number of case histories is reported by Van Nieuwenhuizen with emphasis on corrosion problems occuring during operation [34]. The corrosion performance of the Eilat, Israel, desalination plant was reported by Lev-Er et al. [35] and of the Tijuana, Mexico, distillation plant by Zuccolotto [36].

The corrosion behavior of various nonferrous metals in seawater was surveyed by Fleetwood [124]. The behavior of lead in normal and tropical seawater and chloride solutions is reported by von Fraunhofer, as well as the effect of alloying additives [125]. Zinc and Zn + 0.08% Fe specimens were exposed by Markovich and Suprun under laminar and turbulent flow conditions to Black Sea water and 3% NaCl solution. Iron containing specimens corroded twice as high [126].

A fully airtight apparatus was constructed by Mazo et al. for studying the effect of very pure desalinated water on materials of construction. The apparatus includes a closed circuit water system with either periodical or continuous water purification and a water purity measuring device [165]. A review of the corrosion effects of desalted water was also presented by Crossley and Waters [166]. The quantitative evaluation of the degree of corrosiveness of the distillate produced in a seawater evaporator can be established on the basis of an index describing the ratio of the initial distillate alkalinity to the limiting alkalinity of the solution saturated with $CaCO_3$ [164].

Kohlenstoff-stahl und Eisen-legierungen

2.19.2 Carbon steel and iron-base alloys

The effect of dissolved oxygen, pH, heat transfer, and metal surface on corrosion of carbon steel evaporator tubes was evaluated by Shea and Hollingshad. The importance of deaeration was pointed out. Heat transfer rate and pH have also an effect on corrosion of carbon steel evaporation tubes. The surface condition of the tubes had no appreciable effect [37]. The effect of temperature, salinity, flow rate and reactive anions, chlorides and sulfates, contained in seawater, on the corrosion resistance of steel was investigated by Konstantinova and Semenova with the purpose of determining its application possibilities [38]. It was concluded that mild steel might be used as construction material on condition that the seawater be deaerated and that no contact with electropositive materials exists.

A critical review of data available in the literature on the use of carbon steel in the construction of seawater distillation plants is given by Songa et al. [39]. Experimental results of researches performed in a pilot plant are reported. In a more recent paper researches on the corrosion behavior of carbon steel in desalination plants are reported at temperatures up to 145°C in a modular flash desalination test plant and in a test loop, as well as corrosion tests in various attacking media under static and quasi-static conditions, using laboratory devices [40].

An equation relating the corrosion resistance of carbon steel to the flow rate of seawater was developed by Markovich and Suprun [41].

Grassiani reports on sulfates being more corrosive than chlorides, as well as on the effects of flow rate, aeration and galvanic couples. Cold-drawn tubes are much more attacked than heat laminated tubes [42]. The corrosion rate of mild steel was found by Saito et al. to be much more rapid in moving NaCl solution than in a solution at rest. Almost complete protection was obtained even under moving condition by the application of a cathodic potential [43]. The effect of dissolved oxygen and pH of seawater on the corrosion behavior of mild steel was investigated by Legault [44].

A model for the corrosion resistance of low-alloy steel was presented by Shibata et al. A liquid diffusion layer and a solid oxide film act together as a barrier to the transport of oxygen [45]. The development of a low alloy steel for use in seawater desalination plants is reported by Kowaka and Nagano. Its resistance to corrosion is approximately twice that of mild steel and from 5 to 10 times that of mild steel in deaerated hot seawater. The chemical composition of this alloy is Mn 1%, Cu 0.3%, Cr 2% and Mo 0.15%. It has good mechanical properties and weldability [46].

Low cost iron-base alloys, ranging from binary alloys to multi-component alloys, containing as many as eleven different components, have been prepared by levitation melting. Dynamic loop-tests at 121°C (250°F) in deaerated seawater and studies of anodic polarization curves at room temperature in deaerated 0.1 N NaCl solution have been used to evaluate the relative corrosion behavior of the alloys. Future development of new corrosion resistant alloys is likely to be based on multi-component single phase alloys containing chromium and/or aluminum as major alloying components. Additions of Co, Mo, Ni, Nb, Si, W, Ag, Zr, V, and Ti have all shown beneficial effects as alloying elements [47].

In further studies the relative resistance of iron-base alloys has been evaluated to pitting corrosion in deaerated synthetic seawater at temperatures up to 121°C. Most promising steels have been found in the Fe-Cr-Mo-Co alloy system. Multi-component ferritic steels, containing Cr and Mo as major additives in combination with small concentrations of Mn, Co, Ni, and/or Cu will have exceptionally high resistance to pitting attack in aqueous chloride media. The more highly resistant base alloys were found Fe-Cr-W, Fe-Cr-Mo, Fe-Cr-Mo-Re, and Fe-Cr-Mo-Co. In order to reduce the cost of these alloys, while retaining the corrosion resistance, combinations of elements have been determined, which can be substituted, wholly or partly, for the chromium, molybdenum, and cobalt components. It was concluded that, by judicious selection of such substitutes, it should be possible to develop an economical ferritic steel with adequate mechanical and corrosion properties for use in hot seawater environments [48]. Investigating the effect of alloy components on the corrosion of steel in seawater Sataka and co-workers have found that Mn, Ni, Ti and Al have no influence, carbon is harmful, Cr, P, in some cases as well Cu and Si are favorable components [49].

Takamura et al. examined the effects of alloying elements as a function of the frequency of the dry-and-wet cycle in the seawater splash zone. In relatively dry environments, the addition of Mn, Cu, P, and Cr improved the corrosion resistance of steel. In wet environments, the addition of P improved the corrosion resistance considerably, whereas Cr addition affected the corrosion resistance adversely [50].

Steels containing P, Cu, Cr, Ni, and Al were exposed to the atmosphere and tidal and submerged zones for about five years. The Cr and Al components were effective to increase the corrosion resistance in the marine atmosphere and P, Cr, Ni, and Al were effective in the tidal zone. It was found that Cu-Cr-Al steels have a good corrosion resistance in marine environments [51].

A low alloy, low cost Cu-bearing steel shows as much as 30% improved performance over mild steel in typical desalination environments. The corrosion rate of mild steel doubles over the temperature range of 82 to 121°C (180 to 250°F). The 4% and 8% Ni and 3.5% Cr steels are unsatisfactory in hot seawater containing 125 ppb (mg/m³) dissolved oxygen [9].

A systematic evaluation of seawater corrosion resistance of three iron-aluminum base alloy systems: Alfenol (8, 10, 12, 14 and 16% Al), Thermenol (3.25 Mo, 8, 10, 12, 14 and 16% Al) and modified Thermenol (3.25 Mo, 0.2 Zr, 0.05 C, 12, 14 and 16% Al) was made. Neither baseline nor Fe-Al alloys were severely attacked in the autoclave at 138°C (280°F), where the most severe attack occurred with a modified Thermenol alloy with 16% Al. This alloy was found to be the most corrosion resistant Fe-Al alloy in moving seawater. The Mo-bearing Fe-Al alloys with 14 and 16% Al were equal to or superior to Cu-Ni in velocity tests at room temperature. However, only the modified Thermenol alloy with 16% Al approached the corrosion resistance of Cupronickel at 88°C (190°F) [52].

In further attempts to improve the seawater corrosion resistance of the iron-aluminum base alloys by alloying additions, molybdenum was found to have the most pronounced effect on improving corrosion resistance, chromium, copper, and nickel having only second order effects. The best ternary composition appeared to be 10 Al-4 Mo and of the quaternary compositions 10 Al-4 Mo-2 Cu appeared best. Crystallographic ordering of high aluminum alloys (14 to 16%), induced by appropriate heat treatments, was found to improve corrosion resistance at levels of 25 to 33%. Chemical polishing improved all of the alloys investigated except 10 Al-4 Mo-2 Cu. The 10 Al-4 Mo composition showed the most significant improvement in corrosion resistance after chemical polishing, with corrosion resistance superior to cupro-nickel [53].

In a basic study of ferrous materials for desalination equipment, the material used was Glidden iron, melted in an induction furnace under vacuum. Weight loss tests were conducted in deaerated 10% NaCl solution. The corrosion rate of 99.94% pure Glidden iron after 34 days was 0.66 mg·dm^{-2}·day^{-1}. Corrosion rates decreased with time [54].

Accelerated corrosion tests on welded steel coupons in seawater were conducted by Beckert and Heinecke and the ratio of the dissolution of the weld to that of the base material was determined. The tendency to corrode increased with decreasing Mn/Si ratio [127].

Seawater corrosion on ship hull steel plate welds indicated that steels with high manganese and low carbon content have low corrosion resistance in the transition zone [128]. The potential-probe curve of welds in steels was evaluated as a tool for the determination of the corrosion susceptibility of the welds. Tests with ship steels in seawater indicate that less-noble potential regions in the potential-probe curves are a necessary but not sufficient condition for selective corrosion [129].

2.19.3 Stainless steels

Nicht-rostender Stahl

Austenitic stainless steels are highly susceptible to chloride stress corrosion cracking; steels of the types 18% Cr, 10% Ni and of 17% Cr, 13% Ni, 2.5 to 3% Mo both with C $\leq$ 0.03%, were investigated. It was found that the kind of cation associated with the chloride ion has a pronounced effect of cracking. The aggressiveness of the cations tested can be expressed as: $Mg^{2+} > Ca^{2+} > Na^+$. Oxygen induces a marked increase in the aggressiveness of environment. Steels to which molybdenum has been added appear to be less susceptible to stress corrosion cracking. Results obtained with usual testing method in 42% $MgCl_2$ solution are not necessarily reliable, as far as corrosion resistance in seawater is concerned [55]. The pitting behavior of stainless steel in solutions is dependent on both sulfate and chloride ions concentration. The formation of deep pits was attributed to an overlap in pitting and active dissolution potentials of the repassivated steel [56]. The susceptibility to pitting and crevice attack of wrought stainless steels 304, 316, 409 and 430 were inconsistent within the variables of temperature and dissolved oxygen. Stainless steel cast alloys 304, 316 and 410 showed unsatisfactory performance at temperatures 82 and 107°C (180 and 225°F) in seawater containing 125 ppb (mg/m^3) dissolved oxygen [9].

Pitting and crevice corrosion of stainless steels in seawater environment was also investigated and is reported by Ohte et al. [57], by Henrikson et al. [58], by Degerbeck [59], by Kohl [60], by Tanno et al. [61], by Wilde [62], and by Ulanovskii and Korovin [63].

The effect of molybdenum as an alloying component of stainless steels to their resistance to seawater corrosion is outlined by Mizuno. He has presented a new stainless steel with composition 25% Cr, 5% Ni, 1.5% Mo and low carbon content, claimed to be resistant to seawater [64]. Degerbeck reports that the corrosion of stainless steels might be inhibited by appropriate alloying additions, namely chromium and molybdenum [65]. Further improvements due to the presence of molybdenum are reported by Bäumel et al. [66] and by Kügler et al. [67].

Welded and stressed samples of chromium-free austenitic manganese steels and of chromium manganese steels have been tested for a period from six weeks to two years. Chromium-free manganese steels showed neither intergranular nor transgranular stress corrosion cracking [68].

The application of stainless steels in desalination equipment was evaluated by full-scale plant tests and by specimen-exposure tests [69]. A program for testing the applicability of stainless steels to desalination plant service, involving four plant sections, was described by Harkins and Lawson [70]. In evaluating the possibility of austenitic stainless steel for condenser tube nest construction, it was found that both the molybdenum bearing steels and the straight 18-10 Cr-Ni steels suffered crevice and pitting corrosion on exposure to sea and estuarine cooling waters and can not be recommended for service [71].

Laboratory studies of ferritic stainless steels, based on ternary alloys of Fe-Cr-Mo, have been made for use as condenser tube material in desalination plants. The preparation and characterization of test coupons, representing eight compositions which exhibit a wide range of corrosion resistance, were described. Electrochemical methods were used to determine the relative corrosion resistances of the alloys. It was concluded that this class of material offers both excellent resistance to corrosion in hot seawater and adequate mechanical properties for fabrication into tubing. The final selection of any specific stainless steel, however, will have to be based on the degree of oxygen control maintained in the desalination plant and the correlation between electrochemical test methods and direct exposure of the test coupons to desalination plant conditions [72].

2.19.4 Copper and copper-base alloys

Kupfer und Kupfer-legierungen

Plain copper is not used as tubing material. A corrosion loss from 0.043 to 0.062 mm/y was observed by Yandushkin by flowing rates of 1 to 5 m/s in 2000 h tests [73].

Nonferrous alloys, such as 70/30 cupro-nickel, 90/10 cupro-nickel and aluminum brass, are the most commonly used materials for condenser tubing in desalination plants. In clean seawater, all three materials form a protective film, which stifles corrosion attack so long as the water velocity across it is within certain limits. For most practical purposes, the velocity limit is about 4.5 m/s for 70/30 Cu-Ni, 3.7 m/s for 90/10 Cu-Ni and 3 m/s for aluminum brass, but usually the velocity does not exceed 2.5 m/s [1].

A survey of service data on condenser tube life experienced by a significant number of coastal power plants in the United States, which use seawater for cooling, and from other tube users covered a large number of sets of tubes composed of 8 different alloys. Consideration of tube costs against probable life in polluted seawater indicated that 90/10 Cu-Ni alloy represents the most economic choice. Very low failure rates were shown by 70/30 Cu-Ni tubes and aluminum brass [74].

In a pilot test program to study the behavior of structural metals in the various environments encountered in a desalination plant, six copper alloys showed, regardless of composition, extremely low weight loss under low dissolved oxygen conditions. At higher oxygen concentrations, 20 to 100 ppb (mg/m^3), the alloys containing nickel showed decreased weight loss over the non-nickel alloys. As oxygen is the single most important variable in determining corrosion rate, an oxygen level below 5 mg/m^3 should be maintained in the seawater. At low oxygen levels, water velocity and water turbulence, as well as temperatures in the range of 82 to 121°C (180 to 250°F), are minor factors on the influence of corrosion rate [75].

Evaluation of the behavior of copper alloys under conditions similar to those prevailing in desalination plants are reported by Cohen and Rice in hot, flowing and degassed seawater [76], by De Santis for Cu-Ni alloy combined with iron and manganese [77], by Cohen and Rice under conditions simulating the brine heater section of a MEMS flash desalting plant [78], by Yandushkin et al. under various velocities, whereas it was found that water velocities of 7 to 10 m/s markedly increase the corrosion rate [79], by Anderson confirming that 90/10 Cu-Ni and arsenical aluminum brass exhibit optimum performance [80], by Lucio classifying the resistance to corrosion of various Cu-Ni alloys in a 3% NaCl solution as increasing in the order Admiralty brass < 90/10 Cu-Ni < 70/30 Cu-Ni [81], by Konstantinova et al. investigating the effect of velocity and geometry of seawater flow at inlet regions of heat exchange tubes, consisting of Cu-Ni alloy, Admiralty brass, aluminum brass and aluminum bronze [82], and by Popplewell et al. examining the effect of iron on the corrosion characteristics of 90/10 Cu-Ni in quiescent NaCl solution [177].

Electrochemical studies were made of the corrosion of copper and copper-base alloys under conditions of boiling heat transfer in flowing NaCl solutions to determine the relative contribution of thermogalvanic and local corrosion effects. The studies included Admiralty brass, Al-brass, 90/10 and 90/30 Cu-Ni alloys [83].

Results of long time immersion tests of copper-nickel alloys were reported by Southwell and Alexander for 16 years [84], and by Yandushkin and Korkosh at various velocities for 6000 h [85]. Parameters affecting the corrosion of copper-nickel-iron alloys (Fe-10% Cu-1.5% Ni) applied in piping systems aboard ships are discussed by Kievits et al. [86]. A testing program reported by Kievits and Ijsseling on the corrosion behavior of CuNi10Fe in seawater included exposition to flowing seawater, tests aimed at determining the erosion behavior, electrochemical measurement of polarization behavior, corrosion potential and polarization resistance, as well as electron microscopic investigations into the structure and composition of protective layers [87].

The state of the art of fabricating copper-nickel and titanium alloy tubing, including methods, limitations and costs, was investigated and reported by Moran [88]. The production of bimetallic tubing for flash-distillation plants is reported by Agricola [89]. A corrosion-resistant alloy of the Cu-Zn-Ni-Al family has been developed for desalination systems. Modified by the addition of iron and manganese, the alloy can be fabricated soft and thermally treated to impart high strength. In a 131°C (268°F) saline solution an improvement in corrosion resistance over arsenical aluminum brass by a factor of three was noted. Also studied were means to protect the alloy and other commercially available alloys by anodic protection techniques [90].

Nickel-
legierungen

2.19.5 Nickel-base alloys

In evaluating nickel base alloys, 18Ni180 and 18Ni200 maraging steels have shown high resistance to stress corrosion cracking in seawater. U-bends had no cracks after exposure for over 3 years. 18Ni180 welds also had good resistance, but welds in the 18Ni200 grade suffered stress corrosion cracking. Cathodic protection of U-bends was beneficial. Fracture toughness decreased when the precracked cantilever bars were coupled to zinc. General corrosion rates of maraging steels in seawater and marine atmospheres were approximately half those of HY-80 and 4340 low alloy steels [91].

The relative corrosion behavior of 22 Ni-alloy samples in quiet and low velocity seawater was investigated by Niederberger et al. The Ni-Cr-Mo alloys have superior resistance to pitting and crevice corrosion. The Ni-Cr alloys are susceptible to localized attack. On the basis of performance in U-bend tests in seawater, none of the alloys appears to be susceptible to stress corrosion [92].

Aluminium
und
Aluminium-
legierungen

2.19.6 Aluminum and aluminum-base alloys

Condensers in desalting plants are actually made of copper alloys as material of construction. The use of aluminum alloys would considerably reduce the investment cost. Several aluminum alloys particularly resistant for this use are available. A prototype plant to produce 1500 m³/day was constructed to prove the applicability of aluminum alloys in desalination plant construction [101].

After two years of almost continuous operation, the aluminum desalination test plant in Free-port, Texas, remained in good condition. Aluminum piping, heat exchanger tubes and process equipment have exhibited a high resistance to the desalination plant exposures. Heat exchangers operated satisfactorily at flow rates not exceeding 1.5 m/s (5 ft/s) in the brine heaters and showed no evidence of anhydrite scale even at 127°C (260°F). Some localized attack occurred at dissimilar metal joints between stainless steel pumps and the aluminum alloy flanges. This was mitigated by inserting short lengths of plastic pipe [102]. The use of aluminum tubes in flash evaporator plants and in horizontal film type vapor compressor plants, developed in Israel, is described by Pachter et al. A horizontal multiple-effect aluminum tube evaporator pilot plant for seawater distillation with a capacity of 150 m³/day (50000 gpd) started operation in January 1971 [103].

Several aluminum alloy tubes were tested for 21 months in multi-stage flash distillation plant having carbon steel vessels at a terminal brine temperature of 121°C (250°F). Corrosion problems occurred when electrical insulation between dissimilar metals proved unreliable. Steel corrosion products deposited on the aluminum tubes causing pitting attack. Alloy 5052 exhibited excellent corrosion resistance to hot seawater brine. Other alloys showed promise under certain conditions. The solution potential of aluminum was found to be temperature dependent [93].

The electrochemical behavior of a number of candidate aluminum alloys was studied under the conditions of water chemistry and temperature found in the multi-stage flash distillation process. The study was conducted in several basic areas which included time-potential-temperature behavior, effect of velocity, weldment performance and polarization effect of reduced oxygen level [95]. Of five aluminum alloys tested by Legault, only one 5052 aluminum demonstrated a high level of resistance to a desalination environment [44]. Protection by inhibitor systems were also investigated. The optimum initial immersion conditions for aluminum appears to be 50 to 100 mg/m³ dissolved oxygen. Aluminum alloy performance was significantly more velocity dependent at 5 mg/m³ than at 30 to 125 mg/m³ dissolved oxygen and the general corrosion rate increased with temperature. The Al alloys 3003 and 5454 demonstrated the highest level of performance of the seven aluminum compositions tested [9].

Aluminum and several of its alloys have been tested by Wanklyn et al. in flowing, acid-treated brine at 120 and 140°C. All materials showed similar rates of uniform corrosion and even after relatively short times these had fallen to around 0.025 mm per year. However localized corrosion in the form of both pitting and crevice corrosion occurred in many tests. The pH of the brine and also its retention time in the test apparatus had an important influence on localized corrosion [96].

Stress corrosion cracking of 7079-T6 aluminum alloy in seawater was investigated by bend tests of smooth and precracked specimens. It was found that there was a general agreement between results of the two methods. The susceptibility of this alloy to intergranular cracking was shown to depend on its directional characteristics [97]. Five aluminum alloys, 1100, 3003, 5052, 5554 and 6061, were tested in deaerated seawater at 121°C (250°F) for two weeks. Only 5052 aluminum performed so as to suggest its use in desalination applications [98].

Corrosion films formed on aluminum in hot salt water after short time tests of thirty seconds to one hour were studied by Fraker and Ruff by electron transmission microscopy and diffraction. It was concluded from observations of 6061, 5454, 2024, Al-1.52 Mn and Al-1.66 Ni alloys that one controlling factor in the initiation and growth of the outer film of the duplex oxide film is the alloy composition. The outer film formed earlier on alloys containing copper and the film growth was also more rapid on these alloys. Additions of magnesium, manganese, and nickel retarded this outer film formation, but with nickel additions pitting attack was more severe and deposition of copper from solution was increased. The formation and growth of the outer film was slowest on the 5454 alloy which was also the most resistant alloy tested [99].

A large and representative test loop was used in order to compare and test the behavior of various alloys over a long time period. The test loop, with heat exchangers of light alloys, has been working on the Mediterranean Coast for two years with acid-treated seawater. Various clad and unclad alloys were tested. Some showed an excellent corrosion behavior in cold or hot seawater and their use can be considered in desalting installations [100].

Evaluation of the behavior of aluminum alloys under conditions imitating those prevailing in desalination plants are also reported by Suezawa et al. on aluminum-bronze pipes in a circulating 3% NaCl solution with emphasis on its pH [104], by Verink, testing the applicability of materials of construction in desalination plants [105], by Gladkii and Chmyrev as a function of pH [106], by Konstantinova et al. as a function of temperature, velocity and pH [107], and by Bairamov and Abasov under Caspian Sea water conditions [108].

Selective oxidation, which would yield a coherent corrosion-resistant surface layer on aluminum brass, was suggested as a very promising way of reducing corrosion rate and improving the life of aluminum brass in seawater converters [94].

On bolted stainless steel-Al alloy couples the crevice and surface corrosion of the Al alloy in seawater did not increase at steel to Al alloy area ratios 0.08 to 0.10. At a steel to Al alloy area ratio of 15 corrosion rates of the Al alloy increased considerably [130].

2.19.7 Titanium and titanium-base alloys

Extensive research work has been conducted establishing the technical superiority of titanium for seawater flash distillation equipment. The areas of immunity for commercially pure titanium, Ti-0.2 Pd and the Ti-2Ni formulations are outlined. The economic position of titanium to competitive materials is discussed based upon both operating experience and engineering studies [109]. Because of its compatibility with both primary water or steam and hot concentrated brine, titanium is considered to reduce the probability of tube failure in heat exchangers. It appears to suffer no corrosion attack other than the build-up of an adherent oxide tarnish film. In the hot brine environment pitting or crevice corrosion limits a commercially pure alloy as heat-exchanger tubing to service below 121°C (250°F). Corrosion evaluation studies and subsequent alloy development program, which has resulted in a series of titanium base alloys suitable for service up to 177°C (350°F), are reviewed [110].

Oxide films formed on five titanium alloys were characterized after corrosion in 3.5% NaCl solutions over the temperature range of 100 to 200°C. Electron transmission microscopy and electron diffraction of corroded thin foils showed the oxide film to be TiO and Ti_2O_3 in the early corrosion stages and to change in the 150 to 200°C temperature range to a final surface oxide, TiO_2 (anatase) [114].

Electrochemical aspects of the corrosion of titanium and a number of its alloys were studied in flowing, high-temperature salt solutions at temperatures up to 104°C (220°F) in a titanium loop facility. Polarization curves showed that a pitting potential exists at sufficiently high potentials for all the alloys. Increase of temperature greatly affects the pitting potentials of titanium alloys. Plots of pitting potential as a function of temperature suggest an upper temperature limit of usefulness for many titanium alloys without risk cf severe localized attack. Alloys of titanium which contain molybdenum show promise for application in high temperature saline waters [111].

Deposition of titanium coatings on low-carbon steels, accomplished by both electrolytic and nonelectrolytic techniques, is reported by Warnock and Stetson. Electrolytic coatings were applied at 843°C (1550°F) at 0.15 A/cm² (1 A/in²) and nonelectrolytic coatings were deposited at 927°C (1700°F). Boiling seawater, aerated room temperature seawater, autoclave (138°C or 280°F) and velocity effects corrosion tests showed the nonelectrolytic 0.076 mm (0.003 in) coating to possess a corrosion resistance essentially equal to that of pure titanium. The rate of coating deposition was found to be of prime importance in determining corrosion resistance, because the deposition rate largely controls the coating composition [112].

Galvanically induced hydriding of commercially pure titanium and titanium-2 nickel alloy was investigated by Charlot. Commercially pure titanium shows a degree of immunity to hydrogen pick-up when coupled to metals such as mild steel, copper, nickel or stainless steel. All titanium coupons, of both the nickel and the commercial purity alloys, demonstrated excellent corrosion resistance, apart from hydriding, to the environments. Aluminum and mild steel, of the dissimilar metals coupled to titanium, were aggressively attacked [113].

Evaluation of the corrosion resistance of titanium and titanium alloys in saline environment were also reported by Sato and Sagisaka, indicating that the general corrosion resistance to seawater decreased in the order of Ti > Al brass > cupronickel tubes [115], by Brown presenting the existence of a threshold stress for corrosion cracking of titanium alloys in salt water [116], by Cavallaro and Wilcox on stress corrosion cracking susceptibility of Ti-7Al binary alloy [117], by Shiobara and Morioka on the effects of anions and temperature on the pitting and crevice corrosion of titanium [118], by Shalaby on the effects of galvanic coupling of titanium with Admiralty brass, aluminum-brass, Cu-Ni and Al-Mg alloys, indicating that titanium is a very corrosive resistant material in 33 g/l NaCl solutions and that this behavior was not affected by coupling with the other alloys [119], by Litvin and Hill on the effect of pH cn seawater stress corrosion cracking of the Ti-7Al-2Nb-1Ta alloy [120], by Chu investigating the fracture behavior of selected titanium alloys by bend tests of notching specimens [121], by Jackson and Boyd describing metallurgical factors that appear to play a role in the stress corrosion cracking mechanisms in salt and seawater [122], by Nakajia describing the use of thin-walled titanium tubes for seawater desalination plants [123], and by Posey et al. investigating kinetics of pitting attack of titanium and titanium alloys in chloride and saline solutions [178].

2.19.8 Concrete in seawater environment

In the design studies of large scale multi-stage flash distillation plants, prestressed concrete vessels were used in the vacuum to medium pressure range up to temperatures of 115°C at moderate savings of capital cost.

The behavior of cement and concrete in seawater and in mineral-free distilled water was extensively investigated by the U.S. Bureau of Reclamation. In general, concretes containing high quality aggregates have not been detrimentally affected by exposure to synthetic seawater brine at temperatures between 95 and 121°C (203 to 250°F). No significant corrosion of the steel reinforcing bars or the steel pretensioning rods has occurred. Conventional portland cement concrete, uncoated and untreated, will not withstand the leaching effects of warm to hot flowing mineral-free distilled water. Various coating materials have also been tested and evaluated [167].

Studies in sulfate, chloride, magnesium salts and in seawater attack on portland cement and concrete are reported by Ben-Yair and the problem of building concrete structures for desalination plants is discussed [168]. Mortar and concrete, impregnated with polymers, has improved structural and durability properties. The high strength and anticipated resistance to brine and distilled water make this system a promising candidate for flash distillation units in desalination plants [169].

Experiments have been conducted to determine the feasibility and techniques for impregnation and in situ polymerization of liquid monomers in preformed concrete and the inclusion of monomers directly into the fresh concrete mix, followed by polymerization. It appears that the cost of the treatment will be reasonable. All monomers evaluated to date produce improved properties, their effects in concrete varying only in degree. For high-temperature desalting plant applications, monomers which have high-temperature stability were selected for screening tests. Compressive and tensile strengths and elastic properties measurements performed to date on four monomer systems at temperatures up to 143°C (290°F) indicate that 60% styrene plus 40% trimethylolpropane trimethacrylate produces the best strength properties, followed by 90% diallyl phthalate plus 10% methyl methacrylate [170].

Further investigations on the durability of cement and concrete in seawater are reported by Kalousek and Benton on the mechanisms of seawater attack [171], by Neville on the mechanisms of sulfate and chloride attack [172], by Gjorv examining the effects of 18 different kinds of cement, addition of trass, precuring, cement content, water-cement ratio and seawater versus fresh water as mixing water [173], by Mchedlov-Petrosian examining the effect of sulfate containing media [174], by Kaneko et al. reporting that the corrosion resistance of cement mortar and concrete against hot seawater was more affected by the water to cement ratio rather than by the type of cement used [175], and by Kasai and Shibata reporting that the amount of salts infiltrated into mortar and concrete soaked in seawater was increased gradually with increased period of soaking, but it was slightly affected by the types of cement used and the water to cement ratio [176].

2.19.9 Corrosion inhibition and protection

*Korrosions-
hemmung
und -schutz*

A low oxygen level plus a low level chromate-phosphate inhibitor concentration promises to control corrosion of mild steel in seawater at 121°C (250°F) at costs that make the environmental modification well worth pursuing. Aluminum alloy corrosion rates can be lowered significantly and pitting suppressed by the addition of sodium chromate plus sodium hydrogen carbonate. Since 3 mg/l chromate was effective in dynamic testing, the cost of added protection for aluminum would be sufficiently low to make the use of inhibiting agents an attractive possibility [135].

The effect of inhibitors on corrosion of various metals in seawater, by means of phosphate with either dichromate or chromate for steel, was also reported by Legault and Bettin [136], by Mor and Bonino with calcium gluconate for steel [137], and by Mor and Beccaria with acrylonitrile for copper [138] and copper-zinc alloy [139].

Protective coatings

Diffusion coatings were applied on metallic alloys using the molten salt electrolytic process to determine their erosion-corrosion resistance. Erosion testing was done using an ultrasonic horn. Of the diffusion coating-substrate combinations, those with highest erosion-corrosion resistance were beryllium diffused into copper-base alloys. The resistance was comparable to Inconel 718 and considerably better than 316 stainless steel. Without the beryllide diffusion coating, the copper base alloys showed poor erosion-corrosion resistance [131].

Fourteen paint and varnish coatings were tested to determine their applicability for the protection of marine equipment and desalination plants [132].

The best coating for using carbon steel instead of stainless steel in seawater evaporators was chlorinated chloroprene rubber primer covered with chlorosulfonated polyethylene [133]. Epoxy-pitch type coatings were found to have the highest durability in nuclear desalination plants [134].

Cathodic protection

For steel structures in natural seawater there is an approximate equality between the corrosion rate and the amount of cathodic protection current required to prevent corrosion. The average corrosion rate of steel and iron specimens continuously immersed in natural seawaters world wide is surprisingly uniform. Any rates much higher than the 0.13 mm/y (5 mils/y) result from peculiar environmental corrosion-accelerating factors. Variations in a natural seawater environment that affect the design of cathodic protection systems are water velocity, electrical resistivity, temperature and turbulence. For cathodic protection of structures in natural seawater, galvanic anode systems employing high-efficiency Al alloy anodes are recommended [140]. Improvement is needed in techniques of cathodically protecting marine heat exchangers, condensers and similar structures. Studies to determine design criteria, cathode current density, anode geometry, reference electrode control location and component requirements for cathodic protection of these structures are discussed by Byrne [141].

The application of cathodic protection to protect steel structures in seawater environments is also reported by Wijngaard recommending the use of zinc and aluminum anodes [142], by Ulanovskii recommending that the potential of the protected structure be as much as possible close to the protecting potential in order to have the lowest current density [143], and by Fitzgerald assuming that the life of marine structures could be extended at least by 20 years by means of cathodic protection [144].

Protection of stainless steel structures is reported by Lennox et al. [145], by Fujii et al. [146], and by Vreeland and Bedford [147]. The protection of aluminum and aluminum alloys is reported by Groover et al. [148], by Ulanovskii [149], and by Lyublinskii and Bibikov [150]. General views on the cathodic protection method were published by Robilliard [151], by Lehmann [152], by Doremus and Pass [153], and by Morgan [154].

The corrosion behavior of 5086-H34 and 6061-T6 structural aluminum alloys has been studied by Lennox et al. to determine if corrosion products from aluminum galvanic anodes containing mercury or tin would accelerate the corrosion of the structural alloys. There were no indications that corrosion products from either the Al-Hg-Zn or the Al-Sn-Zn galvanic anodes caused increased corrosion. Cathodic protection from either the Al-Hg-Zn or the Al-Sn-Zn anodes virtually eliminated corrosion on the 5086-H34 and 6061-T6 aluminum specimens that were either continuously or alternately immersed in seawater, except for slight corrosion under some of the anodes [155].

The suitability of various metals as anodic sacrificial material is also reported by Lennox et al. for aluminum compared with zinc [156], by Scott and Craig [157] and by Groover et al. for aluminum [158], by Lyublinskii for zinc [159], and by Lyublinskii et al. for manganese base alloys [160].

Literature to 2.19

[1] B. Todd (Desalination **3** [1967] 106/117). — [2] P. D. Miller, A. B. Tripler, J. J. Ward, F. H. Hanie, W. K. Boyd (Off. Saline Water Res. Develop. Progr. Rept. No. 174 [1966]). — [3] F. W. Fink (Off. Saline Water Res. Develop. Progr. Rept. No. 46 [1960]). — [4] F. W. Fink, E. L. White, W. K. Boyd (Off. Saline Water Res. Develop. Progr. Rept. No. 225 [1966]). — [5] W. F. Dietrich (Off. Saline Water Res. Develop. Progr. Rept. No. 244 [1967]).

[6] W. E. Bell, H. C. Bramer (Off. Saline Water Res. Develop. Progr. Rept. No. 308 [1967]). — [7] V. F. Fitzpatrick (Off. Saline Water Res. Develop. Progr. Rept. No. 346 [1968]). — [8] H. C. Behrens, F. D. Martin, O. Osborn, L. Rice, W. B. Russell, C. F. Schrieber, J. C. Williams (Off. Saline Water Res. Develop. Progr. Rept. No. 417 [1969]). — [9] H. C. Behrens, O. Osborn, L. Rice, C. F. Schrieber, B. P. Webb, A. L. Whitted (Off. Saline Water Res. Develop. Progr. Rept. No. 623 [1970]). — [10] H. C. Bramer, W. E. Bell, R. E. Moore, D. R. Surdock, D. R. Brenneman, J. M. McCrossin (Off. Saline Water Res. Develop. Progr. Rept. No. 535 [1970]).

[11] A. L. Whitted, C. F. Schrieber (AIChE [Am. Inst. Chem. Engrs.] Symp. Ser. **69** No. 129 [1973] 508/14). — [12] F. Bulda. K. R. Fröhner, F. Schmitt (Proc. 4th Intern. Symp. Fresh Water Sea, Heidelberg 1973, Vol. 2, p. 229/37). — [13] E. A. Newton, J. D. Birkett (Desalination **6** [1969] 229/37; Off. Saline Water Res. Develop. Progr. Rept. No. 512 [1969]). — [14] C. R. Southwell (Corrosion Sci. **9** [1969] 179/83). — [15] E. Heitz (Chem. Ingr.-Tech. **41** [1969] 96/9).

[16] G. Boisdé, H. Coriou, L. Grall, C. Mahieu, M. Pelras (Proc. Symp. Nucl. Desalination, Madrid 1968 [1969], p. 895/914). G. Boisdé, H. Coriou, L. Grall, C. Mahieu, R. Mahut, M. Pelras (Proc. 3rd Intern. Symp. Fresh Water Sea, Dubrovnik 1970, Vol. 1, p. 579/90). — [17] K. Yamashita, A. Ishida, S. Kiyooka (Boshoku Gijutsu **18** [1969] 206/11; C.A. **71** [1969] No. 94658). — [18] K. G. Compton (Corrosion **26** [1970] 448/9). — [19] F. W. Fink, G. A. Di Boari, E. L. White, W. K. Boyd, F. H. Haynie (Mater. Prot. Performance **10** No. 7 [1971] 17/22). — [20] S. P. Sukach, Yu. D. Klovatskii, I. B. Shenderovich, N. S. Vinarskii (Vodosnabzh. Kanaliz. Gidrotekhn. Sooruzheniya No. 12 [1971] 3/7; C.A. **75** [1971] No. 132796).

[21] R. Cigna (Rass. Chim. **23** [1971] 137/40). — [22] F. M. Reinhart, J. F. Jenkins (U.S. Natl. Tech. Inform. Serv. AD 743872 and AD 743875 [1972]). — [23] H. Coriou, L. Grall, C. Mahieu, R. Mahut, R. Yvet (Proc. 4th Intern. Symp. Fresh Water Sea, Heidelberg 1973, Vol. 2, p. 239/47). — [24] R. E. Bailie, A. H. Tuthill, B. Todd (Proc. 3rd Intern. Symp. Fresh Water Sea, Dubrovnik 1970, Vol. 1, p. 549/78). — [25] O. Osborn, C. F. Schrieber, H. G. Smith (Chem. Eng. Progr. **66** No. 7 [1970] 74/9).

[26] M. G. J. Desensy (Chem. Eng. **77** No. 13 [1970] 182/9). — [27] R. Bom (Brit. Corros. J. **5** [1970] 258/63). — [28] K. Wangnick (Tech. Mitt. Krupp Werksber. **30** No. 4 [1972] 175/82). — [29] H. Westphal (Meerestechnik **3** [1972] 23/9). — [30] International Nickel Ltd. (Copper-Nickel and other Alloys for Desalination Plant, London 1973).

[31] M. E. Mattson, R. M. Fuller (Off. Saline Water Res. Develop. Progr. Rept. No. 163 [1965]). — [32] S. Yamada, Y. Kibe (Ishikawajima-Harima Giho **11** [1971] 455/62; C. A. **76** [1972] No. 158174). — [33] D. Hirschfeld, O. Mahlstedt, K. Wangnick (Chemiker-Ztg. **95** [1971] 952/8). — [34] D. H. Van Nieuwenhuizen (Ingenieur [Utrecht] **84** No. 36 [1972] W71/W75). — [35] J. Lev-Er, M. Adler, M. Berent, A. C. Friedland (Proc. 4th Intern. Symp. Fresh Water Sea, Heidelberg 1973, Vol. 2, p. 259/67).

[36] H. M. Zuccolotto (Proc. 4th Intern. Symp. Fresh Water Sea, Heidelberg 1973, Vol. 2, p. 307/20). — [37] E. P. Shea, W. R. Hollingshad (Off. Saline Water Res. Develop. Progr. Rept. No. 166 [1965]). — [38] E. V. Konstantinova, L. S. Semenova (Desalination **5** [1968] 90/4; Zashchita Metal. **5** [1969] 130/3; C. A. **70** [1969] No. 108535). — [39] T. Songa, G. Casarini (Termotecnica [Milan] **23** [1969] 623/30), T. Songa, G. Careri (Termotecnica [Milan] **23** [1969] 617/22). — [40] T. Songa, G. Careri, G. Casarini, R. De Santis (Proc. 4th Intern. Symp. Fresh Water Sea, Heidelberg 1973, Vol. 2, p. 285/94).

[41] R. A. Markovich, L. A. Suprun (Zashchita Metal. **6** [1970] 557/61; C. A. **74** [1971] No. 8899). — [42] M. Grassiani (Trib. CEBEDEAU [Centre Belg. Etude Doc. Eaux] **23** [1970] 547/50). — [43] A. Saito, H. Shigeno, T. Kumagai (Boshoku Gijutsu **19** [1970] 302/6; C. A. **74** [1971] No. 106495). — [44] R. A. Legault (Off. Saline Water Res. Develop. Progr. Rept. No. 438 [1969]). — [45] T. Shibata, G. Okamoto, A. Muzao, T. Tsuchida (Trans. Iron Steel Inst. Japan **9** [1969] 239/44; C. A. **71** [1969] No. 104329).

[46] M. Kowaka, H. Nagano (Chem. Econ. Eng. Rev. **4** No. 6 [1972] 51/6; C. A. **77** [1972] No. 92676). — [47] N. Pessall, F. C. Hull, N. Michael, C. Liu (Off. Saline Water Res. Develop. Progr. Rept. No. 394 [1969]). — [48] N. Pessall, F. C. Hull, C. Liu (Off. Saline Water Res. Develop. Progr. Rept. No. 478 [1969], No. 627 [1970]). — [49] Y. Sataka, T. Moroishi, T. Nakajima (Boshoku Gijutsu **18** [1969] 355/60; C. A. **72** [1970] No. 58132). — [50] A. Takamura, K. Arakawa, K. Fujiwara, H. Hirose (Boshoku Gijutsu **19** [1970] 294/301; C. A. **74** [1971] No. 129416).

[51] A. Tamada, M. Tanimura (Boshoku Gijutsu **21** [1972] 513/22; C. A. **79** [1973] No. 22376). — [52] E. R. Duffy, J. F. Nachman (Off. Saline Water Res. Develop. Progr. Rept. No. 626 [1970]). — [53] E. R. Duffy. J. F. Nachman (Off. Saline Water Res. Develop. Progr. Rept. No. 633 [1971]). — [54] A. R. Troiano, R. F. Heheman (Off. Saline Water Res. Develop. Progr. Rept. No. 637 [1970]). — [55] S. Brunet, H. Coriou, L. Grall, C. Mahieu, M. Pelras (Desalination **3** [1967] 118/24).

[56] V. K. Zhuravlev, M. K. Kurpetov, M. N. Fokin, V. I. Oreshkin (Korroziya i Zashchita Metal. **1970** 117/21; C. A. **74** [1971] No. 60074). — [57] S. Ohte, M. Muramatsu, A. Ishida (Zairyo **18** [1969] 418/23; C. A. **71** [1969] No. 24007). — [58] S. Henrikson, J. Olsson, J. Arvesen (Proc. 6th Scand. Corr. Congr., Gothenburg 1971, Paper 35, p. 1/8). — [59] J. Degerbeck (Proc. 6th Scand. Corr. Congr., Gothenburg 1971, Paper 36, p. 1/11). — [60] H. Kohl (Werkstoffe Korrosion **23** [1972] 984/93).

[61] K. Tanno, K. Makabe, Y. Furutani, N. Kawashima (Boshoku Gijutsu **21** [1972] 219/23; C. A. **77** [1972] No. 117098). — [62] B. E. Wilde (Corrosion **28** [1972] 283/91). — [63] I. B. Ulanovskii, Yu. M. Korovin (Zashchita Metal. **8** [1972] 321/3). — [64] M. Mizuno (Chem. Econ. Eng. Rev. **2** No. 4 [1970] 39/42; C. A. **72** [1970] No. 136220). — [65] J. Degerbeck (Chem. Proc. Eng. **52** No. 12 [1971] 47/50).

[66] A. Bäumel, K. Bohnenkamp, K. Schäfer, G. Lennartz (Corrosion Fouling Metals Sea Brackish Waters, Travemünde 1972). — [67] A. Kügler, K. Bohnenkamp, G. Lennartz, K. Schäfer (Stahl Eisen **92** [1972] 1026/30). — [68] A. Bäumel, F. Bachmann (Arch. Eisenhüttenw. **43** [1972] 631/7). — [69] E. H. Phelps, R. T. Jones, H. P. Leckie (J. Electrochem. Soc. **116** [1969] 213c/217c). — [70] T. R. Harkins, H. H. Lawson (Water Wastes Eng. **7** No. 1 [1970] A13/A16).

[71] D. W. C. Baker, W. E. Heaton, B. C. Patient (Corrosion Sci. **12** [1972] 247/64). — [72] N. Pessall, J. Nurminen (Off. Saline Water Res. Develop. Progr. Rept. No. 741 [1972]). — [73] K. N. Yandushkin (Zashchita Metal. **6** No. 1 [1970] 46/8; C. A. **72** [1970] No. 103546). — [74] E. H. Newton, J. D. Birkett (Off. Saline Water Res. Develop. Progr. Rept. No. 278 [1967]). — [75] H. C. Behrens, F. D. Martin, O. Osborn, L. Rice, W. B. Russell, C. F. Schrieber, J. C. Williams (Off. Saline Water Res. Develop. Progr. Rept. No. 417 [1969]).

[76] A. Cohen, L. Rice (Mater. Prot. Performance **8** No. 12 [1969] 67/9). — [77] R. De Santis (Rame **7** No. 25 [1969] 22/30). — [78] A. Cohen, L. Rice (Mater. Prot. Performance **9** No. 11 [1970] 29/35). — [79] K. N. Yandushkin, V. P. Kuris, V. P. Baranik (Zashchita Metal. **7** [1971] 83/7). — [80] D. B. Anderson (Mater. Prot. Performance **10** No. 11 [1971] 26/9).

[81] G. Lucio (Quad. Ric. Sci. No. 58 [1969] 489/507). — [82] E. V. Konstantinova, V. I. Gebelev, L. A. Minuhin (Proc. 4th Intern. Symp. Fresh Water Sea, Heidelberg 1973, Vol. 2, p. 249/57). — [83] P. J. Boden (Corrosion Sci. **11** [1971] 353/62, 363/70). — [84] C. R. Southwell, A. L. Alexander (Mater. Prot. Performance **8** No. 3 [1969] 39/44). — [85] K. N. Yandushkin, S. V. Korkosh (Zashchita Metal. **6** [1970] 429/33).

[86] F. J. Kievits, F. P. Ijsseling, P. J. A. van de Berg, W. Wisse (Inst. Metals Monograph Rept. Ser. No. 34 [1970] 345/52). — [87] F. J. Kievits, F. P. Ijsseling (Werkstoffe Korrosion **23** [1972] 1084/96). — [88] F. J. Moran (Off. Saline Water Res. Develop. Progr. Rept. No. 540 [1970]). — [89] K. R. Agricola (Off. Saline Water Res. Develop. Progr. Rept. No. 686 [1971]). — [90] F. H. Cocks, A. H. Taylor, S. B. Brummer (Off. Saline Water Res. Develop. Progr. Rept. No. 801 [1972]).

[91] N. Kenyon, W. W. Kirk, D. van Rooyen (Corrosion **27** [1971] 390). — [92] R. B. Nieder-berger, R. J. Ferrara, F. A. Plummer (Mater. Prot. Performance **9** No. 8 [1970] 18/22). — [93] D. A. Fauth, R. I. Lindberg (Off. Saline Water Res. Develop. Progr. Rept. No. 583 [1970]). — [94] M. A. H. Howes (Off. Saline Water Res. Develop. Progr. Rept. No. 628 [1970]). — [95] W. H. French, J. S. Snodgrass (Off. Saline Water Res. Develop. Progr. Rept. No. 688 [1971]).

[96] J. N. Wanklyn, N. J. M. Wilkins, D. R. V. Silvester, C. E. Austing, P. F. Lawrence (Proc. 3rd Intern. Symp. Fresh Water Sea, Dubrovnik 1970, Vol. 1, p. 617/29; Desalination **9** [1971] 245/58). — [97] H. P. Chu, G. A. Wacker (J. Basic Eng. **91** [1969] 565/9). — [98] R. A. Legault, W. J. Bettin (Mater. Prot. Performance **10** No. 3 [1971] 9/12). — [99] A. C. Fraker, A. W. Ruff (Corrosion **27** [1971] 151/6). — [100] C. Vargel, M. Pelletier (Proc. 3rd Intern. Symp. Fresh Water Sea, Dubrovnik 1970, Vol. 1, p. 601/15).

[101] C. Vargel, M. Mirabel (Proc. 4th Intern. Symp. Fresh Water Sea, Heidelberg 1973, Vol. 2, p. 295/306). — [102] E. D. Verink (Mater. Prot. Performance **12** No. 2 (Pt. 1) [1973] 34/6), E. D. Verink, P. F. George (Mater. Prot. Performance **12** No. 5 [1973] 26/30) — [103] M. Pachter, A. Barak, J. Weinberg (Proc. 4th Intern. Symp. Fresh Water Sea, Heidelberg 1973, Vol. 1, p. 439/50). — [104] Y. Suezawa, T. Shinohara, T. Terao, H. Ogiyama (Kagaku Kogaku **33** [1969] 1030/7; C.A. **72** [1970] No. 58951). — [105] E. D. Verink (Mater. Prot. Performance **8** No. 11 [1969] 13/5).

[106] I. N. Gladkii, Yu. P. Chmyrev (Khim. Neft. Mashinostr. No. 8 [1969] 18/9; C.A. **71** [1969] No. 115930). — [107] E. V. Konstantinova, S. I. Spector, L. S. Semenova, B. I. Kursanova (Proc. 3rd Intern. Symp. Fresh Water Sea, Dubrovnik 1970, Vol. 1, p. 591/600). — [108] M. M. Bairamov, T. A. Abasov (Vodosnabzh. Sanit. Tekhn. **1970** No. 5, p. 15/8; C.A. **74** [1971] No. 24922). — [109] N. G. Feige, R. L. Kane (Desalination **7** [1969] 17/22). — [110] N. G. Feige (Proc. Symp. Nucl. Desalination, Madrid 1968 [1969], p. 815/25).

[111] F. A. Posey, E. G. Bohlmann (Desalination **3** [1967] 269/79). — [112] R. V. Warnock, A. R. Stetson (Off. Saline Water Res. Develop. Progr. Rept. No. 395 [1969]). — [113] L. A. Charlot (Off. Saline Water Res. Develop. Progr. Rept. No. 624 [1970]). — [114] A. C. Fraker, A. W. Ruff (Corrosion Sci. **11** [1971] 763/5). — [115] S. Sato, K. Sagisawa (Sumitomo Keikinzoku Giho **11** [1970] 114/32; C.A. **73** [1970] No. 58715).

[116] B. F. Brown (J. Mater. Sci. **5** [1970] 786/91). — [117] J. L. Cavallaro, R. C. Wilcox (Corrosion **27** [1971] 157/63). — [118] K. Shiobara, S. Morioka (J. Japan Inst. Metals [Sendai] **35** [1971] 812/20, 980/4). — [119] L. A. Shalaby (Corrosion Sci. **11** [1971] 767/78). — [120] D. A. Litvin, B. Hill (Corrosion **26** [1970] 89/94).

[121] H. P. Chu (Eng. Fract. Mech. **4** [1972] 107/17). — [122] J. D. Jackson, W. K. Boyd (Proc. Intern. Conf. Sci. Technol. Appl. Titanium, 1968 [1970], p. 257/81). — [123] H. Nakajia (Chem. Econ. Eng. Rev. **4** No. 6 [1972] 45/50; C.A. **77** [1972] No. 92623). — [124] N. J. Fleet-wood (Werkstoffe Korrosion **2** [1970] 267/73). — [125] J. A. von Fraunhofer (Anti-Corrosion Methods Mater. **16** No. 5 [1969] 21).

[126] R. A. Markovich, L. A. Suprun (Tr. Tsentr. Nauchn. Issled. Inst. Morsk. Flota **1971** No. 139, p. 32/6; C.A. **77** [1972] No. 38991). — [127] M. Beckert, J. Heinecke (Schweißtechnik **20** [1970] 536/41). — [128] S. Haagenrud (Korrosjons Nytt **9** No. 1 [1970] 6/11; C.A. **76** [1972] No. 29993). — [129] A. Bäumel (Werkstoffe Korrosion **23** [1972] 546/54). — [130] I. B. Ulanovskii, Yu. M. Korovin (Korroziya i Zashchita Metal. **1970** 177/9; C.A. **74** [1971] No. 70770).

[131] J. L. O'Brien (Off. Saline Water Res. Develop. Progr. Rept. No. 716 [1971]). — [132] V. G. Shigorin, Yu. M. Khovanskii, B. N. Egorov (Lakokrasoch. Mater. Ikh Primen. **1969** No. 4, p. 33/5; C. A. **72** [1970] No. 4373). — [133] V. G. Shigorin, I. A. Gavritenkov, B. N. Egorov (Lakokrasoch. Mater. Ikh Primen. **1969** No. 5, p. 19/21; C. A. **72** [1970] No. 13855). — [134] V. G. Shigorin, B. N. Egorov (Proc. 3rd Intern. Symp. Fresh Water Sea, Dubrovnik 1970, Vol. 4, p. 169/86). — [135] B. D. Oakes, J. S. Wilson (Off. Saline Water Res. Develop. Progr. Rept. No. 649 [1971]).

[136] R. A. Legault, W. J. Bettin (Mater. Prot. Performance **9** No. 9 [1970] 35/9). — [137] E. Mor, G. Bonino (Ann. Univ. Ferrara Sez. 5, Suppl. No. 5 [1970] 659/70). — [138] E. D. Mor, A. M. Beccaria (Ann. Chim. [Rome] **62** [1972] 229/38). — [139] E. D. Mor, A. M. Beccaria (Corrosion Traitements Protect. Finition **20** [1972] 460/6). — [140] J. G. Davis, G. L. Doremus, F. W. Graham (J. Petrol. Technol. **24** [1972] 323/8).

[141] P. B. Byrne (Mater. Prot. Performance **10** No. 3 [1971] 21/3). — [142] B. H. Wijngaard (TNO Nieuws **25** [1970] 450/5). — [143] I. B. Ulanovskii (Zashchita Metal. **8** [1972] 213/6). — [144] J. H. Fitzgerald (Mater. Prot. Performance **11** No. 5 [1972] 23/7). — [145] T. J. Lennox, R. E. Groover, M. H. Peterson (Mater. Prot. Performance **8** No. 5 [1969] 41/8).

[146] M. Fujii, S. Saeki, M. Kamada (Nippon Kinzoku Gakkaishi **34** [1970] 534/8; C. A. **73** [1970] No. 30980). — [147] D. C. Vreeland, G. T. Bredford (Mater. Prot. Performance **9** No. 8 [1970] 31/4). — [148] R. E. Groover, T. J. Lennox, M. H. Peterson (Mater. Prot. Performance **8** No. 11 [1969] 25/30). — [149] I. B. Ulanovskii (Zashchita Metal. **6** [1970] 600/2; C. A. **74** [1971] No. 8891). — [150] E. Ya. Lyublinskii, N. N. Bibikov (Zashchita Metal. **8** [1972] 36/9).

[151] C. B. Robilliard (Australian Chem. Proc. Eng. **22** No. 4 [1969] 18/20). — [152] J. A. Lehmann (J. Metals **22** No. 3 [1970] 56/63). — [153] E. P. Doremus, R. B. Pass (Mater. Prot. Performance **10** No. 5 [1971] 23/7). — [154] J. H. Morgan (Brit. Corrosion J. **5** [1970] 237/8). — [155] T. J. Lennox, R. E. Groover, M. H. Peterson (U.S. Naval Res. Lab. Rept. 7648 [1973]).

[156] T. J. Lennox, R. E. Groover, M. H. Peterson (Mater. Prot. Performance **10** No. 9 [1971] 39/44). — [157] J. R. Scott, H. L. Craig (Proc. 26th Conf. Natl. Assos. Corrosion Eng., Philadelphia 1970 [1971], p. 179/87). — [158] R. E. Groover, J. A. Smith, T. J. Lennox (Corrosion **28** [1972] 101/4). — [159] E. Ya. Lyublinskii (Zashchita Metal. **6** [1970] 436/9). — [160] E. Ya. Lyublinskii, R. I. Agladze, K. G. Makharadze, G. Sh. Mamporiya (Elektrokhim. Margantsa **4** [1969] 223/31; C. A. **73** [1970] No. 104818).

[161] I. Ehlert, M. Pantke (Werkstoffe Korrosion **23** [1972] 196/205). — [162] K. G. Compton (Off. Saline Water Res. Develop. Progr. Rept. No. 662 [1971]), C. A. Smith, G. K. Compton, F. H. Coley (Corrosion Sci. **13** [1973] 677/85). — [163] H. P. Vind, M. J. Noonan (U.S. Clearinghouse Fed. Sci. Tech. Inform. AD 684423 [1969]). — [164] L. S. Alekseev (Tr. Vses. Nauchn. Issled. Inst. Vodosnabzh. Kanaliz. Gidrotekhn. Sooruzh. Inzh. Gidrogeol. **1970** No. 25, p. 50/4; C. A. **73** [1970] No. 80377). — [165] A. A. Mazo, I. K. Marshakov, V. K. Filina, A. A. Obraztsov (Tr. Voronezhsk. Gos. Univ. **72** [1969] 85/6; C. A. **78** [1973] No. 7710).

[166] E. I. Crossley, F. O. Waters (J. Am. Water Works Assoc. **62** [1970] 188/94). — [167] U.S. Bureau of Reclamation (Off. Saline Water Res. Develop. Progr. Rept. No. 345 and No. 390 [1968]). — [168] M. Ben-Yair (Desalination **3** [1967] 147/54). — [169] B. Manowitz, M. Steinberg, L. Kukacka, P. Colombo (BNL-13732 [1969]). — [170] J. E. Backstrom, J. T. Dikeou (Desalination **9** [1971] 97/125).

[171] G. L. Kalousek, E. J. Benton (J. Am. Concrete Inst. **67** [1970] 187/92). — [172] A. M. Neville (J. Mater. **4** [1969] 781/816). — [173] O. E. Gjorv (J. Am. Concrete Inst. **68** [1971] 60/7). — [174] O. P. Mchedlov-Petrosian (Silikattechnik **23** [1972] 93/6). — [175] M. Kaneko, W. Kondo, K. Fujii (Semento Konkuriito **1972** No. 306, p. 12/8; C. A. **77** [1972] No. 143389).

[176] Y. Kasai, O. Shibata (Semento Gijutsu Nempo **25** [1971] 274/7; C. A. **77** [1972] No. 143409). — [177] J. M. Popplewell, R. J. Hart, J. A. Ford (Corrosion Sci. **13** [1973] 295/309). — [178] F. A. Posey, E. G. Bohlmann, S. S. Misry, D. V. Subrahmanyam, F. Nelson (Off. Saline Water Res. Develop. Progr. Rept. No. 852 [1973].

2.20 Disposal of effluents from desalination plants

Beseitigung der Sole von Entsalzungs-anlagen

Desalination plants located on the sea shore will discharge their effluents into the marine environment. Two effluent streams will probably be used and each will have different properties: The brine blowdown and the cooling water from either the desalination and the power plant. The blowdown, under all combinations of temperature and salinity, will have a greater specific gravity than seawater and will sink to the ocean floor. The cooling water with increased temperature will be less dense than seawater and stratify at the ocean surface. Mixing of the two makes a stream of intermediate properties.

In addition to increased temperature and salinity, the effluent streams will be decreased in dissolved oxygen and alkalinity, will probably be chlorinated and will contain small amounts of heavy metals, especially copper from corrosion and erosion, eventually phosphates or other chemical additives, as a result of scale control and of marine biofouling, which might have a detectable impact upon the local marine environment.

The concentration of copper in the mixed effluent stream is estimated to be many times that encountered in the open ocean, as well as sufficiently high to be toxic to many organisms and create problems to the environment [1].

Ökologische
Wirkungen

2.20.1 Ecological effects

The presence of certain heavy metals in the brine will depend upon the materials of construction used. Desalination wastes present therefore the problem of ascertaining the possible effect on the various members of marine plankton. Phytoplankton can tolerate slightly elevated temperatures and salinities, but the growth rates will decline markedly. Temperature tolerance will be closely related to the nutritional capacity of the water and to the light intensity to which the plant cells are exposed.

The presence of copper ions in desalination effluents in quantities above the metabolic needs of the marine plankton algae introduces an additional environmental stress which must be carefully considered. Static toxicity experiments for copper ions on marine plankton indicated a broad range of tolerance, from 0.05 to 0.5 mg/l. Other marine organisms are more sensible to the presence of copper [2].

Increased concentrations of copper in the aquatic ecosystem will probably result in a higher accumulation and storage of this element at different trophic levels above their normal content. Copper is concentrated by phytoplankton from the aquatic environment as high as 5000 : 1. Zooplankton will concentrate the element even more, three to four times higher than the phytoplankton which is consumed as food. The concentration pattern of copper in marine organisms can be followed from phytoplankton to zooplankton to favorable sites of concentration in fishes, such as the liver, visceral masses and intestines.

Zinc and nickel ions produce the same type of general effect on both marine plants and animals as found with copper, but the concentrations necessary to produce the effect to the same extent are considerably higher. Copper uptake by certain benthic forms can be doubled by an increase in the temperature of only 10°C [6]. Long-term multivariate, seasonal experiments were conducted on juvenile and adult specimens of benthic fauna in order to evaluate the effects of the seawater-brine mixtures, as well as on the eggs and larval stages of these marine organisms, using short-term bioassay tests [18].

A survey and review of literature has been made in an effort to determine the possible effect of introducing large volumes of brine effluent on the marine environment of the Gulf of California. Heat budget calculations assuming complete vertical mixing show annual mean temperature increases above ambient values to be only 1 to 2°C for any 130 km² area. Salinity increases resulting from the brine effluent could create a hypersaline environment at the northern end of the Gulf. Other physical and chemical effects, such as oxygen, pH and turbidity, were considered to be minor in comparison to temperature and salinity increases [8].

The effect of waste heat and therefore water temperature on the plants and animals in a body of water was also examined by Hemers [9], by Brown indicating that addition of heat to water bodies may produce adverse effects on aquatic organisms [10], by Polk et al. presenting analytical models to allow prediction of the temperature regime for three particular distribution classifications [11], by El Mahgary discussing the thermal diffusion of warm water from power plants into a sea basin [12], by Ross examining the effects on aquatic life of warm water discharges into rivers and the sea [13], and by Elwood reporting on four years of experience in determining standards and control procedures for thermal discharges from power plants [14].

The term of "thermal pollution" is used to indicate that the uncontrolled flow of waste quantities of heat into a river or lake could have detrimental effects on aquatic life. The U.S. Atomic Energy

Commission has promoted the term "thermal beneficiation" on the basis that addition of warm water to a certain extent might be beneficial to lakes and streams. The point is that waste heat need not necessarily be thermal pollution and can eventually be converted to an asset by proper utilization [15].

There are many suggestions about beneficial use of waste heat, which can be found in the technical literature. The importance of the problem might be emphasized by a worldwide idea contest sponsored in early 1970 by the Swedish Association of Engineers and Architects for the proper uses of the enormous quantity of residual energy rejected back to water. The concepts honored by the technical jury are outlined in a paper by Lusby and Somers [16].

2.20.2 Removal of copper

Beseitigung des Kupfers

The average copper content of seawater is about 0.003 mg/l in open ocean. Coastal and inshore waters may contain a total copper concentration higher than that found in open ocean seawater. Analysis of copper in process streams from three desalination plants gave the following copper contents in mg/l:

	Freeport	Clair Engle	Wrightsville Beach
Intake	0.02	0.054	0.135
Cooling water	0.057	0.020	0.019
Blowdown	0.342	0.648	0.605

The estimated copper content of the mixed effluent is 4 to 7 times as high as the recommended water quality criteria of 0.02 mg/l of copper and 25 to 45 times the copper concentration in the receiving environment.

Copper was found to be more soluble in seawater concentrates than in seawater, with a minimum solubility at pH 7.5. It is thus advisable to operate desalination plants with the circulated brine being maintained at a pH around 7.5, where copper solubility is at a minimum.

The removal of copper from brine blowdown by plating on aluminum was found to be technically infeasible. The addition of copper chelating agents is successful in rendering the copper non-toxic, but appears to be too expensive. The use of chelating resins is efficient but also not economical [2].

Carbon sorbates, produced by sorption of chelating agents on active carbon, act as ion exchangers in the hydrogen form and have been found effective in removing copper ions from brines. Six carbons were evaluated and salicylaldoxime, 8-hydroxyquinoline, benzoylacetone, dibenzoylmethane, anthranilic acid and mercaptobenzothiazole reagents appear useful. Extraction capacity, relative rate of extraction, reagent wash-off rates and performance in beds have been evaluated [17].

2.20.3 Dispersion of the effluent

Verteilung der Sole

The physical and chemical properties of the effluents from a distillation plant will depend on the design. The estimated effluent properties from two distillation plants, one using the multi-stage flash evaporation process (MSF) and the other the multiple-effect long tube vertical falling film evaporation process (VTE), both dual-purpose plants with a capacity of about 570 000 m³/day (150 Mgd) and 1800 MW$_e$ electric power are given in Table 2/13, assuming that copper base alloys are used for condenser tubing, as far as the copper content in the effluents is concerned. The effluent, after mixing of the blowdown with the cooling water, will always be more dense than the receiving ocean water and therefore will tend to sink to the ocean bottom. A good dispersion and mixing of the effluents in the receiving waters is required. The theoretical aspects of a vertical jet of dense fluid were investigated by an approximate analytical solution of the differential equations governing the flow and density distribution for an upwardly projected dense jet, as well as by a numerical integration technique [2]. In a third phase the dispersion of the dense effluent into seawater was studied in a laboratory model. The proportionality of the average height of dense jets to the flow conditions and density differences was established experimentally. The distribution of dye concentration across the jet was found to be Gaussian, with two distinctly different slopes identifying a middle region of upward flow, surrounded by an outer region of downward flow. Turbulent diffusion coefficients inside the jet have been calculated. The information obtained was used for the design of outfall systems from desalination plants [3].

Table 2/13

Properties of effluents from distillation plants.

		Flow rate in t/h	in million lbs/h	Temperature in °C	in°F	Salinity in °/oo	Density in g/cm³	Oxygen in mg/l	Copper in mg/l
MSF:	Feed	47174	104	15.6	60	34.0	1.025	7.2	0.003
	Blowdown	23587	52	22.2	72	68.0	1.048	0	0.50
	Cooling water	140616	310	22.2	72	34.0	1.024	7.2	0.01
	Mixed effluent	164203	362	22.2	72	38.8	1.028	6.3	0.08
VTE:	Feed	35381	78	15.6	60	34.0	1.025	7.2	0.003
	Blowdown	11794	26	29.4	85	168.0	1.121	0	4.8
	Cooling water	404611	892	22.8	73	34.0	1.024	7.2	0.01
	Mixed effluent	416405	918	25.3	77.6	37.6	1.026	7.0	0.14

Model studies to evaluate the effects of port diameter, brine flow rate, density differential, and ambient velocity on the geometry and mixing characteristics of a dense jet discharged vertically through a single port were conducted by Holly and Grace. A temperature differential of 10°C between the brine and ambient fluid has no significant effect on the flume mixing characteristics. A multiple-port diffuser has significant advantages over a simple outfall pipe [4]. The data obtained in the flume study were used to calibrate several existing models for the jet regime. The model chosen gives a reasonable simulation of the geometry of the jet axis and the dilution of the peak of the jet, but was found to oversimulate the lateral spreading factor. A numerical simulation of the dispersion of a dense effluent discharge into a homogeneous stream was derived [5].

Further studies have investigated the disposal of effluents from desalination plants into estuarine waters [6]. The dispersion rates of the brine waste and the dynamic equilibrium of the waste were studied on comprehensive models of the San Diego Bay, the Galveston Bay, and the Delaware River in the United States, these models being considered typical of the estuaries on which salt water conversion facilities are likely to be located [7].

Injektion in Tief-bohrungen

2.20.4 Deep well injection

Large volumes of brine from producing oil fields are re-injected in deep wells. This method has been proposed for disposal of brine effluents from inland desalting plants. Deep well injection has been also used for disposal of industrial wastes.

A suitable site for deep well injection requires a permeable sedimentary formation, such as sandstone or limestone, capped by an impermeable formation, such as shale, to prevent pollution of neighboring potable waters. Geologic and hydrologic investigations will be required to assure that the candidate site is satisfactory and to provide data to be used as the basis for designing an injection system [19].

The properties of a good desalination waste brine disposal formation are adequate thickness, areal extent, pore size, porosity and permeability. Variations in geology, geography and plant size will affect the cost of brine disposal by means of deep wells, which might develop to a substantial part of the total cost of desalination. The cost can often be lowered by reducing the volume of waste to be disposed of, by employing higher concentration ratios in the desalination process. Except in unusual circumstances, preconditioning of the waste prior to injection is not necessary [20].

Another technique of subsurface disposal of waste effluents consists of in situ mixing of waste water and indigenous ground water. using either single- and two-well recharge-discharge techniques or pit recharge. The two-well method appears to be the most favorable well technique for in situ mixing. Reasonably good mixing also occurred during pause and pulse type single-well tests [21].

Disposal of liquid wastes by injection underground was also investigated by Piper, emphasizing the problems and hazards of this process [22], by Rouston and Wildung, discussing the disposal of wastes to soil [23], by Garcia-Bençochea and Vernon, reporting on experience, results and potential for waste disposal in the so-called Boulder zone [24], and by Visser, indicating that underground disposal of wastes can be technically and economically attractive [25]. An annotated bibliography including 692 selective references on waste disposal through wells was compiled by Rima et al. [26].

2.20.5 Solar evaporation of waste brines

Solar evaporation as a disposal process has been used in an industrial scale in many arid and semi-arid regions. Studies of the effect of salinity and dyes on solar evaporation of brine have been conducted at the effluent evaporation ponds of the saline water conversion plant at Roswell, New Mexico. Several dyes were tested to evaluate the efficiency of increased evaporation and the economic feasibility of the addition of dyes. Naphthol green dye was found to increase the solar evaporation of brine more than any of the other dyes [27].

A bibliography including 875 publications was compiled to provide references on evaporation ponds, pond linings and soil treatments for watertight ponds, as well as waste disposal, properties of saline water and salt water pollution [28].

Various aspects of the problem of disposal of brine by solar evaporation are considered in a series of reports, sponsored by the U.S. Office of Saline Water. These reports include design data and a computer program for the evaluation of evaporation [29], laboratory studies of the effect of salinity, effect of dyes and design models [30], and results of field experiments [31]. Design criteria were investigated by using a heat transfer and a diffusion model [32]. A brine disposal pond manual, intended for designers and planners of brine effluent disposal from desalting plants, was prepared, which includes design criteria, construction and cost estimates, as well as operation and maintenance of disposal ponds [33]. Various pond lining materials and soil sealants were evaluated in laboratory and field tests. Recommendations on the selection of these materials and a monitoring system for measuring seepage losses from brine disposal ponds are given [34, 35]. Spray systems, as a method of increasing water evaporation rates, were also tested [36]. The technical and economic performance of spray evaporation ponds was compared with conventional solar evaporation. Evaluations are based on five years of hourly weather observations in Phoenix, Arizona, Roswell, New Mexico, Midland, Texas, and Aberdeen, South Dakota [37].

Further investigations related to waste disposal ponds are reported by Shulman, presenting a nomogram and an equation for sizing depth and evaporation rate [38], and by Kumer and Jedlicka, giving a review on materials for pond linings, their physical and chemical properties, as well as information on the performance of lining materials as far as available [39]. The cost of disposal of brines from hypothetical municipal water renovation schemes were investigated for El Paso, Tucson, and Denver. Disposal methods considered were evaporating ponds, deep wells, pipelines, distillation and appropriate combinations. Costs of disposal depended on climate and location [40].

Verdunstung von Sole mittels Sonnen- energie

2.20.6 Conversion of brine effluents to solids

Problems related to the disposal of brine effluents from inland desalination plants would be considerably reduced by converting the effluents in solid state. A combination of multiple-effect evaporation and direct-contact drying was chosen as the lowest cost and most practicable process to convert to dry solids the effluent from a 9500 m³/day (2.5 Mgd) multi-stage flash distillation plant [41].

Where waste disposal is a problem, it would be advantageous to treat the raw water to remove as much calcium as reasonably practical, permitting maximum product water recovery. Reduction of calcium content to 14 mg/l in the raw water permits roughly a five-fold reduction in waste brine volume to about 4% of the product water volume. Crystallization of the salts in a multiple-effect evaporator adds less than 10% to water cost, after credit for the additional water recovered. The crystallizing evaporator is very small. The tonnages and characteristics of the salts produced will usually not justify byproduct recovery for sale. Capital investment is much lower than for solar ponds but total annual costs are similar where solar conditions are favorable. A direct contact multi-stage flash process is proposed as a less expensive means of producing potable water and solids directly from some brackish water sources, characterized by having impurities mainly of calcium and magnesium sulfates [42].

The technical and economic feasibility of the crystallization/evaporation was evaluated for treating brine effluents from inland desalting plants. The results show that the unit cost of water will be increased 10 to 15%, depending on the salt concentration and composition [43].

In another report facility designs, capital costs and annual expenses were presented for systems of disposal of brine effluents produced by desalination plants operating on brackish waters at inland

Umwandlung von Sole in feste Stoffe

locations. The disposal systems reported on include solar ponds and multi-stage flash evaporators in conjunction with either conventional crystallizers or solar ponds. The brines considered were those that would result from desalination of four typical brackish waters found in the Western United States. Geographic applicability of the solar pond system is discussed and the results are displayed on a map of the area considered. A computer program, used to calculate pond size from brine and climate variables, is included [44].

Within legal, ecological and environmental constraints and using a representative range of brackish water compositions, geographical distribution, geological formations and other such factors, a technical and economic evaluation of all practical methods of final disposal was made using a systems approach. This was illustrated by s x case studies. Generalized cost ranges for the component subsystems were prepared for use in approximating systems costs and identifying the least-cost disposal method for use with other specific cases. The study showed that: subsurface injection of waste brines is generally less costly than other methods of final disposal; discharge of waste brines into the sea preceded by pipelining more than 112 km (70 miles) is not economically sound; electrodialysis is the most economical method of concentrating the effluent brine to 20% total dissolved solids at 15 000 m³/day (4 Mgd) product water rate, while solar evaporation is the most economical at the 1500 m³/day (0.4 Mgd) rate and is the most economical method of reducing a 20% brine to bulk dry solids [45].

Weitere Sole- beseitigungs- verfahren

2.20.7 Other effluent disposal methods

A new process for the disposal of waste brines from inland desalting plants was described by LaMont. A fluid-bed evaporator containing a bed of salt particles is used. Brine is fed into the bed and evaporates, while the salts are deposited on the particles. Steam leaving the fluidized vessel is condensed for energy and fresh water recovery. Incremental cost for brine disposal ranged from 3.6 to 10 cents per kgal (0.95 to 2.64 cents/m³) of product when coupled to a vertical tube evaporator plant or 19 cents per kgal when coupled to a reverse osmosis plant [46]. Evaluation of the process showed that it is technically and economically feasible [43].

Another process was suggested by Grinstead and Lingafelter, by which brines from the desalination of brackish water can be converted to either useful products or waste materials, which can be readily disposed of without environmental harm. The method involves precipitation of the sulfate with calcium chloride and exchange of chloride for hydrogen carbonate by means of a liquid-liquid anion exchange extraction system [47].

A freezing process for brine disposal was proposed by Stepakoff et al. At the eutectic point, both ice and salt crystallize out of salt water and brine volume is minimized. Thus, continuous eutectic freezing has major potential for brine disposal and concentration of industrial wastes. A laboratory study of the principal unit operations: crystallization, separation, and washing was described. The results prove technical feasibility and underlie a conceptual design of a two stage eutectic freezing process. Process economics appear to be favorable [48].

Literature to 2.20

[1] P. G. Le Gros, E. F. Mandelli, W. F. McIlhenny, D. E. Winthrode, M. A. Zeitoun, W. E. Pequegnat, W. G. Blanton, T. J. Bright, K. S. Bottoms (Off. Saline Water Res. Develop. Progr. Rept. No. 316 [1968]). — [2] M. A. Zeitoun, E. F. Mandelli, W. F. McIlhenny, R. O. Reid (Off. Saline Water Res. Develop. Progr. Rept. No. 437 [1969]). — [3] M. A. Zeitoun, W. F. McIlhenny, R. O. Reid (Off. Saline Water Res. Develop. Frogr. Rept. No. 550 [1970]; Chem. Eng. Progr. Symp. Ser. **65** No. 97 [1969] 156/66). — [4] F. M. Holly, J. L. Grace (Off. Saline Water Res. Develop. Progr. Rept. No. 714 [1971]). — [5] M. A. Zeitoun, W. F. McIlhenny (Off. Saline Water Res. Develop. Progr. Rept. No. 804 [1972]).

[6] M. A. Zeitoun, E. F. Mandelli, W. F. McIlhenny (Off. Saline Water Res. Develop. Progr. Rept. No. 415 [1969]). — [7] W. H. Bobb, R. A. Boland, F. A. Hermann (Off. Saline Water Res. Develop. Progr. Rept. No. 736 [1971]). — [8] D. A. Thomson, A. R. Mead, J. R. Schreiber (Off. Saline Water Res. Develop. Progr. Rept. No. 387 [1969]). — [9] J. Hemens (Chem. Eng. Progr. Symp. Ser. **65** No. 97 [1969] 47/50). — [10] F. S. Brown (Trans. ASCE J. Power Div. **96** No. PO-3 [1970] 277/86).

[11] E. M. Polk, B. A. Benedict, F. L. Parker (Chem. Eng. Progr. Symp. Ser. **67** No. 119 [1971] 111/9). — [12] Y. S. El Mahgary (J. Environm. Sci. **14** No. 6 [1971] 20/3). — [13] F. F. Ross (Effluent Water Treat. J. **11** [1971] 497/502). — [14] J. E. Elwood (J. New Engl. Water Poll. Contr. Assoc.

6 No. 1 [1972] 18/30). — [15] J. I. Bregman (Chem. Eng. Progr. Symp. Ser. **67** No. 107 [1971] 43/6).

[16] W. S. Lusby, E. V. Somers (Mech. Eng. **94** No. 6 [1972] 12/6). — [17] R. H. Moore (Off. Saline Water Res. Develop. Progr. Rept. No. 651 [1971]). — [18] E. F. Mandelli, W. F. McIlhenny (Off. Saline Water Res. Develop. Progr. Rept. No. 803 [1971]). — [19] W. J. Boegly, D. G. Jacobs, T. F. Lomenick, O. M. Sealand (Off. Saline Water Res. Develop. Progr. Rept. No. 432 [1969]). — [20] P. G. Le Gros, C. E. Gustafson, G. L. Nevill, E. C. Majeske, R. D. Mathews, J. S. Talbot, W. F. McIlhenny (Off. Saline Water Res. Develop. Progr. Rept. No. 456 [1969]).

[21] L. G. Wilson (Off. Saline Water Res. Develop. Progr. Rept. No. 650 [1971]). — [22] A. M. Piper (Chem. Eng. Progr. Symp. Ser. **65** No. 97 [1969] 5/18). — [23] R. C. Rouston, R. E. Wildung (Chem. Eng. Progr. Symp. Ser. **65** No. 97 [1969] 19/25). — [24] J. I. Garcia-Bengochea, R. O. Vernon (Water Resour. Res. **6** [1970] 1464/70). — [25] W. A. Visser (Ingenieur [Utrecht] **84** [1972] A 891/A 897).

[26] D. R. Rima, E. B. Chase, B. K. Myers (U.S. Geol. Surv. Water Supply Paper 2020 [1971]). — [27] N. N. Gunaji, C. G. Keyes (Off. Saline Water Res. Develop. Progr. Rept. No. 351 [1968]). — [28] G. F. De Puy (Off. Saline Water Res. Develop. Progr. Rept. No. 454 [1969]). — [29] C. G. Keyes, N. N. Gunaji, J. V. Lunsford, J. G. Seckler, O. D. Richard (Off. Saline Water Res. Develop. Progr. Rept. No. 547 [1970]). — [30] C. G. Keyes, D. C. Winans, C. Morales, H. R. Pritchett, W. S. Gregory (Off. Saline Water Res. Develop. Progr. Rept. No. 548 [1970]).

[31] C. G. Keyes, N. N. Gunaji, J. V. Lunsford, W. D. Loth (Off. Saline Water Res. Develop. Progr. Rept. No. 563 [1970]). — [32] C. G. Keyes, W. S. Gregory, N. N. Gunaji, J. V. Lunsford (Off. Saline Water Res. Develop. Progr. Rept. No. 564 [1970]). — [33] M. E. Day, E. L. Armstrong (Off. Saline Water Res. Develop. Progr. Rept. No. 588 [1970]). — [34] W. R. Morrison, R. A. Dodge, J. Merriman, L. M. Ellsperman (Off. Saline Water Res. Develop. Progr. Rept. No. 602 [1970]). — [35] W. R. Morrison, R. A. Dodge, J. Merriman (Off. Saline Water Res. Develop. Progr. Rept. No. 734 [1971]; U.S. Natl. Tech. Inform. Serv. Rept. PB 206454 [1971]).

[36] W. G. Smoak (Off. Saline Water Res. Develop. Progr. Rept. No. 480 [1969]). — [37] G. O. G. Löf, J. C. Ward, S. Karaki, A. Dellah (Off. Saline Water Res. Develop. Progr. Rept. No. 764 [1972]). — [38] W. Shulman (Chem. Eng. **79** No. 6 [1972] 134/6). — [39] J. Kumar, J. A. Jedlicka (Chem. Eng. **80** No. 3 [1973] 67/70). — [40] P. M. Rapier (Chem. Eng. Progr. Symp. Ser. **67** No. 107 [1971] 340/51).

[41] P. E. Muhlberg, P. G. Le Gros, P. P. Shepherd, W. F. McIlhenny (Off. Saline Water Res. Develop. Progr. Rept. No. 603 [1970]). — [42] F. C. Standiford (Off. Saline Water Res. Develop. Progr. Rept. No. 636 [1970]). — [43] N. Ganiaris, H. Wightman, R. Glasser (Off. Saline Water Res. Develop. Progr. Rept. No. 639 [1970]). — [44] R. D. Ridley, S. Sack, R. N. Jacobson, E. Chemtob (Off. Saline Water Res. Develop. Progr. Rept. No. 640 [1970]). — [45] J. R. Booth, B. P. Shepherd, W. F. McIlhenny (Off. Saline Water Res. Develop. Progr. Rept. No. 817 [1972]).

[46] P. E. La Mont (Off. Saline Water Res. Develop. Progr. Rept. No. 743 [1971]). — [47] R. R. Grinstead, T. E. Lingafelter (Off. Saline Water Res. Develop. Progr. Rept. No. 810 [1972]). — [48] G. L. Stepakoff, D. Siegelman, R. Johnson, W. Gibson (Proc. 4th Intern. Symp. Fresh Water Sea, Heidelberg 1973, Vol. 3, p. 421/33).

3 Ionic Processes

Processes based on the properties of ions, encountered in brackish and seawater, are described and discussed in this chapter. Whereas in distillation the amount and kind of salts contained in the raw feed water are of no importance to the process and do not affect the economics, in all ionic processes the amount of dissolved salts is of primary importance. This is true independently if salts have to be removed from the solution or if a dilute solution is to be separated from the feed, leaving as waste a more concentrated solution. The amount of salts to be removed affects in ion exchange the amount and cost of regenerant, in electrodialysis the consumption of electrical energy and in reverse osmosis the required pressure. Increased pressure affects membrane life and energy requirements. Thus, ionic processes are in first instance indicated for brackish waters with more or less reduced salinity.

3.1 Ion exchange techniques

Ion exchange processes have been used extensively for the complete demineralization of feed water in boilers and in some cases for improving municipal water supplies. Conventional regeneration of ion exchangers reduced their applicability to low salinity feed waters.

Ion exchange plants can be operated in either small or large units and offer few problems in up-scaling. The process is basically simple and easy to operate. Capital investment is comparatively low. Labor and maintenance costs compare favorably with other existing systems. Corrosion is minimal. Product water quality is comparable to if not better than that from many systems ranging up to complete demineralization. Ion exchange is less affected by iron and manganese contaminations than membrane processes and does not require pH adjustment of the feed water.

However, costly chemicals are required in the conventional ion exchange process for the regeneration of the resins and the amount of regenerants is proportional to the salts removed, which is the main reason for limiting the total salinity that can economically be removed from a brackish water.

Another disadvantage of the process is that it produces a greater waste disposal problem than other desalting processes, because the regenerant chemicals have also to be disposed of, in addition to the salts eliminated from the feed water. Hence the total salt load in the waste water may be double.

Development of improved resins, such as the weak-base anion exchange resins, provided greater flexibility in the choice of regenerant and increased the efficiency. The amount of required regenerant could be reduced to about 110% instead of 300% of the stoichiometric amounts of regenerant previously required.

Novel processes have been developed aiming either to improve the regeneration system in fixed bed operation or to introduce continuous moving bed operation and more efficient regeneration. In each case the overall economics were considerably improved that raw waters with salinities as high as 2500 to 3000 mg/l may now be considered for desalting by means of ion exchange, whereas previously the economics restricted the possibility of ion exchange application only to waters with less than 1000 mg/l total dissolved solids [1].

3.1.1 Fundamental studies

A theoretical analysis of multicomponent ion exchange in fixed beds was derived by Klein et al. in terms of dimensionless parameters. Uniform presaturation and constant feed composition were assumed. Rules for outlining overall concentration profiles predict the number of zones of constant composition (plateau zones), the dependence of the sign of the slope of concentration-profile curves on the position of each component in the relative-affinity series, the order in which the concentrations of various components can become zero and indicate the proper selection of roots obtained in calculating a throughput parameter [2].

In further investigations a nonequilibrium theory for ion exchange column operations involving more than two exchanging ionic species has been developed. Relations between ternary and binary diffusion coefficients based on the kinetic-theory model were converted to a fundamental form and were applied to multicomponent ion exchange with the binary mass-transfer coefficients expressed as functions of composition. Also, the Nernst-Planck diffusion model was extended to multicomponent ion exchange modified to allow the single-component diffusion coefficients to vary with composition. A computer program based on the method of characteristics has been developed to describe ternary ion exchange column operations and used successfully with different rate models, diffusion models and equilibrium behavior. For constant-pattern transitions in ternary systems a second computer program was derived with a graduated time scale and used to describe a single transition zone [3].

The design of an ion exchanger column normally involves the determination of the breakthrough curve of the column effluent, that is the solute concentration in the effluent as a function of time. Engineering design of ion exchange and adsorption columns for large scale use has been hindered by the absence of precalculated breakthrough curves of a general nature. Attention has been given at the University of California, Berkeley, to certain unsolved problems in two-component ion exchange and one-component adsorption in which the solutions were still needed for engineering purposes. Among such problems was the effect of longitudinal mixing or axial dispersion in beds of packed spheres. The combined effects of axial dispersion and mass transfer have been studied for four controlling exchange mechanisms or models: external mass transfer, solid diffusion, pore diffusion and reaction kinetics [4].

Optimal design of large-scale ion exchange systems involves predictions of the effluent concentration history under repeated cycles in which incomplete exhaustion is alternated with incomplete regeneration. The exploratory study of a way to make such predictions and of their use in optimization procedures was outlined. A computer program was described which utilizes a ridge-climbing technique to find the economically optimum design. Input to the program includes cost factors for resin, regenerant apparatus, power and correlation constants obtained from a mass-transfer model of the system. Output comprises the optimum combination of bed height and exhaustion flow rate [5].

Cation exchange resins were compared with reference to the decalcification of seawater. The isomeric bis-sulfonated types demonstrate a quite different selectivity behavior. The partially ortho bis-sulfonic acid type is characterized by a very high selectivity coefficient for calcium ions over magnesium ions. This characteristic will make the product very well suited for the decalcification of seawater [6].

The development of ion exchange technology, both experimental and commercially available, within the framework of the applicability to inland brackish water desalting was reviewed. The most economic fixed-bed and continuous processes were subjected to cost-modelling and analysis exercises. Parameters for design considerations were given, together with basic estimating assumptions. A brief market survey and a complete bibliography of ion exchange references were provided [7]. Recent developments of various ion exchange processes are reviewed by Iammartino [8].

Literature to 3.1

[1] M. A. Lynch, M. S. Mintz (J. Am. Water Works Assoc. 64 [1972] 711/25). — [2] G. Klein, D. Tondeur, T. Vermeulen (Off. Saline Water Res. Develop. Progr. Rept. No. 172 and No. 173 [1966]). — [3] T. Vermeulen (Off. Saline Water Res. Develop. Progr. Rept. No. 326 [1968]). — [4] R. E. Quilici, T. Vermeulen (Off. Saline Water Res. Develop. Progr. Rept. No. 476 [1969]). — [5] S. Pancharatnam, G. Klein, T. Vermeulen (Off. Saline Water Res. Develop. Rept. No. 477 [1969]).

[6] G. J. De Jong (Proc. 3rd Intern. Symp. Fresh Water Sea, Dubrovnik 1970, Vol. 2, p. 73/84). — [7] Control Systems Research Inc. (Off. Saline Water Res. Develop. Progr. Rept. No. 616 [1970]). — [8] N. R. Iammartino (Chem. Eng. 80 No. 1 [1973] 60/2).

3.2 Fixed bed ion exchange processes

In the earlier, conventional regeneration of ion exchange resins, sulfuric acid was normally used as regenerant for the cation exchange bed and sodium hydroxide served for the anion exchange bed. Sulfuric acid was relatively inexpensive. Caustic soda was substantially more expensive and constituted the major part of the regeneration cost. The elimination of expensive anion regenerants by using either less expensive chemicals, such as calcium hydroxide, or the feed water itself as the regenerant for the anion exchanger, opened the applicability of the ion exchange process to water not suitable with older conventional systems.

3.2.1 Desal process

The Desal process is a fixed bed ion exchange process employing three columns in series [1]. In the basic process, the first unit contains a weak-base anion exchange resin (Amberlite IRA-68) in the hydrogen carbonate form. As the influent passes through this unit, neutral salts are converted

*Festbett-
Ionen-
austausch-
Verfahren*

*Das Desal-
Verfahren*

to hydrogen carbonates. The first column acts as alkalization unit (**Fig. 3–1**). The second unit contains a weak-acid cation exchange resin (Amberlite IRC-84) in the hydrogen form and acts as the dealkalization unit. In this unit the hydrogen carbonate salts are converted to carbonic acid. The third unit contains Amberlite IRA-68, but in the free-base form. The resin can absorb carbonic acid from the effluent of the second unit, converting the resin to the hydrogen carbonate form. Hence the third column acts as carbonation column. Because there are losses of carbonic acid during service, supplemental carbon dioxide must be introduced into the carbonation unit. Regeneration of the alkalization unit to the free-base form is made with ammonium hydroxide and that of the dealkalization unit to the hydrogen form with sulfuric acid. Upon completion of the service cycle, the third unit is in the hydrogen carbonate form and the first unit is in the free-base form. By reversing the flow cycle, the third unit becomes the initial anion exchange column and acts as the alkalization unit and the previously first unit acts as the carbonation unit. The principal advantage of this process is the simultaneous removal and recovery of carbonic acid by weak-base anion exchange resin in the third position.

Fig. 3–1

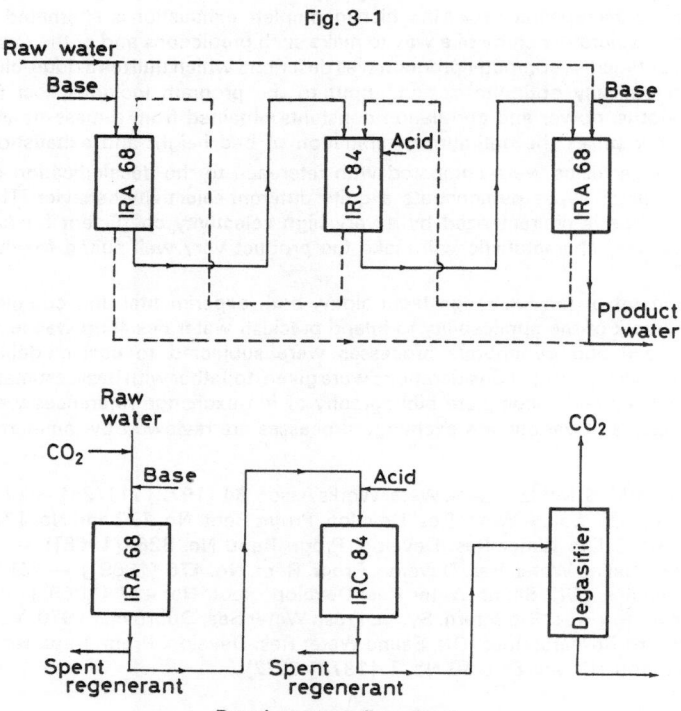

Desal-process flowsheet.

In a modification of the basic process a two-unit design is used and the carbonation column is replaced by a decarbonator. The effluent from the dealkalization unit is passed through the decarbonator to remove carbon dioxide and minimize corrosion.

Cold lime softening may be an economical method for achieving both hardness reduction and partial desalination, when the raw water contains a relatively high percentage of carbonate hardness. Non-carbonate hardness, calcium sulfate and magnesium chloride, which are converted in the dealkalization unit to calcium and magnesium hydrogen carbonates, will then be precipitated in the cold lime softener as calcium carbonate and magnesium hydroxide.

The basic Desal process was tested in pilot plant scale by Sturla [2] and in an Aconex apparatus for continuous operation by Vassiliou and Sturla [3].

A field test program was conducted to determine design criteria for the Three-Bed Desal Process, the Two-Bed Process, the Alkalization plus Cold Lime Softening Process and the Desal plus Cold

Lime Softening Process. The tests were conducted at four sites. Data were obtained at three sites. The ion content before and the minimum and maximum limits of ion content after treatment under various conditions were as follows [4], values in mg/l:

	Roswell		Dalpra Farm		Webster	
	before	after	before	after	before	after
Hydrogen carbonates	218 to 215	0 to 203	470	189 to 287	340	120 to 170
Sulfates	590 to 424	0 to 103	1958	0 to 120	769	1 to 190
Chloride	518 to 685	11 to 174	135	14 to 21	14	1 to 8
Calcium	207 to 232	0 to 62	107	6 to 26	192	4 to 76
Magnesium	62 to 55	0 to 13	65	0 to 4	83	0 to 45
Sodium	344 to 359	6 to 99	936	58 to 133	145	1 to 68
Minimum and maximum limits	1939 to 1970	17 to 496	3671	282 to 500	1543	172 to 490

The field test data were used to determine the total treatment cost for a 3785 m³/day (1 Mgd) plant using the three basic varieties of the Desal process. Reported costs were as follows:

	Three-bed Desal process		Two-bed Desal process		Desal plus Cold Lime Softening	
	cent/kgal	cent/m³	cent/kgal	cent/m³	cent/kgal	cent/m³
Roswell	123.0	32.5	106.6	28.2	—	—
Dalpra Farm	201.3	53.2	—	—	—	—
Webster	75.9	20.1	73.6	19.4	74.2	19.6

The economic projections for plants of 1 Mgd (3785 m³/day) capacity indicate considerable sensitivity to the salinity of the feed waters and the Desal process appears to be competitive with other processes in salinity ranges up to 2000 mg/l, based on the technology developed at the time of the field tests.

3.2.2 Sul-biSul process

Das Sul-biSul-Verfahren

The Sul-biSul process employs two columns in fixed beds for cation and anion exchange. The feed water passes first the cation bed, which contains a strong acid resin, e.g. Dowex HCR. The hydrogen ions on the resin are exchanged for calcium, magnesium, and sodium in solution and the dissolved salts are converted to their corresponding acids. The effluent stream from the cation exchange is, therefore, a stream of dilute acid. The unusual feature of the Sul-biSul process is in the anion exchanger, which uses a strong base resin, e.g. Dowex SBR, in the sulfate form. When the acidic effluent from the cation exchanger passes through this resin, the resin sites holding the sulfate ions are converted to the monovalent hydrogen sulfates. A vacant resin site is created which can absorb anions such as chlorides, sulfates or nitrates. Thus, the acid in the solution from the cation exchanger effluent is absorbed into the anion resin and the effluent from this exchanger is mineral free water. It is in this feature and the manner of regeneration which makes the Sul-biSul process attractive for demineralization. Regeneration of the cation exchanger is made in a conventional manner by means of sulfuric acid. The simplest method of anion exchanger regeneration is in the use of the feed water itself as regenerant, provided it has some alkalinity.

Ions other than sulfates are eluted from the bed by a process of selective ion exchange. The strong base anion exchange resin will selectively remove and hold sulfate ions in preference to chloride ions from dilute solutions. Both the demineralization and regeneration steps are favored by a high sulfate-to-chloride ratio. This ratio is indicative for the brackish waters for which the process is best adapted. The total cost of anion regeneration in the Sul-biSul process employing the rinse regeneration technique is equal to the cost of the available water, which in many cases is the cost of pumping water. It appears that a minimum requirement for this process is a water of both high alkalinity and sulfate content [5].

A three bed Sul-biSul system, with the addition of a carboxylic cation exchanger, is advantageously employed on water supplies containing a substantial percentage of alkalinity. Waste acid from the strong cation unit effluent is used for the regeneration of the week acid exchanger.

Two problems may arise with respect to the regenerant stream. The effluent will be large and it will contain dilute acids. Regeneration by means of a lime slurry will reduce both the volume of the rinse stream and eliminate the acid in the effluent. Field testing of the Sul-biSul process with a mobile unit at several brackish water sites was sponsored by the Office of Saline Water, as for the Desal process. The technical feasibility of the process had already been established and the objectives were to determine the effect of different brackish waters on the performance and to obtain sufficient data to predict the capital and operating cost of both a 1900 and a 3785 m³/day (0.5 and 1 Mgd) plant desalting brackish water [5].

Operating costs vary directly with the mineral content of the raw water. The amount of feed water used to regenerate the anion exchanger is essentially the same as the volume of product water. Water supplies high in hardness and low in sodium content are most suitable for the process, the total dissolved solids not exceeding 2000 mg/l. Sulfate-to-chloride ion ratio should be approximately 9:1 or more.

A municipal water plant using the Sul-biSul process with a capacity of 1900 m³/day (0.5 Mgd) is in operation in the United States at Burgettstown, Pennsylvania, on a high-sulfate water of 1500 to 2000 mg/l total dissolved solids.

Das IRSA-
Verfahren

3.2.3 IRSA process

A combined fixed bed process, which employs four resins in two main columns was developed at the Istituto di Ricerca sulle Acque in Italy. A third column and a degasifier may be used, if required, for chemical mass balance in the system. The resins occupy separate layers in the same column. The feed water is introduced first into the cation exchange column, containing the weak and the strong acid resins. The weak acid resin exchanges hydrogen for calcium and magnesium in the feed. The strong acid resin absorbs sodium; it may be bypassed if sodium is not to be removed. The effluent from this column is highly acidic (**Fig. 3–2**).

Fig. 3–2

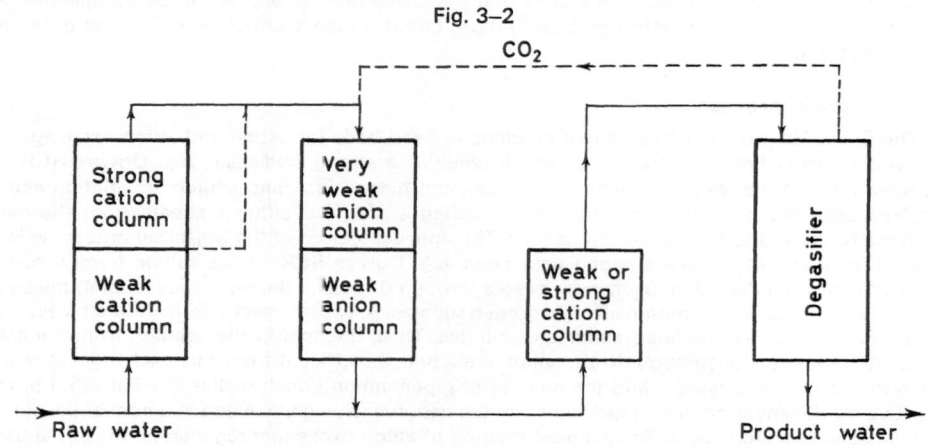

IRSA-process flowsheet.

The anion exchange column contains a very weak and a weak anion exchange resin, which absorb chlorides and sulfates. The feed from the weak acid cation exchanger can be converted to hydrogen carbonates. Carbon dioxide removed by the degasifier may be recycled to the head of the anion exchanger to maintain the hydrogen carbonate content in the feed to the anion exchanger. Provision is also made to recycle liquid from the anion exchanger to the head of the cation exchanger for maintaining a proper ratio of hydrogen carbonates to calcium hardness to increase the efficiency of cation exchange with the weak acid exchanger. Regeneration of the cation exchange resins is made with dilute sulfuric acid. As it passes through the bed, sodium sulfate is produced together with some calcium and magnesium sulfates. However, the concentration of the calcium sulfate is not high enough to cause precipitation in the resin bed. If lime slurry is used to regenerate the anion resin, calcium leakage during the subsequent exhaustion cycle will result in a content of about 100 mg/l, which is not objectionable if the product water is used as potable water [6].

By using carbon dioxide as a buffering agent, a 20% increase in the exchange capacity of the anionic resin was achieved and the consumption of regenerants was appreciably lowered [7]. The regeneration process of the ion exchange resins by means of sulfuric acid and $Ca(OH)_2$-slurry is discussed in two more papers [8]. The addition of an anionic resin column after the degasifier made it possible to desalinate completely a brackish water containing 2500 mg/l dissolved salts [9].

A variety of this process aimed at the pretreatment of the feed water to improve the performance of other desalination processes. Almost complete removal of hydrogen carbonate and sulfate ions in the feed of seawater distillation plants was obtained by means of weak anion exchange resins (see also section 2.18.14, p. 161). It is also possible to eliminate scaling or substitute low transport ions. such as Ca^{2+}, Mg^{2+} and HCO_3^-, in brackish waters, feeding electrodialysis plants, by an ion exchange process of proper design. The regeneration is accomplished with the rejected brine and acid. Scaling problems are eliminated even at severe operating conditions and high concentration factors [10].

The selectivity coefficients, the free exchange energies and the enthalpies between 5 and 45°C of different types of anion resins with functional groups of the amino type were determined for the SO_4^{2-}/Cl^- system in a range of bulk salinities lying between 0.006 and 2.4 N. The experimental results showed that the selectivity for the divalent ion depends strongly on the basicity of the resin, the affinities of the various amino functional groups following the order primary > secondary > tertiary > quarternary. Furthermore, the affinity of every resin for the sulfate ion increases with the dilution of the aqueous phase and with the equilibrium temperature. These results are in full agreement with the theoretical forecasts and show that in heterovalent exchange processes a fundamental role is played not only by the phenomena arising from hydrostatic interactions, but also by those connected with the electrostatic interactions [11].

3.2.4 DUCOL process

The DUCOL process is a strong acid and weak base ion exchange system, which was studied in pilot plant scale. During demineralization raw water passes first through a bed of strong acid resin and then through a bed of weak base resin. Regeneration of the acid resin is made by hydrochloric acid, because it was locally available at low cost. Lime was used as regenerant for the spent weak base resin. Despite contradictory experience in other operations, a 2% lime slurry was successfully used for regeneration, it is claimed. The process featured cost advantages associated with the use of lime slurry regenerant. Fifty cycles were carried out during which over 50 m³ of water containing 50 mg/l total dissolved solids were produced. There was no fall-off in operational capacity over the range of product flow rates of 75 to 325 l/h or with the number of cycles carried out [12].

3.2.5 Other fixed bed ion exchange systems

The TriplEx process employs three ion exchange beds in series, a weak acid resin, a strong acid resin in hydrogen form and a weak base resin. All the feed water is passed through the first bed, where the resin absorbs an amount of cations substantially equal to the alkalinity ions present in the feedwater, exchanging each absorbed cation for a hydrogen ion. The hydrogen ions react with the hydrogen carbonate ions to form water and carbon dioxide. A portion of the stream leaving the first resin bed is sent to the second bed containing the strong-acid resin in the hydrogen form. Most of the remaining cations are exchanged for hydrogen. The effluent from this bed is passed over a weak-base resin in the third bed, where anions and associated hydrogen ions are absorbed. The effluent from this bed is a purer water than is needed to meet municipal water specifications. The remaining portion of the effluent from the first resin bed is mixed with the effluent from the weak-base resin bed and with a small amount of feed water. The combined product water stream is neutral and meets the required water quality specification. The acid resins are regenerated with sulfuric acid and the basic resin is regenerated with ammonia. The amount of acidity left in the waste regenerant stream from the weak-acid resin is just sufficient to react with the ammonia left in the regenerant stream from the weak base resin to yield a neutral brine [13].

The QuadruplEx process is a variety of the TriplEx process, which utilizes a parallel bed containing a strong acid resin in the sodium form. When the feedwater contains a relatively large proportion of permanent hardness ions, a three bed ion exchange process will produce a water with a lower than needed total dissolved solids content. Costs can then be reduced by employing a fourth ion exchange bed that merely softens a portion of the feed water. This fourth bed may be installed in parallel with the three other beds or may be placed after the weak-acid resin and parallel with the last two beds. Both methods produce approximately the same results, although the latter may be somewhat more economical [13].

As the amount of dissolved solids to be removed from the feed water increases, processes using membranes become significantly more economical than all ion exchange processes. The cross-over point appears to be below a total dissolved solids removal of 2500 mg/l under usual process parameters. At higher total dissolved solids levels, a combined system may prove most advantageous. An example of a combined system consists of a weak acid resin bed placed before the membrane process, either electrodialysis (IX-ED) or reverse osmosis (IX-RO). The sodium salts in the brine from the reverse osmosis unit are used to regenerate the spent resin bed [13].

Application assessments in several municipalities of the United States in Colorado [14], in New Mexico [15], and in Montana [16] suggest that water quality improvement systems employing ion exchange techniques are, in certain instances, competitive with the economics of water quality improvement systems based on electrodialysis and reverse osmosis. In case of high salinity and high-hardness water, a combination of ion exchange and either electrodialysis or reverse osmosis may offer greater economic benefits than single systems [17].

Vajna has described a counter-current multi-column ion exchange process, named Intensive Fractionating Process. The process enables the almost stoichiometric utilization of the regenerants. Waste brine, lime, and carbon dioxide are used as regenerants [18]. In a variety of this process descaling of seawater reduces the calcium content to 19 to 20% of the original concentration. By using two pairs of ion exchange beds, the total hardness was reduced to 5.3% and with four beds to 0.6% of the incoming hardness. It was found that an optional relation exists between the volumes of the solution and the exchange beds [19]. Another multi-column process of seawater demineralization by ion exchange for calcium and hydrogen carbonate ions and their subsequent removal by thermal decomposition was described by Ranck [20].

An ion exchange process, combined with two other processes, precipitation and complex formation, was described by Glueck. Fresh water is produced from seawater by passing the brine through a cationic exchanger presaturated with silver ion. The precipitated silver chloride is dissolved with ammonia, forming the diaminosilver complex, which is subsequently taken up by the resin. The effluent stream from this step of the process consists the waste brine [21].

Literature to 3.2

[1] R. Kunin (Proc. 1st Intern. Symp. Water Desalination, Washington, D.C., 1965 [1967], Vol. 2, p. 69/85; Desalination 4 [1968] 38/44), R. Kunin, D. G. Downing (Chem. Eng. 78 No. 14 [1971] 67/9; Chem. Eng. Progr. Symp. Ser. 67 No. 107 [1971] 575/80). — [2] P. Sturla (Proc. 1st Intern. Symp. Water Desalination, Washington, D.C., 1965 [1967], Vol. 1, p. 416/28). — [3] B. Vassiliou, P. Sturla (Desalination 5 [1968] 40/8). — [4] A. C. Epstein, M. B. Yeligar (Off. Saline Water Res. Develop. Progr. Rept. No. 631 [1970]). — [5] K. Schmidt, D. Senger, D. Schwark, H. Dalaly (Off. Saline Water Res. Develop. Progr. Rept. No. 446 [1969]).

[6] G. Boari, L. Liberti, C. Merli, R. Passino (Proc. 3rd Intern. Symp. Fresh Water Sea, Dubrovnik 1970, Vol. 2, p. 63/72). — [7] G. Boari, M. Derchi, C. Merli, R. Passino (Quad. Ric. Sci. No. 58 [1969] 237/51). — [8] G. Boari, L. Liberti, C. Merli, R. Passino (Quad. Ric. Sci. No. 58 [1969] 252/74, 275/92). — [9] G. Agostinelli, G. Boari, E. Gadda, L. Lorenzo, C. Merli, R. Pascali (Termotecnica [Milan] 26 [1972] 215/21). — [10] G. Boari, L. Liberti, C. Merli, R. Passino, M. Santori, G. Tiravanti (Proc. 3rd Intern. Symp. Fresh Water Sea, Dubrovnik 1970, Vol. 2, p. 47/62).

[11] G. Boari, L. Liberti, C. Merli, R. Passino (Proc. 4th Intern. Symp. Fresh Water Sea, Heidelberg 1973, Vol. 3, p. 25/48). — [12] S. Evans, N. C. Daltrophe (Proc. 4th Intern. Symp. Fresh Water Sea, Heidelberg 1973, Vol. 3, p. 49/56). — [13] S. A. Bresler (Off. Saline Water Res. Develop. Progr. Rept. No. 781 [1972]). — [14] F. S. Agardy, H. Daubert (Off. Saline Water Res. Develop. Progr. Rept. No. 702 [1971]). — [15] D. E. Morris, W. L. Prehn (Off. Saline Water Res. Develop. Progr. Rept. No. 767 [1971]).

[16] I. C. Watson, F. M. Heider (Off. Saline Water Res. Develop. Progr. Rept. No. 783 [1972]). — [17] S. A. Bresler, E. F. Miller (J. Am. Water Works Assoc. 64 [1972] 764/72). — [18] S. Vajna (Proc. 3rd Intern. Symp. Fresh Water Sea, Dubrovnik 1970, Vol. 2, p. 107/19). — [19] M. Vajna (Proc. 3rd Intern. Symp. Fresh Water Sea, Dubrovnik 1970, Vol. 2, p. 121/8). — [20] J. P. Ranck (Desalination 6 [1969] 75/85).

[21] A. R. Glueck (Desalination 4 [1968] 32/7).

3.3 Continuous ion exchange processes

Some continuous ion exchange processes have already been described briefly in section 2.18.14, p. 161, for the removal of scale forming compounds prior to introducing seawater to distillation systems. The continuous processes are aimed to reduce labor and regeneration costs to competitive levels. It appears that unforeseen troubles have made commercial operation of the continuous processes less attractive than originally projected. Frequent operation of valves has required correspondingly frequent maintenance and replacement. Valves and screens both have plugged in operation with resin or precipitate. Problems were also encountered in scale-up of the hydraulic resin-transfer mechanisms [1].

3.3.1 Asahi process

The essential feature of the Asahi process is the continuous or near continuous treatment of impure waters by ion exchange resins. This process has been successfully employed in the low-cost treatment of boiler feed waters, the removal of undesirable ions from sugar liquors and the recovery of copper from industrial waste waters. The design is oriented towards moving resins and fluids in the proper relationship throughout the processing equipment. In a continuous system, ion absorption, resin regeneration and rinsing require the timely and automatic control of valves during the movement of the resins through the system (**Fig. 3–3**). Feed water flows into the cation exchanger in which the cations are exchanged for hydrogen ions. The effluent from this column is acidic. The effluent water may move into a degasifier to remove CO_2 by the breakdown of carbonic acid. This loss of CO_2 is equivalent to removing acid from the solution. Next, the feed moves to the anion exchanger to remove chlorides, sulfates and other anionic constituents. Demineralized water is the effluent from the anion exchanger.

Fig. 3–3

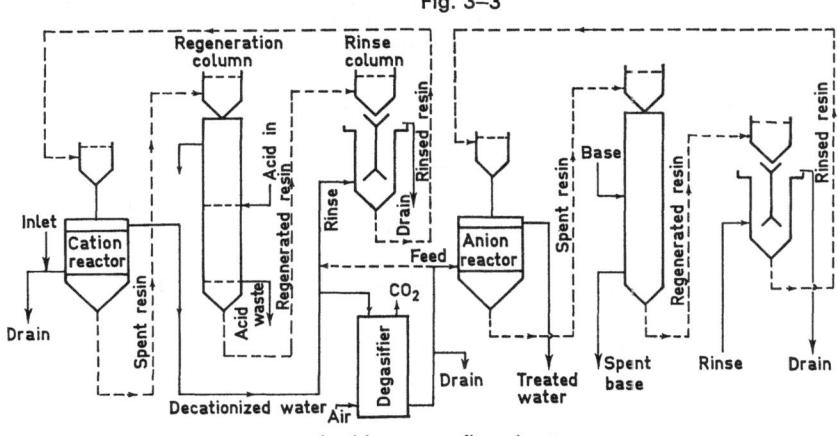

Asahi-process flowsheet.

The movement of the resins in the Asahi process is by hydraulic action. The spent resin is moved into the regeneration column by means of service water and after regeneration the resin is moved hydraulically into the rinse column to remove excess regenerant. Finally, the resins are conveyed hydraulically to the metering hoppers above the cation or anion exchanger from which a measured amount of the newly regenerated resin is returned into the exchanger. During the absorption cycle, the raw feed waters to the system flow upward through the exchangers. This tends to expand the resin bed and, to promote more efficient absorption of the ions in solution, resin retainer screens are employed to keep the resin particles in a compact bed.

There are several notable advantages of the Asahi process. Resins and fluids flow countercurrently, the resin movement being downward. The absorption, regeneration and rinsing columns can be calculated separately and each operation can be designed according to the flow rate, salinity and regeneration efficiency.

The choice of resins for the Asahi process depends upon the kind of waters to be treated and the specification of the final product water itself. Since resin costs are a large part of the cost of the operation of the processes, equipment designs can be manipulated to take advantage of resin properties and resin costs. The most serious disadvantage of the Asahi process is in the instrumentation of the system so that all operations are properly timed to assure continuous operation. A variation of this design may be the substitution of a mixed bed ion exchanger. However, this substitution requires a separate vessel in the plant design to effect a separation of the spent cation and anion resins prior to regeneration. After the separation, the resins are moved into the proper vessels for regeneration and rinsing.

Experience and the latest state of development, particularly in water treatment, of the Asahi process and the modified Asahi-Degremont monobed and mixed continuous ion exchange process are reviewed by Bucher [2]. The review includes comparison with the Permutit CCIX system, the Fluicon process and the underneath described ChemSeps process.

*Das
ChemSeps-
Verfahren*

3.3.2 ChemSeps process

The ChemSeps system is also a continuous, countercurrent ion exchange process developed by Higgins [3], which employs a loop for performing the three operations of ion exchange demineralization. Two loops are required, one for the cation and one for the anion exchange operations. The resins in the plant are moved around the loop by hydraulic ram action at precise intervals. All solution flows are countercurrent to the flow of the resin and it is this mode of operation which permits the optimum achievement of sorption, regeneration and rinsing. The contaminants are removed during the run cycle, while the resins are moved during the pulse cycle. The operation of the ChemSeps cycle depends upon the proper sensing of the effluent streams, whether they are in the loading section, the regenerant section or the rinsing section. In the loading section, the raw water flows through fresh resins and the appropriate ions are sorbed. At the same time, a regenerating fluid flows upward through the spent resins and a similar operation occurs in the rinsing section. The duration of a process or run cycle may be from 4 to 10 minutes depending upon the contaminants to be removed, the resin characteristics and the process parameters.

The operation of the ChemSeps system is illustrated in **Fig. 3–4**. Valve A opens and a slug of spent

Fig. 3–4

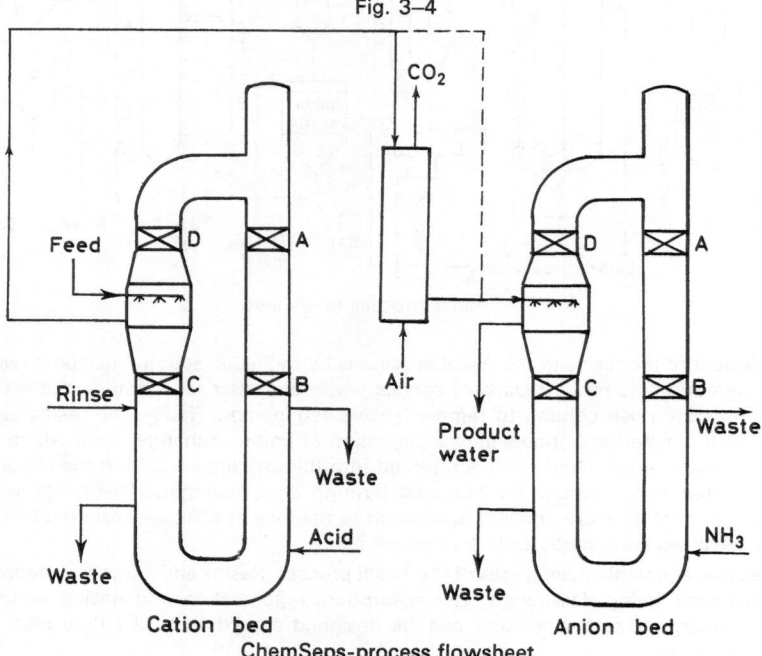

ChemSeps-process flowsheet.

resin drops into the regenerant section. A pulse of raw water is admitted to wash out any suspended solids brought over with the spent resin. Next, valves B, C and D open in a timed sequence and, by a pulse of water, the resins slide around the loop to their respective positions. When the indicated valves close, other valves controlling the feed water, the regenerant and rinse water open and the appropriate operations are performed. When a run cycle is complete as detected by a sodium or a chloride detector, for example, the system then moves into the pulse cycle for the movement of the resins around the loop. The entire process is then completed [3].

The resin inventory for a ChemSeps plant is much less, being on the order of 10 to 30% of the fixed bed resin requirements. Alternatively, the resin attrition rate may be greater than for a fixed-bed system in which the resin is not subjected to as much mechanical stress.

3.3.3 Avco process   *Das Avco-*
                                                                          Verfahren

A continuous ion exchange process was developed by Avco Systems with a moving bed of ion exchange resin. The column consists basically of an ion exchange zone and a regeneration zone, where the liquid moves countercurrent to the resin bed, and two driver zones where the liquid, either raw feed for the secondary zone or treated water for the primary zone, moves concurrent to the resin (**Fig. 3–5**). In the exchange zone, the calcium ions in the feed water are exchanged with sodium ions. In the regeneration zone the spent resin is treated with a strong sodium chloride solution. The zones where the chemical exchanges take place are maintained in a fixed position by moving the resin countercurrent to the flow at a rate equal to the exchange zone speed in a fixed bed operation. The length of these zones is selected according to the desired degree of exchange and regeneration.

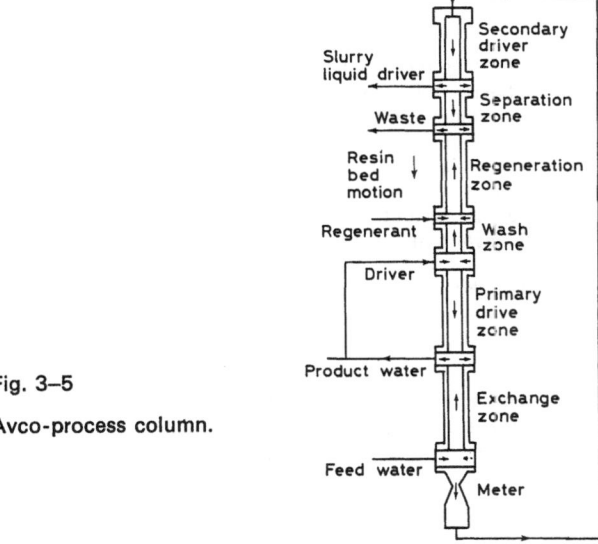

Fig. 3–5

Avco-process column.

Separation and washing zones are inserted between the main exchange and driver zones in order to prevent the mixing of treated water, feed water and regenerant liquid and, where necessary, to wash one liquid off the resin before it is passed into a zone containing another liquid. These separation zones constitute the primary difference between the proposed system and other continuous contercurrent systems. For a softening operation a single column, as shown in Fig. 3–5, is sufficient, while for brackish water desalting two such columns would be used in series. The first column would contain a cation resin which has free hydrogen ions and the second column would contain an anion resin which has free hydroxyl ions [4].

The process has successfully been tested in a 57 m ³/day (15000 gpd) pilot plant. The system is claimed to be particularly suitable for large scale membrane softening pretreatment and brackish water desalting because of its inherently smaller plant size and almost stoichiometric chemical usage.

Literature to 3.3

[1] N. R. Iammartino (Chem. Eng. **80** No. 1 [1973] 60/2). — [2] H. Bucher (Dechema Monograph. **64** [1971] 159/72; Chem. Anlagen Verfahren **1971** No. 7, p. 55/6).—[3] I. R. Higgins (Chem. Eng. Progr. **65** No. 6 [1969] 59/62), I. R. Higgins, R. C. Chopra (Ion Exch. Process Ind., Papers Conf., London 1969 [1970], p. 121/6). — [4] H. Gold, A. Todisco, A. A. Sonin, R. F. Probstein (Proc. 4th Intern. Symp. Fresh Water Sea, Heidelberg 1973, Vol. 3, p. 57/65).

Nicht-chemische Verfahren für die Regenerierung von Ionenaustauscherharzen

Thermische Regenerierung von Ionenaustauscherharzen

3.4 Non-chemical methods of ion exchange resin regeneration

Whereas the preceeding processes are based on the regeneration of the ion exchange resins by chemical means, there have been published in recent years reports of methods for the regeneration of ion exchange resins by the use of heat, pressure or electricity.

3.4.1 Thermal regeneration of ion exchange resins

Sirotherm is a demineralization process in which regeneration of ion exchange resins is accomplished by the use of low grade heat rather than by chemicals. The process depends on the marked influence of temperature on the equilibrium between a solution of sodium chloride and a mixed bed of weak base R_B and weak acid R_AH ion exchange resins: $R_AH + R_B + NaCl \rightleftharpoons R_ANa + R_BHCl$. Adsorption of ions occurs in the cold; when fully loaded, the resins may be stripped by passing hot water, which reduces the adsorption of salt and displaces the equilibrium to the left. The equilibrium depends on many variables besides temperature, such as the basicity and acidity of the resins, pH value and ionic strength of the solution, resin affinities for different ions and the ratio of equivalents of acid to basic resin [1].

An extensive study of the complex equilibria in a mixed bed of resins has been made, in which the properties of weakly basic and weakly acidic ion exchange resins were examined [2]. The titration curve of a resin containing tertiary amino groups of the same type exhibits a remarkable plateau. On raising the temperature from 20 to 80°C the plateau becomes displaced to lower pH values. It has been shown that it is possible to reduce the salinity of a salt solution from 1000 to about 250 mg/l and to produce an effluent with a peak concentration of 5000 mg/l by regenerating with the feed water at 80°C. Divalent as well as monovalent ions are reversibly desorbed [3]. It has been further found advantageous to use a resin which exhibits a pronounced plateau on titration and for which the level of basicity decreases markedly at higher temperatures [4]. The process operates at neutral pH; the absorption of sodium chloride is slow. Even with the most porous resins available, the rates obtained were inadequate for the practical operation of the process if resin beads of normal size are to be used [5].

Advances in the process include the synthesis of new, thermally regenerable, ion exchange resins having practical working capacities and fast adsorption rates. Field trials have shown considerable promise. These resins have better abrasion resistance than most commercial resins and are chemically stable under the thermal regeneration conditions, provided that oxygen is removed from the feed water [6].

A significant enhancement in the rates of salt uptake by thermally regenerated mixtures of weakly basic and weakly acidic ion exchangers has been achieved after studies of many different physical forms of the resins. A new resin system has been devised, which consists of composite beads prepared by embedding micro particles of both types of exchangers in a water- and salt-permeable matrix, to produce a "plum pudding" structure. Being of normal size, the resin can be handled by conventional means. Its adsorption rate is comparable with that of strong electrolyte systems. Nevertheless, improvements in both capacity and mechanical strength are essential before this approach could be considered a solution to the economic operation of a thermally regenerable desalting process [17].

The state of the art of the Sirotherm process was recently presented by Weiss [7]. Pilot plant studies were also reported. Product yields were 77 to 91%. Well water of Perth with 1190 mg/l solids gave a product with an average of 550 mg/l. Municipal water of Adelaide with 280 to 415 mg/l was reduced to 100 mg/l [8, 9]. A theoretical model has been developed to describe the thermal regeneration of ion exchange resins in the Sirotherm process [10].

Another ion exchange process using thermally regenerated reagents utilizes the causticization of ammonium chloride by magnesium oxide and the hydrolysis of magnesium chloride in a recycle loop. By the application of thermal energy, almost all the ammonia and hydrochloric acid regenerants are recovered, thereby eliminating the need to use additional chemicals. Solid wastes, rather than a brine stream, result from this procedure and scale forming compounds are readily handled. Operating parameters for the process have been determined experimentally and a plant design based on these studies was presented. An economic analysis of the process indicates that it compares most favorably with other ion exchange desalting technology and is competitive with present reverse osmosis designs [11].

An apparatus in which desalination was achieved on ion exchange resins by periodic change in temperature synchronized with an alternation in the course of liquid flow was described by Ginde and Chu. Experiments showed that higher separation can be obtained with an increase in temperature difference, pressure, rate of liquid flow or resin-to-liquid ratio [12].

3.4.2 Electrical regeneration of ion exchange resins

Mixed beds of ion exchange resins that had been used to de-ionize a salt solution were electro-regenerated in situ in the depleting compartments of an electrodialysis cell. The technique showed promise for de-ionizing water for use in hemodialysis. Feed solutions of NaCl were demineralized easily. Heavy metals could also be removed. The presence of Ca^{2+} and HCO_3^- ions in the feed caused problems in electro-regeneration but these ions can be removed by pretreatment [13]. A pilot plant ion exchange filter was developed in Russia with electro-chemical regeneration of the resins. The anode is located inside the cation exchange layer and the cathode inside the anion exchange layer. Regeneration is effected by the formation of acids and bases on the electrodes. Portable units of 3 to 12 m^3/day capacity and initial salt content 4 g/kg in the feed are economical when water is used as the electrolyte [14].

Elektrische Regenerierung von Ionen-austauscher-harzen

3.4.3 Pressure regeneration of ion exchange materials

The application of mechanical pressure to ion exchange materials equilibrated with saline solution produces an effluent with considerably reduced salt concentration. Using the most efficient and a final pressure of 440 kg/cm² (6255 psi), the concentration of 0.7 molar sodium chloride was reduced to 0.01 molar in four cycles. Most of the imbibed salt appears in the first half of the pressure eluted solution, its concentration dropping rapidly thereafter [15]. In order to produce water of a specified salinity, it is necessary to discard the earlier fractions of the exudate and save some of the later ones. It is possible to minimize the mechanical work by correctly choosing how much of the imbibed water shall be squeezed out of the resin and what portions of it shall be kept as product. Minimum work requirements have been estimated as a function of the feedwater salt concentration and the extent of demineralization desired. For slightly brackish waters, the mechanical work requirements are low enough to make the process interesting [16].

Regenerierung von Ionenaus-tauschern durch Druck

Literature to 3.4

[1] D. E. Weiss, B. A. Bolto, R. McNeill, A. S. Macpherson, R. Siudak, E. A. Swinton, D. Willis (Proc. 1st Intern. Symp. Water Desalination, Washington, D.C., 1965 [1967], Vol. 2, p. 103/37). — [2] D. E. Weiss, B. A. Bolto, R. McNeill, A. S. Macpherson, R. Siudak, E. A. Swinton, D. Willis (Australian J. Chem. **19** [1966] 561/87, 589/608, 765/89, 791/9). — [3] D. E. Weiss (Desalination **1** [1966] 107/28). — [4] B. A. Bolto, R. McNeill, A. S. Macpherson, R. Siudak, D. E. Weiss, D. Willis (Australian J. Chem. **21** [1968] 2703/10). — [5] B. A. Bolto, R. E. Warner (Desalination **8** [1970] 21/34).

[6] H. A. J. Battaerd, N. V. Blesing, B. A. Bolto, A. F. G. Cope, G. K. Stephens, D. E. Weiss, D. Willis, J. C. Worboys (Desalination **12** [1973] 217/37). — [7] H. A. J. Battaerd, A. F. Cope, G. K. Stephens, J. C. Worboys, N. V. Blesing, B. A. Bolto, D. E. Weiss, D. Willis (Proc. 4th Intern. Symp. Fresh Water Sea, Heidelberg 1973, Vol. 2, p. 13/23). — [8] H. A. J. Battaerd, N. V. Blesing, B. A. Bolto, A. F. G. Cope, G. K. Stephens, D. E. Weiss, D. Willis, J. C. Worboys (Australian Chem. Process. Eng. **25** No. 8 [1972] 19/21). — [9] D. E. Weiss (Ion Exch. Membranes **1** [1972] 109/11). — [10] S. D. Hamann (Australian J. Chem. **24** [1971] 1979/92, 2439/47).

[11] G. A. Busch, S. Lynn (Univ. of California, Berkeley, Saline Water Conversion Lab. Rept. No. 72 [1972]). — [12] V. R. Ginde, C. Chu (Desalination **10** [1972] 309/17). — [13] T. A. Davis,

R. E. Lacey (U.S. Natl. Tech. Inform. Serv. PB-210163 [1972]). — [14] R. Ya. Lapkes, Z. Ya. Yaroslavskii, G. I. Nikoladze (Opresnenie Solen. Vod Ispol'z. ikh Vodosnabzh. **1972** 30/7; C.A. **79** [1973] No. 9687). — [15] Western Independent Research Laboratories Inc. (Off. Saline Water Res. Develop. Progr. Rept. No. 75 [1963]).

[16] L. Dresner (Off. Saline Water Res. Develop. Progr. Rept. No. 508 [1970] 218/22). — [17] B. A. Bolto, K. Eppinger, A. S. Macpherson, R. Siudak, D. E. Weiss, D. Willis (Desalination **13** [1973] 269/85).

*Ionen-
selektive
Membranen*

3.5 Ion selective membranes

A bibliography on membrane technology, pertaining to saline water desalination and ranging from 1908 to 1965, has been compiled by the Office of Saline Water. The bibliography contains publications on membrane preparation, properties, and application in electrodialysis and includes patents in the principal countries [1].

Ion selective membranes contain ion exchange material in sheet form. Reviews on ion selective membrane processes were also published in several books and technical journals. Helfferich deals with ion exchange membranes in chapter 8 of his book on Ion Exchange [2]. A pioneer book on electrodialysis describes the process and the use of ion exchange membranes in water desalting [3]. Chapters in books by Tuwiner [4] and by Spiegler [5] also describe the application of ion selective membranes in electrodialysis as a desalination process. Lacey treats the physical chemistry of membrane phenomena in his book Membrane Processes for Industry [6]. Reviews on transport phenomena in artificial membranes were published by Lakshminarayanaiah [7] and on membrane separation processes by Friedlander and Rickles [8]. From several other reviews in various languages, only two of more general character are mentioned in the following. Sollner has recently reviewed the nature and classes of membranes, the preparation and electrochemical properties of ion exchange membranes, as well as transport mechanisms and models [9]. Reboiras reporting on electrodialysis as a desalination process describes the properties of artificial membranes and their characterization [10].

Two more books were recently published: by Lakshminarayanaiah on Transport Phenomena in Membranes [11] and by Lacey and Loeb on Industrial Processing with Membranes [12]. In view of this large bibliographic work, only the more recent publications and some fundamental earlier papers will be considered in this and the following sections.

The various phenomena, underlying membrane processes, are shown in **Fig. 3—6**. The driving forces which cause a flux or a flow of ionic or molecular species through a membrane separating two solutions are: difference of electric potential ΔE, difference of pressure Δp, difference of

Fig. 3—6

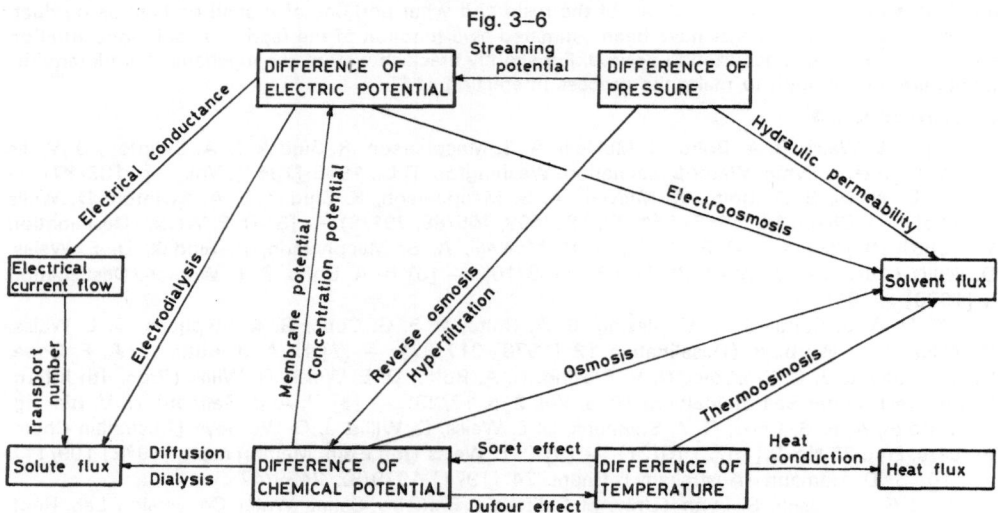

Transport phenomena arising across membranes.

chemical potential $\Delta\mu$ and difference of temperature Δt. They are arranged in the centre of the diagram. Resulting fluxes or flows are placed at the two sides. A prerequisite for the successful operations indicated in the figure is the availability of suitable membranes. Not all these operations are pertaining to desalisation applications.

Ion selective membranes contain ion exchange material as the ion selecting material and might be classified heterogeneous or homogeneous, depending on their visible structure.

Heterogeneous ion selective membranes have been prepared by incorporating ion exchange particles into film forming resins by dry molding or calendering mixtures of the ion exchange and film-forming materials, by casting films from a solution of ion exchange materials in a film-forming polymer and then evaporating the solvent, and by casting films from a mixture of the ion exchange material with a partially polymerized film-forming material and completing the polymerization.

Homogeneous membranes contain the ion exchange component in a continuous phase through-out the resin matrix. Homogeneous ion exchange membranes have been prepared by condensation polymerization of mixtures of reactants, one of which at least can be made anionic or cationic; by additional polymerization of mixtures of reactants, one of which at least contains anionic or ca-tionic groups and one of which is usually a cross-linking agent; by introduction of anionic or cationic groups into preformed films using various techniques; by casting films from a solution of a mixture of a linear film-forming polymer and a linear polyelectrolyte and then evaporating the solvent.

Membranes made by any of the above methods may be cast or formed around reinforcing materials to improve their strength and dimensional stability. Table 3/1, p. 200, shows the properties of some representative commercially available ion exchange membranes. Most of the given properties are as reported by the manufacturers [12].

Preparation, chemical composition, characteristics and properties of improved ion selective membranes are reported in a large number of papers. Research sponsored by the Office of Saline Water on improved ion selective membranes for electrodialysis is reported by Sheehan [13] and by Salyer et al. [15]. Plummer and coworkers evaluated newly developed membranes for electrodialysis and/or transport depletion [14]. Anion exchange membranes specific for Na^+/K^+ were prepared by Varoqui et al. [16]. Preparation of membranes from various resins is reported by Chiolle et al. [21], by Laskorin et al. [23] and by Jendrychowska-Bonamour [25]. Preparation and characterization of membranes based on cellulosic and other polymeric materials is reported by Riande and Guzman [17], by Sueszer [18], by Martin et al. [19], by Grot [20] and by Caplan and Sollner [26]. Properties of improved membranes, as well as membranes resistant to fouling were reported by Korngold [24], by Kusumoto et al. [22] and by Hodgdon et al. [27]. Kelly reported on new boron-nitrogen polyions and related compounds as components of ion exchange membranes [28].

Literature to 3.5

[1] G. F. Mangan, J. M. Shackelford, K. A. Kase (Off. Saline Water Res. Develop. Progr. Rept. No. 83 [1963]), J. M. Shackelford, K. A. Kase (Off. Saline Water Res. Develop. Progr. Rept. No. 153 [1965]). — [2] F. Helfferich (Ionenaustauscher, Weinheim/Bergstr. 1959; Ion Exchange, New York 1962). — [3] J. R. Wilson (Demineralization by Electrodialysis, London 1960). — [4] S. B. Tuwiner (Diffusion and Membrane Technology, New York 1962). — [5] K. S. Spiegler (Principles of Desalination, New York 1966).

[6] R. E. Lacey (Membrane Processes for Industry, Southern Research Institute, Birmingham, Ala., 1966). — [7] N. Lakshminarayanaiah (Chem. Rev. **65** [1965] 491/565). — [8] H. Z. Friedlander, R. N. Rickles (Chem. Eng. **73** [1966] No. 5, p. 111/6, No. 7, p. 121/4, No. 11, p. 153/6), R. N. Rickles, H. Z. Friedlander (Chem. Eng. **73** [1966] No. 9, p. 163/8, No. 13, p. 217/24). — [9] K. Sollner (J. Macromol. Sci. Chem. **3** [1969] 1/86). — [10] M. D. Reboiras (Rev. Plast. Mod. **23** [1972] No. 191, p. 667/73, 682, No. 192, p. 843/51, No. 193, p. 76/86, 100).

[11] N. Lakshminarayanaiah (Transport Phenomena in Membranes, New York 1969). — [12] R. E. Lacey, S. Loeb (Industrial Processing with Membranes, New York 1972). — [13] W. C. Sheehan (Off. Saline Water Res. Develop. Progr. Rept. No. 202 [1966]). — [14] C. W. Plummer, J. Enos, A. B. La Conti, J. R. Boyack (Off. Saline Water Res. Develop. Progr. Rept. No. 481 [1969]). — [15] I. O. Salyer, E. V. Kirkland, P. H. Wilken (Off. Saline Water Res. Develop. Progr. Rept. No. 483 [1969]).

[16] R. Varoqui, C. Pusineri, A. Schmitt, H. Benoit (Compt. Rend. **269** [1969] 905/6). — [17] E. Riande, G. M. Guzman (Anales Quim. **65** [1969] 483/91). — [18] A. Sueszer (Negev Institute

Table 3/1

Properties of some commercially available ion exchange membranes.

Designation	Type	Resistance Ω · cm²	Transference No. of counterion	Thickness mm	mils	Available cm	size inches	Manufacturer
		(0.6 N KCl)						
C-60	Cat	5 ± 2	0.80 (0.5/1.0 N KCl)	0.30	12	112	44 wide rolls	[1]
C-100	Cat	7 ± 2	0.90 (0.5/1.0 N KCl)	0.22	8.5	112	44 wide rolls	[1]
A-60	An	6 ± 2	0.80 (0.5/1.0 N KCl)	0.30	12	112	44 wide rolls	[1]
A-100	An	8 ± 2	0.90 (0.5/1.0 N KCl)	0.23	9	112	44 wide rolls	[1]
		(0.5 N NaCl)						
CK-1	Cat	1.4	0.85 (0.25/0.5 N NaCl)	0.23	9	112 × 112	44 × 44	[2]
DK-1	Cat	1.8	0.85 (0.25/0.5 N NaCl)	0.23	9	112 × 112	44 × 44	[2]
CA-1	An	2.1	0.92 (0.25/0.5 N NaCl)	0.23	9	112 × 112	44 × 44	[2]
DA-1	An	3.5	0.92 (0.25/0.5 N NaCl)	0.23	9	112 × 112	44 × 44	[2]
		(0.5 N NaCl)						
CMV	Cat	3	0.93 (0.5/1.0 N NaCl)	0.15	6	112	44 wide rolls	[3]
CSV	Cat	10	0.92 (0.5/1.0 N NaCl)	0.30	12	112	44 wide rolls	[3]
AMV	An	4	0.95 (0.5/1.0 N NaCl)	0.15	6	112	44 wide rolls	[3]
ASV	An	5	0.95 (0.5/1.0 N NaCl)	0.15	6	112	44 wide rolls	[3]
		(0.1 N NaCl)						
MC-3142	Cat	12	0.94 (0.5/1.0 N NaCl)	0.20	8	102 × 305	40 × 120	[4]
MC-3235	Cat	18	0.95 (0.1/0.2 N NaCl)	0.30	12	102 × 305	40 × 120	[4]
MC-3470	Cat	35	0.98 (0.1/0.2 N NaCl)	0.20	8	102 × 305	40 × 120	[4]
MA-3148	An	20	0.90 (0.5/1.0 N NaCl)	0.20	8	102 × 305	40 × 120	[4]
MA-3236	An	120	0.93 (0.5/1.0 N NaCl)	0.30	12	102 × 305	40 × 120	[4]
IM-12[1])	An	12	0.96 (0.1/0.2 N NaCl)	0.15	6			[4]
MA-3475 R	An	11	0.99 (0.5/1.0 N NaCl)	0.36	14			[4]
		(0.1 N NaCl)						
CR-61	Cat	11	0.93 (0.1/0.2 NaCl)		23	46 × 102	18 × 40	[5]
AR-111 A	An	11	0.93 (0.1/0.2 NaCl)		24	46 × 102	18 × 40	[5]
		(0.5 N NaCl)						
CL-2.5 T	Cat	3	0.98		6	102 × 127	40 × 50	[6]
CLS-25 T[2])	Cat	3	0.98		6	102 × 127	40 × 50	[6]
AV-4 T	An	4	0.98		7	102 × 127	40 × 50	[6]
AVS-4 T[2])	An	5	0.98		7	102 × 127	40 × 50	[6]

Manufacturers:

[1] American Machine and Foundry Co., Stamford, Connecticut, U.S.A.
[2] Asahi Chemical Industry Co., Ltd., Tokyo, Japan
[3] Asahi Glass Co., Ltd., Tokyo, Japan
[4] Ionac Chemical Co., Birmingham, New Jersey, U.S.A.
[5] Ionics Inc., Cambridge, Massachusetts, U.S.A.
[6] Tokuyama Soda Co., Ltd., Tokyo, Japan

[1]) Membrane high diffusive to acids. — [2]) Univalent selective membranes.

Arid Zone Research, Beer Sheva 1971). — [19] E. C. Martin, R. D. Dietert, H. F. Hamil (Separ. Sci. **6** [1971] 637/44). — [20] W. Grot (Chem.-Ingr. Tech. **44** [1972] 167/9).

[21] A. Chiolle, L. Credali, P. Parrini (Proc. 4th Intern. Symp. Fresh Water Sea, Heidelberg 1973, Vol. 3, p. 81/90). — [22] K. Kusumoto, T. Sata, Y. Mizutani (Proc. 4th Intern. Symp. Fresh Water Sea, Heidelberg 1973, Vol. 3, p. 111/8). — [23] B. N. Laskorin, N. M. Smirnova, E. I. Semenova, T. A. S'yanova, T. I. Savel'eva, Yu. N. Banin (Proc. 4th Intern. Symp. Fresh Water Sea, Heidelberg 1973, Vol. 3, p. 119/27). — [24] E. Korngold (Proc. 4th Intern. Symp. Fresh Water Sea, Heidelberg 1973, Vol. 3, p. 99/109). — [25] A. M. Jendrychowska-Bonamour (J. Chim. Phys. **70** [1973] 12/9, 20/6).

[26] S. R. Caplan, K. Sollner (J. Colloid Interface Sci. **46** [1974] 46/66, 67/76, 77/84). — [27] R. B. Hodgdon, E. Witt, S. S. Alexander (Desalination **13** [1973] 105/27). — [28] H. C. Kelly (Off. Saline Water Res. Develop. Progr. Rept. No. 794 [1972]).

3.5.1 Characterization of ion selective membranes

A Test Manual for Permselective Membranes has been prepared by the Office of Saline Water in cooperation with the U.S. Bureau of Reclamation. The membrane test methods included in the Manual have been gathered from a number of sources consisting mainly of organizations, that are producers or users of electrodialysis membranes. The Manual covers specified methods for sampling, testing, storing and shipping permselective membrane materials of both cationic and anionic types [1].

3.5.2 Selectivity

Ion exchange membranes are selective for either cations or anions. Cation selective membranes contain anionic groups fixed in the resin matrix. Anion selective membranes contain cationic groups fixed in the resin matrix.

When a cation selective membrane is immersed in an electrolyte solution the cations in solution will enter in the resin matrix and replace the cations present, but anions are prevented from entering the matrix by the repulsion of the anions affixed to the resin. The opposite phenomenon takes place when an anion selective membrane is immersed in an electrolyte solution. Because of the selective permeability, they are also termed permselective membranes.

Counter-ions within an ion exchange resin or membrane are ions with a charge opposite to the charges affixed to the membrane matrix. Co-ions are ions with a charge as the fixed charges of the matrix. Hence ion selective membranes are selectively permeable to counter-ions and selectively not permeable to co-ions. The selectivity, which might be expressed in terms of the transference number of counter-ions in the membranes, is not generally restricted to all ions of one charge. Membranes may as well be more selective to some ionic species than to others. There are commercially available membranes, as indicated in table 3/1, that will selectively transport univalent ions in electrodialysis. With these membranes not only the concentrations but also the composition of electrolyte solutions can be altered. A membrane possessing specific selectivity between divalent and univalent ions would be useful for removing sulfate from a solution of chloride and sulfate, or to fractionate a mixture of the ions into two solutions, each one containing only one of the ions.

The preparation of membranes of improved specificity was attempted by Gregor, by employing compounds which were presumed to be specific to sulfate. It was demonstrated that the relative transport numbers for sulfate and chloride may vary from 0.75 to 1.12. This means that a significantly higher or lower concentration of sulfate ions at the membranes surface can be achieved, if brackish waters are demineralized by membrane electrodialysis [2]. Membrane pore size was found to have a pronounced effect on chloride-sulfate ion exchange equilibrium and consequently on the transport ratio of these ions [3].

An extensive research work on the permeability of divalent to univalent ions in cation exchange transfer is reported as part of a general study aiming at the use of electrodialysis for seawater desalination and salt recovery. The permselectivities of magnesium and calcium ions against sodium ions increased with the decrease in current density. The decrease in concentration of sulfate was very much less than that of chloride and the molar fraction of sulfate in desalinated water increased with the progress of desalination. The permselectivity coefficient of sulfate against chloride increased with the decrease in concentration of chloride, but the effect of current density on the coefficient was very small [4].

Charakterisierung von ionenselektiven Membranen

Selektivität

The permeation rate of magnesium ions was found to decrease with increasing total concentration of magnesium ions. Permeation cf Na ions increased under similar conditions. At constant NaCl concentration, the permeation rate of Na ions decreased with increasing $MgCl_2$ concentration [5]. Similar work was reported for the selective permeability of magnesium and hydrogen ions, as well as of calcium and magnesium ions against sodium ions [6]. For univalent cations permselectivity was found to decrease in the order $NH_4 > K > Na > Li$ [7], and for anions in the order $Cl > Br > J$ [8]. The permselectivity of fluoride against chloride ions was determined by Takemoto et al. The results indicated that fluoride in seawater forms complexes such as MgF^+, which is not permeable through ion exchange membranes [10]. Permselectivity for divalent cations to sodium was found to increase in the order of $Mg < Ca < Ba < Sr$ [9]. The permselectivity between two ions having the same electric charge was also investigated by Azechi [11].

The selectivity between chlorides and nitrates was studied experimentally as function of the degree of ionization of the membrane, of ionic strength and of the ionic equivalent fraction of chloride in the solution. A simple expression of selectivity between chlorides and nitrates was found to apply with constant interaction coefficients [12].

An investigation of the length of the heat treatment of ion exchange membranes on some of the electrochemical properties indicated that at higher temperatures the selective ionic permeability decreases and the electrical conductivity increases. These changes are caused by breakdowns in the structure of the organic matrix of the ion exchange and binding material [13].

The preferential ion transport through ion exchange membranes of a certain ionic species to another was described by Kitamoto and Takashima [14] by using two dimensionless parameters. An equation for the flux of a trace counter-ion between two solutions separated by an ion exchange membrane was derived and tested for systems containing singly and multiply charged trace and bulk ions. Experimental evidence was presented by Blaedel et al. [15]. Transference numbers of counter-ions in ion exchange membranes derived from the electromotive force data are lower than those derived by the Hittorf method. The difference was attributed to transport of water across the membrane [16]. EMF measurements for selectivity determinations of highly selective membranes, using a new flow-cell method and a membrane-skip method, were reported by Kwak [17].

*Ionen- und
Wasser-
transport*

3.5.3 Ion and water transport

Several transport processes occur simultaneously during electrodialysis of saline solutions. Counter-ion transport is the major electrical ion movement and a certain quantity of water is transported with the counter-ions by electro-osmosis. Co-ion transport is small and depends upon the selective properties of the membrane and the brine concentration. Water transport from the low concentration compartments into the higher concentration brine compartments is taking place by osmosis. Diffusion of electrolyte occurs from the brine compartment to the low concentration compartment. Water transport is also associated with electrolyte diffusion. The total quantity of water moved across various membranes has been measured by several investigators and has been reviewed by Lakshminarayanaiah [18] and more recently by Läuger [19].

The influence of counter-ions on permselective membrane performance has been examined by Kertesz et al. [20], coupling between chemical reactions and transport phenomena was investigated by Bailey and Luss [29], and transport of ions through ion exchange membranes was studied by Woerman [30], by Sata [33], by Rogers and Sternberg [37] and by Kamo et al. [38 to 41]. Water transport accompanying the passage of ions was investigated by George and coworkers [21 to 23], by Lakshminarayanaiah et al. [24 to 26], by Foley and Meares [27] and by Wallace and Ampaya [28]. Transport properties of cation exchange membranes were reported by Gardner and Paterson [31, 32, 36], by Fedotov et al. [34] and by Tombalakian [35].

Mosaic membranes are produced by pressing grains of different polymers in sheet form. Properties of mosaic membranes were investigated by Weinstein, Caplan et al. [42, 43]. Transport characteristics of a family of ion exchange membranes composed of contiguous layers of anion and cation exchange materials were investigated by Sonin and Grossman [44].

*Elektro-
osmose*

3.5.4 Electroosmosis

Water transport by osmosis and electroosmosis accompanies electrodialysis and generally occurs in the same direction as the salt transfer. Both are effects which limit the effectiveness of electrodialysis as a method of concentrating electrolyte solutions.

Water transference numbers rise with temperature in cation exchange membranes in the H-form and slightly decrease in the K-form [45]. Membranes with low water content, less than 14%, produce electroosmotic flows independent of current density, while membranes with higher water content show current dependence. At external concentration exceeding 0.1 N the current dependence of water flow disappears and the flow decreases with increase of external concentration [46]. The effects of counter-ions and co-ions on the electroosmotic stream were investigated by McHardy et al. [47].

Measurements of membrane properties have been made on both cation and anion exchange membranes in a variety of ionic states. Calculated electroosmotic coefficients were found to agree with most experimentally measured values. The results indicate that membrane pore structure is well approximated by a model employing parallel-plate pores and that ions are partially stripped off their hydration sheath while within the membrane. A hydrodynamic model for electroosmosis in ion exchange membranes has been developed [48]. In the case of fine pores the electrokinetic phenomena can be described based on true electroosmotic transfer. The cause of electroosmosis is the effect of the electric field on the excess ions. The electroosmotic transfer decreases with increasing average concentration of the excess ions. This was confirmed by experiment. With increasing total exchange capacity the true electroosmotic transfer decreases [49].

3.5.5 Concentration polarization

The study of membrane phenomena and the understanding of the detailed structure and function of ion exchange membranes, as suggested by Gregor, would lead to the fabrication of membranes for specific applications [50]. Cooke [51] and Solt [52] have drawn attention to the fact that even in the most idealized electrodialysis compartment, the local conditions of mass transfer are not uniform.

Transport numbers of ions in the ion exchange membrane are different from those in the solutions on both sides of the membrane. Because of the lower transport number of ions in the solution, the number of ions transported to the membrane surface by the electrical current is in deficiency to the ions removed from that surface and transferred through the membrane. The opposite phenomenon occurs on the other side of the membrane. A greater number of ions are transferred from the entering to the outgoing membrane surface than can be carried away by the electrical current. Two boundary layers with opposite concentration gradients are formed at both sides of the membrane. This tendency for concentration and depletion is opposed by diffusion and physical mixing. Hence the thickness of the boundary layers depends on hydrodynamic conditions and on the degree of turbulence. However there remain layers adjacent to the membrane in which the solutions are completely static.

Increasing the current density has the effect of increasing the concentration gradients at both surfaces of the membrane and the point may be reached at which the concentration of ions at the entering side of the membrane approaches zero. This is the limiting current density. When the limiting current density is exceeded, hydrogen and hydroxyl ions are transported through the solution and the membrane, causing changes of pH inside the membrane and at the boundary layers of the solutions, as well as an increase in the overall electrical resistance. The desired ions participate with only a small amount in the transport. The phenomenon is termed concentration polarization and is the major limitation of the production rates achievable by electrodialysis.

An apparatus was described by Kooistra for the automatic registration of potentiodynamic polarization curves of ion exchange membranes. A strict time schedule of stirring and non-stirring periods was applied. The parts of the potential–current density curves registered during the non-stirring periods clearly showed the limiting current density, due to the building up of a depleted layer on one side of the membrane. The continuation of the polarization curve beyond the limiting current region supplies valuable information about ion generation and loss of selectivity at higher polarization voltages [53].

From studies on the effect of the boundary layer on the limiting current density, it was concluded that the limiting current density in the laminar flow is approximately proportional to the 1st power of the concentration, the $1/2$ power of the flow velocity of the solution and the $1/2$ power of the membrane length in the direction of the flow. In natural convection it is proportional to the $5/4$ power of the concentration of the solution [54].

The limiting current densities of the ion exchange membranes were measured in the systems formed with various membranes and solutions. Cation exchange membranes give lower values of limiting current densities and a lesser effect on the disturbance of the pH value of the solution than the anion exchange membrane in NaCl solutions [55]. Investigating the pH changes at anion selective membranes, concentration of OH-ions was measured as a function of current density on the receiving side of the polarizing anion selective membrane [56].

The role of water dissociation during electrodialysis was also studied. The curves which describe dependence of water dissociation on the current density and concentration of the electrolyte show that an increase of electrolyte concentration in the polarization layer decreased participation of decomposition products of water in the dialysis. An increase of NaCl concentration and a decrease in current density acted in the same direction. A quantitative study was made of the rate of formation of H^+ and OH^- ions during the electrodialysis of aqueous NaCl solutions under polarizing conditions. A new method was proposed for determining the limiting current based on the recording of the transfer numbers of the dissociation products of water as a function of the current density [57].

The rate of ionic mass transfer at the limiting current density was measured in two-dimensional flows, both laminar and turbulent, through a channel between a pair of ion exchange membranes. For laminar flow, the experimental results showed good agreement with those of numerical analysis. For turbulent flow, the analytical results were also brought into good agreement with the experimental results by introducing a new concept of the eddy migration coefficient. The local electrical conductivity in an intensely agitated fluid may be greater by 5 to 10% than that in a stationary solution [58].

The potential difference between the two sides of the membrane is determined by the concentration difference between the two membrane-solution interfaces and the potential caused by the occurrence of electrical dipoles developed in the membrane: the electret potential. Electret potential and increase in membrane resistance during continued transfer of direct current were measured on various types of ion exchange membranes. Comparison of the results showed a qualitative correlation between the polymeric skeleton of the membrane and its polarization performance [59]. An expression was derived for the current—voltage curve characteristic of an ion exchange membrane system with an electrolyte on each side. It was concluded that upon increasing the voltage, current plateaus should be reached, both in the absence and in the presence of a polyelectrolyte, which cannot permeate the membrane. The so-called limiting currents are rarely plateaus of this kind [60].

In an attempt to investigate polarization phenomena, including substantial participation of hydrogen and hydroxyl ions in conduction, current—voltage curves were measured in different sections along the tortuous flow path of an electrodialysis apparatus containing a single anion exchange membrane. While the measured current—voltage curves exhibited true limiting current plateaus only in the last section, all sections yielded plateau currents after correction for the measured variation of coulombic efficiency. These corrected curves were in fair agreement with those computed from the Nernst-Planck equations for ion migration, as developed to take due account of the variation of salt concentration along the flow path [61]. An experimental investigation indicated that the Nernst model, involving a limiting diffusion layer of constant thickness, is applicable to a precise case, from which conclusions can be reached to detect limiting currents in the electrochemical polarization at the interface of ion exchange membranes and aqueous saline solution [62].

The concentration polarization on cation and anion exchange membranes was investigated in dilute NaCl, NaOH, and HCl solutions. Current—voltage and current—resistance curves were plotted both to define values of limiting current intensities and to examine the properties of the membranes. The experimental values of limiting current agreed with calculated values for dilute NaCl solutions by use of cation exchangers, whereas both for NaOH solution with cation exchanger and for HCl solutions with anion exchanger membranes the experimental values appeared to be two to three times lower than the calculated values [63].

Measurements of persistent electrical polarization as a function of composition, temperature, applied field, molecular weight of the polarizing component, and time were made on a series of membranes containing sodium polystyrenesulfonate in matrices of poly(vinylalcohol), polyacrylamide and poly(vinylpyrrolidone). A model for the process was developed. The model is compatible with experimental data and explains observed saturation effects and the effects of such variables as membrane composition and polarizing temperature [64].

The validity of a correlation, which directly involves the principal operative variables of electro-dialyzers and the resulting desalination, has been proved by the results of a large number of runs carried out with a pilot plant. The experimental data have supplied precise indications on the behavior of the apparatus in presence of polarization [65].

The steady-state transport characteristics of a family of ion exchange membranes composed of contiguous layers of anion and cation exchange materials was described in terms of a simple model. Membranes consisting of up to four layers were considered. Explicit analytic expressions were derived, which relate the current—voltage characteristics and transport numbers to membrane structure and to the concentration of the bounding solutions. An analysis was given for the general performance characteristics, including the effects of concentration polarization in the bounding solutions, of a three-layered membrane consisting of a thick central ion exchange layer sandwiched between two extremely thin ion exchange layers of opposite sign. This combination may serve as a model for the effects of certain types of membrane fouling in practical applications such as electrodialysis. It was shown that even very thin surface layers can reduce the limiting current to a value significantly below the diffusion-controlled one, which is expected in the absence of the surface films [66].

An improved method for calculating the saline concentration at the solution membrane inter-face in a system consisting of two electrolyte solutions separated by a charged membrane makes use of potential measurements carried out both while the current is passing through the system and after it has been switched off. The method is valid for natural convection systems. A comparison of the two series of measurements makes it possible to identify a critical value of the current density char-acterized by the fact that the current is being transported by species other than the counter-ion. It occurs at values of the interfacial concentration corresponding to about 25% of the bulk concentra-tion of the electrolyte [67]. The concentration polarization occurring at the membrane—solution inter-face was analyzed on the basis of the continuity condition of the steady flux of movable components. The equation obtained for the degree of polarization is represented in terms of the thickness of the stagnant layers remaining at the membrane surface and of the difference between mass fixed trans-ference numbers of counter-ions in the membrane and in the bulk solution [68].

3.5.6 Electrical properties

Elektrische
Eigenschaf-
ten

Ion exchange membranes should have a high electrical conductivity, when in equilibrium with a dilute solution, as membrane conductivity affects the ohmic resistance of the electrodialysis unit.

Electrical conductivities of cation- and anion-permeable electrodialysis membranes in a number of monovalent ionic forms have been determined under precisely controlled conditions of tempera-ture, membrane water content, and direct current voltage gradient. Measurements were carried out in the absence of external electrolyte. A new semitheoretical mobility equation was developed and experimentally verified for the passage of ions in common electrodialysis membranes. Calculated mobilities of hydrogen, sodium, and potassium ions in typical cationic electrodialysis membranes were presented as a function of the entire membrane water-content range at room temperature. The calculated mobilities of hydroxyl and chloride ions in typical anionic membranes were given over the total membrane water-content range [69].

Membrane conductivity obeys the Arrhenius equation with apparent activation energy values ranging from 11 to 16 kcal/mole in the glassy membrane. Computed diffusion coefficients of hydro-gen ions increase with sulfuric acid content and are appreciably higher in the rubbery than in the glassy membrane states [70]. Conductivity of various membranes was measured, the temperature was raised to 50 to 60°C and then lowered to the original. The activation energy of conductivity re-mained the same in both periods [71]. Electrical conductivity and the coefficient of self-diffusion of ions in various membranes were determined and the mobility of counter-ions was calculated [72]. Conductivity measurements of ion exchange membranes were carried out with a newly developed measuring system. The concentration dependence of the electrolytic conductivity of the pore-solution was described quantitatively by an equation, taking into account the conductivity coefficient [73].

The usual electrical conductivity-electrolyte concentration relation is inapplicable for the ion exchange membranes. An empirical relation between the specific electrical conductivity and the exchange capacity of the membrane was suggested. This relation appears to be applicable for all known membranes and for any polymer ion exchange material [74].

The influence of the fixed ion concentration on the conductivity of an ion exchange membrane was investigated and a new classification of ion exchange membranes was proposed, which distinguishes between homogeneous and heterogeneous systems with respect to the matrix and the fixed ion concentration [75]. Measurements of the resistance of ion exchange membranes swollen in water by a contact method involving electrodes directly adjacent to the membrane surface lead to a formula for determining the conductivity of the membranes from the contact resistance [76]. The electrical resistances of electrodialyzer chambers and of swollen grains of an ion exchanger in a nonconducting medium were calculated in the same manner. The relative electrical conductivity of swollen ion exchange resins in a nonconducting medium depends not only on the specific electrical conductivity of the ion exchanger material but also on the geometry of the system, especially the diameter of the electrical contact among the grains [77].

The electrical conductivity of the hydrogen-, lithium-, sodium-, and potassium-forms of cation exchange membranes in water has been determined between 30 and 180°C in a specially made cell. The sodium−potassium ion exchange equilibrium has been investigated under static conditions over the same temperature range. The variation in selectivity of sorption corresponds to the change in salt forms of the membrane [78].

Specific and transport properties of a cation and an anion exchange membrane in Na^+ and Cl^- form respectively have been measured in water and dioxane−water mixtures. A greater electrolyte absorption and a lesser specific conductivity and counter-ion transport number in solvents at higher dioxane content indicate strong electrostatic interactions between fixed groups in the membrane and counter-ions [79].

The dielectric behavior of the hydrogen-, sodium-, cholin-, and triethylbenzylammonium-forms of sulfonic acid cation exchange resin membranes with different specific degrees of hydration has been investigated. A region of dielectric dispersion due to the movement of bound counter-ions has been observed. The critical dispersion frequency and the low-frequency dielectric increment increase with the moisture content and depend on the nature of the cations. For all the forms of the cation exchanger, the high-frequency dielectric constant increases linearly with moisture content, showing, however, a negative deviation from the additive line [80].

Electrical potentials arising across three cation-selective membranes, separating NaCl solutions of different concentrations, have been measured and electrolyte uptake of these membranes has been estimated. These data together with the apparent membrane transport number have been introduced in a simple relation [81]. Membrane potentials were measured by two methods and activity gradients obtained. The potentials predicted from the transference numbers agreed well with those measured. The osmotic water flux appeared to be a more serious cause of polarization than the salt diffusion flux [82]. The membrane potential is a direct measurement of the permselectivity of an ion exchange membrane. The behavior of a given membrane in various conditions may be anticipated provided that the membrane phenomena would be theoretically explained [83].

A theory was proposed on the transmembrane potential caused by a membrane with asymmetrical distribution with respect to the surface charges of the membrane [84]. Theoretical equations were derived for the bi-ionic potential between two univalent electrolyte solutions separated by a membrane and for the membrane potential between two solutions of the electrolyte of different concentrations [85]. Measurements have been made of the selectivity coefficients, cation-interchange fluxes and two-cationic cell potentials for the exchange between cadmium, calcium, cobalt, copper, nickel, and barium ions across an ion exchange membrane separating 0.1 N solutions of divalent chloride salts. The experimental data have been used to estimate the magnitudes of the ionic mobility ratios, interdiffusion and effective single-ion diffusion coefficients of the counter-diffusing cations in the membrane. The dependence of the membrane interdiffusion coefficient on the ionic composition of the membrane for the various divalent cation exchange systems has been determined [86].

3.5.7 Scaling and fouling of membranes

Krusten-bildung und Faulen von Membranen

An increase in the pH of the solution associated with polarization promotes the formation of alkaline precipitates, such as calcium carbonate and magnesium hydroxide, on the membrane surface. Scaling of membranes causes additional electrical and flow resistance, a decrease in electrodialysis efficiency and an increase in pumping power requirements. Mechanical damage to the membranes may also occur.

Scale formation is particularly pronounced at high current densities, but it also occurs, to a lesser extent, at low current densities. Calcium sulfate scale may also deposit on the membrane surface, when the feed solution has a high sulfate content. The allowable level of supersaturation of calcium sulfate in the waste stream of electrodialysis demineralizers was investigated in an early study. It was concluded that supersaturated solutions cannot be handled in electrodialysis units without stabilizing additives, such as sodium hexametaphosphate [87].

Experience with demineralization of brackish natural water by electrodialysis has shown that the electrical resistivity of the membranes increases during extended operation. This increase occurs in two phases. A primary resistance increase develops rapidly during the first several hours of operation after which a secondary resistance increase occurs at a slower rate. The major cause of the resistance increase was an accumulation of an iron-rich gel on membrane surfaces. Sulfuric acid injected into the dilution stream eliminated the secondary resistance increase when the product acidity was lowered to pH 5.1. Although acidification adds to the direct cost of desalting, the added cost was considered to be partly offset by savings in energy consumption due to a lower overall electrical resistance [88]. Problems related to the acidification of brines during the electrodialysis of water containing calcium ions for prevention of precipitation on membranes were also investigated [89]. Electrodialytic desalination with pH adjustment of the concentrated solution by the addition of mineral acid in order to prevent scale deposition has recently been studied. Limiting current density and current efficiency were examined in an experimental cell and the results were confirmed in field tests with a commercial stack [90].

Chronopotentiometric measurements were used to study the polarization of cation exchange membranes in the absence of convective mixing of an electrolyte solution as a function of the ionic form of the membrane, concentration and nature of the electrolyte, taking into account possible deposition of difficulty soluble products on the membrane surface. The formation of deposits usually decreased the transit time. Such change in the transit time is presumed to be determined mainly by the specifics of adhesion of a precipitate and its density and was not associated with the value of the solubility product of the depositing compound [91].

Anionic membranes foul owing to the deposition of insoluble acidic colloids on the membrane surface facing the diluate. The fouling is caused by hydrogen ions generated by even minimal polarization at the membrane surface. Fouling increases under polarizing conditions, such as decreasing salt concentration and flow velocity and increasing current density and colloid concentration. It is much stronger on rough surfaced than on glossy surfaced membranes and stronger on membranes which show high dynamic polarizability. An alkaline pH decreases the humate type of fouling. Fouled stacks can be regenerated successfully by rinsing with alkali [92]. The effect of chlorination on the colloidal-fouling of electrodialysis membranes by solutions of sodium alginate and sodium humate was determined. Chlorination was found to be more efficient for the prevention of fouling for the alginate. At higher levels of chlorination, fouling by both substances was substantially reduced [93]. To prevent the formation of deposits in electrodialysis, methods such as inserting local resistances into the chambers, preliminary acidification of water, use of frequent impulses of the reversible current, addition of inhibitors and use of less selective ion exchange membranes were described [94]. Deposition of $CaCO_3$ is inhibited by the presence of Mg^{2+} and PO_4^{3-}. The former increases the solubility of $CaCO_3$ and the latter is proved to favor precipitation of calcite rather than aragonite [95]. The addition of sulfuric acid to keep pH below 5.0 for prevention of $CaCO_3$ scaling in concentrated seawater is also reported [96].

The effects of the acid concentration of seawater to prevent carbonate scale formation on three types of conventional strongly acidic and basic membranes and a univalent cation-selective membrane were studied [98].

The effect of ions in seawater to aragonite formation was studied in a solution containing the same ion concentration as seawater for about 100 h in an electrodialysis apparatus. On the basis of the X-ray diffraction analysis and microscopic observation on the $CaCO_3$ scale deposited, conclusions were obtained, as to the conditions of formation of either aragonite or calcite in various forms [99].

States of the interfacial layer at the desalting side of an ion exchange membrane under water decomposition conditions in NaCl solution were observed. The limiting current densities estimated from current–voltage curves agreed with those at which the concentration at the membrane surface reached zero. In the process of the interfacial layer formation, the concentration decrease at the surface of the cation exchanger was more rapid than that of the anion exchanger. At the anion exchanger the concentration distribution remained stable even under water decomposition conditions. Under water decomposition conditions, the concentration distribution in the interfacial layer of cation exchanger showed complicated configuration consisting of about 4 parts. Phenomena in the interfacial layer on the desalting side of an anion exchange membrane in $NaHCO_3$ and $KHCO_3$ electrodialysis were also studied, as well as differences of the state of the interfacial layer between the water decomposition and the carbonate decomposition layers. For a membrane without treatment for giving low permeability to divalent anions a break point in the current–voltage relation was observed at the beginning of the water decomposition. For a membrane with this treatment two break points corresponding to both the carbonate and the water decomposition were observed [100].

Experimental data were presented by Grossman and Sonin for the performance of a multi-compartment electrodialysis system with laminar flow in plane, unobstructed channels. The results are shown to be in good agreement with the theory of Sonin and Probstein (see section 3.7.5, p. 219). Fouling of the membrane surfaces is shown to bring about a reduction of the limiting current by as much as a factor of two below the theoretical value. The magnitude of the decrease depends on flow speed and other operating conditions [101].

In electrodialysis systems, membrane fouling may be caused by deposits of either neutral matter or colloidal matter with ion exchange properties. A model for membrane fouling was developed and expressions were derived for the reduction in limiting current caused by both neutral fouling films and films having ion exchange properties. The effect of fouling was shown to depend not only on the properties of the fouling film but also on the hydrodynamic conditions in the channels and in the case of ion exchange fouling films on the salt concentration in the dialysate channel. Ion exchange fouling films were shown to be much more effective in reducing the limiting current than neutral films. Experimental data confirm the trend of the theory for fouling with ion exchange films [102].

A very extensive research work on scale formation in concentrating seawater by electrodialysis was performed by Takamoto. This process was given serious consideration in Japan for the extraction of common salt. Electrodialysis was carried out using 0.5 N NaCl solution and concentration change in the interfaces from 0.5 N to equilibrated concentration was shown by photographs. Concentration difference between anion and cation exchange membrane increased with current density. The concentration at membrane surface C_i is a linear function of that of the bulk solution C_b, as $C_i = \beta \cdot C_b + C_s$. Both constants β and C_s were characteristic of the membrane species and particularly were independent of current density [97].

Vergiftung von Membranen

3.5.8 Poisoning of membranes

Membranes which lose their characteristic properties because of the effect of a chemical agent are said to be poisoned. For a series of permselective membranes, the influence of counter-ions, such as hexacyanoferrate(II), hexacyanocobaltate(III), pentacyanonitrosoferrate, and hexaamine-cobalt(III) ions, on resistance, permselectivity, swelling and co-ion content was examined. The size and charge of the counter-ions were respectively varied. Characteristic differences between membranes of similar initial properties were observed and correlated with differences in the structure. Membranes were found to show two extremes of behavior. One group of membranes changed resistance less during poisoning but the sign of their permselectivity changed more easily. The other group increased their resistance by several orders of magnitude, but needs poisoning ions of higher valency and size to change the sign of their permselectivity. Small, hydrated ions affect the membrane properties less than large ions, which are presumably less hydrated [103].

The poisoning by surface active ions, particularly anions, was described and the mechanism of poisoning elucidated. Protection for resins and membranes from poisoning could be derived by providing for a high degree of crosslinking. Surface active agents to neutralize the poisoning effect of surfactants were examined [104]. The transport properties of ion selective membranes, which absorbed or exchanged a surface active agent, were expected to change when used in electrodialysis. However, no change was observed by the absorption of polyoxyethylene laureate on the surface of the cation or the anion exchange membrane [105]. A similar investigation was also made in the presence of other surface active agents [106].

The current–voltage curves of ion exchange membranes in dilute sodium chloride solutions were determined in the absence and in the presence of detergents of different structure. The effect of small concentrations of detergents is negligible [107].

The poisoning of commercial anion selective membranes by sodium dodecylsulfate was investigated. During this poisoning the volume of the membranes increased and the water content decreased. Moreover the membranes showed a very strong increase of the resistance and a decrease of the permselectivity tending towards zero. During electrodialysis of a NaCl solution containing 5×10^{-4} M sodium dodecylsulfate a "soap layer" is formed at the diluate side in the membranes. The chloride ions are transported through this layer by diffusion of sodium chloride [108]. The poisoning of three anion-selective membranes by a commercial humic substance was further investigated. It was shown that small humic ions with a mean equivalent weight of 70 to 80 g/eq and a mean apparent equivalent volume of 40 to 50 ml/eq can be sorbed by the tested membranes in a stoichiometric ion exchange process. As a result of this poisoning the permselectivity of the membranes, measured between 0.01 and 0.02 M KCl solutions, does not change. However, the volume, the water content, and the a. c. resistance increase [109]. The effects of humic acid and sodium dodecylbenzenesulfonate on the current–voltage curves of various anion exchange membranes have been investigated and essential differences between these two poisoning agents were discovered, indicating different types of inhibition of ion transport. The results of this study are compared with those of an investigation of membranes from a pilot plant, that had processed for two years chemically purified surface water to which traces of a detergent had been added [110].

The manufacture and electrochemistry of ion exchange membranes for electrodialysis and their application for the desalination of raw water, seawater and in various other fields, as waste water purification, was described and membrane polarization, membrane poisoning and factors affecting membrane improvement were discussed [111].

Literature to 3.5.1 to 3.5.8

[1] Bureau of Reclamation (Off. Saline Water Res. Develop. Progr. Rept. No. 77 [1964]). — [2] Polytechnic Institute of Brooklyn (Off. Saline Water Res. Develop. Progr. Rept. No. 37 [1960]). — [3] H. Schweigart (Desalination 2 [1967] 154/60). — [4] T. Azumi, R. Dohno, T. Hakushi, S. Takashima, T. Ishino (Nippon Kaisui Gakkai-Shi 22 [1969] 383/91; C.A. 74 [1971] No. 34509). — [5] R. Dohno, T. Azumi, S. Takashima (Himeji Kogyo Daigaku Kenkyn Hokoku A No. 22 [1969] 97/102; C.A. 72 [1970] No. 70954).

[6] R. Dohno, T. Azumi, S. Takashima (Nippon Kaisui Gakkai-Shi 23 [1969] 21/5, 26/30; C.A. 74 [1971] No. 25365, No. 15654; Nippon Kagaku Zasshi 91 [1970] 1133/6; C.A. 75 [1971] No. 6761). — [7] R. Dohno, T. Azumi, S. Takashima (Nippon Kagaku Zasshi 91 [1970] 131/4; C.A. 73 [1970] No. 18843). — [8] R. Dohno, S. Morita, T. Azumi, S. Takashima (Nippon Kagaku Zasshi 91 [1970] 521/5; C.A. 73 [1970] No. 80971). — [9] R. Dohno, T. Azumi, S. Takashima (Nippon Kagaku Zasshi 92 [1971] 136/9; C.A. 76 [1972] No. 50479). — [10] N. Takemoto, K. Mashiko, M. Setoguchi (Nippon Kaisui Gakkai-Shi 23 [1970] 271/5; C.A. 73 [1970] No. 101879).

[11] S. Azechi (Nippon Kaisui Gakkai-Shi 24 [1970] 25/36, 54/67; C.A. 76 [1972] No. 10671, 74 [1971] No. 25365, No. 15654; Nippon Kagaku Zasshi 91 [1970] 1133/6; C.A. 75 [1971] 633/52). — [13] V. A. Makarova, M. V. Pevnitskaya, V. D. Grebenyuk (Izv. Sibirsk. Otd. Akad. Nauk SSSR Ser. Khim. Nauk 1970 No. 3, p. 151/4; C.A. 74 [1971] No. 42968). — [14] A. Kitamoto, Y. Takashima (J. Chem. Eng. Japan 3 [1970] 54/62; C.A. 73 [1970] No. 38871). — [15] W. J. Blaedel, T. J. Haupert, M. A. Evenson (Anal. Chem. 41 [1969] 583/90).

[16] N. Lakshminarayanaiah (J. Phys. Chem. 73 [1969] 97/102). — [17] J. C. T. Kwak (Desalination 11 [1972] 61/9). — [18] N. Lakshminarayanaiah (Chem. Rev. 65 [1965] 491/565). — [19] P. Läuger (Angew. Chem. 81 [1969] 56/67; Angew. Chem. Intern. Ed. Engl. 8 [1969] 42/54). — [20] D. Kertesz, F. de Körösy, E. Zeigerson (Desalination 2 [1967] 161/9).

[21] J. H. B. George (Off. Saline Water Res. Develop. Progr. Rept. No. 159 [1965]). — [22] J. H. B. George, C. R. Schlaikjer (Off. Saline Water Res. Develop. Progr. Rept. No. 203 [1966]). — [23] J. H. B. George, R. A. Horne, C. R. Schlaikjer (Off. Saline Water Res. Develop. Progr. Rept. No. 321 [1968]; J. Electrochem. Soc. 117 [1970] 892/8). — [24] N. Lakshminarayanaiah (Desalination 3 [1967] 97/105). — [25] N. Lakshminarayanaiah (J. Macromol. Sci. Phys. 5 [1971] 159/65).

[26] N. Lakshminarayanaiah, F. A. Siddiqi (Z. Physik Chem. [Frankfurt] 78 [1972] 150/64). — [27] T. Foley, P. Meares (Experientia Suppl. No. 18 [1971] 313/9). — [28] R. A. Wallace, J. P.

Ampaya (Desalination **14** [1974] 121/34). — [29] J. E. Bailey, D. Luss (Proc. Natl. Acad. Sci. U.S. **69** [1972] 1460/3). — [30] D. Woerman (Chem. Ingr.–Tech. **44** [1972] 158/63).

[31] H. Ferguson, C. R. Gardner, R. Paterson (J. Chem. Soc. Faraday Trans. II **68** [1972] 2021/9). — [32] C. R. Gardner, R. Paterson (J. Chem. Soc. Faraday Trans. II **68** [1972] 2030/40). — [33] T. Sata (J. Colloid Interface Sci. **44** [1973] 393/406). — [34] N. A. Fedotov, K. Kh. Urusov, Ya. B. Skuratnik (Zh. Fiz. Khim. **46** [1972] 2842/5; Russ. J. Phys. Chem. **46** [1972] 1617/8). — [35] A. S. Tombalakian (Can. J. Chem. Eng. **50** [1972] 203/6).

[36] R. Paterson, C. R. Gardner (J. Chem. Soc. A **1971** 2254/61). — [37] C. E. Rogers, S. Sternberg (J. Macromol. Sci. **5** [1971] 189/206). — [38] N. Kamo, Y. Toyoshima, H. Nozaki, Y. Kobatake (Kolloid-Z. Z. Polymere **248** [1971] 914/21). — [39] N. Kamo, Y. Toyoshima, Y. Kobatake (Kolloid-Z. Z. Polymere **249** [1971] 1061/8). — [40] N. Kamo, Y. Kobatake (Kolloid-Z. Z. Polymere **249** [1971] 1069/76).

[41] T. Ueda, N. Kamo, N. Ishida, Y. Kobatake (J. Phys. Chem. **76** [1972] 2447/52). — [42] J. N. Weinstein, B. J. Bunow, S. R. Caplan (Desalination **11** [1972] 341/77). — [43] J. N. Weinstein, B. M. Misra, D. Kalif, S. R. Caplan (Desalination **12** [1973] 1/17). — [44] A. A. Sonin, G. Grossman (Off. Saline Water Res. Develop. Progr. Rept. No. 814 [1972]). — [45] R. Arnold (J. Phys. Chem. **73** [1969] 1414/20).

[46] N. Lakshminarayanaiah (J. Electrochem. Soc. **116** [1969] 338/43). — [47] W. J. McHardy, P. Meares, A. H. Sutton, J. F. Thain (J. Colloid Interface Sci. **29** [1969] 116/28). — [48] B. R. Breslau, I. F. Miller (Ind. Eng. Chem. Fundamentals **10** [1971] 554/65). — [49] O. L. Alekseev, F. D. Ovcharenko (Kolloidn. Zh. **33** [1971] 3/5; C.A. **74** [1971] No. 91 618). — [50] H. P. Gregor (Proc. 1st Intern. Symp. Water Desalination, Washington, D. C., 1965 [1967], Vol. 1, p. 452/72; Off. Saline Water Res. Develop. Progr. Rept. No. 193 [1966]).

[51] B. A. Cooke (Proc. 1st Intern. Symp. Water Desalination, Washington, D. C., 1965 [1967], Vol. 1, p. 219/37). — [52] G. S. Solt (Proc. 1st Intern. Symp. Water Desalination, Washington, D. C., 1965 [1967], Vol. 1, p. 13/23). — [53] W. Kooistra (Desalination **2** [1967] 139/47). — [54] R. Yamane, T. Sata, Y. Mizutani, Y. Onoue (Bull. Chem. Soc. Japan **42** [1969] 2741/8). — [55] T. Yamabe, M. Seno (Desalination **2** [1967] 148/53).

[56] T. R. E. Kressman, F. L. Tye (J. Electrochem. Soc. **116** [1969] 25/31; Comments by A. H. Heit, R. Prober and reply ibid. p. 1714). — [57] Yu. A. Kononov, B. M. Vrevskii (Zh. Prikl. Khim. **44** [1971] 927/9, 929/32; C.A. **75** [1971] No. 10684, No. 40187). — [58] A. Kitamoto, Y. Takashima (J. Chem. Eng. Japan **3** [1970] 182/91; Comments by A. A. Sonin, R. F. Probstein and reply ibid. **4** [1971] 283/5). — [59] C. Forgacs (Desalination **7** [1969] 111/21). — [60] K. S. Spiegler (Desalination **9** [1971] 367/85).

[61] C. Forgacs, N. Ishibashi, J. Leibovitz, J. Sinkovic, K. S. Spiegler (Desalination **10** [1972] 181/214). — [62] G. Khedr, R. Varoqui (Compt. Rend. C **275** [1972] 1185/8). — [63] M. V. Pevnitskaya, V. K. Varentsov (Izv. Sibirsk. Otd. Akad. Nauk SSSR Ser. Khim. Nauk **1970**, No. 6 p. 8/13, 13/8; C.A. **74** [1971] No. 130787, No. 91 602). — [64] C. Linder, I. F. Miller (J. Phys. Chem. **76** [1972] 3434/45; J. Electrochem. Soc. **120** [1973] 498/502). — [65] D. Barba, R. Cigna, S. Di Cave (Quad. Ing. Chim. Ital. **7** [1971] 156/9).

[66] A. A. Sonin, G. Grossman (J. Phys. Chem. **76** [1972] 3996/4006). — [67] G. Boari, G. Lacava, C. Merli, R. Passino, G. Tiravanti (Proc. 4th Intern. Symp. Fresh Water Sea, Heidelberg 1973, Vol. 3, p. 169/80). — [68] Y. Kobatake, N. Kamo (Proc. 4th Intern. Symp. Fresh Water Sea, Heidelberg 1973, Vol. 3, p. 91/8). — [69] R. A. Wallace (Off. Saline Water Res. Develop. Progr. Rept. No. 407 [1969]). — [70] R. A. Wallace, B. K. Jindal (J. Electrochem. Soc. **118** [1971] 707/9).

[71] F. de Körösy, M. F. Zevulun (Israel J. Chem. **7** [1969] 117/26). — [72] A. M. Filimonova, N. N. Rodin, N. I. Nikolaeva (Zh. Fiz. Khim. **43** [1969] 1292/4; C.A. **71** [1969] No. 105656). — [73] C. Steymans (Ber. Bunsenges. Physik. Chem. **74** [1970] 46/53). — [74] V. P. Greben, P. E. Tulupov, A. I. Kasperovich, V. N. Zaitsev (Elektrokhimiya **7** [1971] 939/44; C.A. **75** [1971] No. 101613). — [75] P. Groll, F. Grass (Electrochim. Acta **16** [1971] 31/40).

[76] V. V. Kryuchenkov, L. A. Gubareva, V. S. Musinova (Zh. Fiz. Khim. **46** [1972] 936/40; Russ. J. Phys. Chem. **46** [1972] 540/2). — [77] V. D. Grebenyuk, N. P. Gnusin, V. D. Bolotova (Vodopodgotovka Ochistka Prom. Stokov **1972** No. 9, p. 174/82; C.A. **78** [1973] No. 88449). — [78] A. T. Davydov, I. I. Ignatov, Yu. I. Ignatov (Zh. Fiz. Khim. **47** [1973] 426/7; Russ. J. Phys. Chem. **47** [1973] 238/9). — [79] S. D'Alessandro, A. Tantillo (Desalination **9** [1971] 225/34). —

[80] V. I. Frolov, B. V. Moskvichev, G. V. Samsonov (Zh. Fiz. Khim. **46** [1972] 1180/5; Russ. J. Phys. Chem. **46** [1972] 680/2).

[81] N. Lakshminarayanaiah, F. A. Siddiqi (J. Polymer Sci. A I **8** [1970] 2949/55). — [82] D. G. Dawson, P. Meares (J. Colloid Interfance Sci. **33** [1970] 117/23). — [83] R. Jerome (Ind. Chim. Belge **36** [1971] 578/88). — [84] S. Ohki (J. Colloid Interface Sci. **37** [1971] 318/24). — [85] Y. Toyoshima, H. Nosaki (J. Phys. Chem. **74** [1970] 2704/10).

[86] A. S. Tombalakian, G. K. Markarian (Can. J. Chem. Eng. **51** [1973] 124/7). — [87] R. M. Lurie, M. E. Berg, A. Giuffrida (Off. Saline Water Res. Develop. Progr. Rept. No. 48 [1961]). — [88] D. H. Furukawa (Off. Saline Water Res. Develop. Progr. Rept. No. 285 [1968]). — [89] G. A. Lebediskaya, N. I. Isaev, N. N. Zubets (Tr. Voronezh. Univ. **1971** No. 82, p. 145/8; C.A. **76** [1972] No. 158156). — [90] T. Asawa, I. Nakamura, T. Kawahara, H. Hani (Proc. 4th Intern. Symp. Fresh Water Sea, Heidelberg 1973, Vol. 3, p. 143/50).

[91] E. I. Ivakina (Tr. Voronezh. Univ. **1971** No. 94, p. 65/70). — [92] F. de Körösy, A. Süszer, E. Korngold, M. F. Taboch, M. Flitman, E. Bandel, R. Rahav (Off. Saline Water Res. Develop. Progr. Rept. No. 605 [1970]), E. Korngold, F. de Körösy, R. Rahav, M. F. Taboch (Desalination **8** [1970] 195/220). — [93] E. Korngold (Desalination **9** [1971] 213/6). — [94] V. A. Kirdun (Tr. Vses. Nauchn. Issled. Inst. Vodosnabzh. Kanaliz. Gidrotekhn. Sooruzh. Inzh. Gidrogeol. **1970** No. 25, p. 31/6; C.A. **73** [1970] No. 80360). — [95] H. Yamamoto, N. Yugi (Nippon Kaisui Gakkai-Shi **26** [1972] 66/73; C.A. **78** [1973] No. 151498).

[96] T. Watanabe, H. Yamamoto, M. Akiyama, N. Yugi (Nippon Kaisui Gakkai-Shi **26** [1972] 83/90; C.A. **78** [1973] No. 163885). — [97] N. Takemoto (Nippon Kaisui Gakkai-Shi **23** [1969] 54/9; C.A. **74** [1971] No. 45441). — [98] N. Takemoto (Nippon Kaisui Gakkai-Shi **26** [1972] 21/6, 26/32, 32/7; C.A. **78** [1973] No. 163882, No. 163881, No. 163883). — [99] N. Takemoto (Nippon Kagaku Kaishi **1972** No. 10, p. 1832/7, No. 11, p. 2053/8; C.A. **78** [1973] No. 33736, No. 33737). — [100] N. Takemoto (Nippon Kagaku Kaishi **1973** No. 1, p. 44/9, No. 3, p. 482/6; C.A. **78** [1973] No. 88471, No. 151477).

[101] G. Grossman, A. A. Sonin (Off. Saline Water Res. Develop. Progr. Rept. No. 742 [1971]; Desalination **10** [1972] 157/80). — [102] G. Grossman, A. A. Sonin (Off. Saline Water Res. Develop. Progr. Rept. No. 813 [1972]; Desalination **12** [1973] 107/25). — [103] F. de Körösy, E. Zeigerson (Desalination **5** [1968] 185/99), F. de Körösy, E. Zeigerson, M. Zevulun (Off. Saline Water Res. Develop. Progr. Rept. No. 380 [1968]). — [104] H. Small, R. C. Gardner (Off. Saline Water Res. Develop. Progr. Rept. No. 565 [1970]). — [105] T. Sata (Kolloid.-Z. **243** [1971] 157/9).

[106] T. Sata, R. Izuo, Y. Mizutani, R. Yamane (J. Colloid Interface Sci. **40** [1972] 317/28). — [107] P. J. Van Duin (Proc. 3rd Intern. Symp. Fresh Water Sea, Dubrovnik 1970, Vol. 2, p. 141/53). — [108] E. J. M. Kobus, P. M. Heertjes (Desalination **10** [1972] 383/401). — [109] E. J. M. Kobus, P. M. Heertjes (Desalination **12** [1973] 333/42). — [110] P. J. Van Duin (Proc. 4th Intern. Symp. Fresh Water Sea, Heidelberg 1973, Vol. 3, p. 253/9).

[111] P. Meares (Chem. Ind. [London] **1973** 1103/7).

3.6 Inorganic ion exchange membranes

Anorganische Ionenaustauschermembranen

Inorganic cation exchange membranes have been successfully used in fuel cells. Inorganic materials, due to their comparative thermal stability and lack of carbonaceous ingredients, could presumably overcome disadvantages of organic membranes, such as degradation at elevated temperatures and fouling, if fabricated into a suitable ion exchange membrane. Preliminary work in preparing synthetic inorganic cation and anion exchange membranes was promising in respect with their stability and electrochemical properties [1].

Aluminum vanadate precipitated from acid solution demonstrated reversible cation exchange properties and, if precipitated from a basic solution, it demonstrated reversible anion exchange properties. Rigid membranes formed by pressing and sintering techniques had high cation and anion selective properties. Desalting experiments with these membranes in a small electrodialysis cell gave current efficiencies of 90%. Flexible membranes formed from aluminum vanadate exchangers using a number of organic binders and curing methods demonstrated only cation selectivity. Electrodialysis experiments with these cation selective membranes demonstrated them to be 80% as efficient as commercial organic membranes. The transport number was 0.85 to 0.90 and the specific resistivity 2000 ohm · cm (40 ohm · cm²). Tested at 60°C (140°F) they demonstrated excellent thermal stability [2].

14*

Cation and anion permselective inorganic ion exchange membranes were also prepared based on zirconium phosphate and hydrous thorium oxide, respectively. Electrical, chemical and physical properties of the membranes were determined and compared to those of organic ion exchange membranes. Inorganic cation permselective membranes exhibited resistances in the 10 to 40 ohm $\cdot$ cm^2 range and transference numbers near 0.9. Anion permselective membranes exhibited resistances in the 2 to 40 ohm $\cdot$ cm^2 range and transference numbers near 0.9. Electrodialysis cell operation indicated that the inorganic membranes exhibit higher solids removal and higher current density but lower current efficiency than the organic membranes [3].

Other membranes were prepared by utilizing inorganic ion exchangers such as phosphates and pure hydroxides of zirconium, titanium, and tin as well as mixed hydrous oxides of thorium with several di-, tri-, and tetravalent cations. Cation exchange membranes of zirconium phosphate and anion exchange membranes of thorium hydrous oxide exhibited ion transference numbers of 0.90 to 0.96 and electrical resistances of 2 to 10 ohm $\cdot$ cm^2. When tested in a multicompartment electrodialysis unit, salt removal of 57.5 and 48.6% for NaCl and KCl brines, respectively, was achieved [4].

The electrical conductivity of heterogeneous membranes from phosphate cation exchangers was measured. The ratio of equivalent electrical conductivity of the free solution and the calculated value for the membrane in a solution of the same concentration was much higher for solutions of barium and magnesium ions than for solutions of univalent ions. The conductivity decreased in the order H > Na > Li > Ba > Mg [5].

Possible variations in the performance characteristics of inorganic ion exchange membranes in the electrodialysis of chloride brines have been investigated as a function of some important variables such as hydraulic flow rates, ratios of throughput rates, concentration of feed solution, and the liquid flow pattern. The electrodialysis unit consisted of six cell pairs with an effective flow area of 39 cm^2 (6 in^2) per membrane. Results of complete current density studies which were carried out in order to examine the variables have indicated that linear flow rates in the range 3.3 to 6.0 cm/s and a 1:1 throughput ratio of diluting to concentrating streams were optimum. By employing a combined parallel-series liquid feed system, a current efficiency of > 99% and a demineralization per pass of 66% were obtained. For an intermediate feed concentration of 2000 mg/l total dissolved solids, a satisfactory combination of 97% current efficiency and 65.5% salt removal was achieved [6].

The moisture content, ion exchange capacity, electrical resistance, and transference number of the zirconium phosphate cation exchange membrane were 18%, 1.0 to 1.2 meq/g, 8 ohm $\cdot$ cm^2 and 0.92 to 0.94 respectively, and those of the thorium hydrous oxide anion exchange membrane were 44%, 0.9 meq/g, 2 ohm $\cdot$ cm^2 and 0.91 to 0.96 respectively. Transference numbers of these membranes approached 1.0 at low concentration of the surrounding electrolyte solution. Salt leak coefficients of 9.4×10^{-9} and 1.6×10^{-8} eq/cm$^2 \cdot$ s were obtained for the cation and anion exchange membranes respectively. Bi-ionic potentials across the cation exchange membrane showed the following decreasing order for the mobilities of the cations: H$^+$, NH$_4^+$, K$^+$, Na$^+$, Li$^+$, (CH$_3$)$_4$N$^+$ and (C$_2$H$_5$)$_4$N$^+$. The ion mobilities of the anions Cl$^-$, Br$^-$, I$^-$, NO$_3^-$ and CNS$^-$ did not show marked differences [7].

The structure and morphology of many inorganic membranes were studied by isobaric dehydration, X-ray and electron diffraction, low angle electron diffraction, infrared, and Mössbauer spectroscopy. The pore structure of heavy metal cyanide complexes consists of the interstices between spheroidal or platelike particles. The morphology is more significant than the particle size. Hydrated copper(II) hexacyanoferrate(II) is unique in forming sheets, assumed to be the active semi-permeable structure, water being transported through the channel-like interstices, regular in this case, but irregular in membranes composed of spheroidal particles [8]. In continuation of the previous work, some of the heavy metal hexacyanoferrates(II), hexacyanoferrates(III), and hexacyanocobaltates(III), of which copper(II)hexacyanoferrate(II) represents the classical example, were examined. Most of the heavy-metal complex cyanides form a cubic isomorphous series, with lattice constants slightly larger than 10 Å. The cyanide complexes of the lanthanides and other cations with a crystal radius larger than 1 Å, tend to form a second isomorphous series that is hexagonal rather than cubic [9].

Ion exchange membranes based on zirconium phosphate, hydrous thorium oxide and mixed hydrous oxide of aluminum and thorium were prepared by casting and sheeting techniques using polyvinyl chloride, polystyrene and cellulose acetate as binders. The quantity of anions diffusing into the cation exchange membranes increased with the decrease in the ionic radius and increase in the valency of the ion. Desalting experiments were carried out in an electrodialysis cell using 3 pairs of membranes. For a solution of KCl containing 5000 mg/l total dissolved solids, a maximum salt removal of 61.8% could be achieved [10].

The permeability of various electrolytes through parchment-supported silver chloride, silver phosphate and silver tungstate membranes increased as Li < Na < K for univalent cations and as Ba < Sr < Mg < Ca for divalent cations; Al had the lowest permeability. Membrane potentials arising across these membranes have been used to derive values for the quantity of charge present on the membranes [11]. The permeability of various electrolytes through hexacyanoferrate (II) membranes, at a given temperature, was $Cl^- > NO_3^- > NCS^- > CH_3CO_2^- > SO_4^{2-}$ for both monovalent and divalent cations. For any given anion, the cations followed the sequence $NH_4^+ > Li^+ > Ba^{2+} > Ca^{2+} > Mg^{2+} > Al^{3+}$. This sequence was correlated with the size of the hydrated ion. Membrane potentials arising across hexacyanoferrate (II) membranes were used to derive values for the quantity of charge present on the membranes [12].

In order to obtain a better interpretation of the observed inversion in the sign of permselectivity over heterogeneous zeolitic membranes, experiments were carried out with synthetic sodium-analcime membranes. Electroosmotic flow of water, as well as the transport number of the sodium ion in the membranes were measured for different sodium chloride concentrations and zeolite percentages. The ionic-conduction inversion seems to be connected to the existence in the analcime crystalline structure of some preferential sites for sodium ions [13].

Literature to 3.6

[1] C. Berger, R. Hubata, M. Plizga (Off. Saline Water Res. Develop. Progr. Rept. No. 138 [1965]). — [2] G. A. Guter, H. K. Bishop (Off. Saline Water Res. Develop. Progr. Rept. No. 279 [1967]); H. K. Bishop, J. A. Bittles, G. A. Guter (Desalination 6 [1969] 369/80). — [3] J. I. Bregman (Off. Saline Water Res. Develop. Progr. Rept. No. 148 [1966]). — [4] K. S. Rajan (Off. Saline Water Res. Develop. Progr. Rept. No. 222 [1966]), K. S. Rajan, D. B. Boies, A. J. Casolo, J. I. Bregman (Desalination 1 [1966] 231/46). — [5] E. A. Materova, S. S. Mikhailova (Vestn. Leningrad. Univ. Fiz. Khim. 1971 No. 1, p. 46/8; C.A. 75 [1971] No. 53695).

[6] K. S. Rajan, D. B. Bois, A. J. Casolo, J. I. Bregman (Desalination 5 [1968] 371/90). — [7] K. S. Rajan (Off. Saline Water Res. Develop. Progr. Rept. No. 328 [1968]). — [8] W. O. Milligan, M. Uda, M. L. Beasley, D. R. Dillin, W. E. Bailey, J. J. McCoy (Off. Saline Water Res. Develop. Progr. Rept. No. 594 [1970]). — [9] W. O. Milligan, M. Uda, D. R. Dillin, W. E. Bailey, R. J. Williams (Off. Saline Water Res. Develop. Progr. Rept. No. 723 [1971]). — [10] K. R. Pai, N. Krishnaswamy (Ind. J. Technol. 10 [1972] 229/32).

[11] F. A. Siddiqi, N. Lakshminarayanaiah, S .K. Saxena (Z. Physik. Chem. [Frankfurt] 72 [1970] 298/306, 307/15). — [12] F. A. Siddiqi, N. Lakshminarayanaiah, M. N. Beg (J. Polymer Sci. A I 9 [1971] 2853/67, 2869/75). — [13] E. Drioli, F. Alfani (Proc. 3rd Intern. Symp. Fresh Water Sea, Dubrovnik 1970, Vol. 2, p. 131/9).

3.7 Electrodialysis

Elektrodialyse

Electrodialysis is the transport of ions through ion-selective membranes as a result of an electrical driving force. The process takes advantage of the ability of these membranes to discriminate between differently charged ions, allowing for free passage to either cations or anions and being not permeable to ions of the opposite charge. Next to distillation, electrodialysis is the second completely developed and leading process for desalination of brackish water and, under certain circumstances, for seawater as well.

3.7.1 History of electrodialysis

Geschichte der Elektrodialyse

Early work on electrodialysis and associated phenomena has been extensively reviewed by Prausnitz and Reitstötter [1]. In a comparative study of capillary systems, Manegold and Kalauch stressed the importance of developing ion-selective membranes for the electrodialysis process [2]. Examining the passage of the electric current through selective membranes, Meyer and Strauss suggested the use of a multicompartment arrangement [3]. Wyllie and Patnode developed membranes from artificial cation exchange materials, which were used for ion activity measurements [4]. These membranes still had a high electrical resistance to be useful in electrodialysis. Juda and McRae have, about the same time, reported on ion exchange gels and membranes [5], and Kressman on ion exchange resin membranes and res n impregnated filter paper [6]. Indications were given for the preparation of ion selective membranes with low electrical resistance.

From 1952 frequent publications appear in the technical journals and patents are issued in several countries on ion-selective membranes and the electrodialysis process. Membranes were then developed, which appeared to be of practical application. In the same year the Organization for European Economic Cooperation (O.E.E.C.) considered the demineralization of brackish water as a subject of international concern and the Organization for Applied Scientific Research (T.N.O.) in the Netherlands was appointed to work on the development of the electrodialysis process. Similar work was performed in the United States and it appears interesting to note that the first Research and Development Progress Report, published by the Office of Saline Water, was on the results of selected laboratory tests of an Ionics demineralizer by electrodialysis [7].

In the following years the preparation of ion exchange selective membranes and the electrodialysis process were fully developed to commercial scale and a large number of electrodialysis plants was erected and successfully operated. The largest plant which was erected up to now is that of Benghazi, in Libya, with a capacity of 19200 m³/day (5.07 Mgd).

Das Elektro-
dialyse-
Verfahren

3.7.2 The electrodialysis process

The electrodialysis process is performed in cells consisting of many, at least three, compartments formed alternatively by an anion and a cation exchange membrane placed between an anode and a cathode. Multicompartment electrodialysis cells are usually termed as electrodialysis stacks. The principle of operation is shown in **Fig. 3–7.**

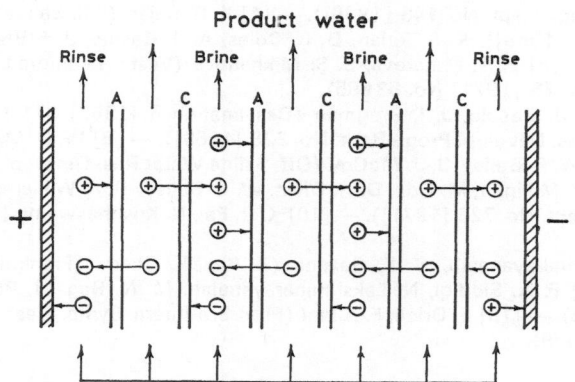

Fig. 3–7

Electrodialysis
principle of operation.

The saline water feed is pumped through the compartments of the membrane stack and, when a direct current potential is applied, cations pass easily through the cation-permeable membrane and are stopped when they reach an anion-permeable membrane. Similarly anions have free passage through the anion-permeable membrane and are stopped at the cation-permeable membrane. The ion concentration increases in alternate compartments. Simultaneously, the compartments between them become depleted of ions. Hence two streams of water are extracted from the electrodialysis stack: one stream with low ion concentration, which is the product water, and one stream with high salt concentration, which is the reject brine.

If the feed water has a low salinity, it is possible to obtain an acceptable potable water in a single pass through the electrodialysis stack. If the feed water is high in salinity this is normally not practical. Feed velocities have to be maintained above a certain minimum value and this requirement is handicaped by the necessity to provide also for a minimum residence time of the fluid in the compartment. An increase of the current density may lead to concentration polarization, as discussed in section 3.5.5, p. 203. On the other hand, the required length of the flow passage increases as a function of the salinity in the feed water. Engineering considerations and economic reasons make it preferable to use a number of stacks, depending on feed salinity and polarization characteristics [8].

   Fig. 3–8 is a view of an electrodialysis stack before assembly, that shows the main components. Typically the design is based on the configuration used in the plate and frame filter press. The end frames have provisions for holding the anode or cathode and are usually made relatively thick and rigid that pressure can be applied to hold the stack components together. The inside surfaces of the end frames are recessed to form an electrode-rinse compartment and provisions are made for introducing and withdrawing the solutions. Spacer frames with gaskets at the edges and ends are placed between membranes to form the solution compartments when ion exchange membranes and spacer frames are clamped together.

Fig. 3–8

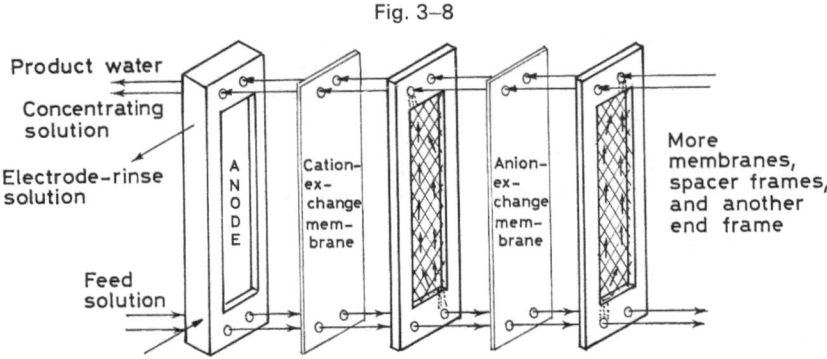

Electrodialysis stack before assembly.

   Fig. 3–9 shows a typical tortuous-path spacer for an electrodialysis stack.
   Usually the supply ducts for the various solutions are formed by matching holes in the spacer frames, membranes, gaskets, and end frames. Each spacer frame is provided with solution channels, as shown in Fig. 3–8, connecting the supply ducts with the solution compartments. The spacer frames have mesh spacers, or some other adequate device, in the liquid compartment space to support the ion exchange membranes from collapsing when a differential pressure develops between two compartments [9].

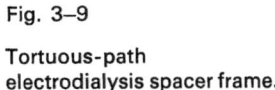

Fig. 3–9

Tortuous-path
electrodialysis spacer frame.

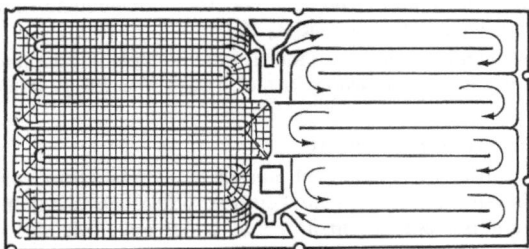

   A standard design consists of stacked membranes separated by one piece plate-like gaskets. The perimeter of these gaskets maintains the membrane spacing and forms the liquid compartment. The inner area of the gaskets consists of flow paths which connect with the appropriate ducts to allow either brine or product water to flow over the membrane surface. Baffles cause turbulence, which breaks up salinity gradients perpendicular to the flow path.
   Another design provides for separate outer gaskets with membrane separation being accomplished by use of a polymer mesh. Flow distribution to the individual compartments is effected by using soft ring gaskets to prevent flow into the compartments where appropriate.
   Finally a third design consists of membranes held at their perimeter by peripheral gaskets formed of natural rubber. Membrane separators are made of a woven synthetic cloth. Multiple entry and exit holes in the outer frame control flow distributions to the individual compartments.

Various materials have been used to frabricate electrodes, including graphite, nickel alloys, stainless steel, as well as platinum coated electrodes because of their high level of corrosion resistance.

Membranes may have an area of 0.2 to 1.3 m² (300 to 2000 in²) depending on the design. Typical stack heights are in the range of 0.45 to 1.22 m (1.5 to 4 ft) and contain in the range of 100 to several hundred membranes. Desired product salinity is achieved by passing the feed liquid through several stacks in series, while desired plant throughput is obtained by operating several electrodialysis paths in parallel.

Two alternative types of an electrodialysis membrane assembly modular unit were presented by Schechter and Forgacs. These units utilize two concepts, namely the cross flow of the two process streams and the thin and short diluate compartments. As main feature is claimed that they enable single stage desalination of almost any type of brackish water. Approximate economic calculations showed that brackish water can be desalted in large electrodialysis plants, based on these modular units, at a total production cost of 7.4 to 8.5 cents per m³ (28 to 32 cents/kgal), depending on membrane price, in plants with a capacity of 3785 m³/day (1 Mgd) and at 4.2 to 5 cents per m³ (16 to 19 cents/kgal) in plants with a capacity of 190000 m³/day (50 Mgd), excluding scale control. The total capital cost per m³ per daily production capacity would be 12.4 to 14 cents (47 to 53 cents per gal of daily capacity) for 1 Mgd plants and 6.1 to 7.7 cents (23 to 29 cents per gal of daily capacity) for 50 Mgd plants [10].

Vorbehand-
lung des
Rohwassers

3.7.3 Pretreatment of feed water

The presence of certain ions dissolved in the feedwater can lead to the precipitation of scale on the membrane surface. Scaling and fouling of the membrane surface and means to prevent it were discussed in section 3.5.7, p. 206.

Excessive ionic concentrations of iron and manganese in the feedwater also cause the formation of a scale on the membrane surfaces. Removal of these impurities is accomplished by chemical pretreatment of the feed with magnesium oxide and potassium permanganate, respectively, and subsequent filtration of the precipitates. Dissolved organic impurities are removed by passing the feedwater through activated carbon filters.

An engineering study has been made to determine the cost effectiveness of pretreatment systems for membrane desalting processes. Four brackish waters of different characteristics were treated by ion exchange and chemical precipitation to hardness levels of 50 and 200 mg/l, expressed as $CaCO_3$, and a calcium level to avoid precipitation of calcium sulfate in the concentrate stream. Chemical precipitation by the cold process was recommended to reduce the hardness to a calcium content preventing precipitation of $CaSO_4$ [67].

The size and complexity of the pretreatment plant is dependent upon the nature and the abundance of impurities in the feedwater (see also section 3.7.5, p. 218).

Energie-
bedarf

3.7.4 Energy requirements

The ratio of the theoretical current requirements for achieving a given salinity reduction in an electrodialysis plant to the actual current requirements is termed as the current efficiency. It is usually expressed in percent. The main factors which adversely affect the current efficiency are imperfect ion selectivity of the membranes, water transport through the membrane and current leakages.

Ion selectivity as a membrane property was discussed in section 3.5.2, p. 201.

The ion transport number t is the parameter, which describes the selectivity ψ of a particular either cation permeable or anion permeable membrane, and is defined by:

$$\psi_c = \frac{\bar{t}_+ - t_+}{t_-} \quad \text{or} \quad \psi_a = \frac{\bar{t}_- - t_-}{t_+}$$

where $\bar{t}_+$ and $\bar{t}_-$ are the transport numbers of cations or anions in the respective membrane, t_+ and t_- the corresponding transport numbers in the solution and the subscripts c and a relate to the cation or anion permeable membrane.

Transport numbers are dimensionless and represent the fraction of current carried by an ion. For a perfect cation permeable membrane $\psi_c = 1.0$ and $\psi_a = 0$. A less perfect membrane might exhibit transport numbers 0.9 and 0.1, respectively.

The efficiency η_s with which current transports salt through a set of membranes may be defined by:

$$\eta_s = \frac{n_c\,t_-\,\psi_c + n_a\,t_+\,\psi_a}{n_c\,t_- + n_a\,t_+}$$

where n_c and n_a are the numbers of cation and anion permeable membranes in the electrodialysis stack [8].

Water transfer as a membrane property is discussed in section 3.5.3, p. 202. Some water is transported through the membranes along with the electrolytes due to electroosmosis (see section 3.5.4, p. 202). The amount of electroosmotic water transport varies with membrane type, ionic species, and concentration of solution. When the feed is a low salinity brackish water, transport becomes seldom a problem. In systems involving a high degree of desalination, water transport can reduce the current efficiency. Hence, the overall current efficiency η_w is defined by:

$$\eta_w = 1 - n\bar{\tau}_w \left(0.018 \, \frac{m}{\eta_s}\right)$$

where $\bar{\tau}_w$ is the transference number for water and m the molality of the feedwater. The transference number τ is defined as the number of moles water transferred per Faraday and is related to the dimensionless transport number t according to:

$$t = \tau Z$$

where Z is the valence, with sign, in equivalents per mole. The valence of water, as a compound, is zero and therefore the transport number is also zero. However, the transference number of water is finite and may be either positive or negative [8].

Current leakage through the stack manifolds causes a reduction in current efficiency of usually less than 5%. Therefore, the efficiency η_m associated with the bypass of current through the manifolds is approximately very close to 1.0.

The Faraday efficiency η_F of an electrodialysis stack is defined as the ratio of the theoretical current required to transport a given quantity of salt to the actual current required. This is expressed by:

$$\eta_F = \eta_s \cdot \eta_m$$

The overall current efficiency η is defined as the ratio of the theoretical current required to produce a given quantity of product water at a particular salinity to the actual current required to perform the same function. Since this includes the effects of water transfer, η is given by:

$$\eta = \eta_s \cdot \eta_m \cdot \eta_w$$

where the three terms on the right hand side account for the effects of membrane permselectivity, current bypass through the manifolds and water transfer respectively [8].

The amount of salt removed from the product water stream depends on the number of membranes in the system and on the transport numbers of the ions in the membranes and solutions. The net removal of salt from the product stream is given by:

$$\Delta\,(N \cdot V) = \frac{I \cdot \xi_F}{F}$$

where $\Delta\,(N \cdot V)$ is the change in product stream normality and flow rate on passing through the stack, F is the Faraday Constant (96494 C/eq) and I is the membrane current in Ampere. ξ_F is the Faraday utilization factor, which represents the amount of salt removed per unit of current and is given by:

$$\xi_F = \eta_m\{n_c\,([\bar{t}_+]_c - t_+) + n_a\,([\bar{t}_-]_a - t_-)\}$$

where the subscripts c and a refer to cation permeable and anion permeable membranes, respectively. Taking into account the effects of water transfer, an overall current utilization factor

$$\xi_I = \eta_w \cdot \xi_F$$

may be defined [8].

The voltage drop per cell pair of an electrodialysis stack is proportional to the current density and the cell pair resistance R in $\Omega \cdot$ cm. An electrodialysis cell pair is the repeating unit of an anion exchange membrane, a cation exchange membrane, an enriching-solution compartment, and a depleting-solution compartment. Typical electrodialysis stacks operate with voltage drops of about 2 Volts per cell pair.

The total energy consumption can be expressed as:

$$C = \frac{i \cdot R \cdot F}{\eta_F} \quad W \cdot sec \text{ per eq removed}$$

where i is the electric current density. The energy consumption increases with current density and cell pair resistance and decreases with improved Faraday efficiency. Typical energy consumption in electrodialysis is about 50 Wh per equivalent removed.

In concentrated solutions the bulk of the individual cell resistance is due to the membrane. Therefore, most of the voltage drop across an electrodialysis stack is spent in heating membranes. The resulting membrane temperature raise depends on the solution temperature, on the thickness, thermal conductivity, and electrical resistance of the membrane and on the coefficient of heat transfer between the solution and the membranes. The current density must be kept below that value, which would result in overheating and deteriorating of the membrane.

3.7.5 Electrodialysis process analysis and optimization

A first attempt to perform an analysis of electrodialysis by considering the process as an electrical network composed of resistive elements representative of various electrochemical subprocesses was made by Berger et al. [11]. The total effect of all subprocesses was unified into a single mathematical equation for the network. The engineering equations developed can be used to quantitatively analyse the electrical resistance of the stages in an electrodialysis plant. The treatment gives a breakdown of the various factors that contribute to electrical resistance with emphasis on those factors that must be improved to make technological advances in the process. Application of the analysis to two commercial plants concluded that the major resistive factors in the operation of these plants were electrolyte resistance and ohmic polarization due primarily to scale and membrane potentials. The minor resistive elements were found to be membrane resistance, electrode polarization, and parasitic duct losses [11].

A process design optimization technique was developed, which is quite general in character, for low cost electrodialysis conversion of brackish waters. Electrodialysis spacers should be much thinner than those used at the time of the study. Improved spacer design would result in immediate process cost reduction. Further process cost reductions could be achieved by reducing the unit costs of membranes, the direct current energy costs and the membrane resistance [12].

The increase of the feedwater flow velocity in the compartments of electrodialysis units is a means to prevent polarization of the membranes. In this respect the hydraulic peculiarities of the liquid passage in electrodialysis compartments with zig-zag flow and with spacers were studied. Corrugated and perforated PVC sheets, expanded PVC sheets and several types of nylon network were used as spacer material. The dependence of the hydraulic resistant coefficient of zig-zag type compartments and compartments with various spacer types to increase turbulence on Reynolds number was determined as well as the dependence of the critical current density on saline water velocity and its salt content. Taking these dependences into account, optimal conditions of brackish water desalination in various types of electrodialysis units were suggested [13]. The hydrodynamic effectiveness of various spacer screens for electrodialysis was studied with the aid of a photographic technique. The technique produced a visual record of the membrane-solution interfacial flow patterns in electrodialysis cells. Additional tests were conducted and criteria established for the characterization of each screen with respect to porosity, dead flow area, drag coefficient, and influence on electrical resistance. The experiments demonstrated that some screens are considerably more effective than others in improving electrodialysis performance, especially at flow rates that give Reynolds numbers greater than 180. Seven out of 21 of the more promising screens were further evaluated in electrodialysis batch tests to determine how current efficiency, membrane stack pressure drop and total power requirements vary with flow rate [14].

The effect of turbulence promoters on local mass transfer has been studied. An instrument has been developed to measure local mass transfer in a channel typical of an electrodialysis cell and to provide a detailed profile of mass transport downstream from a turbulence promoter. The instrument works by analogy between a redox electrode and a polarized membrane. Data were taken under conditions where the mass transfer could be calculated from diffusion theory. Agreement between the data and the calculation was exceptionally good [15]. Mass transport promoters are commonly used to increase the usable current density. The development of techniques for optimization of such promoters was investigated. Runs were made under conditions, which matched as closely as possible the conditions for which a mathematical solution to the mass transfer equations could be obtained. The data yielded a diffusion constant which correlated closely to values reported in the literature [16].

Hydrostatic phenomena in the diluting and the concentrating compartments of an electro-dialysis apparatus of the filter-press type were studied and an equation was derived to calculate the height of the point ϑ, at which hydrostatic pressures in the diluting and the concentrating compartments balance. To prevent maldistribution of flow in the compartments, a proper value of ϑ should be chosen [17].

Three systems of feeding flow distribution in an electrodialysis apparatus were studied while the apparatus was in operation for seawater concentration. Pressure drops across several membrane cells were measured and were proportional to the 1.3 power of the feed rate. By using this relation, the liquids velocity in the cells was evaluated. Conduits, which were designed to equalize the pressures in the internal conduits of dialysate and to improve the distribution of flow in the diluting compartments, gave only a small effect [18].

Sonin and Probstein have developed a hydrodynamic theory of electrodialysis for a multichannel system with steady laminar flow between plane, parallel membranes. The modelling of the system was found to be governed by four basic similarity parameters: a dimensionless applied potential, the product of the channel aspect ratio and the inverse Peclet number, the ratio of brine and dialysate inlet concentrations and a parameter measuring membrane resistance [19]. The hydrodynamic theory of desalination by electrodialysis was extended for a parallel channel system with unobstructed fully developed turbulent flow. From the equations and the boundary conditions it was shown that there are five parameters which govern the system performance. Four are the same as for the corresponding laminar case, while the fifth measures the thickness of the concentration diffusion sublayer [20]. An investigation was made subsequently and a method was presented by Sonin and Isaacson for determining whether the hydrodynamic design, that is the choice of channel geometry and flow speed, is a costing factor in an electrochemical process and for determining the optimum design and flow conditions for product cost minimization. This method was applied to assess the state of the art in the hydrodynamic design of electrodialysis systems used for the desalting brackish water. The performances of various possible channel designs were reviewed and it was shown that, for brackish waters of low salt concentration, a further reduction in product cost of the order of 30% is in principle possible, if the hydrodynamic design can be significantly improved from the present state of the art. For feed concentrations above about 6000 mg/l, however, the hydrodynamic design is not a cost-limiting factor at the present state of the art and further development in this direction is unnecessary [21].

Various mathematical and analytical models have been developed to describe the flow in electrodialysis cells. An analytical model was presented for the flow between two parallel plates in the presence of a net-like material [22]. A mathematical model was also proposed for the con-centration field between two membranes in an electrodialysis desalination cell [23]. The mean velocity in an electrodialysis cell was shown to satisfy the potential flow equations. A method of design was suggested to eliminate almost-stagnation regions [24].

Using models of both the cost procedures and the physical plant design, an optimization routine was constructed [25]. A mathematical model was developed for engineering design of electro-dialysis plants [26].

Two analytical models for mass transfer in an electrodialysis cell with spacers of finite mesh were also proposed. Representative graphs and a method for evaluating spacer characteristics, applicable to the scale-up of electrodialysis units, were presented [27]. The mass transfer rate in a narrow channel in the presence of turbulence promoters, similar to those in electrodialysis, was measured and the results were compared with the analytical model previously proposed [28]. The com-plexity of the optimum problem of electrodialysis was explained by the quantities of parameters of processes, their implicit interaction and free unknown quantities in the system of equations between technological and design parameters. The most important stage of research was the creation of the economic and mathematical model [29].

A parametric engineering and economic evaluation of the electrodialysis process was conducted, in which a comprehensive computational algorithm for electrodialysis process analysis, design and costing was developed. It permitted systematic investigation of variations in process design and economic parameters [30].

Auswertung des Elektro-dialyse-Verfahrens

3.7.6 Electrodialysis process evaluation

In an early report of the Office of Saline Water the design, construction, field testing, and cost analysis of an experimental electrodialysis demineralizer for brackish waters were evaluated [31]. Mathematical relationships to predict costs of the electrodialysis process for the conversion of a number of selected saline waters to potable water were proposed. A computer program was developed from these mathematical relationships. The method, although based on the state of the art of electrodialysis at the time of study, can readily be adapted to make use of future developments related to the many variables [32]. Another parametric, economic, and engineering analysis of the electrodialysis process focused on plants of 3785, 37850, and 190000 m³/day (1, 10, and 50 Mgd) capacity, treating each of three reference brackish waters. Plants using both reinforced and unreinforced membranes were used and a total of 18 cases resulted. A computer program performs necessary process and plant design calculations in order to provide capital and water costs for each plant. Prime variables used in the computer analysis were the number of plant stages, number of cell pairs per stack and solution velocity in a stack. For each of the 18 cases, a set of operating conditions was specified which delivered the lowest-cost water [33].

A demonstration plant for the electrodialysis was constructed by the Office of Saline Water at Webster, South Dakota, with a capacity of about 950 m³/day (250000 gpd). The feed is a hard water containing approximately 1700 mg/l total dissolved solids, principally calcium and magnesium sulfate and hydrogen carbonate. Of this total, about 4 to 5 mg/l represent iron and manganese, which exceeds the acceptable limits for the process [34]. In a series of reports on the operation of the Webster plant, the electrodialysis process was extensively evaluated and valuable information is given [35]. Treatment of the feedwater was necessary to provide an improved quality feed to the electrodialysis equipment. A pilot plant was constructed to test the pretreatment process and a final reduction of iron to 10 µg/l (ppb) or less and manganese to 50 µg/l or less was obtained [36]. The Webster plant is the only Office of Saline Water Test Bed Plant for converting brackish water to potable water by the electrodialysis process and its production capacity reached 1230 m³/day (325000 gpd). Since May 1969, the feed water has been pretreated using lime softening to reduce the level of iron, manganese, and hardness. The performance of the Test Bed Plant was also evaluated on the basis of anion membranes subject to organic fouling, using carbon filtration pretreatment for feed water organic removal to compensate for the anion fouling [37].

A comparative field test evaluation of reverse osmosis and electrodialysis desalting pilot plants was made, using a 21 m³/day (5000 gpd) electrodialysis pilot plant. The tests were run with feed waters having a salinity of 2000 and 5000 mg/l at ambient temperatures [38].

A study was also made on the economics and the use of large scale electrodialysis plants on selected tidal estuarine waters and a computer optimization program was developed. The data and considerations that must be predetermined for any similar electrodialysis plant are given. The study determined the design characteristics and optimized the electrodialysis type water desalination plant on the basis of the sum of annual operating costs over the life of the plant of 95000, 189000, 378500, and 1135600 m³/day (25, 50, 100, and 300 Mgd) average capacity at four different sites, five different operating modes and product water delivered to the plant boundary and to selected existing reservoirs [39].

A combination of the electrodialysis and ion exchange processes in treating boiler feed water enables appreciable reduction in the consumption of reagents. A power consumption of 0.3 to 0.5 kWh is required to remove 200 to 250 mg/l salt. With allowance for pumping water and brine through the electrodialysis plant, total power requirements are 1.5 to 2.0 kWh/m³ treated water [40].

Elektro-dialyse bei erhöhter Temperatur

3.7.7 High temperature electrodialysis

The electrical resistance of an electrodialysis system will be by about 2% lower for each 1°C rise in temperature. This makes the operation of the electrodialysis process at elevated temperatures attractive. However, the main problem is that the membranes to be used must preserve their physicochemical properties under high temperature conditions. Several other engineering problems may also be associated with operation at elevated temperatures, but they might be considered of minor importance in respect with the main problem of the stability of the membranes. At least four commercially available membranes were found to behave satisfactorily at the time of an early study [41].

High temperature transference numbers were measured for 0.1 N sodium chloride solution at temperatures from 15 to 125°C [42].

Research conducted in Israel utilizes the Joule heat evolved in the process to increase the temperature of operation thus reducing the energy consumption. Initial calculations indicated that by suitable increase of temperature the energy consumption may be reduced as much as 60 to 70%. The economics of the process seem sufficiently favorable to justify its application to seawater desalination. More extensive investigation of the changes in membrane resistance with temperature has shown that it decreases much more markedly below the range 70 to 80°C than above it. It was also found that electroosmotic water transfer through membranes increases, especially in the higher temperature region (70 to 100°C). This behavior of membranes was reflected in the results obtained from laboratory scale high temperature electrodialysis experiments. The saving of energy consumption by increasing the temperature of operation from 30 to 70°C was 60 to 70%, with different types of membranes. Further increase in temperature contributed less substantially to the decrease in energy consumption. During these experiments it became evident that electrodialysis apparatus built from conventional materials of construction using general purpose ion exchange membranes may be successfully operated for long periods, up to at least 70°C [43].

A bench scale high temperature electrodialysis unit has been operated successfully on natural seawater up to 70°C. Effects governing scale formation were investigated and conditions of operation under which no scale is formed were determined. Power consumption, desalination efficiency and water transport have been measured. Differences between these results and those obtained with NaCl solution have been shown to be due to the presence of Mg^{2+} ion in natural seawater [44].

The capacity, water content, resistance and burst strength of various anion and cation selective membranes were determined and membranes showing no detectable changes in appearance, resistance or ion exchange capacity were indicated. Using an activation energy of 27 kcal, calculated from the decomposition of a commercial anion exchange resin, a half-life of 200 000 h for thermal decomposition was estimated for these membranes at 80°C. It was concluded that the lifetime in the field will be determined by factors other than thermal decomposition. The temperature coefficient of resistance is somewhat greater than 2% per 1°C. With one exception, the transport number is invariant with temperature. The water transfer decreases by roughly 15% on passing from room temperature to 82°C (180°F), again with one, but a different, exception. Potentials at least as high as 3 V per cell pair can be applied at 82°C without polarization. The pumping energy is substantially reduced and it is possible to accomplish a degree of demineralization in one stage, which would take two or more stages at 27°C (80°F) at the same potential and linear velocity. The cost of demineralizing seawater at 82°C was estimated for a 900 l/s state-of-the-art facility, including a captive electric generator. The investment is about 12.5 million $ and total water costs are about 10 cents/m³ (40 cents/kgal) which compares quite favorably with projections for very much larger dual-purpose distillation plants. The economics of producing 7500 m³/day (2 Mgd) of some 500 mg/l water from a 10°C (50°F) 2500 mg/l feed were estimated for various operating temperatures. A 15% saving was estimated for operation at 38°C (100°F) as compared to 10°C and a 25% saving at 66°C (150°F). With free waste heat the cost savings are about 25 and 35% respectively [45].

A commercial unit, the Stackpack, was evaluated, using lime-softened water at the Webster Test Plant of the Office of Saline Water at operating temperatures of 10, 20, 27, 43 and 60°C (50, 68, 80, 110 and 140°F). The major benefit of operation at 60°C as opposed to operation at 10 or 20°C is in decreased power requirements, because of the decrease in electrical resistance with increasing temperature. For example, power expenditure was reduced to about 36% at 60°C for a salt reduction from about 1290 to 760 mg/l, compared with operation at 10°C for the same salt reduction and the same amount of product water. The major contribution to the cell pair resistance was from the dilution stream rather than from the membranes, regardless of temperature. No special problems or limitations occurred as a result of high temperature electrodialysis, up to 60°C, provided that polyethylene spacers are used below 43°C and block polymer spacers are used in the 43 to 60°C range, and the flow channel linear velocity does not exceed 50 cm/s. Prolonged use of polyethelene spacers at 60°C wi l cause deformation of the spacer straps and low limiting current values [45].

New components, spacers and membranes have been developed for electrodialysis at temperatures up to 80°C and the performance of a seawater plant using these components was evaluated. Total water costs depend on membrane production costs and membrane life. Even with conservative assumptions with regard to membrane cost and life, it appears possible to desalt water for less than 26 cents/m³ [47].

A 10 000 gpd (38 m³/day) electrodialysis unit was operated at 38°C (100°F) for a total of 5000 h at the Webster Test Facility. Preliminary economics based on the operation of this unit were projected for a 1 Mgd (3785 m³/day) facility [48]. Investigations on high temperature electrodialysis, covering a three years research work, were reported by Leitz et al. In Phase I the evaluation of properties of standard membranes and the development of thin membranes is reported; the effect of spacer design variables was investigated and the operation of a stack at temperatures of 150°F (66°C) was evaluated. In Phase II, 20 mil spacers and 6 mil membranes were fabricated and endurance in long term tests was investigated. In Phase III furtrer testing of the above unit and design of a field test unit is reported. The reports on three Phases are condensed in a summary report [49].

Betriebs-
erfahrungen

3.7.8 Operating experience

A large number of electrodia ysis plants have been built in the past two decades all over the world. Most of them are small units, not included in the Inventory, published by the Office of Saline Water. Table 3/2 summarizes the number and the daily total capacity of these electrodialysis plants, having a capacity exceeding 95 m³/day (25 000 gpd), as reported in the mentioned inventory [50]. Figures for 1972 may not be final.

Reports on details of construction and on operating experience have been published for various plants. Only the most recent publications will be mentioned here.

An account of the experience acquired by French manufacturers of electrodialysis apparatus was given by Gomella with special emphasis on treatment of very hard and ferruginous waters [51]. Solt gave a description of the water situation in Benghasi and described the problem of the water treatment plant for this town [52]. Nebbia has described the Vieste plant in Italy [53]. There are several papers published on the Mashabei Sadeh plant, a joint venture of the Israel Government and the U.N. Development program [54]. Shishlyannikov et al. reported on the three years operation of the Mointy electrodialysis plant [55]. Balice et al. reported on the combined ion-exchange electrodialysis plant at Brindisi [56]. Botteri gave an account of experience gained in Argentina [57] and Harkare at al. [58] as well as Govindan et al. [59] on progress on electrodialysis techniques in India, and Smagin et al. in the Soviet Union [60].

Table 3/2

World capacity of electrodialysis plants.

Year	Plants	Capacity		Year	Plants	Capacity	
		kgd	m³/day			kgd	m³/day
1956	1	56	212	1966	3	184	696
1959	1	28	106	1967	7	351	1 329
1960	1	70	265	1968	4	367	1 389
1961	3	170	644	1969	11	8096	30 647
1962	1	650	2460	1970	7	1106	4 187
1963	3	390	1 476	1971	12	6142	23 250
1964	2	215	814	1972	8	6287	23 799
1965	5	861	3 259	Total	69	24 973	94 533

Hygienische
Überprüfung
des erzeug-
ten Wassers

3.7.9 Hygienic evaluation of product water

Biomedical aspects of water desalination, especially when using ion exchange materials, were investigated by Shtannikov [61]. Desalted water from two electrodialysis prototype plants were tested with animals for sanitary and toxicological effects. No adverse effect on the organism could be detected [62]. Similar physiological studies on living organisms were carried out by El'piner and Shafirov [63]. Seawater reduced in salt content to about 300 to 350 mg/l by electrodialysis may contain up to 4.5 mg/l B and 1.5 to 1.8 mg/l Br. It was considered unsuitable for use as drinking water [64]. A model seawater was demineralized by electrodialysis. It was found that the product water would not be suitable for use, as it contained 4 mg/l B and 2 mg/l bromide [65].

In a hygienic assessment of water desalted by electrodialysis the reduction of inorganic constituents was found satisfactory, but the content of organic materials in the product did not conform to the requirements [66].

Literature to 3.7

[1] P. Prausnitz, J. Reitstötter (Elektrophorese, Elektroosmose, Elektrodialyse in Flüssigkeiten, Dresden 1931). — [2] E. Manegold, K. Kalauch (Kolloid-Z. **86** [1939] 93/101, 186/205, 257/73, 313/9). — [3] K. H. Meyer, W. Strauss (Helv. Chim. Acta **23** [1940] 795/800). — [4] M. R. J. Wyllie, H. W. Patnode (J. Phys. Chem. **54** [1950] 204/27). — [5] W. Juda, W. A. McRae (J. Am. Chem. Soc. **72** [1950] 1044).

[6] T. R. E. Kressman (Nature **165** [1950] 568). — [7] N. W. Rosenberg, T. A. Kirkham, C. E. Tirrell, N. E. Saliba (Off. Saline Water Res. Develop. Progr. Rept. No. 1 [1954]). — [8] Hittman Associates Inc. (Off. Saline Water Res. Develop. Progr. Rept. No. 610 [1970]). — [9] R. E. Lacey, S. Loeb (Industrial Processing with Membranes, New York 1972). — [10] J. Schechter, C. Forgacs (Off. Saline Water Res. Develop. Progr. Rept. No. 681 [1971]).

[11] C. Berger, G. A. Guter, G. Belfort (Off. Saline Water Res. Develop. Progr. Rept. No. 238 [1967]), G. Belfort, G. A. Guter (Desalination 5 [1968] 267/91). — [12] Process Research Inc. (Off. Saline Water Res. Develop. Progr. Rept. No. 325 [1968]). — [13] V. A. Klyachko, L. D. Ushakov (Desalination 2 [1967] 279/82). — [14] G. Belfort, G. A. Guter (Off. Saline Water Res. Develop. Progr. Rept. No. 459 [1969]; Desalination 10 [1972] 221/62). — [15] P. H. Bradley, J. L. Greatorex, F. B. Leitz (Off. Saline Water Res. Develop. Progr. Rept. No. 597 [1970]).

[16] L. Marincic, F. B. Leitz (Off. Saline Water Res. Develop. Progr. Rept. No. 793 [1972]). — [17] S. Azechi (Nippon Kaisui Gakkai-Shi **23** [1970] 276/81; C.A. **74** [1971] No. 77811). — [18] S. Azechi, Y. Fujimoto (Nippon Kaisui Gakkai-Shi **23** [1970] 134/47; C.A. **74** [1971] No. 79399). — [19] A. A. Sonin, R. F. Probstein (Off. Saline Water Res. Develop. Progr. Rept. No. 375 [1968]; Desalination 5 [1968] 293/329). — [20] R. F. Probstein, A. A. Sonin, E. Gur-Arie (Off. Saline Water Res. Develop. Progr. Rept. No. 771 [1972]; Desalination 11 [1972] 165/87).

[21] A. A. Sonin, M. Isaacson (Proc. 4th Intern. Symp. Fresh Water Sea, Heidelberg 1973, Vol. 3, p. 237/51). — [22] D. Pnueli, D. Grossman (Desalination 6 [1969] 303/8). — [23] D. Pnueli, D. Grossman (Desalination 7 [1970] 297/308). — [24] D. Pnueli (Desalination 7 [1970] 309/22). — [25] G. Belfort, J. A. Daly (Desalination 8 [1970] 153/66).

[26] M. Avriel, N. Zelingher, J. R. Olie, A. Raz (Proc. 3rd Intern. Symp. Fresh Water Sea, Dubrovnik 1970, Vol. 2, p. 181/94), M. Avriel, N. Zelingher (Desalination 10 [1972] 113/46). — [27] A. Solan, Y. Winograd, U. Katz (Proc. 3rd Intern. Symp. Fresh Water Sea, Dubrovnik 1970, Vol. 2, p. 259/66). — [28] Y. Winograd, A. Solan, M. Toren (Proc. 4th Intern. Symp. Fresh Water Sea, Heidelberg 1973, Vol. 3, p. 261/6). — [29] V. N. Smagin, D. A. Yaroshevsky (Proc. 4th Intern. Symp. Fresh Water Sea, Heidelberg 1973, Vol. 3, p. 229/35). — [30] G. R. Olsson, A. P. Christodoulou (Proc. 3rd Intern. Symp. Fresh Water Sea, Dubrovnik 1970, Vol. 2, p. 215/39), A. P. Christodoulou, G. R. Olsson, W. P. Sommers (Proc. 3rd Intern. Symp. Fresh Water Sea, Dubrovnik 1970, Vol. 2, p. 240/58), A. P. Christodoulou, G. R. Olsson, H. J. Monnik (Off. Saline Water Res. Develop. Progr. Rept. No. 488 [1969]).

[31] J. P. Dankese, T. A. Kirkham, G. Maheras, M. S. Mintz, J. H. Powell, N. W. Rosenberg (Off. Saline Water Res. Develop. Progr. Rept. No. 11 [1956]). — [32] Mason-Rust (Off. Saline Water Res. Develop. Progr. Rept. No. 134 [1965]). — [33] J. W. Porter, S. Cherney (Off. Saline Water Res. Develop. Progr. Rept. No. 470 [1969]). — [34] B. W. Calvit, J. J. Sloan (Proc. 1st Intern. Symp. Water Desalination, Washington 1965 [1967], Vol. 3, p. 11/22). — [35] Mason-Rust (Off Saline Water Res. Develop. Progr. Rept. No. 101 [1964], No. 132 [1964], No. 164 [1965], No. 241 [1967], No. 296 [1967], No. 567 [1970], No. 568 [1970], No. 692 [1971]).

[36] G. R. Bell (Off. Saline Water Res. Develop. Progr. Rept. No. 201 [1966]). — [37] J. S. Nordin, N. Call, R. A. Ackerman, D. R. Bogue, J. E. Gegeler (Off. Saline Water Res. Develop. Progr. Rept. No. 805 [1972]). — [38] E. G. Kaup (Off. Saline Water Res. Develop. Progr. Rept. No. 899 [1973]). — [39] D. H. Guild (Off. Saline Water Res. Develop. Progr. Rept. No. 777 [1971]). —

[40] V. N. Smagin, P. D. Shchekotov (Teploenergetika **20** [1973] 17/20; Therm. Eng. [USSR] **20** [1974] 22/5).

[41] C. Forgacs (Proc. 1st Intern. Symp. Water Desalination, Washington 1965 [1967], Vol. 3, p. 155/72). — [42] E. B. Dismukes (Off. Saline Water Res. Develop. Progr. Rept. No. 87 [1964]). — [43] T. Bejerano, C. Forgacs, J. Rabinowitz (Desalination **3** [1967] 129/34). — [44] C. Forgacs, L. Koslowsky, J. Rabinowitz (Desalination **5** [1968] 349/58). — [45] W. A. McRae, W. Glass, F. B. Leitz, J. T. Clarke, S. S. Alexander (Desalination **4** [1968] 236/47).

[46] Mason-Rust (Off. Saline Water Res. Develop. Progr. Rept. No. 342 [1968]). — [47] F. B. Leitz, M. A. Accomazzo, W. A. McRae (Proc. 4th Intern. Symp. Fresh Water Sea, Heidelberg 1973, Vol. 3, p. 195/203). — [48] R. G. Parent, J. W. Arnold (Off. Saline Water Res. Develop. Progr. Rept No. 900 [1974]). — [49] F. B. Leitz (Phase I: Off. Saline Water Res. Develop. Progr. Rept. No. 912 [1974]), F. B. Leitz, H. I. Viklund (Phase II: Off. Saline Water Res. Develop. Progr. Rept. No. 913 [1974]), F. B. Leitz, M. A. Accomazzo, H. I. Viklund (Phase III: Off. Saline Water Res. Develop. Progr. Rept. No. 914 [1974], Summary Rept. No. 915 [1974]). — [50] F. O'Shaughnessy (Desalting Plants Inventory Report No. 4, Department of the Interior, Washington, D.C., 1973).

[51] C. Gomella (Desalination **2** [1967] 283/6). — [52] G. S. Solt (Proc. 3rd Intern. Symp. Fresh Water Sea, Dubrovnik 1970, Vol. 2, p. 267/80). — [53] G. Nebbia (Acqua Ind. Inquinamento **11** No. 7 [1969] 18/21). — [54] Anonymous (Ingenieur [Utrecht] **83** No. 35 [1971] Ch 65/Ch 67); J. R. Olie, H. Weiss (Develop. Desalination Technol. Israel **1972** 62/9), J. Maoz (Proc. 4th Intern. Symp. Fresh Water Sea, Heidelberg 1973, Vol. 3, p. 205/18), B. M. Halpern, J. Olie (J. Am. Water Works Assoc. **64** [1972] 735/40). — [55] L. A. Shishlyannikov, F. T. Shostak, E. E. Ergozhin (Vestn. Akad. Nauk Kaz.SSR **28** No. 2 [1972] 44/51; C.A. **76** [1972] No. 117364).

[56] V. Balice, G. Boari, R. Passino, M. Santori, G. Tiravanti (Proc. 4th Intern. Symp. Fresh Water Sea, Heidelberg 1973, Vol. 3, p. 151/68). — [57] A. B. Botteri (Saneamiento **54** [1970] 157/89, 246; C.A. **77** [1972] No. 24687). — [58] W. P. Harkare, V. K. Indusekhar, P. K. Narayanan, M. N. Prajapati (Proc. 3rd Intern. Symp. Fresh Water Sea, Dubrovnik 1970, Vol. 2, p. 195/203). — [59] K. P. Govindan, V. K. Indusekhar, N. Krishnaswamy (Proc. 4th Intern. Symp. Fresh Water Sea, Heidelberg 1973, Vol. 3, p. 181/8). — [60] V. N. Smagin, V. L. Demkin, V. K. Egorov, O. V. Evdokimov (Proc. 4th Intern. Symp. Fresh Water Sea, Heidelberg 1973, Vol. 3, p. 219/27).

[61] E. V. Shtannikov (Proc. 1st Intern. Symp. Water Desalination, Washington 1965 [1967], Vol. 1, p. 429/31). — [62] A. F. Aksyuk, E. E. Gorshkova, G. R. Tsyplakova, T. V. Yudina, A. P. Sergeev, K. M. Saldadze, S. N. Gvozdeva, G. A. Bobkova (Ionoobmen. Membrany Elektrodialize **1970** 267/71; C.A. **74** [1971] No. 15653). — [63] I. I. El'piner, Yu. B. Shafirov (Ionoobmen. Membrany Elektrodialize **1970** 177/82; C.A. **74** [1971] No. 45466). — [64] T. A. Nikolaeva, A. I. Bokina, Yu. A. Rakhmanin, V. P. Plugin, T. S. Khachatryan, G. V. Verbitskaya, L. I. El'piner, Yu. B. Shafirov, I. M. Khovakh, L. D. Shamrova (Gigiena i Sanit. **35** No. 11 [1970] 11/4; C.A. **74** [1971] No. 45467). — [65] A. F. Aksyuk, Yu. V. Novikov, Z. A. Anisimova, T. K. Parkhomchuk, E. F. Gorshkova, E. M Oleinik, A. N. Sergeev, K. M. Saldadze, S. N. Gvozdeva (Gigiena i Sanit. **37** No. 4 [1972] 19/23; C.A. **76** [1972] No. 158147).

[66] N. I. Anan'ev, N. A. Demin (Gigiena i Sanit. **36** No. 2 [1971] 89/90; C.A. **74** [1971] No. 130233). — [67] J. S. Kneale, E. M. Kelley (Off. Saline Water Res. Develop. Progr. Rept. No. 425 [1969]).

Varianten der
Elektro-
dialyse

3.8 Variants of electrodialysis

The basic electrodialysis process has been further developed to several variants with the aim to overcome some of the problems associated with the basic process.

Transport-
verarmung

3.8.1 Transport depletion

The transport depletion process is a variant of electrodialysis in which an array of alternate cation exchange and nonselective membranes is used. The process is based on the concentration gradients established at the faces of the membranes, because of the different transference numbers of the ions in the membrane and in the solution. When direct electric current is passed, depleted and concentrated boundary layers form at the two sides of the cation exchange membranes, but not at the nonselective neutral membranes. The latter serve only to separate the depleted and concentrated boundary layers and create alternate diluting and concentrating compartments [1].

Research on the transport of electrolytes through near-neutral cellophane and parchment membranes and through highly selective membranes was undertaken to obtain data on the characteristics of actual use and to develop theoretical expressions that would describe the performance of all electrically driven membrane processes. Data from electrical transport experiments and from dialysis experiments conducted under similar hydrodynamic conditions have permitted the calculation of transference numbers for both ions and water that are characteristic of each membrane and the associated boundary layer in a given system. The data were also used to develop flux equations that combine the effects of electrical dialytic and osmotic transport through membranes and to calculate the phenomenological admittance coefficients that appear in the flux equations derived from irreversible thermodynamics [2].

A new experimental method was developed that permitted the measurement of the osmotic permeabilities of membrane systems in the absence of any influence from counter-diffusing ions. The osmotic permeability data obtained permitted the calculation of values of the admittance coefficients in phenomenological equations that are characteristic of a given membrane. The characterization under dynamic conditions of membranes was extended to broader ranges of current densities in NaCl solutions. Preliminary studies were made of two different methods of arranging cation-selective and nonselective membranes for possible use in demineralization devices. In one of these arrangements a solution-permeable nonselective membrane is spaced only a small distance from each cation-selective membrane and solution flow is directed towards the surfaces of the cation-selective membrane. The other arrangement of membranes combines the features of transport depletion and electrogravitation and appears to offer the possibility of high product-to-waste flow ratios [3]. Experimental and mathematical techniques were developed to determine admittance coefficients and friction factors for phenomenological equations that describe transport of electrolytes and water in membrane systems. The experimentally determined admittance coefficients were shown to be descriptive of transport by comparisons of calculated net fluxes of electrolyte with fluxes that were determined in earlier studies of the transport depletion process [4].

The transport depletion process was investigated in a wide range of solution velocities, current densities and types of feed water. The coulomb efficiency ranged from 0.35 to 0.55. With water containing a low concentration of sulfates, transport depletion can be carried out without encountering precipitation problems associated with conventional electrodialysis. With high-sulfate water the problems of $CaSO_4$ precipitation were similar. Energy requirements were 1.7 to 3.0 times that of conventional electrodialysis. Cost of membranes and capital costs are lower [5].

For testing in pilot plant scale, a forty-cell-pair plate-and-frame design with four separate channels was selected. Each channel had separate electrodes, electrode rinse streams and manifolding. Membrane spacing was 0.75 mm (0.03 in). Tests were made with a feed solution of 3000 mg/l NaCl under varying operating parameters. Current efficiencies ranged from 36 to 70%. Product concentrations as low as 120 mg/l were produced. A study of a transport-depletion plant of 1900 m³/day (500000 gpd) showed a cost of 1.4 million $ and a product water cost of $ 0.48 per m³ (1.81 per kgal). This cost would be reduced to $ 0.335 and 0.214 per m³ ($ 1.27 and $ 0.811 per kgal) for an 1 Mgd and 10 Mgd plant respectively [6].

A radial-flow round stack electrodialysis unit was tested to assess its design features and operating characteristics. The stack, containing 20 cell pairs (cation/neutral) of 13.56 dm² (1.46 ft²) per cell pair, was capable of demineralizing a 3000 mg/l water to 500 mg/l water at a capacity of 3000 gpd (11.3 m³/day), while a 2000 mg/l feed was demineralized to 500 mg/l at a capacity of 5400 gpd (20.4 m³/day), all at high coulombic efficiencies. Round stack power costs were about three times higher than expected and can probably be attributed to the Dacron separators and/or the neutral membranes. Stack performance was improved by changing the separator material. High current efficiencies noted in a few test runs suggest anomalous selectivity of the neutral membranes is being encountered [7]. A new type of neutral membrane has been developed in Israel for the transport depletion process. The membrane can be used at a high current density and is claimed to be superior to the cellulosic membrane [8].

The principal advantages of the transport depletion process over conventional electrodialysis result from the elimination of anion exchange membranes and the associated polarization and fouling problems.

3.8.2 Electrogravitational demineralization

Electrogravitational demineralization may be achieved in a cell, in which only cation exchange membranes are used. When a direct electric current is passed, depleted and concentrated boundary layers form at each side of the membrane, but the solution in the depleted boundary layer rises and collects at the top of the depleted compartment because of its lower density than that of the bulk of the solution. Similarly the solution in the concentrated boundary layer slides downward and collects at the bottom of the enriched compartment. Additional density differences are obtained by electrical heating of the depleted solution, because of higher electrical resistance. Extreme simplicity in construction and operation are advantages of the process and very high concentration gradients can be achieved. However, the process is not competitive with other desalination processes [5].

3.8.3 The osmionic process

The osmionic process was conceived by Murphy, developed to a small laboratory unit [9] and was then evaluated for practical application [10].

The process resembles the electrodialysis process in several ways, but it differs in the source of energy used. The osmionic process makes use of the difference in concentration between two solutions as a source of energy, instead of the external electromotive force used in electrodialysis. The unit is formed by two cation-permeable membranes and two anion-permeable membranes and is immersed in concentrated brine, thus forming a three compartment cell. The edges of the unit are closed on each side so that passage of ions can occur only through the membranes or around the assembly. Because of the difference in concentration between the brine and dilute solutions, sodium ions tend to flow from the concentrated brine through the cation-permeable membrane to the first compartment and chloride ions from the concentrated brine through the anion-permeable membrane to the third compartment. Electrical neutrality is maintained by migration from the middle compartment of chloride ions to the first compartment. The first and third compartments thus become more concentrated in sodium chloride and the middle compartment becomes less concentrated. The combination of brine and water of lower salinity may be considered as the power source.

In examining the practicality of the osmionic process it was estimated that even under favorable conditions, the cost of demineralization would be about 2 to 3 times as much as the cost of demineralization by electrodialysis [11]. In concluding, a theoretical study was made by Murphy and Matthews of the behavior of ions in aqueous solutions of mixed electrolytes with respect to the osmionic cell operation [12]. The osmionic effect for demineralization of water was also studied by Moldau et al. in a five chamber cell [13].

3.8.4 Electrosorption and desorption

The electrosorption process is similar in some respects to electrodialysis and in some respects to the electrochemical process of adsorbing ions on carbon electrodes (see section 3.10.1, p. 232). Special ion exchange membranes are required for the process, which comprises an inner layer that is permeable to ions and solution, sandwiched between an anion-permeable layer at one surface and a cation-permeable layer at the opposite surface.

An electrosorption demineralizer consists of many electrosorption membranes arranged, between a pair of electrodes, in a manner that solution compartments are formed between the parallel membrane surfaces. When a voltage is applied to the electrodes in the proper direction, cations are transported from the solution through the cation-permeable layer into the inner layer, but are blocked by the anion-permeable layer on the opposite side. Simultaneously, the anions are transported in the opposite direction through the anion-permeable membrane surface into the inner layer, but are blocked by the cation-permeable layer. Concentrated electrolyte solution is trapped at the inside membrane layer and the external solution is demineralized as it flows through the spaces between the membranes. From time to time the voltage applied to the electrodes is reversed and the ions accumulated in the middle layer of the membrane are driven back to the external solution. The membranes are regenerated and the external solution is sent to waste during discharge. The regeneration period requires appreciably less time than the demineralization part of the cycle [14].

When the water being treated contains calcium sulfate, precipitates form within the electrosorption membranes but these are removed during the desorption period. A flattened tube with solution inside and ion exchange membranes for walls is a method of providing an electrosorption membrane with a conductive inner layer [15]. At current densities less than the limiting value, the Faraday efficiencies obtained in the electrosorption process are as high as in electrodialysis and resistances per cell pair are about 30% lower. Calculations, based on the experimentally measured values of Faraday efficiencies, resistances, cycle times, and degree of desalination have indicated that an appreciable reduction in the total cost of desalination might be possible, mainly because of the low capital cost of the electrosorption stacks. An improved design for electrosorption stacks and methods of making acceptable membranes have been subsequently developed. The stack was used in the field to demineralize two types of natural brackish waters and electrodialysis experiments were conducted with the same waters. The cost of demineralization by electrosorption was estimated to be about 17% lower for one of the waters and about 6% lower for the other than that for demineralization by electrodialysis [16].

A novel cyclic electrochemical membrane separation process, which uses periodic flow reversal, applied to an electrically driven absorption - desorption process, was described by Bass and Thompson. Absorption takes place in a stack of multi-layer ion-selective membranes. Two bench-scale designs of the stack have been investigated and the effect of nine system parameters on the separation of aqueous sodium chloride solutions in batch operation has been analyzed. The process is said to have potential application in continuous systems [17].

Another tested system is the interposition of granulated ion exchange resins between the membranes of an electrodialyzer, which increases the overall electrical conductivity of the cell at low ion concentrations. The proper selection of the filler permits the selective removal of one ion at a time. This may find applications in the recovery of valuable materials from industrial effluents [18].

The effect of an electric current on a system anion membrane/ion exchange resin/cation membrane was studied by Grebenyuk et al. An equation was derived to express the concentration of the solution corresponding to the steady state of both systems [19]. The electrical conductivity of the solution in an electrodialysis cell, filled with a mixture of ion exchangers in the salt form, will increase as salt is transferred to the solution. The increase in salt concentration of the circulating medium is proportional to the quantity of beads of ion exchanger contacted and to the current density [20].

3.8.5 Forced-flow electrodesalination

Forced-flow electrodesalination utilizes high-shear flow through thin solution compartments to reduce concentration polarization. Hence high values of limiting current density can be obtained with high solution velocities. Cell-pair resistance is also lower with the thin solution compartments. With the ability to operate at the economic optimum current density and the low resistances attainable, the costs of treatment of brackish waters by forced-flow electrodesalination were estimated to be about 25% less than costs of treatment by conventional electrodialysis. A method of manifolding solutions entering and leaving the thin compartments was developed that results in good distribution of solution across the entire width of each compartment and equal distribution to all compartments. A system of equations was developed to describe the operation of the forced-flow electrodesalination stack, with which the operating conditions, energy consumption and production rate of a given stack can be predicted. A normalized current density, the polarization parameter, which is a function only of the solution velocity in the compartments, and a normalized cell-pair resistance, which is a function only of the membranes and spacer materials used, were employed to simplify calculation procedures. The process was evaluated with two natural brackish waters, one with high $Ca(HCO_3)_2$ and NaCl content and the other with high $CaSO_4$ content. A forced-flow electrodesalination stack and an electrodialysis stack were used. Both stacks were operated for approximately 3600 h at each well site. Precipitates formed in both stacks when limiting current densities were exceeded. Limiting values of current density and degree of demineralization with the forced-flow electrodesalination stack were approximately twice the values attainable with the electrodialysis stack. Cost estimates indicated significantly lower costs in large-scale electrodialysis plants if forced-flow electrodesalination stacks are used. Graphite anodes deteriorated during field operations, so they were replaced with platinized titanium anodes which performed satisfactorily [21].

Elektro-entsalzung unter erzwungener Strömung

Literature to 3.8

[1] R. E. Lacey (Off. Saline Water Res. Develop. Progr. Rept. No. 80 [1963]). — [2] M. S. Mintz (Off. Saline Water Res. Develop. Progr. Rept. No. 182 [1966]). — [3] M. S. Mintz, R. E. Lacey

(Off. Saline Water Res. Develop. Progr. Rept. No. 259 [1967]). — [4] R. E. Lacey (Off. Saline Water Res. Develop. Progr. Rept. No. 343 [1968]). — [5] E. L. Huffman (Off. Saline Water Res. Develop. Progr. Rept. No. 439 [1969]).

[6] R. Redman (Off. Saline Water Res. Develop. Progr. Rept. No. 683 [1971]). — [7] R. N. Smith, R. A. Knight (Off. Saline Water Res. Develop. Progr. Rept. No. 669 [1971]). — [8] E. Korngold (Develop. Desalination Technol. Israel **1972** 70/3). — [9] G. W. Murphy (Off. Saline Water Res. Develop. Progr. Rept. No. 14 [1957]). — [10] R. E. Lacey, E. W. Lang, C. E. Feazel (Off. Saline Water Res. Develop. Progr. Rept. No. 38 [1960]).

[11] Southern Research Institute (Off. Saline Water Res. Develop. Progr. Rept. No. 64 [1962]). — [12] G. W. Murphy, R. R. Matthews (Off. Saline Water Res. Develop. Progr. Rept. No. 76 [1963]). — [13] M. Moldau, A. Kh. Suit, L. R. Suit (Tr. Voronezhsk. Gos. Univ. **72** [1969] 116/20; C.A. **77** [1972] No. 168476). — [14] R. E. Lacey, E. W. Lang (Off. Saline Water Res. Develop. Progr. Rept. No. 106 [1964]; Desalination **2** [1967] 387/93). — [15] R. E. Lacey (Off. Saline Water Res. Develop. Progr. Rept. No. 135 [1965]).

[16] R. E. Lacey (Off. Saline Water Res. Develop. Progr. Rept. No. 195 [1965], No. 228 [1967], No. 398 [1969]). — [17] D. Bass, D. W. Thompson (Proc. 4th Intern. Symp. Fresh Water Sea, Heidelberg 1973, Vol. 3, p. 269/77). — [18] N. I. Nikolaev, A. M. Filimonova, M. D. Kalinina, G. G. Chuvileva (Ionoobmen. Membrany Elektrodialize **1970** 160/5; C.A. **74** [1971] No. 25364). — [19] V. D. Grebenyuk, T. Z. Sotskova, N. P. Gnusin (Zh. Fiz. Khim. **45** [1971] 730/1; C.A. **74** [1971] No. 146711). — [20] V. D. Grebenyuk, T. A. Sotskova, N. P. Gnusin (Zh. Fiz. Khim. **45** [1971] 1744/7; C.A. **75** [1971] No. 133387).

[21] T. A. Davis, R. E. Lacey (Off. Saline Water Res. Develop. Progr. Rept. No. 557 [1970], No. 710 [1971]).

3.9 Specific electrodialysis applications

Spezielle Anwen-dungen der Elektro-dialyse

In several other applications of the electrodialysis process emphasis is given, besides desalting of seawater, to the treatment of various wastes and the recuperation of solid materials rather than the production of potable water. Some of these applications are briefly reviewed.

3.9.1 Seawater and brine

Meerwasser und Sole

The main interest for the development of the electrodialysis process in Japan was the production of common salt. Combination of flash distillation and electrodialysis lead to more concentrated brines for the recovery of mineral by-products. About 13%, amounting to 120000 t, of the total annual salt consumption were produced in Japan by electrodialysis [1]. A more recent review on salt production in Japan was presented by Nomiyama, including a comprehensive utilization of the use of electrodialysis in the process [2]. A demonstration plant producing concentrated brine from seawater to yield 3000 t per year of common salt was described by Yamane et al. [3].

Another important application of electrodialysis is the use of compact stacks on board ships. Such units producing about 600 l/day of fresh water with 300 to 400 mg/l total dissolved solids consume about 25 kWh/m³ [4]. Electrodialysis was found to be more economical than the eva-poration process, when used in a small capacity plant in special environments such as on ships and islands [5]. A compact electrodialysis apparatus with a capacity of either 0.5 or 2 m³/day potable water had been developed with a view to using it for desalting seawater on large fishing boats. By using wire electrodes, seawater could be used as the electrode rinse solution without acidification which contributed in making the equipment compact and easy to operate. The cost of product water is fairly tolerable for use on fishing boats [6].

Seawater was desalinated by electrodialysis in a pilot plant, working at the Atlantic Ocean for two months. It was probably carried on an expedition vessel. Energy consumption was 30 kWh/m³ by 20 mA/cm² current density. The product water contained 520 to 590 mg/l total dissolved solids. A rather high content of B, Br and F ions was observed in the product water [7]. Desalination of seawater was also studied under laboratory conditions. A decrease of salt content from 25.6 to 1 g/l was obtained [8]. Electrodialytic desalination of Black Sea water in pilot plant scale reduced the salt content from 18 g/l to 500 to 700 mg/l. No scale formation was observed in either the diluate or brine compartments [9].

A study was made on electrodialytic concentration of waste brine for salt production from effluents discharged from a multi-stage flash seawater desalting plant. The results showed that, in utilizing waste brine, several advantages such as an important decrease in electric energy consumption, a lessening of the problem of scale deposition in the concentrate compartments, an increase in the productivity of the apparatus and a decrease in the cost of salt production could be expected, as compared with the utilization of seawater [10]. The most favorable conditions were sought for electrodialytic concentration of either the waste brine coming from a seawater desalination plant by the multi-stage flash evaporation process or the warm seawater coming from the condenser of a steam-power station. In the electrodialytic process with warm feed solutions a lower operating cell voltage and less electrical energy consumption may be expected and less scale may be formed. More NaCl may be produced than with conventional processes [11]. Subsequently electrodialysis of waste brine was carried out on a full scale to evaluate the effectiveness of concentrating the waste brine discharged from the multi-stage flash evaporators. The concentration of the solution obtained and the amount of NaCl manufactured were almost the same as those of seawater at the same current density. The consumed electric power was less than for seawater [12].

A review on electrodialytic concentration of seawater is given by T. Nishiwaki in chapter 6 of the book on Industrial Processing with Membranes, edited by Lacey and Loeb [13].

3.9.2 Treatment of wastes

Electrodialysis was proposed as an experimental unit process for the removal of phosphates from waste water. Control experiments with NaCl, $NaNO_3$, and Na_3PO_4 singly and in combination were used to show the extent of removal possible. A comparison was made to complex waste waters that can potentially foul and poison the permselective membranes [14].

Tests were made for the removal of sodium ions from Na_2SO_4 solutions in a multichamber electrodialyzer. The best results were obtained with countercurrent flow in the compartments. The concentration of sodium ions could be reduced to less than 0.065 M [15].

A more important field of application for electrodialysis is the regeneration of acid wastes. Three electrodialysis schemes were tested for the recuperation of acid suitable for use in regeneration of cation exchange resin beds. Two of them were highly efficient in HCl production with an energy consumption of about 0.265 kWh/g-eq. [16]. Sulfuric acid could be concentrated from 20 to 34 g/l by electrodialysis and following repeated dialysis up to 100 g/l [17]. Dilute sulfuric acid solutions were partially demineralized using an electrodialysis cell. Graphs were given of product concentration versus current density with superimposed isothermal lines. A graph was also given for concentration versus flow rate of solution at a fixed current density [18].

Electrodialysis of etching solutions containing titanium, ammonium, and sodium salts was investigated by Kochergin et al. [19]. Spent sulfuric acid containing iron and titanium was treated in an electrodialysis cell for the recovery of sulfuric acid [20]. The electrodialytic treatment of spent liquors from pulp mills was investigated by Santiago with neutral, cationic, and anionic membranes. Cationic membranes gave the best separation of chemicals and were relatively inert to the medium [21]. An electromembrane process for treating spent sulfuric acid pickle liquor to produce usable acid was investigated by Lacey and found to be technically feasible [22]. Electrodialysis was used for the recovery of magnesium hydrogen sulfite in reprocessing of spent sulfite liquors. The recovery reached 80%. It depended mostly on the current density and not on the liquor temperature [23]. The use of electromembrane processes for recovery of constituents from pulping liquors is described by Ahlgren in chapter 5 of the book on Industrial Processing with Membranes, edited by Lacey and Loeb [13].

3.9.3 Separation of inorganic constituents

Electrodialysis was used for the separation of various inorganic constituents from solutions of electrolytes. The separation of complex chlorides of platinum, iridium, palladium, and rhodium from chlorides of copper, nickel, and iron was investigated by Ezerskaya and Solovykh [24]. Electrodialysis was studied by Shivrin et al. as a possible process for recovering gold and silver from cyanide solutions. Copper and zinc were removed from the solution by precipitation with aluminum [25].

Behandlung von Abwässern

Abtrennung von anorganischen Bestandteilen

The process was also used for the preparation of carrier-free ^{90}Y from ^{90}Sr-^{90}Y equilibrium mixtures [26]. Electrodialytic behavior of ^{60}Co, ^{106}Ru, and ^{144}Ce in seawater was investigated by Honda et al. [27].

Investigating the conditions for the separation of amphoteric compounds by electrodialysis, Isaev and Malinovskaya separated zinc and lead from aluminum solutions [28]. The kinetics of the de-ionization of aluminum hydroxide gels by electrodialysis and dialysis were investigated as a function of pH. It was found that the control of the de-ionization by means of the pH is possible [29].

Electrodialysis of certain salt solutions in the presence of ion exchange membranes is frequently accompanied by the formation of metallic precipitates and the evolution of oxygen on the membrane. The phenomenon was investigated during the electrolysis of alkaline solutions of Cu^{II}, Sn^{II}, and Ag^{I} in a cell equipped with an anion exchange resin [30].

The removal of calcium ions from boric acid by electrodialysis was investigated; at 20 mA/cm^2 during 30 h of electrodialysis the calcium concentration in the saturated H_3BO_3 solution decreased from 0.75 to 0.001 g/l. The transfer rate of the calcium ions increased with current density. The concentration of H_3BO_3 remained practically constant [31]. Sodium carbonate and hydrogen carbonate solutions were purified using highly basic ion selective membranes. A 95% degree of desalination of the solution was obtained [32].

Changes of current density and flow rate had no significant effect on the separation of calcium ions from sodium ions. The yields of the electrodialysis reached only 90% and decreased with increasing current density [33].

The feasibility of recovering nickel from plating wastes by electrodialysis was investigated by Trivedi and Prober [34]. A small electrodialyzer in conjunction with a strongly acidic cation exchange membrane with univalent permselectivity were used to produce NaH_2PO_4 by a substitution reaction between Na_2HPO_4 and raw wet-process phosphoric acid. A series of pilot-scale tests were also made in a medium-size commercial plant [35].

A 36-chamber electrodialyzer was used to investigate the demineralization of recycled water from non-ferrous metallurgy plants. The following operation characteristics were obtained: Current density 8.25 mA/cm^2, flow rate 5 cm/s, temperature 23°C and energy consumption 3 to 10 kWh/m^3 [36]. A similar study was made for the desalination of industrial waters. The salt content was reduced to 25 mg/l with a current density of 0.5 to 1 A/dm^2 and an energy consumption of 10 kWh/m^3 [37].

A review was presented by Katz for the preparation of high purity boiler feed water either by electrodialysis alone or in series with ion exchange final treatment [38].

Abtrennung von organischen Bestand- teilen

3.9.4 Separation of organic constituents

An experimental investigation was made on the electrodialysis of acetate or butyrate ions with chloride ions and the effects on the transport of the individual anions were studied [39].

The preparation of L-amino acids was investigated by Khleborodova and Ryazanov [40]. Electrodialysis of casein hydrolyzate was carried out to remove hydrochloric acid from amino acid mixtures. A review of amino acid separation methods based on electrodialysis was also given [41].

The transport of organic acid anions in electrodialysis under dynamic conditions was investigated by Kotov et al. [42].

Thin sugar juices were demineralized by electrodialysis to reduce their ash content. A decrease of 25% in calcium and magnesium ions was obtained [43]. Semitechnical experiments to demineralize and decolorize sugar juice were also performed by Perschak [44]. Demineralization of milk and whey by means of the Morinaga process, which was developed in Japan, is reported by Huwe [45]. A review on electromembrane processing of cheese whey is given by Ahlgren in chapter 4 of the book on Industrial Processing with Membranes, edited by Lacey and Loeb [13].

Literature to 3.9

[1] M. Kaho, T. Watanabe (Ind. Water Eng. 6 No. 11 [1969] 30/2; C.A. 72 [1970] No. 58 936). — [2] Y. Nomiyama (Kagaku Kogyo 23 [1972] 251/7; C.A. 76 [1972] No. 142 934). — [3] R. Yamane, M. Ichikawa, Y. Mizutani, Y. Onoue (Ind. Eng. Chem. Process Design Develop. 8 [1969] 159/65). — [4] S. Itoi (Desalination 2 [1967] 378/86). — [5] T. Matsuda, S. Ogawa, Y. Onoue (Desalination 3 [1967] 295/303).

[6] Y. Tsunoda, M. Kato (Desalination 3 [1967] 66/81). — [7] V. B. Gaidaymov, M. M. Senyavin, A. A. Zaborskii, M. N. Kukovkina, Yu. G. Ponomarev (Ionoobmen. Membrany Elektrodialize 1970 171/6; C. A. 74 [1971] No. 45442). — [8] G. A. Bedyukh, N. N. Zubets, I. A. Anishchenko, Z. D. Lavrova, G. A. Lebedinskaya, L. D. Maysheva (Tr. Voronezhsk. Univ. 1971 No. 94, p. 110/3; C. A. 77 [1972] No. 118061). — [9] B. N. Laskorin, N. M. Smirnova, Y. I. Tisov, A. V. Borisov, T. I. Savelyeva (Proc. 4th Intern. Symp. Fresh Water Sea, Heidelberg 1973, Vol. 3, p. 189/94). — [10] T. Watanabe, M. Kaho (Proc. 3rd Intern. Symp. Fresh Water Sea, Dubrovnik 1970, Vol. 2, p. 281/9).

[11] T. Watanabe, S. Azechi, Y. Tanaka, S. Nagatsuka, N. Yugi (Nippon Kaisui Gakkai-Shi 24 [1970] 104/28; C. A. 75 [1971] No. 25154). — [12] Y. Tanaka, T. Yuyama, N. Yugi, T. Yuzurihara (Nippon Kaisui Gakkai-Shi 25 [1972] 189/97; C. A. 77 [1972] No. 92517). — [13] R. E. Lacey, S. Loeb (Industrial Processing with Membranes, New York 1972). — [14] T. Helfgott, J. V. Hunter (Chem. Eng. Progr. Symp. Ser. 65 No. 97 [1969] 218/31). — [15] V. T. Kunin, A. V. Kuz'mina (Izv. Vysshikh Uchebn. Zavedenii Khim. i Khim. Tekhnol. 14 [1971] 100/3; C. A. 74 [1971] No. 102806).

[16] N. I. Isaev, M. N. Romanov (Izv. Vysshikh Uchebn. Zavedenii Khim. i Khim. Technol. 12 [1969] 924/7; C. A. 72 [1970] No. 15592). — [17] F. Wolf, R. Bachmann (Uniw. Adama Mickiewicza Poznania Pr. Wydz. Mat. Fiz. Chem. Ser. Chim. No. 10 [1969] 1/13; C. A. 72 [1970] No. 124875). — [18] N. P. Gnusin, E. Kh. Ignatenko, A. S. Gumen (Ukr. Khim. Zh. 38 [1972] 659/62; C. A. 77 [1972] No. 118667). — [19] V. P. Kochergin, T. F. Moiseeva, G. I. Krokhin (Izv. Vysshikh Uchebn. Zavedenii Khim. i Khim. Tekhnol. 14 [1971] 110/2; C. A. 74 [1971] No. 116334). — [20] H. Tamura, Y. Matsuda, S. Tsujino (Kogyo Kagaku Zasshi 74 [1971] 2003/5; C. A. 76 [1972] No. 9707).

[21] E. Santiago (Tappi 54 [1971] 1641/5). — [22] R. E. Lacey (Water Pollut. Contr. Res. Ser. Rept. 12010-EQF [1971]). — [23] K. A. Beinov, L. A. Mazitov (Bumazhn. Proml. 1972 No. 5, p. 7/9; C. A. 77 [1972] No. 63651). — [24] N. A. Ezerskaya, T. P. Solovykh (Izv. Akad. Nauk SSSR Ser. Khim. 1969 No. 5, p. 993/9; C. A. 71 [1969] No. 42681). — [25] G. N. Shivrin, B. N. Laskorin, E. M. Shivrina (Tsvetn. Metal. 43 [1970] 89/93; C. A. 72 [1970] No. 123482).

[26] P. Groll, F. Grass, K. Buchtela (Radiochim. Acta 12 [1969] 152/6). — [27] Y. Honda, Y. Kimura, N. Tsurii (Radioisotopes [Tokyo] 21 [1972] 269/75). — [28] N. I. Isaev, E. M. Malinovskaya (Zh. Fiz. Khim. 43 [1969] 1742/6; C. A. 71 [1969] No. 116896). — [29] H. Saisse, J. Cohen, R. Kern (J. Chim. Phys. 69 [1972] 428/35). — [30] N. Ya. Kovarskii, M. V. Pevnitskaya, L. G. Kolzunova (Ionnyi Obmen Ionity 1970 199/203; C. A. 74 [1971] No. 150293).

[31] N. M. Smirnova, B. N. Laskorin (Ionnyi Obmen Ionity 1970 194/6; C. A. 74 [1971] No. 113848). — [32] N. M. Smirnova, B. N. Laskorin (Ionnyi Obmen Ionity 1970 190/4; C. A. 74 [1971] No. 130708). — [33] J. Lindeman, H. Czarczynska, W. Trochimczuk, H. Galina (Polimery 16 [1971] 266/9; C. A. 76 [1972] No. 76910). — [34] D. S. Trivedi, R. Prober (Ion Exch. Membranes 1 [1972] 37/46). — [35] T. Nishiwaki, H. Hani, S. Itoi (Membrane Sci. Technol. Ind. Biol. Waste Treat. Processes Proc. Symp., 1969 [1970], p. 150/70; C. A. 77 [1972] No. 37094).

[36] N. Ya. Lyubman, A. I. Uskov, O. M. Gruzo (Tr. Gos. Nauchn. Issled. Proekt. Inst. Obogashch. Rud Tsvet. Metal. „Kazmekhanobr" 1970 No. 3, p. 247/61; C. A. 77 [1972] No. 105376). — [37] N. Ya. Lyubman, O. M. Gruzo, I. A. Sondar (Tr. Gos. Nauchn. Issled. Proekt. Inst. Obogashch. Rud Tsvet. Metal. "Kazmekhanobr" 1970 No. 3, p. 262/71; C. A. 77 [1972] No. 105380). — [38] W. E. Katz (Proc. Am. Power Conf. 33 [1971] 830/41). — [39] J. T. Schrodt, P. J. Yonikas, R. B. Grieves (Can. J. Chem. Eng. 47 [1969] 398/402). — [40] R. T. Khleborodova, A. I. Ryazanov (Zh. Prikl. Khim. 42 [1969] 1053/8; C. A. 71 [1969] No. 74482).

[41] A. M. Gostomczyk, T. Winnick (Pr. Nauk. Inst. Inz. Chem. Urzadzen Cieplnych Politech. Wroslaw 3 No. 2 [1970] 64/99; C. A. 78 [1973] No. 76242). — [42] V. V. Kotov, N. I. Isaev, V. A. Shaposhnik (Zh. Fiz. Khim. 46 [1972] 529/40; Russ. J. Phys. Chem. 46 [1972] 314/5). — [43] K. Ciz, V. Cejkova (Listy Cukrov. 85 [1969] 230/2; C. A. 72 [1970] No. 102073). — [44] F. Perschak (Zucker 26 [1973] 239/45). — [45] K. H. Huwe (Deut. Molkerei-Ztg. 93 [1972] 1500/2).

3.10 Electrochemical and physicochemical methods of desalination

Several attempts were also made to develop electrochemical and physicochemical methods for the demineralization of saline waters. However, none of them has reached a degree of practical application. The principles of these methods are outlined in the following.

Elektro-
chemische
und
physiko-
chemische
Entsalzungs-
verfahren

3.10.1 Adsorption on porous carbon electrodes

The feasibility of water demineralization by adsorption of ions on pairs of cation-responsive and anion-responsive electrodes was demonstrated by Murphy and his coworkers. The chemical groups that are formed in the manufacture of carbons, especially blacks and chars, are cation-responsive. Important factors that influence the cation capacity of carbons are surface area and oxygen content. Anion capacity depends on the amount of electrochemically active nitrogen containing groups and the amount of cation-responsive groups remaining on the surface, which compete with them. The process is based on pairs of electrodes, both made from carbon, one of which specifically adsorbs cations and the other anions upon application of an appropriate low voltage. In a second phase of the process the electrode polarity is reversed and the ions are then desorbed to a rejected solution. The process was at first successfully tested in the laboratory [1].

In continuing research work improved ion-responsive electrodes were prepared [2], various types of experimental cells were constructed and an electrochemical theory of demineralization previously developed was completed [3]. A bench-scale unit was for some years in continuing satisfactory operation and the theory of electrochemical demineralization was extended [4]. The adsorption capacity of a carbon electrode is a function of the applied potential over the range 0.3 to 1.0 V. For a pair of carbon electrodes the capacity increase with applied potential diminishes at approximately 0.6 V. Severe oxidation of the carbon yields a large capacity and high selectivity for cations, exceeding 1 meq/g, high temperature treatment in an ammonia atmosphere reduces the cation capacity of the carbon. The relatively low anion capacity of many commercial carbons can be improved by incorporating additives such as a cationic polyelectrolyte, a quaternary ammonium salt or polyethylenimine [5]. Further research work on electrode development was reported by Murphy and Cooper and the removal of Na, Ca, Mg, Cl, SO_4, HCO_3, NO_3, and PO_4 ions from brackish water was achieved [6].

The process was further tested in a 75 l/day (20 gpd) pilot plant unit and an engineering evaluation has been made [7]. The unit was made from a commercial electrodialysis apparatus by eliminating the concentrating chambers. Saline water was made to flow between the anion and cation selective membranes of the diluting chambers. Under the influence of a direct current, anions and cations were absorbed on the respective membranes. After saturation the polarity was reversed and the ions desorbed into the water stream and this was rejected to waste. It was predicted that 20 membrane pairs would yield 17.9 gpd (68 l) of product water, when used as an ion-adsorption process. The same unit would yield 500 gpd (1893 l), when used as an electrodialysis apparatus on the same water. It was concluded that the process is not competitive with the electrodialysis process [8].

Polymers containing oxidizable or reducible organic groups were synthesized and tested [9]. The electrically induced demineralization of water by using electrodes, built-up from an ion exchanger mixed with carbon powder was tested and a bench-test apparatus for desalting water by using porous electrodes was designed. The cell contained 20 electrodes. Anodes were aminated and the cathodes sulfonated. The initial salt concentration of 5 g/l could be reduced by 15 to 20% during a 15 min half-cycle of the process at 5 V. The average current output of the process was 40% [10].

Demineralization electrodes were studied in half-cells and in complete cells to establish the electrochemical characteristics of individual electrodes and to determine overall polarization characteristics [11]. Extensive coulometric and mass-balance experiments were conducted with cation-responsive paste electrodes. The mechanism of ion uptake and release at such electrodes was presented and the exchange characteristics were evaluated [12].

3.10.2 Environmentally modulated adsorption

Reversible adsorption reactions, which can be modulated by changes in the temperature, pressure, and/or concentration of saline species in the adsorption system have been studied. By virtue of the large working areas and short ion transport distances that can be attained in adsorption beds consisting of fine granular particles, high desalting rates are possible even when thermodynamic driving forces are small. A thermally modulated system making use of suitably pretreated granular carbon adsorbents seems to be especially attractive. The technical feasibility of such a process has been demonstrated and the useable salt exchange capacity has been found to be sufficient for practical use with respect to energy consumption and to initial cost. Reversible adsorption of salt in a thermally modulated adsorption process for desalting water would be accomplished in beds composed of porous, granular, electrically conducting carbon by suitable cycling the temperature within the range of ambient to 100°C [13].

Development of electrochemical ion exchange materials and techniques for desalting water by means of the Thermosorb and the Electrosorb processes were reported by Venolia and Johnson. Carbon sorbents are cyclically regenerated by the use of either thermal or electrical energy [14]. Thermally cycled carbon beds were shown to be capable of desalting aqueous sodium chloride in a reproducible way by the Thermosorb process. Inexpensive raw materials, straightforward preparative methods, and moderate temperatures were used. Rates were favorable and they were capable of being manipulated. Rate optimization was not attempted. No evidence of a durability problem was encountered in an uninterrupted desalting run of more than two thousand hours duration. Beds of salt sorptive carbons appear to be competitive with beds of ion exchange resins on the basis of effective capacity per weight of sorbent. It appeared likely that carbons will also outperform ion exchange resins on the basis of initial cost and durability [15]. In developing the Electrosorb process, experimental work has shown the effectiveness of the system when operating on feedwaters composed of the principal ionic species that are commonly found in natural waters. Preliminary cost studies have been made which indicate that attractive costs can be expected for the improvement of waters of low salinity. The experimental program has not demonstrated degradation-free performance. A variety of non-conflicting routes toward possible attainment of a durable system have been defined [16].

A porous electrode model was developed and analyzed for ionic adsorption on porous carbon. In the absence of concentration variations, the system behaves like a distributed network of resistances and capacitances. Experimental results supported the basic model and showed a preferential adsorption of divalent ions [17].

3.10.3 Electrochemically controlled ion exchange

Undissociated fixed groups, which are ionized as a result of electrochemical reactions, participate in the electrochemically controlled ion exchange process. The properties of weak-acid and weak-base ion exchange resins are incorporated into ion-responsive electrodes. An anion-responsive electrode is coupled with a cation-responsive electrode. During demineralization, sodium and chloride ions are removed from the solution and released during regeneration by polarity reversal. Cell potentials for demineralization and regeneration at 1.0 mA were less than 1 V [18]. The mechanism of electrochemically controlled ion exchange was described by Evans et al. [19].

Ion exchange kinetics at anion- and cation-responsive electrodes were used to refine a previously [20] developed model to represent the transport processes associated with electrochemically controlled ion exchange. Electrochemical experiments were carried out to obtain information as to the limiting currents that could be associated with this process, to evaluate new combinations of binder systems, ion exchange resins and carbons and to determine the suitability of electrodes to ions other than sodium and chloride [21].

Mixed beds of ion exchange resins that had been used to demineralize a salt solution were electroregenerated in situ in the depleting compartments of an electrodialysis cell. Feed solutions of sodium chloride were demineralized easily. The presence of calcium and hydrogen carbonate ions in the feed caused problems in electroregeneration and these ions should be removed by pretreatment [22].

3.10.4 Donnan softening as a pretreatment process

Donnan membrane water softening is a technique, which exploits the Donnan exclusion properties of ion exchange membranes to selectively remove divalent hardness ions from the feed waters to desalination systems. Laboratory scale test systems were constructed, process design examples have been developed and costs of water softening have been estimated [23]. For a simplified economic analysis of Donnan softening only the costs of the membrane and added salt were considered and four specific examples discussed, based on using electrodialysis membranes to soften the feed waters at Yema, Arizona, Webster, South Dakota, Roswell, New Mexico, and a synthetic seawater. With present-day electrodialysis membranes, Donnan softening is not interesting economically. A reduction in membrane cost, achieved e.g. by making the membranes thinner, would make Donnan softening an attractive candidate for further development [24].

Elektro-chemisch kontrollierter Ionen-austausch

Donnan-Enthärtung als Vor-behand-lungs-verfahren

Schaum-
fraktionie-
rung

3.10.5 Foam fractionation

To explore the possibility of separation of aqueous solutions of inorganic salts into their components, water and salt, by means of foam fractionation the persistence of single bubbles in aqueous solutions of sodium chloride, sodium bromide, sodium sulfate, potassium chloride, magnesium chloride, and two mixtures of sodium chloride and magnesium chloride was measured. This persistence usually was between 0.08 and 1 second, was little affected by the nature of salt, it slightly increased with the salt concentration and was almost independent of the bubble diameter. Calcium, iron, and manganese can be partially removed from the aqueous solutions of their salts by adding a foaming agent to the solution and bubbling air through. The foam obtained contains a higher concentration of the metal than the initial or the residual solution. The accumulation ratio is greater for iron and manganese than for calcium. Long-chain sulfates and poly(oxyethylene) sulfates appeared most promising as foaming agents. None of the non-ionic surfactants tested was satisfactory. The accumulation ratio for the metal generally increases when the drainage time of the foam and the weight ratio surfactant to metal increase. The accumulation ratio for the surfactant increases with the duration of drainage [25].

Einfluß von
magneti-
schen
Feldern

3.10.6 Effect of magnetic fields

The degree of concentration of ions in an electrolyte solution, subjected to electric and magnetic fields, was investigated; in none of the experiments a concentration gradient large enough to be measured could be observed [26].

Several systems have been proposed in which the ionic solution is flowed through a magnetic field. Velocities of the ions relative to the magnetic field of the order of 100 to 1000 m/s are necessary to obtain an adequate resultant force and this appears to be achievable only by moving the magnetic field relative to the ions. Such systems do not appear to be competitive with other desalination methods [27].

The potential difference produced by electric induction in an electrolyte, streaming through a magnetic field, was used to drive the ions through cation- and ion-selective membranes. This method may be considered as an electrodialysis without electrical power supply [28]. Samples of reduced salinity could be obtained from a flowing sodium chloride solution subjected to crossed electrical and magnetic fields [29].

Behandlung
des Wassers
durch
Mikro-
flotation

3.10.7 Water treatment by microflotation

The removal of organic color from water by the process of microflotation appears promising. It has several advantages over the conventional sedimentation processes. The effects of chemical conditions on the performance of microflotation and settling were established and the microflotation process was compared to settling [30]. Microflotation has been successfully applied to the separation of bacteria, algae and organic color from water. It produces a thin, relatively dry foam on which the colloids are collected. Effects of pH and aluminum salts, as coagulants, on the foam separation by microflotation of several species of bacteria and materials, which cause organic color in water, were determined [31]. The conditions for the most efficient separations were related to the stability of the colloidal sols as a function of pH in the presence of aluminum salts [32].

The microflotation process is a non-selective separation method also capable of efficiently removing colloidal particles of varying chemical and physical properties. The microflotation of colloidal silica was carried out in the presence of aluminum salts at various pH values. The colloidal properties of the silica-alumina system as a function of pH have been established and were discussed in detail by Mangravite et al. [33]. In continuation of previous work, the study was extended to cover the range of aluminum hydroxide precipitation and the "aluminum salt-pH" domain was expanded to include the silica–aluminum system [34].

Literature to 3.10

[1] University of Oklahoma (Off. Saline Water Res. Develop. Progr. Rept. No. 45 [1960], No. 58 [1962]). — [2] G. W. Murphy, J. J. Bloomfield, F. W. Smith, W. E. Neptune, J. O. Purdue, D. Candle, A. L. Stevens, J. Tucker, E. N. Wood, L. Tague, M. Lawson, R. Rose, D. James (Off. Saline Water Res. Develop. Progr. Rept. No. 92 [1964]). — [3] G. W. Murphy, J. J. Bloomfield, F. W. Smith, W. E. Neptune, D. Candle, A. L. Stevens, J. Tucker, E. N. Wood, L. Tague, B. A. Arnold,

J. W. Blair, A. Satter (Off. Saline Water Res. Develop. Progr. Rept. No. 93 [1964]). — [4] D. D. Candle, J. H. Tucker, J. L. Cooper, B. B. Arnold, A. Papastamataki, E. N. Wood, R. Hock, G. W. Murphy (Off. Saline Water Res. Develop. Progr. Rept. No. 188 [1966]). — [5] G. W. Murphy, J. H. Tucker (Desalination **1** [1966] 247/59).

[6] G. W. Murphy, J. L. Cooper (Off. Saline Water Res. Develop. Progr. Rept. No. 399 [1969]). — [7] G. W. Reid, F. M. Townsend, A. M. Stevens, A. Al-Awady, A. Hu, J. P. Abichandani, A. Abbas (Off. Saline Water Res. Develop. Progr. Rept. No. 293 [1968]). — [8] G. W. Reid, F. M. Townsend, E. H. Klehr, L. E. Streebin, A. Hu, S. W. Swafford, D. A. Dodson, C. Chieu (Off. Saline Water Res. Develop. Progr. Rept. No. 498 [1969]). — [9] T. O. Rouse, A. Factor (Off. Saline Water Res. Develop. Progr. Rept. No. 524 [1970]). — [10] Yu. N. Efimov, V. K. Solyakov (Tr. Vses. Nauchn. Issled. Inst. Vodosnabzh. Kanaliz. Gidrotekhn. Sooruzh. Inzh. Gidrogeol. **1970** No. 25, p. 11/9; C.A. **73** [1970] No. 80373), Yu. N. Efimov (Tr. Vses. Nauchn. Issled. Inst. Vodosnabzh. Kanaliz. Gidrotekhn. Sooruzh. Inzh. Gidrogeol. **1971** No. 29, p. 57/60; C.A. **78** [1973] No. 7668).

[11] J. Farrar, S. Evans, G. D. Seele (Off. Saline Water Res. Develop. Progr. Rept. No. 126 [1965]). — [12] S. Evans, W. S. Hamilton (Off. Saline Water Res. Develop. Progr. Rept. No. 156 [1966]). — [13] A. M. Johnson (Off. Saline Water Res. Develop. Progr. Rept. No. 155 and 218 [1966]). — [14] A. W. Venolia, A. M. Johnson (Off. Saline Water Res. Develop. Progr. Rept. No. 300 [1968]). — [15] A. W. Venolia (Off. Saline Water Res. Develop. Prcgr. Rept. No. 502 [1969]).

[16] A. M. Johnson, A. W. Venolia, R. G. Wilbourne, J. Newman (Off. Saline Water Res. Develop. Progr. Rept. No. 516 [1970]). — [17] A. M. Johnson, J. Newman (J. Electrochem. Soc. **118** [1971] 510/7). — [18] S. Evans, M. A. Accomazzo, W. S. Hamilton, J. E. Lewis (Off. Saline Water Res. Develop. Progr. Rept. No. 284 [1968]). — [19] S. Evans, M. A. Accomazzo, J. E. Accomazzo (J. Electrochem. Soc. **116** [1969] 307/9). — [20] S. Evans, M. A. Accomazzo, J. E. Accomazzo, M. Ladacki, K. A. Lossett (Off. Saline Water Res. Develop. Progr. Rept. No. 409 [1969]).

[21] S. Evans, M. A. Accomazzo, M. Ladacki, K. A. Lossett (Off. Saline Water Res. Develop. Progr. Rept. No. 598 [1970]). — [22] T. A. Davis, R. E. Lacey (U.S. Natl. Tech. Inform. Serv. PB 210163 [1972]). — [23] J. L. Eisenmann, J. D. Smith (Off. Saline Water Res. Develop. Progr. Rept. No. 506 [1970]), J. D. Smith, J. L. Eisenmann (Ind. Water Eng. **7** No. 9 [1970] 38/9). — [24] L. Dresner (AIChE [Am. Inst. Chem. Engrs.] J. **12** [1973] 148/59). — [25] J. J. Bikerman (Off. Saline Water Res. Develop. Progr. Rept. No. 248 [1967], No. 510 [1970]).

[26] B. K. Cooper (Off. Saline Water Res. Develop. Progr. Rept. No. 187 [1966]). — [27] E. R. Gilliland (Off. Saline Water Res. Develop. Progr. Rept. No. 231 [1967]). — [28] W. Schäfer (Desalination **3** [1967] 174/82). — [29] M. Khalifa, A. A. Abdel-Hamid, M. M. S. Abdel-Salam (Z. Physik. Chem. [Leipzig] **247** [1971] 273/81, 333/9). — [30] T. Buzzell, E. A. Cassell (Off. Saline Water Res. Develop. Progr. Rept. No. 757 [1972]).

[31] E. A. Cassell, A. J. Rubin, H. B. LaFever, E. Matijevic (Off. Saline Water Res. Develop. Progr. Rept. No. 758 [1972]). — [32] E. A. Cassell, E. Matijevic, F. Mangravite, T. Buzzell, S. B. Blabac (Off. Saline Water Res. Develop. Progr. Rept. No. 760 [1972]). — [33] F. J. Mangravite, E. A. Cassell, E. Matijevic (Off. Saline Water Res. Develop. Progr. Rept. No. 756 [1972]). — [34] E. Matijevic, F. J. Mangravite, E. A. Cassell (Off. Saline Water Res. Develop. Progr. Rept. No. 755 [1972]).

3.11 Reverse osmosis

Umgekehrte Osmose

Osmotic flow, direct or reversed, depends on the selective property of some membranes to allow certain components of a solution, usually the solvent, to pass through the membrane. This intrinsic property of the membrane is termed as semipermeability. If two solutions of different concentration, or a pure solvent and a solution, are separated by a semipermeable membrane, the solvent will flow under normal conditions from the less concentrated department through the membrane into the concentrated solution with the tendency that both solutions reach the same concentration. This flow is known as osmosis (**Fig. 3–10**, p. 236). Osmotic flow through the membrane will stop, when the concentrated solution reaches a sufficiently higher pressure than prevailing in the less concentrated solution or the solvent compartment. The equilibrium pressure difference between solvent and solution, or the two solutions, known as osmotic pressure, is a property of the solution. Equilibrium can also be reached by applying an external pressure to the concentrated salt solution equal to the osmotic pressure (**Fig. 3–11**, p. 236). Further increase of the pressure on the concentrated solution, beyond the osmotic pressure, causes reversal of the osmotic flow. Pure solvent passes from the

solution, through the membrane, in opposite direction into the solvent compartment (**Fig.** 3-12).
In fact only the difference in osmotic pressure between the feed solution and the product has to be
overcome. The reversal of the osmotic phenomenon is the basis of the reverse osmosis process of
desalination.

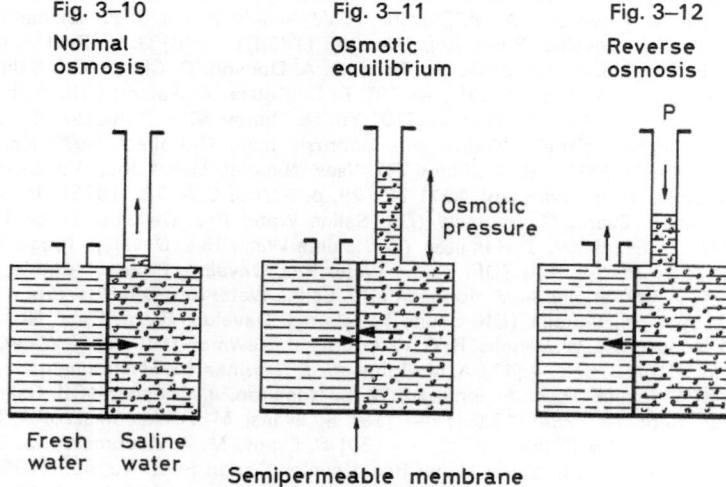

Fig. 3-10   Fig. 3-11   Fig. 3-12
Normal   Osmotic   Reverse
osmosis   equilibrium   osmosis

Reverse osmosis principle of operation.

Reverse osmosis has some analogies with filtration, in that both remove other substances from
a liquid. As the matter removed by reverse osmosis is in solution, the process is sometimes termed
"hyperfiltration", when membranes are used which are highly semipermeable with respect to low-
molecular weight solutes. On the other hand "ultrafiltration" uses membranes with little or no
selectivity for such solutes and is mainly applied for the separation of colloids or macromolecules
from low molecular weight solutes.

Another differentiation is that in reverse osmosis or hyperfiltration there is usually a significant
osmotic pressure to overcome, whereas in the ultrafiltration processes the osmotic pressure is
insignificant. Hence ultrafiltration is performed at low pressures, usually less than 5 kg/cm², whereas
reverse osmosis is carried out at high pressures, typically 50 kg/cm².

The major attractions of the reverse osmosis process from an economic point of view are its
simplicity and the relatively low energy consumption. However, an increased resistance requirement
is imposed upon the material due to the pressure differential on both sides of the membranes.

Geschichte
der umge-
kehrten
Osmose

3.11.1 History of reverse osmosis

Knowledge of osmotic phenomena dates back more than two centuries and was concentrated
to direct osmotic flow. The prospect that reversal of osmotic flow may prove a commercially feasible
method of obtaining fresh water from saline sources gave an important impulse in the study of
osmotic phenomena, the development of improved osmotic membranes being the main target.
Membrane phenomena have extensively been studied and an important body of theory has gradually
evolved to explain and correlate these phenomena.

In 1953 Reid, of the University of Florida, suggested to the Office of Saline Water that a serious
investigation of reverse osmosis, as a potential desalination method, be undertaken [1]. Screening
programs were performed to determine the most promising film-forming materials, which would
exhibit semipermeability to seawater salts. Cellulose acetate appeared to have the most promise in
respect with high degree of osmotic semipermeability and flux. Dobry had already suggested,
in her studies of ultrafiltration membrane preparation, the use of saturated aqueous magnesium
perchlorate solutions as a solvent for cellulose acetate [2]. By incorporating acetone as a fourth

component in the casting solution, Loeb and Sourirajan found it possible to fabricate asymmetric cellulose acetate membranes having greatly improved performance [3]. This achievement and the results of other parallel research on membrane development, together with the expectancy that reverse osmosis is the desalting process of the future, caused an intensive research effort in many countries and a flood of publications. Materials other than cellulose acetate and membranes other than the Loeb-Sourirajan type were thoroughly investigated. This tendency is best illustrated by the large and increasing number of reports on reverse osmosis, published by the Office of Saline Water in the series of Research and Development Progress Reports on desalting, as well as by other institutions and individuals.

The worldwide interest in reverse osmosis is also illustrated by the large number of review papers published in scientific and technical journals of many countries. A selection of such general reviews may include papers by Karelin on principles of water desalination by hyperfiltration [4], by Agrawal and Sourirajan on reverse osmosis with particular reference to Loeb-Sourirajan membranes [5], by Sharples presenting an introduction to reverse osmosis [6, 11], by Shenouda treating reverse osmosis as a process for water desalting, waste water treatment and concentration of aqueous solutions [7], by Goodall and Lorch describing the four types of reverse osmosis concepts [8], by Lonsdale outlining separation and purification processes by reverse osmosis [9], by Strathmann describing material separation by pressure filtration with semipermeable membranes [10], by Leightell reviewing the operation and the costs [12], by Kimura describing the desalination by reverse osmosis [13], as well as by Suk Ho Kang [19], Rice [20], Kimura [22], by Peri outlining the basic principles of the process [14], by Moore on water purification by reverse osmosis [15], by Nusbaum et al. on the potential of reverse osmosis for municipal water supply and waste treatment [16], by Cruver on the same subject [17], by Lynch on membrane and ion exchange processes [18], by Lacey on membrane separation processes [21], by Shields on reverse osmosis for municipal water supply [23], by Walch on the reverse osmosis mechanism and membrane development [24], by Baujard wondering if reverse osmosis will keep its promises [25], by Dejmek et al. on ultrafiltration and reverse osmosis [26], by Ardabili on reverse osmosis in the service of mankind [27], by Murkes reviewing the available apparatus [28], by Steiner and Zibinski on reverse osmosis as unit operation [29], by Haase on new principles in chemical engineering on reverse osmosis and ultrafiltration [30], by Kreutzinger on water desalination by means of reverse osmosis [31], by Pepper on reverse osmosis and desalination techniques [32], etc.

Reviews have also been published with special reference to reverse osmosis membranes, including a review by Podall on recent advances on reverse osmosis membranes [33], by Kesting on the preparation, structure and applications of cellulose membranes [34], by Staude on the preparation and properties of reverse osmosis membranes [35], by Hochscherf and Mandre on the preparation of semipermeable membranes and their use in reverse osmosis and ultrafiltration [36], by Yamabe on reverse osmosis membranes and their performance [37], and recently by Lonsdale on recent advances in reverse osmosis membranes [38].

Books and monographs were also published on reverse osmosis and membrane processes in general, such as a pioneer book by Schlögl on "Mass Transfer through Membranes" [61], "Desalination by Reverse Osmosis" edited by Merten [39], "Membrane Processes for Industry" edited by Lacey [40], "Transport Phenomena in Membranes" by Lakshminarayanaiah [41, especially chapters 6, 8 and 9], "Reverse Osmosis" by Sourirajan [42], a review on patent literature in the United States edited by McDermott [43], and "Industrial Processing with Membranes" edited by Lacey and Loeb [44, especially part II, chapters 7 to 12]. The Office of Saline Water has also issued a report on the State of the Art [1969] of reverse osmosis desalting [45]. A monograph containing papers presented at a "Symposium on Reverse Osmosis and Membrane Research" was edited by Lonsdale [46].

Despite an important research and development effort, the practical achievements of reverse osmosis as a desalination process, compared with distillation and electrodialysis, are rather restricted. Pilot plants and some more or less small capacity commercial plants have been put in operation or are under construction and a larger participation of reverse osmosis in the field of water desalination may be recorded in the years to come. On the other hand reverse osmosis has found a promising field of application in other industrial treatments, such as in the pulp and paper industry, in the food industry, in industrial and municipal waste treatment, etc.

3.11.2 Mechanism of reverse osmosis

Early treatments of pressure-driven permeation processes were based on a porous model of the membrane. It was assumed that flow occurs through pores which have a characteristic size distribution. Various other models have also been suggested.

Sourirajan has admitted that the desalination effect arising when saline water is forced through a porous membrane is caused by the absence of ions from the interface between the water and the membrane [48]. Glueckauf explained the behavior of porous membranes by the repulsion of ions which occurs when an aqueous solution is enclosed in a narrow pore of material of low dielectric constant [49]. Another mechanism to account for the fact that cellulose acetate membranes are capable of rejecting salt from aqueous solution was proposed by Banks and Sharples. This mechanism assumes that the membranes are effectively free from pores in the surface layer and that both water and salt are transmitted through this layer by a true diffusion process [50]. Merten has developed the finely porous transport model [39], which combines the concept that flow within the membrane pores occurs by both viscous flow and diffusion with the concept of frictional interactions within the pores of ionic membranes, as proposed by Spiegler [51].

A theory of reverse osmosis, based on non-equilibrium thermodynamics, was presented by Spiegler and Kedem [52]. A mathematical analysis based on boundary layer theory was presented by Srinivasan and Chi Tien for the case of reverse osmosis in multicomponent systems [53]. A finite difference solution for reverse osmosis in turbulent flow was also presented by the same authors [54].

The effect of hydrostatic pressure, applied to a solution on one side of a membrane, on the steady state concentration distribution of permeating solvent inside the membrane has been examined by Rosenbaum and Cotton to distinguish between different mechanisms of permeation. The determined concentration gradients support the view that permeation takes place by diffusion in nonporous membranes [55].

Values for the permeabilities and rejection coefficients and effective pore area-length ratios, measured by Klein et al., generally agreed with the assumed structural differences between the membranes. The capillary-pore theory and the irreversible thermodynamics theory were used to calculate Staverman coefficients [56]. The transport properties of reverse osmosis membranes can be described by the thermodynamics of irreversible processes, if appropriate models are used. Normal cellulose acetate membranes can satisfactorily be described by the solution-diffusion model and modified Loeb-type membranes by the same model if the effective thickness of the membrane is taken into consideration, was concluded by Pusch [57]. Treatment of membrane processes by nonequilibrium thermodynamics was reviewed by Vink and the application of the method to osmosis and ultra-filtration was considered [58].

The selectivity of cellulose acetate membranes in the desalination of water was explained by Dytnerskii et al. on the basis of the capillarity model of semipermeability [59]. The theory of liquid diffusive flow through graded membrane under applied pressure difference was developed and demonstrated on four simple models by Peterlin and Williams [60].

The reactions between membrane forming monomers, water and sodium chloride are expected to give new information on the mechanism of reverse osmosis. Hydrogen bonding and hydrate formation can be studied independently of transport effects. Hydrogen bonding as well as miscibility increase with pressure [47].

Literature to 3.11

[1] E. J. Breton (Off. Saline Water Res. Develop. Progr. Rept. No. 16 [1957]). — [2] A. Dobry (Bull. Soc. Chim. France [5] **3** [1936] 312/8). — [3] S. Loeb, S. Sourirajan (Advan. Chem. Ser. **38** [1963] 117/32). — [4] F. N. Karelin (Desalination 3 [1967] 207/12). — [5] J. P. Agrawal, S. Sourirajan (Ind. Eng. Chem. **61** No. 11 [1969] 62/89).

[6] A. Sharples (Chem. Ind. [London] **1970** 322/4). — [7] F. Shenouda (Rheinstahl-Technik **4** [1970] 178/89). — [8] J. B. Goodall, W. F. Lorch (Elga Progr. **9** No. 2 [1970] 5/12). — [9] H. K. Lonsdale (Progr. Separ. Purif. **3** [1970] 191/232). — [10] H. Strathmann (Chem. Ingr.-Tech. **42** [1970] 1095/102).

[11] A. Sharples (Chem. Ingr.-Tech. **43** [1971] 1253/7). — [12] B. Leightell (Filtr. Separ. **8** [1971] 715/20). — [13] S. Kimura (Hyomen **9** [1971] 237/48; C. A. **77** [1972] No. 168431). —

[14] C. Peri (Ind. Aliment. Agr. [Paris] **88** [1971] 1323/8). — [15] R. E. Moore (Australian Chem. Process. Eng. **25** No. 2 [1972] 9/12).

[16] I. Nusbaum, R. E. Cruver, J. H. Sleigh (Chem. Eng. Progr. **68** No. 1 [1972] 69/70; AIChE [Am. Inst. Chem. Engrs.] Symp. Ser. **68** No. 124 [1972] 270/82). — [17] J. E. Cruver (Mar. Technol. **9** [1972] 216/22). — [18] M. A. Lynch (J. Am. Water Works Assoc. **64** [1972] 711/25). — [19] Suk Ho Kang (Hwahak Konghak **10** [1972] 93/100). — [20] P. C. Rice (Ion Exch. Progr. **11** No. 1 [1972] 1/3).

[21] R. E. Lacey (Chem. Eng. **79** No. 19 [1972] 56/74). — [22] N. Kimura (Shokuhin Kogyo **15** No. 18 [1972] 63/72). — [23] C. P. Shields (Water Sewage Works **119** No. 1 [1972] 64/70). — [24] A. Walch (Chemiker-Ztg. Chem.-Tech. **2** [1973] 7/11). — [25] J. Baujard (Traitement Surface **14** No. 119 [1973] 21/5).

[26] P. Dejmek, P. Eriksson, B. Hallstrom, A. Klima (Kem. Tidskr. **85** No. 3 [1973] 46/8). — [27] M. Ardabili (Oberflächentechnik **1973** 217/20). — [28] J. Murkes (Chem. Tech. [Leipzig] **25** [1973] 601/4). — [29] R. Steiner, E. Zibinski (Verfahrenstechnik [Mainz] **7** [1973] 337/42). — [30] H. Haase (Chem. Anlagen Verfahren **1973** No. 10, p. 61/6).

[31] H. P. Kreutzinger (Chem. Rundsch. [Solothurn] **26** No. 42 [1973] 9/11). — [32] D. Pepper (Effluent Water Treat. J. **13** [1973] 779/84). — [33] H. E. Podall (Chem. Eng. Progr. Symp. Ser. **67** No. 107 [1971] 260/6). — [34] R. E. Kesting (High Polymer **5** [1971] 1233/60). — [35] E. Staude (Chemiker-Ztg. **96** [1972] 27/32).

[36] E. Hochscherf, G. Mandre (Chem. Ingr.-Tech. **44** [1972] 1152/60). — [37] T. Yamabe (Seisan-Kenkyu **24** [1972] 415/20). — [38] H. K. Lonsdale (Desalination **13** [1973] 317/33). — [39] U. Merten (Desalination by Reverse Osmosis, Cambridge, Mass., 1966). — [40] R. E. Lacey (Membrane Processes for Industry, Birmingham, Ala., 1966).

[41] N. Lakshminarayanaiah (Transport Phenomena in Membranes, New York 1969). — [42] S. Sourirajan (Reverse Osmosis, London 1970). — [43] J. McDermott (Desalination by Reverse Osmosis, Park Ridge, N. J., 1970). — [44] R. E. Lacey, S. Loeb (Industrial Processing with Membranes, New York 1972). — [45] Hittman Associates, Inc. (Off. Saline Water Res. Develop. Progr. Rept. No. 611 [1970]).

[46] H. K. Lonsdale (Reverse Osmosis Membrane Research, New York 1972). — [47] W. Schneider, K. Fischbeck (Proc. 4th Intern. Symp. Fresh Water Sea, Heidelberg 1973, Vol. 3, p. 139/40). — [48] S. Sourirajan (Ind. Eng. Chem. Fundamentals **2** [1963] 51/5). — [49] E. Glueckauf (Proc. 1st Intern. Symp. Water Desalination, Washington, D.C., 1965 [1967], Vol. 1, p. 143/56). — [50] W. Banks, A. Sharples (Off. Saline Water Res. Develop. Progr. Rept. No. 143 [1965]).

[51] K. S. Spiegler (Trans. Faraday Soc. **54** [1958] 1408/28). — [52] K. S. Spiegler, O. Kedem (Desalination **1** [1966] 311/26), K. S. Spiegler (Off. Saline Water Res. Develop. Progr. Rept. No. 353 [1968], No. 613 [1970]). — [53] S. Srinivasan, Chi Tien (Desalination **5** [1968] 139/56). — [54] S. Srinivasan, Chi Tien (Desalination **7** [1969] 51/74). — [55] S. Rosenbaum, O. Cotton (J. Polymer Sci. A I **7** [1969] 101/9).

[56] E. Klein, J. K. Smith, R. P. Wendt (J. Polymer Sci. C No. 28 [1969] 209/23). — [57] W. Pusch (Chem. Ingr.-Tech. **41** [1969] 127/8). — [58] H. Vink (Z. Physik. Chem. [Frankfurt] **71** [1970] 51/64). — [59] Yu. I. Dytnerskii, G. V. Polyakov, L. S. Lukavyi (Teor. Osn. Khim. Tekhnol. **6** [1972] 628/31; C.A. **77** [1972] No. 118045). — [60] A. Peterlin, J. L. Williams (J. Appl. Polymer Sci. **15** [1971] 1493/505).

[61] R. Schlögl (Stofftransport durch Membranen, Darmstadt 1964).

3.12 Reverse osmosis membranes

Membranen für die umgekehrte Osmose

The flux capability of a reverse osmosis membrane is characterized by the membrane constant A, which at a given temperature is determined by the membrane properties alone and is defined as

$$A = \frac{J_1}{\Delta p - \Delta \Pi} \text{ in } cm^3/cm^2 \cdot s \cdot atm$$

where J_1 is the water flux, expressed in $cm^3/cm^2 \cdot s$ (gal/ft$^2 \cdot$ day), Δp is the pressure difference between the feed stream and the product stream and $\Delta \Pi$ the osmotic pressure difference between the two streams, both expressed in atm (psi). In the anglo-saxon systems of units A is expressed in gal/ft$^2 \cdot$ day $\cdot$ psi.

Hence, the water flux through the membrane is approximately given by the expression:

$$J_1 = A(\Delta p - \Delta\Pi) = A \cdot p_{eff}$$

This equation illustrates the significance of pressure on the water production rate. Osmotic pressures associated with typical brackish water feeds are in the range of 2 to 10 kg/cm² (30 to 150 psi). The osmotic pressure to be overcome when desalting seawater may be as high as 31 to 42 kg/cm² (450 to 600 psi). To assure a reasonable flow of product water through current high salt rejecting membranes, fluid system pressures in the range of 42 to 56 kg/cm² (600 to 800 psi) are required for brackish water feeds and up to 105 kg/cm² (1500 psi) for seawater feeds.

Other membrane coefficients in common use may be defined as follows:

The salt transfer flux J_2, in cm³/s or lb/h, expressed by the difference of the salt concentrations C_F and C_P in the feed water and the product water, respectively.

The equilibrium distribution coefficients K_1 and K_2 at each membrane interface, respectively, for the distribution of water and of salt between the membrane and the external solution. K_i is the distribution coefficient for component i.

The salt rejection coefficient defined as the fraction of the salt in the feed stream which is rejected and does not pass through the membrane.

The water permeability coefficient expressed in cm³/cm² · s or in m³/m² · day. When expressed in m/day the water permeability coefficient illustrates the osmotic stream as a function of time.

The solute permeability coefficient expressed in mole/cm² · s · atm.

The requirements and testing of membranes for use in desalination by reverse osmosis have been described by Podall in a series of reports. The intrinsic water to salt permeability requirements are quantitatively defined in terms of the principal process parameters. Process parameters include composition and concentration of salts in the feed and product water, pH of feed and product water, temperature of feed and product water, circulation velocity of feed stream over the membrane, applied pressure and effective pressure drop across the membrane from the feed water to the product water side, percentage of feed recovered as product water, and concentration of salts in the rejected brine stream. Characteristic properties of the membrane are the salt rejection coefficient and the flux of the effluent product water through the membrane. A detailed theoretical description of the reverse osmosis process was given with the aim to enable the proper characterization of a candidate reverse osmosis membrane and recommended procedures were suggested in a third report for the testing and evaluation of reverse osmosis membranes [1].

An ideal reverse osmosis membrane would reject all salts contained in the feed water and be highly permeable to the flow of water. However, from the standpoint of both theory and practice these criteria can not be obtained and membranes, which are commercially marketed, may have the following characteristics:

Water permeability in m³/m² · day	0.5 to 0.8	maximum 1.4
Water permeability in gal/ft² · day	12 to 20	maximum 35
Salt rejection ability in %	95 to 97	maximum 99.5
Lifetime	up to 2 years	

The water permeability of a membrane degrades with time because of membrane fouling by scale, contaminants etc., of membrane compaction and compression, due to maintained high pressure which deteriorates the porous membrane structure, and eventually of hydrolytic deterioration of the membrane material, especially cellulose acetate material.

The structure of cellulose acetate membranes, investigated by scanning electron microscopy, is predominantly determined by the nature of the casting solvent. Tubular membranes are more susceptible than sheet membranes to occlude voids in their matrix. The formation of void and large cavities in internal flow tubular reverse osmosis membranes can be eliminated by increasing the acetone-evaporation period or by precipitating the membranes in salt solution rather than in pure water [5]. The density of cellulose acetate membranes precipitated in aqueous solutions increases with the decrease of the water activity of the leaching bath [6]. Large cellular cavities, "bubbles", in membranes have been shown to be aqueous intrusions, which grow during the gelling stage of the membrane formation. Their formation can be modified and sometimes inhibited. By a technique involving the removal of successive layers of an anisotropic cellulose acetate membrane, it is possible to calculate some of the properties of each decrement which characterizes its structure. These are: average pore diameter, area ratio of pores to total surface and pore volume fraction [7]. The formation

of large voids in membranes can be eliminated by lowering the tendency of the nonsolvent to penetrate into the casting solution or increasing the viscosity of the cast solution or creating a thick gel
layer on top of this cast solution [8].

Diffusion and salt permeability measurements by Saltonstall et al. have demonstrated that electrical conductivity is a rapid and reliable tool for evaluating polymeric films for use in reverse osmosis.
Drying conditions were found to exert a marked effect on the value of the salt diffusion coefficient in
cellulose acetate and to have a lesser but still significant influence on the distribution coefficient of
salt between solution and film. The electrical resistance method proved to be particularly useful in
the study of the active layer of swollen membranes [2].

In examining the effects of liquid membranes at the cellulose acetate membrane saline solution
interface, Kesting et al. showed by means of spectrophotometric studies that the effectiveness of
liquid membranes on the semipermeability of underlying gelatinous membrane layers depends in
part on the existence of surfactant-water interactions. It was concluded that semipermeability depends
on the chemical and physical structure of liquid membranes [3]. Swelling, asymmetry, void size, pore
size, rugosity and morphological changes on hydrolysis were also examined by Kesting et al. by means
of scanning electron microscopy, using various types of membranes [4].

Water permeabilities, equilibrium water sorption levels and rates of approach to sorption equilibrium were measured by Vieth et al. for membranes of various materials [9].

Theoretical equations were derived by Toyoshima and Nozaki for the streaming potential, when
a pressure difference is applied across the membrane [102]. A phenomenological expression for the
reverse osmosis streaming potential was presented by Tanny and Kedem. Criteria for membrane
homogeneity were defined, utilizing the equivalent pore radius calculated from the electroosmotic
coefficient and the hydraulic permeability [103]. Streaming potential across charged membranes is
also reported by Tanny et al. [104], and as a probe of water structure in membranes by Tanny [105].

A method was developed by Raridon et al. for rapid measurement of the salt rejecting characteristics of reverse osmosis membranes, based on the theory of concentration polarization in an unstirred
system [10]. An apparatus for the characterization of reverse osmosis membranes, described by
McKinney, provides for simultaneous measurements of product flux and rejection, together with a
subsequent test for the presence of pinhole defects [11]. A new procedure was described by Steymans and Fischer for the measurement of the hydrodynamic permeability of membranes. The measurement of the volume flow through the membrane is refined, so that small volume displacement can be
accurately measured. The method permits the use of low pressures, thus avoiding the need for a
membrane support which may lead to errors [12]. A method of measurement of rejection coefficients,
described by Greco and Astarita, is based on easy measurements of flow rates and pressures. Polarization effects are avoided [13]. An apparatus was presented by Matz and Elata, which has been used
for the direct determination of transport coefficients and compaction characteristics of reverse osmosis
membranes. Results were given for a range of membranes of different characteristics and the effect
of membrane preparation techniques on their performance coefficients is described [14].

Membrane permeabilities were measured by Wendt et al., using a rotating batch dialyzer. An
empirical expression was obtained for calculating boundary layer permeabilities from known values
for the rotational speed of the cell, the density and viscosity of the solutions and the diffusion coefficient of the solute in solution. Data obtained were confirmed by comparing the results with those
determined on a flat plate dialyzer [15]. An experimental reverse osmosis apparatus was described
by Korngold and de Körösy, in which ceramic tubes carry a salt rejecting layer. Inorganic membranes
were prepared by filling the surface layer of the tubes with material containing pores of nearly molecular dimensions [16].

An apparatus for studying the mechanism of membrane formation has been constructed by
Rosenthal et al. The apparatus enables the preparation of polymeric membranes by the "casting-
leaching" method, the control of the physical parameters which affect the structure and properties
of these membranes and the study of the processes taking place during membrane formation. The
measuring system enables, furthermore, direct measurement of the rate of solvent flow out of the
concentrated polymer solution as the membrane is precipitated [17]. A rapid and simple method
employing the laboratory centrifuge was developed by Hollahan et al. for the evaluation of membrane
performance during reverse osmosis. Results were presented for cellulose acetate membranes for
rejection of salt and urea as dissolved solids [18].

The determination of the transport and compaction coefficients of cellulose acetate membranes, as well as of the reflection and the solute transport coefficients were reported and discussed by Matz and Elata [110]. Further work on reverse osmosis was reported by Anderson and Hamilton on the impact of reverse osmosis and progressive mode ion exchange on water pollution [111], and by Loeb et al. on the use of osmotic sink osmosis for dewatering of concentrated brines [112].

In a recent review on the developments of the past several years in the area of reverse osmosis membranes, Lonsdale concludes that the progress made has been substantial, although largely empirical. These developments have occurred along two lines: new membrane materials and new membrane fabrication methods [118].

The most important characteristic properties, materials of preparation and methods of fabrication of reverse osmosis membranes are reviewed in the following sections.

Wasser-transport und Salz-rückhalte-vermögen

3.12.1 Water transport and salt rejection

The water permeability and the rejection of various solutes by anisotropic cellulose acetate membranes has been investigated by a number of groups. Criteria of separation have been sought by Blunk that would enable the prediction of semipermeability from the properties of the solute. Three criteria of separation have been studied: the application of surface tension versus solute concentration characteristics to the Gibbs adsorption equation, the hydrogen bonding ability of the solute, and the permselective nature of the membrane. The Gibbs adsorption equation did not serve as an adequate criterion for predicting the semipermeability of cellulose acetate membranes to aqueous solutions. The qualitative correlation of experimental data indicated that, when hydrogen bonding can occur between the ionic species of a uni-univalent electrolyte, the rejections of electrolytes having a common ion tend to decrease with increase in strength of the hydrogen bonds between the species. When hydrogen bonding cannot occur between the ionic species of a uni-univalent electrolyte, the rejections for a series of electrolytes having a common ion will decrease with increase in the size of the unhydrated counterion. Electrolytes other than uni-univalent will be strongly rejected regardless of hydrogen-bonding characteristics [108].

Transport phenomena involving aqueous solutions of organic compounds have been reported by Kesting and Eberlin [19], and for phenol by Lonsdale et al. [20]. The selective properties of cellulose acetate membranes toward ions found in natural waters were investigated by Erickson et al. [21], as well as by Govindan and Sourirajan [109]. A scale of membrane selectivity was presented by Agrawal and Sourirajan for twelve inorganic and two organic solutes. The performance data for different membranes, solution systems and operating conditions have been calculated [22].

Flow equations for salt and water in double-layer membranes were suggested by Jagur-Grodzinski and Kedem and salt rejection was derived as a function of transport coefficients. Maximum salt rejection observed with modified cellulose acetate membranes ranged from 69 to 99% [23].

General equations describing the transport of water and salt through non-ideal membranes were critically reviewed and further developed by Elata. The equations are integrated for steady state flow across homogeneous, composite and imperfect membranes. The constants of these equations may be transformed to the transport coefficients, which evolve naturally without additional assumptions. Methods are also listed by which the relevant intrinsic characteristics of membranes may be determined [24].

The transport property requirements of membranes for use in the reverse osmosis process were defined by Podall in terms of three phenomenological coefficients: the hydraulic permeability, the solute permeability, and the reflection factor, derived by application of irreversible thermodynamics to the reverse osmosis process. These coefficients were, in turn, quantitatively related to various process parameters, such as concentration of the saline water feed, its flow rate over the membrane surface, the membrane area, pressure and membrane type, as well as to performance parameters, such as concentration of the product water, production rate of the product water, the volume fraction of the saline water feed recovered as product water and the concentration of the brine reject stream [25].

An analysis describing the reverse osmosis process in laminar and turbulent systems was also presented by Gill et al. [26]. Equations describing the transport of water and of binary electrolytes through semipermeable membranes were derived by Bennion and Rhee and the resulting six independent transport parameters were experimentally determined [27]. Flow equations derived by

Osborn and Bennion gave good correlation of cellulose acetate membrane performance in aqueous sodium chloride solutions. The membrane diffusion parameter in the electrolyte flux equation demonstrates a small concentration dependence and was correlated by a simple empirical equation. The membrane diffusion parameter in the water flux equation is concentration independent. Although the fraction of total transport through the membrane is generally small, the coupling terms aid in correlation of membrane transport data. The use of activity differences in place of concentration differences as driving forces contributed to better data correlation [28].

With the help of a simple theoretical model, it was shown by Hodgson that there exists a unique relationship between the rejection of one salt and that of another for a given membrane tightness. If the rejection of a particular salt by a given membrane is known, it is possible to predict the rejections of other salts by the same membrane. The manufacture of membranes to reject a given ion in preference to other ions is not possible. With more than one anion and one cation present in an aqueous solution, each ion is rejected to a different extent. If an ion finds itself in an environment of more permeable ions of like charge, its rejection is increased. If it finds itself in an environment of less permeable ions of like charge, its rejection is decreased. Using this model, the rejections of individual ions in multi-ion solutions can be predicted [29].

The mathematical modeling of a system of coupled mass transfer of binary and ternary solutions across porous membranes in the presence of pressure and concentration gradients was presented by Ramirez. The developed mathematical model was experimentally tested on binary und ternary aqueous solutions of sodium chloride and the simulation results were found in agreement with the experimental data [30].

Transport equations developed by Spiegler and Kedem were used by Kimura to obtain three coefficients, namely hydraulic permeability, solute permeability and reflection coefficient only from reverse osmosis data. To elucidate properties of membrane dense layer more clearly, dense homogeneous membranes were cast and such properties, as solute permeability, diffusion coefficient, water content and distribution coefficient, were determined and used to obtain the relation between hydraulic permeability, reflection coefficient and friction coefficients [31].

Using the linear relations of thermodynamics of irreversible processes, the transport coefficients of mechanical permeability, the osmotic membrane coefficient, the coupling coefficient and the reflection coefficient were measured by Pusch for an asymmetric cellulose acetate membrane with various electrolytes. The experimental findings have shown a strong dependence of the three transport coefficients on solute concentration, which can be attributed to a concentration gradient within the porous sub-layer of the asymmetric membrane. The transport coefficients of an asymmetric membrane depend on the solute concentration on both sides of the membrane rather than on the mean concentration, as would be the case for a homogeneous membrane [32].

Experiments were carried out by Boari et al. to determine the thermodynamic transport coefficients of cellulose acetate membranes, as well as the relationship between these coefficients and the conditions of membrane preparation. The experiments confirmed that the evaporation phase and the heating phase during membrane preparation have a marked effect on the transport coefficients of cellulose acetate membranes [33].

Investigating the variation of the rejection coefficient, Alfani and Drioli have studied the behavior of various electrolytes using cellulose acetate membranes. A detailed analysis of the obtained data was made on the basis of the theories available in the literature and with a model proposed by the authors [34].

Data were collected by Lonsdale et al. and interpreted on the rejection of various salts, including sodium chloride, nitrates, phosphates, boric acid, carbonates and low molecular weight organic compounds. Measures were developed for improving the rejection of certain solutes, to which cellulose acetate is relatively permeable [35]. A noncoupled flow model of solute and water transport was used to compare intrinsic water and solute permabilites and reverse osmosis semipermeability data. Flow coupling was generally not important and some factors affecting the transport properties of cellulose acetate films were examined [36].

A numerical solution was presented by Diez Roche to the unsteady one-dimensional diffusional flow towards a membrane. By proper selection of the dimensionless variables entering in the problem, the solution may lead to a form, that allows for the determination of the rejection coefficient and the permeability constant by a single measurement of the amounts of solvent and solute, that have traversed the unit area of the membrane during a measured time interval [37].

Membranes prepared by Sono et al. were annealed at 70 to 80°C and examined as to their salt rejection capacity. The size and charge of the solute was the most important parameter affecting rejection, which was found to follow the order: Na < Fe < Ca < Mg [38], as well as the order NaCl < MgCl$_2$ < sucrose, consistent with the lyotropic order [100]. The mobilities of alkali and alkaline earth metal chlorides in asymmetric cellulose acetate membranes can be ordered as Li < Na < K > Rb > Cs for alkali metal chlorides and Be > Mg < Ca < Sr > Ba for alkaline metal chlorides [101]. Certain additions to the feed solution of reverse osmosis with cellulose acetate membranes were found to enhance boric acid rejection by complex formation. Results obtained with mannitol additions are consistent with the known complex-formation equilibria [106]. The rejection of carbon dioxide by asymmetric cellulose acetate membranes falls in the order CO_3^{2-} > HCO_3^- > CO_2, implying that above pH about 6 the pH of the product water is lower than that of the feed water. Below pH 5.5 there is a reversal in this tendency, presumably because the membrane is permeable to the hydrogen ion [107].

A model has been proposed by Kiørboe et al. to determine the optimal membrane for the separation of two dissolved substances by means of reverse osmosis. The model is based on certain idealized assumptions, but is nevertheless useful in giving an approximation of separation effectiveness for different membranes from flux and retention data. The model was verified on four different types of cellulose acetate membranes. In addition to the proposed model, the example illustrates how the membrane type, as determined by the annealing temperature and pressure, influence the separation efficiency [39].

A theory was presented by Probstein and coworkers for the salt rejecting characteristics of porous membranes, whose pore size is large compared with molecular dimensions under reverse osmosis conditions. The salt-rejection mechanism was assumed to be an electrokinetic mechanism resulting from charge built-up on the interior surfaces of the material, when in contact with the salt solution. The performance of the porous material was shown to depend on three parameters: the ratio of Debye length of effective pore radius, a dimensionless wall potential related to the zeta potential, and a filtration Peclet number based on the filtration velocity through the pore, the membrane thickness and the diffusion coefficient of the salt in the water. An universal correlation was given for the fractional salt rejection in terms of the three membrane parameters. Experiments were carried out on the salt rejecting properties of compacted clay, through which saline solutions were forced under high pressure [40]. Numerical solutions for the rejections are given for large Peclet numbers and analytical solutions when the Debye length is also large. Good agreement was shown with reverse osmosis data on dynamic membranes for the limiting case of single salts which dissociate into divalent and univalent ions [41].

Konzentra-
tions-
polarisation

3.12.2 Concentration polarization

Salts are carried by the flowing solution to the membrane and, since they are not passing through the membrane, a salt concentration builds up at its surface, which exceeds the concentration in the saline solution. Back diffusion of salt, produced by this concentration gradient, is counterbalanced by the convection of salt to the membrane surface, thus creating a concentration polarization. Besides various other effects, which influence adversely the desalination process, concentration polarization may cause precipitation at the membrane surface of such salts, the solubility limit of which is exceeded.

The boundary layer effects in reverse osmosis have been discussed by Merten [42], by Merten et al. [43], and a theoretical approach for concentration polarization effects was presented by Sherwood et al. [44]. A finite-difference solution was obtained by Brian for salt concentration polarization with laminar flow of saline water between parallel flat membranes. The analysis accounts for the permeation flux falling off as the salt concentration at membrane surface builds up. Average polarization over the length of the membrane is very nearly the same with the constant flux solution [45]. Concentration polarization ratios have been presented for reverse osmosis desalination systems operating at a variety of conditions including laminar and turbulent flow in tubular membranes. For turbulent flow, the polarization ratio decreases as the brine inlet velocity increases. This reduces the extra osmotic pressure due to concentration polarization, but the frictional pressure drop increases correspondingly. For laminar flow the polarization ratio is independent of brine velocity if a given fractional water removal is accomplished in a single pass [46].

A mathematical analysis for the simultaneous development of the velocity and concentration profiles in a reverse osmosis system consisting of two parallel flat membranes was presented by Srinivasan et al. The initial velocity and concentration profiles were assumed to be uniform and the analysis was restricted to the two-dimensional, steady state laminar flows with constant physical properties. The momentum and diffusion equations were solved simultaneously using the approximate integral method. The analysis is applicable to both the entrance region and the fully developed region at large distances from the conduit inlet. Data were presented on the concentration build-up at the wall for a wide range of parameters and typical results for the water flux produced and productive capacity were given. It was shown that the extent of concentration polarization may be significantly different depending on wether the velocity profile in the stream entering the membrane section is parabolic or flat. An explanation was given for the fact that the water flux is greater and the polarization is less in fully developed flow than when the velocity profile is flat at the system entrance [47].

Concentration polarization data for sodium chloride solutions of various concentrations and cellulose acetate membranes were determined by Strathmann in a thin channel laminar flow cell, as well as at a polyelectrolyte resin membrane and tetrabutylammonium chloride solutions of various concentrations. Spacers to modify the flow distribution in the channel and moderate irregularities in channel depth had no significant effect on the concentration polarization in laminar flow. In the analysis of the experimental data, an extrapolation method related to the particular experimental situation has been developed for the determination of membrane characterization data and of the concentration polarization [48]. It has been further concluded that concentration polarization can be controlled with small expenditure of energy in laminar flow and thin channels for membrane permeation fluxes up to 2.4 m³/m² · day (60 gal/ft² · day) and a fractional feed water recovery of more than 50 % [49].

Polarization in turbulent flow was studied by Bixler and Cross, in particular with turbulence induced by membrane surface roughness and by partial obstruction of the channel. Neither of these was found to be significant [50].

The problem of concentration build-up at the surface of a tubular membrane in the mass transfer entry length in fully developed turbulent flow was formulated by Winograd and Solan, taking into account incomplete salt rejection at the membrane and axial variation of the osmotic pressure. A limiting closed-form similarity solution was also presented and used together with known results for the fully developed concentration profile to derive an approximate expression for the development length [51].

A simplified procedure was presented by Srinivasan and Chi Tien for the prediction of concentration polarization in reverse osmosis operation for multicomponent systems by using the information obtained from one-solute system. The method, which is based on the observation that the local Sherwood number for a given species of solute is approximately proportional to the cubic root of its Schmidt number and relatively independent of other parameters, involves the simultaneous solution of n coupled integral equations. The method was tested for two- and three-solute systems and the results were in good agreement compared with those obtained from the more rigorous method [52].

The salt concentration polarization at the phase boundary of external reverse osmosis membranes supported by horizontal and vertical tubes has been studied by Tsao. For horizontal tubes it was found that the salt concentration at the upper stagnation point, the radius and the membrane constant are the factors which affect the salt concentration and water flux distributions. For vertical tubes the membrane constant is the main factor for salt concentration and water flux distributions [53]

Build-up of excess salt concentration in brine adjacent to a reverse osmosis membrane was studied experimentally and theoretically by Williams et al. Rejection loss and diffusion wave phenomena were identified theoretically. For the cell, experiments showed each theory holds within theoretically anticipated range of applicability. In an intermediate range of time and of ratio of osmotic to applied pressure, no theory agrees well with experiment. For the channel, membrane salt concentrations agree well with predictions of an integral theory. The new integral theory can be applied easily in design calculations. Experimental observations of buoyant instability revealed convection patterns. Observations were correlated by convection-jet model. Buoyant convection greatly reduces average excess concentration and increases membrane flow rates [54].

The time of complete polarization in a reverse osmosis process was measured by Drioli and Alfani for an unstirred batch system at constant pressure and the influence of the membrane constant on the complete polarization was determined [55]. A salt flux decay and a salt rejection increase during the first two or three hours of the runs at high applied pressure was recorded. The runs with low applied pressure presented, on the contrary, rejection decline during the first hours of the experiment. The rejection decline along the channel, at constant time, seems to be greatly affected by the inlet-zone disturbance [56]. Experimental results and theoretical analysis were reported by Liu and Williams for a model of an unstirred batch cell, that aids in understanding the performance of desalination equipment. Transient flow in the device is governed by a salt-diffusion equation [57].

The problem of concentration polarization in cases of concentration dependent diffusivity was analyzed by Shah. A numerical solution of the mass transport equation for the laminar flow through two equally permeating, flat, parallel membranes was obtained. Three types of diffusivity-concentration relationships and two values of solute rejections were investigated. The decrease in diffusivity with an increase in concentration was found to increase the value of concentration polarization modulus over that obtained in the case of constant diffusivity under the same system conditions. The increment was found to be larger for the case of a stronger diffusivity-concentration relationship and for a larger magnitude of membrane wall concentration. A method was proposed by which the effect of variable diffusivity on the value of concentration polarization modulus can be calculated for a wide range of practical conditions using the existing theoretical results for the case of constant diffusivity [58]. The utility of drawing off the concentrated boundary layer near the membrane to reduce concentration polarization and thereby decrease product salinity and increase system productivity was demontrated qualitatively by Shaw et al. Concentration polarization and product salinity were reduced most significantly when the ratio of the osmotic pressure of the feed solution to the pressure drop across the membrane approaches zero as its lower limit. System productivity was increased most profoundly as this ratio approaches its upper limit of unity. The technique becomes increasingly effective as the fraction of the feed to be removed in a single pass is increased [59].

The effect of concentration dependent viscosity and diffusivity on concentration polarization was studied theoretically by Doshi et al. using the following three methods: integral method, Leveque solution, and the method of rapidly varying boundary condition. Both the integral method and the method of rapidly varying boundary conditions proved to be satisfactory. The sensitivity of the concentration polarization, productivity and velocity profile to the variation in viscosity and diffusivity increases as the ratio of the osmotic pressure of the feed to the pressure drop across the membrane decreases [60].

A mathematical expression was proposed by Strathmann for superconcentration occurring in reverse osmosis of various salt solutions in a stirred cylindrical cell and for laminar and turbulant flow in a rectangular channel [61].

Concentration dependence of intrinsic rejection of ion-exclusion reverse osmosis membranes was included by Thomas in turbulent flow concentration polarization equations [62]. Markedly reduced concentration polarization appears to occur at Schmidt numbers of about 600 in fully developed flow, when the product water flux (transpiration velocity) exceeds about 0.3 cm/min. Analysis in terms of a simple eddy diffusion model indicates that, under certain experimental conditions, the thickness of the concentration boundary layer is appreciably reduced [63].

A mathematical model for the simulation of concentration polarization was also proposed by Alfani and Drioli [64], and an analytical model was described for the calculation of the concentration profiles in reverse osmosis processes. The model is based on the hypothesis of a reaction, which is variable in time [65].

A test cell for measuring the concentration polarization boundary layer in a reverse osmosis system was devised by Goldsmith and Lolachi. Arrangements were made permitting access to the boundary layer that is formed on the membrane surface and governing equations were given. In continuing research the main results achieved were the development of Ag-AgCl electrodes that respond to chloride ions in a Nernstian fashion which were found to be insensitive to pressure and velocity, the development of a test cell for simultaneous and direct measurements of concentration at the surface of the membrane, product water salinity, product water flow rate and the membrane compaction in an one-dimensional reverse osmosis desalinating system and a study and direct measurement of concentration polarization boundary layer profile in a practical two-dimensional desalinating module [66]. Research concerning concentration polarization and fouling reduction control was continued by Lolachi. It was shown that fluidized beds were compatible with reverse osmosis membranes and did not damage the active membrane surface. Application of a fluidized bed diminishes the concentration polarization to such extent that at low Reynolds number, i. e. 50 to 250, potable water better in quality than in some systems operating with turbulent flow was obtained. High water recoveries were obtained, for example, from an 1.5 m (5 ft) long tube with $3/4$ inch outer diameter; 30 to 50% product water recoveries were obtained at pressures ranging between 42 and 105 kg/cm² (600 to 1500 psi) [67].

A study by Srinivasan and Chi Tien of the two-dimensional boundary layer equations describing reverse osmosis in a flat membrane duct revealed the existence of maximum concentration polarization phenomenon corresponding to a particular value of salt rejection ratio, which is contrary to the common belief that the extent of concentration polarization always increases as membranes become more perfect. An one-dimensional model was used to analyze the same problem and explicit relationship was presented between membrane characteristics, operating variables and the presence of maximum concentration polarization [68]. Boundary layer theory was used by Muhlenkamp and Srinivasan to analyze a flat horizontal membrane system with solution flowing below the membrane. Varying wall flux in the longitudinal direction and natural convection of the solute from the membrane surface were considered. The resulting momentum and diffusion equations were then solved simultaneously using the integral method. The results indicate a reduction in concentration polarization as the effect of natural convection increases [69].

Analytical and experimental studies of reverse osmosis system by Bansel et al. include investigations on the effect of concentration dependent viscosity and diffusivity on concentration polarization and the increased productivity by reduction of concentration polarization in laminar flow [113].

A review on concentration polarization in reverse osmosis systems was recently presented by Pusch [70].

3.12.3 Compaction of membranes

When a large pressure driving force is used, the porosity of the membrane layer may decrease with distance from the high pressure side and the water flux through the membrane may decrease as well with time, because of membrane compaction. The effect of pressure and the resulting membrane permeability was studied by Harriot and Michelsen for cellophane and polyvinyl alcohol films [71].

Compaction of cellulose acetate membranes appears to be related to creep of plastics under mechanical stress. Wet compressive creep studies of cross-linked, pigment-reinforced or higher molecular weight cellulose acetate films indicated that only pigment reinforcement was promising. An anhydrous organic-coated aluminum silicate (Burgess KE) was the most effective pigment for reducing wet compressive creep, but affected unfavorably salt rejection. The original value of 98% dropped to 87%. A solution to this problem was found by coating the Burgess KE reinforced membrane with cellulose acetate before water immersion. Long term tests (320 h) gave water fluxes of 0.4 to 0.5 m³/m² · day (10 to 12 gal/ft² · day) and maximum salt rejections ranging from 97 to 98% [72]. The nature and the properties of the fillers to reduce wet compressive creep of cellulose acetate membranes were also discussed [73].

A model of compaction for a composite membrane has been developed by Lonsdale and coworkers. Improvement of membrane compaction resistance by adding filler particles to the casting solution appeared to be of limited value and a technique difficult to perfect [74]. The curves for the permeability of cellulose acetate membranes as a function of pressure showed that the residual deformation is much less after 48 h than after 24 h. This was explained by the ability of the membrane to partially recover its structure with time [75].

Polymeric gels were synthesized by crosslinking cellulose acetate and were used to replace a portion of the linear cellulose acetate in a standard asymmetric membrane, with the aim of reducing long-term flux decline. Testing was made on a 3.5% NaCl solution at 105 kg/cm² (1500 psi) for up to 400 h. Appropriate solution polymerized gels reduced the compaction slope by about a factor of two, without seriously compromising salt rejections. By casting somewhat thinner membranes, the fluxes could be maintained at control level. Experiments on a series of dense cellulose acetate membranes of different thickness revealed that essentially all the hydrodynamic resistance in an asymmetric membrane is in the dense salt-rejecting layer and further indicated that this is the locus of flux decline processes [76].

Several types of filled cellulose acetate membranes were prepared to determine the effect of the filler properties and polymer properties on permeability of the composite materials. Various cellulose acetates and silica gels were used. Casting procedures were chosen to give a dense cellulose acetate phase and a uniform distribution of filler throughout the membrane. The composite membranes were evaluated in a reverse osmosis cell. The permeability to water increased with increasing filler concentration and the effect was greater when the polymer permeability was low. For cellulose acetate with 39.8% acetyl content, 50 volume percent filler gave a 5-fold increase in water flux and 96% salt rejection [77].

The preparation of membranes for demineralizing brackish waters and seawater with retention of at least 80% of their initial flux after one year of operation was studied by Hoernschemeyer et al. Two membranes, one made from a blend of cellulose acetates and one made from crosslinkable cellulose acetate methacrylate, were selected for development to meet these goals. Improved performance of both types of membranes was sought by investigating the relationship between membrane properties, polymer solubility and the solubility parameters of the casting solution components. The cellulose acetate blend membranes gave an initial flux of 1.42 $m^3/m^2 \cdot$ day (35 $gal/ft^2 \cdot$ day) with 97% salt rejection. However, these membranes contained small voids, which were found to rupture at pressures of 42 kg/cm^2 (600 psi) or larger. Attempts to eliminate voids from cellulose acetate methacrylate membranes have been unsuccessful [78].

Alterung und Faulen

3.12.4 Degradation and fouling

A type of cellulose acetate membrane failure has been observed, which is characterized by an exponential increase in both the salt and water permeability of the membrane. The symptoms strongly suggested attack by microorganisms. In an investigation, carried out by Cantor and coworkers, three sources of potentially destructive organisms were cultured and 26 different microbial isolates were obtained. Freshly prepared, sterilized membranes were exposed to each of the 26 isolates and showed significant losses in salt rejection of certain samples after two months. Losses up to 50% of the acetyl content of the membranes were observed. Cellulose triacetate membranes are more resistant to biological attack under conditions identical to those causing degradation of diacetate membranes with a degree of substitution of 2.3 to 2.5 [79]. In a continuing investigation, losses in permeability of cellulose acetate membranes were related with biological degradation [80].

A study was conducted by Kissinger and Willits to develop means to prevent microbial destruction of reverse osmosis membranes, used in the concentration of maple sap [81]. Practical methods were established by Lawrence et al. to protect reverse osmosis units from bacterial attack. Five biocides —formaldehyde, an organic mercury compound, methylene dithiocyanate and two quaternary ammonium salts— were chosen for long term storage tests. The storage resulted in some reduction of the salt rejection but little effect was noted upon the membrane flux. Formaldehyde appears to be the best agent to use. The method of membrane manufacture has no effect on the behavior of membrane in biocide. Temperatures of -18 and $-10°C$ (0 and 14°F) were found as limits to which membranes could be subjected for months, without a change in performance or in physical properties. Dry membranes were tested with a limited amount of success [82].

Cellulose acetate membranes were stored in 20 different solutions for periods of time up to 45 months. Where the pH of the storage solution was controlled to minimize hydrolysis, the samples showed good transport properties after storage. Although growth of microorganisms occurred in some storage solutions, there was no evidence that these organisms changed the transport characteristics of the membrane. Copper(II) sulfate controlled such growth and did not damage the membrane. Benzoic acid attacked the membrane. Chlorine was an effective desinfectant. At low pH, as much as 95% of the chlorine was transported through the membrane. However, in high concentrations chlorine attacked the membrane. Stabilized chlorine dioxide was not readily transported through the cellulose acetate membrane [83].

Fouling of membranes may also occur by the formation of deposits on their surface with an adverse effect on their performance. Investigating the mechanism of formation, Jackson and Landolt studied the rate of fouling by deposition of iron hydroxide on tubular reverse osmosis membranes under closely controlled experimental conditions resembling those of brackish water desalination. Feed brines of approximately 5000 mg/l NaCl were used, to which small amounts of iron hydroxide were added. The effect of brine flow rate, product water flux, pH and oil contamination on the rate of fouling layer build-up was investigated [84].

Effects of fouling in plant performance can be observed in both flux and rejection decline. The effect of axial velocity on fouling characteristics of a commercial cellulose acetate membrane was investigated by Sheppard and Thomas. Feed was demineralized water or untreated river water spiked with magnesium chloride to give solution concentrations of 0.04 to 0.08 M. Substantially no flux decline was observed with demineralized water feed in a 93 h test run at 7.3 m/s (24 ft/s). Even with untreated river water feed, containing 10 to 100 mg/l suspended solids, flux decline was relatively modest at the same circulation velocity. Reduction of circulation velocity to 0.5 m/s (1.64 ft/s) resulted in a marked decline in flux with time [85]. An attempt was also made to predict quantitatively the rejection of reverse osmosis membranes, which are covered by uniform layers of fouling material. Uniform permeable plastic films were used as a substitute for fouling layers [86]. The rate of flux decline varied with feed velocity, when tubular reverse osmosis modules were operated with the same river feed water, but with different feed velocities. Operation at feed velocities in excess of the threshold velocity prevented accumulation of particulates on the membrane and the rate of flux decline was significantly less than predicted by extrapolation based on the half-power relationship [87].

The development of regenerative methods to recover the flux and salt rejection properties of service-deteriorated cellulose acetate membranes was investigated by Cantor et al. Various in situ simplified one-step processes have been successfully developed for both mentioned objectives [88]. Laboratory investigations were carried out by Hamer and Kalish to determine the feasibility of removing scale from cellulose acetate membranes and restoring their performance. Descaling was tested on flat membranes coated with deposits of iron(III)oxide and calcium sulfate. Dissolving the iron component, calcium sulfate was loosened and removed. No chemical degradation of the membrane was apparent. Best regeneration resulted from heat soaking the compacted membranes at a temperature approaching the originally used in their preparation [89].

The ambient temperature regeneration process with compacted or service-deteriorated blend membranes is capable of nearly complete restoration of the osmotic properties. The successful repair of damaged cellulose acetate membranes using colloidal solutions of guar gum has provided another means for extending the service life of reverse-osmosis units which because of leaks might otherwise require dismantling and remembraning [90]. Another way of in situ regeneration of a sealed reverse osmosis system was accomplished by Littman et al. by removal of the spent membrane with a solvent and deposition of a new one from a casting solution. There was no deterioration of performance, demonstrating the usefulness of the in situ deposition concept. Several membrane support materials were investigated for this purpose. Fluxes for different membranes ranged from 0.16 to 1.6 m³/m² · day (4 to 40 gal/ft² · day), salt rejections from 53 to 97%. Parametric studies were carried out to determine the effects of higher pressures, higher salt concentrations and multivalent ions. Membranes cast in situ had a structure similar to that of flat membranes [91].

3.12.5 Diffusion   *Diffusion*

In an investigation by Richardson of thermal diffusion for saline water conversion, mass diffusion properties of seawater were found markedly dependent on temperature. Their variation with concentration was relatively slight. Both theoretical and experimental evidence indicated that ions diffuse at different rates and sometimes in opposite directions, depending upon whether the ion is in a binary or multicomponent ionic solution [114]. The principles and practice of the diaphragm cell technique for the study of the diffusion of electrolytes were outlined by Janz and Mayor [115].

The diffusion coefficient of a salt in homogeneous nonionic membrane may readily be evaluated from the electrical conductivity of the membrane, when it is immersed in an aqueous salt solution. The diffusion coefficient of sodium chloride in dense isotropic cellulose acetate membranes has been measured by this method by Saltonstall et al. and the activation energy for diffusive salt transport has been obtained from the temperature dependence. From the effect of heat treatment on the electrical conductivity of anisotropic gel membranes, an estimate of the thickness of the active layer was obtained [92].

The system $H_2O-MgCl_2-NaCl$ was examined by Wendt and Shamim and values for diffusion coefficients were reported. These can be compared with values predicted from theory for multi-ion systems containing a divalent ion. A method was described for determining diffusion coefficients with moderate accuracy [116].

Equations were derived by Karelin to determine the permeability coefficients for particular ions through cellulose acetate membranes. The investigated ions can be classified in three groups on the basis of an accepted membrane porosity [93]. The diffusion and permeation constants for water and sodium chloride were measured and reported for cellulose acetate and nylon-6 membranes [94]. Experimental data on a small laboratory two-chambered diffusion cell for studying the permeability and the performance of different types of reverse osmosis membranes were also presented by the same group of authors [95]. Various techniques were used by Williams to obtain solutions to the diffusion equation of flow through a semipermeable membrane under hydrostatic pressure difference [96].

The effect of natural convection on polarization and flow patterns in liquid phase convective diffusion in a vertical duct with semipermeable membrane walls has been investigated theoretically by Ramanadhan and Gill. At low flow rates, gravitational fields can play a significant role in distorting the velocity profiles and natural convection also affects mass transfer rates. Results were presented for both momentum and mass transfer in upward and downward flows for different wall Peclet numbers. The hydrodynamic stability of the system has been investigated and critical values of the buoyancy parameters were reported. The analysis is of practical interest in reverse osmosis and other membrane separation processes [97]. Equations governing the effect of longitudinal diffusion in specified reverse osmosis systems were also presented by Ohya and Sourirajan [98].

Henderyckx has described a modified heat exchanger with a vapor permeable and a vapor not permeable membrane and with a vapor space between them. A vapor gradient is then formed, instead of a pressure gradient like in reverse osmosis. Evaporation is effected by diffusion through the membranes [117].

Examining the diffusion-layer structure in reverse osmosis channel flow, Hendricks and Williams have measured salt concentration profiles in brine adjacent to the membrane with electrical conductivity microprobes for fully developed two-dimensional channel flow in a closed-return water tunnel. Cellulose acetate membranes were employed with various solutions at hydrostatic pressures from 10 to 40 atm and with Reynolds numbers from 137 to 1365. The excess concentration was found to vary exponentially with distance normal to the membrane. An integral technique, developed on the basis of an exponential-profile assumption, was found to provide reasonable agreement between theoretical and experimental concentration profiles. This integral theory gave good values for the wall concentration [99].

Literature to 3.12

[1] H. E. Podall (Off. Saline Water Res. Develop. Progr. Rept. No. 255, No. 267, No. 268 [1967]). — [2] C. W. Saltonstall, F. C. Burnett, W. S. Higley, W. M. King, A. L. Vincent (Off. Saline Water Res. Develop. Progr. Rept. No. 232 [1967]). — [3] R. E. Kesting, W. J. Subcasky, J. D. Paton (J. Colloid Interface Sci. 28 [1968] 156/60), R. E. Kesting, W. J. Subcasky (J. Macromol. Sci. Chem. 3 [1969] 151/5). — [4] R. E. Kesting, M. Engdahl, W. Stone (J. Macromol. Sci. Chem. 3 [1969] 157/67). — [5] M. A. Frommer, D. Lancet (Off. Saline Water Res. Develop. Progr. Rept. No 774 (Pt. 1) [1972] 1/31).

[6] M. A. Frommer, R. Matz, U. Rosenthal (Ind. Eng. Chem. Prod. Res. Develop. 10 [1971] 193/6; Off. Saline Water Res. Develop. Progr. Rept. No. 774 (Pt. 4) [1972] 68/82). — [7] R. Matz (Desalination 10 [1972] 1/15, 11 [1972] 207/15; Off. Saline Water Res. Develop. Progr. Rept. No. 774 (Pt. 5/6) [1972] 83/116). — [8] M. A. Frommer, R. M. Messalem (Ind. Eng. Chem. Prod. Res. Develop. 12 [1973] 328/33). — [9] W. R. Vieth, A. S. Douglas, R. Block (J. Macromol. Sci. Phys. 3 [1969] 737/49). — [10] R. J. Raridon, L. Dresner, K. A. Kraus (Desalination 1 [1966] 210/24).

[11] R. McKinney (Anal. Chem. 41 [1969] 1513/6). — [12] C. Steymans, H. Fischer (Desalination 6 [1969] 203/14). — [13] G. Greco, G. Astarita (Quad. Ing. Chim. Ital. 6 No. 3 [1970] 49/53). — [14] R. Matz, C. Elata (Proc. 3rd Intern. Symp. Fresh Water Sea, Dubrovnik 1970, Vol. 2, p. 511/34). — [15] R. P. Wendt, R. J. Toups, J. K. Smith, N. Leger, E. Klein (Ind. Eng. Chem. Fundamentals 10 [1971] 406/11).

[16] E. Korngold, F. de Körösy (Desalination 11 [1972] 125/7). — [17] U. Rosenthal, J. Nechushtan, A. Kedem, D. Lancet, M. A. Frommer (Desalination 9 [1971] 193/200). — [18]

J. Hollahan, T. Wydeven, R. P. MacGullough (J. Appl. Chem. Biotechnol. **23** [1973] 669/74). —
[19] R. E. Kesting, J. Eberlin (J. Appl. Polymer Sci. **10** [1966] 961/7). — [20] H. K. Lonsdale,
U. Merten, M. Tagami (J. Appl. Polymer Sci. **11** [1967] 1807/20).

[21] D. L. Erickson, J. Glater, J. W. McCutchan (Ind. Eng. Chem. Prod. Res. Develop. **5** [1966]
205/11). — [22] J. P. Agrawal, S. Sourirajan (Ind. Eng. Chem. Process Design Develop. **8** [1969]
439/49). — [23] J. Jagur-Grodzinski, O. Kedem (Desalination **1** [1966] 327/41). — [24] C. Elata
(Off. Saline Water Res. Develop. Progr. Rept. No. 291 [1968]; Desalination **6** [1969] 1/12). —
[25] H. E. Podall (Off. Saline Water Res. Develop. Progr. Rept. No. 303 [1968]).

[26] W. N. Gill, L. J. Derzansky, M. R. Doshi (Off. Saline Water Res. Develop. Progr. Rept.
No. 403 [1969]). — [27] D. N. Bennion, B. W. Rhee (Ind. Eng. Chem. Fundamentals **8** [1969]
36/48). — [28] J. C. Osborn, D. N. Bennion (Univ. of California School Eng. Appl. Sci., Los Angeles
Rept. No. 69-49 [1969]; Ind. Eng. Chem. Fundamentals **10** [1971] 273/80). — [29] T. D. Hodgson
(Desalination **8** [1970] 99/138). — [30] W. F. Ramirez (Ind. Eng. Chem. Process Design Develop.
10 [1971] 210/4).

[31] S. Kimura (Proc. 4th Intern. Symp. Fresh Water Sea, Heidelberg 1973, Vol. 4, p. 197/206). —
[32] W. Pusch (Proc. 4th Intern. Symp. Fresh Water Sea, Heidelberg 1973, Vol. 4, p. 321/32; Chem.
Ingr.-Tech. **45** [1973] 1216/22). — [33] G. Boari, C. Merli, G. Mossa, R. Passino (Proc. 4th Intern.
Symp. Fresh Water Sea, Heidelberg 1973, Vol. 4, p. 49/64). — [34] F. Alfani, E. Drioli (Proc. 4th
Intern. Symp. Fresh Water Sea, Heidelberg 1973, Vol. 4, p. 13/23). — [35] H. K. Lonsdale, C. E.
Milstead, B. P. Cross, F. M. Graber (Off. Saline Water Res. Develop. Progr. Rept. No. 447 [1969]).

[36] H. K. Lonsdale, B. P. Gross, F. M. Graber, C. E. Milstead (J. Macromol. Sci. Phys. **5** [1971]
167/87). — [37] J. T. Diez Roche (Proc. 3rd Intern. Symp. Fresh Water Sea, Dubrovnik 1970,
Vol. 2, p. 361/70). — [38] M. Sono, Y. Kurokawa, N. Yui (Technol. Rept. Tohoku Univ. **37** [1972]
181/92; C.A. **79** [1973] No. 9685). — [39] L. Kiørboe, C. E. Boesen, G. Jonsson (Desalination
12 [1973] 35/43). — [40] G. Jacacio, R. F. Probstein, A. A. Sonin, F. Yung (Off. Saline Water Res.
Develop. Progr. Rept. No. 809 [1972]; J. Phys. Chem. **76** [1972] 4015/23).

[41] R. F. Probstein, A. A. Sonin, D. Yung (Proc. 4th Intern. Symp. Fresh Water Sea, Heidelberg
1973, Vol. 4, p. 309/20). — [42] U. Merten (Ind. Eng. Chem. Fundamentals **2** [1963] 229/32). —
[43] U. Merten, H. K. Lonsdale, R. L. Riley (Ind. Eng. Chem. Fundamentals **3** [1964] 210/3). —
[44] T. K. Sherwood, P. L. T. Brian, R. E. Fisher, L. Dresner (Ind. Eng. Chem. Fundamentals **4** [1965]
113/8; Off. Saline Water Res. Develop. Progr. Rept. No. 508 [1970] 115/20). — [45] P. L. T. Brian
(Off. Saline Water Res. Develop. Progr. Rept. No. 145 [1965]; Ind. Eng. Chem. Fundamentals **4**
[1965] 439/45).

[46] P. L. T. Brian (Proc. 1st Intern. Symp. Water Desalination, Washington, D.C., 1965 [1967],
Vol. 1, p. 349/70). — [47] S. Srinivasan, Chi Tien, W. N. Gill (Off. Saline Water Res. Develop.
Progr. Rept. No. 243 [1967]). — [48] H. Strathmann (Off. Saline Water Res. Develop. Progr. Rept.
No. 336 [1968]). — [49] H. Strathmann, B. Keilin (Desalination **6** [1969] 179/201). — [50] H. J.
Bixler, R. A. Cross (Off. Saline Water Res. Develop. Progr. Rept. No. 469 [1969]).

[51] Y. Winograd, A. Solan (Desalination **7** [1969] 97/109). — [52] S. Srinivasan, Chi Tien
(Desalination **7** [1970] 133/45). — [53] C. K. Tsao (Desalination **8** [1970] 243/58). — [54] F. A.
Williams, T. J. Hendricks, M. K. Liu (Off. Saline Water Res. Develop. Progr. Rept. No. 622 [1970]). —
[55] E. Drioli, F. Alfani (Quad. Ing. Chim. Ital. **6** [1970] 179/85).

[56] E. Drioli, F. Alfani, G. Iorio (Proc. 4th Intern. Symp. Fresh Water Sea, Heidelberg 1973,
Vol. 4, p. 115/24). — [57] M. K. Liu, F. A. Williams (Intern. J. Heat Mass Transfer **13** [1970] 1441/57).
— [58] Y. T. Shah (Intern. J. Heat Mass Transfer **14** [1971] 921/30). — [59] R. A. Shaw, R. Deluca,
W. N. Gill (Desalination **11** [1972] 189/205). — [60] M. R. Doshi, A. K. Dewan, W. N. Gill (AIChE
[Am. Inst. Chem. Engrs.] Symp. Ser. **68** No. 124 [1972] 323/39).

[61] H. Strathmann (Chem. Ingr.-Tech. **44** [1972] 1160/7). — [62] D. G. Thomas (Ind. Eng.
Chem. Fundamentals **11** [1972] 302/7). — [63] D. G. Thomas (Ind. Eng. Chem. Fundamentals **12**
[1973] 395/405). — [64] F. Alfani, E. Drioli (Chim. Ind. [Milan] **54** [1972] 235/40). — [65]
E. Drioli, F. Alfani, G. Iorio (Chim. Ind. [Milan] **53** [1971] 674/5).

[66] H. Goldsmith, H. Lolachi (Off. Saline Water Res. Develop. Progr. Rept. No. 527 [1970],
No. 727 [1971]). — [67] H. Lolachi (Off. Saline Water Res. Develop. Progr. Rept. No. 843 [1973]). —
[68] S. Srinivasan, Chi Tien (Desalination **13** [1973] 287/301). — [69] S. P. Muhlenkamp, S. Srinivasan (Proc. 4th Intern. Symp. Fresh Water Sea, Heidelberg 1973, Vol. 4, p. 251/8). — [70] W. Pusch
(in: H. K. Lonsdale, Reverse Osmosis Membrane Research, New York 1972, p. 43/58).

[71] P. Harriott, D. L. Michelsen (Off. Saline Water Res. Develop. Progr. Rept. No. 330 [1968]). —
[72] B. Baum, S. A. Margosiak, W. H. Holley (Off. Saline Water Res. Develop. Progr. Rept. No. 467
[1969]; Ind. Eng. Chem. Prod. Res. Develop. **11** [1972] 195/9). — [73] D. R. Deanin, B. Baum,
S. A. Margosiak, W. H. Holley (Ind. Eng. Chem. Prod. Res. Develop. **9** [1970] 172/5). — [74] H. K.
Lonsdale, R. L. Riley, L. D. La Grange, C. R. Lyons, A. S. Douglas, U. Merten (Off. Saline Water
Res. Develop. Progr. Rept. No. 484 [1969] 71/84). — [75] Yu. I. Dytnerskii, G. V. Polyakov, S. L.
Zakharov (Khim. Prom. **48** [1972] 504/5; C.A. **77** [1972] No. 118066).

[76] S. L. Rosen, C. Irani, L. Baayens (Off. Saline Water Res. Develop. Progr. Rept. No. 791
[1972]), L. Baayens, S. L. Rosen (J. Appl. Polymer Sci. **16** [1972] 663/70). — [77] F. Klunker,
P. Harriot (AIChE [Am. Inst. Chem. Engrs.] Symp. Ser. **68** No. 124 [1972] 340/8), P. Harriot, J. Wu,
F. Klunker (Off. Saline Water Res. Develop. Progr. Rept. No. 846 [1973]). — [78] D. L. Hoernsche-
meyer, C. W. Saltonstall, O. S. Schaeffler, A. J. Secchi (Off. Saline Water Res. Develop. Progr. Rept.
No. 849 [1973]). — [79] P. A. Cantor, B. J. Mechalas, O. S. Schaeffler, P. H. Allen (Off. Saline
Water Res. Develop. Progr. Rept. No. 340 [1968]). — [80] P. A. Cantor, B. J. Mechalas (J. Polymer
Sci. C No. 28 [1969] 225/41).

[81] J. C. Kissinger, C. O. Willits (Food Technol. **24** [1970] 481/4). — [82] R. W. Lawrence,
B. J. Mechalas, P. H. Allen, R. R. Jay (Off. Saline Water Res. Develop. Progr. Rept. No. 673 [1971]). —
[83] K. D. Vos, I. Nusbaum, A. P. Hatcher, F. O. Burris (Desalination **5** [1968] 157/66). — [84]
J. M. Jackson, D. Landolt (Univ. of California, Los Angeles, UCLA-ENG-7266 [1972]; Desalination
12 [1973] 361/78). — [85] J. D. Sheppard, D. G. Thomas (Desalination **8** [1970] 1/12).

[86] J. D. Sheppard, D. G. Thomas (AIChE [Am. Inst. Chem. Engrs.] J. **17** [1971] 910/5). —
[87] J. D. Sheppard, D. G. Thomas (Desalination **11** [1972] 385/98). — [88] P. A. Cantor, W. S.
Highley, C. W. Saltonstall (Off. Saline Water Res. Develop. Progr. Rept. No. 601 [1970]). — [89]
E. A. G. Hamer, R. L. Kalish (Off. Saline Water Res. Develop. Progr. Rept. No. 471 [1969]). — [90]
W. S. Highley, C. W. Saltonstall (Off. Saline Water Res. Develop. Progr. Rept. No. 694 [1971]).

[91] F. E. Littman, H. K. Bishop, R. McMillan (Off. Saline Water Res. Develop. Progr. Rept. No. 701
[1971]), F. E. Littman, H. K. Bishop, G. Belfort (Desalination **11** [1972] 17/30). — [92] C. W.
Saltonstall, W. M. King, D. L. Hoernschemeyer (Desalination **4** [1968] 309/27). — [93] F. N.
Karelin (Proc. 3rd Intern. Symp. Fresh Water Sea, Dubrovnik 1970, Vol. 2, 467/86). — [94] G. R. G.
Steiner, R. F. Eaton, T. K. Kwei, A. V. Tobolsky (U.S. Clearinghouse Fed. Sci. Tech. Inform. AD 714 899
[1970]). — [95] G. R. Garbarini, R. F. Eaton, T. K. Kwei, A. V. Tobolsky (J. Chem. Educ. **48** [1971]
226/30).

[96] F. A. Williams (SIAM [Soc. Ind. Appl. Math.] J. Appl. Math. **17** No. 1 [1969] 59/73). —
[97] K. Ramanadhan, W. N. Gill (AIChE [Am. Inst. Chem. Engrs.] J. **15** [1969] 872/84). — [98]
H. Ohya, S. Sourirajan (AIChE [Am. Inst. Chem. Engrs.] J. **15** [1969] 780/2). — [99] T. J.
Hendricks, F. A. Williams (Desalination **9** [1971] 155/80). — [100] N. Nakagawa, Y. Kurokawa,
N. Yui (Technol. Rept. Tohoku Univ. **37** [1972] 207/17; C.A. **79** [1973] No. 9688).

[101] K. W. Choi, D. N. Bennion (Univ. of California, Los Angeles, Rept. Eng-7370 [1973]). —
[102] Y. Toyoshima, H. Nozaki (J. Phys. Chem. **73** [1969] 2134/41). — [103] G. Tanny, O. Kedem
(Proc. 4th Intern. Symp. Fresh Water Sea, Heidelberg 1973, Vol. 4, p. 395/405). — [104] G. Tanny,
E. Hoffer, O. Kedem (Experimentia Suppl. No. 18 [1971] 619/30). — [105] G. Tanny (Nature **242**
[1973] 474/5).

[106] F. M. Graber, H. K. Lonsdale, C. E. Milstead, B. P. Cross (Desalination **7** [1970] 249/58). —
[107] C. E. Milstead, A. B. Riedinger, H. K. Lonsdale (Desalination **9** [1971] 217/23). — [108]
R. Blunk (Univ. of California, Los Angeles, Dept. Eng. Rept. No. 64-28 [1964]). — [109] T. S.
Govindan, S. Sourirajan (Ind. Eng. Chem. Process Design Develop. **5** [1966] 422/9). — [110]
R. Matz, C. Elata (Off. Saline Water Res. Develop. Progr. Rept. No. 543 [1970]).

[111] J. R. Anderson, R. S. Hamilton (AIChE [Am. Inst. Chem. Engrs.] Symp. Ser. **68** No. 124
[1972] 349/55). — [112] S. Loeb, N. Zelinger, M. R. Bloch (Proc. 4th Intern. Symp. Fresh Water
Sea, Heidelberg 1973, Vol. 4, p. 235/41), S. Loeb, M. R. Bloch (Desalination **13** [1973] 207/15). —
[113] B. Bansel, R. DeLuca, L. Derzansky, A. K. Dewan, M. R. Doshi (Off. Saline Water Res. Develop.
Progr. Rept. No. 854 [1973]). — [114] J. L. Richardson (Off. Saline Water Res. Develop. Progr.
Rept. No. 107 [1964]). — [115] G. J. Janz, G. E. Mayor (Off. Saline Water Res. Develop. Progr.
Rept. No. 196 [1966]).

[116] R. P. Wendt, M. Shamim (Off. Saline Water Res. Develop. Progr. Rept. No. 504 [1969]). —
[117] Y. Henderyckx (Desalination **3** [1967] 237/42; Chem. Ingr.-Tech. **41** [1969] 124/7). —
[118] H. K. Lonsdale (Desalination **13** [1973] 317/32).

3.13 Cellulose acetate membranes

A reverse osmosis desalination membrane must have a high degree of semipermeability for the solvent, an adequate water flux at reasonable pressures and must keep the above characteristics for a time as long as possible.

The original development of the anisotropic cellulose acetate membrane has been described by Loeb and Sourirajan. In the fabrication technique developed, the casting solution was a mixture (by weight) of 22.2% cellulose acetate, 1.1% magnesium perchlorate, 10.0% water and 66.7% acetone. The membrane was cast on a glass plate at a cast thickness of 0.25 mm. The casting solution and all casting components were kept between -5 and $-10°C$. After an acetone evaporation period of 3 minutes at this temperature, the membrane-plate assembly was immersed in ice water for at least an hour. The membrane was then removed from the glass plate and heated by immersion in water at 75 to 85°C. The performance of the membrane, fabricated in this manner, was 0.2 to 0.45 $m^3/m^2 \cdot$ day (5 to 11 gal/ft$^2 \cdot$ day) water under a pressure differential of 105 to 140 kg/cm^2 (1500 to 2000 psi) from a brine containing 5.25% sodium chloride [1].

Loeb and Manjikian have reported improvements by the addition of small quantities of a fifth component, such as hydrochloric acid [2]. Other cellulose esters, including cellulose acetate-butyrate and cellulose propionate, were used with some success in place of cellulose acetate, but the membranes obtained did not exhibit better performances than that of cellulose acetate [3]. A number of useful membrane casting solutions containing only nonelectrolyte components was described. The ternary mixture (by weight) 25% cellulose acetate, 30% formamide and 45% acetone was found to be the most useful. Membranes made from this mixture perform better than those fabricated from solutions containing electrolytes and casting can be accomplished at room temperature [4].

Membranes prepared by these techniques are said to be anisotropic or to have a dense surface layer (skin), which might have a thickness of 5 to 12% of the total membrane thickness. This layer is the active layer of the membrane. It was admitted that all water present in this layer is bound to the hydroxyl groups of cellulose acetate. Salt is insolube in such water and is rejected. The bulk of the membrane contains large quantities of capillary water, which permits the movement of water through the body of the membrane at rapid rates [5].

The quantities of bound water contained in desalination membranes, before and after heat and pressurization treatments, vary in a manner similar to that of the total water content. The permeabilities of a number of inorganic salts vary inversely with the extent of hydration of the component ions. The permeabilities of a number of sterically similar compounds with different functional groups vary with the hydrogen-bonding capacities of the solute. The decrease in desalination characteristics, which occurs upon storage in saline solution, has been interpreted on the basis of microbial decomposition rather than hydrolysis due to the brine itself [5].

An extensive research program was carried out by Merten and coworkers on the flow relationships and boundary-layer effects in reverse osmosis, the theories of membrane transport, the properties of cellulose acetate membranes, the preparation and properties of alternate membrane materials and the economics of the reverse osmosis process [6]. Cellulose acetate membranes and their permeability to both water and salts, the structure and method of formation of modified membranes, the variables affecting the lifetime of these membranes, membrane compaction and the preparation of thin films have been studied in depth. Membrane materials which might be used as alternates for cellulose acetate have also been searched by the same group of authors [7].

Earlier work had indicated that pyridine, as a casting solution ingredient, increased the strength of semipermeable membranes cast from cellulose acetate solutions. Pyridine was most effective in the range of 5 to 25% by weight, with cellulose acetate, acetone and formamide as the other casting solution ingredients. Besides strength, toughness and elongation of the membrane were improved. Formulation, evaporation time and curing temperature were found to be more critical for solutions containing pyridine. Aging of the casting solution for about ten days before casting seemed to improve membrane performance. The most valuable and practical result of pyridine addition appears to be a significantly superior membrane performance at higher operating pressures. A casting solution employing a combination of glycerol and n-propanol as a flux promoter was also developed [60]. Data on the activity of acetone for the binary system acetone–formamide and on a specific ternary system cellulose acetate–acetone–formamide were given by Ohya and Sourirajan [61].

An analysis of existing data indicated that a significant portion of salt leakage may take place through regions of very low polymer density or defects in the matrix. Experiments with three types of cellulose acetate membranes, a high salt rejection low flow rate, a low rejection high flow rate and an intermediate rejection intermediate flow, showed that modified cellulose acetate membranes become progressively more leak to salt as the water permeability increases and that the pressure dependence of the salt flux increases with increasing water permeability. The observed salt flux behavior is qualitatively consistent with three transport mechanisms: purely diffusive transport without coupling to the water flux, diffusive transport with coupling, and hydraulic defect or pore flow combined with diffusive flow. Several techniques were used to improve the membrane salt rejection characteristics. Precipitation of insoluble salts inside the membrane matrix was not so successful. The addition of water soluble polymers to the saine feed stream, such as polyvinyl methyl ether and copolymers of polyethylene oxide and propylene oxide, was extremely effective in increasing the salt rejection efficiency. It appears that the additive tends to precipitate or absorb on the membrane, where it forms a second barrier membrane acting in series with the barrier layer of the cellulose acetate [8].

A generalized barrier model for the separation of dissolved components was presented by Podall. The mechanism of membrane separation processes via selective permeation by means of naturally formed or artificially introduced separation barriers was discussed and the basic morphological and structural requirements of reverse osmosis membranes were outlined [9]. The significance of the membrane barrier effect for the permeation of water through semipermeable membranes was also discussed by Fischbeck [10].

By varying the concentration of formamide a series of membranes having a wide range of performance characteristics was obtained by Kesting and Manefee [11]. Thickness of the salt rejecting region of a cellulose acetate membrane can be obtained by measuring the electrical capacitance. The salt rejecting layer becomes more efficient and thicker on heat treatment [12].

The asymmetric behavior of air-dried cellulose acetate membranes was investigated by Gröpl and Pusch. The membranes were effective in desalination only if the air-dried surface was juxtaposed with the brine. The experiments proved that the asymmetric behavior of modified cellulose acetate membranes is the result of concentration polarization. It was also proved that the film-diffusion-theory model predicts salt rejection behavior [13].

As it might be concluded from the reported work, the principal parameters in the fabrication of useful desalination membranes are: composition of the casting solution; temperatures at which casting, drying and immersion of the membrane are effected; thermal soak as a post-treatment of the cast membrane. A major effort in early work was devoted to understanding the events, which occur in the preparation of membranes. The presence or absence of the membrane salt appears to have some effect on the flux. The temperature of membrane treatment is important to the optimum membrane capability.

Modified cellulose acetate membranes contain large quantities of water in their pores. If this water is allowed to evaporate under ambient conditions, the membranes suffer an irreversible loss in desalination properties. In order to avoid deterioration, a method was conceived, by which cellulose acetate membranes are soaked in a surface active agent and then dried at either ambient or elevated temperatures, up to 110°C. Upon being rewet, the membranes did not show any loss in desalination or physical properties as compared with the undried control membrane. All types of surface active agents will function well. Because the dry membranes are somewhat brittle, glycerin and ethylene glycol have been used as plasticiziers for the dry membrane [14].

Experimental data support the hypothesis that the surface layer of the asymmetric Loeb-Sourirajan type membranes has a heterogeneous microporous structure. A general method was proposed by Kopecek and Sourirajan for improving the performance by pumping pure water past the back side of the membrane. The membrane is then used in the reverse osmosis experiments in the normal manner with the active surface layer facing the feed solution. Smaller pores on the surface layer are opened more than the bigger pores during the back side operation, involving structural changes in the film during backpressure treatment [15].

A simple method suggested by Agrawal and Sourirajan for predicting the performance of Loeb-Sourirajan type porous cellulose acetate membranes for low concentrations of mixed solutes in aqueous feed solution systems involves two or more inorganic salts with a common ion. The method requires only data on membrane specifications and the applicable mass transfer coefficient correlation for the corresponding single solute systems [16]. A methodology to determine the performance characteristics of flat asymmetric noncellulosic membranes has also been developed. A mathematical analysis to model the expected radial flow and to predict the concentration polarization has been performed. Experiments defined the validity limits of the model [17].

The effects of casting solution composition and evaporation period on the performance of resulting porous cellulose acetate membranes have been discussed by Kunst and Sourirajan in terms of casting solution structure, solvent evaporation rate during film formation and the film shrinkage temperature profile [69]. Several sets of porous cellulose acetate membranes were made at various casting solution temperatures and solvent evaporation conditions. The productivity of the films for brackish water desalination was 100 to 150% higher than that of other porous cellulose acetate membranes at 18 kg/cm² (250 psi) or less. The obtained results are of practical importance in low pressure reverse osmosis applications [18]. It is emphasized in a further paper, that even small changes in casting solution temperature can bring about significant changes in solution structure and hence membrane performance. A quantitative confirmation is presented of the governing significance of the solution structure-evaporation rate concept, relating to the mechanism of phase separation and pore formation at the membrane surface [19]. In further development membranes for low pressure reverse osmosis applications were cast from cellulose acetate-acetone-dioxane-aqueous magnesium perchlorate solutions. The membrane obtained showed slightly better performance than that previously prepared. As they were made on the basis of the solution structure-evaporation rate concept on the mechanism of reverse osmosis membrane formation, their development and good performance appears to confirm the validity of this concept [20]. Changes in reverse osmosis membrane performances, caused by including dioxane as a substitution for a part of acetone in a casting dope, were studied and discussed by the same group of authors [21]. Preparation of cellulose acetate membranes by the asymmetric film making technique and details on the casting conditions were also reported by Kunst and Floreani [22].

Membranes cast from acetone and dioxane are dense and structurally homogeneous. Membranes cast from acetic acid, dimethyl formamide or dimethyl sulfoxide are all skinned. They consist of a dense cellulose acetate layer, which is supported by a porous substructure. The location of the skin depends upon the nature of the solvent. The porosity of cellulose acetate membranes decreases considerably when annealed at temperatures higher than 90°C. A considerable reduction in water permeability and increase in selectivity after annealing is the case. A mechanism based on the assumption that the density of each layer of the membrane is determined by the concentration of the corresponding layer of the cast polymer solution at the precipitation point has been suggested. The volume concentration of the polymer at the precipitation point has been correlated to the concentration of nonsolvent required for precipitating cellulose acetate and to the direction and magnitude of osmotic flows of liquids into or out of the cast solution during leaching [23].

The study of a Loeb-Sourirajan type membrane leads to the conclusion that the dense layer of the asymmetric membrane contains two types of porosity. Water and NaCl fluxes are predominantly governed by the molecular porosity. Annealing causes all types of pores to shrink. The activation energies of the permeation of water and all the electrolytes remain independent of the annealing treatment [24]. Membranes made from binary mixtures of cellulose acetate and cyclohexanone consistently gave porosity factor values of 2.9. If a ternary mixture was used, containing a swelling agent such as formamide, a lower porosity factor value was obtained which first increased on annealing. The result was interpreted by Glueckauf and Russell by the disappearance of a vast number of very fine interstices during annealing, so that the mean pore size shows an increase in spite of the fact that all pores get narrower during the annealing [25].

The optimum performance of cellulose acetate membranes, cast from a dioxane based dope, is associated with subtle changes in the morphology of the upper layers, in contact with the salt solution, of the three-layer structure. The formation of the membrane is affected from the casting conditions and the heat treatment. The non-reproducibility of the system is compatible with the instability of the dope, the complexity of the morphology and its sensitivity to evaporation conditions during casting [26]. In a study of membranes, cast from an acetone-formamide dope, a three-layer membrane structure was also observed and was thought to have similar origins in the casting process. Both transmission and scanning microscopy are required for a detailed analysis of all aspects of membrane morphology. Transmission microscopy reveals more information on the active surface layer, while scanning microscopy was shown to reveal more of the texture of the bulk of the membrane [27]. The effect of pH and ionic strength on the crystallization and morphology of polymer cellulose acetate membranes was investigated by Mandelkern [28].

The water in a membrane plays a significant role in the efficiency of the salt rejection and the chemical structure of the membrane exerts some influence over the water structure. Solvent structure enforcement can be associated with increases in activity coefficients and decreases in solubility. Cellulose acetate has a stronger structure making influence than cellulose or partially acetylated cellulose. The mechanism of salt rejection postulates the formation of a salt free water layer immediately adjacent to the membrane surface. The thickness of water layer in relation to the diameter of the pores, through which water passes, determines the salt rejection of the membrane [29]. A combination of adsorption and calorimetric measurements was made to study structuring of water at cellulose acetate surfaces of varying polarity. Cellulose acetate powders with varying degrees of acetylation were used. The results obtained support experimental evidence from diverse sources of a relation between solvent structuring at interfaces and solute rejection [30].

The quaternary phase diagram of the system cellulose acetate–acetone–formamide–water has proved to be a useful tool in the discussion of membrane structures and properties. A mechanism has been suggested to explain the observed asymmetry in the membrane structure. X-ray diffraction and electron microscopy were used to supplement flux and retention data of the prepared membranes [31]. A phase diagram of the cellulose acetate–formamide–acetone casting system has also been determined by Fahey and Grethlein, which gives the region of natural solubility of this system. By varying casting solution composition and holding time in a systematic way, improvement in water flux at a given level of salt rejection has been obtained. Statistically designed experiments have been helpful in optimizing these two variables [32].

The performance and shrinking temperature profile of cellulose acetate-acetone-formamide prepared membranes were affected by the casting solution composition, acetone concentration in the casting atmosphere, evaporation time, evaporation rate constant and remoteness of the casting solution from corresponding phase boundary composition [33].

Investigating the structure of cellulose acetate membranes, Kepinski and Chlubek measured their electrical potential and electrical resistance. The potentials were found to depend on the position of the membranes and the potential curves show extreme values for membranes heated at about 80°C or more. This may indicate a discontinuity in the structure of water contained in the membranes [34].

In a study of the chemical and physical variables, associated with the preparation of Loeb-type cellulose acetate membranes, and the effect of these variables on membrane performance, a primary-to-secondary free hydroxyl ratio in cellulose acetate polymer of about 0.65 was found optimal. Salt rejection rates of 90% or better were obtained. Molecular weights around 40 000 are desirable. Casting speed and evaporation time were found to be particularly important [35].

Loeb-type reverse osmosis membranes can also be formed without the necessity of heating the membranes, if the immersion of the membrane is conducted in special electrolyte solutions rather than in water. Membrane performance was highly dependent on electrolyte concentration, on anion and cation size and valence. However, membranes made by this method were not as good as standard cellulose acetate membranes [36].

By using differential scanning calorimetry in determining the relative amounts of frozen and nonfrozen water in cellulose acetate membrane of different structures, a large fraction of the water remained in the unfrozen state at a temperature of −60°C. The amounts of nonfreezing water decrease with the increase in the packing density of polymer within the membrane. Nuclear magnetic resonance techniques proved to be a powerfull tool for studying freezing of water in membranes [37].

Conformational properties of the cellulose acetate molecule and aggregation behavior of cellulose acetate in concentrated solutions may have appreciable influence on the properties of a cast film. The thermodynamics and conformational properties of various cellulose esters in dilute solution were reviewed and data, including the temperature dependence of the second virial coefficient, the mean square radius of gyration and the intrinsic viscosity, obtained on fractions of cellulose acetate with a 2.45 degree of substitution were reported by Tanner et al. Cellulose acetate is strongly associated in dilute solution in most of the solvents examined. Light scattering measurements on concentrated solutions of cellulose acetate showed that interchain aggregation becomes more pronounced as the polymer concentration is increased [38].

The melting temperatures and heats of fusion have been determined by Kokta et al. for cellulose acetate in the form of a powder, dense film, ultra-thin film and reverse osmosis membrane. Powdered samples showed broad melting peaks with low heats of fusion. Higher heats of fusion and sharper peaks were obtained from films prepared by the dipping technique. Varying the sample techniques did not change the value for the heats of fusion. Still higher heats of fusion were obtained when the cellulose acetate was cast from acetone. Some increase in the heats of fusion resulted when the latter samples were heat treated at 85°C. The heats of fusion of commercial membranes were independent of the reverse osmosis properties. It was suggested that the rejection layer of Loeb-type membranes has the same degree of crystallinity as the porous substrate [39].

A time–temperature equivalence exists for the ultimate elongation and birefringence of 2,5-acetyl cellulose acetate. This equivalence can be extended below the glass transition temperature. An empirical method has been developed to obtain a master curve of reduced birefringence as a function of reduced elongation, independent of elongation ratio, rate and temperature of stretching [40]. A time–temperature equivalence exists also for the transport characteristics of non-asymmetric cellulose acetate membranes. The desalinated water flux through oriented low flux membranes shows a decrease with increasing rate of stretching and degree of orientation. The performance characteristics of unstretched and stretched membranes indicate that the membrane properties depend on the chemical nature of the treatment media, temperature and, in the case of stretched membranes, also on the degree of elongation and on the rate of stretching. Desalination studies of low flux cellulose acetate membranes heat treated in water or in glycerine-water mixtures revealed that the membrane characteristics depend on the chemical nature of the treatment medium and its temperature. The water uptake and the effect of quenching temperature on the desalination characteristics of these membranes showed consistent trends. Apparently, the multiple secondary transitions of cellulose acetate seem to justify the experimental observations [41].

Neutron spectroscopic investigations were carried out by Safford and Leung to systematically study and determine, at a molecular level, the factors and properties that control the diffusion and bonding of water in dense skin cellulose acetate membranes. Data were taken both to determine the functional form of the diffusive kinetics, that characterizes the transport of water molecules in membranes, and then study the influence on the bonding and kinetics of the individual water of extent of membrane hydration, number of hydrophilic groups, variations in molecular weight, temperature, the presence of all included salt, heat treatment and compaction, as well as their relationship to the casting process. To complement the neutron scattering results, differential scanning calorimetric measurements were made to determine the transitions characteristics of the dense layer materials [42].

Statistically designed experiments were used by Grethlein for optimizing the casting variables and improve the performance of cellulose acetate membranes. By adjusting the heat treatment, membranes cast at the optimum conditions gave 94 to 96% rejection and 1.6 to 1.2 $m^3/m^2 \cdot$ day (40 to 30 $gal/ft^2 \cdot$ day) product water [43]. The variation in flux and rejection of replicate cellulose acetate membranes was measured within a membrane sheet as well as between membranes. Although the casting conditions were carefully controlled and various influences were studied, the source of variation in replicates was not identified. It was shown that the cause of variation is not due to hand or machine casting technique, casting solution composition, heat treatment, dust, overall film thickness or pressure history, and it is still not known what causes variations within and between replicates when all the variables are fixed [44].

A dynamically formed membrane was prepared by Wakakuwa et al. from a cellulose acetate and formamide solution at an annealing temperature of 70 to 80°C. Additives of Zr, Th, Al, Si, and polyethyleneimine had marked effects on salt rejection, but results were not reproducible [52].

Experimental studies on a series of homogeneous cellulose acetate membranes of graded porosity have indicated that predicted values of membrane pore size form a reliable indicator of membrane functional properties, as regards solute rejection, diffusive versus bulk flow, membrane compaction, the presence or absence of a surface layer, etc. [45]. Average pore radius in cellophane and cellulose acetate Loeb-type membranes has been estimated by Volgin by comparing water filtration and evaporation rates through the membranes. The results have been verified by forcing air through wetted membranes [46].

The selectivity and the mechanical properties of cellulose acetate membranes depend on the degree of substitution of cellulose acetate, the cellulose acetate concentration in solution, the solvent concentration in the precipitating bath and the annealing time of the membranes. Increasing the degree of substitution of the cellulose acetate, elastic modulus, tensile strength and breaking elongation were increased. Moisture content decreased. Membrane selectivity decreased by increasing alcohol concentration in the bath [47].

The degree of substitution of cellulose acetate has an effect on the membrane performance. An attempt was made to elucidate the possible inter-relationship of the substituted acetyl groups in connection with the rejection phenomenon [49].

A two-stage method of making semipermeable high flux reverse osmosis membranes was developed by Carter et al. using water-ethanol mixtures to precipitate the cellulose acetate. This eliminated the need for heat treatment and produced membranes with fluxes up to 5 $m^3/m^2 \cdot$ day and sodium chloride rejections up to 85%. Their properties are compared with membranes made by the three-stage method involving heat treatment and show more sensitivity to the effects of pressure [50].

Micellar morphology, which may constitute a definition of asymmetry in a common class of reverse osmosis membranes, was suggested by Panar et al. The morphology of the permeate side of the membrane, in contrast to that of the functioning feed side, can vary with composition and preparation [51].

Reverse osmosis membranes exhibit NaCl rejection even before heat treatment. This would seem to indicate the presence of an incipient dense layer, which is subsequently consolidated by the heating process [48].

Optimum conditions for the preparation of cellulose acetate membranes from dimethylformamide solutions were investigated by Gulyamov et al. The heat treatment conditions depend on the parameters of the preceding stages of preparation. The duration of heat treatment does not affect membrane properties, but temperature does. With increasing temperature water flux decreases and salt rejection increases to a maximum and then drops. Membranes prepared from 20% cellulose acetate and 80% dimethylformamide exhibited a water flux of 0.2 to 0.3 $m^3/m^2 \cdot$ day at a salt rejection less than 95%, measured with 10 g/l NaCl solution at 50 kg/cm^2. Varying the casting conditions a membrane was obtained exhibiting a water flux of 0.525 $m^3/m^2 \cdot$ day at a salt rejection of 94% [53].

Asymmetric microporous membranes of both wet and dry process types were considered by Kesting from the standpoint of the effect of process variables upon structural and functional performance characteristics and ease to handling. In the dry process, developed by the author, the solution is simply cast and allowed to dry to completion. It yields membranes, which are wet-dry reversible and amenable to direct study, even in their formative phases. The polymer solutions, from which the membranes are frabricated, were investigated by freeze-quenching, lyophilization and scanning electron microscopy. It has been demonstrated both that the emulsoid nature of the membrane gel originates in its sol precursors and that differences between skin and substructure are quantitative rather than qualitative in nature [54].

A study has been undertaken by Noël and Monnerie of the physicochemical phenomena taking place during evaporation, coagulation and heat treatment in the fabrication procedure of cellulose diacetate membranes from casting solution containing inorganic salts. Requirements of experimental conditions (casting solution composition, temperature and time of acetone evaporation, nature of immersion bath) in order to produce membranes having good desalination capability were determined [55]. The effect of the preparation conditions on the water content of Loeb-type membranes was also studied. A correlation between swelling and yield was established [56]. In further work the water concentration required for precipitating the polymer from concentrated solutions and the time necessary for coagulation were determined [57].

Further work on the behavior of cellulose acetate membranes is reported by Brun et al. [58], on swelling of these membranes by Trudelle and Nicolas [59], by Mason et al. on cellulose acetate dialysis membranes [62], by Re and Bennion on relative mobilities of alkali cations in modified cellulose acetate membranes [63], by Boeddeker and Kaschemekat on the preparation and characterization of cellulose acetate membranes [64], by Serafini et al. discussing the effect of poly(styrenesulfonic acid) on the electret effect in cellulose acetate membranes [65], by Nakagaki and Miyata on the potential and permeability coefficient of cellulose acetate membranes [66], by Wakakuwa et al. on the membrane potential of these membranes used for desalination by reverse osmosis [67], and by Furuta et al. on the sorption of water vapor by cellulose acetate membranes used for reverse osmosis [68].

Literature to 3.13

[1] S. Loeb, S. Sourirajan (Univ. of California, Los Angeles, Dept. Eng. Rept. No. 60-60 [1960]; Advan. Chem. Ser. **38** [1963] 117/32). — [2] S. Loeb, S. Manjikian (Univ. of California, Los Angeles, Dept. Eng. Rept. No. 63-22 [1963]). — [3] S. Loeb, F. Milstein (Dechema Monograph. **47** [1962] 707/33), S. Loeb (Univ. of California, Los Angeles, Dept. Eng. Rept. No. 62-26 [1962]). — [4] S. Manjikian, S. Loeb, J. W. McCutchan (Proc. 1st Intern. Symp. Water Desalination, Washington, D.C., 1965 [1967], Vol. 2, p. 159/79). — [5] B. Keilin (Off. Saline Water Res. Develop. Progr. Rept. No. 84 [1963], No. 117 [1964]).

[6] U. Merten (Proc. 1st Intern. Symp. Water Desalination, Washington, D.C., 1965 [1967], Vol. 1, p. 275/91), H. K. Lonsdale, U. Merten, R. L. Riley, K. D. Vos, J. C. Westmoreland (Off. Saline Water Res. Develop. Progr. Rept. No. 111 [1964]), H. K. Lonsdale, U. Merten, R. L. Riley, K. D. Vos (Off. Saline Water Res. Develop. Progr. Rept. No. 150 [1965]). — [7] U. Merten, H. K. Lonsdale, R. L. Riley, K. D. Vos (Off. Saline Water Res. Develop. Progr. Rept. No. 208 [1966]). — [8] L. L. Markley, R. A. Cross, H. J. Bixler (Off. Saline Water Res. Develop. Progr. Rept. No. 281 [1967]). — [9] H. E. Podall (Advan. Colloid Interface Sci. **2** [1969] 331/46). — [10] K. Fischbeck (Chem. Ingr.-Tech. **41** [1969] 121/4).

[11] R. E. Kesting, A. Manefee (Kolloid-Z. Z. Polymere **230** [1969] 341/6). — [12] C. P. Bean (Off. Saline Water Res. Develop. Progr. Rept. No. 465 [1969]). — [13] R. Gröpl, W. Pusch (Desalination **8** [1970] 277/92). — [14] K. D. Vos, F. O. Burris (Off. Saline Water Res. Develop. Progr. Rept. No. 348 [1967]; Ind Eng. Chem. Prod. Res. Develop. **8** [1969] 84/9). — [15] J. Kopecek, S. Sourirajan (J. Appl. Polymer Sci. **13** [1969] 637/57).

[16] J. P. Agrawal, S. Sourirajan (Ind. Eng. Chem. Process Design Develop. **9** [1970] 12/8). — [17] J. P. Agrawal, C. R. Antonson, N. W. Rosenblatt (Desalination **11** [1972] 71/90). — [18] B. Kunst, S. Sourirajan (Desalination **8** [1970] 139/52; J. Appl. Polymer Sci. **14** [1970] 723/33). — [19] B. Kunst, S. Sourirajan (Natl. Res. Council Can. Rept. No. 11 523 [1970]; J. Appl. Polymer Sci. **14** [1970] 1983/96). — [20] B. Kunst, A. M. Basnec, G. Arneri (Proc. 4th Intern. Symp. Fresh Water Sea, Heidelberg 1973, Vol. 4, p. 217/25).

[21] B. Kunst, G. Arneri, P. Goran, A. M. Basnec (Off. Saline Water Res. Develop. Progr. Rept. No. 840 [1973]). — [22] B. Kunst, B. Floreani (Kolloid-Z. Z. Polymere **251** [1973] 600/2). — [23] R. Bloch, M. A. Frommer (Desalination **7** [1970] 259/64), M. A. Frommer, I. Feiner, O. Kedem, R. Bloch (Desalination **7** [1970] 393/402), R. Bloch, M. A. Frommer, O. Kedem, I. Feiner, A. Lancet (Off. Saline Water Res. Develop. Progr. Rept. No. 499 [1969] 1/41). — [24] E. Glueckauf, D. C. Sammon (Proc. 3rd Intern. Symp. Fresh Water Sea, Dubrovnik 1970, Vol. 2, p. 397/422). — [25] E. Glueckauf, P. J. Russell (Desalination **8** [1970] 351/7).

[26] G. J. Gittens, P. A. Hitchcock, D. C. Sammon, G. E. Wakley (Desalination **8** [1970] 369/91). — [27] G. J. Gittens, P. A. Hitchcock, G. E. Wakley (Desalination **12** [1973] 315/22). — [28] L. Mandelkern (Off. Saline Water Res. Develop. Progr. Rept. No. 718 [1971]). — [29] W. V. Johnston, J. Greyson (Off. Saline Water Res. Develop. Progr. Rept. No. 707 [1971]). — [30] M. Korsch, W. V. Johnston (Off. Saline Water Res. Develop. Progr. Rept. No. 837 [1972]).

[31] H. Strathmann, P. Scheible, R. W. Baker (J. Appl. Polymer Sci. **15** [1971] 811/28). — [32] P. M. Fahey, H. E. Grethlein (Desalination **9** [1971] 297/313). — [33] R. Pilon, B. Kunst, S. Sourirajan (J. Appl. Polymer Sci. **15** [1971] 1317/34). — [34] J. Kepinski, N. Chlubek (Proc. 3rd Intern. Symp. Fresh Water Sea, Dubrovnik 1970, Vol. 2, p. 487/95). — [35] F. S. Model, H. J. Davis, A. A. Boom, C. Helfgott, L. A. Lee (Off. Saline Water Res. Develop. Progr. Rept. No. 657 [1971]).

[36] E. Zisner, S. Loeb (Proc. 3rd Intern. Symp. Fresh Water Sea, Dubrovnik 1970, Vol. 2, p. 615/28). — [37] M. A. Frommer, D. Lancet, M. Shporer (U.S. Natl. Tech. Inform. Serv. AD 733 467 [1971]), M. A. Frommer, D. Lancet (J. Appl. Polymer Sci. **16** [1972] 1295/303). — [38] D. W. Tanner, G. C. Berry, J. Borch, T. G. Fox (Off. Saline Water Res. Develop. Progr. Rept. No. 737 [1971]). — [39] B. Kokta, F. Luner, R. Suen (Off. Saline Water Res. Develop. Progr. Rept. No. 829 [1972]). — [40] E. A. Meinecke, D. V. Mehta (Desalination **10** [1972] 341/52; Reverse Osmosis Membrane Research, New York 1972, p. 111/29).

[41] D. V. Mehta, E. A. Meinecke (Desalination **10** [1972] 353/68, 369/82). — [42] G. J. Safford, P. S. Leung (Off. Saline Water Res. Develop. Progr. Rept. No. 834 [1972]). — [43] H. E. Grethlein (Proc. 4th Intern. Symp. Fresh Water Sea, Heidelberg 1973, Vol. 4, p. 147/57). — [44] H. E. Grethlein (Desalination **12** [1973] 45/60). — [45] M. N. Sarbolouki, I. F. Miller (Desalination **12** [1973] 343/59).

[46] V. D. Volgin (Kolloidn. Zh. **35** [1973] 546/9; Colloid J. [USSR] **35** [1973] 506/9). — [47] L. P. Perepechkin, V. P. Dubyaga, N. I. Naimark, B. A. Fomenko (Vysokomol. Soedin. A **15** [1973] 1173/7; C.A. **79** [1973] No. 93619). — [48] R. Matz (Desalination **12** [1973] 273/5; Off. Saline Water Res. Develop. Progr. Rept. No. 774 (Pt. 7) [1972] 117/25). — [49] A. Yi Yan (Proc. 4th Intern. Symp. Fresh Water Sea, Heidelberg 1973, Vol. 4, p. 443/8). — [50] J. W. Carter, G. Psaras, M. T. Price (Desalination **12** [1973] 177/88).

[51] M. Panar, H. H. Hoehn, R. R. Hebert (Macromolecules **6** [1973] 777/80). — [52] M. Waka-kuwa, Y. Kurokawa, N. Yui (Technol. Rept. Tohoku Univ. **37** [1972] 193/206; C.A. **79** [1973] No. 9686). — [53] T. Gulyamov, B. R. Rashidov, E. Maksudov, B. I. Aikhodzhaev (Uzbeksk. Khim. Zh. **17** No. 3 [1973] 39/41; C.A. **79** [1973] No. 83327). — [54] R. E. Kesting (Proc. 4th Intern. Symp. Fresh Water Sea, Heidelberg 1973, Vol. 4, p. 185/96). — [55] C. Noël, L. Monnerie (Proc. 4th Intern. Symp. Fresh Water Sea, Heidelberg 1973, Vol. 4, p. 275/83).

[56] C. Lemoyne, C. Noël, L. Monnerie (Compt. Rend. **276** [1973] 1163/5). — [57] J. L. Halary, C. Noël, L. Monnerie (Desalination **13** [1973] 251/67). — [58] R. Brun, E. Dubois, Y. Duriau (Desalination **4** [1968] 119/30). — [59] Y. Trudelle, L. Nicolas (Desalination **4** [1968] 198/208). — [60] S. Manjikian, C. Allen (Off. Saline Water Res. Develop. Progr. Rept. No. 378 [1968]).

[61] H. Ohya, S. Sourirajan (J. Appl. Polymer Sci. **15** [1971] 705/13). — [62] N. S. Mason, O. Lindan, R. E. Sparks (Chem. Eng. Progr. Symp. Ser. **67** No. 114 [1971] 139/50). — [63] M. F. Re, D. N. Bennion (Ind. Eng. Chem. Fundamentals **12** [1973] 69/75). — [64] K. W. Boeddeker, J. Kaschemekat (GKSS-71/E/29 [1971]). — [65] T. T. Serafini, P. Delvigs, R. D. Vannucci (J. Appl. Polymer Sci. **17** [1973] 3519/52).

[66] M. Nakagaki, K. Miyata (Yakugaku Zasshi **93** [1973] 1105/11; C.A. **80** [1974] No. 16664). — [67] K. Wakakuwa, Y. Kurokawa, N. Yui (Technol. Rept. Tohoku Univ. **38** [1973] 181/6; C.A. **80** [1974] No. 30521). — [68] T. Furuta, Y. Kurokawa, N. Yui (Technol. Rept. Tohoku Univ. **38** [1973] 207/12; C.A. **80** [1974] No. 19323). — [69] B. Kunst, S. Sourirajan (Proc. 3rd Intern. Symp. Fresh Water Sea, Dubrovnik 1970, Vol. 2, p. 497/509; J. Appl. Polymer Sci. **14** [1970] 2559/68).

Modifizierte
Cellulose-
acetat-
Membranen

3.14 Modified cellulose acetate membranes

Blends of cellulose diacetate and cellulose triacetate have been found to afford membranes with salt rejections in excess of 99%. Hence, cellulose esters were prepared having a degree of substitution intermediate between those of commercial cellulose diacetate and cellulose triacetate. Starting with two commercial cellulose acetate esters, two series of mixed esters were synthesized, the second ester group being propionate, butyrate, benzoate, methacrylate or crotonate. Cellulose acetate methacrylate, with a degree of substitution of methacrylate of 0.1 to 0.2, has yielded membranes of good osmotic properties. In 200 h tests, both crosslinked and uncrosslinked cellulose acetate methacrylate membranes exhibited sodium chloride rejections of 99.7%, fluxes of 0.32 to 0.37 m^3/$m^2 \cdot$ day (8 to 9 gal/ft$^2 \cdot$ day) and extremely low flux decline slopes [1].

Improved membranes have been fabricated from cellulose diacetate and from blends of cellulose diacetate and triacetate. High fluxes and salt rejections were obtained, while the flux decline slope could be reduced. Crosslinking of cellulose acetate methacrylate membranes to complete acetone insolubility increased the flux stability, as well as reduced the flux decline slope. Fluxes of 1.7 m^3/$m^2 \cdot$ day (41 gal/ft$^2 \cdot$ day) at 90% salt rejection were obtained. A more open membrane of this type exhibited a flux of 3.34 m^3/$m^2 \cdot$ day (82 gal/ft$^2 \cdot$ day) and a sodium sulfate rejection of 97% [2]. Membranes, prepared from cellulose diacetate and cellulose triacetate with degrees of substitution between 2.5 and 2.7, appeared particularly suited to seawater desalination exhibiting fluxes of 0.4 m^3/$m^2 \cdot$ day. Similar membranes were suitable for the desalination of brackish water with fluxes up to 2 m^3/$m^2 \cdot$ day. Membranes prepared from cellulose acetate methacrylate exhibited fluxes approaching 4 m^3/$m^2 \cdot$ day at 56 kg/cm^2 operating pressure [3].

Considerable emphasis was given in further work to characterizing the flux stability of the mem-branes. The two best types demonstrated the capability of a projected mean first-year flux of 1.42 m^3/$m^2 \cdot$ day (35 gal/ft$^2 \cdot$ day) with 91% rejection of sodium chloride at 56 kg/cm^2 (800 psi). Cellulose acetate methacrylate membranes can be crosslinked sufficiently to result in essentially complete elimination of flux decline [4].

Asymmetric membranes made from a blend of cellulose diacetate and cellulose triacetate exhibited initial fluxes of 0.45 to 0.53 m^3/$m^2 \cdot$ day (11 to 13 gal/ft$^2 \cdot$ day) and salt rejections of 99.3 to 99.6%

at a pressure of 105 kg/cm² (1500 psi) when demineralizing seawater. Projected fluxes after one year of operation were 0.33 to 0.40 m³/m² · day (8 to 10 gal/ft² · day). For the purification of an 1% NaCl feed water, fluxes as high as 1.83 m³/m² · day (45 gal/ft² · day) were obtained at 90% salt rejection and at 56 kg/cm² (800 psi) pressure [5].

Cellulose films were prepared by Smith and coworkers with a high degree of orientation from xanthate solutions of cellulose. The purpose of inducing orientation was to increase the tensile strengths of the films when wet and to supplant the usual cellulosic pore structure with an essentially homogeneous surface. Films have been prepared by two modifications of the viscose process to retard regeneration and thus to facilitate orientation imposed by stretching. Chemical modification of oriented cellulose membranes was carried out. Acetylation of the oriented membranes was found to be rapid and relatively simple. The combination of wet spinning and chemical modification offers a complete spectrum of membranes with various mechanical strength, wettability, water of imbibition and permeability. Mixed ether esters of cellulose offer the potential of selectivity with better mechanical and biological stability [6]. The potential for the use in reverse osmosis of asymmetric membranes from ethyl cellulose has been investigated by the same group of authors. Asymmetric membranes cast from methyl acetate-methanol-formamide-water solutions exhibited fluxes of 0.49 m³/m² · day (12 gal/ft² · day) with 99.5% rejection of Na_2SO_4 at 105 kg/cm² (1500 psi). Other casting formulations yielded membranes with 98 to 99% rejection of sodium chloride and fluxes of 0.2 to 0.28 m³/m² · day (5 to 7 gal/ft² · day) at 105 kg/cm². These membranes were all cast at room temperature and not heat treated. Satisfactory compaction resistance has been observed in 200 h performance tests [7]. A solution prediction model has been developed which allows the formation of asymmetric membranes from a larger number of solvent combinations. The data from both flat sheet and tubular asymmetric ethyl cellulose membranes reveal an outstanding stability to alkaline solutions, good rejections for most solutes but relatively low flux. The compaction of these membranes is of the same degree as cellulose acetate [25].

Various formulations have been tested by Manjikian and coworkers, employing cellulose acetate butyrate as the film base instead of the conventional cellulose acetate. Ultraselective membranes have been produced without heat treatment. Salt rejection of 99.8% at a flux of 0.33 to 0.4 m³/m² · day (8 to 10 gal/ft² · day) has been achieved treating feed water of 5000 mg/l NaCl at 42 kg/cm² (600 psi). Using seawater as feed and an operating pressure of 70 kg/cm² (1000 psi), a 99.7% rejection of total dissolved solids was achieved at a flux of 0.2 m³/m² · day (5 gal/ft² · day) [8]. The characterization of cellulose acetate butyrate membranes, the optimization of their performance and the production of tubular butyrate membranes were further investigated. In general, butyrate membranes exhibit rejection properties substantially better than acetate membranes for a variety of inorganic and organic solutes and seem to be more resistant to hydrolysis. The increased separation of nitrates, borates, urea and hexachlorophene appears to be of particular interest. For urea concentrations in the range of 1000 to 2000 mg/l, butyrate membrane selectivity is about three times as high as that of acetate membranes and approaches quantitative rejection. However, the rate of flux decline is somewhat greater than that of acetate membranes. Experimental cellulose acetate butyrate tubular membranes have also been produced. A glyoxal modified tubular butyrate membrane exhibited a flux of 0.49 m³/m² · day (12 gal/ft² · day) and 99.4% salt rejection when operated at 105 kg/cm² (1500 psi) [9].

Regenerated cellulose films treated in alkaline aqueous solutions of 1-cyclohexyl-3-(2-morpholinoethyl)-carboimide retard the passage of dilute calcium salt solutions compared to sodium. Infrared spectra showed that a substituted urea had been formed in the film, presumably with a quaternary ammonium group attached. Films treated in a 67% carbodiimide solution showed an average salt rejection value of 97.4% for 0.001 M calcium over the range 17.5 to 70 kg/cm² (250 to 1000 psi) [10].

Characterization of the active layer of modified cellulose acetate membranes was made by Yasuda and Lamaze, utilizing asymmetric saponification from the active side of the membrane. The study of the effect of heat treatment indicates that the structure of modified cellulose acetate membranes is not a discrete composite of an active and a porous supporting layer, but a vectored density gradient membrane. The heat treatment seems to decrease the permeability of the active layer. The transport resistance and the resistance of the active layer to saponification increase nearly exponentially with increasing heat treatment temperature [11].

A nuclear magnetic resonance method for determining the distribution of acetyl groups in cellulose acetates was developed by Goodlett et al. The method was applied to studying the hydrolysis of cellulose triacetate with ammonia, the acetylation of cellulose acetate with acetyl chloride and with acetic anhydride [12].

Asymmetric cellulose acetate membranes may be prepared using anhydrous ammonia, incorporated by absorption at room temperature, as modifying additive. Membrane performance is comparable to that of formamide-modified membranes, at a considerably lower level of initial concentration of the additive [13].

The effect of twenty different secondary additives in the casting solution was studied by Johnston and Sourirajan. Porous cellulose acetate membranes, capable of giving 5 to 20% increase in productivity at a 90% level of solute separation for a 3500 mg/l aqueous NaCl solution at 17.5 kg/cm² (250 psi), were produced by using 5% by weight ethyl ether as the secondary additive in the casting solution. The use of secondary additives offers a flexibility in the choice of film-casting conditions [14].

Studies were performed by Subcasky and Segovia to determine the effect of additives upon the performance of flat sheet and tubular membranes. Poly(vinyl methyl ether) was found to be an effective additive for feed solutions. Cellulose acetate membranes treated with additives had higher product fluxes and high salt reduction factors than could be obtained by heat treatment alone. Some of the polyoxyethylenenonylphenols, when used as additives, also gave membrane performance better than could be achieved by heat treatment alone. Membrane tightness, hydrophilic to hydrophobic ratio and concentration of additive were all factors in determining the effectiveness of the additive [15].

Surface pressure-area isotherms of cellulose triacetate and amylose triacetate at the air-water interface were determined by Hittmeier et al. on a vertical and horizontal film balance. Both polymers show anomalous behavior on the vertical balance. The behavior of the cellulose triacetate film is attributed to heterogeneous gelation of the monolayer at low pressures. A mechanism involving the lateral tilting of pyranose units at the interface and subsequent stabilization of the monolayer by van der Waal's interactions accounts for the properties of this film. A helical conformation for amylose triacetate at the air-water interface was proposed to explain the low compressibility of this monolayer [16]. Short term reverse osmosis experiments were made on thin films of cellulose diacetate, cellulose triacetate and amylose diacetate. The experimental relationship between water and salt permeability was discussed in terms of the effect of changes in these parameters with changes in operating pressure, feed concentration and required salt rejection [17].

A method for measuring the resistance change of a film with time was used by Barnes et al. to determine the diffusion and apparent permeability coefficients of sodium chloride in cellulose acetate and amylose acetate polymeric films. The results with cellulose diacetate were comparable to previous results, including reverse osmosis. Amylose diacetate was shown to have a lower diffusion coefficient and apparent permeability to sodium chloride than the cellulose diacetate films. Heating amylose acetate films greatly reduced the derived coefficients. These results are accounted for on the basis of the ease of crystallization of amylose acetate [18]. The transport of water through untreated and heated cellulose diacetate membranes and films and through amylose diacetate films was studied at 25°C by means of both steady-state and unsteady-state methods. The coefficients obtained by steady-state measurements were confined to the high water vapor activity regions. The unsteady-state measurements were made in small steps over the range from 0.1 to 0.9 relative vapor pressure for comparison with the steady-state results. The permeability and diffusion coefficients obtained by both techniques increased with increasing film thickness and the true diffusion coefficient was calculated where possible by correcting for film thickness. The latter coefficients decreased exponentially with increasing mean relative vapor pressure. The steady-state permeability and diffusion coefficients for untreated cellulose and amylose diacetates were of the same order of magnitude. Heating reduced the coefficients for the cellulose diacetate films, but not those of amylose diacetate. The experimental water adsorption isotherms in cellulose and amylose acetates were discussed in terms of three theories. The Hailwood-Horrobin theory was shown to fit the sorption isotherm throughout most of the relative vapor pressure region while the B.E.T. theory fitted the experimental isotherm up to 0.6 relative vapor pressure. Heating cellulose acetate was shown to have little effect on sorption, but increasing the acetyl content reduced both the hydrated or surface bound water and the water of solution. Amylose acetate sorbed less of both hydrated water and water of solution and heating reduced the water of solution considerably. Clustering in all polymers was shown by evaluation of the Zimm-Lundberg cluster function. Heating cellulose acetate or increasing the acetyl content yielded some initial clustering. Clustering was lower in the amylose acetates [19].

Styrene grafting to cellulose acetate was studied by Hopfenberg et al. in an attempt to synthesize a membrane with improved resistance to compaction [20]. Reduction of product flux and/or rejection with time can be reduced by controlled radiation-induced graft copolymerization of polystyrene to the cellulose acetate backbone. The most attractive method for the preparation of grafted asymmetric membranes appears to be radiation of a styrene-containing ternary casting solution, followed by casting according to a modified Loeb-Manjikian technique [21]. The predominant effect of short chain grafting is destructuring of the cellulose acetate. Long chain grafting tends to form domains which act predominantly as secondary valence crosslinks. The long chain domains are rather water insensitive, resulting in decreased water sorption and a structuring rather than destructuring of the surrounding cellulose acetate. The short chain grafting yielded a markedly more open membrane resulting in a monotonic increase in flux with increasing percent graft and a companion decrease in rejection at high grafting levels [22]. A number of relevant parameters of the grafting system cellulose acetate-styrene was correlated with the resultant product flux, salt rejection, compaction and tensile creep properties of the various grafted membranes. The parameters studied include cellulose acetate state of aggregation, film thickness, film asymmetry, radiation dose, dose rate, swelling agent type and concentration, multifunctional monomer concentration and order of processing steps [23].

A large number of monomers were grafted and chemically attached to the cellulose acetate molecules of Loeb membranes with subsequent growth of polymeric side chains of the monomer. Grafting of nonpolar momomers gave no improvement in either water flux or salt rejection efficiency. Strongly polar monomers, especially positively charged quaternaries, substantially increased the water flux without loss in salt rejection efficiency as compared to an ungrafted control. Grafting on completely prepared membranes gave results that were more reproducible and uniform and permitted in turn a more meaningful comparison with reference membranes [24]. Reverse osmosis properties of cellulose acetate and methylacrylate grafted to cellulose acetate membranes were also investigated by Kasai et al. [26].

Literature to 3.14

[1] C. W. Saltonstall (Off. Saline Water Res. Develop. Progr. Rept. No. 434 [1969]). — [2] D. L. Hoernschemeyer, C. W. Saltonstall, O. S. Schaeffler, L. W. Schollenbach, A. J. Secchi, A. L. Vincent (Off. Saline Water Res. Develop. Progr. Rept. No. 556 [1970]). — [3] C. W. Saltonstall (Proc. 3rd Intern. Symp. Fresh Water Sea, Dubrovnik 1970, Vol. 2, p. 579/86). — [4] D. L. Hoernschemeyer, L. W. Lawrence, C. W. Saltonstall, O. S. Schaeffler, A. J. Secchi (Off. Saline Water Res. Develop. Progr. Rept. No. 700 [1971]), D. L. Hoernschemeyer, R. W. Lawrence, C. W. Saltonstall, O. S. Schaeffler (Reverse Osmosis Membrane Research, New York 1972, p. 163/75). — [5] W. M. Kind, D. L. Hoernschemeyer, C. W. Saltonstall (Reverse Osmosis Membrane Research, New York 1972, p. 131/61).

[6] J. K. Smith, F. Morton, E. Klein (Off. Saline Water Res. Develop. Progr. Rept. No. 507 [1970]). — [7] J. K. Smith, E. Klein, F. Morton, T. Casebonne (Off. Saline Water Res. Develop. Progr. Rept. No. 630 [1970]). — [8] S. Manjikian, S. Liu, M. Foley, C. Allen, B. Fabrick (Off. Saline Water Res. Develop. Progr. Rept. No. 534 [1970]). — [9] S. Manjikian, M. I. Foley (Off. Saline Water Res. Develop. Progr. Rept. No. 612 [1970], No. 654 [1971]). — [10] J. Steigman, B. Luftig (Off. Saline Water Res. Develop. Progr. Rept. No. 704 [1971]).

[11] H. Yasuda, C. E. Lamaze (in: A. F. Turbak, Membranes from Cellulose and Cellulose Derivatives, Appl. Polymer Symp. No. 13 [1970] 157/67). — [12] V. W. Goodlett, J. T. Dougherty, H. W. Patton (J. Polymer Sci. A I 9 [1971] 155/61). — [13] K. W. Böddeker, J. Kaschemekat, H. Woldmann (Proc. 4th Intern. Symp. Fresh Water Sea, Heidelberg 1973, Vol. 4, p. 65/71; Z. Physik. Chem. [Frankfurt] 87 [1973] 60/3). — [14] H. K. Johnston, S. Sourirajan (J. App. Polymer Sci. 17 [1973] 2485/99). — [15] W. J. Subcasky, G. Segovia (Off. Saline Water Res. Develop. Progr. Rept. No. 466 [1969]).

[16] M. Hittmeier, L. S. Sandel, P. Luner (Off. Saline Water Res. Develop. Progr. Rept. No. 827 [1972]). — [17] W. W. Carey, P. Luner (Off. Saline Water Res. Develop. Progr. Rept. No. 828 [1972]). — [18] M. Barnes, C. Skaar, P. Luner (Off. Saline Water Res. Develop. Progr. Rept. No. 825 [1972]. — [19] M. Barnes, C. Skaar, P. Luner (Off. Saline Water Res. Develop. Progr. Rept. No. 831 [1972], No. 832 [1972]). — [20] H. B. Hopfenberg, F. Kimura, P. T. Rigney, V. Stannett (J. Polymer Sci. C No. 28 [1969] 243/60).

[21] H. B. Hopfenberg, V. Stannett, F. Kimura, P. T. Rigney (in: A. F. Turbak, Membranes from Cellulose and Cellulose Derivatives, Appl. Polymer Symp. No. 13 [1970] 139/55). — [22] F. Kimura, H. B. Hopfenberg, V. Stannett (Reverse Osmosis Membrane Research, New York 1972, p. 177/203). — [23] V. Stannett, H. B. Hopfenberg, F. Kimura (Off. Saline Water Res. Develop. Progr. Rept. No. 788 [1971]). — [24] C. Horowitz (Off. Saline Water Res. Develop. Progr. Rept. No. 545 [1970]). — [25] J. K. Smith, E. Klein (Off. Saline Water Res. Develop. Progr. Rept. No. 863 [1973]).

[26] K. Kasai, Y. Kurokawa, N. Yui (Technol. Rept. Tohoku Univ. **38** [1973] 187/96; C. A. **80** [1974] No. 30522).

Polymer-
Membranen

3.15 Polymer film membranes

A vast research work has been conducted, since the potential of reverse osmosis as a desalination process was established, to develop high performance membranes other than cellulose acetate. Commercially available polymers were evaluated as membranes and attempts were made to improve the performance of the most promising polymeric films. Novel polymeric membrane material was also synthesized. To a certain degree part of these attempts were successful in producing reverse osmosis membranes with improved characteristics. However, cellulose acetate and its derivatives still keep a leading role as membranes for reverse osmosis.

Salt filtering layers can be formed on porous bodies by circulating pressurized solutions containing small concentrations of organic polyelectrolytes. These films reject substantial fractions of solute from dilute solutions, frequently with high transmission rates through the membranes, and the possibilities of their application in reverse osmosis processes was discussed [1]. It has been found in further studies that salt-rejecting interfacial layers or membranes with remarkably high permeation rates can be formed dynamically on porous bodies, when certain additives are present in the pressurized feed solutions. Rejecting layers have been formed on substrates with pore diameters as high as 5 μ. The additives which so far have given the most interesting results include organic polyelectrolytes, colloidal dispersions of hydrous oxides, solutions of hydrolyzable ions, ground-up low-cross-linked ion-exchange beads etc. The chemical nature of the porous substrate is of no primary importance. With the dynamic membranes, so far developed, salt rejection is less than with cellulose acetate and, for most of them, rejection decreases with increasing salt concentration [2]. Yasuda and Lamaze concluded that many polymers in the form of dense films showed quite good salt rejecting property, but not, at the time of the study, quite as good as cellulose acetate [3].

An attempt was made to analyze the relationship between salt rejection and water flux of nonionic polymer membranes on the basis of the movement of water in the membranes. Good agreement was found between the experimental data and the calculated curve. Excessive swelling of membranes results in bulk flow of water with coupled transport of salt. Hence, the salt rejection decreases quickly as water flux increases beyond a treshold value above which water flux can be characterized as bulk flow [4]. The water permeability of homogeneous ionic polymer membranes and their salt rejection were examined with cationic and anionic membranes of block and graft copolymers. The principle of salt rejection by ionic membranes was explained by the difference in the transport volumes for mobile co-ions and water. The repulsive force between a fixed ion and a mobile co-ion decreases the transport volume of the latter, thus creating a transport depletion of salt flux relative to water transport. This transport depletion is governed by the amount of water sorbed by a fixed ionic site, which also determines the water flux. The relation of salt rejection and water permeability significantly differs from that found between salt rejection and water permeability for nonionic polymer membranes. The decline of salt rejection with increasing water permeability is much less in ionic membranes than in nonionic membranes. However, in the high salt rejection region, water permeability for both ionic and nonionic membranes becomes similar as the dominant mode of water transport changes from flow to diffusion [5]. Membranes of graft copolymers of polyethylene with poly(sodium styrene sulfonate), poly(4-vinylpyridinium methyl bromide) and poly(sodium acrylate) were prepared and examined as a function of grafting yield. The overall reverse osmosis characteristic is dependent on the number of ionic sites introduced and also on the swelling capability of the membrane. The salt rejection depends on the conditions of the grafting reaction [6].

The pressure driven transport of liquids is attributed by Paul and Ebra-Lima to a solution-diffusion mechanism in highly swollen polymer membranes. A theory based on this mechanism correlated successfully permeation fluxes for such membranes. Confirmation of this theory was provided by direct measurement of the proposed concentration gradient. The study of the temperature dependence of the liquid diffusion coefficient in the polymer membrane has provided additional evidence of a hydrodynamic regime of diffusion in highly swollen membranes. It was also shown that the proposed ceiling flux in reverse osmosis is equal to the pervaporation flux [7].

Methods to measure the permeability of polymer membranes to solvating liquids and vapors have been reviewed by Blackadder and Keniry. Experiments indicated that the various methods are not necessarily equivalent. The medium over which the membrane is supported and more particularly the nonpermeant pressure differential during permeation are shown to influence the permeation rate [8]. A general equation developed for the flow of water through hydrated polymer membranes is expected to be extendable to more general permeants in swollen polymer membranes. The difference between the hydraulic and diffusive permeabilities under pressure and concentration gradients, respectively, are explained by the incipient viscous flow at high degree of swelling [9].

The development of reverse osmosis membranes from various polymer materials is reviewed in the following, with the group of polymers classified more or less in chronological sequence of development.

Early work on polymer membranes. Chemical crosslinking of polyvinyl alcohol and selective gelatinization of cellophane films were found to give improved salt rejection, while maintaining attractive transport of water [10]. Novel techniques for preparing semipermeable membranes have been developed at the Massachusetts Institute of Technology. Crosslinking and polar group content of a polymer were considered to be important factors in determining membrane semipermeabilty. Membranes were prepared from copolymers of hydroxyethyl methacrylate and ethylene glycol dimethacrylate as crosslinking agent. The degree of salt rejection rose as the membrane was more tightly crosslinked and then dropped off sharply [11]. By changing to trimethylol propane, as crosslinking agent, films without support material and with thickness ranging down to 0.07 mm (2.7 mils) could be prepared. Up to 93% salt rejection was obtained [12]. However, the water flux was about two-thirds that of homogeneous cellulose acetate films. Studies of hydroxyethyl methacrylate and ethyl methacrylate copolymers have shown that it is possible to control the permeability of the film by varying the ratio of the two components and an optimum membrane composition at a 5:2 ratio was found. Water and salt fluxes were somewhat smaller than for cellulose acetate and the mechanical strength was inferior. Polyurethane membranes were prepared as a reaction product of toluene diisocyanate and polyethylene glycol with trimethylol propane as the crosslinking agent. These membranes exhibited moderate flux and moderate rejection characteristics [13].

Homogeneous polysalt films prepared from poly(sodium styrene sulfonate) and poly(vinylbenzyl trimethylammonium chloride) exhibited salt rejection of the order of 50% for a feed containing 5% NaCl [14]. Polyacrylonitrile and polymethylacrylonitrile membranes exhibited interesting water fluxes but unsufficient salt retentions. Membranes prepared by casting from a solution of a high-molecular-weight poly(vinylene carbonate) in dimethylformamide exhibited salt retentions of about 83% at low flux rates [15]. Ultrathin films of the polymer showed rapid deterioration followed by an increase in flux and salt permeation [16].

Polyvinylpyrrolidone polymers. The polyvinylpyrrolidone-polyisocyanate interpolymer system forms a series of hydrophilic materials, the water permeability and selectivity of which may be varied over a wide range. Membranes of these materials have excellent physical properties and appear to be free of imperfections. The permeability to both water and solutes is strongly dependent on the equivalent ratio of isocyanate to vinylpyrrolidone. Water sorption data indicate a largely random distribution of water at low sorptions, with some clustering at high water contents. Permeability studies of high water content membranes indicate that coupling occurs between water and solute flows [17]. The transport properties of crosslinked polyvinylpyrrolidone membranes have been described in detail by Riley, Lyons and Merten [18]. It has further been shown that the promising interpolymer system based on polyvinylpyrrolidone and polyisocyanates does not form good thin films. High water fluxes could not be combined with high salt rejections [19].

Membranes were also prepared from water insoluble complexes of poly(vinyl pyrrolidone) homo and copolymers with carboxyl group containing reaction products of poly(alkyl vinyl ether/maleic anhydrides). The best combination of mechanical properties, relative insensitivity to pH change above 5 and salt rejection properties was obtained using a capped diisocyanate to covalently crosslink complexes from poly(vinyl pyrrolidone) and hydrolyzed poly(butyl vinyl ether/maleic anhydride). Salt rejection as high as 89% at 42 kg/cm² (600 psi) and trivalent iron ions rejection up to 88.5% were obtained at low water flux [20].

Derivatives of acrylic acid. A copolymer system based on galactose methacrylate and methyl methacrylate has been used for the preparation of reverse osmosis membranes. Their resistance to pressure in the homogeneous state was considered to be better than for cellulose acetate. The membranes had the ability to reject divalent ion salt, even when sodium chloride rejection was quite low [21].

Criteria have been established by Maconochie et al. for the synthesis of new polymer systems suitable for use in connection with the preparation of reverse osmosis membranes. Inherent behavior for a range of chemically different polymers was established and data obtained on standard physical structures, utilizing homogeneous porefree membranes, were given [22]. Membranes were prepared from the galactose methacrylate-methyl methacrylate and glyceryl methacrylate-methyl methacrylate copolymers using various solvents. By addition of a swelling agent to the solvent mixture and immersion before completion of drying, Loeb-type heterogeneous films were cast. Examples of performance, for a film of 1:9 mole ratio galactose copolymer cast from chloroform-ethanol and formamide, are rejection 83% of an 1% sodium chloride feed and flow rate 0.02 m³/m² · day (0.56 gal/ft² · day) at 56 kg/cm² (800 psi); for a film of the same copolymer, cast from aqueous acetone and zinc chloride, rejection 78.4% and flow rate about the same [23].

A new class of membranes has been developed by polymerizing mixtures of N-methylol acrylamide, acrylic acid, ethyl acrylate and trimethylol propane trimethylacrylate, followed by heat treatment. Performance capabilities were slightly better than dense cellulose acetate membranes under comparable conditions [24]. Tested on dilute salt solutions, the membranes exhibited salt rejections greater than 98%. Considering the effect of membrane composition on product flux and salt rejection it was concluded that improved membranes should have as high as possible a concentration of hydrophilic groups, distributed randomly through a lightly crosslinked rubbery polymer matrix [25]. A series of crosslinked poly(hydroxyethyl methacrylate) membranes has been prepared, which exhibited salt rejection increases from 78% to a maximum of 94% [26].

The effect of the hydrophilic nature of membranes prepared by the copolymerization of a hydrophilic and hydrophobic monomer upon the transport of NaCl and water was studied by Kopecek and Vacik. The 2-(2-hydroxyethoxy)ethyl methacrylate-methyl methacrylate copolymer crosslinked with ethylene dimethylacrylate has a higher permeability for the above compounds than 2-hydroxyethyl methacrylate-methyl methacrylate copolymer crosslinked with ethylene dimethacrylate at the same degree of hydration [27].

Poly(glutamic acid) and poly(acrylic acid) dynamically formed membranes exhibited rejections, which were high in comparison with those of organic polyelectrolytes previously studied. Transmission rates through cast poly(glutamic acid) films were much lower than through those dynamically formed of the same polyelectrolyte. Cast polyacrylate membranes, however, had extremely high permeation rates for films having comparable rejection made of any material. Fluxes were in many cases higher than those of polyacrylate membranes formed dynamically [28].

Polyacrylic acid composite membrane properties as a function of pH, water flux and salt rejection rates as a function of feed concentration and change of the counterion were studied by Milstead and Tagami on various desalination and sewage treatment operations. Optimum performance was obtained with a support material produced from polysulfone and methyl cellulose [29]. Membranes consisting of a layer of polyacrylic acid on hydrous Zr^{IV} oxide are reported by Johnson et al. and show promise for reverse osmosis of low-salinity waters [30].

Ethylene oxide polymers. Block copolymers have been synthesized from peroxycarbamate-terminated poly(ethylene oxide) and the vinyl monomers styrene, methyl methacrylate and acrylonitrile. Membranes fabricated from these block copolymers appear to be relatively insensitive to the structure of the hydrophobic cosegments. A substantial increase in the crystallinity of the poly(ethylene oxide) phase is accompanied by a substantial decrease in water permeability. Compaction of the membrane resulted in a significant increase in selectivity with only a slight decrease in water permeability. One of the systems studied gave membrane films having osmotic transport characteristics comparable in all respects to dense cellulose acetate. Syndiotactic poly(vinyl alcohol) membranes showed rather poor selectivity for aqueous sodium chloride feed solutions, but have shown promise for hard water applications, based on experience with magnesium sulfate solutions. The effect of grafting with acrylonitrile appeared to be in the decrease of water permeability and the increase in compaction resistance. Unsignificant improvement in selectivity was observed [31].

The use of inorganic swelling salts and non-solvent additives, such as H_2O_2, in attempts to prepare anisotropic films and therefore upgrade the performance of poly(ethylene terephthalate)-b-poly(ethylene oxide) was not successful. However, the block copolymer poly(bisphenol-A carbonate)-b-poly(ethylene oxide) gave membranes of excellent mechanical properties. Flux rates could be controlled by varying solvents, casting time and evaporation temperature. The flux range appeared to be as wide as 0.14 to 8.14 m³/m² · day (0.1 to over 200 gal/ft² · day) without materially affecting selectivity [32].

Polyphenylene oxide polymers. Sulfonated 2,6-dimethyl polyphenylene oxide membranes were prepared by Plummer et al. for reverse osmosis application and the effects of membrane and water feed variables on flux and salt rejection were extensively investigated. The membranes exhibited an ion exchange capacity of 2.0 to 3.0 meq of H^+ per g of dry polymer, water content of 80 to 200% based on dry polymer and 4 to 12.7 μ (0.15 to 0.5 mil) thickness. Testing was made at a pressure of 14 to 84 kg/cm² (200 to 1200 psi) at temperatures of 21 to 54°C (70 to 130°F) and various water compositions. Fluxes and salt rejections for the various feeds were generally in the broad range of 0.4 to 1.63 m³/m² · day (10 to 40 gal/ft² · day) and 98 to 70%, respectively, depending upon experimental conditions and membrane properties. The life test on 1% NaCl feed was 220 h and that on Webster water was 524 h. The membranes were considered to be best for applications involving mixed feeds primarily containing salts with divalent ions, such as brackish water. Rejections are substantially decreased as monovalent ion concentrations increase in feeds containing divalent ions [33].

A polymer and membrane manufacturing procedure, based on the effects of various process variables on membrane performance, was devised by Chludzinski et al. The membranes were found to be reproducible in flux to about ±25% and in rejection to ±5%. Almost all of the membranes showed flux better than 1 m³/m² · day (25 gal/ft² · day) and 85% salt rejection when operated on natural Webster water at 77 kg/cm² (1100 psi) and 54°C (130°F). It was concluded that the sulfonated polyphenylene oxide membrane is showing high flux and rejection, good life, resistance to compaction, chemical resistance and tolerance to high temperatures [34].

A wide range of salt rejections and water permeabilities can be obtained with sulfonated poly(2,6-dimethylphenylene ether) cation exchange membranes. They posses 8 to 10 times the water permeability of homogeneous cellulose acetate membranes at the same salt rejection [35]. The ion exchange capacity of these membranes is strongly dependent on the degree of sulfonation. The membranes proved to be exceptionally stable mechanically and chemically in tests with brackish water, caustic feed, secondary sewage and synthetic wash water [36].

Polyolefin polymers. Polyethylene was modified by grafting with acidic monomers such as acrylic and methacrylic acids, basic monomers such as aminoalkyl methacrylates and vinylpyridines, and neutral monomers such as hydroxyalkyl methacrylates. Crosslinking during grafting was accomplished by use of difunctional monomers in the grafting solution. The membranes so produced maintained much of the tensile strength and elongation of the original polyethylene. The hydrophilicity of the membranes was reduced by crosslinking or by cografting with a hydrophobic monomer. The acrylic acid grafted membranes were thermally sensitive. Heating in air above 100°C decreased water flux and increased salt rejection, while heating in water above 100°C led to increases in water flux and decreases in salt rejection [37].

Crude mosaic membranes were also prepared by selective areal grafting of acidic and basic monomers onto polyethylene film. Preparative procedures were evaluated, which included direct grafting, indirect grafting and peroxidation grafting using γ irradiation, β irradiation and both high and low energy electrons. The peroxidation technique using electron irradiation was most successful. A better procedure for mosaic membrane preparation involves chemical addition of cationic and anionic sites to a styrene grafted polyethylene film using a masking technique to obtain the desired control of charge domain size and spacing. Preparation of random polysalt membranes by sequential grafting of polyethylene with acid and basic monomers afforded membranes, which displayed selective transport properties for different ions when evaluated with mixed ionic species in the feed [38].

Aromatic polyamide polymers. Membranes prepared by McKinney and Rhodes from aromatic polyamides exhibited a selectivity of over 99.4% for dissolved salts at fluxes of 0.04 to

1.22 m³/m² · day (1 to 30 gal/ft² · day) with excellent chemical stability and mechanical properties. The membrane structure consists of a thin dense surface layer supported on a porous matrix, similar to that reported for cellulose acetate membranes [39]. In continuing work, polyterephthalamide of p-aminobenzhydrazide and polyterephthalamide of 1,3-bis-(3-amino)-benzamide were prepared by low temperature polymerization in dimethylacetamide. Anisotropic membranes were cast and reverse osmosis experiments were performed using 0.5% NaCl at 42 kg/cm² (600 psi), 3.5 and 6.0% NaCl at 105 kg/cm² (1500 psi) with a surface linear velocity over 200 cm/s. The unannealed membrane had a flux of 1.22 m³/m² · day (30 gal/ft² · day) at 60% rejection using 0.5% NaCl and 42 kg/cm². Membranes annealed in boiling water had a flux decrease of 0.08 to 0.16 m³/m² · day (2 to 4 gal/ft² · day) at 99.7% rejection [40]. The use of crosslinking based on ionic bonds has been successfully employed in membranes prepared from aromatic polyamides for the improvement of long-term flux stability. Preliminary data taken for seamless tubular membranes indicated a performance level of 0.28 m³/m² · day (7 gal/ft² · day) and 99.3% rejection at a low compaction rate under comparable conditions of reverse osmosis [41].

Reverse osmosis properties of poly(trans-2,5-dimethylpiperazine fumaramide) films have been found to be particulary interesting for practical reverse osmosis applications. A high permeability to water makes the material suitable for use in preparation of composite membranes or hollow fibers [42]. Anisotropic membranes were prepared from poly(trans-2,5-dimethylpiperazine-thiofurazanamide) and showed a water permeability that may vary from 0.09 to 2.3 μg/cm · s depending on the structure. Their properties very closely approach those of the corresponding cellulose acetate membranes [43].

The intrinsic permeability characteristics of a fully aromatic polyamide film has been determined by Frommer et al. and its suitability for desalination by reverse osmosis was compared with that of the commonly used cellulose acetate. Although the solubility of sodium chloride in the polyamide film is higher than that in the cellulose acetate film, polyamide membranes will reject salt better than cellulose acetate membranes having identical structure and morphology. This is because the diffusivity in NaCl through the polyamide film is substantially lower and the permeability of water through it, as well as the solubility and diffusivity of water in it, are higher than the comparable values for cellulose acetate films [44].

Polypropylene polymers. The morphology of peroxidized and γ irradiated polypropylene film grafted with acrylic acid was investigated by Alessandrini et al. Electron microscopy examination showed that the laminar structure in the oriented sample was progressively disrupted giving irregular surface pits which increased in size on grafting. The graft film had a high water vapor permeability and was tested as a reverse osmosis membrane [45]. Isotactic polypropylene membranes respectively grafted with acrylic polyacid, poly-β-hydroxyethylmethacrylate and polyacrylamide were prepared by Pegoraro and coworkers. The membrane structure has been studied either at the electron scanning microscope or by transmission. Thermal treatments scarcely influenced the structure. The water transport depended on the resistance of the membrane-vapor limit layer. The desalination properties depended on the membrane thickness. Increased thickness decreased flux, but salt rejections increased to more than 80% [46]. The obtained desalting results were interpreted on the basis of the membrane structure, as observed with the electron microscope [47].

The methods for the preparation of three types of membranes of polypropylene grafted with polyacrylic acid were described by Pegoraro and Penati. These methods refer to extraction membranes, grafted bioriented films and grafted not oriented films. The desalting properties of the membranes have been determined by reverse osmosis and, of the three types of membranes, the grafted not oriented film has been found to be the most effective [48].

Other polymeric membrane materials. Block copolymers of 2-vinylpyridine and various methacrylic acid esters were synthesized for use as desalination membranes by living anionic polymerization at low temperature. Quaternization of the 2-vinylpyridine block followed by dehydrohalogenation yields a polymer containing anionic and cationic blocks. Salt rejection and flux rate of the membranes were found to be generally good [49]. 2-vinylpyridine-trimethylsilyl methacrylate block copolymer, styrene-2-vinyl-pyridine block copolymer and styrene-trimethylsilyl methacrylate block copolymer were synthesized. The copolymers were hydrolyzed to convert the acrylate block to the free acid and then the pyridine segment was quaternized. Desalination membranes of the first block copolymer formed from sodium sulfate solution showed better salt rejection at pH 4 in dual-layer membranes with zirconium sublayer than in single-layer membranes. About 80% salt rejection was observed [50].

Quaternized 2-polyvinylpyridine and polyvinylamine hydrochloride were deposited on partially cured cellulose acetate membranes. Reverse osmosis measurements revealed that a prolonged exposure of such membranes to diluted polyelectrolyte solutions cause a substantial increase in their salt rejection properties [51]. Membranes were also obtained by radiation induced grafting, using 4-vinylpyridine on a polytetrafluoroethylene matrix. Subsequent quaternization of the pyridine groups improved greatly the reverse osmosis properties of the membranes, considering both flux and rejection [52].

Asymmetric membranes, prepared from polyvinyl alcohol and its derivatives, had an average salt rejection for chloride and sulphate ions. The separation properties of the membranes increase when the polarization of the organic component is reduced [53].

Polyimidazopyrrolone polymers were reported to have high water permeability and salt rejection properties and might compare favorably with those of cellulose acetate [54].

The polymerization of allylamine in a radio-frequency electrodeless plasma to form thin polymer films on microporous filter media provided very effective dry composite membranes for reverse osmosis. Salt and urea rejections as high as 98% and 46%, respectively, have been achieved from a solution containing 10 g/l NaCl and 10 g/l urea [55].

The synthesis and characterization of hydrophilic tri-block polymers prepared by the hydroxylation of diene polymers was investigated by McIntyre and coworkers, but salt rejection and permeation by these membranes was considered not to be sufficient [56].

A high flow asymmetric membrane was obtained by Chapurlat by sulfonation of an aromatic polysulfone polymer and tested under reverse osmosis conditions. The permeation coefficients to water and to salt were determined and their ratio was found to be 1470 g/cm^3. Under the same conditions a cellulose acetate film exhibited a ratio of 1200 g/cm^3. The membranes are reported to be suitable for desalting brackish water, resistant in pH variations and have a good chemical stability and resistance to bacterial attack [57].

Three classes of hydrophilic polymers, which include cellulosics, vinyl copolymers and polyethers, were modified to effect water insolubility. Films of these materials were successfully evaluated for water and salt permeability and performance in reverse osmosis [58].

Heteroaromatic membranes were obtained from polymeric precursors of polyimidazopyrrolones and from the sodium-salt form of polyamide imides. The films were not damaged by compaction or by chemical attack from a variety of compounds and exhibited over 99% NaCl rejection and more than 0.81 $m^3/m^2 \cdot day$ (20 $gal/ft^2 \cdot day$) water flux, when tested at 105 kg/cm^2 (1500 psi) against 3.5% NaCl. They also gave excellent rejections against $MgCl_2$, Na_2SO_4 and $MgSO_4$. Low concentrations of phenol, urea and boric acid were as well rejected at various levels [59].

Polyacrylonitrile copolymers, vinyl acetate copolymers and their partial hydrolysis products were selected by Wilken et al. as study materials for new membrane compositions. Dense membranes of acrylonitrile-methylvinylpyridine compositions had excellent handling and processing characteristics and rejected 73% of sodium chloride. Asymmetric membranes prepared from solutions of acrylonitrile-methylvinylpyridine copolymer in dimethylacetamide-glycerine were also tested. The best salt rejection and flux was obtained in vinyl acetate-vinyl alcohol copolymers containing from 10 to 25% vinyl alcohol [60].

The modification of poly(aryl ethers) by incorporation of heterocyclic hydrophilic monomers or amide containing monomers was found to improve their water uptake. Low water fluxes under reverse osmosis conditions were generally observed for all novel poly(aryl ethers). Attempts to increase performance with ethoxylation were not successful. Either water flux was low or, when water flux was increased, the salt rejection decreased [61].

In a continuing search to prepare asymmetric membranes that exhibit intrinsic chemical and reverse osmosis transport properties superior to those of cellulose acetate, polymers containing ether and sulfone groups appeared to be most promising candidates. Sulfonation of bakelite polysulfone yielded a product possessing intrinsic osmotic properties superior to those of cellulose acetate. An investigation was also carried out relating to the modification of Penton, in film form, through the radiation grafting of selected hydrophilic monomers [62].

In a series of polymers based on poly(epichlorohydrin) or on the copolymer of epichlorohydrin and ethylene oxide, a pronounced increase in the equilibrium water sorption was observed with increasing number of polar groups. Extremely high water clustering was determined in all polymers indicating a very heterogeneous distribution of the water in the polymer. The water permeability increased drastically with increasing water sorptivity, but was in all cases very low. All polymers gave poor films, which were suited for reverse osmosis tests only when crosslinked [63].

Asymmetric skin reverse osmosis membranes were prepared by Strathmann et al. from various polymers such as polyacrylonitrile, polyamide and polyimide. The membrane transport properties were determined in reverse osmosis tests and their structures were shown in scanning electron micrographs. The various parameters such as casting solution composition, polymer concentration, membrane post treatment procedures and their effect on membrane structure and performance were discussed [64].

A series of asymmetric membranes were made from the same basic polymer, the reaction adduct of a polyanion (polysodium styrene sulfonate) and a polycation (polyvinyl benzyltrimethylammonium chloride). Compaction and fouling of the membranes were evaluated using brackish and seawater feeds. The product flux of the membranes increased linearly with applied hydrostatic pressure up to 42 kg/cm² (600 psi). Long term evaluation indicated that compaction was of the same order of magnitude as that observed with cellulose acetate membranes. The rejection of monovalent and divalent ions did not change significantly with time [65].

Crazes similar to true cracks can be nucleated, under stress, in many glassy thermoplastic polymers. These crazes are structured regions analogous to a porous sponge. The presence of particles in the polymeric matrix enhances the rate of craze formation and equalizes the craze distribution. On the basis of these considerations, composite membranes were prepared, using a matrix of acrylonitrile-butadienestyrene and glass beads as a filler. Permeability tests were performed under applied pressures of 6 and 40 atm [66].

Collagen (gelatine) membranes were prepared by Kepinski and Chlubek and tested for reverse osmosis. The obtained results were comparable to those of Loeb-type cellulose acetate membranes [67].

Literature to 3.15

[1] K. A. Kraus, H. O. Phillips, A. E. Marcinkowsky, J. S. Johnson (Desalination 1 [1966] 225/30). — [2] K. A. Kraus, A. J. Shor, J. S. Johnson (Desalination 2 [1967] 243/6). — [3] H. Yasuda, C. E. Lamaze (Off. Saline Water Res. Develop. Progr. Rept. No. 473 [1969]). — [4] H. Yasuda, C. E. Lamaze (J. Polymer Sci. A II 9 [1971] 1537/51). — [5] H. Yasuda, C. E. Lamaze, A. Schindler (J. Polymer Sci. A II 9 [1971] 1579/90).

[6] C. E. Lamaze, H. Yasuda (J. Appl. Polymer Sci. 15 [1971] 1665/77). — [7] D. R. Paul, O. M. Ebra-Lima (J. Appl. Polymer Sci. 15 [1971] 2199/210). — [8] D. A. Blackadder, J. S. Keniry (J. Appl. Polymer Sci. 16 [1972] 2141/52). — [9] A. Peterlin, H. Yasuda, H. G. Olf (J. Appl. Polymer Sci. 16 [1972] 865/70). — [10] Monsanto Research Corp. (Off. Saline Water Res. Develop. Progr. Rept. No. 61 [1962], No. 69 [1962]).

[11] R. F. Baddour, W. R. Vieth, A. S. Douglas (Off. Saline Water Res. Develop. Progr. Rept. No. 144 [1965]; J. Colloid Interface Sci. 22 [1966] 588/98). — [12] R. F. Baddour, W. R. Vieth, A. S. Douglas, A. S. Hoffman (Off. Saline Water Res. Develop. Progr. Rept. No. 274 [1967]). — [13] W. R. Vieth, A. S. Douglas, R. Bloch (Off. Saline Water Res. Develop. Progr. Rept. No. 352 [1968]), R. Bloch, W. R. Vieth (J. Appl. Polymer Sci. 13 [1969] 193/204). — [14] A. S. Michaels, H. J. Bixler, R. W. Hausslein, S. M. Fleming (Off. Saline Water Res. Develop. Progr. Rept. No. 149 [1965]). — [15] C. W. Saltonstall, W. S. Higley, R. E. Kesting (Off. Saline Water Res. Develop. Progr. Rept. No. 167 [1966]), C. W. Saltonstall, W. S. Higley, W. M. King (Off. Saline Water Res. Develop. Progr. Rept. No. 220 [1966]).

[16] C. W. Saltonstall, W. S. Higley (Off. Saline Water Res. Develop. Progr. Rept. No. 360 [1968]). — [17] U. Merten, H. K. Lonsdale, R. L. Riley, K. D. Vos (Off. Saline Water Res. Develop. Progr. Rept. No. 369 [1968] 60/78). — [18] R. L. Riley, C. R. Lyons, U. Merten (Off. Saline Water Res. Develop. Progr. Rept. No. 484 [1969] 87/120; Desalination 8 [1970] 177/93). — [19] H. K. Lonsdale, R. L. Riley, C. E. Milstead, L. D. LaGrange, A. S. Douglas, S. B. Sachs (Off. Saline Water Res. Develop. Progr. Rept. No. 577 [1970] 64/75). — [20] H. S. Schultz, N. D. Field (Off. Saline Water Res. Develop. Progr. Rept. No. 609 [1970]).

[21] A. Sharples, G. Thomson (Off. Saline Water Res. Develop. Progr. Rept. No. 329 [1967]). —
[22] G. Maconochie, A. Sharples, G. Thomson (European Polymer J. **7** [1971] 499/511). — [23]
G. Maconochie, G. Thomson (Off. Saline Water Res. Develop. Progr. Rept. No. 533 [1970]; European
Polymer J. **7** [1971] 513/21). — [24] A. S. Hoffman, M. Modell (Off. Saline Water Res. Develop.
Progr. Rept. No. 374 [1968]). — [25] A. S. Hoffman, M. Modell, P. Pan (Chem. Eng. Progr. Symp.
Ser. **64** No. 90 [1968] 324/5; J. Appl. Polymer Sci. **14** [1970] 285/301).

[26] T. A. Jadwin, A. S. Hoffman, W. R. Vieth (J. Appl. Polymer Sci. **14** [1970] 1339/59). —
[27] J. Kopecek, J. Vacik (Collection Czech. Chem. Commun. **28** [1973] 854/60). — [28] S. B.
Sachs, W. H. Baldwin, J. S. Johnson (Desalination **6** [1969] 215/28). — [29] C. E. Milstead,
M. Tagami (Reverse Osmosis Membrane Research, New York 1972, p. 405/18). — [30] J. S. Johnson,
R. E. Minturn, P. H. Wadia (J. Electroanal. Chem. **37** [1972] 267/81).

[31] T. W. Brooks, C. L. Daffin (Off. Saline Water Res. Develop. Progr. Rept. No. 385 [1969],
No. 460 [1969]). — [32] T. W. Brooks, D. A. Warner, C. L. Daffin, J. Patton (Off. Saline Water Res.
Develop. Progr. Rept. No. 660 [1971]). — [33] C. W. Plummer, G. Kimura, A. B. LaConti (Off.
Saline Water Res. Develop. Progr. Rept. No. 551 [1970]). — [34] P. Chludzinski, J. F. Austin, J. Enos
(Off. Saline Water Res. Develop. Progr. Rept. No. 697 [1971]). — [35] G. Kimura (Ind. Eng. Chem.
Prod. Res. Develop. **10** [1971] 335/9).

[36] A. B. LaConti, P. J. Chludzinski, A. P. Fickett (Reverse Osmosis Membrane Research,
New York 1972, p. 263/84). — [37] H. F. Hamil, W. W. Harlowe (Off. Saline Water Res. Develop.
Progr. Rept. No. 530 [1970]). — [38] H. F. Hamil (Off. Saline Water Res. Develop. Progr. Rept.
No. 800 [1972]). — [39] R. McKinney, J. H. Rhodes (Macromolecules **4** [1971] 633/7). — [40]
R. Mc Kinney (Reverse Osmosis Membrane Research, New York 1972, p. 253/61).

[41] R. McKinney, W. L. Hofferbert, J. A. Carden (Off. Saline Water Res. Develop. Progr. Rept.
No. 886 [1973]). — [42] L. Credali, A. Chiolle, P. Parrini (Polymer **14** [1972] 503/6). — [43]
L. Credali, A. Chiolle, P. Parrini (Proc. 4th Intern. Symp. Fresh Water Sea, Heidelberg 1973, Vol. 4,
p. 95/105; Desalination **14** [1974] 137/50), L. Credali, P. Parrini, E. Leonelli, A. Chiolle (Chim.
Ind. [Milan] **56** [1974] 19/24). — [44] M. A. Frommer, J. S. Murday, R. M. Messalem (European
Polymer J. **9** [1973] 367/73). — [45] G. Alessandrini, M. Pegoraro, A. Penati, G. Mossa (Chim.
Ind. [Milan] **54** [1972] 105/11).

[46] M. Pegoraro, A. Penati, G. Alessandrini (Chim. Ind. [Milan] **54** [1972] 505/13). — [47]
M. Pegoraro (Pure Appl. Chem. **30** [1972] 199/215). — [48] M. Pegoraro, A. Penati (Proc. 4th
Intern. Symp. Fresh Water Sea, Heidelberg 1973, Vol. 3, p. 129/38). — [49] J. K. Stille, M. Kamachi,
M. Kurihara (Off. Saline Water Res. Develop. Progr. Rept. No. 792 [1972]), M. Kamachi, M. Kurihara,
J. K. Stille (Macromolecules **5** [1972] 161/7). — [50] M. Kurihara, M. Kamachi, J. K. Stille (J.
Polymer Sci. Polymer Chem. **11** [1973] 587/610).

[51] G. Tanny, J. Jagur-Grodzinski (Desalination **13** [1973] 53/62). — [52] G. Canepa, S. Munari,
C. Rossi, F. Vigo (Desalination **13** [1973] 159/70). — [53] D. Mittelstädt, S. Peter (Proc. 4th Intern.
Symp. Fresh Water Sea, Heidelberg 1973, Vol. 4, p. 243/9), S. Peter, D. Mittelstädt (Kolloid Z.-Z.
Polymere **251** [1973] 225/31). — [54] H. Scott, F. L. Serafin, P. L. Kronick (J. Polymer Sci. Polymer
Letters B **8** [1970] 563/71). — [55] J. R. Hollahan, T. Wydeven (Science [2] **179** [1973] 500/1).

[56] D. McIntyre, S. Krishnamurthy, B. H. Meyer (Off. Saline Water Res. Develop. Progr. Rept.
No. 766 [1972]). — [57] R. Chapurlat (Proc. 4th Intern. Symp. Fresh Water Sea, Heidelberg 1973,
Vol. 4, p. 83/93). — [58] R. V. Cartwright, M. A. Grable, B. M. Riggleman (Off. Saline Water Res.
Develop. Progr. Rept. No. 857 [1973]). — [59] L. C. Scala, L. W. Frost, P. K. Lee, D. F. Ciliberti,
G. D. Dixon (Off. Saline Water Res. Develop. Progr. Rept. No. 860 [1973]). — [60] P. H. Wilken,
A. J. Blardinelli, J. L. Schwendeman, I. O. Salyer, L. E. Erbaugh (Off. Saline Water Res. Develop.
Progr. Rept. No. 862 [1973]).

[61] L. A. Pilato, L. M. Litz, J. E. McGrath, R. N. Johnson, I. E. Kochevas (Off. Saline Water
Res. Develop. Progr. Rept. No. 867 [1973]). — [62] A. F. Graefe, W. J. Schell, C. W. Saltonstall,
V. F. Stannet, H. E. Hopfenberg (Off. Saline Water Res. Develop. Progr. Rept. No. 870 [1973]). —
[63] H. Strathmann, R. Devarakonda (Off. Saline Water Res. Develop. Progr. Rept. No. 726 [1971]).
— [64] H. Strathmann, H. D. Saier, R. W. Baker (Proc. 4th Intern. Symp. Fresh Water Sea, Heidelberg
1973, Vol. 4, p. 381/94). — [65] H. Strathmann, R. A. Cross, T. R. Rich, B. E. Carroll (Off. Saline
Water Res. Develop. Progr. Rept. No. 894 [1973]).

[66] E. Drioli, L. Nicolais, A. Ciferri (J. Polymer Sci. **11** [1973] 3327/9), L. Nicolais, E. Drioli
(Quad. Ing. Chim. Ital. **10** [1974] 11/4). — [67] J. Kepinski, N. Chlubek (Proc. 4th Intern. Symp.
Fresh Water Sea, Heidelberg 1973, Vol. 4, p. 181/4).

3.16 Ultrathin and composite membranes

Asymmetric membranes have an active layer on a porous material of the same composition. In preparing composite membranes the active layer of a salt rejecting substance is cast on a porous substrate, which is usually of different chemical composition. The active layer might be very thin, defined as ultrathin, or simply thin in respect with the thickness of the porous substrate.

Ultrathin membranes on porous support films were prepared by Francis from polymers having a broad range of chemical structures and an estimated thickness between 1400 and 5000 Å. The polymeric structures used were containing the following groups, either as pendant groups or linking groups: ester, ether and acetal, nitrate esters, nitrile, urethane, aromatic and hydroxyl. All of these structures were permeated by water and rejected significant proportions of salt in the saline feed. Water fluxes were generally low, except for cellulose acetate (ester, acetal and hydroxyl groups). Several membrane treatments, including annealing, solvent vapor exposure, membrane drying and partial fractionation, were investigated to increase permeability of membranes or to increase their capability for salt rejection. Cellulose acetate membranes when subjected to those treatments showed a wide range of permeability and salt rejection. However, as one property was improved, the other was found to suffer [1].

Cellulose derivatives were made either by direct substitution of cellulose or by using a cellulose diacetate or dimethyl cellulose and substituting the remaining hydroxyl groups with the desired groups. Cellulose triacetate gave very favorable desalination results when evaluated as an ultrathin membrane. Fluxes about equal to secondary cellulose acetate membranes were obtained at salt rejections exceeding 99%. Cellulose triacetate is much more resistant to hydrolysis than cellulose diacetate. Addition of substituents containing any alkyl group with more than one carbon atom decreased the water flux. Methyl carbonate esters, as active substituents, appear to offer the most promise for higher flux. Methyl sulfonate esters gave better water flux but did not give high salt rejection [2].

The use of an ultrathin film of cellulose acetate together with a separate porous supporting membrane appeared to be a promising approach to circumvent the selectivity and permeability problems, encountered with the existing membranes in desalination of seawater [3]. Ultrathin membranes of β-glucan acetate appear to be as effective in reverse osmosis desalination as cellulose acetate membranes and may surpass them in some properties, such as flux decline. β-glucan acetate has also shown promise as a new polymer for anisotropic membranes. Polysulfone has some advantages over cellulose acetate as a material for supporting ultrathin membranes [4].

A process of casting ultrathin membranes directly onto the inside surface of tubes has been developed, involving the use of a two-phase cellulose acetate casting solution to cast the active ultrathin layer on a preformed tubular polysulfone support. The most effective type of module contained a precast tubular polysulfone liner, sealed into a commercially available porous resin-reinforced fiber glass tube. In brackish water tests, these modules with membranes approximately 1500 Å in thickness have provided over 95% salt rejection at up to 0.85 m³/m² · day (21 gal/ft² · day) water flux. Ultrathin membranes were removed and replaced in situ without manual manipulation of the tubes. The membrane was removed by hydrolizing it with 5.4% ammonium hydroxide and the residue floated away from the tube wall in a few minutes, without affecting the polysulfone support. After washing a new membrane could be formed on the tube by the same casting procedure [5]. In continuing work, composites containing cellulose acetate membranes approximately 400 Å thick exhibited water fluxes above 0.81 m³/m² · day (20 gal/ft² · day) throughout the entire testing period of 3000 h. During testing the fluxes declined 0.08 to 0.12 m³/m² · day (2 to 3 gal/ft² · day), after approximately 900 h, but were restored to slightly less than the initial values by cleaning with a citric acid solution. The salt rejections remained above 96% during the entire testing period. Tube composites containing membranes less than 200 Å in thickness exhibited a flux of 1.26 m³/m² · day (31 gal/ft² · day) at 94.4% salt rejection [6].

Two tubular modules containing ultrathin cellulose acetate membranes underwent a reverse osmosis test sequence. The test consisted of three runs of 100 h each with 0.5% NaCl at 42 kg/cm² (600 psi). The modules were drained after each run and held for at least 50 h in a chamber maintained at 49°C (120°F) or higher and at less than 5% relative humidity. A decline in the water flux between service cycles was observed and attributed to the hydrophobicity of the polysulfone support. The flux decline after the second drying was significantly less than that after the first drying. The flux was expected to become essentially constant after further dry cycles, probably at above 0.41 m³/m² · day (10 gal/

ft^2 · day). The incorporation of hydrophilic additives into the polysulfone support to help offset this hydrophobicity appeared to be successful [7]. In a second phase, investigations were made on a proper adhesive system for this composite membrane system to resist extreme flushing conditions while maintaining adequate flux, rejection and wet-dry properties. An adhesive system was developed that looked promising in two-feet tubes. However, it was detrimental to the membrane in five-foot tubes for the modules, but could be eliminated in another procedure used for producing a wet-dry membrane [26].

Composite membranes may have an active layer of varying thickness. Support membranes were prepared by casting mixtures of cellulose acetate and cellulose nitrate. The thin films was casted directly on the membrane with an effective thickness of 3000 to 20000 Å. A flux decline over time resulted either from changes in the water permeability of the thin film or from decreased porosity of the support membrane [8]. In further development work on thin-film composite membranes, methods were developed for the semicontinuous production of porous cellulose ester membranes and cellulose nitrate-cellulose acetate membranes were produced in pieces that were used for preparing composites. The importance of the membrane casting temperature has been demonstrated and methods for coating the pores of the support membrane were investigated. The technique studied in most detail was airless spraying of polyacrylic acid from a dilute ethanol-water solution [9]. However, the cellulose nitrate-cellulose acetate support does not possess the highly and finely porous surface structure that is necessary to combine maximum salt rejection with the high fluxes achievable only with very thin films. Membranes prepared by depositing a film of polyacrylic acid on the surface of a finely porous support appear to have potential for the desalination of brackish waters [10].

The properties of composite membranes, prepared by casting polyacrylic acid on or by dip-coating the glossy surface of a porous support, were described by Sachs and Lonsdale. These membranes exhibited reverse osmosis performance for dilute solutions containing a single salt and also a mixture of salts markedly better than those of polyacrylic acid membranes. The water flux is linear in the pressure range up to 100 kg/cm^2. Salt rejection is a function of pressure, of the concentration of the feed solution and of the charge of the counterion [11].

Salt-rejecting membranes can be formed from a variety of additives on woven hoses coated with filter aid. Intrinsic salt rejections as high as 85% with production rates of 3.87 m^3/m^2 · day (95 gal/ft^2 · day) were observed at pressures of 19 kg/cm^2 (275 psi). Production rates were proportional to circulation velocity and temperature [12].

Cellulose acetate membranes, cast from a ternary casting solution of cellulose acetate-formamide-acetone under a variety of conditions on various support materials, were tested with the objective to examine the possibility of casting thin membranes. Good quality membranes can be cast by either dip casting or extrusion techniques [13].

Impregnated felt has acceptable mechanical and flux transmission properties. A Loeb-type membrane cast on such felt operated normally during a 30 days test [14].

The effect of porous substrate characteristics on the overall performance of a composite reverse osmosis membrane assembly was investigated by Dagan and Gollan for the two most representative concepts, the parallel and the radial flow pattern systems. Pressure drop and flux decline curves were determined as functions of physical parameters of the system [15].

A composite membrane having a 400 Å semipermeable barrier has been developed by Riley et al. for seawater desalination. A thin film of a polymer, generally cellulose triacetate, is formed directly upon the finely porous surface of a supporting membrane. The membrane was considered capable of desalting seawater in a single pass [16]. A major improvement in the desalination properties of polyacrylic acid composite membranes has been achieved by using porous polysulfone support, in replacement of the previously used cellulose nitrate-cellulose acetate support. The polysulfone support has the advantage of greater chemical resistance and, thus, the polyacrylic acid membrane can resist hydrolysis over a wide range of pH. Polyacrylic acid-polysulfone membranes have exhibited flux and rejection characteristics superior to the high-flux asymmetric cellulose acetate membranes under brackish water desalination operating pressures. They have shown water fluxes greater than 1.83 m^3/m^2 · day (45 gal/ft^2 · day) with rejections of about 75% from a 0.05 M sodium chloride feed at 56 kg/cm^2 (800 psi). However, the membranes have shown a decided tendency to foul, both by organic and inorganic contaminants in the feed, but they also have a great capability for recovery and withstand drastic cleansing procedures [17].

The design, fabrication and assembly of a prototype unit for the continuous production of thin-film composite membrane support was reported by Riley et al. The release properties of the membrane from a number of stainless steel casting surfaces were investigated and a method was devised for maintaining a constant level of solution within the casting knife reservoir during casting. The formation of 800 Å thin-films simulated continuous casting conditions and the casting of the membrane directly on fabric for improved strength were also examined [18]. In continuing research the cause of aging of the cellulose nitrate-cellulose acetate support and possible solutions for the effect of this aging on the physical and transport properties of the thin-film composite membrane were investigated. A composite membrane spiral-wound module (see chapter 3.20, p. 297) was developed and a study of the wet-dry stability of the composite membrane module was conducted in an effort to develop a module capable of undergoing repeated wet and dry cycles without a loss in water flux or an increase in salt permeability. Crosslinking of the support prior to application of the thin film was also examined [19].

Scaling-up of the preparation of the thin-film composite membrane, to be used for single-stage seawater desalination in the spiral-wound configuration, using the new continuous prototype casting machine, was undertaken. Methods for optimizing the transport properties of the composite membrane were investigated, as well as testing with seawater as function of pressure, pH and water recovery under simulated spiral-wound module operating conditions. Porous supporting membranes were prepared from copolymers of polyacrylonitrile and polyvinylchloride. Novel noncellulosic materials, primarily of the aromatic heterocyclic types, were synthesized and evaluated [20]. Advances in the continuing development of the thin-film composite membrane have made it possible to produce potable water from seawater in a single stage at water recovery rates greater than 50%. Stable long-term water fluxes of 1.02 $m^3/m^2 \cdot$ day (25 gal/ft$^2 \cdot$ day) and NaCl rejections of greater than 99.5% were attained with seawater at an applied pressure of 102 kg/cm^2 (1500 psi) [21].

Composite membranes have also been prepared by Yasuda by the vapor phase polymerization and crosslinking of heteroaromatic nitrogen compounds, aromatic amines, aliphatic amines and derivatives of pyrrolidone in an electrodeless glow discharge tube. The plasma polymerized semipermeable membranes are highly crosslinked and are intimately bonded to the substrate. The ultrathin membranes have exhibited salt rejections greater than 99.5% at high water flux levels. Membrane performance was found to be stable after 20 days with no visible decline in water flux using 3.5% NaCl at 105 kg/cm^2 (1500 psi). It was anticipated that proper selection of membrane and substrate would lead to an excellent composite membrane and that the plasma polymerization technique can be applied to porous substrates in planar, tubular and hollow fiber form [22].

An improved porous support was developed by Smith and Sirkar, consisting of a porous polyethylene tube with an external braid of stainless steel wire. After application of end fittings, a membrane is inserted to form the complete desalination unit. The composite support containing a cellulose acetate membrane, heat treated in situ, was subjected to desalination testing. It was concluded that there was no gross damage to the membrane in the test of the reinforced support [23].

Model glass fiber reinforced epoxy resin tubes, produced by a novel winding machine, are considered suitable for reverse osmosis membrane supports showing water permeabilities in the range 2.0 to 20 $m^3/m^2 \cdot$ day and high strength with ultimate tensile stresses. The technique involves glass melting, fiber production, resin impregnation and winding in one operation followed by the appropriate resin cure [24].

The preparation of a composite membrane is reported by Lefebvre and Van Haute, using an impregnated synthetic paper with pores of an average radius of 12000 Å as a support. A phenol-formaldehyde solution has been used for impregnating the paper up to 42 g/m^2. The products used for the active layer are cellulose triacetate and cellulose acetate. To obtain a convenient adherence to the impregnated paper, it was necessary to apply an intermediate layer of polyacrylic acid. With a salt solution of 5000 mg/l rejections of 30 to 90% and quantities of desalted water of 0.65 to 0.02 $m^3/m^2 \cdot$ day were obtained [25].

Literature to 3.16

[1] P. S. Francis (Off. Saline Water Res. Develop. Progr. Rept. No. 177 [1966]). — [2] P. S. Francis, J. E. Cadotte (Off. Saline Water Res. Develop. Progr. Rept. No. 247 [1967]). — [3] U. Merten, H. K. Lonsdale, R. L. Riley, K. D. Vos (Desalination 3 [1967] 353/8). — [4] L. T. Rozelle, J. E. Cadotte, R. D. Corneliussen, E. E. Erickson (Off. Saline Water Res. Develop. Progr. Rept. No. 359

[1968]). — [5] L. T. Rozelle, J. E. Cadotte, D. J. McClure (Off. Saline Water Res. Develop. Progr. Rept. No. 531 [1970]).

[6] L. T. Rozelle, J. E. Cadotte, W. L. King, A. J. Senechal, B. R. Nelson (Off. Saline Water Res. Develop. Prog. Rept. No. 659 [1971]; Reverse Osmosis Membrane Research, New York 1972, p. 419/35). — [7] L. T. Rozelle, J. E. Cadotte, B. R. Nelson (Off. Saline Water Res. Develop. Progr. Rept. No. 725 [1971]). — [8] R. L. Riley, H. K. Lonsdale, L. D. LaGrange, C. R. Lyons (Off. Saline Water Res. Develop. Progr. Rept. No. 386 [1969]). — [9] H. K. Lonsdale, R. L. Riley, L. D. LaGrange, A. S. Douglas, U. Merten (Off. Saline Water Res. Develop. Progr. Rept. No. 484 [1969]). — [10] H. K. Lonsdale, R. L. Riley, C. E. Milstead, L. D. LaGrange, A. S. Douglas, S. B. Sachs (Off. Saline Water Res. Develop. Progr. Rept. No. 577 [1970]).

[11] S. B. Sachs, H. K. Lonsdale (Proc. 3rd Intern. Symp. Fresh Water Sea, Dubrovnik 1970, Vol. 2, p. 561/78; J. Appl. Polymer Sci. **15** [1971] 797/809). — [12] J. A. Dahlheimer, D. G. Thomas, K. A. Kraus (Ind. Eng. Chem. Process Design Develop. **9** [1970] 566/9). — [13] R. L. Nickelson, E. A. Birkhimer, D. E. Coverdell, Y. Y. Lai, D. G. J. Wang (Off. Saline Water Res. Develop. Progr. Rept. No. 520 [1970]). — [14] A. C. Wrotnowski, E. G. Bernard (Off. Saline Water Res. Develop. Progr. Rept. No. 599 [1970]). — [15] G. Dagan, A. Gollan (Off. Saline Water Res. Develop. Progr. Rept. No. 614 [1970]), A. Gollan, G. Dagan (Desalination **8** [1970] 261/75).

[16] R. L. Riley, H. K. Lonsdale, C. R. Lyons (Proc. 3rd Intern. Symp. Fresh Water Sea, Dubrovnik 1970, Vol. 2, p. 551/60). — [17] R. L. Riley, C. E. Milstead, H. K. Lonsdale, K. J. Mysels (Off. Saline Water Res. Develop. Progr. Rept. No. 729 [1971]). — [18] R. L. Riley, G. R. Hightower, H. K. Lonsdale, J. F. Loos, C. R. Lyons (Off. Saline Water Res. Develop. Progr. Rept. No. 799 [1972]). — [19] C. E. Milstaed, G. R. Hightower, C. R. Lyons, K. J. Mysels, R. L. Riley (Off. Saline Water Res. Develop. Progr. Rept. No. 847 [1973]). — [20] R. L. Riley, C. E. Milstead, W. J. Wrasildo, R. L. Grabowsky, G. R. Hightower (Off. Saline Water Res. Develop. Progr. Rept. No. 851 [1973]).

[21] R. L. Riley, G. R. Hightower, C. R. Lyons, M. Tagami (Proc. 4th Intern. Symp. Fresh Water Sea, Heidelberg 1973, Vol. 4, p. 333/47), R. L. Riley, G. R. Hightower, C. R. Lyons (Appl. Polymer Symp. No. 22 [1973] 255/67). — [22] H. Yasuda (Off. Saline Water Res. Develop. Progr. Rept. No. 811 [1972]; Appl. Polymer Symp. No. 22 [1973] 241/53). — [23] H. P. Smith, K. K. Sirkar (Off. Saline Water Res. Develop. Progr. Rept. No. 807 [1972]). — [24] T. R. Bott, K. D. Walford (Proc. 4th Intern. Symp. Fresh Water Sea, Heidelberg 1973, Vol. 4, p. 73/82). — [25] C. Lefebvre, A. Van Haute (Proc. 4th Intern. Symp. Fresh Water Sea, Heidelberg 1973, Vol. 4, p. 227/34).

[26] J. E. Cadotte, K. E. Cobian, L. T. Rozelle (Off. Saline Water Res. Develop. Progr. Rept. No. 874 [1973]).

3.17 Other types of membranes

Andere Membran-typen

Biological and protein based membranes. Membranes perform very important functions in living organisms. Biological membranes possess carrier mechanisms, which transport ions against chemical and electropotential gradients through the expenditure of metabolic energy. Ion transport is carried out with a high degree of selectivity. Kidneys cleanse the blood of chemical wastes and regulate the acid–base balance of the body. Sweat and secretory glands exhibit a salt-conserving action. Concentration of sodium chloride is effected by intestinal mucosa, gills of fish and the skin of amphibians. Analogies were investigated in natural phenomena with the aim of drawing conclusions for practical application in membrane processes.

The common ability of animals and cells to accumulate and maintain ions against concentration gradients is termed active transport. Sodium chloride transport through the isolated frog skin was extensively studied by many investigators. A method seeking to utilize physiological transport of sodium chloride by living organisms aims at the use of unicellular algae, which, possessing large surface area per unit volume, would accumulate sodium chloride in an optimum environment of saline water. Transferred to an unfavorable environment, the algae should dump sodium chloride in approaching equilibrium with the environment. After this they would be removed to the original suspension to accumulate more sodium chloride, gradually depleting the salinity of the water as the cycling continues. An experimental investigation indicated that the process might be feasible as well as that algae might well be used for protein production [1].

In further studies of ion transport across microbial membranes, the mechanisms involved in the transport of inorganic ions across the cytoplasmic membranes of Serratia marcescens and Serratia marinorubra were investigated. The cells of the latter were markedly more impermeable to monovalent cations than Serratia marcescens, when grown in a medium lacking sodium chloride and examined under essentially identical conditions [2]. Both the microorganisms tend to accumulate potassium ions during growth and to exclude sodium ions against concentration gradients. The amount of potassium ions accumulated within the cells is increased, as the ionic strength is raised in the medium in which the organisms are grown [3].

The mechanism of ion transport in plant tissues was investigated by Epstein [4]. The physical chemistry of complex acidic lipics, which are compounds of cell membranes and have important functions in regard to the movement of ions across these membranes, has been studied by Katzman [5]. Sodium transport across living membranes was investigated by Bricker, using as experimental model the isolated urinary bladder of the fresh water turtle [6]. The influence of lipid phase transitions on the diffusion characteristics of sodium and potassium in the polar regions of condensed polar lipid-like systems and the bearing of these effects on ionic regulatory mechanisms of biological membranes is reported by Nelson and Blei [7].

Salient mechanistic and structural characteristics of selective ion transport by avian salt glands were investigated with the objective of application on a desalination process. The salt glands of White China geese have been studied both in vivo and in vitro by Usdin et al. [8]. Considerable progress was made by Bean et al. in developing an understandig of the nature of ion movement across biological membranes. Reasonable extrapolations can be made to the formulation of model systems for desalination using biological principles and materials [9].

Several models have been proposed to describe the structure of biological membranes. Many theories differ fundamentally because the properties of this biological entity are very poorly defined. The nature of the lipid-protein interaction has not been established. The organization of the protein in the membrane is not clearly understood and the manner in which the membrane effects selective permeability is not known. An attempt was made by Zull to improve understanding of some of these fundamental problems [10].

The permeability properties of natural lipoprotein films derived from red cell membranes were investigated by Meriwether et al. The presence of lipoprotein from red cell membranes in various porous membrane supports decreases the salt flux and increases the water flux through these filters. The properties of cellulosic membranes cast from formulations containing lipoproteins and their components have proven interesting. Investigation of the isolated membrane components supports the dominating role of the protein in these systems. It appears that the protein portion serves as the superstructure of the cell membrane, determining the three dimensional organization, whereas the lipid components fill in the pores of this superstructure in order to control and select molecules that may penetrate [11].

In continuation of previous work the energetics of sodium transport in frog skin were investigated, especially oxygen consumption in the short-circuited state and the effects of electrical potential on oxygen consumption. An analysis was given of nonequilibrium thermodynamics of ion transport and membrane metabolism. Flux ratios and isotope interaction in an ion exchange membrane were investigated [12].

Work was also described by Ginzburg and Ginzburg on two species of unicellular organisms, Halobacterium and Dunaliella parva, isolated from the Dead Sea. Both species have been found to be highly permeable to the hydrogen ion and to the ions of the alkali metals. The cells of these organisms are not surrounded by any functional membrane. It appears that the mechanism for salt control in Halobacterium lies in the structure of the cell proteins and in the form taken by potassium within the cell. Changes in either of these two parameters lead to changes in the concentration of cell sodium. Thus, the site of control is within the cell interior rather than in the cell membrane, as is the case in other organisms [13].

The role of water structure, especially the structure of water near interfaces, in the properties and functioning of membranes, particularly biological membranes, was investigated by Drost-Hansen. It was found conceptually important to study membranes ranging from the most simple types to complex, living membranes. It was suspected and, in turn, supported by the results of the experiments, that the structuring of water in or near membranes appears to play an important and at times dominating role in determining functional properties of membranes, regardless of the level of morphological complexity or the operational intricacies of the membranes [14].

A review of biological desalination phenomena was presented by Katchalsky with emphasis on the advantage of evolution in the geological time span, which permitted the organisms extensive experimentation and ultimately resulted in the "Wisdom of the organism". It may also be of profit to the student of technical desalination [15].

A program of exploratory research was undertaken by Fisher and Hsiao to determine the intrinsic transport properties of the protein zein, the prolamine fraction of corn gluten, in homogeneous film form toward water, sodium chloride and other solutes. Crosslinked zein films exhibited mechanical and permeability properties in practical levels comparable to other desalination membranes. Reflection coefficients to NaCl in the range of 0.90 to 0.99 were obtained with intrinsic water flux values similar to those of cellulose acetate. Incorporation of 10 to 15% of poly(methacrylamide) into crosslinked zein films, followed by post-crosslinking, resulted in a 25% improvement of their intrinsic water permeability and a 5% increase in their selectivity to sodium chloride. Acetylation of zein markedly reduced its hydrophilic character. Films of acetyl zein exhibited very high selectivity toward sodium chloride, but low intrinsic water permeability. The relative permeability of crosslinked zein films to various solutes appeared to be governed mainly by solute size. The permeabilities decreased in the order urea, NaCl, $CaCl_2$, $MgCl_2$, Na_2SO_4, $MgSO_4$ and sucrose. Selectivity toward sulfates and sucrose was very high and indicates potential for zein as a barrier to desalination of high sulfate-content brackish waters [16].

In continuation of previous work, zein was prepared in the form of hollow fibers, as ultrathin film on porous substrates and as flat films, each crosslinked and cured to achieve insolubility in water and withstand high pressures. Hollow fibers were prepared in a variety of wall thicknesses by spinning zein dissolved in isopropyl alcohol/water or chloroform/methyl alcohol mixtures to which were added a formaldehyde-type crosslinking agent and catalyst. Similar zein solutions were used to prepare ultrathin zein films on a polyacrylonitrile-polyvinylchloride filter paper of uniform pore size. In reverse osmosis tests, the best results were obtained with filters precoated lightly with a polyacrylic acid/glycerol mixture to provide a base for the zein/formaldehyde applications. Zein solutions containing a formaldehyde-type crosslinking agent and fortified with a polymer additive such as polyacrylic acid or polyethylene glycol were cast into sheets, which became ultrathin when dried and cured. Rewet samples exhibited good rejection and permselective properties in reverse osmosis tests using 0.5% NaCl solutions [17].

Water permeability coefficients, as well as their temperature dependence, were determined by H. T. Tien for black lipid membranes formed from oxidized cholesterol, cholesterol-dodecyl acid phosphate, cholesterol-hexadecyl trimethylammonium bromide and chloroplast extracts. It has been demonstrated that photoelectric effects are possible with the chloroplast black lipid membranes and that transport of water can be initiated by light. The most important conclusion is that black lipid membranes constituted from photoactive pigments are a unique model system for study, which could aid in the understanding and exploitation of the photolysis of water [18].

Ion and water permeability of lipid phases was studied by Bean with lipids incorporated into membranes or dispersed in a bulk phase. Lipid-loaded membranes generally had a low resistance at temperatures below a transition temperature near the melting point of the lipid. At this point an abrupt and important increase in resistance occurred in all cases, except with primary alkylamines. At temperatures below the liquefying transitions, some membranes had excellent ion selective properties [19].

Charged membranes. Salt rejection in charged membranes was evaluated by Hoffer and Kedem according to the fixed charge model as described by Teorell [20] and by Meyer and Sievers [21]. Taking into account the concentration dependence of the transport coefficients, it was found that salt rejection at given volume flow is a function of the ratio between membrane charge and salt concentration, the mobilities of the ions and the effective thickness of the membrane. The limiting salt rejection at high flow rates is given by the reflection coefficient, which is completely determined by the ratio between membrane charge and feed concentration, the valencies of the ions and the transport number of the counter-ion in free solution. For divalent counter-ions salt rejection was lower. High salt rejection was obtained for divalent co-ions [22]. Charged membranes were also prepared by crosslinking albumin in a collodion matrix. High water flow was obtained. Reverse osmosis of hydrochloric acid and of sulfuric acid in this membrane gave negative rejection. Positive rejection of copper ions and negative rejection of negative ions were obtained in mixtures of copper sulfate and sulfuric acid and of copper chloride and hydrochloric acid [23]. Polylysine and polyvinylamine were used as polyelectrolytes to prepare porous charged membranes. High rate fluxes were obtained, but low salt rejections. Good separation between hydrochloric acid and copper chloride was achieved. The acid concentrates in the product and the salt remains in the feed [24].

In further investigations the same group of authors calculated salt rejections of copper chloride and hydrochloric acid, using the fixed charge membrane model described by Teorell and by Meyer and Sievers. Concentration gradients of these species were numerically integrated across the membrane and a numerical optimization method was used to evaluate rejections. A considerable separation between the two cations was to be expected and, in many cases, separation was larger than the gap between the rejections during reverse osmosis of separate solutions. This difference can be explained in terms of different ion distributions at the feed boundary and of the interdependence of streaming potentials. Appreciable separations between divalent or trivalent salt solutions and corresponding mineral acids can also be achieved by reverse osmosis of mixtures through ion exchange membranes. Enrichment of acid in the product and concentration of salt in the feed were obtained. The gap between rejections was larger in reverse osmosis of mixtures than in that of separate solutions [25]. In continuation of previous work on water transport through various types of charged membranes, Vofsi and Kedem have prepared asymmetric membranes from maleinated acetyl cellulose and have examined their performance under reverse osmosis conditions. Prediction of salt rejection from equilibrium measurements and streaming potential was made [26].

Ion distribution between polybase-collodion membranes and aqueous salt solutions could be described by an effective charge density, following the association model for polyelectrolyte solutions. The effective charge density resulting from the equilibrium measurements was introduced into the equation for salt rejection, derived previously for the Teorell-Meyer-Sievers model [27]. The reverse osmosis properties of fixed charge membranes containing pores of various widths have been examined under conditions where flows are large and restricted to the pores. A generalized expression for the salt rejection coefficient has been derived, indicating that the relation between salt rejection and the salt distribution coefficient simplifies to be of the same form as that which applies for the fixed charge membrane described by Teorell, Meyer and Sievers. The absolute values of the rejection coefficient are smaller, however, for porous membranes [28]. A procedure for the measurement of ionic transport numbers, suitable for adsorbed polymer layers and dynamic membranes, was suggested by Kedem. For membranes adequately described by the fixed charge model, allowing for ion association, a simple relation between transport numbers and salt rejection is derived [29].

Assuming the Donnan equilibrium as boundary condition, a differential equation was numerically solved by Nakagaki and Kobayashi. The equation expresses the ionic flux inside the membrane, when aqueous electrolyte solutions are separated by a membrane with fixed charge [30].

Ion exchange membranes. Ion exchange membranes have been considered for desalination by the reverse osmosis process, because of their ion excluding ability. Reasonable salt rejection can be achieved with ion selective membranes. However, the selective property of the membranes decreases with increasing salt concentration and most of the conventional ion exchange membranes have been found to exhibit uninterestingly low water permeability [31].

Methods have been presented by Merten et al. for calculating the performance of ion exchange membranes in reverse osmosis as a function of the feed concentration, fixed-charge density, pore radius and porosity. The applied pressure, required to achieve a given water flux and salt rejection, has been defined in terms of the physical structure of the membrane and the mobility of ions within the pore fluid. The membrane properties predicted to be close to optimum have been determined and it was concluded that, in principle, ion exchange membranes of this type should be useful for brackish water recovery [32]. Membranes, 30 to 40 μ thick, demonstrated either high flow and no salt rejection, indicating cracks, or unmeasurably low flow. An investigation of the pore radii showed that the pore radii were exceedingly small, 2.5 to 5 Å, indicating either that many of the pores were blind or that the pores were very small with the membrane preparation conditions used [33].

Investigations by Pusch on reverse osmosis using ion exchange membranes include the determination of the mechanical permeability, the reflection coefficient, the solute rejection efficiency and the volume flow density for various aqueous solutions. The dependence of the salt rejection and the volume flow density on pressure was measured up to 400 atm using solutions of hydrochloric and sulfuric acids. The salt rejection and the volume flow density were determined with cation exchange membranes as function of concentration with aqueous solutions of sodium hydroxide and sodium chloride. The measured dependence of the salt rejection and volume flow density on pressure yields that above 10 atm there exists no linear relation between pressure and either salt rejection or volume flow density [34].

Polyelectrolytes were grafted onto cellophane, utilizing both strong-electrolyte cationic and anionic types. Anion exchange films were prepared by appropriate reactions with poly(vinyl chloride) films. Membranes were also cast from mixtures of polyelectrolytes and cellulose acetate and tested under reverse osmosis conditions [35]. Experiments on transport of salts and water across ion exchange membranes are also reported by Meares and Foley [36].

Dynamically formed hydrous zirconium oxide and aluminum oxide membranes were prepared by Hoornaert et al. and tested under reverse osmosis conditions. The membranes were much more suitable for divalent co-ions than for monovalent co-ions and were very poor for polyvalent counter-ions [37]. Pilot plant tests with composite zirconium oxide-polyacrylic acid dynamic membranes showed an average rejection of total dissolved solids in the range of 90 to 91%, sulfate rejection 93 to 95%, and chloride rejection 83 to 94%. The flux was over 4 $m^3/m^2 \cdot$ day (100 $gal/ft^2 \cdot$ day) after two weeks of continuous operation [38]. Membranes from inorganic hydrous oxide gels were also dynamically prepared and tested by Kuppers [39].

Wolf and Abicht reported that anion exchange membranes exhibit a lower water permeation than membranes based on cellulose acetate. However, salt rejection might exceed 90% [40].

An investigation was also conducted into the behavior of a dynamically formed membrane consisting of small particles of ion exchange material. The determined retention values appeared to agree quantitatively with Donnan's exclusion theory if a certain degree of leakage of the membrane is assumed. The compressibility of the membrane material explains the behavior of product flux, leakage and capacity of the ion exchanger as a function of the pressure. The temperature was found to have considerable influence on the permeate flux, but not on the actual retention [41].

Thin-wall tubes of high ion exchange capacity were prepared by chlorosulfonation of polyethylene films swollen with carbon tetrachloride and casting the chlorosulfonated material from its solutions in chlorobenzene. The chlorosulfonated polyethylene is converted by hydrolysis into cation exchange membranes, while amination followed by quaternization converts it into an anion exchange material. The water uptake of the ionically charged material may be reduced by chemical or thermal crosslinking of the chlorosulfonated polyethylene. The performance of such membranes under reverse osmosis conditions and in dialytic Donnan water softening was investigated by Jagur-Grodzinski et al. [51].

Membranes containing graphitic oxide. Graphitic oxide layers deposited on porous supporting bases are able to function as semipermeable membranes in separating sodium chloride from its solution in water. The water flux rates ranged from 0.03 to 0.24 $m^3/m^2 \cdot$ day (0.7 to 5.8 $gal/ft^2 \cdot$ day) and the salt rejections from 65 to 83%. Both membrane characteristics are lower than those given for cellulose acetate membranes [42].

Graphitic oxide membranes and membranes formed from bentonite and vermiculite were also tested for reverse osmosis performance. The performance of the graphitic oxide membranes was characteristic of a weak-acid cation exchange membrane. A general change in performance properties was observed and aging of the original graphitic oxide preparations was suspected. Sodium bentonite membranes were fabricated, containing additions of polyanions such as oxalate, polyphosphates or tannic acid as particle edge modifier, and showed a salt rejection efficiency of 57 and 60%. In some membranes fluxes as high as 1.14 $m^3/m^2 \cdot$ day (28 $gal/ft^2 \cdot$ day) were measured, however the salt rejection efficiency of the membrane was only 28%. Vermiculite membranes gave rejection efficiencies of only 10% [43].

The mechanical strength of graphitic oxide membranes was significantly improved by overcoating them with a layer of microcrystalline collagen. Use of collagen layers to sandwich the graphitic oxide membrane affected adversely desalination performance, while further improving mechanical strength. Standard six layer membranes exhibited salt rejection values of 94.2% and water flux of 0.022 $m^3/m^2 \cdot$ day (0.55 $gal/ft^2 \cdot$ day) when used with 0.5% NaCl solution at feed pressures of 42 kg/cm^2 (600 psi). Ammonium bentonite membranes could achieve salt rejection efficiency values varying from 60 to 65%, according to the mode of operation [44].

Porous glass membranes. The use of porous glass as membrane material has been explored using the glass in the form of tublets, which gave sodium chloride rejections as high as 97% from a 10000 mg/l salt feed. It has been demonstrated that porous glass membranes can reject a wide variety of materials, such as urea, phenol, nitrates, borates and magnesium salts, to about the same extent as sodium chloride and have therefore a potential for use in the reverse osmosis process as a hollow fiber material. The relationships between manufacturing variables and performance as a reverse osmosis membrane were determined by exploring the effect of composition of the parent glass, such as silica content, type of alkali and other constituents. The conditions of the phase separation were optimized for each composition and a novel leaching process worked out. The data suggest that useful combinations of salt rejection and flux can be obtained. The major problem is the stabilization of the silica matrix to maintain high performance over long periods of time. Very promising results were obtained in the application of porous glass membranes to life support problems, such as urine and waste water recovery. High quality water was obtained from waste water and the ability of porous glass to handle raw urine without pretreatment was established [45].

In a continuing research new materials were developed having vastly superior properties. Salt rejection was improved from 20 to 40% to well over 90%, flux from 0.04 to 0.08 $m^3/m^2 \cdot$ day (0.1 to 0.2 gal/ft$^2 \cdot$ day) up to 0.1 $m^3/m^2 \cdot$ day (2.5 gal/ft$^2 \cdot$ day) and service life from less than 200 h to at least 3000 h. The latter result war obtained with 3% zirconium oxide as stabilizing agent [46].

Treatment by reverse osmosis of sodium chloride with porous glass membranes, at 40 to 120 kg/cm^2 pressure, exhibited rejection data which were consistent with the functioning of the porous glass as a low capacity ion exchange membrane [47]. The pore size distributions of porous glass reverse osmosis membranes were calculated from nitrogen desorption isotherm data. Pore volume distributions of unused membranes peaked sharply at 19 to 22 Å pore radius. Membranes from batches with higher solute rejection appeared to have sharp cut-offs of the pore volume distribution tail for larger pores and median pore radius less than 20 Å. A permeability equation with a tortuosity term was used to relate porous glass membrane permeabilities to the pore size distributions. Membranes that lost solute rejection capability during life tests showed a marked broadening of the pore volume distribution and a new pore volume distribution peak at 60 to 80 Å radius [48].

Hollow fiber membranes suitable for the two-stage desalination of seawater have been prepared from two-phase-separable Na_2O-SiO_2-B_2O_3 glasses. The hollow fibers had outside diameters in the range 20 to 200 microns and corresponding wall thicknesses of 5 to 40 microns. Using a 3.5% NaCl solution at 120 bar, product fluxes of about 1.0 $m^3/m^2 \cdot$ day at 88% rejection were obtained with fibers of one glass and fibers of the other glass withstood hydraulic compressive pressures over 200 bar. There was no evidence of compaction or hydrolytic degradation [49].

The role of water in porous glass desalination was investigated by Belfort and Scherfig, as well as the question whether the state of water inside the pores is similar to bulk water or not. The investigation was extended by two complementary studies. The first study is macroscopic in nature and concerns the determination of transport coefficients of the glass membranes. The second study is microscopic in nature and involves the proton relaxation phenomena of the interfacial water inside the pores using pulse nuclear magnetic resonance. The results of these independent studies corroborate the existence of motionally restricted water adsorbed on the pores of porous glass [50].

Literature to 3.17

[1] Resources Research Inc. (Off. Saline Water Res. Develop. Progr. Rept. No. 52 [1961]). — [2] B. H. Goldner (Off. Saline Water Res. Develop. Progr. Rept. No. 160 [1965]). — [3] N. L. Gale, B. H. Goldner (Off. Saline Water Res. Develop. Progr. Rept. No. 323 [1968]). — [4] E. Epstein (Off. Saline Water Res. Develop. Progr. Rept. No. 161 [1965]). — [5] R. Katzman (Off. Saline Water Res. Develop. Progr. Rept. No. 178 [1966]).

[6] N. S. Bricker (Off. Saline Water Res. Develop. Progr. Rept. No. 206 [1966]). — [7] S. S. Nelson, I. Blei (Off. Saline Water Res. Develop. Progr. Rept. No. 221 [1966]). — [8] E. Usdin, J. M. Spurlok, J. A. Simmons (Off. Saline Water Res. Develop. Progr. Rept. No. 433 [1969]). — [9] R. C. Bean, R. W. Albers, S. R. Caplan, E. Epstein, R. Katzman (Off. Saline Water Res. Develop. Progr. Rept. No. 513 [1970]). — [10] J. E. Zull (Off. Saline Water Res. Develop. Progr. Rept. No. 652 [1971]).

[11] L. S. Meriwether, L. K. Bjornson, H. F. Jaillet (Off. Saline Water Res. Develop. Progr. Rept. No. 667 [1971]). — [12] A. Essig (Off. Saline Water Res. Develop. Progr. Rept. No. 752 [1972]). — [13] B. Z. Ginzburg, M. Ginzburg (Off. Saline Water Res. Develop. Progr. Rept. No. 751 [1972]). — [14] W. Drost-Hansen (Off. Saline Water Res. Develop. Progr. Rept. No. 790 [1972]). — [15]

A. Katchalsky, B. Z. Ginzburg, M. Ginzburg (Proc. 1st Intern. Symp. Water Desalination, Washington, D.C., 1965 [1967], Vol. 1, p. 441/51).

[16] B. S. Fisher, H. Y. Hsiao (Off. Saline Water Res. Develop. Progr. Rept. No. 482 [1969]). — [17] N. R. S. Hollies, H. Y. Hsiao, B. S. Watson (Off. Saline Water Res. Develop. Progr. Rept. No. 663 [1971]). — [18] H. T. Tien (Off. Saline Water Res. Develop. Progr. Rept. No. 696 [1971]). — [19] R. C. Bean (Off. Saline Water Res. Develop. Progr. Rept. No. 812 [1971]). — [20] T. Teorell (Proc. Soc. Exptl. Biol. Med. 33 [1935] 282/5).

[21] K. H. Meyer, J. F. Sievers (Helv. Chim. Acta 19 [1936] 649/77). — [22] E. Hoffer, O. Kedem (Desalination 2 [1967] 25/39). — [23] E. Hoffer, O. Kedem (Desalination 5 [1968] 167/72). — [24] E. Hoffer, J. Sinnreich (Off. Saline Water Res. Develop. Progr. Rept. No. 591 [1970] 21/6). — [25] E. Hoffer, O. Kedem (Ind. Eng. Chem. Process Design Develop. 11 [1972] 221/5, 227/8).

[26] D. Vofsi, O. Kedem (Off. Saline Water Res. Develop. Progr. Rept. No. 324 [1968], No. 401 [1969], No. 591 [1970], No. 653 [1971], No. 787 [1972], No. 850 [1973]). — [27] E. Hoffer O. Kedem (J. Phys. Chem. 76 [1972] 3638/41). — [28] R. Simons, O. Kedem (Desalination 13 [1973] 1/16). — [29] O. Kedem (Israel J. Chem. 11 [1973] 313/4). — [30] M. Nakagaki, M. Kobayashi (Yakugaku Zasshi 93 [1973] 864/74; C.A. 79 [1973] No. 86813).

[31] J. G. McKelvey, K. S. Spiegler, M. R. Wyllie (Chem. Eng. Progr. Symp. Ser. 55 No. 24 [1959] 199/208). — [32] U. Merten, H. K. Lonsdale, R. L. Riley, K. D. Vos (Off. Saline Water Res. Develop. Progr. Rept. No. 265 [1967] 122/36). — [33] U. Merten, H. K. Lonsdale, R. L. Riley, K. D. Vos (Off. Saline Water Res. Develop. Progr. Rept. No. 369 [1968] 78/81). — [34] W. Pusch (Ber. Bunsenges. Physik. Chem. 74 [1970] 444/9; Proc. 3rd Intern. Symp. Fresh Water Sea, Dubrovnik 1970, Vol. 2, p. 535/49). — [35] W. H. Baldwin, C. E. Higgins, J. Csurny (Off. Saline Water Res. Develop. Progr. Rept. No. 508 [1970] 163/8).

[36] P. Meares, T. Foley (Off. Saline Water Res. Develop. Progr. Rept. No. 584 [1970]). — [37] P. Hoornaert, C. Lefebvre, A. Van Haute (Desalination 11 [1972] 315/27). — [38] D. G. Thomas (Proc. 4th Intern. Symp. Fresh Water Sea, Heidelberg 1973, Vol. 4, p. 407/16). — [39] J. R. Kuppers (J. Colloid Interface Sci. 31 [1969] 577/8). — [40] F. Wolf, K. Abicht (Plaste Kautschuk 19 [1972] 20/6).

[41] J. W. van Heuven, R. K. Bloebaum (Proc. 4th Intern. Symp. Fresh Water Sea, Heidelberg 1973, Vol. 4, p. 435/42). — [42] L. C. Flowers, D. E. Sestrich, D. Berg (Off. Saline Water Res. Develop. Progr. Rept. No. 224 [1966]). — [43] L. C. Flowers, P. K. Lee, E. S. Bober, M. H. Gjertsen, D. Berg (Off. Saline Water Res. Develop. Progr. Rept. No. 418 [1969]). — [44] E. S. Bober, L. C. Flowers, P. K. Lee, D. E. Sestrich (Off. Saline Water Res. Develop. Progr. Rept. No. 544 [1970]). — [45] F. E. Littman, G. A. Guter (Off. Saline Water Res. Develop. Progr. Rept. No. 379 [1968], No. 505 [1970].

[46] F. E. Littman, F. D. Kleist, G. A. Croopnick (Off. Saline Water Res. Develop. Progr. Rept. No. 720 [1971]). — [47] E. V. Ballou, T. Wydeven, M. I. Leban (Environ. Sci. Technol. 5 [1971] 1032/8). — [48] E. V. Ballou, T. Wydeven (J. Colloid Interface Sci. 41 [1972] 198/207). — [49] D. C. Crozier, P. W. McMillan, S. V. Phillips, J. McC. Taylor (Proc. 4th Intern. Symp. Fresh Water Sea, Heidelberg 1973, Vol. 4, p. 107/14). — [50] G. Belfort, J. Scherfig (Proc. 4th Intern. Symp. Fresh Water Sea, Heidelberg 1973, Vol. 3, p. 69/79).

[51] J. Jagur-Grodzinski, B. Bikson, D. Vofsi (Proc. 4th Intern. Symp. Fresh Water Sea, Heidelberg 1973, Vol. 4, p. 171/80).

3.18 Reverse osmosis process development

Entwicklung des Verfahrens der umgekehrten Osmose

The vapor gap osmotic distillation process is a forerunner of the reverse osmosis process and was conceived by Hassler and McCutchan at the University of California, Los Angeles. It is based upon the difference between the water vapor pressures over a saline solution and that over pure water. Osmotic membranes affect a separation of pure water from saline solutions with a diffusion of the liquid through the filtering membrane. The process was later further developed by Hutkin et al. [1].

The reverse osmosis process has been conceived and first developed with the plate and frame concept, using the filter-press principle. It has then been replaced by the tubular design, the spiral wound module and the process using hollow fibers as membranes. The fine tubules concept is an intermediate design combining the tubular and hollow fiber concepts. The plate and frame design makes use of a rigid plate with the membranes mounted on opposite sides and sealed to the plate. Salt water under pressure is flowing on the outer side of the membrane and the product water is forced through the membranes into the interior of the porous plate. The plate, instead of being porous, can also have hollow channels, through which the product water flows to a collecting point. **Fig. 3–13** illustrates the principle of the plate and frame design with circular plates inside a cylindrical vessel.

Fig. 3–13

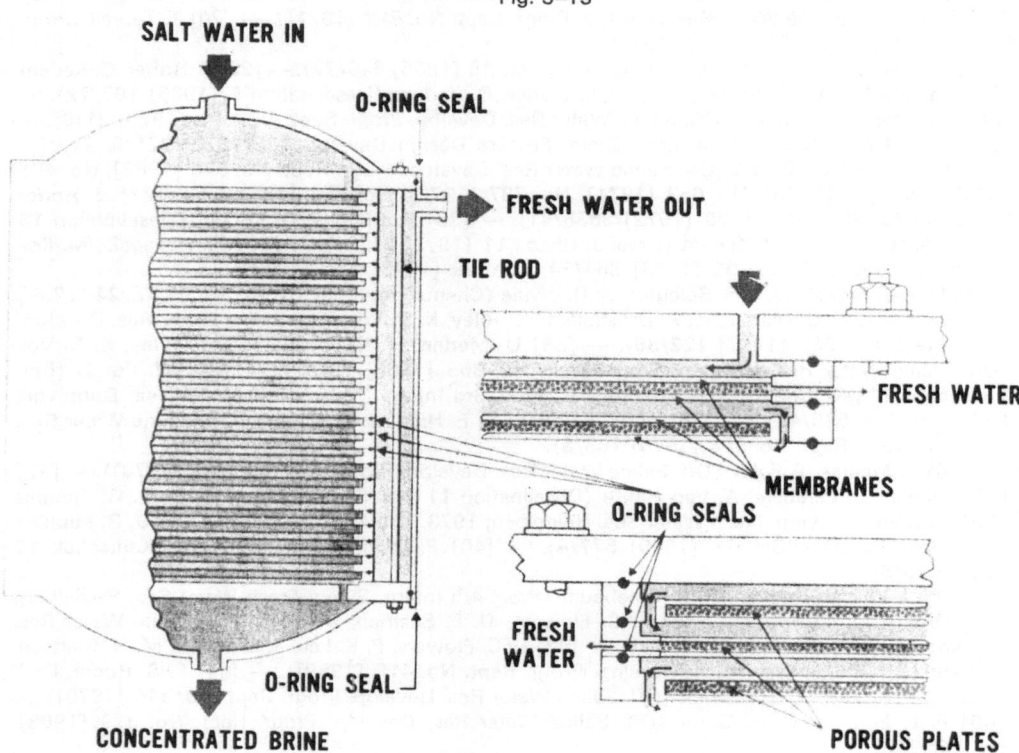

Principle of plate and frame reverse osmosis configuration.

A pilot plant has been designed and erected by Aerojet-General Corporation to evaluate the technical feasibility of the reverse osmosis process for desalination of brackish and seawater at a nominal production capacity of about 3.8 m³/day (1000 gpd) of potable water [2]. An engineering design was consequently made of a 50 000 gpd (190 m³/day) pilot planf to operate on seawater feed [3]. Results of the operation of these pilot plants were given by Aerojet-General Corp. [4] and by DeHaven et al. [5]. A parametric study of a 1 Mgd (3785 m³/day) reverse osmosis plant was also made to define system pressure, flow ratio between product water and brine stream, as well as the general flow arrangement best suited for larger plant sizes [3].

Design criteria for reverse osmosis desalination plants were presented by Keilin and DeHaven [6]. An analysis of continuous constant pressure two-dimensional flow system, such that system performance is predicted explicitly in terms of operating variables, was given by Gill et al. [7]. The plant designer has a choice between high flow rate and low salt rejection membranes or high salt rejection and low flow rate membranes. Geometrical shape, shear in the brine passing the membrane surface, the possibility of multi-stage operation, the choice of flow patterns and reduction of required membrane area at the downstream end of flow channels were quantitatively examined by Johnson et al. Multi-stage operation might be more advantageous for large plants [8].

A test cell has been constructed by Miller and Spatz to produce 5 gpd (19 l/day) as a household appliance using low pressure already present in the water system [9]. Membranes were produced by extrusion of the Sourirajan-Loeb solution onto a continuous sheet of tough polyester filter paper, serving as membrane support. The characteristics of membranes are reported for brackish water feed at 1.75 to 3.5 kg/cm² (25 to 50 psi) and compared to the results obtained by Sourirajan for 42 to 105 kg/cm² (600 to 1500 psi) [10].

Two compact cartridges for reverse osmosis applications were fabricated and tested by Ellington. The first cartridge consisted of three membranes (2.4 ft²), was operated at a recovery factor of 8% on a 10000 mg/l NaCl feed and has produced 23 gal/ft² · day (0.94 m³/m² · day) at 93% salt rejection. The second unit (10 membranes, 8 ft²), operated at a recovery factor of 50% on a 5000 mg/l NaCl feed, performed 30 gal/ft² · day (1.22 m³/m² · day) at 95% salt rejection. A process was also developed for manufacturing reverse osmosis membranes reinforced with nylon cloth [11]. A 1000 gpd (3.8 m³/day) reverse osmosis pilot plant, based on the Union Carbide flat-plate sandwich type permeator cartridge, was also constructed. Product flux of 1150 gpd (4.4 m³/day) at 94% rejection was obtained with 52 ft² of permeator surface. No problems in operation at pressures up to 1500 psi (105 kg/cm²) were encountered [12]. The development, design, fabrication and test by Amicon Corporation of a small scale (5.5 ft² of effective membrane area) cartridge to demonstrate a practical, compact, disposable cartridge design for reverse osmosis desalination using thin channel laminar flow feed stream management was described by Michaels et al. In addition, a 1500 psi pressure vessel for containing the cartridge was designed, fabricated and tested together with the cartridge [13].

Materials have been identified that provide adequate permeation of product water parallel to a structural member plate, thereby simplifying plate design by eliminating product collection grooves. This concept was demonstrated using a high sulfate, low chloride retention membrane with synthetic seawater. Flux equivalency with the standard design required higher operating pressure [16].

The plate and frame design, as already said, was the first to be used in early reverse osmosis applications. A plate and frame unit was used at the 1966 International Trade Fair in Bari, Italy, to demonstrate the reverse osmosis process. It was then shipped to the Brackish Water Conversion Demonstration Plant at Webster, South Dakota, where brackish water was successfully converted to potable water for 5 months. Product water flow rate declined sharply during the first week of operation. Various types of feedwater pretreatment were employed to minimize membrane fouling, but no procedure was devised to restore the original product rate. Mechanical performance was troublefree. Time for maintenance was less than 1% of operating time [17]. A filter press apparatus in improved design for the desalination of brackish water was also described by Karelin and Lishnevskii [18].

Turbulence promoters for reverse osmosis were prepared, studied and economically evaluated by Thomas et al. The use of turbulence promoters greatly affects optimum plant geometry. Higher fluxes may result to appreciable reduction in water cost [19].

Building of a laminar-flow narrow channel desalination device that minimizes concentration polarization was considered by Davies and Canamare to be feasible. This approach has engineering advantages such as short feed-side paths, large membrane area per unit volume, ease of utilization on a modular basis and control of concentration polarization [20]. The feasibility of a thin-channel membrane unit was demonstrated by 1812 total hours of pilot plant operation at channel spacings from 7 to 22 mils (0.2 to 0.6 mm). Economic evaluation of thin-channel systems showed them to be competitive with other reverse osmosis methods. A conceptual design study for reverse osmosis plants with capacities of 1, 10 and 50 Mgd considered scale-up of the current 1000 gpd unit without modification, plant design based on improvements in the present unit. The parametric economic study established comparative costs for the conceptual designs [21].

Performance of a reverse osmosis system for water purification depends on several engineering factors: osmotic pressure, membrane properties, temperature, degree of fouling and concentration polarization. A review was given by Kamp on the various parameters affecting the process, including osmotic pressure. Characteristics of semipermeable membranes are given and the properties of cellulose acetate membranes, such as water flux, the effect of temperature, flux decline, causes and cures of fouling, the prediction of flux, salt permeability and rejection factor, were outlined. The effects of concentration polarization were also described. [22].

Reverse osmosis is a general separation process, which can be designed and operated in stages. Application of the design equations was illustrated by Kimura et al. by a set of calculations [23].

A generalized approach to reveres osmosis process design was also presented by Ohya and Sourirajan. A reverse osmosis system for water desalination can be specified in terms of the properties of the solution, specifications of the membrane and the operating conditions. Analytical expressions were derived for the change of volume of solution, concentration of bulk solution, concentrated boundary solution on the pressure side of the membrane, change in the permeating velocity of solvent water through the membrane, solute separation and other related quantities. The application of these equations indicate that operating pressures of about 150 kg/cm^2 or more, high mass transfer coefficients and least membrane compaction during continuous operation are the most favorable conditions for the economic desalination of seawater [24].

A Fortran computer program, HYPFIL, was presented by Griffith et al. for the investigation of process parameters in single-stage reverse osmosis processes. The model was developed to calculate relative water costs under various operational parameters in rectangular or circular tubular systems. The inlet velocity and pressure, water recovery and flow conditions can be optimized to obtain minimum water costs. The optimized solution has an upper limit in the salt concentration of the product water and a lower limit in the Reynolds number [25].

A computer program was developed by Perona and Dillon, which finds optimum inlet flow velocities and inlet pressures for reverse osmosis plants having as many enriching stages as necessary to produce water meeting product specifications. Stripping stages are used if high water recoveries are specified. Membranes with intrinsic rejections of 40 to 95% were studied with feed concentrations of 1000 to 5000 mg/l. Optimum membrane permeabilities and water recoveries were found. Flow velocities, pressures and water recoveries which result in a minimum cost for a given membrane often do not minimize the number of stages required. Optimum permeabilities did not coincide with minimum stages. Therefore, a reduction in stage requirements is not in itself a valid means of cost reduction [26].

A continuous reverse osmosis desalination system with a perfect or imperfect membrane was considered by Liu. For high Schmidt number, the velocity field is approximated by the first term of the velocity expansion near the wall. The species equation, after coordinate transformations, has been solved by an iterative technique. For the first-order iteration, the reverse osmosis boundary condition leads to a nonlinear integral equation. Series solution of the integral equation for small distances from the leading edge have been obtained in terms of the operating parameters and the transverse coordinate. Although the present iterative solution is expected to be best for large axial distance, the series solutions give results accurate within 5%. Calculations for imperfect membranes have also been carried out. The results show that finite salt rejection has profound effect on the wall concentration and cannot be neglected in real analysis [27].

A mathematical analysis based on boundary layer theory was presented by Srinivasan and Chi Tien to describe the effect of natural convection in reverse osmosis. It was shown that the effect of including the buoyancy term is to decrease the concentration polarization. Numerical results were presented for a variety of cases [28].

General assumptions in the computer model are valid at seawater concentrations with two-stage operation. Accelerated deterioration of the membrane under the severe operating conditions in the first stage produces difficulty in relating computed data to the time dependent characteristics of the membranes. The second stage performance, which is equivalent to a brackish water unit, was predicted with considerable accuracy for a variety of module configurations [29]. Simple analytical expressions for concentrations of permeate and retentate and for apparatus capacity necessary to reach these were derived by Dejmek for solutions with appreciable osmotic pressure [30].

Truck mounted reverse osmosis desalting systems are both technically and economically sound for several services such as: to substitute for several small municipal plants for brackish water purification, to substitute for several small resort hotel plants desalting seawater and as a rapidly transportable emergency water supply in disaster areas where normal water supply has been interrupted by floods, earthquakes, etc. [31].

A commercial reverse osmosis system capable of delivering 1.4 Mgd (5300 m^3/day) of high purity water for boiler feedwater make-up is in operation. Large reverse osmosis systems also are in use for desalination of brackish water and the preparation of ultra-pure water for the production of semiconductor devices. The range of potential uses for reverse osmosis systems is very broad and the ultimate capacity of the systems is practically unlimited, has been concluded by Lepper [32].

The following table shows the existing reverse osmosis plants in the world and their total capacity. Data are taken from the Desalting Plant Inventory Report No. 4 [33] and include plants with a capacity over 95 m^3/day (25000 gpd). Figures referring to 1972 might be incomplete.

Year	Plants	Capacity	
		m³/day	Mgd
1967	2	378	0.10
1969	2	416	0.11
1970	11	4921	1.30
1971	7	7609	2.01
1972	24	14422	3.81

Further work on reverse osmosis development and applications is reported by Wutschel on the operation of a pilot plant with a production of 6 m³/day [34], by Connelley on the operating experience with five reverse osmosis plants in industry [35], by Marquardt on possible applications and limitations of the process [36], by Lising and Alward on unsteady state operation of a reverse osmosis unit [37], by Alward et al. on the effect of a variable power input on the performance of a reverse osmosis unit [38], by Gillam on recent developments of the reverse osmosis process [39], by Hendricks et al. on observations on buoyant convection in reverse osmosis [40].

Process optimization. Analysis and optimization studies have been carried out by Fan et al. in order to develop procedures that can be used to design optimal desalination systems. A boundary layer flow model is used to relate water production rate to the operating pressure, Reynolds number and membrane area. Equations relating the capital and operating costs to the design variables were also developed. These relations were then used to determine design and operating variables of the system, which minimizes the cost of water production. Several multi-stage reverse osmosis systems were considered. In the optimization study, the recirculation rate in each stage, the brine composition leaving each stage, the ratio of membrane area to feed at each stage and the operating pressure in each stage were controlled to arrive at a minimum water production cost. The effect of a flow work exchanger on water production costs has also been investigated [41]. In a further study a pattern search method was used in the optimization of a single process, while dynamic programming was employed for the multi-stage processes. The results were compared with those of previous work, which assumes a uniform bulk concentration within the membrane unit. A parametric search with respect to the exit concentration of a two-stage process has been carried out [42].

A computer program was developed by Perona et al. for parametric analysis of reverse osmosis plants. Provisions are made in the code to permit repressurization between sections through booster pumps and to allow for recirculation of solution within each section with a separate pump. With the code, optimum costs can be obtained at fixed water recoveries. The code was applied to cost analysis of reverse osmosis of solutions with 1000 to 5000 mg/l of salt. Conditions were found, where two-stage plants produced cheaper water than one-stage plants. Two-stage plants can use membranes with higher permeability than one-stage plants. For feed concentrations up to 4000 mg/l, lower costs were obtained with the moderately rejecting high flux membranes than with the cellulose acetate type, if plant costs per unit membrane area are equal [43].

A mathematical model of a reverse osmosis system for desalination of brackish water was also developed by Hatfield and Graves. A nonlinear programming problem was formulated with the objectives of maximizing product flux and determining the optimal arrangement of assemblies. The solution of this problem pointed out that by using optimization techniques, substantial gains can be made in reducing the size of reverse osmosis systems and thus reducing the cost of the water produced [44].

Numerical calculations were performed by Thomas et al. to obtain process parameters for reverse osmosis with membranes having characteristics similar to cellulose acetate. The concentration polarization and friction loss equations used in the calculation were modified to include provision for a wide range of turbulence promoter characteristics, based on potential performance curves derived from an extensive literature survey. The presence of turbulence promoters causes marked reduction of optimum entrance velocity and tube length. The channel pressure drop for tubes with turbulence promoters is less under optimum conditions than the value for tubes with no promoters. There is a cost advantage for turbulence promoters which minimize fouling, thereby giving higher average fluxes for longer periods of time. Only where the membrane is strained to produce water of required quality is there a direct effect of turbulence promoters on product water cost, because of this beneficial effect on concentration polarization [45].

The primary mechanism by which tubular reverse osmosis plants were designed and optimized is a generalized computer program known as SALT, developed by Braunheim during the study. Four feed waters, which embrace a broad spectrum of brackish waters, were specified. A detailed cost model has been developed for a 37850 m³/day (10 Mgd) plant and integrated into the computer program. The cost model showed that the installed capital costs varied between $ 37.7 and 43 per m² ($ 3.50 and 4.— per ft²) of membrane. The total cost of water production varied between 5.55 and 19 cents per m³ (21 and 72 cents per kgal) for the four feed waters, depending on the feed water composition, remembraning cost, power cost and brine disposal cost. The operating and maintenance costs were by far the most significant costs in terms of total production costs [46].

Two types of two-stage reverse osmosis units were considered by Apeltsin et al. For each of these units the relations were shown between the rate of flows and the concentration of salts ensuring the operation of the unit with minimum power consumption at certain salt rejection of membranes. The two types of units were compared from the point of view of specific power consumption and required area of membranes. The problems of cellulose acetate membranes permeability for various substances were considered [47].

Further work is reported by Müller on optimization problems in the preparation of fresh water with the reverse osmosis process [48], by Rautenbach et al. on multi-stage membrane processes for desalting seawater, system analysis and optimization [49], etc.

Power recovery. The reject brine stream in reverse osmosis remains at the high pressure and has to be depressurized before being removed from the system. Power recovered from the reject brine stream could be used to pressurize some of the incoming feed stream to reduce the overall energy requirement of the process. A flow-work exchanger is a unified piece of equipment, which uses two displacement vessels to form closed loops with a high pressure processing system. Each of the displacement vessels is alternately filled by a low-pressure feed and a high-pressure product, both pressurized and depressurized respectively, by substantially non-flow processes [50]. The hydraulic operation of pilot plant units has been generally satisfactory and the technical feasibility of such units has been demonstrated. Efficiencies as high as 90 to 95% at rated capacities were obtained [51]. Two types of flow work exchangers, the piston and the bladder-type units, were tested at pressures up to 105 kg/cm² (1500 psi) and flow rates up to 75 l/min (20 gpm). It was concluded that units in sizes up to 1400 m³/day (365000 gpd) of fresh water can be built with a 25 % energy recovery factor [52]. A survey of existing means for high pressure pumping and energy recovery was carried out by Hickman et al. and the best suited pumps and hydraulic turbines have been indicated [53].

A new type of pump has been developed by Sanchez Tarifa especially suitable for reverse osmosis applications. The experimental performances of a number of working models are reported and the feasibility of utilizing a combined turbo-pump group of this type for energy recovery in reverse osmosis plants is outlined [54]. The principal conclusions of a parametric study of energy recovery in reverse osmosis plants using turbo-pump groups were discussed in a further paper. Energy recovery appears to be economical, even for very small capacity plants [55].

The hydrostatic pressure existing in several hundred meters depth may be used for desalination by reverse osmosis, if the pressure at the fresh water side of the membranes is maintained near atmospheric. In that case the work to be performed consists merely of the energy required for lifting the product water up to the level of a nearby shore. The membranes will by placed at the outside of a pressure vessel and no salt concentration build-up at the membrane surface is expected, because of natural currents [56]. In evaluating the principle of submarine units Drude and Klapp concluded, that it might be possible to arrive at water costs comparing favorably with those of other processes under the assumption of careful optimizing calculations, comprising a wide range of parameters related to oceanography, design, mode of operation and energy consumption. The applicability of submarine units will depend greatly on their reliability and maintenance requirements and on the effects of reduced energy demand upon possibly higher investment costs [57].

Process evaluation. A parametric engineering and economic analysis of the reverse osmosis process was performed by Harris et al., in which water costs were examined for demineralizing feedwaters with salinities of 2000, 3000 and 5000 mg/l in plant sizes of 3785, 37850 and 189250 m³/day (1, 10 and 50 Mgd) capacity, product water recoveries of 50, 60 and 80% and operating pressures of 28, 42 and 56 kg/cm² (400, 600 and 800 psi). Capital and operating costs were prepared using state-of-the-art technology for tubular, plate-and-frame, spiral wound and hollow fine fiber reverse osmosis plant design concepts. Mathematical models were prepared to represent the hydrodynamic characteristics of each design concept. These models, in conjunction with computer codes, were used to optimize plant design and operating parameters [58].

Estimated investment costs for brackish water reverse osmosis plants are in the range of 15.8 to 34.3 cents per liter or 0.6 to 1.3 $ per gal of daily capacity. Membrane equipment constitutes the major capital cost component of the plant. Since membrane characteristics greatly influence investment costs, development of high flux membranes holds promise for reduction of reverse osmosis costs. Fixed charges, depreciation and interest account for the major portion of the operating costs of reverse osmosis plants. Energy costs and membrane replacement contribute about equally to the operating costs. A comparison of reverse osmosis with other desalination methods shows that investment and operating costs for small reverse osmosis plants are in the same range as those of much larger distillation plants [59]. Capital and operating costs were also discussed by Banfield, using a 3785 m^3/day (1 Mgd) scheme. Breakdown of capital costs and the effect of increase in the level of membrane design flux on the capital required are discussed and the effect of water salinity and changes in process variables are outlined [60].

For a practical comparison of the two processes, the capital requirements were estimated by Clark for 1 and 10 Mgd reverse osmosis and electrodialysis plants for three brackish feed waters. Economic scaling factors were applied to extend the plant investment estimates to 0.5 and 50 Mgd capacities. The cost estimates for producing fresh water indicate that the electrodialysis process has a slight advantage over the reverse osmosis process on feedwaters in the 1000 to 2500 mg/l total dissolved solids range, whereas the reverse osmosis process has an advantage for feedwaters above the 3000 mg/l level. Membrane replacement and power costs were the most important operating costs. With capital costs they accounted for 66 to 78% of the product water cost [61].

Performance of semipermeable membranes in reverse osmosis systems of various configurations has been investigated by Manjikian et al. Membranes tested in very small flat test cells have in general shown consistent and effective operating characteristics. Membranes cast separately and then inserted in tubular structures were found to be subjected to both compressive and tensile stresses in operation, resulting in various degrees of deformation. In directly cast tubular systems, this type of deformation is held to a minimum. Membrane performance was found to be superior in precast tubular units. Work on spiral wound modules was accomplished on commercially available units. Brine concentration and pressure effects might be critical in their effect on membrane performance. Specific casting solutions are needed to produce optimum performing membranes in each configuration. Conditions affecting membrane performance in various configurations during casting, curing, assembly and operating stages were reported [62].

The performance characteristics of reverse osmosis and electrodialysis plants as well as of experimental freeze desalination units were evaluated by Denton et al. The data given on the quality of product water obtained are useful in the optimization of the process. An electrodialysis plant provided a continuous output of 9 m^3/h water over a 4 year period at an annual membrane replacement of 1% [63].

Reverse osmosis, electrodialysis and ion exchange processes were compared by Dryden as to their overall efficiency in reducing 500 to 1500 mg/l total dissolved solids by about 34% in plants producing up to 20 Mgd (75000 m^3/day) purified water from seawater or waste water. The overall efficiency was lowest for the reverse osmosis process, even though the spiral wound membrane modular unit operated for 13 months before a significant membrane failure was observed [64].

A modular digital computer system has been developed by Gembicki et al. for the analysis and evaluation of specific community brackish water needs based on reverse osmosis desalination. The system provides an analytical tool to facilitate individual communities to determine the cost effective total water system best suited for their specific requirements. It also provides current technology and cost information, as well as a highly flexible format which readily facilitates incorporation of new and projected desalting technology. The heart of the system is a set of unit programs, which perform all engineering and financial calculations for the individual basic components of a water supply system: well, pretreat, pump, reverse osmosis plant, pipe, pond, etc. Other unit programs perform cost summaries and community financial analysis which compute bond repayment schedules and their relation to projected per capita income, population, and real estate values [65].

Hollow-fiber and spiral-wound modules were operated by Minturn in reverse osmosis of synthetic waters in order to establish a correlation between the practical solubility limits of calcium sulfate as a function of operating conditions and the equilibrium solubility limits that were determined in concurrent work. The results of the experimental program are given, thermodynamic data and procedures for their use and a theoretical discussion of concentration polarization in multicomponent systems are presented [66].

A mobile brackish water test facility was established by the Office of Saline Water to demonstrate the ability of reverse osmosis and electrodialysis processes to produce potable water from brackish water in the Southwest United States. The reverse osmosis units consisted of tubular, spiral wound and hollow fine fiber modules with capacities from 1.14 to 19 m³/day (300 to 5000 gpd). Tests were made to determine problems of operation at high recovery or conditions under which scaling might occur. Pretreatment consisted of removing calcium and magnesium, sediment filtration, as well as iron, manganese and organic material removal. The flux decreases observed were due largely to fouling and scaling. Flushing the unit restored the flux except where calcium sulfate was encountered. Electrodialysis unit was operated successfully on two low salinity wells. Iron and calcium sulfate build-up gave some problems [67]. Further similar tests were reported by Kaup on 2000 and 5000 mg/l salinity feed waters at ambient temperature [68].

Operating testing of hollow fine fiber, spiral wound and helical wound reverse osmosis pilot plants at the O.S.W. Webster Test Bed Plant was reported by Ackerman et al. The plants were operated at a set product to feed conversion ratio and at constant brine discharge rates to determine flux declination. The chief cause of flux decline was found to be iron. Various methods for controlling this condition, including citric acid, flushing and hexametaphosphate feed, were studied [69].

Evaluations of the reverse osmosis process for various practical applications of water desalting include the possibility of reducing the salinity of the Potomac river [70], the effectiveness of the process in reclaiming drainage containing acid and iron from coal mines [71], desalting of water from brackish water sources [72], treatment of Rhine river water [73], the desalination of high sulfate water in the island of Corfu [74], and the potential application in the United Kingdom [75]. Reviews on the application of reverse osmosis and ultrafiltration were also presented by Henderyckx with emphasis on the advantages and disadvantages of the two processes [76], by Madsen reporting on experience with commercial ultrafiltration and reverse osmosis plants [77], by Oesterle on membrane separation processes with emphasis on water desalination [78], by Marquardt on fresh and waste water treatment by means of reverse osmosis and ultrafiltration compared with or as supplement of ion exchange techniques [79], etc.

Disposal of effluents. Reverse osmosis plants, when operating on seawater feed, will use the standard methods of effluent disposal, as outlined in chapter 2.20, p. 179. When located at inland sites reverse osmosis, as well as electrodialysis, effluents might need special consideration in order to minimize the impact of the cost of effluent disposal on the product water cost.

Reverse osmosis desalination systems were examined by LeGros et al. for six inland locations in the United States to appraise the effects of disposal method on total system costs. Communities were selected which would represent a range of feedwater types, evaporation rates and underground geology. It was concluded that the lowest total system costs are found at the highest practicable plant recovery rates in every case; evaporation ponds are costly due to the high cost of the required lining material; the cost of disposal by deep well injection ranges widely due to the wide variation in geological properties of the disposal horizon [80].

A study was also made of unconventional methods for concentrating the waste effluents from plants at inland locations. Two solvent extraction methods and three processes involving ion exchange with production of chemical regenerants from the dissolved salts appear feasible. Preliminary flowsheets, material balances and capital and operating cost estimates were prepared for these processes. Several of these should find application as primary desalting processes, with the unique feature that only very concentrated waste effluents are produced [81].

Parametric studies of reverse osmosis were carried out by Griffith et al. to determine optimal flow conditions and relative water costs when waste brine disposal costs are included in the economic model and to further investigate tapered plant geometries. Relative water costs were computed at optimum flow conditions by use of a Fortran computer program, HYPFIL (see p. 284), which deals with a single-stage reverse osmosis system with tubular membranes under turbulent conditions. It was found that the optimum permeability of an ion exchange type membrane which gave the minimum relative water cost was lower with increasing brine disposal costs. The optimum water recovery increased with increased disposal cost. The effect of brine disposal costs on relative water cost and the characteristics of cellulose acetate resulted in higher optimum water recoveries and increased the relative cost of water by a much high percentage when the feed concentration was near the inherent rejection ability of the membrane to yield a product with satisfactory salinity [82].

Literature to 3.18

[1] I. J. Hutkin, M. H. Dawson, R. Wintersdorff (Off. Saline Water Res. Develop. Progr. Rept. No. 212 [1966]). — [2] Aerojet-General Corporation (Off. Saline Water Res. Develop. Progr. Rept. No. 86 [1964]). — [3] D. T. Bray, H. F. Menzel (Off. Saline Water Res. Develop. Progr. Rept. No. 176 [1966]). — [4] Aerojet-General Corporation (Off. Saline Water Res. Develop. Progr. Rept. No. 213 [1966]). — [5] C. G. DeHaven, M. A. Jarvis, C. R. Wunderlich (Off. Saline Water Res. Develop. Progr. Rept. No. 356 [1968]).

[6] B. Keilin, C. G. DeHaven (Proc. 1st Intern. Symp. Water Desalin., Washington, D.C., 1965 [1967], Vol. 2, p. 367/88). — [7] W. N. Gill, Chi Tien, D. W. Zeh (Off. Saline Water Res. Develop. Progr. Rept. No. 185 [1966]). — [8] K. D. B. Johnson, J. R. Grover, D. Pepper (Desalination 2 [1967] 40/55). — [9] C. S. Miller, D. D. Spatz (Off. Saline Water Res. Develop. Progr. Rept. No. 339 [1968]). — [10] J. L. Brock, P. M. Fahey, C. S. Miller, D. D. Spatz (Off. Saline Water Res. Develop. Progr. Rept. No. 441 [1969]).

[11] R. C. Ellington (Off. Saline Water Res. Develop. Progr. Rept. No. 410 [1969]). — [12] H. W. McRobbie, L. M. Litz (Off. Saline Water Res. Develop. Progr. Rept. No. 423 [1969]). — [13] A. S. Michaels, H. J. Bixler, R. A. Cross, D. S. Cleveland, B. Carrol (Off. Saline Water Res. Develop. Progr. Rept. No. 795 [1972]). — [14] C. G. DeHaven, E. R. Watson (Off. Saline Water Res. Develop. Progr. Rept. No. 422 [1969]). — [15] E. R. Watson (Off. Saline Water Res. Develop. Progr. Rept. No. 429 [1969], No. 430 [1969] and No. 431 [1969]).

[16] G. A. Fluke (Off. Saline Water Res. Develop. Progr. Rept. No. 428 [1969]). — [17] W. H. Bossert, G. A. Fluke (Off. Saline Water Res. Develop. Progr. Rept. No. 427 [1969]). — [18] F. N. Karelin, V. A. Lishnevskii (Tr. Vses. Nauchn. Issled. Inst. Vodosnabzh. Kanaliz. Gidrotekhn. Sooruzh. Inzh. Gidrogeol. 1970 No. 25, p. 19/24; C.A. 73 [1970] No. 80367). — [19] D. G. Thomas, P. H. Hayes, W. R. Mixon, J. D. Sheppard, W. L. Griffith, R. M. Keller (Environ. Sci. Technol. 4 [1970] 1129/36). — [20] D. S. Davies, J. G. Canamare (Off. Saline Water Res. Develop. Progr. Rept. No. 420 [1969]).

[21] D. S. Davies, I. Bemberis (Off. Saline Water Res. Develop. Progr. Rept. No. 671 [1971]). — [22] E. C. Kaup (Chem. Eng. 80 No. 8 [1973] 46/55). — [23] S. Kimura, S. Sourirajan, H. Ohya (Ind. Eng. Chem. Process Design Develop. 8 [1969] 79/89). — [24] H. Ohya, S. Sourirajan (Desalination 6 [1969] 153/78; AIChE [Am. Inst. Chem. Engrs.] J. 15 [1969] 826/36). — [25] W. L. Griffith, R. M. Keller, R. A. Ebel, R. S. Dillon (ORNL-Y-1699 [1970]).

[26] J. J. Perona, R. S. Dillon (Desalination 11 [1972] 149/63). — [27] Mei-Kao Liu (Desalination 9 [1971] 181/91). — [28] S. Srinivasan, Chi Tien (Desalination 10 [1972] 273/86). — [29] A. S. Hodgson (Desalination 11 [1972] 113/23). — [30] P. Dejmek (Chem. Eng. Sci. 27 [1972] 1577/83).

[31] R. A. Tidball, S. C. May (Proc. 4th Intern. Symp. Fresh Water Sea, Heidelberg 1973, Vol. 4, p. 417/24). — [32] F. R. Lepper (Power Eng. 77 No. 5 [1973] 56/9). — [33] F. O'Shaghnessy (U.S. Dept. Interior Desalting Plant Inventory Rept. No. 4 [1973]). — [34] A. Wutschel (Oberfläche-Surface 12 No. 3 [1972] 42/6). — [35] E. J. Connelley (Proc. Amer. Power Conf. 34 [1972] 716/24).

[36] K. Marquardt (Chemiker-Ztg. - Chem. Tech. 2 [1973] 245/53). — [37] E. R. Lising, R. Alward (Desalination 11 [1972] 261/8). — [38] R. Alward, E. R. Lising, T. Lawand (Proc. 4th Intern. Symp. Fresh Water Sea, Heidelberg 1973, Vol. 4, p. 25/34). — [39] W. S. Gillam, H. E. Podall (Proc. 3rd Intern. Symp. Fresh Water Sea, Dubrovnik 1970, Vol. 2, p. 385/95; Desalination 9 [1971] 201/11). — [40] T. J. Hendricks, J. F. Macquin, F. A. Williams (Ind. Eng. Chem. Fundamentals 11 [1972] 276/9).

[41] L. T. Fan, C. Y. Cheng, L. E. Erickson, C. L. Hwang (Desalination 3 [1967] 225/36); L. T. Fan, C. Y. Cheng, L. Y. S. Ho, C. L. Hwang, L. E. Erickson (Off. Saline Water Res. Develop. Progr. Rept. No. 332 [1968]; Desalination 5 [1968] 237/65, 6 [1969] 131/52). — [42] L. T. Fan, J. T. Tseng, C. L. Hwang, L. E. Erickson (Proc. 3rd Intern. Symp. Fresh Water Sea, Dubrovnik 1970, Vol. 2, p. 371/84). — [43] J. J. Perona, T. W. Pickel, K. A. Kraus (Desalination 8 [1970] 73/91). — [44]

G. B. Hatfield, G. W. Graves (Desalination **7** [1970] 147/77). — [45] D. G. Thomas, W. L. Griffith, R. M. Keller (Desalination **9** [1971] 33/50).

[46] S. T. Braunheim (Off. Saline Water Res. Develop. Progr. Rept. No. 684 [1971]). — [47] I. E. Apeltsin, F. N. Karelin, V. A. Lishnevskii, A. B. Kosminskii (Proc. 4th Intern. Symp. Fresh Water Sea, Heidelberg 1973, Vol. 4, p. 35/47). — [48] H. U. Müller (Chem. Ingr.-Tech. **45** [1973] 431). — [49] R. Rautenbach, A. Pasternack (Verfahrenstechnik [Mainz] **6** [1972] 19/22), R. Rautenbach, A. Pasternack, K. Rauch (Verfahrenstechnik [Mainz] **6** [1972] 399/405). — [50] C. Y. Cheng, S.W. Cheng, L. T. Fan (AIChE [Am. Inst. Chem. Engrs.] J. **13** [1967] 438/42).

[51] C. Y. Cheng, L. T. Fan (Off. Saline Water Res. Develop. Progr. Rept. No. 357 [1968]). — [52] G. B. Gilbert (Off. Saline Water Res. Develop. Progr. Rept. No. 672 [1971], No. 680 [1971]). — [53] K. E. Hickman, J. P. Tramoni, J. T. Ganley, E. J. Fahiminian (Off. Saline Water Res. Develop. Progr. Rept. No. 457 [1969]). — [54] C. Sanchez Tarifa (Proc. 3rd Intern. Symp. Fresh Water Sea, Dubrovnik 1970, Vol. 3, p. 475/90). — [55] C. Sanchez Tarifa (Proc. 4th Intern. Symp. Fresh Water Sea, Heidelberg 1973, Vol. 4, p. 369/80).

[56] B. C. Drude (Desalination **2** [1967] 325/8). — [57] B. C. Drude, E. Klapp (Proc. 4th Intern. Symp. Fresh Water Sea, Heidelberg 1973, Vol. 4, p. 125/34). — [58] F. L. Harris, G. B. Humphreys, H. Isakari, G. Reynolds (Off. Saline Water Res. Develop. Progr. Rept. No. 509 [1969]). — [59] H. Alcalay, R. Matz (Proc. 3rd Intern. Symp. Fresh Water Sea, Dubrovnik 1970, Vol. 2, p. 311/27). — [60] D. L. Banfield (Chem. Ind. [London] **1970** 348/51).

[61] C. F. Clark (Off. Saline Water Res. Develop. Progr. Rept. No. 638 [1970]). — [62] S. Manjikian, M. Foley, C. Allen (Off. Saline Water Res. Develop. Progr. Rept. No. 670 [1971]). — [63] W. H. Denton, W. H. Hardwick, K. D. B. Johnson (Water Treat. Seventies Proc. Symp., Reading, Engl., 1970, p. 145/59). — [64] F. D. Dryden (Ind. Water Eng. **8** No. 7 [1971] 24/6). — [65] S. A. Gembicki, R. L. Brewster, P. W. Weaver, T. T. Shanon (Off. Saline Water Res. Develop. Progr. Rept. No. 876 [1973]).

[66] R. E. Minturn (Off. Saline Water Res. Develop. Progr. Rept. No. 897 [1973]). — [67] E. I. Ewoldson, M. H. Jhawar, P. Mothes (Off. Saline Water Res. Develop. Progr. Rept. No. 885 [1973]). — [68] E. G. Kaup (Off. Saline Water Res. Develop. Progr. Rept. No. 899 [1973]). — [69] R. A. Ackerman, D. R. Bogue, J. E. Gugele (Off. Saline Water Res. Develop. Progr. Rept. No. 895 [1973]). — [70] J. Schulz (Off. Saline Water Res. Develop. Progr. Rept. No. 216 [1966]).

[71] A. Riedinger, J. Schulz (Off. Saline Water Res. Develop. Progr. Rept. No. 217 [1966]), S. S. Kremer, A. Riedinger, J. H. Sleigh, R. L. Truby (Off. Saline Water Res. Develop. Progr. Rept. No. 586 [1970]). — [72] J. Schulz, A. Riedinger, H. McCracken (Off. Saline Water Res. Develop. Progr. Rept. No. 237 [1967]). — [73] D. Kuiper, C. A. Bom, J. L. van Hezel, J. Verdouw (Proc. 4th Intern. Symp. Fresh Water Sea, Heidelberg 1973, Vol. 4, p. 207/14). — [74] D. Pepper, A. I. Rogan, C. Tanner (Proc. 4th Intern. Symp. Fresh Water Sea, Heidelberg 1973, Vol. 4, p. 297/308). — [75] W. H. Hardwick (Chem. Ind. [London] **1970** 297/301).

[76] Y. Henderyckx (Tech. Eau Assainissement No. 304 [1972] 29/37, No. 305 [1972] 39/45, No. 309 [1972] 33/47). — [77] R. F. Madsen, O. J. Olsen (Chemiker-Ztg. - Chem. Tech. **3** [1974] 81/4). — [78] K. M. Oesterle (Chem. Rundschau [Solothurn] **26** [1973] 14, 17/23). — [79] K. Marquardt (Oberfläche-Surface **14** [1973] 271/5, 303/6, 338/41; Metalloberfläche-Angew. Elektrochem. **27** [1973] 169/82; Chemiker-Ztg. - Chem. Tech. **2** [1973] 245/53). — [80] P. G. LeGros, C. E. Gustafson, B. P. Shepherd, W. F. McIlhenny (Off. Saline Water Res. Develop. Progr. Rept. No. 587 [1970]).

[81] R. F. Battey, R. N. Jacobson, S. Foster (Off. Saline Water Res. Develop. Progr. Rept. No. 641 [1971]). — [82] W. L. Griffith, R. M. Keller, D. G. Thomas, W. J. Boegly (Desalination **11** [1972] 91/112).

3.19 Tubular reverse osmosis assembly

Vorrichtung für die umgekehrte Osmose mit schlauchförmiger Membran

The tubular design uses the surface of a tube as a support for the membrane and the tube wall as a pressure vessel. Usually the membrane is placed on the inner wall of the tube and the salt water flows, under pressure, inside the tube. Product water passes through the membrane to the tube wall and is conducted, under low pressure, to the outside of the tube. This may be done by using either a porous tube, permitting direct flow of the product water to the outer tube surface, or a solid tube with small holes. In the latter case a porous fabric material is placed between the membrane and the inner tube wall to provide a path for the product water to the tube holes. **Fig. 3–14** illustrates the principle of operation of the tubular reverse osmosis process.

Fig. 3–14

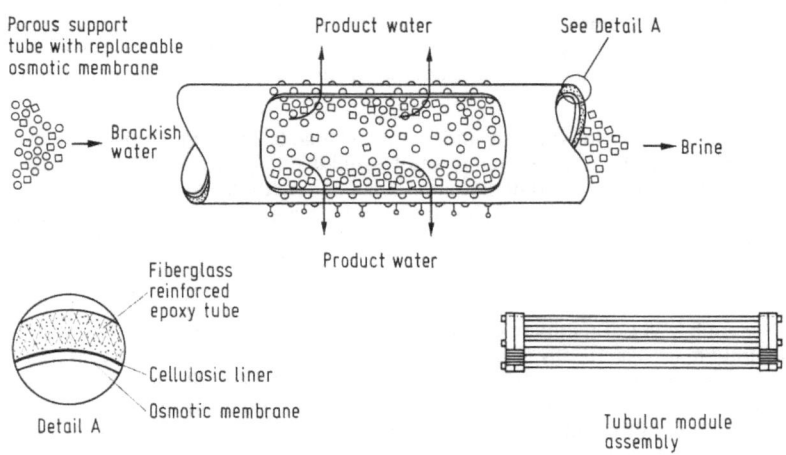

Tubular module configuration.

The methods for producing flat sheet cellulose acetate membranes, developed by Loeb and co-workers, were incorporated into the fabrication of tubular assemblies, with modifications required for the tubular geometry. The casting solution was a 25:30:45 mixture by weight of cellulose acetate, formamide and acetone. Casting was accomplished at room temperature. The membrane tube was wrapped with several layers of a porous material, such as Nylon parchment, to provide for a support and the ends were treated to secure tightness in the assembly. Tubular assemblies fabricated in this manner have been found to give performances as good and as reproducible as those obtained from flat sheets. Performance data were largely obtained from various tubes in a pilot plant, producing fresh water for the city of Coalinga, California, since 17 June 1965 [1]. The plant had a capacity of 5000 gpd (19 m³/day) and utilized 112 composite tubular assemblies arranged in series for the containment of brine passing through in turbulent flow. Each assembly consisted of a tubular cellulose acetate membrane, retained within a perforated copper support tube one inch (2.54 cm) in diameter, 10 ft (3 m) long and having flare-type end fittings. The plant operated at 600 psi (42 kg/cm²), at which pressure 50% of the feed brine, containing 2500 mg/l dissolved solids, was recovered as fresh water [2]. The plant has been on stream 98.5% of the time till 1966 and the salt content of the product water was 200 to 350 mg/l. Declines in production rate due to deposition of iron (III) hydroxide were reduced by eliminating dissolved oxygen with sodium sulfite addition to the feed water. A better antifouling method has been found in scrubbing the membrane tubes in place every 3 to 4 days by passing a foam ball through the array [3]. A report on the operating experiences of the Coalinga plant has also been given, after the plant has been for three and one-half years in operation [4].

An analysis was made by Fisher et al. to express the salt concentration field as a function of the tube diameter, tube length, axial saline water flow velocity and water flux through the membrane. In order to keep less than a two-fold build-up of the salt concentration at the membrane surface over the initial feed concentration, small diameter tubes, short channel lengths or multiple pass operations are required at the higher water fluxes [5].

A mathematical analysis, based on boundary layer theory, was presented by Srinivasan and Tien for the case of a tubular reverse osmosis duct. The analysis considers the simultaneous development of velocity and concentration profiles. The momentum and diffusion equations were solved by the integral method, which also takes into account the non-linear effects created by the varying water flux produced. The analysis is applicable to both the entrance region and the fully developed region.

Typical results were presented for the concentration build-up at the membrane wall, the flux of water produced and the productive capacity of the system [6]. A parametric study was carried out by Griffith et al., using the HYPFIL Fortran computer program (see chapter 3.18, p. 284). The parametric study dealt largely with variables, such as membrane permeability and rejection, feed concentration, plant geometry, temperature, and economic parameters. An advantage of high-flux membranes is that they could permit matching of membrane characteristics to optimum pressure operation [7].

Analytical studies in which the independent variables were membrane properties, tube dimensions and arrangements, feed water quality and Reynolds numbers were also made by Fluke and computer programs were developed. Materials and fabrication methods were investigated and a tubular support structure for the membrane was developed, which consists of a paper substrate on the inside surface with 20 mils of fiber glass resin composite. Tubular assemblies, free from defects, have produced up to 0.8 $m^3/m^2 \cdot$ day (20 gal/ft$^2 \cdot$ day) with rejection rates of 80% NaCl and 97% Na_2SO_4 when operating at 800 psi (56 kg/cm^2) [8].

A complete set of mathematical equations, describing the whole process of reverse osmosis and ultrafiltration for electrically uncharged membranes, was presented by Murkes and Bohman. The model refers to a module of tubular membranes. The equations were used as a basis for modelling programs for single or arranged in parallel membrane tubes of arbitrary length and diameter, for an arbitrary number of membrane tubes connected serially, where the concentrate from one tube constitutes the feed to the following and for a batch process where the initial liquid is recirculated an arbitrary number of times with or without drain-off. The process parameters can be calculated both for laminar and turbulent flow conditions. The model covers most practical cases of interest for the tubular membrane design [9].

A surface favorable for dynamic formation of tubular reverse osmosis membranes can be attained by thinly coating a coarsely porous material with particulate or fibrous substances. The coating is deposited by circulating a suspension of filter aids under pressure [10]. Optimal perforation spacing in tubular reverse osmosis elements was studied and a criterion for optimum spacing was formulated by Winograd and Solan [11]. Membranes prepared by depositing ultrathin cellulose acetate discs in situ upon a polysulfone porous support in tubes have exhibited rejections of 93 to 96% of sodium sulfate and 31.5 to 33.5% sodium chloride at fluxes up to 0.7 $m^3/m^2 \cdot$ day (17 gal/ft$^2 \cdot$ day), when evaluated on 3000 mg/l test solutions at 56 kg/cm^2 (800 psi) [12].

Replacement of membranes may be facilitated by the use of in situ cast regenerable membranes. The economic advantage of the developed technique is to eliminate labor of replacing the membranes, by simply manipulating the solvents and the casting and curing solutions within a sealed pressure shell. A 50 000 gpd reverse osmosis plant was designed, using in situ regenerable membranes [13].

The use of a static mixer, a form of twisted tape insert, for convection promotion within a tubular desalination membrane, was studied by Pitera and Middleman for Reynolds numbers in the range 10 to 1500. The polarization ratio was significantly reduced, leading to improvement of permeate quality and an increase in permeate flux [14].

The effectiveness of surface active agents as feed additives to improve membrane performance was studied by Subcasky. Poly(vinyl methyl ether) improved salt rejection from 4 to 34% or, in product flux, from 8 to 18% for NaCl-containing feed solutions. Other additives used were a series of polyoxyethylene-nonylphenols with an improvement in salt reduction 14 to 36% or, in product flux, 4 to 32%, block copolymers of ethylene and propylene oxides with ethylenediamine with improvement in salt rejection 3 to 14%, or, in product flux, 2 to 35%, and a cationic surfactant of the quaternary ammonium salt type with salt rejection over 400% or, in product flow, more than 140% [15].

Variables which affect permeation properties of tubular cellulose acetate membranes were also examined by Richardson et al. The effects of membrane fabrication conditions, saline feed solution type and concentration level, inlet pressure, inlet temperature, inlet flow rate, surface-active feed additive concentration and time were measured. The permeation property data were used to formulate a model of tubular module system performance. Calculations have shown that maximum performance is obtained with the smallest diameter tubes. Increases in integrated performance of about 20% can result in a reduction of total operating cost of more than 10% [16]. Membrane fabrication variables were studied by the same group of authors for the design of a continuous casting machine. Flexible braided fiberglass sleeves were obtained. A small demonstration unit was designed, constructed and tested [17]. Basic elements of a new tubular reverse osmosis module design approach have been derived from a research investigation of continuous casting and flexible braided sleeve membrane pressure supports. Techniques were developed for batch casting membranes and the results were used to define the principal characteristics of a continuous casting apparatus. This device was developed and used to produce tubular membrane. The apparatus proved to be capable of producing tubular membranes whose properties are essentially constant in the longitudinal direction. The performance of the continuous casting apparatus at casting rates up to 0.66 ft/min was demonstrated [18]. Continuous tubular membrane casting offers the potential of minimizing labor costs in the casting process, while at the same time maximizing the tendency for uniformity in wall thickness and freedom from imperfections. Performance data were obtained for single, batch-cast 6 m (20 ft) long nominal 0.5 inch tubular membranes, cast with machine caster. Cellulose acetate 25, formamide 30 and acetone 45% were used in the test batches [19].

The hydrocasting method was developed to provide a means for the in situ formation of a tubular membrane within a preassembled plant, flushing out the membrane and regenerating fresh membrane surfaces, without dismantling the assembly. Cellulose acetate membrane is produced within a porous support tube, filled with casting solution and forced through the system under pressure. A column of water leaches the solvent from the casting solution and gels the cellulose acetate into a tubular membrane [20]. Membrane thicknesses for vertical and inclined draining orientations were discussed, preliminary experiences and performance data of in situ cast membranes were also given [21]. The contraction phenomenon of the forming bubble was explained by Gollan and Tulin by gas being dissolved from the bubble into the casting solution. Exact and approximate analytical solutions of the unsteady mass transfer problem were presented. Experimental results with nitrogen and helium gas bubbles at various operating conditions agree very well with the theoretical predictions [22]. Hydrocasting in single non-porous tubes using modified Loeb-Sourirajan casting solutions has been further perfected and extended to casting in multiple non-porous tubes. Various means were found effective in suppressing the undesirable macropores formed during membrane tubule fabrication by hydrocasting. The way was thus opened for relatively high pressure testing of short supported sections as well as low pressure characterization of unsupported membranes of almost full produced length of about 77 cm (30 inch). A process for prefabrication of porous nylon tubes of high water permeability was developed [23]. Radii optimization of internally skinned membrane tubules, convective flows and large void formation during membrane precipitation as well as the structure of cellulose acetate membranes were also discussed. The surface tension of polymer solutions and preliminary production cost estimate of supported and unsupported hydrocast modules were outlined [24].

An apparatus was described by Saldadze et al. for preparing tubular, semipermeable cellulose acetate membranes. A solution of cellulose acetate is passed on the inner wall of a glass runoff tube, using a resilient molder to mold the runoff solution into shape. Limiting the diameter shrinkage by using pressure during heat treatment did not change the water to NaCl permeability of the membrane [25]. A tubular membrane with high osmotic properties was made with this elastic molder at 20 to 22°C. Water rate was 1.7 to 2.3 cm/s, air motion rate inside the tube 0.2 to 0.4 m/s, evaporation time 3 to 5 s, immersion time 15 min and heat treatment temperature 85 to 87°C. The film thickness can be regulated by changing the diameter ratio of the molder and the watering tube [26].

A variant of the tubular reverse osmosis module is to be found in the preparation by extrusion techniques of rigid tubules. These fine tubes may have an outside diameter of 0.75 to 0.9 mm (0.03 to 0.035 inch) and a wall thickness of about 0.2 mm (8 to 10 mils). Several thermoplastic materials were evaluated for this purpose, the best of which was a polyvinyl chloride homopolymer. Water fluxes up to 1.22 $m^3/m^2 \cdot day$ (30 gal/$ft^2 \cdot day$) were obtained with material having a pore size of about 2000 Å [27]. Coating of the tubules with effective high flux desalination barriers was the second step in developing this process. The tubules were coated individually at first, but coating in bundles was also attempted. The target performance for the coated tubules was at least 95% salt rejection. Coating work has emphasized high flux systems such as asymmetric and ultrathin cellulosics, polyacrylic acid and dynamically formed zirconium oxide/polyacrylic acid [28]. Further work involved improvement in the tubules formulation, membrane improvements and investigation of coating and fabrication parameters to enable construction of a small experimental brackish water bundle element [29].

The application of the hydrocasting method for the production of small reverse osmosis membrane tubules with an inside skin for unsupported use at low pressures was described by Gollan et al. Investigated hydrocasting parameters included: gas exposure time, velocity of the gelling water inside the tubules with and without the addition of surfactants, salt concentration in the gelling water, and annealing temperature. The most important parameters to be considered during hydrocasting appears to be the gelling water velocity inside the casting tube, which must be kept between 2.5 and 5 cm/s (1 and 2 inch/s) to obtain good membranes [30].

An analytical study on the mass transfer characteristics of the reverse osmosis process in curved tubular membrane duct was made by Srinivasan and Chi Tien. Based on the velocity profile, the concentration distribution of the binary salt solution to be processed in a curved tubular membrane duct was obtained. It was found that a significant increase in mass transfer rate can be achieved due to the Coriolis force. This increase in mass transfer resulting a substantial reduction in concentration polarization is considered significant not only in terms of the increase of system capacity, but also the possible prolonging of the life span of the membrane duct as well [31]. This new design for reverse osmosis in a curved tube system takes advantage of the scouring-silting nature of the secondary flow pattern by utilizing a split membrane over the circumference of the tube. The wall concentration and concentration boundary layer thickness at the inside of the bend grow rapidly unless salt is allowed to escape there, whereas at the outside of the bend the wall concentration and concentration boundary layer thickness remain small compared to straight tubes, thus indicating improved performance [32].

Methods for fabricating composite support elements of braided fiberglass and for casting high flux membranes were developed by Blevitt et al. Tubular membranes of cellulose diacetate and cellulose triacetate gave initial flux of 1.2 to 1.4 m³/m²·day (29.6 to 35.1 gal/ft²·day) with 90.9 to 87.9% rejection from 1% NaCl feed at 56 kg/cm² (800 psi) [33].

The increase of the lifetime of braided pressure supports used in reverse osmosis tubular modules by means of selection of the most appropriate fiberglass yarns and yarn coatings was also investigated. A mathematical model has been developed to allow a more analytical approach to yarn selection and braided pressure support design. An extensive body of yarn test data has been generated in the evaluation of potential yarns and coatings. The most promising yarns appear to be those, which are composite strands of polyester and fluorocarbon coated glass. Thermosetting acrylic impregnants coupled with fluorocarbon top coats were found to be the most effective in preserving glass yarn strength in a wet environment [34].

Tubular elements composed of polyphenylene oxide sulfonate membranes cast onto porous substrates and held in standard porous fiberglass backing tubes were developed by Chludzinski. Results showed the feasibility of using a hydrophobic grade of porous polypropylene as substrate. The tubular membrane on substrates demonstrated integral films of polymer on the substrate which were nigged and easily handled. Tests with flat sheets of the meniscus-coated membrane on porous polypropylene using secondary sewage showed the ability to regenerate the system after fouling. It was concluded that because of the polyphenylene oxide sulfonate membranes resistance to high temperature and corrosive chemicals, it would be a good candidate for use on secondary sewage [43].

A brackish water pilot plant incorporating 36 series-connected helical tubular segments, each formed from an one inch diameter, 15 m (50 ft) long tubular membrane was designed and fabricated. The assembled pilot plant was tested in the laboratory to verify design integrity and performance at various operating pressures, fractional recoveries and feedwater velocities. Subsequently, the pilot plant was installed and operated [35]. **Fig. 3–15** shows a helical tubular segment containing 100 ft of 1 inch tubing.

A tubular pilot plant with a capacity of 1000 gpd (3.8 m³/day) was designed, fabricated and tested. The design was based on 10 modules, each comprising 14 tubular membrane elements 12.7 mm diameter (0.5 inch) and 1.22 m (4 ft) length mounted between circular tube sheets and each containing 0.68 m² (7.3 ft²) of replaceable cellulose acetate membrane. Turbulence promoters were incorporated in 8 modules. Well water of 3500 mg/l was desalted under a pressure of 42 kg/cm²

(600 psi) with 70% recovery. The output was 1000 gal/day by over 6 month operation. Product water had 300 mg/l salts. Drop in membrane water flux of about 10% was observed in the initial 3 months [36].

A 23 m³/day (5000 Imperial gpd) pilot plant for the development and evaluation of membranes has been erected in Appleby Parva, Leicestershire, United Kingdom, and has been operated on two borehole waters. Initially, this plant contained prototype modules with membranes cast on the outside of tubes. Over 3 million module hours of operation have been achieved and the satisfactory performance of both membranes and membrane supports has confirmed that the principle of a membrane cast on the outside of a suitable tubular support is sound. This led to the design of the "spaghetti"

Fig. 3–15

Helical tubular segment.

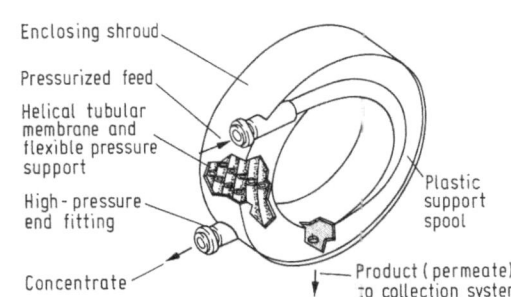

module system. Operating experience with these modules has nearly reached 2 million module hours on a number of brackish waters and on secondary sewage effluent [37]. An illustration of the "spaghetti" module is given in **Fig. 3–16**, p. 296.

The development of the reverse osmosis process has raised commercial interest and a great deal of manufactures in many countries already offer reverse osmosis equipment. Guy has reported on the Havens reverse osmosis design, which utilizes 0.5 inch porous fiber glass tubes lined with 5 mil thick cellulose acetate membrane [38]. The tubular reverse osmosis module assembly, developed by Aerojet-General Corporation, was described by Fluke [39]. Three reverse osmosis concepts, developed by the General American Transportation Corporation and based on porous support tubes with external membranes, arranged in a shell-and-tube configuration, were evaluated by Harris and Humphreys. It was concluded that this concept does not offer economic advantages over the large scale tubular design [40].

A large scale tubular reverse osmosis plant was put into operation at Yotvata in the Negev Desert of Israel in September 1968. The plant, intended to represent a prototype for possible future installations, was designed to produce 200 m³ per day of fresh water from brackish water containing 2400 mg/l salt, of which about half are divalent ions. The plant represents a departure from previous experience in the United States, because of the doubled length and diameter of the tubular assemblies, new materials used for support tubes and the vertical orientation of the tubular train. Assembly fabrication, design and operating data were presented by Cohen and Loeb [41]. Field experience with the Yotvata plant was also reported by Cohen, including evaluation of a 1 Mgd reverse osmosis project and operation parameters of three more plants at Gilat, Abu Rodeis and Timna [42].

Fig. 3–16

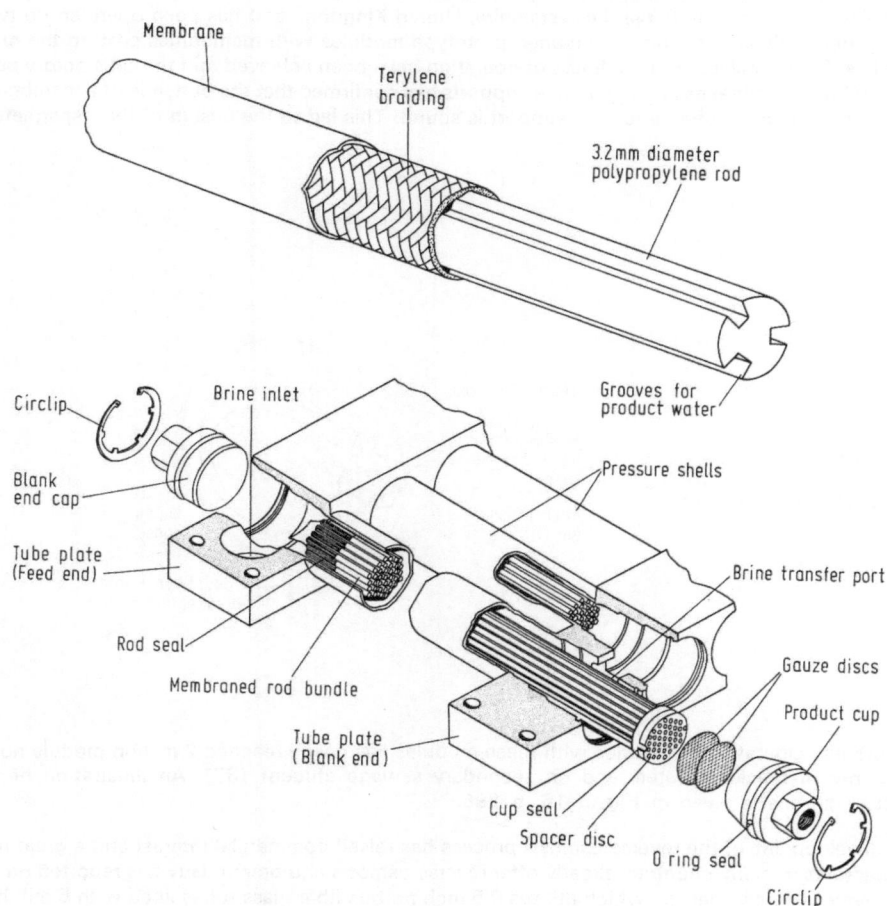

The "spaghetti" module.

Literature to 3.19

[1] S. Loeb (Desalination **1** [1966] 35/49). — [2] D. Stevens, S. Loeb (Desalination **2** [1967] 56/74). — [3] S. Loeb, E. Selover (Desalination **2** [1967] 75/80). — [4] J. S. Johnson, J. W. McCutchan, D. N. Bennion (Univ. of California, Los Angeles, School Eng. Appl. Sci. Rept. No. 69-45 [1969]), J. W. McCutchan, J. S. Johnson (J. Am. Water Works Assoc. **62** [1970] 346/53). — [5] R. E. Fisher, T. K. Sherwood, P. L. T. Brian (Off. Saline Water Res. Develop. Progr. Rept. No. 141 [1965]).

[6] S. Srinivasan, C. Tien (Desalination **3** [1967] 5/16). — [7] W. L. Griffith, R. M. Keller, K. A. Kraus (Desalination **4** [1968] 283/308), Corrigendum by W. L. Griffith, R. M. Keller, D. G. Thomas, W. J. Boegly (Desalination **11** [1972] 129/32). — [8] G. A. Fluke (Off. Saline Water Res. Develop.

Progr. Rept. No. 426 [1969]). — [9] J. Murkes, H. Bohman (Desalination 11 [1972] 269/301). — [10] J. S. Johnson, K. A. Kraus, S. M. Fleming, H. D. Cochran, J. J. Perona (Desalination 5 [1968] 359/69).

[11] Y. Winograd, A. Solan (Desalination 8 [1970] 237/41). — [12] C. W. Saltonstall, H. F. Shuey (Off. Saline Water Res. Develop. Progr. Rept. No. 730 [1971]). — [13] H. K. Bishop, G. Belfort, G. A. Guter (Off. Saline Water Res. Develop. Progr. Rept. No. 464 [1969]). — [14] E. W. Pitera, S. Middleman (Ind. Eng. Chem. Process Design Develop. 12 [1973] 52/6). — [15] W. J. Subcasky (Off. Saline Water Res. Develop. Progr. Rept. No. 532 [1970]).

[16] J. L. Richardson, G. Segovia, W. Baerg, M. L. Anderson (Off. Saline Water Res. Develop. Progr. Rept. No. 455 [1969]). — [17] J. L. Richardson, G. Segovia, A. O. Brodie (Off. Saline Water Res. Develop. Progr. Rept. No. 576 [1970]). — [18] J. L. Richardson, G. Segovia, J. W. Mason, W. J. Subcasky (Off. Saline Water Res. Develop. Progr. Rept. No. 661 [1971]). — [19] J. L. Richardson, G. Segovia, W. H. Bachle, H. A. Parker-Jones (Reverse Osmosis Membrane Research, New York 1972, p. 205/40). — [20] R. Matz, M. P. Tulin, A. Gollan, H. S. Preiser, H. Alcalay (Off. Saline Water Res. Develop. Progr. Rept. No. 542 [1970]).

[21] A. Gollan (Proc. 3rd Intern. Symp. Fresh Water Sea, Dubrovnik 1970, Vol. 2, p. 443/65). — [22] A. Gollan, M. P. Tulin (Proc. 3rd Intern. Symp. Fresh Water Sea, Dubrovnik 1970, Vol. 2, p. 423/42). — [23] A. Gollan, M. P. Tulin, C. Elata (Off. Saline Water Res. Develop. Progr. Rept. No. 806 [1972]). — [24] A. Gollan, M. Frommer, R. Matz, M. Tulin (Off. Saline Water Res. Develop. Progr. Rept. No. 855 [1973]). — [25] K. M. Saldadze, G. Z. Nefedova, A. A. Yasminov, E. A. Kotyakhov, L. I. Gracheva, N. E. Kozhevnikova (Plast. Massy 1973 No. 3, p. 40/1; C.A. 79 [1973] No. 43321).

[26] A. A. Yasminov, E. A. Kotyakhov, L. I. Gracheva (Opresnenie Solen. Vod Ispol'z. ikh Vodonabzh. 1972 153/8; C.A. 79 [1973] No. 83329). — [27] B. Baum, R. A. White, W. H. Holley, W. R. Diehl, R. C. Trudeau, D. C. Button (Off. Saline Water Res. Develop. Progr. Rept. No. 604 [1970]), B. Baum, R. A. White, W. H. Holley (Reverse Osmosis Membrane Research, New York 1972, p. 475/90). — [28] B. Baum, R. A. White, H. Stiskin, W. H. Holley (Off. Saline Water Res. Develop. Progr. Rept. No. 865 [1973]). — [29] B. Baum, R. A. White, W. H. Holley, H. Stiskin (Off. Saline Water Res. Develop. Progr. Rept. No. 868 [1973]). — [30] A. Gollan, J. Ricklis, M. P. Tulin (Proc. 4th Intern. Symp. Fresh Water Sea, Heidelberg 1973, Vol. 4, p. 135/46).

[31] S. Srinivasan, Chi Tien (Proc. 3rd Intern. Symp. Fresh Water Sea, Dubrovnik 1970, Vol. 2, p. 587/600; Desalination 9 [1971] 127/39, Corrigendum in: Desalination 12 [1973] 277/9). — [32] R. I. Nunge, L. R. Adas (Desalination 13 [1973] 17/36). — [33] R. Blevitt, R. H. Hartupee, M. Marks (Off. Saline Water Res. Develop. Progr. Rept. No. 665 [1971]). — [34] R. H. Williams, A. O. Brodie, C. H. Lewis, J. S. Field, J. C. Britt (Off. Saline Water Res. Develop. Progr. Rept. No. 884 [1973]). — [35] R. H. Williams, G. Segovia, W. H. Bachle, J. C. Britt, J. L. Richardson (Off. Saline Water Res. Develop. Progr. Rept. No. 871 [1973]).

[36] E. A. G. Hamer (Off. Saline Water Res. Develop. Progr. Rept. No. 424 [1969]). — [37] J. R. Grover, M. H. Delve (Chem. Eng. [London] No. 257 [1972] 24/8), J. R. Grover (Chem. Ind. [London] 1973 369/70), J. R. Grover, R. Gayler, M. H. Delve (Proc. 4th Intern. Symp. Fresh Water Sea, Heidelberg 1973, Vol. 4, p. 159/69). — [38] D. B. Guy (Chem. Eng. Progr. Symp. Ser. 65 No. 91 [1969] 43/7). — [39] G. A. Fluke (Off. Saline Water Res. Develop. Progr. Rept. No. 582 [1970]). — [40] F. L. Harris, G. B. Humphreys (Off. Saline Water Res. Develop. Progr. Rept. No. 642 [1971]).

[41] H. Cohen, S. Loeb (Proc. 3rd Intern. Symp. Fresh Water Sea, Dubrovnik 1970, Vol. 2, p. 339/49). — [42] H. Cohen (Proc. Symp. Develop. Desalination Technol. Israel, Jerusalem 1972, p. 22/41). — [43] P. J. Chludzinski (Off. Saline Water Res. Develop. Progr. Rept. No. 881 [1973]).

3.20 Spiral-wound membrane module

Modul in Wickelform

In this design the membrane is wound into a spiral unit before being placed into the pressure vessel. Hence, the membrane and the support between the high- and the low-pressure sides of the membrane can be formed into an ordinary pipe, as shown in **Fig. 3–17**, without the need of a special pressure vessel. Three sides of the membrane are sealed, forming a kind of envelope, and one end is left open to transfer the product water to a collecting tube attached at the open end. A plastic or glass fabric material, capable of withstanding the high pressures, is placed within the envelope and is sufficiently porous to carry the product water to the collector tube. To form the spiral, the envelope is wrapped into a roll around the central product collecting tube. A coarsely woven fabric is rolled with the membrane envelope to form a path for the salt water to flow through the module over the membrane surface. **Fig. 3–18** illustrates the principle of operation of the reverse osmosis spiral-wound module. The spiral-wound concept was developed by Gulf General Atomic Corp. and described by Merten [1].

Fig. 3–17   Fig. 3–18

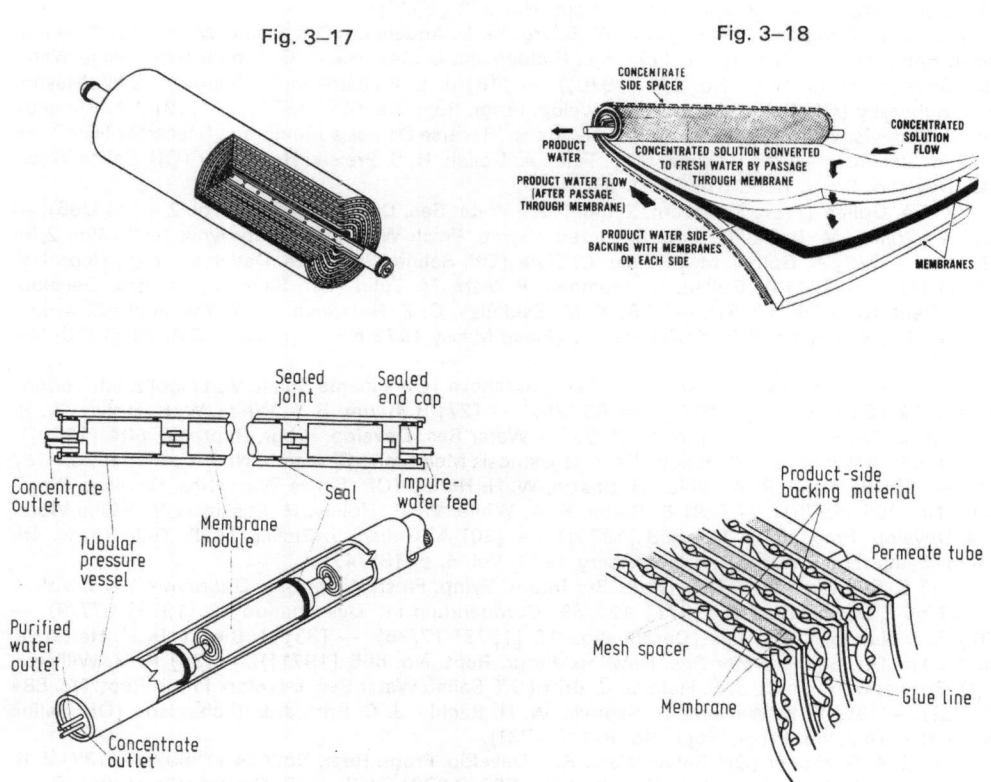

Spiral-wound reverse osmosis module.

A parametric study was first made by Menzel of an 1 Mgd (3785 m³/day) brackish water reverse osmosis plant, using the spiral-wound module concept [2]. Module development, feedwater pretreatment and module cleaning, module sterilization and storage, as well as parametric studies were reported by Larson et al. [3]. The primary problem during the development period was the synthesis of a composite backing material that would provide a support for the membrane and a porous flow path for the water and allow the assembly of the module into a compact configuration, as well as the development of a technique for sealing the membrane to this backing material. Based on the module performance data and the boundary layer predictions, parametric studies were conducted for an 1 Mgd brackish water desalination plant [4]. The development of larger spiral-wound modules for economical application in large reverse osmosis systems was then studied. Several larger module concepts were evaluated and a 10000 gpd (38 m³/day) test pilot plant was designed, fabricated and tested [5]. Problems encountered during the operation of the pilot plant were reported by Larson and Gibbons and techniques were developed for the fabrication of reproducible, high-performance spiral-wound modules [6]. A description of the pilot plant, as well as a report on its operation and model testing was given by Sudak [7]. Larsen has also reported on operating experience with a number of units ranging from 5000 to 50000 gpd (19 to 190 m³/day) capacity [8]. Further work to improve and develop spiral membrane modules was reported by Foreman et al. [9].

The spiral module configuration provides the opportunity to attain high packing density in terms of effective membrane area per unit of pressure vessel volume. An improved product channel material was developed, having less than one-half the thickness of the initial product channel assembly. The packing density is in excess of 490 m²/m³ (150 ft²/ft³) [10]. Long term tests verified the performance of larger modules and improvement in the cost [11].

Development and demonstration of a second generation spiral-wound element were successfully undertaken. Techniques to reuse expensive components in the element and to restore adequate salt rejection by in situ reheat treatment of the membrane were investigated. A low cost filter paper was used as a replacement for part of the woven polyester sailcloth support used in the element. Work was also directed toward attaining the capability to produce on a volume basis large spiral-wound membrane elements with membrane cast directly on the support fabric. The main objective of this effort was to develop high-rejection, compaction-resistant membrane elements capable of producing a minimum water flux of 0.65 m³/m² · day (16 gal/ft² · day) at 42 kg/cm² (600 psi) from a natural feedwater after 10000 h of operation. Salt rejection capability throughout the test was approximately 96.5% [12]. In continuing work the feasibility of a spiral-wound module measuring 30 cm (12 inch) in diameter and 9.15 m (30 ft) in length to produce about 300 m³/day (100000 gpd) of purified water was demonstrated [13].

Literature to 3.20

[1] U. Merten (Off. Saline Water Res. Develop. Progr. Rept. No. 165 [1966]). — [2] H. F. Menzel (Off. Saline Water Res. Develop. Progr. Rept. No. 236 [1967]). — [3] T. J. Larson, S. S. Kremen, L. Nusbaum, A. B. Riedinger (Off. Saline Water Res. Develop. Progr. Rept. No. 313 [1968]). — [4] T. J. Larson, L. Nusbaum, A. B. Riedinger, J. Astl (Off. Saline Water Res. Develop. Progr. Rept. No. 338 [1968]). — [5] A. B. Riedinger, J. K. Laughlin, R. G. Sudak (Off. Saline Water Res. Develop. Progr. Rept. No. 341 [1968]).

[6] T. J. Larson, B. C. Gibbons (Off. Saline Water Res. Develop. Progr. Rept. No. 442 [1969]). — [7] R. G. Sudak (Off. Saline Water Res. Develop. Progr. Rept. No. 453 [1969]). — [8] T. J. Larson (Desalination 7 [1970] 187/99). — [9] G. E. Foreman, S. S. Kremen, A. B. Riedinger, R. L. Truby, W. W. Wight (Off. Saline Water Res. Develop. Progr. Rept. No. 675 [1971]). — [10] S. S. Kremen, A. B. Riedinger (Off. Saline Water Res. Develop. Progr. Rept. No. 676 [1971]).

[11] G. E. Foreman, S. S. Kremen, J. F. Loos (Off. Saline Water Res. Develop. Progr. Rept. No. 677 [1971]). — [12] G. E. Foreman, S. S. Kremen, W. W. Wight, T. D. Wolfe (Off. Saline Water Res. Develop. Progr. Rept. No. 875 [1973], No. 880 [1973]). — [13] J. M. Chirrick, G. E. Foreman, A. B. Riedinger, W. W. Wight (Off. Saline Water Res. Develop. Progr. Rept. No. 901 [1973]).

3.21 Hollow fiber module

Hohlfaser-Modul

The third design of commercial application of reverse osmosis uses hollow fine fibers, which were prepared from membrane materials in outer diameters as fine as 50 to 250 μ (2 to 10 mils) and about half of it as inside diameter. In such very small diameters, the microscopic tubules can withstand enormous pressures and do not need any pressure support. Fibers of this type, closely packed, in millions per unit, provide a very large membrane surface area per unit of pressure vessel volume. With a fiber outer diameter of 50 μ and a packing density of 50%, a hollow fiber device can achieve a surface area 29500 m²/m³ (9000 ft²/ft³) of vessel volume. Product flow per unit of fiber surface is usually less than for the same area of conventional membranes. However, the difference in area per volume compensates for the difference in flow rate. The fibers are placed in a pressure vessel either with one end sealed and the open end embedded in a plastic seal plate or are U-bended and both ends are embedded in the seal plate. Generally, the design is based on the shell and tube heat exchange configuration. The salt water, under pressure, is on the outside of the fibers and the product water is forced through the fiber walls to the inside and flows to the open end outside of the pressure vessel. Fiber failure under external pressure results in collapse but not rupture. The principle of operation of the hollow fiber reverse osmosis process is shown in **Fig. 3–19**.

The hollow fiber module was first developed by Du Pont, using aromatic polyamide fibers, with the trade name "Permasep". Commercial permeators were described and the advantages of hollow fibers as semipermeable membranes were reviewed by Cooke [1]. Advantages of hollow fibers, membrane evaluation, field test results, as well as capital and operating costs were also reviewed by Mattson and Tomsic [2]. A similar review was presented by Mahon et al. Flux rates of 0.83 to 1.66 m³/day per kg of fiber (100 to 200 gal/day per lb of fiber) are reported with 50% product water recovery [3].

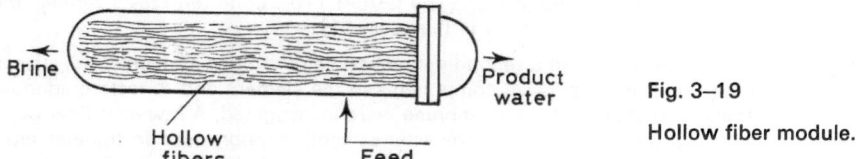

Brine

Product water

Hollow fibers   Feed

Fig. 3–19

Hollow fiber module.

Composite membranes of crosslinked cellulose substrate and a thin film of cellulose triacetate were developed by Cohen and Riggleman for hollow fiber application. Salt rejections up to 99.6% were achieved in some membranes. Composite membranes on spontaneously gelled nitrocellulose substrates gave the same salt rejections [4]. Hollow fiber asymmetric membranes of cellulose acetate were further prepared and the characteristics given. They offer a good potential for utilization in reverse osmosis modules, exhibiting water flux rates of 0.24 and 0.16 m³/m² · day (6 and 4 gal/ft² · day) at respectively 42 and 28 kg/cm² (600 and 400 psi), while rejecting 95 to 97% of the NaCl from a 5000 mg/l NaCl feed solution. The optimum hollow fiber module should be short to minimize frictional pressure losses. The presence of polyethylene glycol in the ultrathin selective layer of composite membranes resulted in an increase in membrane water flux, while maintaining a high selectivity in rejecting 99% of a 5000 mg/l NaCl feed [5]. The duration and the temperature of the annealing process were varied to yield asymmetric hollow fiber membranes with higher flux rates. Composite hollow fiber membranes on cellulose nitrate substrates produced up to 0.72 m³/m² · day (17.8 gal/ft² · day) at 14 kg/cm² (200 psi), but collapsed at higher pressure. Preparation of stronger fiber membranes resulted in lower fluxes [6]. 3,3'-Diaminobenzidine-diphenyl isophthalate copolymer reverse osmosis membranes have approximately the same flux as cellulose acetate membranes at ambient temperature, but when the temperature was raised to 75 to 90°C the flux of the membrane increased, while that of cellulose acetate decreased to zero. A wide range of flux and rejection was obtained with hollow fibers from this polymer [7].

A model based in that the pressure difference causes concentration gradients in the membrane and the transport of liquid particles is made by diffusion was presented by Paul. The model was applied to hollow fiber configurations [8]. A predictive model of hollow fiber reverse osmosis systems was developed by Gill and Bansal, using the equivalent annulus assumption. The fraction of feed recovered depends on five parameters. Optimum values exist for the ratio of the inside to outside fiber radius and for the outside fiber radius. For dilute systems, a simple closed form expression is obtained for recovered feed, which enables to determine optimum values easily. The effects of pressure, temperature, flow rate, concentration, viscosity of the feed, system length, membrane rejection parameter and number of fibers were studied. Countercurrent was found superior to concurrent operation [9].

In further development of the hollow fiber process, the preparation of hollow fibers from cellulose acetate, other cellulose derivatives, poly(vinyl alcohol) compositions and aliphatic polyamides was reported by Orofino. An unique solution-spinning technique for cellulose acetate systems was developed, which resulted in the successful translation of the Loeb skin-matrix composition to hollow fiber geometry. Performance of 0.16 m³/m² · day (4 gal/ft² · day), rejection 97% with Webster brackish water at 17.5 kg/cm² (250 psi) and 50% recovery were obtained [10]. The performance of low-pressure cellulose acetate hollow fiber system was considerably improved and translation of the preparative technique to multi-orifice spinning was demonstrated. Typical performance indices for members of this hollow fiber system are 0.2 m³/m² · day (5 gal/ft² · day) with 90% rejection and 0.12 m³/m² · day (3 gal/ft² · day) with 97% rejection, obtained with 3000 mg/l sodium chloride feed at 250 psi external pressure. Appropriate techniques for assembly of large fiber bundles and fabrication of high pressure seals were developed [11].

Additional development of asymmetric cellulose acetate hollow fibers resulted in a brackish water desalting system for hardness removal and partial demineralization, exhibiting fluxes of 0.35 m³/m² · day (8.5 gal/ft² · day) at 7 kg/cm² (100 psi) operating pressure. The dioxane-formamide doped fiber systems tested showed good stability with flux modules in excess of 5000 gpd capacity [12]. Use of alternate solvents, coagulation media and annealing conditions provided hollow fibers varying in univalent/divalent salt passage ratio and with markedly improved product water flux as compared with state-of-the-art systems. Laboratory performance of a hollow fiber system especially useful for divalent ion rejection exceeded 0.4 m³/m² · day (10 gal/ft² · day) and 95% rejection of magnesium sulfate at 10.5 kg/cm² (150 psi), after seven days of uninterrupted test [13].

A nominal 76 m³/day (20000 gpd) hollow fiber module of single-cartridge, radial flow design, was constructed and field-tested for three months. Preparatory work included detailed engineering and economic evaluation of two alternate axial flow modules, also of nominal 20000 gpd productivity, and fabrication of an instrumented test loop to accommodate the unit finally built. The module showed good flux stability throughout the test period, exceeding target performance with regard to module output and volumetric efficiency. Salt rejection stabilized at 85% for the remainder of the test [14].

The technique of membrane surface modification, developed by Hillman, was applied to hollow fibers in order to improve system efficiency by yielding higher total flux rates per unit volume of membrane. New polymers and copolymers were synthesized for the purpose of surface modification and hollow fiber formation. The polyamides and carbonate-ester amides produced good salt rejection (86.6%) and good flux, but the membranes were subject to compaction. Polymerized polyamide showed low salt rejection and suffer from compaction. Blends of polyamide and epoxy resins show that flux is enhanced, while rejection is maintained. Polysulfonamides of various compositions produce 70 to 72% salt rejection, while the flux must be enhanced by asymmetric membrane preparation techniques [15].

Laboratory evaluations have shown that polybenzimidazole hollow fiber membranes possess good desalination properties. Water flux as high as 0.4 m³/m² · day (10 gal/ft² · day) and salt rejections as high as 99% have been achieved. Structure-desalination performance correlations were established in morphological examinations employing thermomechanical analysis, optical microscopy and scanning electron microscopy. Radial void structure, void concentration and shrinkage behavior were found to be related at least in a semiquantitative way to desalination performance based on a Loeb-type membrane structure. Membrane fouling was established as the major cause of the observed decline in flux and rejection, with compaction playing a minor role. Fouling was usually eliminated by installation of a highly efficient filtration system [16].

Asymmetric hollow fiber membranes with the discriminating skin on the inside of the fiber were prepared from aromatic polyamides and cellulose acetate. The fibers gave water flux rates of 0.8 m³/ m² · day (20 gal/ft² · day) at 40% magnesium sulfate and 20% sodium chloride rejection levels when tested with a 1500 mg/l brackish feed brine. The burst pressure of these fibers averaged 10.5 kg/cm² (150 psi). A single fiber testing device was constructed, which allowed screening of candidate fibers immediately after spinning, without having to wait for a module to be constructed [17].

A small melt extruder was designed and constructed for the production of low flux hollow fibers from cellulose acetate. The screw driven extruder, equipped with special stretching and take-up mechanisms, was used to produce fibers having diameters ranging from 0.076 to 1.27 mm (0.003 to 0.05 inch), with a variable wall thickness to diameter ratio. Hollow fiber bundles were prepared by casting fiber ends in epoxy blocks for insertion into high pressure test cells. Hollow fibers were also produced by solution extrusion. Desalination characteristics of unstretched and stretched membranes were studied as a function of membrane thickness. The applicability of time-temperature superposition to the ultimate elongation of cellulose acetate was also established in these studies. Model studies were made to determine the influence of fiber geometry to the various mechanisms of fiber failure [18].

Two semipermeable hollow fiber modules undergoing reverse osmosis were compared. The inherent liminations of a constant seepage assumption were deduced from a more general approach, which includes the simultaneous coupling between the transport across the fiber wall and the flow through the fiber bore. For the coupled model, the critical radii corresponding to a maximum axial mass flux for symmetric and asymmetric membranes were shown to be comparable under certain limiting conditions. A distribution of membrane permeabilities was employed to account for the random variation of fiber properties within a bundle. Rigorous lower bounds on an effective bundle permeability were derived and compared with experiments [19].

Reverse osmosis modules using cellulose triacetate hollow fiber membranes were developed and tested in two demonstration plants. After 1460 h of operation, the unit at Webster had a flux of 0.06 m³/m² · day (1.49 gal/ft² · day), a salt rejection of 98% at 52% recovery (600 psi) and a total output of about 3 m³/day (800 gpd). At Yuma after 1040 h, the hollow fiber module had a flux of 0.07 m³/m² · day (1.66 gal/ft² · day) and a salt rejection of 94.6% at 48% recovery (500 psi). The production rate was about 2.2 m³/day (570 gpd). External feed water flow is more advantageous. A high speed melt spinning process was developed for the production of cellulose triacetate hollow fibers and a machine for high speed winding of shell-and-tube cartridge [20]. A minor modification resulted in a flow rate of 0.11 m³/m² · day (2.6 gal/ft² · day) and 96% rejection using brackish water [21]. In further development modules were designed and tested for seawater desalination (see chapter 3.22).

A 570 m³/day (150000 gpd) reverse osmosis hollow fiber unit was constructed at Greenfield, Iowa, for municipal service [22]. A 30000 gpd (114 m³/day) cellulose triacetate hollow fiber module, for potential use in large scale brackish water desalting plants, was fabricated and tested on a brackish water with 3650 mg/l total dissolved solids for 60 days. Capital and operating cost estimates were made for large brackish water desalting plants, using the 30000 gpd module [23].

A diffusion solubility model for reverse osmosis was developed by Machacek and Osburn. The sorption of a dye by hollow fiber nylon membranes was measured. Equations for the unsteady state reverse osmosis separation of the dye from the water were solved by numerical analysis and were shown to fit data on this process. The assumed model is that of solution of dye in the membrane followed by diffusion through it [24]. A review on naturally occurring and synthetic hollow fibers, including applications and future possibilities, was given by Tanquary [25]. Three types of hollow fiber processes were outlined in another review: low pressure for low salinity waters, medium pressure for brackish water and high pressure for one-pass seawater desalting [26].

Literature to 3.21

[1] W. P. Cooke (Desalination 7 [1969] 31/46). — [2] R. J. Mattson, V. J. Tomsic (Chem. Eng. Progr. 65 [1969] 62/8). — [3] H. I. Mahon, E. A. McLain, E. W. Skiens, B. J. Green, T. E. Davis (Chem. Eng. Progr. Symp. Ser. 65 No. 91 [1969] 48/51). — [4] M. E. Cohen, B. M. Riggleman (Off. Saline Water Res. Develop. Progr. Rept. No. 400 [1969]). — [5] M. E. Cohen, M. A. Grable, B. M. Riggleman (Off. Saline Water Res. Develop. Progr. Rept. No. 518 [1970], No. 608 [1970]).

[6] M. B. Harbert, M. E. Cohen, M. A. Grable, J. W. Morton, B. M. Riggleman (Off. Saline Water Res. Develop. Progr. Rept. No. 745 [1972]). — [7] F. S. Model, L. A. Lee (Reverse Osmosis Membrane Research, New York 1972, p. 285/97). — [8] D. R. Paul (J. Appl. Polymer Sci. 16 [1972] 771/83). — [9] W. N. Gill, B. Bansal (AIChE [Am. Inst. Chem. Engrs.] J. 19 [1973] 823/31). — [10] T. A. Orofino (Off. Saline Water Res. Develop. Progr. Rept. No. 549 [1970]).

[11] T. A. Orofino (Off. Saline Water Res. Develop. Progr. Rept. No. 798 [1972]). — [12] R. L. Leonard (Off. Saline Water Res. Develop. Progr. Rept. No. 872 [1973]). — [13] R. L. Leonard, T. A. Orofino (Off. Saline Water Res. Develop. Progr. Rept. No. 887 [1973]). — [14] T. A. Orofino, R. L. Leonard, J. C. Berry, T. E. Britt, J. A. Garden (Off. Saline Water Res. Develop. Progr. Rept. No. 893 [1973]). — [15] A. M. Stake, C. Giori, J. J. Hillman (Off. Saline Water Res. Develop. Progr. Rept. No. 768 [1972]).

[16] A. A. Boom, H. J. Davis, M. Jaffee, L. A. Lee, F. S. Model (Off. Saline Water Res. Develop. Progr. Rept. No. 818 [1972]). — [17] D. S. Cleveland, M. Rambeau, A. Czernicki, T. R. Rich (Off. Saline Water Res. Develop. Progr. Rept. No. 856 [1973]). — [18] E. A. Meinecke, D. V. Mehta, S. Grume (Off. Saline Water Res. Develop. Progr. Rept. No. 845 [1973]). — [19] C. Chen, C. A. Petty (Desalination 12 [1973] 281/93). — [20] E. L. Dance, T. E. Davis, H. I. Mahon, E. A. McLain, W. E. Skiens, J. O. Spano (Off. Saline Water Res. Develop. Progr. Rept. No. 763 [1971]).

[21] T. E. Davis, W. E. Skiens (Polymer Prepr. Am. Chem. Soc. Div. Polymer Chem. 12 [1971] 378/84). — [22] N. P. Chopey (Chem. Eng. 78 No. 3 [1971] 28/30). — [23] H. I. Mahon, M. D. Bearden (Off. Saline Water Res. Develop. Progr. Rept. No. 923 [1974]). — [24] R. F. Machacek, J. O. Osburn (J. Polymer Sci. Polymer Symp. No. 41 [1973] 109/16). — [25] A. C. Tanquary (Bull. S. Res. Inst. 26 No. 1 [1973] 10/3).

[26] Anonymous (Sci. Am. 230 [1974] 10/1).

3.22 Desalting of seawater and wastes by reverse osmosis and ultrafiltration

*Entsalzung
von Meer-
wasser und
Ablaugen
durch
umgekehrte
Osmose und
Ultra-
filtration*

Seawater. In developing the reverse osmosis process, both brackish water and seawater were considered as feed solutions. It appeared that brackish water could be desalted in a single stage, whereas two stages were required for seawater to reach the desired degree of desalination.

A 9.5 m³/day (2500 gpd) pilot plant was designed by Manjikian and constructed by the U.S. Navy to produce potable water of salt content less than 500 mg/l from seawater in two stages. First stage, operated at 70 to 140 kg/cm² (1000 to 2000 psi), reduced salt content to 3000 to 5000 mg/l and the second stage, operated at 45 kg/cm² (650 psi), reduced further the salt content to less than 500 mg/l. Details on the fabrication of most effective tubular membranes are given. During one month test period the flux remained nearly constant at about 0.4 m³/m² · day (10 gal/ft² · day) with average rejection of about 90% [1]. A complete report on the operation of this unit was given by Hodgson and Pal. The integrity of the unit for operation with natural seawater was doubtful. The first stage suffered failures of almost all operating components within 1000 h and constant repairs and maintenance was required to keep the unit operational [2].

Due to the high sodium chloride rejection (99.2%) of the tubular cellulose acetate-cellulose triacetate membranes, they were considered to be suitable for the single-pass desalination of seawater. Their predicted average flux and total water product over the first year of operation was 12% greater than that of a membrane with an initial flux of 0.4 m³/m² · day (10 gal/ft² · day) and a flux decline parameter of 0.04 [3]. However, when extruded into tubes, membranes from cellulose acetate-triacetate blend formulations showed high salt permeation. Three types of defects, wavemarks, voids and fiber penetrations, have been identified and means for improvement were developed. The membranes were then successfully evaluated with seawater in the laboratory [4].

In evaluating the single and multiple pass seawater desalination, Envirogenics Co. concluded that the two-pass concept is the least expensive configuration when using membranes available at the time of the study at a pressure up to 105 kg/cm² (1500 psi). With membranes potentially available in the future the one-pass configuration can become cost effective at 1500 psi. Break-even point in the water production cost from the one-and two-pass configurations was calculated. At higher pressures the one-pass design becomes cost effective with lesser membrane properties. The membrane replacement and tube costs were investigated in respect with their impact on the product water cost. Plant capacity did not influence design selection. A hybrid combination of reverse osmosis and electrodialysis would produce water at nearly the same cost as an one-pass reverse osmosis plant (84 cents per kgal) [5].

Continuous experimentation has shown, that cellulose acetate membranes were not affected by hydrolysis. After 10 months operation in natural seawater, with a single-stage equipment, the characteristics of the obtained product water remained good. With two-stage equipment, water with less than 500 mg/l total salinity could be obtained [6]. A vertical tube single-stage reverse osmosis plant, producing 8 m³/day potable water of slightly over 1000 mg/l total dissolved salts, was erected on the island of Cavallo, near Corsica [7]. A two-stage vertical tube plant with a capacity of 50 m³/day (13 200 gpd) has also been constructed at the french island of Houat, on the Atlantic Ocean [8]. An apparatus with baglike packing of cellulose acetate membranes was described by Kul'skii and Kucheruk. Although the apparatus operated successfully on simulated seawater, it was concluded that it is impossible to obtain potable water from seawater in a single-step desalination in the tested apparatus [9].

Data which show the effect of concentration polarization on the product rate and product quality for reverse osmosis system operating on seawater feed were presented by Johnson and McCutchan. The tests were conducted on asymmetric cellulose acetate membranes of the Loeb-Sourirajan type. Membranes cured at temperatures of 83, 88 and 92°C were tested at 70 kg/cm² (1000 psi). The applicability of the Brian equation for use in the turbulent flow regime was checked and the possibility of adapting the same equation in the laminar region by use of a correction term was discussed [10]. The effects of operating pressure, membrane curing temperature, time in service, seawater pH and flow rate, membrane compaction and feed seawater temperature on desalination performance were studied by Johnson and McCutchan on a test unit containing 28 tubular assemblies equipped with modified cellulose acetate membranes. The data indicated that one-pass desalination of normal seawater is possible with these membranes [11]. A dynamic programming was developed and solved to arrive at optimum operating parameters for a possible two-stage and three-stage seawater tubular reverse osmosis desalting plant with cellulose acetate membranes [12].

In a single step, it was concluded by Pasternack et al., cellulose acetate type membranes are only capable of producing fresh water from brackish water with a salt concentration up to 1%. With higher concentrated solutions as feed, either the flux or the rejection of the membranes are too low. Multi-stage processes may possibly eliminate this problem. Multi-stage membrane processes require membranes with different characteristics be present at each stage. The influence of important parameters such as tube-diameter, load-factor and module-design on production costs as well as capital costs was discussed. It was assumed that the membranes in tubular form are placed within support-tubes and that flow conditions are turbulent. Seawater can be processed in a two-stage development. The production costs of two-stage-processes are relatively high and this should limit the process to plant capacities up to 1000 m³/day [13].

Stable, high productivity membranes made from a blend of cellulose diacetate and cellulose triacetate have been incorporated into 10 cm diameter by 56 cm long spiral-wound elements for use with brackish water. Depending on the salt concentration in the feed water, such elements produce up to 5 m³/day of product water with 90% removal of salt or up to 4 m³/day with 95% salt removal at 42 kg/cm² operating pressure. Elements made with a more retentive membrane yielded a final product containing 300 mg/l or less of salts in the two-pass desalination of seawater. Structural modifications together with a membrane of the maximum retention provided elements capable of single-pass desalination of seawater at 105 kg/cm². These elements produced 0.6 m³/day of water with less than 200 mg/l of dissolved salts [14].

The construction and operation of a first pass seawater reverse osmosis tubular assembly, as used in a two-pass system, and the desalting performance evaluation of cellulose acetate blend membranes under actual field service conditions were described by May and Tidball. The system is composed of 1240 tubular elements operating at 70 kg/cm² (1000 psi) on pretreated seawater. Membrane performance was satisfactory with an initial flux of 0.65 m³/m² · day (16 gal/ft² · day) and a product salinity of 2000 mg/l. A flux decline coefficient of 0.02 was measured for the initial testing period [13].

Seawater pretreatment methods which could be used in conjunction with reverse osmosis desalination to prolong membrane effectiveness were investigated. The investigation consisted of: bench-scale tests to determine optimum application of coagulants, chlorine and possible need for pH adjustments; pilot plant test to determine the effect of hydraulic loading; pilot plant test to determine the effects of chlorination, aluminum coagulation, settling, sand, manganese-zeolite, granular activated carbon and diatomaceous earth pressure filters; and a two-week continuous run to determine the performance of the above mentioned filters under prolonged operation. The study showed that chlorination, sand pressure filtration and activated carbon filtration were sufficient to obtain the desired water quality in respect with turbidity at all times during the test period. The bench-scale tests indicated that alum coagulation and settling provide satisfactory clarification. A pilot plant investigation at the Office of Saline Water, Wrightsville Beach, test facility evaluated the effectiveness of five pretreatment systems in improving the quality of seawater prior to reverse osmosis. Two identical filtration trains provided data for the following pretreatment systems: chlorination, rapid sand, manganese-zeolite and activated charcoal filtration [16].

An engineering study was also made to determine the optimum minimum size pilot plant to provide useful scale-up data for hollow fiber reverse osmosis seawater desalting plants in the 1890 to 3780 m³/day (0.5 to 1.0 Mgd) range. A conceptual design based on reverse osmosis modules of 2500 gpd capacity, a cost estimate and a construction schedule were developed. A test agenda and time schedule for testing were also formulated. Major test areas were marine fouling control, filtration, residual iron control and removal, high pressure piping, and materials of construction [17].

An amide hydrazide film membrane developed by Du Pont is considered as an outstanding candidate for seawater desalination among the family of non-cellulosic barriers. Characteristic properties are: one-pass desalting capability of 99% chloride rejection with initial fluxes of 0.4 m³/m² · day (10 gal/ft² · day) and better, at 70 kg/cm² (1000 psi) and integrity in seawater for at least 500 days. Tests have been conducted in cells and tubular devices. Fouling studies defined pretreatment as the most effective control measure [18].

Small scale, 100 gpd (0.38 m³/day), reverse osmosis modules using cellulose triacetate hollow fiber membranes were developed for single-stage desalination of seawater. The modules were tested in an once through system for periods of up to 3600 h. The small experimental cartridges were designed for radial flow of the seawater over the outer surface of the fibers [19]. These modules were further developed from a 4 inch diameter (100 gpd) to an 8 inch (20 cm) diameter module with a target capacity of 9.5 m³/day (2500 gpd) for the desalination of seawater in one-pass [20].

Washwater recovery in spacecrafts. Manned space flight in extended missions requires that the washwater for laundry and body hygiene must be recycled and be acceptable from a health and comfort standpoint. Tests were run on both synthetic and real washwaters and rejection efficiencies were determined [21]. Four selected reverse osmosis pumps were evaluated during a 30-day test under spacecraft washwater recovery system conditions. The test data include pump power requirements, efficiency and other design parameters [22]. Necessary modification and development of sulfonated 2,6-dimethyl polyphenylene oxide membrane was conducted to obtain optimum reverse osmosis performance on synthetic and real washwater at sterilization temperatures of 74°C (165°F) and more, and maximum pressure of 56 kg/cm² (800 psi). The goal was to obtain a reverse osmosis permeate containing less than 10 microorganisms/cm³ and to achieve a stable flux of 0.2 m³/m² · day (5 gal/ft² · day) with rejections of NaCl (85.0%), urea (99.3%), lactic acid (99.0%) and detergent or soap (99.9%) with a synthetic wash water feed containing 1500, 1000, 7000 and 10 000 mg/l, respectively, of the washwater species. It was demonstrated with real shower water that reverse osmosis operation with these membranes at 165°F yields a permeate free of microorganisms [23]. An aerospace-configured gasketed plate-and-frame reverse osmosis hardware was designed and fabricated. A description of the components and their function was presented. Parametric life testing was conducted with 9.5 l/day (2.5 gpd) units. Synthetic and real washwater were used in the study [24].

Three membrane modules were evaluated for concentration of washwater at 165°F. The membranes tested were: polybenzimidazole (hollow fiber configuration); cellulose acetate blend (spiral wound configuration); and sulfonated polyphenylene oxide (plate-and-frame configuration). Tests were conducted both with a synthetic washwater which contained sodium dodecyl benzene sulfonate, urea, lactic acid and sodium chloride, and with simulated washwater [25].

A spiral-wound module containing blend cellulose di- and triacetate membrane was also evaluated at 21kg/cm² (300 psi) in real laundry water for 17 days at 74°C (165°F). Throughout the test period, the rejections of detergent, lactic acid, ionic species and total dissolved solids were above 98% at product recoveries ranging from 50 to 85%. A prototype system processing about 0.1 m³ (27 gal) of wash water in 8 h or less was constructed and tested. A carbon column to remove traces of detergent and an ion exchange column to insure that ammonia nitrogen would remain below 5 mg/l were included in the prototype. The addition of urease enzyme to laundry water at 74°C was shown to be a feasible method for reducing urea content [26].

Waste water and sewage. The application of pressure driven membrane processes in the treatment of industrial wastes is described by Okey in chapter 12 of the monograph on Industrial Processing with Membranes, edited by Lacey and Loeb [27].

Reverse osmosis and ultrafiltration are gaining an increasing interest for application in such processes as the reclamation of waste water and sewage for reuse, the treatment of waste liquids in the pulp and paper industry, the purification of liquid foods and the treatment of various industrial wastes either to reclaim water for reuse or materials otherwise wasted.

All grades of municipal waste water, was concluded in a report by Aerojet General Corp., may be significantly improved by the reverse osmosis process. High removals of dissolved minerals, organic substances and suspended matter have been achieved. The relative effects of reverse osmosis test-cell geometry on solids deposition and membrane performance were presented. A phenomenological model was postulated describing the role of undissolved solids and organic substances in producing product water flux decline and the subsequent maintenance of constant product water fluxes [28].

Review papers on ultrafiltration having a more or less general interest were published by Stavenger [29], by van Altena [30], by Goldsmith [31], by Porter et al. [32], by Blatt et al. [33], by Madsen et al. [34], by Forbes on optimization of ultrafiltration [35], by Porter on concentration polarization with membrane ultrafiltration [36], by Palmer et al. on the same subject [37], by McDonald [38], etc. Reviews and papers were also published by van Oss on ultrafiltration membranes [39], by Cross on asymmetric hollow fibers for ultrafiltration and dialysis [40], by Harriott on the mechanism of partial rejection by ultrafiltration membranes [41], etc.

Other work on the application of reverse osmosis and ultrafiltration in waste water treatment includes papers by Hindin and Bennett on water reclamation by reverse osmosis [42], by Channabasappa on the application of reverse osmosis process for water reuse [43], by Savage et al. on effluents from primary treatment of sewage [44], by Hauck and Sourirajan on the applicability of reverse osmosis for the treatment of sewage waters [45], by Feuerstein and Bursztynsky on renovation of municipial waste water by reverse osmosis [46], by Bregman on tertiary treatment of sewage and water reclamation [47], by Leitner on waste water treatment [48], by Golomb and Besik giving a review on the application of reverse osmosis to waste treatment [49], by Douglas et al. on membrane materials for waste water reclamation [50], by Kraus on the application of reverse osmosis to treatment of municipal sewage effluents [51], by Bishop on the use of improved membranes in tertiary treatment by reverse osmosis [52], by Short et al. on low-pressure ultrafiltration systems for waste water contaminant removal [53], by Fisher and Lowell on the treatment of waste water by reverse osmosis [54], by Dytnerskii et al. on the purification of waste waters [55], by Feuerstein and Bursztynsky on design considerations for treatment of solids-loaden waste waters by reverse osmosis [56], by Iwai on decoloring of waste water [57], by Grigoropoulos and Jennet on the treatment of wastes produced during waste water reclamation by reverse osmosis [58], by Shelef et al. on the recovery and recycling of sewage effluent by means of membrane separation processes [59], by Conn on direct reclamation and phosphate removal from raw sewage by reverse osmosis [60], by James on reverse osmosis treatment of water and waste water [61], by Cocconi on treatment of industrial and municipal waste waters by reverse osmosis [62], by Matsuura and Sourirajan on studies for water pollution control [63], by Nefedova et al. on the possible use of reverse osmosis and ultrafiltration for purifying waste and mine waters [64], by Quinn et al. on the use of reverse osmosis and ion exchange for the treatment of water and waste water [65], by Thomas and Mixon on the effect of axial velocity and flux decline of cellulose acetate membranes in reverse osmosis of primary sewage effluents [66], by Middleton and Stenburg on techniques for advanced waste treatment [67], by DeBussy and Whitmore on industrial water and waste treatment by reverse osmosis [68], by Leitner on reverse osmosis for waste water treatment [69], by Rodziller and Golovenkov on purification of waste water by reverse osmosis [70], by Ishizaka et al. on the treatment of secondary sewage effluent by reverse osmosis [71], by Merten on reverse osmosis for waste water treatment [72], by Newkirk and Schroeder on waste treatment by reverse osmosis [73], by Besik on reverse osmosis in treatment of domestic sewage [74], by Itoi on waste water treatment by ion exchange membranes [75], by Heist on the selection of membrane equipment for water recycle systems in Army mobile hospitals [76], by Shelef et al. on ultrafiltration and microfiltration membrane processes for treatment and reclamation of pond effluents in Israel [77], by Bashaw et al. on hollow fiber technology for advanced waste treatment [78], by Leitner on reverse osmosis for water recovery and reuse [79], by Rozelle et al. on ultrathin membranes for treatment of municipal and metal finishing waste waters by reverse osmosis [80], by Belfort et al. on membrane regeneration for waste water reclamation using reverse osmosis [81], by Porter on membrane ultrafiltration for pollution abatement and by-product recovery [82], by Hoppenberg on water treatment by membrane ultrafiltration [83], etc.

Pilot scale tubular and spiral-wound reverse osmosis units were used by Rex Chainbelt Inc. to establish the feasibility of renovating municipal sewage. Prior treatment with ferric chloride and alum will be necessary even when using the tubular reverse osmosis system. 92% water recovery rates, with no precipitation problems in the reverse osmosis concentrate, were achieved, but reverse osmosis product waters would require further treatment for organic removal prior to discharge. The feasibility of biodegrading reverse osmosis concentrates was also established [84].

Fouling characteristics of reverse osmosis membranes of sewage contaminated surface waters were studied by Beckman et al. and techniques for prevention and/or cleaning were developed. Laboratory experiments were conducted to characterize sewage-contaminated surface waters and to screen various feed additives. Field tests were run to determine the effects of operating variables such as brine velocity, pressure, brine-to-product flow ratio and brine channel turbulence. Samples of membranes from these tests were used for laboratory screening of chemical cleaners. Sodium perborate, "Biz" enzyme detergent and ethylenediaminetetraacetic acid restored the product water flux to 80 to 85% of the initial value [85].

Fouling of reverse osmosis modules was also studied in pilot plant units. Surface waters were simulated by mixing various combinations of humic acid, salts and clay. Membrane of three different permeabilities was used. Pressure adjustment gave constant permeate flow. Measurement of apparent streaming current provided data on neutralization of charge and consequently on coagulation as distinct from precipitation. It was thus a measure of colloidal stability. Two different fouling mechanisms were postulated and two different methods for prevention of fouling were studied [86].

Pulp and paper industry. Applications for reverse osmosis in the pulp and paper industry were reviewed by Wiley, Ammerlaan and Dubey in chapter 11 of the monograph on Industrial Processing with Membranes, edited by Lacey and Loeb [87].

Further work on this subject may be found in papers published by Ammerlaan et al. on membrane processing of dilute pulping wastes by reverse osmosis [88], by Beder and Gillespie on the removal of solutes from paper mill effluents [89], by Wiley et al. on the concentration of dilute pulping wastes by reverse osmosis and ultrafiltration [90], by Leszczynski and Zielinski on the same subject [91], by Morris et al. on the treatment of waste water from a pulp and paperboard mill [92], by Dytnerskii et al. on the use of reverse osmosis and ultrafiltration for the purification of wastes from a pulp and paper mill [93], etc.

In developing a mathematical model for optimizing the design of reverse osmosis systems for concentration of dilute pulping effluents, Bansal and Wiley have compared experimental data with the values calculated using a computerized model. The model was sensitive to variation of the NaCl flux rate and the osmotic pressure of the liquors being processed. For the design of a manifolding and piping system of a reverse osmosis unit, the number of modules operated in series was more important than the number of stages and the recovery ratios of water in each stage [94].

Food industry. The use of reverse osmosis in the food industry was reviewed by Merson and Ginette in chapter 10 of the monograph on Industrial Processing with Membranes, edited by Lacey and Loeb [95]. Reverse osmosis and ultrafiltration are processes of special interest for heat sensitive food liquids, as they can replace evaporation and operate at ambient temperature and also can accomplish separations that no other process can do.

Reports on the application of reverse osmosis and ultrafiltration in the food industry were given by Merson et al. [96], by Morse comparing the relative costs of reverse osmosis, electrodialysis and ion exchange for food processing [97], by Porter et al. on using reverse osmosis to treat the waste effluents of potato starch factories [98], by Pompei on food industry wastes [99], by Porter and Michaels on the application of ultrafiltration in food processing [100], by Huyman and Müller on the preparation of brewing water [101], by Hess et al. on the recovery of chocolate waste [102], by Spatz on reclamation and reuse of waste products from food processing [103], by Underwood and Willits on partial concentration of maple sap [104], by Fortunato on the concentration of fruit juices [105], by Vernois on applications in the sugar industry [106], by Madsen on the same subject [107] and by Baker et al. [108].

Another application field of reverse osmosis and ultrafiltration is in the treatment of dairy products and whey. Papers were presented on the treatment and concentration of cheese whey by McDonough and Mattingly [109], by Fenton-May et al. [110], by Goldsmith and Horton [111], by Suzuki [112], by Furukawa [113], by Pepper and Marquardt [114], by Gross et al. [115]. Goldsmith et al. discussed the application of reverse osmosis and ultrafiltration to concentrate, fractionate or purify a molecular or colloidal solution [116], Lawhon et al. reported on recycling of effluent from membrane processing of cotton seed wheys [117], Kennedy et al. on concentrating liquid foods [118], etc.

Metals and inorganics. Reverse osmosis has been used for the recovery of various metals from dilute solutions and wastes. Concentration of low-level radioactive Co solutions is reported by Goff and Gloyna [119]. The application of reverse osmosis to electroplating waste treatment is discussed by Golomb for recovery of nickel, copper and gold [120], by Hauck and Sourirajan [121], by Weiner [122], by Spatz [123] and by Götzelmann [124] on the treatment of diluted plating solutions, by Rozelle on treating metal finishing effluents [125], by Kay et al. on processing of photographic effluents [126], etc.

Sword has investigated the technical feasibility of desalting agricultural tile drainage [127], Antoniuk and McCutchan of desalting of irrigation field drainage water which is rich in $CaSO_4$ [128], Rex Chainbelt Inc. of demineralization of acid mine drainage [129]. Desalination of boiler feed water for power plants was reported by Hohenhinnebusch [130], as well as by Rowland et al. [131], by Kemp [132], etc. Separation of inorganic salts from solutions were reported by Nakane and

Ishizaka for sulfates of heavy metals [133], by Dytnerskii and Zakharov for separation of KNO_3 from
KCl solutions [134], by Magniikii et al. for the separation of NaCl from $K_2Cr_2O_7$ solutions [135], etc.
The removal of toxic metals, such as Ba, Cd, Zn, Cu, Cr and Pb, from water was also reported by
Mixon [136].

Organics. Reverse osmosis has also been used for the separation of various organic constituents from solutions. Papers were published by Ohya and Sourirajan on the separation of urea
in aqueous solutions [137], by Hindin et al. on the removal of organic compounds [138], by Kamizawa and Ishizaka on the transport of carboxylic acids [139], by Kopecek and Sourirajan on separation of mixtures of organic liquids [140], by Feberwee and Evers on aqueous solutions of organic
solutes [141], by Matsuura and Sourirajan on the separation of organic solutes [142], of alcohols,
penols and monocarboxylic acids, of aldehydes, ketones, ethers, esters and amines, of phenol, cresol
and chlorophenol, of organic acids and of hydrocarbons [143], by Anderson et al. on the rejection
of organic solutes [144], by Kamizawa et al. on the permeation behavior of amino acid solutions
[145], by Hamoda et al. on organics removal by low pressure reverse osmosis [146], etc.

Other applications of reverse osmosis are reported by van Roosen in the concentration of water
soluble paints [147], by Aurich et al. in the treatment of textile dyeing wastes [148], by Staude on
the same subject [149], by Sorber et al. on virus rejection [150], by Abron and Osburn on the removal of DDT and aldrin from water [151], by Desai et al. on the refinement of petrochemical plant
effluents [152].

Further applications of ultrafiltration are reported by Porter on ultrafiltration of colloidal suspen-
sions [153], by Kozinski and Lightfoot on protein ultrafiltration [154], by Grieves et al. on membrane
ultrafiltration of a nonionic surfactant [155], etc.

Water pollution. An investigation was made by Marynowski et al. on the future applications
of desalting processes for reduction of industrial water pollution. The major water-using industries
were examined to identify several specific attractive applications of desalting processes and to
evaluate the technical and economic feasibility of the selected illustrative applications. Three applications analyzed in detail are relatively novel: ion exchange treatment of ammonium nitrate plant
waste, evaporation of steel mill waste ammoniacal liquor and the treatment of power plant stack gas
scrubber blowdown by electrodialysis and evaporation. Estimates were made of the capacity of
waste water treatment equipment that will be needed for effluent demineralization by selected
industries and the corresponding investment and operating expenses. Recommendations are made
on the direction of research efforts needed to aid and implement the use of desalting techniques for
reduction of industrial water pollution [156].

Literature to 3.22

[1] S. Manjikian (Off. Saline Water Res. Develop. Progr. Rept. No. 448 [1969]). — [2] A. S.
Hodgson, D. Pal (Off. Saline Water Res. Develop. Progr. Rept. No. 674 [1971]). — [3] W. M. King
(Off. Saline Water Res. Develop. Progr. Rept. No. 682 [1971]). — [4] W. M. King, M. L. O'Hair
(Off. Saline Water Res. Develop. Progr. Rept. No. 694 [1971]). — [5] Envirogenics Co. (Off. Saline
Water Res. Develop. Progr. Rept. No. 762 [1971]).

[6] R. Brun, Y. Duriau, P. Dussaussoy (Proc. 3rd Intern. Symp. Fresh Water Sea, Dubrovnik 1970,
Vol. 2, p. 329/38; Chim. Ind. Genie Chim. 103 [1970] 1639/42). — [7] P. Treille (Proc. 3rd Intern.
Symp. Fresh Water Sea, Dubrovnik 1970, Vol. 2, p. 601/13). — [8] P. Treille, J. M. Rovel (Proc.
4th Intern. Symp. Fresh Water Sea, Heidelberg 1973, Vol. 4, p. 425/34). — [9] L. A. Kul'skii, D. D.
Kucheruk (Vodopodgotovka Ochistka Prom. Stokov 1972 No. 9, p. 171/3: C. A. 78 [1973] No. 62021).
— [10] J. S. Johnson, J. W. McCutchan (Desalination 10 [1972] 147/56).

[11] J. S. Johnson, J. W. McCutchan (AIChE [Am. Inst. Chem. Engrs.] Symp. Ser. 69 No. 129
[1973] 493/507). — [12] J. W. McCutchan, V. Goel (Proc. 4th Intern. Symp. Fresh Water Sea,
Heidelberg 1973, Vol. 4, p. 259/73). — [13] A. Pasternack, K. Rauch, R. Rautenbach (Proc. 4th
Intern. Symp. Fresh Water Sea, Heidelberg 1973, Vol. 4, p. 285/95). — [14] C. W. Saltonstall (Proc.
4th Intern. Symp. Fresh Water Sea, Heidelberg 1973, Vol. 4, p. 363/8). — [15] S. C. May, R. A.
Tidball (Off. Saline Water Res. Develop. Progr. Rept. No. 878 [1973]).

[16] Ralph Stone Co. (Off. Saline Water Res. Develop. Progr. Rept. No. 890 [1973], No. 891
[1973]). — [17] B. P. Shepard, H. C. Behrens, J. R. Massie, W. F. McIlhenny (Off. Saline Water
Res. Develop. Progr. Rept. No. 873 [1973]). — [18] N. W. Rosenblatt (Proc. 4th Intern. Symp. Fresh
Water Sea, Heidelberg 1973, Vol. 4, p. 349/61). — [19] E. L. Dance, T. E. Davis, H. I. Mahon (Off.

Saline Water Res. Develop. Progr. Rept. No. 922 [1974]). — [20] R. D. Ammons, H. I. Mahon (Off. Saline Water Res. Develop. Progr. Rept. No. 924 [1974]).

[21] R. M. Lawrence (Off. Saline Water Res. Develop. Progr. Rept. No. 848 [1973]). — [22] M. S. Bonura, G. M. Wells (Off. Saline Water Res. Develop. Progr. Rept. No. 892 [1973]). — [23] J. M. Amore, J. F. Enos, A. B. LaConti (Off. Saline Water Res. Develop. Progr. Rept. No. 815 [1973]). — [24] J. M. Amore, J. F. Enos, A. B. LaConti (Off. Saline Water Res. Develop. Progr. Rept. No. 866 [1973]). — [25] R. L. Goldsmith, S. Hossain, M. Tan (Off. Saline Water Res. Develop. Progr. Rept. No. 877 [1973]).

[26] R. Lawrence, R. A. Tidball, C. W. Saltonstall (Off. Saline Water Res. Develop. Progr. Rept. No. 905 [1974]). — [27] R. W. Okey (in: R. E. Lacey, S. Loeb, Industrial Processing with Membranes, New York 1972, p. 249/77). — [28] Aerojet General Corp. (U.S. Natl. Tech. Inform. Serv. PB 197659 [1969]). — [29] P. L. Stavenger (Chem. Eng. Progr. 67 No. 3 [1971] 30/6). — [30] C. van Altena (Chem. Anlagen Verfahren No. 8 1971 52, 57/8).

[31] R. L. Goldsmith (Ind. Eng. Chem. Fundamentals 10 [1971] 113/20). — [32] M. C. Porter, P. Schratter, P. N. Rigopoulos (Ind. Water Eng. 8 No. 6 [1971] 18/24). — [33] W. F. Blatt, L. Nelsen, E. M. Zipilivan, M. C. Porter (Separ. Sci. 7 [1972] 271/84). — [34] R. F. Madsen, O. J. Olsen, I. K. Nielsen, W. K. Nielsen (Filtr. Separ. 9 [1972] 567/75). — [35] F. Forbes (Chem. Eng. [London] No. 257 [1972] 29/34).

[36] M. C. Porter (Ind. Eng. Chem. Prod. Res. Develop. 11 [1972] 234/48). — [37] J. A. Palmer, H. B. Hopfenberg, R. M. Felder (J. Colloid Interface Sci. 45 [1973] 223/34). — [38] D. P. McDonald (Process Eng. 1973 No. 1, p. 76/8). — [39] C. J. van Oss (Progr. Separ. Purif. 3 [1970] 97/132). — [40] R. A. Cross (AIChE [Am. Inst. Chem. Engrs.] Symp. Ser. 68 No. 120 [1972] 15/20).

[41] P. Harriott (Separ. Sci. 8 [1973] 291/307). — [42] E. Hindin, P. J. Bennett (Water Sewage Works 116 [1969] 66/73). — [43] K. C. Channabasappa (Chem. Eng. Progr. Symp. Ser. 65 No. 97 [1969] 140/7). — [44] H. C. Savage, N. E. Botton, H. O. Phillips, K. A. Kraus, J. S. Johnson (Water Sewage Works 116 [1973] 102/6). — [45] A. R. Hauck, S. Sourirajan (Environ. Sci. Technol. 3 [1969] 1269/75).

[46] D. L. Feuerstein, T. A. Bursztynsky (Water Pollut. Contr. Res. Ser. ORD-17040-EFQ-12 [1969]). — [47] J. I. Bregman (Environ Sci. Technol. 4 [1970] 296/302). — [48] G. F. Leitner (Metal Progr. 98 No. 6 [1970] 62/3). — [49] A. Golomb, F. Besik (Proc. Ont. Ind. Waste Conf. 17 [1970] 67/97; Ind. Water Eng. 7 No. 10 [1970] 16/9). — [50] A. S. Douglas, M. Tagami, C. E. Milstead (Water Pollut. Contr. Res. Ser. ORD-17040-EFO-06 [1970]).

[51] K. A. Kraus (U.S. Clearinghouse Fed. Sci. Tech. Inform. PB 197671 [1970]). — [52] H. K. Bishop (Water Pollut. Contr. Res. Ser. No. 17020-DHR [1970]). — [53] W. L. Short, R. T. Skrinde, D. G. Newton (Membrane Sci. Technol. Ind. Biol. Waste Treat. Processes Proc. Symp. 1969 [1970], p. 188/95). — [54] B. S. Fisher, J. R. Lowell (Water Pollut. Contr. Res. Ser. No. 17020-DUD-09/70 [1971]). — [55] Yu. I. Dytnerskii, R. G. Kocharov, G. V. Makarov, V. A. Minaev (Khim. Prom. 47 [1971] 895/9; C.A. 76 [1972] No. 49492).

[56] D. L. Feuerstein, T. A. Bursztynsky (Chem. Eng. Progr. Symp. Ser. 67 No. 107 [1971] 568/74). — [57] S. Iwai (Senryo To Yakuhin 17 [1971] 159/65; C.A. 77 [1971] No. 117896). — [58] S. T. Grigoropoulos, J. C. Jennet (Proc. 26th Ind. Waste Conf. Eng. Ext. Ser., Lafayette 1971, No. 140, p. 336/47). — [59] G. Shelef, R. Matz, M. Schwartz, H. Alcalay (Proc. Symp. Develop. Desalination Technol. Israel, Jerusalem 1972, p. 45/52). — [60] W. M. Conn (AIChE [Am. Inst. Chem. Engrs.] Symp. Ser. 68 No. 124 [1972] 294/9).

[61] E. James (Mar. Technol. 9 [1972] 216/22). — [62] L. Cocconi (Nuova Chim. 48 [1972] 71/3). — [63] T. Matsuura, S. Sourirajan (Water Res. 6 [1972] 1073/86). — [64] G. Z. Nefedova, N. E. Kozhevnikova, L. I. Gracheva, S. A. Tverskaya, E. A. Kotyakhov (Tsvetn. Metal. 45 No. 3 [1972] 48/51; C.A. 77 [1972] No. 9398). — [65] R. M. Quinn, R. S. Hamilton, J. R. Anderson, C. O. Weiss (AIChE [Am. Inst. Chem. Engrs.] Symp. Ser. 68 No. 124 [1972] 262/9).

[66] D. G. Thomas, W. R. Mixon (Ind. Eng. Chem. Process Design Develop. 11 [1972] 339/43). — [67] F. M. Middleton, R. L. Stenburg (J. Sanit. Eng. Div. Am. Soc. Civil Engrs. 98 [1972] 515/28). — [68] R. P. DeBussy, H. B. Whitmore (Natl. Eng. 76 No. 2 [1972] 10/4). — [69] G. F. Leitner (Tappi 55 [1972] 258/61). — [70] I. D. Rodziller, Yu. N. Golovenkov (Zh. Vses. Khim. Obshchestva 17 [1972] 184/8; C.A. 77 [1972] No. 65981).

[71] S. Ishizaka, S. Suzuki, K. Kosaka, T. Nakamura (Kogyo Yosui No. 164 [1972] 16/22; C.A. 77 [1972] No. 105353). — [72] U. Merten (Air Water Pollut. Proc. Summer Workshop, 1970 [1972], p. 155/75). — [73] R. W. Newkirk, P. J. Schroeder (Hydrocarbon Process. 51 No. 10 [1972] 103/6).

— [74] F. Besik (Water Sewage Works **119** No. 10 [1972] 76/85). — [75] S. Itoi (Nenryo Oyobi Nensho **39** [1972] 625/39; C.A. **77** [1972] No. 156094).

[76] J. A. Heist (Intern. Conf. Appl. New Concepts Phys.-Chem. Wastewater Treat., Nashville, Tenn., 1972, p. 347/56). — [77] G. Shelef, R. Matz, M. Schwartz (Intern. Conf. Appl. New Concepts Phys.-Chem. Wastewater Treat., Nashville, Tenn., 1972, p. 335/45). — [78] J. D. Bashaw, J. K. Lawson, T. A. Orofino (Environ. Prot. Tech. Ser. No. EPA-R2-72-103 [1972]). — [79] G. F. Leitner (Chem. Eng. Progr. **69** No. 6 [1973] 83/5); AIChE [Am. Inst. Chem. Engrs.] Symp. Ser. **69** No. 133 [1973] 29). — [80] L. T. Rozelle, J. E. Cadotte, B. R. Nelson, C. V. Kopp (Appl. Polymer Symp. No. 22 [1973] 223/39).

[81] G. Belfort, F. E. Littman. H. K. Bishop (Water Res. **7** [1973] 1547/59). — [82] M. C. Porter (AIChE [Am. Inst. Chem. Engrs.] Symp. Ser. **69** No. 129 [1973] 100/22). — [83] H. B. Hoppenberg (U.S. Natl. Tech. Inform. Serv. PB 220685 [1973]). — [84] Rex Chainbelt Inc. (Water Pollut. Contr. Res. Ser. No. EPA-17040-EUE [1972]). — [85] J. E. Beckman, E. Bevege, J. E. Cruver, S. S. Kremen, I. Nusbaum (Off. Saline Water Res. Develop. Progr. Rept. No. 882 [1973]).

[86] E. Bevege, J. E. Cruver, J. G. Kilbride, S. S. Kremen, A. B. Riedinger (Off. Saline Water Res. Develop. Progr. Rept. No. 883 [1973]). — [87] A. J. Wiley, A. C. F. Ammerlaan, G. A. Dubey (in: R. E. Lacey, S. Loeb, Industrial Processing with Membranes, New York 1972, p. 223/47). — [88] A. C. F. Ammerlaan, B. F. Lueck, A. J. Wiley (Tappi **52** [1969] 118/22), A. C. F. Ammerlaan, A. J. Wiley (Chem. Eng. Progr. Symp. Ser. **65** No. 97 [1969] 148/55). — [89] H. Beder, W. J. Gillespie (Tappi **53** [1970] 883/7). — [90] A. J. Wiley, G. A. Dubey, J. M. Holderby, A. C. F. Ammerlaan (J. Water Pollut. Contr. Fed. **42** [1970] R279/R289), A. J. Wiley, G. A. Dubey, I. K. Bansal (Water Pollut. Contr. Res. Ser. No. 12040-EEL [1972]).

[91] C. Leszczynski, J. Zielinski (Przeglad Papier. **26** [1970] 385/9; C.A. **74** [1971] No. 102849). — [92] D. C. Morris, W. R. Nelson, G. O. Walrafen (Water Sewage Works **118** [1971] 124A/146A; Water Pollut. Contr. Res. Ser. No. 12040 [1972]). — [93] Yu. I. Dytnerskii, A. A. Swittsov, Yu. K. Romanenko, Yu. N. Zhilin, V. P. Semenov, O. V. Trupchaninova (Bum. Prom. **1972** No. 7, p. 22/4; C.A. **77** [1972] No. 130337). — [94] I. K. Bansal, A. J. Wiley (Tappi **56** [1973] 112/5). — [95] R. L. Merson, L. F. Ginette (in: R. E. Lacey, S. Loeb, Industrial Processing with Membranes, New York 1972, p. 191/221).

[96] R. L. Merson, L. F. Ginette, A. I. Morgan (Dechema Monograph. **63** [1969] 179/201). — [97] T. Morse (Engineer **230** [1970] 35/7). — [98] W. L. Porter, J. Siciliano, S. Krulick, E. G. Heisler (Membrane Sci. Technol. Ind. Biol. Waste Treat. Processes Proc. Symp., Columbus, Ohio, 1969 [1970], p. 220/30). — [99] C. Pompei (Ind. Aliment. Agr. [Paris] **88** [1971] 1585/91). — [100] M. C. Porter, A. S. Michaels (Chem. Technol. **1** [1971] 56/63, 248/54, **2** [1972] 56/61).

[101] E. Huymann, H. J. Müller (Brauwelt **111** [1971] 387/8). — [102] P. W. Hess, R. H. Thackery, M. C. Morrison (AIChE [Am. Inst. Chem. Engrs.] Symp. Ser. **69** No. 129 [1973] 123/6). — [103] D. D. Spatz (AIChE [Am. Inst. Chem. Engrs.] Symp. Ser. **69** No. 129 [1973] 89/99). — [104] J. C. Underwood, C. O. Willits (Food Technol. **23** [1969] 787/90). — [105] A. D. Fortunato (Ind. Quim. [Buenos Aires] **27** [1969] 185/8).

[106] G. Vernois (Z. Zuckerind. **19** [1969] 214/8, 387/9; Sucrerie Belge **89** [1970] 129/34). — [107] R. F. Madsen (Dechema Monograph. **64** [1971] 325/31; Z. Zuckerind. **21** [1972] 612/4; Intern. Sugar J. **75** [1973] 163/7). — [108] R. W. Baker, F. R. Eirich, H. Strathmann (J. Phys. Chem. **76** [1972] 238/44). — [109] F. E. McDonough, W. A. Mattingly (Food Technol. **24** [1970] 88/91). — [110] R. I. Fenton-May, C. G. Hill, C. H. Anundson, P. D. Auclair (AIChE [Am. Inst. Chem. Engrs.] Symp. Ser. **68** No. 120 [1971] 31/40).

[111] R. L. Goldsmith, B. S. Horton (Water Pollut. Contr. Res. Ser. No. 12040-DXF [1971]), R. L. Goldsmith (Proc. Symp. Memb. Processes Ind. Biomed. 1971, p. 267/300). — [112] A. Suzuki (Shokuhin Kogyo **15** No. 6 [1972] 73/81). — [113] D. H. Furukawa (AIChE [Am. Inst. Chem. Engrs.] Symp. Ser. **68** No. 124 [1972] 104/7). — [114] D. Pepper, K. Marquardt (Deut. Molkerei-Ztg. **93** [1972] 1504/9). — [115] M. C. Gross, J. Markind, R. R. Stana (AIChE [Am. Inst. Chem. Engrs.] Symp. Ser. **69** No. 129 [1973] 81/8).

[116] R. L. Goldsmith, R. P. De Filippi, S. Hossain (AIChE [Am. Inst. Chem. Engrs.] Symp. Series **68** No. 120 [1972] 7/14). — [117] J. T. Lawhon, C. M. Cater, K. F. Mattil (Food Technol. **27** [1973] 26/34). — [118] T. J. Kennedy, L. E. Monge, B. J. McCoy, R. L. Merson (AIChE [Am. Inst. Chem. Engrs.] Symp. Ser. **69** No. 132 [1973] 81/6). — [119] D. L. Goff, E. F. Gloyna (U.S. Clearinghouse Fed. Sci. Tech. Inform. AD 697149 [1969]). — [120] A. Golomb (Plating **57** [1970] 1001/5, **59** [1972] 316/9).

[121] A. R. Hauck, S. Sourirajan (J. Water Pollut. Contr. Fed. **44** [1972] 1372/83). — [122] R. Weiner (Galvanotechnik **63** [1972] 614/9). — [123] D. D. Spatz (Prod. Finishing [Cincinnati] **36** No. 11 [1972] 79/89). — [124] W. Götzelmann (Galvanotechnik **64** [1973] 588/600). — [125] L. T. Rozelle (Water Pollut. Contr. Res. Ser. No. 12010 [1971]).

[126] M. Kay, F. J. Quinn, M. W. Marshall, H. Meikle (J. Soc. Motion Picture Television Engrs. **81** [1972] 461/4). — [127] B. R. Sword (U.S. Natl. Tech. Inf. Serv. PB 213890 [1971]). — [128] D. Antoniuk, J. W. McCutchan (Univ. of California, Los Angeles, Rept. Eng. 7368 [1973]). — [129] Rex Chainbelt Inc. (Water Pollut. Contr. Res. Ser. No. EPA-14010-FQR [1972]). — [130] W. Hohen-hinnebusch (Umschau **69** [1969] 417/8).

[131] H. Rowland, I. Nusbaum, F. J. Jester (Power **115** No. 12 [1971] 47/8). — [132] A. R. Kemp (Proc. Am. Power Conf. **34** [1972] 703/9). — [133] T. Nakane, S. Ishizaka (Nippon Kaisui Gakkai-Shi **36** [1972] 11/5; C.A. **79** [1973] No. 9491). — [134] Yu. I. Dytnerskii, S. L. Zakharov (Zh. Prikl. Khim. **46** [1973] 1455/8; C.A. **79** [1973] No. 118684). — [135] A. L. Magniikii, V. A. Bakunov, G. V. Makarov, N. N. Zav'yalova, I. A. Donetskii, Yu. A. Kosmrov (Khim. Prom. **49** [1973] 315; C.A. **79** [1973] No. 83806).

[136] F. O. Mixon (Off. Saline Water Res. Develop. Progr. Rept. No. 889 [1973]). — [137] H. Ohya, S. Sourirajan (Ind. Eng. Chem. Process Design Develop. **8** [1969] 131/42). — [138] E. Hindin, P. J. Bennett, S. S. Narayanan (Water Sewage Works **116** [1969] 466/70). — [139] C. Kamizawa, S. Ishizaka (Kogyo Kagaku Zasshi **72** [1969] 1227/31; C.A. **72** [1970] No. 36178). — [140] J. Kopecek, S. Sourirajan (Ind. Eng. Chem. Process Design Develop. **9** [1970] 5/12).

[141] A. Feberwee, G. H. Evers (Lebensm.-Wiss. Technol. **13** [1970] 41/3). — [142] T. Matsuura, S. Sourirajan (Ind. Eng. Chem. Process Design Develop. **10** [1971] 102/8). — [143] T. Matsuura, S. Sourirajan (J. Appl. Polymer Sci. **15** [1971] 2905/27, **16** [1972] 1663/86, **16** [1972] 2531/54, **17** [1973] 3661/82, **17** [1973] 3683/708). — [144] J. E. Anderson, S. J. Hoffman, C. R. Peters (J. Phys. Chem. **76** [1972] 4006/11). — [145] C. Kamizawa, H. Masuda, S. Ishizaka (Bull. Chem. Soc. Japan **45** [1972] 2964/6).

[146] M. F. Hamoda, K. T. Brodersen, S. Sourirajan (J. Water Pollut. Contr. Fed. **45** [1973] 2145/6). — [147] D. van Roosen (Paint Varnish Prod. **59** No. 7 [1969] 69/71). — [148] C. Aurich, C. A. Brandon, J. S. Johnson, R. E. Minturn, K. Turner, P. H. Wadia (J. Water Pollut. Contr. Fed. **44** [1972] 1545/51). — [149] E. Staude (Chem. Ingr.-Tech. **45** [1973] 1222/5). — [150] C. A. Sorber, J. F. Malina, B. P. Sagik (U.S. Natl. Tech. Inform. Serv. AD 735750 [1971]; Water Res. **6** [1972] 1377/88).

[151] L. A. Abron, J. O. Osburn (Water Res. **7** [1973] 461/77). — [152] S. V. Desai, J. K. Smith, E. Klein, R. E. C. Weaver (AIChE [Am. Inst. Chem. Engrs.] Symp. Ser. **68** No. 124 [1972] 379/87). — [153] M. C. Porter (AIChE [Am. Inst. Chem. Engrs.] Symp. Ser. **68** No. 120 [1972] 21/30). — [154] A. A. Kozinski, E. N. Lightfoot (AIChE [Am. Inst. Chem. Engrs.] J. **18** [1972] 1030/40). — [155] R. B. Grieves, D. Bhattacharyga, W. G. Schomp, J. L. Bewley (AIChE [Am. Inst. Chem. Engrs.] J. **19** [1973] 766/74).

[156] C. W. Marynowski, C. F. Clark, R. C. Phillips (Off. Saline Water Res. Develop. Progr. Rept. No 898 [1973]).

3.23 Piezodialysis

Piezo-dialyse

The transport property requirements of membranes for use in the "pressure dialysis" process were defined by Podall in terms of the phenomenological coefficients: hydraulic permeability, solute permeability, and reflection coefficient. These coefficients were, in turn, quantitatively related to various process and performance parameters for brackish water desalination, where this process is considered best applicable [1].

The term "pressure osmosis" was used by Merten to describe a process, in which only partial salt removal is achieved or in which the permeate is actually enriched in salt relative to the pressurized solution. Two model membranes, a neutral and a mosaic membrane, suitable for the purpose were suggested and their expected performance was calculated [2]. Coupling phenomena in synthetic mosaic arrays were studied by Caplan with the object of developing basic techniques and principles for use in the design and testing of pressure dialysis equipment [3].

"Piezodialysis", another term for pressure dialysis or pressure osmosis, was considered as a promising coupled-transport process for removing ionic constituents from an aqueous solution by causing them to pass, under the influence of an hydraulic pressure gradient, through membranes which, at equilibrium, tend to sorb salt in preference to water and which exhibit substantial coupling between ion and water transport. In early work ion exchange membranes were used, consisting of small regions of anion selective material interpersed with small regions of cation selective material, each such region extending substantially uniformly from one surface of the membrane to the other. The necessary topological, electrical, hydraulic and electrokinetic properties were determined. The measured transport properties of several ion selective membrane formulations were shown to be promising for piezodialysis. Fine grain mosaic membranes were made from styrenized polyethylene membranes which were capable of withstanding application of 105 kg/cm² (1500 psi) pressure difference. Salt enrichments of 25 to 32% and 37 to 42% were obtained at 68 and 102 atm, respectively. These values were very close to the 27% and 40% predicted for these resins [4]. Equations were derived for relating permeate concentration and flux to the high pressure solution concentration, pressure head and circulating current flow path length for a given set of transport coefficients [5]. An average salt enrichment factor of 5, was concluded, should make the process roughly competitive with reverse osmosis. With a newly developed latex-polyelectrolyte membrane, a salt enrichment factor greater than 5 was obtained with the best example of this membrane [6].

Three different approaches were investigated by Schindler and Yasuda in the preparation of charge mosaic membranes. In the first approach phase separation in block copolymer films composed of styrene and vinyl pyridine was utilized. Piezodialysis effects were not observed with these membranes in their amphoteric state. In the second approach phase separation was utilized in films prepared from blends of random copolymers containing cross-linkable butadiene sequences. This preparation technique remained unsuccessful because of insufficient cross-linking across the phase boundaries. In the third approach blending of block copolymers with homopolymers was investigated. This method seems to be the most promising approach. Under reverse osmosis conditions, these composite films exhibited salt rejections identical to those as obtained from films of the same block copolymer [7].

Three types of ionic block copolymers have been studied by Lopatin et al. as piezodialysis membranes. Their properties and reverse osmosis behavior have been investigated towards their use as piezodialysis components and they were found to be very exceptional in strength, versatility and salt rejection. Two types of charge mosaic membranes were developed sufficiently for piezodialysis testing and substantial salt enrichment of the effluent [8]. The effects of different process variables on the salt transport and solution flux of charge mosaic membranes were evaluated. Fine grained charge mosaic membranes of a type with a well defined and controlled geometry were used to test the effects of temperature, pressure, concentration, thickness ratio and anionic exchange capacity on the salt and solution transport properties of the membranes. Fine grained charge mosaic membranes with high particle exposure at the membrane surfaces were prepared and tested, but no noteworthy piezodialysis activity was observed for these membranes [9].

Based on irreversible thermodynamics, a mathematical description of the piezodialysis process has been developed by Leitz and Shorr, which predicts the kind of performance that might be obtained. Strong coupling between the counterions and the gel water in the resins is required. Several ion-exchange resins have been developed, which have adequate coupling between counterions and water. A variety of techniques has been investigated for using these resins in composite membranes, some of which have shown significant salt enrichment. A laboratory piezodialysis module, which has given useful desalination, is described [10]. A relationship between geometric factors of the membrane and test apparatus and the effective circulating current path length was also presented. Performance of a module which demonstrates continuous desalination by piezodialysis is described [11]. For piezodialysis to become a useful desalination process, the most critical requirement is the development of appropriate membranes. The membrane being sought is a mosaic of cation-passing and anion-passing resins, each resin having a high degree of coupling between the mobile counterions and water. Of a wide variety of possible fabrication techniques, three are of particular interest: phase separation, pattern molding and latex-polyelectrolyte fabrication. The latex-polyelectrolyte fabrication technique has produced membranes, which bring piezodialysis close to practical reality [12].

The concentrations of the sample solutions obtained by using the piezodialysis membrane were determined by Yamabe et al. as functions of the pressure. The degree of the concentration was inversely proportional to the pressure, although the degree of concentration was very small [13]. Styrene-butadiene copolymer membrane was formed with the casting method, crosslinked with anhydrous tin (II) chloride and prepared by sulfonation, by chloromethylation and by quaternary amination. It was found that the total exchange capacity was fairly large and the exchange capacity ratio was not far from unity [14].

A schematic theory of the effect of the heterogeneity of charge-mosaic membranes on piezo-dialysis was given by Dresner. An effect exists because the ions leave the feed bath at different points and arrive in the filtrate bath at different points. The effect is always to reduce the salt flux through the membrane. The amount of reduction has been calculated for two geometric arrangements of the parts of the mosaic membrane, namely, alternating strips and a checkerboard of squares [15].

A nonequilibrium thermodynamic analysis of piezodialysis was presented by Gardner et al. The expressions derived in a previous paper [16] were used to predict the fractional recovery and rate of production of potable water in desalination by piezodialysis. An illustrative calculation is performed for a cylindrical tube made from a hypothetical mosaic membrane. This calculation sets upper limits on the performance of a given membrane by assuming that its properties, rather than the hydrodynamic conditions, are rate controlling [17].

Literature to 3.23

[1] H. E. Podall (Off. Saline Water Res. Develop. Progr. Rept. No. 304 [1968]). — [2] U. Merten (Desalination **1** [1966] 297/310). — [3] S. R. Caplan (Off. Saline Water Res. Develop. Progr. Rept. No. 413 [1969]). — [4] F. B. Leitz, S. S. Alexander, A. S. Douglas (Off. Saline Water Res. Develop. Progr. Rept. No. 452 [1969]). — [5] F. B. Leitz, W. A. McRae (Proc. 3rd Intern. Symp. Fresh Water Sea, Dubrovnik 1970, Vol. 2, p. 293/307; Desalination **10** [1972] 293/307).

[6] F. B. Leitz, S. S. Alexander, D. M. de Winter, C. W. Plummer (Off. Saline Water Res. Develop. Progr. Rept. No. 620 [1970]). — [7] A. Schindler, H. Yasuda (Off. Saline Water Res. Develop. Progr. Rept. No. 689 [1971]). — [8] G. Lopatin, H. A. Newey, E. T. Bishop, W. P. O'Neill, A. B. Krewinghaus (Off. Saline Water Res. Develop. Progr. Rept. No. 690 [1971]). — [9] G. Lopatin, H. A. Newey, O. D. Bergren (Off. Saline Water Res. Develop. Progr. Rept. No. 824 [1973]). — [10] F. B. Leitz, J. Shorr (Off. Saline Water Res. Develop. Progr. Rept. No. 775 [1972]).

[11] F. B. Leitz, J. Shorr (Proc. 4th Intern. Symp. Fresh Water Sea, Heidelberg 1973, Vol. 4, p. 451/63). — [12] J. Shorr, F. B. Leitz (Proc. 4th Intern. Symp. Fresh Water Sea, Heidelberg 1973, Vol. 4, p. 465/74). — [13] T. Yamabe, S. Yoshida, N. Takai (Kogyo Kagaku Zasshi **74** [1971] 2410; C. A. **72** [1972] No. 63070). — [14] T. Yamabe, K. Umezawa, S. Yoshida, N. Takai (Proc. 4th Intern. Symp. Fresh Water Sea, Heidelberg 1973, Vol. 4, p. 475/83). — [15] L. Dresner (Desalination **10** [1972] 47/66).

[16] J. N. Weinstein, B. J. Bunow, S. R. Caplan (Desalination **11** [1972] 341/77). — [17] C. R. Gardner, J. N. Weinstein, S. R. Caplan (Desalination **12** [1973] 19/33).

*Gefrier- und
Hydrat-Ver-
fahren.
Flüssig-
flüssig-
Extraktion*

4 Freezing processes. Hydrate processes. Liquid-liquid extraction

The ice which separates from brine is almost pure. A salt concentration builds up at the interface of the brine and the ice. In freezing processes, as applied in desalting, many small ice crystals are formed which are dispersed in the liquid. Brine is enclosed in the slurry and washing the ice crystals with pure water consumes part of the fresh water produced. Techniques which would allow the ice crystals to be formed in more regular and larger shapes would reduce the washing problem and improve the efficiency of the freezing process.

*Keimbildung
und
Wachstum
der Eis-
kristalle*

4.1 Nucleation and growth of ice crystals

The growth of ice crystals in saline solutions has been found by Farrar and Hamilton to occur more rapidly in the basal plane (a-axis) than perpendicular to it (c-axis). While growth along the c-axis is strongly inhibited by salts and additives, along the a-axis growth remains unaffected by salt or additives and is largely under thermal control. The free energy of ice–water interfaces appears to be of importance in ice nucleation and crystal growth [1]. Nucleation stimuli could be observed by using large water masses. The nucleation rates were determined as functions of the sample size and sample source, the concentration of electrolytes and suspended matter, and the intensity of mechanical agitation [2].

The growth rate of unconfined ice crystals along their basal lane in flowing subcooled water was measured by Fernandez and Barduhn and was found proportional to the $3/2$ power of the applied subcooling temperature difference and to the $1/2$ power of the flow velocity. A model was presented that assumes the growth rate is determined solely by the rate at which the heat of solidification is transferred by convection to the flowing water [3]. Temperature difference as driving force initiates a heat flow through the ice layer and the adjacent aqueous solution. Exothermous ice formation is an additional source of heat, which partly offsets the temperature gradient to be rejected [9]. Further data were presented on ice dendrite growth in salt solutions which flow over the dendrite tip and a theoretical model, based on the boundary-layer energy and diffusion equations, has been formulated to predict ice growth rates in flowing NaCl solutions [4]. The main uncertainty with the obtained ice growth data was the velocity of the flowing subcooled solution at the tip of the growing crystals. A new cell for crystal growth was designed, permitting accurate knowledge of the local velocity at the ice tip. Variation of the flow velocity allows control of the heat transfer away from at least part of the crystal [5]. New crystal growth measurements at lower subcooled water velocities were made and correlated with those obtained previously. It was found that all the data could be represented satisfactorily by the equation:

$$v = 0.0439 \cdot V^{0.5} \cdot \Delta t^{1.56}$$

where v is the growth rate in cm/s, V the water velocity in cm/s and Δt the applied subcooling temperature in °C. Accuracy of the coefficient 0.0439 is ± 0.0015 and of the power 1.56 is 0.03 [6]. The linear growth rate in the basal plane of unconfined ice crystals has been measured in flowing subcooled water and aqueous sodium chloride solutions [7].

Growing ice in a supercooled aqueous sodium chloride solution has usually the shape of needles or dendrites. Under special conditions a film of clear ice grows up, which does not include any brine. A relationship exists from which the maximum growing rate for clear ice can be obtained [8].

Nucleation rates have been calculated by Wood from the frequency with which liquid droplets solidify. The influence of the droplets environment on the nucleation process was found to be ineffective [10]. An experimental investigation into the secondary nucleation characteristics of ice–brine slurries was initiated by Estrin, using a Couette flow crystallizer. The major operational problem, stable deposition of ice upon the cooling surfaces, was found to be more severe in steady state than in batch experiments. The level of agitation appears to be an important variable determining nucleation kinetics [11]. A general size-dependent growth model based upon diffusion-controlled growth mechanism and a simple power law for the nucleation rate involving the supersaturation and total surface area were developed. The results indicated that the heat transfer properties of the crystallizer, the level of agitation, the coolant temperature and the number of seeds sensitively determine the final distribution [12].

A particle size analyzer and a freeze crystallizer set-up were used by Schneider and Farrar for studies of crystal growth and nucleation in ice–brine slurries [13]. The distribution of ice particle size from freeze crystallizers was investigated by means of a forward light-scattering apparatus and a birefringent analyzer. A photographic technique was employed to obtain data on the effect of moderate variations in Δt, residence time, salinity and refrigerant flow rate upon the ice particle size distribution [14]. An automatic particle size analyzer was then assembled. Automatically determined size distributions of ice-brine mixtures were in good agreement with manual determinations using the photographic technique [15]. Ice particle growth velocity was found to be directly proportional to supercooling and to decrease with increasing salt concentration. The results are best fitted by a process controlled by the rate of heat and mass transfer [16]. In a further report a guide is given for the operation and maintenance of the automatic particle size analyzer [17].

An experimental investigation was reported by Simpson et al. of the a-axis and c-axis growth rates in water and brine in a channel immersed in a thermostatically controlled bath. While growth rates in water agree well with those of previous workers, those in brine are lower than any previously reported. A new theory, based on conduction in the ice crystal, is outlined for the growth rate at high Δt [18].

Under certain conditions a smooth solid–liquid interface separating ice and brine becomes morphologically unstable resulting in entrapped brine. An effort to prevent interface instability was made by stirring the liquid in the vicinity of the solid–liquid interface. This is done by causing pressure fluctuations, which give rise to air bubbles composed of air rejected by the freezing ice [19]. The effective distribution of solute between a liquid and a growing solid phase is governed not only by the equilibrium partitioning characteristics, but also by the transport of rejected solute away from the interface. A method was derived for determining the solute distribution in a solid when the partitioning is a function of concentration. Freezing potentials were measured concomitantly with the partitioning studies and a linear voltage–velocity relationship was found for dilute NaCl solutions [20].

Further work on the subject includes papers by Orcutt on the nucleation and growth of ice crystals in secondary refrigerant freezing [21], by Renz on thermodynamic properties of ice [22], by Miksch on solidification of ice dendrites in flowing supercooled water [23], by Boari et al. on nucleation and growth of ice crystals in desalination [24], by Terwilliger and Dizzio on salt rejection phenomena in the freezing of saline solutions [25], by Miyauchi et al. on the rate of ice crystal growth in aqueous salt solutions [26], by Jones and Chadwick on solid–liquid interfacial energies in the ice–water–sodium chloride system [27], by Milonov and Seiitkurbanov on the conditions of ice crystal formation and crystal purification in water desalination [28], by Proshin on size distribution of ice crystals in flow-type crystallizer during the desalination of water [29], by Rozental on problems of the formation of ice in water and solutions [30], by Janzow and Chao on salt entrainment in ice crystallized from brine and on induced crystallization of large free ice crystals in slowly flowing brine [31].

Literature to 4.1

[1] J. Farrar, W. S. Hamilton (Off. Saline Water Res. Develop. Progr. Rept. No. 127 [1965]). — [2] J. Farrar, K. Youel (Off. Saline Water Res. Develop. Progr. Rept. No. 157 [1966]). — [3] R. Fernandez, A. J. Barduhn (Desalination 3 [1967] 330/42; Off. Saline Water Res. Develop. Progr. Rept. No. 229 [1967], No. 230 [1967]). — [4] R. Fernandez, J. B. Pangborn, S. L. Colten, A. J. Barduhn (Off. Saline Water Res. Develop. Progr. Rept. No. 333 [1968]). — [5] A. J. Barduhn, J. M. Poisot, G. M. Roux, T. C. Tsao (Off. Saline Water Res. Develop. Progr. Rept. No. 370 [1968]).

[6] J. G. Vlahakis, T. C. Tsao, H. A. Richard, A. J. Barduhn, S. A. Stern (Off. Saline Water Res. Develop. Progr. Rept. No. 497 [1969]). — [7] J. G. Vlahakis, H. S. Chen, M. S. Suwandi, A. J. Barduhn (Off. Saline Water Res. Develop. Progr. Rept. No. 830 [1972]). — [8] W. Schneider, K. Fischbeck (Desalination 5 [1968] 217/30). — [9] W. Schneider, K. Fischbeck (Chem. Ingr.–Tech. 41 [1969] 116/21). — [10] G. R. Wood (Off. Saline Water Res. Develop. Progr. Rept. No. 500 [1969]), G. R. Wood, A. G. Walton (J. Appl. Phys. 41 [1970] 3027/36).

[11] J. Estrin (Off. Saline Water Res. Develop. Progr. Rept. No. 494 [1970]). — [12] J. S. Wey, J. Estrin (Ind. Eng. Chem. Process Design Develop. 12 [1973] 236/46; Desalination 14 [1974] 103/20). — [13] G. R. Schneider, J. Farrar (Off. Saline Water Res. Develop. Progr. Rept. No. 292 [1968]). — [14] G. R. Schneider, P. R. Newton, D. F. Sheehan (Off. Saline Water Res. Develop. Progr. Rept. No. 408 [1969]). — [15] G. R. Schneider (Off. Saline Water Res. Develop. Progr. Rept. No. 770 [1972]).

[16] G. R. Schneider (Off. Saline Water Res. Develop. Progr. Rept. No. 838 [1973]). — [17] G. R. Schneider (Off. Saline Water Res. Develop. Progr. Rept. No. 839 [1973]). — [18] H. C. Simpson, G. C. Beggs, J. Deans, J. Nakamura (Proc. 4th Intern. Symp. Fresh Water Sea, Heidelberg 1973, Vol. 3, p. 395/407). — [19] R. F. Sekerka, R. G. Seidensticker, D. R. Hamilton, J. D. Harrison (Off. Saline Water Res. Develop. Progr. Rept. No. 319 [1968]). — [20] R. G. Seidensticker, R. F. Sekerka (Off. Saline Water Res. Develop. Progr. Rept. No. 517 [1970]).

[21] J. C. Orcutt (Desalination **7** [1969] 75/96). — [22] U. Renz (Kältetech. Klim. **21** [1969] 266/9). — [23] E. S. Miksch (Trans. AIME **245** [1969] 2069/72). — [24] G. Boari, G. Lacava, C. Merli, R. Passino (Quad. Ric. Sci. No. 58 [1969] 377/400). — [25] J. P. Terwilliger, S. F. Dizzio (Chem. Eng. Sci. **25** [1970] 1331/49).

[26] T. Miyauchi, Y. Tanaka, T. Yoshida (Asahi Garasu Kogyo Gijutsu Shoreikai Kenkyu Hokoku **17** [1970] 353/65; C. A. **75** [1971] No. 101 154). — [27] D. R. H. Jones, G. A. Chadwick (J. Cryst. Growth **11** [1971] 260/4). — [28] V. V. Milonov, S. S. Seiitkurbanov (Tr. Turkm. Politekhn. Inst. **1971** No. 12, p. 181/90; C. A. **78** [1973] No. 101 822). — [29] E. A. Proshin (Tr. Vses. Nauchn. Issled. Inst. Vodosnabzh. Kanaliz Gidrotekhn. Sooruzh. Inzh. Gidrogeol. **1971** No. 32, p. 134/41; C. A. **79** [1973] No. 83 328). — [30] O. M. Rozental (Zh. Fiz. Khim. **46** [1972] 191/2, 657/9, 971/2; Russ. J. Phys. Chem. **46** [1972] 111, 376/7, 559/60).

[31] E. F. Janzow, B. T. Chao (Desalination **12** [1973] 141/61, 163/75).

Gefrier-
verfahren

4.2 Freezing processes

History. A historical review on early experiments on water desalination by freezing was compiled by Nebbia and Nebbia-Menozzi. The review refers mainly to experiments in the second half of the 17th century and later, and describes extensively the purification of seawater by freezing developed by A. M. Lorgna [1]. Contemporary early work includes reports by Rose and Hoover [2] and by Hendrickson and Moulton [3] on research and development of processes for desalting water by freezing.

General. All variants of the freezing processes are based on the well-known phenomenon that, when a saline solution is cooled to its freezing temperature, ice crystals of pure water will form. There are several theoretical advantages inherent to the freezing processes of desalination. The latent heat of phase transition to the solid state is only 79.7 kcal/kg against 538.8 kcal/kg for the heat of vaporization. Operation at low temperatures minimizes scale and corrosion problems. Freezing and melting of ice can be accomplished without the use of heat transfer barriers and the recovery of the heat of fusion is almost complete. The energy consumed in a freezing process is practically limited to mechanical requirements and to energy losses associated with temperature differences between incoming feed water and outgoing product water. However, the major disadvantage of the freezing processes is the necessity of washing the ice crystals from adhering brine, an operation which inevitably consumes part of the product water.

A great deal of basic research for the development of desalination processes by means of freezing has been performed at various Research Centres and Universities. Adams and Rohatgi have reported on the solidification and separation of ice from saline water and especially on the conditions of dendrite ice crystal formation [4]. Sherwood et al. [5] and Sherwood and Brian reported on salt concentration at phase boundaries in desalination processes. The same authors reported also on washing of brine from ice crystals [6]. However, Leinroth et al. presented a more extensive report on washing of brine from ice crystals based on an experimental investigation. The obtained data show that washing is accomplished with the least loss of fresh water if long columns are used, with no agitation of the bed, and that washing is much more effective in flooded than in drained beds, as well as from beds of larger particles [7].

The kinetics of the crystallization of ice from supercooled water was investigated by Sherwood and Brian [6] and by Brian and Sperry examining the temperature distribution in the crystallization of undercooled liquids in cylindrical tubes. Numerical values of interface temperature rise as a function of the various parameters of a capillary crystallization experiment were presented [8]. Data for heat transfer from water to melting ice spheres and mass transfer in the case of dissolving spheres suspended in agitated water were reported by Sherwood and Brian. The transport coefficients were found to depend on agitator power input but not on agitator design. A correlation involving Nusselt and Prandtl or Schmidt numbers together with a dimensionless group involving agitation power was developed. The correlation is essentially independent of solid-liquid density ratio in the range of 0.8 to 1.25 [9].

A hydrodynamic model was developed by Shwartz and Probstein for flooded counterwashers of the type used in washing brine from ice crystals. In the model the brine is displaced by fresh water from the interstices of the ice plug. Similarity parameters which can be used for comparing the performance of different counterwashers were derived. The production rate of fresh water was found to increase with an increase in the ice crystal size, the ice plug length above the screens, the concentration of ice crystals in the slurry and the external mechanical restraining forces on the ice plug. It was also found to increase with a decrease in the ice plug length below the screens [10]. An investigation was presented by the same authors of the characteristics and performance of a laboratory scale counterwasher slurry separator, like those used for the separation of brine from ice crystals. Polyethylene particles were used to simulate the ice crystals. Main purpose was to check out a model and theory of such counterwashers previously derived by the authors. The experiments demonstrated the validity of the theoretical conclusions. The feasibility of obtaining high rates of crystal washing was demonstrated. The highest steady production rate obtained in the experiments is equivalent to 39 t/h $\cdot$ m² (8000 lb/h $\cdot$ ft²) of brine-free ice in the freeze-distillation process [11]. A completely analytical solution for the general sink model of a two-dimensional counterwasher was presented by Kemp. The results are simple formulae for the performance of the counterwasher in terms of the geometric, flow and pressure parameters. The formulae are in good agreement with the numerical solution given by Schwartz and Probstein for their special case [12].

A pilot-scale wash column was operated for the purpose of obtaining data on wall friction and on brine dispersion. Good correlation between the data and simple physical models was obtained [13]. An investigation was also made with the purpose of determining the relationships between the operation of a continuous stirred crystallizer and the suitability of the ice product for washing to remove brine. Photographs of effluent revealed that the ice particles were disk shaped and that the particle size distributions in all cases passed through a maximum. Growth rates correlated well with values predicted from heat and mass transfer rates in the size range 0.6 to 2 mm but increased sharply at the smaller sizes. Permeabilities of ice beds formed from the crystallizer product were found to be a strong function of both the method of bed formation and the size and shape of the crystals. An important practical result of the study was the indication that much larger ice production rates per unit crystallizer volume should be possible without sacrifice of crystal size [14].

Further work on fundamental operations of the freezing process was reported by Barak and Dagan on an analytical investigation of the flow in the saturated zone of ice counterwashers [15], by Kawasaki and Owa on washing of the gravitationally drained bed of ice crystals [16], by Seiitkurbanov and Milonov discussing the factors contributing to the efficient operation of three major components of the freezing processes [17], by Tanaka emphasizing that the most difficult problem is the removal of salts adhering to the ice surface [18], by Proshin on removal of brine from ice crystals by centrifuging [19], by Kolodin et al. presenting equations for calculating the centrifuge parameters [20], by Vorstman and Thijssen reporting on the separation of ice and aqueous solutions in a wash column [21], by Hobson and McGrath describing a continuous column crystallizer [22], etc. A review of the state-of-the-art (1968) of the freezing processes was presented by Barduhn [23].

Direct freezing. In the direct freezing process, the precooled saline water is introduced into a chamber which is at a very low pressure. A portion of the water evaporates reducing the temperature of the remaining brine below the freezing point and about one half of the water is frozen into ice crystals. The slurry is transferred in a separation column, where the ice crystals float to the top, forming a porous bed of ice. The rising ice bed is washed with fresh water obtained in the conversion process. After washing, the ice is separated by a mechanical scraper and conveyed to a melt tank. The feasibility of the process has been demonstrated on a pilot unit capable of producing 50 kg of ice per hour [24]. A pilot plant with a design capacity of 15000 gpd (57 m³/day) was operated at the O.S.W. Test Facility in Wrightsville Beach. The vapor, which is drawn off from the freezer, was absorbed by a concentrated solution of lithium bromide. The diluted LiBr solution was reconcentrated by distillation and the condensate was part of the product water [25].

Vacuum freezing vapor compression process. This process has been developed in Israel, known as Zarchin process, and in the United States by Colt Industries. The incoming seawater feed is filtered, de-aerated, precooled by the two effluent streams, the product water and the reject brine, almost to the freezing point and directed to the freezing section of the hydroconverter. This section is maintained at a vacuum of about 3.5 Torr, and part of the feed, when entering to this reduced pressure, flashes into vapor. The latent heat of vaporization is extracted from the precooled salt water, causing ice crystals to form. The vapor formed in the freezer section is passed through carryover separators and its pressure is increased by a compressor located near the top of the hydroconverter. The vapor is then passed over refrigeration coils, to extract heat added to the system, and channelled to the melter. Ice crystals conveyed from the counterwasher are mixed with the pure water vapor from the hydroconverter condenser. The vapor condenses and melts the ice, both forming the product water. The ice crystals-brine slurry is conveyed from the freezer section of the hydroconverter into the bottom of the counterwasher vessel. An ice plug is formed on top of the brine. Concentrated brine is continually being drained from the counterwasher and the top of the ice pack is being washed with pure water. A rotating mechanical scraper removes from the top of the ice pack the clean ice, which is transferred to the melter chamber of the hydroconverter. **Fig. 4–1** shows a simplified flow sheet of the process. Pachter and Barak have reported on the pilot plant at Tel Aviv and the demonstration plant at Eilat. Zarchin's main contribution was the development of the compressor having a rotor with flexible unmachined stainless steel blades [26]. After several years of experimentation the Eilat plant was shut down.

Fig. 4–1

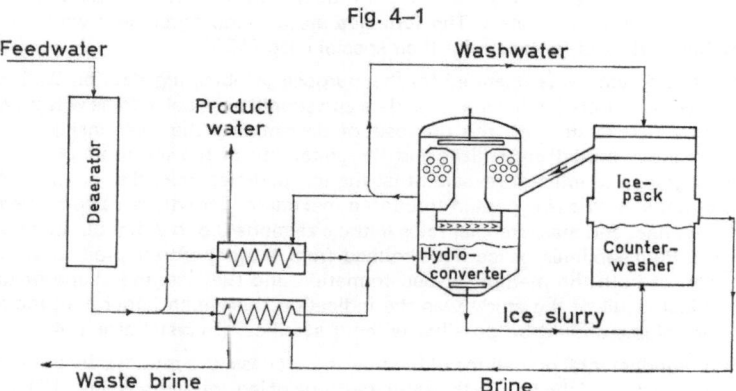

Vacuum freezing vapor compression process flow sheet.

The development made by Colt Industries is described in several papers. Consie et al. reported on a 60000 gpd (227 m³/day) pilot plant, erected first in Beloit and then transferred to Wrightsville Beach [27]. A study was prepared on 1 and 5 Mgd (3785 and 18900 m³/day) desalting plants using the vacuum freezing vapor compression process. These plants utilize 0.5 Mgd factory assembled modules in groupings of 2 and 10 to form the desalting plants [28]. The process was further evaluated on a 15000 gpd (57 m³/day) mobile plant on high salinity brackish water and seawater. A number of tests were carried out with yields of 20 to 80%. No significant evidence of precipitation or scaling was noticed. Product water containing no more than 500 mg salt/l was obtained [29]. Fraser and Emmermann reported on the evaluation of the 60000 gpd pilot plant. The capacity of the plant could be increased to 120000 gpd and the total power consumption reduced from 65 to 44.7 kWh/kgal (11.8 kWh/m³). Various improvements could also be made [30].

A theoretical study was made by Kemper et al. on the use of multiple-phase ejectors in desalination systems instead of conventional vapor compressors and the advantages of the system were outlined [31]. A bench scale study on the vacuum freezing ejector absorption process was made by Koretchko and Hajela. Performance tests on a 6000 gpd equivalent bench size ejector test facility showed the ejector to be free of any operating difficulties; projected mass ratios were obtained. The studies indicated that NaOH was the most suitable absorbent for the process [32]. An economic study was subsequently made. A 1 Mgd (3785 m³/day) reference desalting plant was designed in sufficient detail to demonstrate plant operability and allow accurate estimating of the capital cost and the water cost. A table is given, which shows the cost of desalted water for the reference plant as the

Water Desalting   319

feed salinity and temperature are varied. Nomograms were developed to obtain the minimum water cost. For the assumed parametric ranges, the water cost varied from 53 to 125 cents/kgal (14 to 33 cents/m³) [33].

The productivity of the process depends mainly on the production rate of the separation column, which is determined by the size of the ice crystals produced in the freezer. Another limitation on conversion rate is in the size of the equipment. A practical size for factory assembly and shipment of major components is about 570 m³/day (150000 gpd).

Further work on this process was reported by Uchida on the desalination of seawater by the vacuum freezing process [34], by Bridge et al. reporting on a commercial vacuum freezing vapor compression desalting unit [35]. A review of the state-of-the-art for the vacuum freezing vapor compression desalting process up to 1968 was prepared by Hittman Associates [36].

Secondary refrigerant freezing. This variant of the freezing process is also a direct contact process, but it differs in the way by which freezing is accomplished. It uses an immiscible heat carrier, such as a hydrocarbon. Liquefied butane is bubbled in the freezer department through the precooled seawater. The butane vaporizes, cools further the water and ice crystals are formed. The butane vapor is compressed in the primary compressor and then introduced into the melter where it condenses on the ice, which is melted. The condensed butane and the product water are separated in a decanter. The liquid butane is recycled back to the freezer. Butane vapor not required for ice melting is also compressed by the secondary compressor, condensed by seawater cooling and recycled to the freezer. The ice-brine slurry is treated in a similar way as by direct freezing. **Fig. 4–2** shows a simplified flow sheet of the secondary refrigerant freezing process.

Fig. 4–2

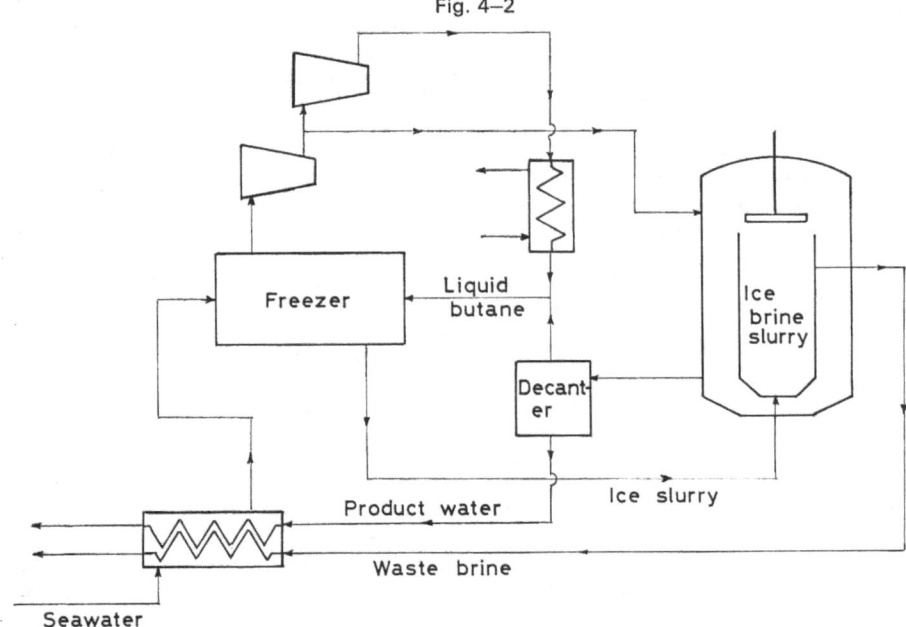

Secondary refrigerant freezing process flow sheet.

The process was originally developed at the Government Chemical Industrial Research Institute in Japan by Umano and coworkers in an attempt to concentrate seawater by means of refrigeration as a first step in salt making. A selected bibliography of the original references of the Umano group is given [37]. The process was further developed by Struthers Scientific and International Corp. [38].

A pilot plant with a capacity of 8 t/h ice (190 m³/day, 50000 gpd) was erected in Japan to study process components. A moving bed column was developed for washing the ice crystals [39].

A pilot plant with a capacity of 57 m³/day (15000 gpd) was also constructed in the United States; a report on its operation was given by Ganiaris et al. [40].

A similar process was developed at the Cornell University by Wiegandt. A rapid direct freeze step, a washing operation in which the ice bed is propelled by a hydraulic piston action, and a melting step in which refrigerant is condensed directly on ice are the characteristics of the process [41]. An engineering study of a plant to produce 37850 m³/day (10 Mgd) of fresh water from seawater was subsequently made [42]. The piston bed is a moving, porous ice bed acting as a free piston, which is constantly cut off at one end and replenished at the other. A drainage port is provided at an intermediate level. Brine flows upward and wash water downward, each toward the drainage port. Washed ice is removed at the top [43]. An electrical analog technique was developed and then applied to several variations of piston bed columns [44]. Mixon examined the behavior of a freely draining rectangular hydraulic piston wash separator and presented a solution, which permits the calculation of the location and shape of the interface between drained and filled portions of the piston [45]. An optimistic study showed a water cost of 7.7 cents/m³ (29 cents/kgal) for a single train secondary refrigerant freezing plant producing 50 Mgd (190000 m³/day). The cost increases to 71 and 176 cents/kgal for 5 and 1 Mgd, respectively [46].

A pilot plant with a capacity of 35000 gpd (132 m³/day) combining the process developed by Wiegandt and a Rotocel-type system was erected in St. Petersburg, Florida. The Rotocel consists of a large number of open-topped cells, in which ice is formed by spraying hydrocarbon refrigerant and seawater into each cell. The small size of the ice crystals formed showed little promise for draining and washing operations. The piston bed was also evaluated.

Scale-up of the freezing process to large sizes and the optimization of the process with respect to temperature driving forces and ice crystal sizes require a better understanding of the ice melting step. Brian et al. have presented a study on vapor flow limitations in a melter-condenser using a secondary refrigerant [47]. The design and scale-up of the melter-condenser must be based upon the surface area of contact between the ice bed and the vapor. Due to the small vapor penetration depth, the minimum bed depth for a melter-condenser is governed by the capillary rise of the ice particles discharged from the washer [48]. For ice beds composed of platelet-shaped particles, the condensation rate is only about one-half of the theoretical value and this discrepancy is apparently due to difficulties in describing the properties of beds produced by anisotropic particles [49].

The effect of particle size distribution on porosity and permeability in porous media was investigated by Barduhn and coworkers and a method for measuring the permeability of beds with a variety of particle sizes was developed. A stripping column and flash evaporation chamber were constructed for investigating the removal of trace amounts of butane and other refrigerants from water in freezing plants [50]. Saturated butane solutions were flashed at reduced pressures and further stripped, using the evolved butane as its own stripping agent. Butane concentrations were measured along the column for different flow rates and pressures, and the overall liquid mass transfer coefficient was derived. The Sherwood number obtained was observed to be independent of pressure indicating that the liquid phase resistance is controlling [51]. Further research on anisotropic ice beds, butane flashing and butane boiling was presented in another report [52].

Butane stripping from the product water or brine is necessary to improve the waters taste and quality, to prevent explosion hazards and to recover its economic value. Rates of butane removed from water in a vacuum spray chamber have been measured for different chamber pressures, spray velocities and at two nozzle heights. Most of the mass transfer is taking place in the liquid sheet attached to the nozzle orifice and not in the spray itself [53]. An investigation was carried out by Simpson et al. into the evaporation of butane droplets in a vertical column of water or brine [54].

A considerable effort has also been made in the United Kingdom for the development of the secondary refrigerant freezing process. The effect on ice quality of variation in the degree of agitation, the refrigerant addition rate, the brine residence time in the crystallizer, and the salinity of the raw brine was tested by Landau and Martindale. Increasing inlet brine concentration and butane rate worsened, increased residence time and turbulence improved product quality [56]. The crystallization and washing of ice has been studied by Denton et al. on the realistic scale of about 3 t/day (7000 lb/day) as part of an extensive experimental investigation of the immiscible refrigerant freezing process. The quality of the crystals produced in the continuous stirred tank crystallizer was independent of the residence time but improved with the production rate, the temperature of the feed butane and the overall temperature difference. This can be explained by the importance of self agitation which was found to govern the heat transfer and which increases with the same variables [57]. It was confirmed in further work that ice salinity is reduced by increased stirring rate, slurry density and residence time. A larger unit functioned satisfactorily without mechanical agitation. Larger overall temperature driving forces are beneficial with this arrangement. A draught tube crystallizer was incorporated in a 45 m³/day (10000 gpd) experimental plant which also included a wash column separator, a direct contact melter and a butane/water decanter. The design and commissioning of this plant was described and preliminary experimental work was discussed by Martindale et al. [58].

An extensive experimental research program using equipment with a throughput of 90 t/day (20000 lb/day) of ice has provided information on the scale-up of the immiscible refrigerant freezing process. Good quality water containing less than 100 mg/l chloride has been produced for 90 h from seawater. The highly rated crystallizer has proved trouble-free and produced ice with a good permeability. The performance of dumped bed melters has been shown to be severely limited by drainage of the immiscible liquids and the nucleation of metastable drops of butane entrained in water to be more important than the stripping of dissolved butane [59]. The development of secondary refrigerant freezing plants in the United Kingdom has reached a stage where the process has been adequately demonstrated and detailed estimates of commercial plant costs shown to be attractive. The approach to meet the stringent design requirements encountered on an actual site in the United Kingdom was outlined by Burton and Lloyd and the design considerations of the primary and secondary compressors were examined in detail. Additionally two other aspects, safety and environment, affecting component specifications, were discussed [60]. A prototype plant, to be built near Ipswich, would incorporate the process development effected in the United Kingdom [61]. The erection of this plant was delayed.

The secondary refrigerant freezing process has been described by Orcutt et al. by mathematical models of the major system components. The mathematical equations were used with the aid of a computer to study the operational-design economics of a freezing process and to predict the best operating conditions. Optimization computations showed that the economics of process operation depend largely on the temperature maintained in the freezer and the overall difference in refrigerant and equilibrium freezing temperature. The effect of departures from optimal design was assessed. An analysis of the linearized freezer dynamic equations showed the freezer to be stable and did not indicate regions of difficult control [62]. Data obtained from transient dye tracer studies of a mixed freezer-crystallizer have been analyzed with the help of a simulation model programmed for analog computer. A parameter representing the extent of a deadwater region was found to depend on the specific power input, while another parameter representing the exchange of material between the deadwater region and the remainder of the vessel contents was strongly affected by the position of the agitator [63].

Further research work on the secondary refrigerant freezing process was reported by Kolodin et al. on desalination of salt water by freezing with butane [64], by Chiu et al. reporting on an experimental study of ice-making in a freezing process [65], by Tamaru and Nagashima on operating performance of a countercolumn in washing brine from ice [66], by Kolodin and Seiitkurbanov on ice formation during contact cooling of a solution [89], by Kolodin and Rutgaizer on thermal loads in contact vaporizers under water crystallization conditions [67], by Kolodin et al. on washing of ice crystals in a washing tower during desalination of brackish waters by freezing [68], by Kawasaki and Owa on the height of saturation zone and permeability of the ice crystal bed in a gravitational field [69], by Kennaway giving a review on freeze desalination [70], by Brian on potential advantages and development problems in water desalination by freezing [71], by Adamski describing a process using a horizontal rotating tube crystallizer [72].

Octafluorocyclobutane (R-C 318) has also been advantageously used in the direct freeze separation process in lieu of butane [55].

Spray freezing. Spray freezing, as opposed to tank freezing, can be accomplished at high capacities in a single pass. Because of the small size of the crystals produced, washing has been difficult. A study was made by Hubbard et al. of how changes in the freezing process affected the washing characteristics of beds of crystals [73]. The effect of butane gas agitation on the crystallization of ice has been investigated by Kawasaki in an experimental crystallizer. The structure of the butane spray nozzle was similar to that of a pneumatic atomizer. This device could promote both the evaporation of butane and the strong agitation of seawater in the crystallization room and, therefore, ice could efficiently be crystallized. The specific production rate of ice in this crystallizer was 0.55 kg/h per kg of solution and the superheating for evaporation of butane was in the range from 0.8 to 1.6°C [74].

Research on a spray freezer was also reported by Johnson et al. High rates of heat transfer between feed and refrigerant were effected in a finely dispersed well-mixed spray of the two fluids. The performance of a triplet nozzle configuration was studied from the viewpoint of production rate and permeability. A theoretical model describes decanting of a stationary, initially uniform, mixture of Freon and water in a vertical column. The analysis provides a simple experimental method to determine droplet size distribution in a liquid-liquid dispersion band and measurement of its decanting rate [75]. The solubility and rate of hydrolysis of Freon 114 ($CCIF_2CCIF_2$) have been measured over a range of conditions relating to secondary refrigerant freeze desalination technology. Results were presented for the solubility of gaseous Freon 114 as a function of pressure in pure and saline water. Semi-empirical relations were presented for solubility trends of fluorocarbon refrigerants as a function of molecular parameters, of molar volume and polarizibility. A theoretical model has been developed for the calculation of the hydrolysis rate constant of Freon refrigerants. The theory includes the effects of hydration of ions as well as the effects of the dielectric constants and ionic strength of the solution. The model can be used to predict the hydrolysis rate constants of other refrigerants which are of interest to freeze or hydrate desalination processes [76]. Spray freezing is used in the Avco-Crystalex freeze desalination process, which is beeing developed. The process flow sheet is shown in **Fig. 4–3**.

Fig. 4–3

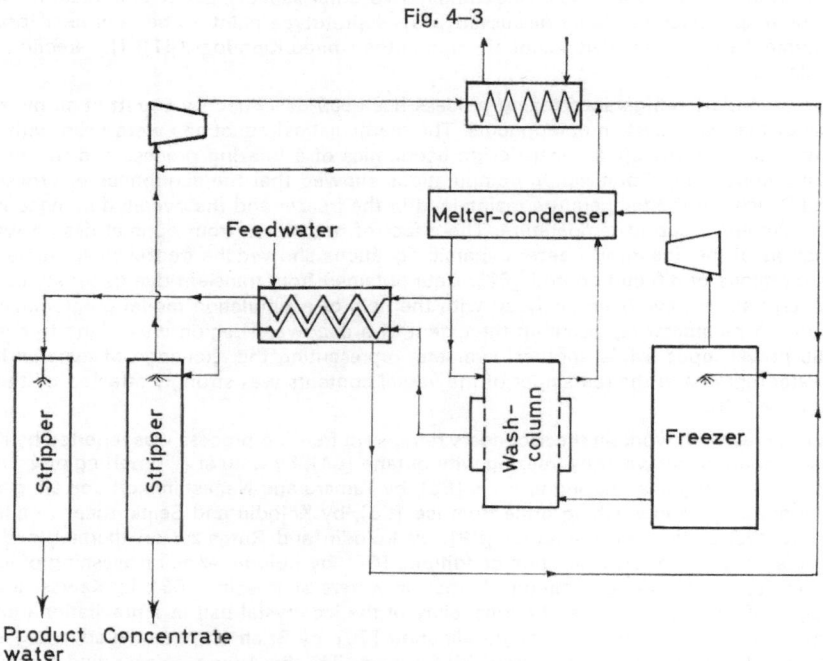

Avco-Crystalex process flow sheet.

Data were obtained in a laboratory loop with a nominal capacity of 5.7 m³/day (1500 gpd). The spray freezer developed to operate at high production involves a triplet injector nozzle through which three jets, two of saline feed and one of refrigerant, are generated and impacted over a small volume element. The resulting liquid sheet becomes unstable and breaks up into fine droplets. Freon evaporation proceeds and ice crystals are grown through a given fall height. Basic information on decanting Freon-water dispersions was obtained. Platelets of 150 μ grow in less than one second without dendrites. The low Freon residue in the slurry is then separated out by a hydrocyclone which provides an enriched ice fraction for the wash column. Pressurized wash column operation exhibited washing rates twenty times that of gravity column rates [77].

Results of an economic optimization and parametric study of the Avco-Crystalex process for water desalination were presented by Fraser and Olsson. The study indicates that this process should desalinate seawater at a cost which is significantly lower than the costs reported for competitive processes. A parametric study indicated water costs ranging from $ 0.53 to $ 1.08 per 1000 gallons (14 to 28.5 cents/m³) of product water for 1 Mgd plants and installed costs, within the plant boundary, ranging from $ 0.82 to $ 0.96 per gallon of daily capacity. For 0.1 Mgd plants, the estimated water costs ranged from $ 1.64 to $ 2.54 per 1000 gallons (43 to 67 cents/m³) of product water, while the installed costs varied from $ 2.38 to $ 2.58 per gallon of daily capacity [78].

The erection of a pilot plant at the O. S. W. Test Facility in Wrightsville Beach was decided in order to test and evaluate the Avco-Crystalex process for a period of twelve months.

Other freezing process. Akins et al. reported on a potential desalting process, in which a properly selected auxiliary system, consisting of a partially frozen organic liquid, is mixed with precooled seawater. The organic slurry melts and the seawater is partially frozen. The ice is separated from the brine, washed and then contacted again with the organic liquid. The mixture is pressurized to melt the ice and partially to freeze the organic liquid. The water and the organic slurry are separated and expanded producing fresh water and organic slurry as recycle [79].

In the batch evaporative freezing process studied by Curran, saline water is sprayed into a rotating cylindrical basket at a pressure below the triple point pressure. Removal of vapor results in the formation of an annular ice—brine semisolid layer on the lateral surface of the basket. The brine is removed by washing and the residual ice is melted. The process is theoretically and technically feasible, but the energy required makes the economic feasibility questionable [80]. An alternate method of freezing was presented by Bustany et al. Seawater as the dispersed immiscible phase moves countercurrently to a cold organic refrigerant. Large agglomerates of ice particles almost the size of the drops were obtained. The crystals were platelets of largest dimension above 1 mm and of thickness-to-diameter ratio generally less than 1/3. A mathematical model with four to eight parallel plates correlates well [81].

The multi-stage secondary refrigerant freezing process has been studied in discontinuous and continuous system by Boari et al. The investigation of the discontinuous plant consisted in the examination of the influence of operational variables on ice crystals dimensions and on the productivity of the freezer [82]. In order to assess actual feasibility of the secondary refrigerant freezing process a 50 kg/h and a 300 kg/h pilot plant were designed and tested. It was found that draught tube crystallizers behave very satisfactorily and give outstanding outputs. Crystallizer and wash column design were based on previous experiences and on mathematical models which, if confirmed experimentally, should be very useful in industrial plant design [83].

The programmed indirect freezing process presented by Cheng et al. is a cyclic process to be conducted in a unified freezer-melter. Each cycle consists of a feeding step, a freezing step, an in situ washing step and an in situ melting step. By an appropriate coordination of the design of a unified freezer-melter and the control of the freezing step, a consolidated ice bed of a structure similar to that of a hydraulic ice washing column used in a direct contact freezing process is obtained at the end of each freezing step. Therefore an efficient washing of ice bed can be accomplished with a small loss of wash water. New ways of accomplishing heat reuse have been incorporated in the process to significantly reduce the energy requirement [84].

A high pressure isothermal freezing process was described by Jellinek. The proposed technique consists in principle of subjecting saline water to pressures of about 10000 atm at room temperature. Part of the water is converted to ice VI. Separating the water substance contained in ice VI from the mother liquor by releasing the pressure yields fresh water. Results obtained indicated that it appears possible to obtain potable water of sufficient low salinity [85].

Fournier et al. reported on a natural freezing process in open tanks, utilizing night radiation to the sky in desert arid areas, at environment temperatures as high as + 5°C. The ice has a salt content 2 to 5 times smaller than the initial water. An experimental station was installed in Atacama desert (Chile) and rectangular juxtaposed tanks with a depth of 20 cm were built. Average water production was 9 litres/m²·day and the cost was estimated at 0.10 dollar/m³ [86]. A similar suggestion has been made by Wankat. Theoretical calculations were given to predict the time required to reach the freezing temperature. The melted ice water from a single freezing without a wash step has three to six times less salt than the feed. Possible suitable locations, a preliminary economic analysis and methods of increasing the product water by washing the ice are discussed [87].

Further applications of freezing processes include the development by Stepakoff et al. of an eutectic freezing process for brine disposal (see section 2.20.7, p. 184) and a process for concentration and recovery of water from plating wastes [88].

Literature to 4.2

[1] G. Nebbia, G. Nebbia-Menozzi (Desalination 5 [1968] 49/54). — [2] A. Rose, T. B. Hoover (Off. Saline Water Res. Develop. Progr. Rept. No. 7 [1955]). — [3] H. M. Hendrickson, R. W. Moulton (Off. Saline Water Res. Develop. Progr. Rept. No. 10 [1956]). — [4] C. M. Adams, P. K. Rohatgi (Off. Saline Water Res. Develop. Progr. Rept. No. 94 [1964], No. 142 [1965]). — [5] T. K. Sherwood, P. L. T. Brian, R. E. Fisher (Off. Saline Water Res. Develop. Progr. Rept. No. 95 [1964]).

[6] T. K. Sherwood, P. L. T. Brian (Off. Saline Water Res. Develop. Progr. Rept. No. 96 [1964]), No. 179 [1966]). — [7] J. P. Leinroth, W. P. White, T. K. Sherwood, P. L. T. Brian (Off. Saline Water Res. Develop. Progr. Rept. No. 128 [1965]). — [8] P. L. T. Brian, P. R. Sperry (Off. Saline Water Res. Develop. Progr. Rept. No. 98 [1964]). — [9] T. K. Sherwood, P. L. T. Brian (Off. Saline Water Res. Develop. Progr. Rept. No. 334 [1968]). — [10] J. Shwartz, R. F. Probstein (Off. Saline Water Res. Develop. Progr. Rept. No. 294 [1968]; Desalination 4 [1968] 5/29), Comments by P. Harriot, J. P. Leinroth, R. L. Von Berg, H. F. Wiegandt (Desalination 6 [1969] 117/9), Reply by J. Shwartz, R. F. Probstein (Desalination 6 [1969] 121/6).

[11] J. Shwartz, R. F. Probstein (Desalination 6 [1969] 239/66). — [12] N. H. Kemp (Desalination 12 [1973] 127/39). — [13] T. K. Sherwood, P. L. T. Brian, A. F. Sarofim, K. A. Smith (Off. Saline Water Res. Develop. Progr. Rept. No. 436 [1969]). — [14] T. K. Sherwood, P. L. T. Brian, A. F. Sarofim (Off. Saline Water Res. Develop. Progr. Rept. No. 474 [1969]), G. Margolis, T. K. Sherwood, P. L. T. Brian, A. F. Sarofim (Ind. Eng. Chem. Fundamentals 10 [1971] 439/52). — [15] A. Barak, G. Dagan (AIChE [Am. Inst. Chem. Engrs.] J. 16 [1970] 9/17).

[16] S. Kawasaki, M. Owa (Nippon Kaisui Gakkai-Shi 23 [1970] 256/63; C. A. 74 [1971] No. 15632). — [17] S. Seiitkurbanov, V. V. Milonov (Izv. Akad. Nauk Turkm.SSR Ser. Fiz. Tekhn. Khim. i Geol. Nauk 1970 No. 6, p. 113/7; C. A. 74 [1971] No. 130236). — [18] S. Tanaka (Sekiyu To Sekiyu Kagaku 15 No. 13 [1971] 53/6; C. A. 76 [1972] No. 103547). — [19] E. A. Proshin (Tr. Vses. Nauchn. Issled. Inst. Vodosnabzh. Kanaliz. Gidrotekhn. Sooruzh. Inzh. Gidrogeol. 1971 No. 29, p. 45/9; C. A. 78 [1973] No. 7672). — [20] M. V. Kolodin, S. Seiitkurbanov, V. V. Milonov (Izv. Akad. Nauk. Turkm.SSR Ser. Fiz. Tekhn. i Geol. Nauk 1971 No. 5, p. 114/8; C. A. 76 [1972] No. 37279).

[21] M. A. Vorstman, H. A. C. Thijssen (Ingenieur [Utrecht] 84 No. 45 [1972] Ch65/Ch69). — [22] M. D. Hobson, L. McGrath (Proc. 4th Intern. Symp. Fresh Water Sea, Heidelberg 1973, Vol. 3, p. 357/69). — [23] A. J. Barduhn (Desalination 5 [1968] 173/84). — [24] C. M. Bosworth, S. S. Carfagno, D. J. Sandell (Off. Saline Water Res. Develop. Progr. Rept. No. 23 [1959]), C. M. Bosworth, S. S. Carfagno, A. J. Barduhn, D. J. Sandell (Off. Saline Water Res. Develop. Progr. Rept. No. 32 [1959]). — [25] W. J. Hahn, R. C. Burns, R. S. Fullerton, D. J. Sandell (Off. Saline Water Res. Develop. Progr. Rept. No. 113 [1964]).

[26] M. Pachter, A. Barak (Desalination 2 [1967] 358/67). — [27] R. Consie, D. Emmermann, J. Fraser, W. B. Johnson, W. E. Johnson (Off. Saline Water Res. Develop. Progr. Rept. No. 295 [1968]). — [28] R. Consie, R. Darling, D. Emmermann, J. Fraser, W. Johnson, J. Koretchko, F. Torvbraten (Off. Saline Water Res. Develop. Progr. Rept. No. 451 [1969]). — [29] M. M. Jhawar, J. H. Fraser (Off. Saline Water Res. Develop. Progr. Rept. No. 541 [1970]). — [30] J. H. Fraser, K. K. Emmermann (Off. Saline Water Res. Develop. Progr. Rept. No. 573 [1970]).

[31] C. A. Kemper, G. F. Harper, S. W. Gouse, J. H. Leigh (Off. Saline Water Res. Develop. Progr. Rept. No. 118 [1964]). — [32] J. Koretchko, G. Hajela (Off. Saline Water Res. Develop. Progr. Rept. No. 744 [1971]). — [33] J. Koretchko (Off. Saline Water Res. Develop. Progr. Rept. No. 833 [1972]). — [34] T. Uchida (Shinku Kagaku 16 No. 2 [1969] 84/9; C. A. 73 [1970] No. 18353). — [35] R. R. Bridge, K. A. Smith, S. Johnson, W. W. Rinne (Chem. Eng. Progr. Symp. Ser. 67 No. 107 [1971] 212/6).

[36] Hittman Associates (Off. Saline Water Res. Develop. Progr. Rept. No. 491 [1969]). — [37] S. Umano, S. Kawasaki (Tokyo Kogyo Shikensho Hokoku 53 [1958] 365/7, 370/3, 387/92; C. A. 1960 20375; Tokyo Kogyo Shikensho Hokoku 54 [1959] 246/9; C. A. 56 [1962] 11371; Tokyo Kogyo Shikensho Hokoku 54 [1959] 299/303, 304/9; C. A. 1961 21698, 21699), S. Umano, Y. Nakano, S. Kawasaki, I. Hayano (Tokyo Kogyo Shikensho Hokoku 53 [1958] 368/9; C. A. 1960 20375), S. Umano, Y. Nakano (Tokyo Kogyo Shikensho Hokoku 53 [1958] 374/8, 379/83, 384/6; C. A.

1960 20375; Tokyo Kogyo Shikensho Hokoku **54** [1959] 27/35, 263/7, 268/83, 284/8, 289/92; C.A. **1961** 21 698; Tokyo Kogyo Shikensho Hokoku **57** [1962] 583/92, 593/8, 599/603; C.A. **62** [1965] 5065), T. Uchida, K. Havada (Tokyo Kogyo Shikensho Hokoku **57** [1962] 409/16, 417/22; C.A. **62** [1965] 5065), I. Hayano, S. Umano (Tokyo Kogyo Shikensho Hokoku **57** [1962] 534/9; C.A. **62** [1965] 5065). — [38] H. Svanoe, W. F. Swiger, J. S. Colton, J. E. Jewett, I. B. Margiloff (Off. Saline Water Res. Develop. Progr. Rept. No. 47 [1961]); J. W. Pike (Proc. 1st Intern. Symp. Water Desalination, Washington, D.C., 1965 [1967], Vol. 3, p. 173/88). — [39] T. Uchida (Proc. 3rd Intern. Symp. Fresh Water Sea, Dubrovnik 1970, Vol. 3, p. 93/100). — [40] N. Ganiaris, J. Lambiris, R. Glasser (Off. Saline Water Res. Develop. Progr. Rept. No. 416 [1969]).

[41] H. F. Wiegandt (Off. Saline Water Res. Develop. Progr. Rept. No. 41 [1960]). — [42] G. Karnofsky, P. F. Steinhoff (Off. Saline Water Res. Develop. Progr. Rept. No. 40 [1960]). — [43] H. F. Wiegandt (Advan. Chem. Ser. No. 27 [1960] 82/9), H. F. Wiegandt, R. L. Von Berg, J. P. Leinroth (Off. Saline Water Res. Develop. Progr. Rept. No. 290 [1968]). — [44] H. F. Wiegandt, R. L. Von Berg, J. P. Leinroth (Ind. Eng. Chem. Process Design Develop. **11** [1972] 404/14). — [45] F. O. Mixon (Desalination **7** [1970] 229/43).

[46] H. F. Wiegandt, P. Harriott, J. P. Leinroth (Off. Saline Water Res. Develop. Progr. Rept. No. 376 [1968]). — [47] P. L. T. Brian, K. A. Smith, L. M. Petri (Off. Saline Water Res. Develop. Progr. Rept. No. 269 [1967]). — [48] K. A. Smith, A. F. Sarofim, G. Margolis (Off. Saline Water Res. Develop. Progr. Rept. No. 621 [1970]). — [49] K. A. Smith, P. L. T. Brian (Off. Saline Water Res. Develop. Progr. Rept. No. 750 [1972]). — [50] D. J. Fontugne, H. Mason, J. E. Bajolle, P. A. Rice, A. J. Barduhn (Off. Saline Water Res. Develop. Progr. Rept. No. 501 [1969]).

[51] J. E. Bajolle, P. A. Rice, A. J. Barduhn (Proc. 3rd Intern. Symp. Fresh Water Sea, Dubrovnik 1970, Vol. 3, p. 13/30, Vol. 4, p. 249/50; Desalination **9** [1971] 351/66), J. E. Bajolle, A. Naimpally, H. L. Mason, J. Missirlis, P. A. Rice, A. J. Barduhn (Off. Saline Water Res. Develop. Progr. Rept. No. 658 [1971]). — [52] J. S. Huang, A. V. Naimpally, J. Missirlis, P. A. Rice, A. J. Barduhn (Off. Saline Water Res. Develop. Progr. Rept. No. 797 [1972]). — [53] Y. S. Cheng, R. Nene, P. A. Rice, A. J. Barduhn (Off. Saline Water Res. Develop. Progr. Rept. No. 823 [1972]). — [54] H. C. Simpson, G. C. Beggs, M. Nazir (Proc. 4th Intern. Symp. Fresh Water Sea, Heidelberg 1973, Vol. 3, p. 409/20). — [55] C. A. Johnson, S. J. Moore, N. D. Wagaman, D. J. Sandell (Off. Saline Water Res. Develop. Progr. Rept. No. 256 [1967]).

[56] M. Landau, A. Martindale (Desalination **3** [1967] 318/29). — [57] W. H. Denton, N. S. Hall Taylor, J. T. Klaschka, M. J. S. Smith, H. R. Diffey, C. H. Rumary (Proc. 3rd Intern. Symp. Fresh Water Sea, Dubrovnik 1970, Vol. 3, p. 51/69). — [58] A. Martindale, B. R. Parr, M. J. S. Smith (Proc. 3rd Intern. Symp. Fresh Water Sea, Dubrovnik 1970, Vol. 3, p. 71/82). — [59] W. H. Denton, M. J. S. Smith, J. T. Klatschka, R. Forgan, H. R. Diffey, C. H. Rumary, R. W. Dawson (Proc. 4th Intern. Symp. Fresh Water Sea, Heidelberg 1973, Vol. 3, p. 291/311). — [60] W. R. Burton, A. J. Lloyd (Proc. 4th Intern. Symp. Fresh Water Sea, Heidelberg 1973, Vol. 3, p. 281/90; Desalination **14** [1974] 151/61).

[61] Anonymous (Energy Intern. **8** No. 5 [1971] 39), J. E. Lock (Chem. Process. [London] **17** No. 6 [1971] 14), H. Vollbrecht (Chem. Anlagen Verfahren **1971** No. 8, p. 59/60). — [62] J. C. Orcutt, F. O. Mixon, F. J. Hale (Off. Saline Water Res. Develop. Progr. Rept. No. 365 [1968]), J. C. Orcutt, F. J. Hale (Desalination **7** [1970] 201/27). — [63] J. C. Orcutt, T. P. Carey (Ind. Eng. Chem. Process Design Develop. **9** [1970] 58/63). — [64] M. V. Kolodin, S. Seiitkurbanov, V. V. Milonov (Vodosnabzh. Sanit. Tekhn. **1969** No. 11, p. 1/4; C.A. **72** [1970] No. 93 256). — [65] S. Y. Chiu, L. T. Fan, R. G. Akins (Ind. Eng. Chem. Process Design Develop. **8** [1969] 347/56).

[66] S. Tamaru, Y. Nagashima (Proc. 3rd. Intern. Symp. Fresh Water Sea, Dubrovnik 1970, Vol. 3, p. 83/91). — [67] M. V. Kolodin, E. M. Rutgaizer (Kholodil'n. Tekhn. **48** No. 4 [1971] 20/3; C.A. **75** [1971] No. 52 650). — [68] M. V. Kolodin, S. Seiitkurbanov, V. V. Milonov (Vodosnabzh. Sanit. Tekhn. **1971** No. 6, p. 20/2; C.A. **75** [1971] No. 80132). — [69] S. Kawasaki, M. Owa (Nippon Kaisui Gakkai-Shi **23** [1970] 263/70; C.A. **74** [1971] No. 57 548). — [70] T. Kennaway (Chem. Process Eng. **52** No. 6 [1971] 91/2).

[71] P. L. T. Brian (Chem. Engineer No. 249 [1971] 191/7). — [72] T. Adamski (Chem. Anlagen Verfahren **1973** No. 10, p. 78/82). — [73] D. S. Hubbard, J. P. Leinroth, H. F. Wiegandt (Off. Saline Water Res. Develop. Progr. Rept. No. 312 [1968]). — [74] S. Kawasaki (Proc. 4th Intern. Symp. Fresh Water Sea, Heidelberg 1973, Vol. 3, p. 383/93). — [75] W. Johnson, J. H. Fraser, W. E. Gibson, A. P. Modica, G. Grossman (Off. Saline Water Res. Develop. Progr. Rept. No. 786 [1972]), G. Grossman (Ind. Eng. Chem. Process Design Develop. **11** [1972] 537/42).

[76] G. L. Stepakoff, A. P. Modica (Desalination **12** [1973] 85/105, 239/50). — [77] W. Gibson, G. Grossman, A. Modica, G. Siegelman, G. Stepakoff (Off. Saline Water Res. Develop. Progr. Rept. No. 816 [1972]), J. H. Fraser, W. E. Gibson (J. Am. Water Works Assoc. **64** [1972] 746/8), W. Johnson, A. Pallone, R. F. Probstein (Proc. 4th Intern. Symp. Fresh Water Sea, Heidelberg 1973, Vol. 3, p. 371/82), W. Gibson, D. Emmermann, G. Grossman, W. Johnson, A. Modica, A. Pallone (Proc. 4th Intern. Symp. Fresh Water Sea, Heidelberg 1973, Vol. 3, p. 343/55), D. K. Emmermann, W. E. Gibson, G. Grossman, A. P. Modica, A. Pallone (AIChE [Am. Inst. Chem. Engrs.] Symp. Ser. **69** No. 129 [1973] 520/6). — [78] J. H. Fraser, T. A. Olsson (Off. Saline Water Res. Develop. Progr. Rept. No. 916 [1973]; Proc. 4th Intern. Symp. Fresh Water Sea, Heidelberg 1973, Vol. 3, p. 331/42). — [79] R. G. Akins, L. T. Fan, L. E. Erickson (Off. Saline Water Res. Develop. Progr. Rept. No. 318 [1968]). — [80] H. M. Curran, C. P. Howard (Off. Saline Water Res. Develop. Progr. Rept. No. 511 [1970]), H. M. Curran (Desalination **7** [1970] 273/84; Proc. 3rd. Intern. Symp. Fresh Water Sea, Dubrovnik 1970, Vol. 3, p. 41/50).

[81] S. Bustany, P. Harriott, R. L. Von Berg, H. F. Wiegandt (Off. Saline Water Res. Develop. Progr. Rept. No. 738 [1971]). — [82] G. Boari, G. Lacava, C. Merli, R. Passino, M. Santori (Proc. 3rd Intern. Symp. Fresh Water Sea, Dubrovnik 1970, Vol. 3, p. 31/9). — [83] A. C. Di Pinto, G. Lacava, R. Passino, A. Rozzi, M. Santori, L. Spinosa (Proc. 4th Intern. Symp. Fresh Water Sea, Heidelberg 1973, Vol. 3, p. 313/9). — [84] C. Y. Cheng, G. Van Riper, V. G. Fox (Off. Saline Water Res. Develop. Progr. Rept. No. 802 [1972]). — [85] H. H. G. Jellinek (Off. Saline Water Res. Develop. Progr. Rept. No. 765 [1972]).

[86] J. Fournier, J. L. Grange, S. Vergara (Proc. 4th Intern. Symp. Fresh Water Sea, Heidelberg 1973, Vol. 3, p. 320/9). — [87] P. C. Wankat (Desalination **13** [1973] 147/57). — [88] R. J. Cambell, D. K. Emmermann (Mech. Eng. **95** No. 7 [1973] 29/32). — [89] M. V. Kolodin, S. Seiit-kurbanov (Kholodil'n. Tekhn. **47** No. 11 [1970] 15/9).

Hydrat-
verfahren

4.3 Hydrate processes

Hydrates, like ice, when formed from a saline solution with a hydrating agent, contain in their structure only pure water, which is delivered by decomposing the hydrate. In the hydrate desalination process, the saline water and hydrating agent, such as a light hydrocarbon or a halogenated derivative, are mixed together under the required conditions of temperature and pressure. The crystals are separated from the brine, washed with fresh water and then melted. The melting is accomplished by temperature or pressure change or both. The resulting two liquids are separated by decantation or centrifugal separation and the hydrating agent is recycled.

Formation and properties. An extensive research has been performed on the identification of the most appropriate hydrating agents, the thermodynamic properties of the hydrates, the kinetics of hydrate formation and the solubilities of hydrate systems. A bibliography of publications on compounds of gas hydrate type was compiled by the University of Pittsburgh [1] and a bibliography on structure of clathrate hydrates was also compiled by the same University and edited by Kase [2].

The properties of gas hydrates and their potential use in demineralizing seawater were reviewed by Barduhn et al. [3]. The same group of authors published also reports on new agents for use in the hydrate process [4], on the solubilities of hydrating agents in water, as well as on the mechanism and kinetics of hydrate formation [5], on chlorine hydrate and the solubility of F-142b (CH_3CClF_2) in water and aqueous sodium chloride solutions [6], on the solubility of four gas hydrate formers ($CHCl_2F$, CH_2ClF, CH_3Br and CH_3CClF_2) in water and NaCl solutions [7], on the hydrates of iso- and n-butane [8], on the kinetics of gas hydrate formation and especially of methyl bromide hydrate [9], on the rate of formation of CH_2ClF hydrate in a continuous stirred reactor and hydrolysis rates of hydrating agents [10], on the rate measurement and economic analysis of hydrolysis losses in the hydrate process of desalination [11], on the properties of the known 51 gas hydrates and on nine hydrating agents having good possiblities, the best being CH_2ClF, as well as on carbon dioxide hydrate [12], and on the properties of the hydrates of chlorine and carbon dioxide [13].

Sugi and Saito reached the conclusion that dichlorofluoromethane is the most profitable hydrating agent [14], Uchida and Hayano confirmed that of the C_4-hydrocarbons only iso-butane seems to be able to form solid hydrate [15], Saito and Iijima reported on the recovery of the hydrating agent by the carrier gas method [16], Tabuchi on the formation of LPG (liquified petroleum gas) hydrate [17], Hayano and Uchida on hydrate formation in propane–n-butane systems [18].

Tester and Wiegandt reported on phase equilibria of methylene chloride and chloroform binary and tertiary hydrate systems [19], Hafeman and Miller on clathrate hydrates of cyclopropane [20], Werezak on the concentration of an aqueous solution by a clathrate type of gas hydrate formation [21], De Graauw and Rutten on the mechanism and rate of hydrate formation [22], Wilms and Van Haute on the thermodynamics and composition of chlorine hydrate [23], the same authors on the determination of a composition of a gas hydrate by the method of Miller and Strong [24], Krasnov and Klimenok on the influence of methanol on the formation of gas hydrates [25], Backhurst and Harker on natural gas and hydrocarbon hydrates [26], Davidson on the motion of guest molecules in clathrate hydrates [27], Morris and Davidson on a clathrate hydrate of cyclobutanone [28], Iskenderov and Musaev on the latent heat of hydrate formation [29], Robinson and Mehta on the hydrates in the propane–carbon dioxide–water system [30], Byk et al. on the calculation of saturated vapor pressure over a crystalline lattice of gas hydrate [31], on the fugacity of water over the crystal lattice of a gas hydrate [32], and on the heat of adsorption in the formation of a gas hydrate [33], Parrish and Prausnitz on dissociation pressures of gas hydrates formed by gas mixtures [34], Daee on the properties of some clathrate-like cluster structures [35], Tester et al. on the thermodynamic properties of water clathrates [36], Gerard and Pernolet on the formation of ethylene and ethane clathrate hydrates [37], Smirnov on the formation of Freon-12 hydrates [38].

Desalting processes. In the Koppers hydrate process, hydrate formation takes place in a reactor vessel which contains boiling liquid hydrate agent and salt water. The reactor produces a hydrate slurry which contains approximately 10% solid material. The slurry is filtered and a hydrate cake is formed, which is first washed with pure water and then decomposed. Decomposition produces a mixture of pure water and liquid hydrate agent, which are separated by decantation. The hydrating agent is returned to the reactor and a large part evaporates due to the heat of hydrate formation. The vapor is compressed and its temperature rises sufficiently to make it usable for decomposition of pure hydrate. The operating temperature of hydrate formation is slightly higher than the freezing point of water. The temperature of decomposition depends on the hydrating agent. Propane hydrate decomposes at 6°C, refrigerant 12 (CF_2Cl_2) forms a hydrate which decomposes at 12°C, refrigerant 31 (CH_2FCl) has a hydrate which is stable up to 18°C. Hydrates are formed only above a minimum pressure, characteristic for the hydrating agent [39].

A 10 000 gpd (38 m³/day) pilot plant was erected at the O.S.W. Test Facility in Wrightsville Beach and tested to demonstrate the Koppers hydrate process. It was concluded that the process is technically feasible and that seawater could be used instead of fresh water for washing purposes [40].

Sweet Water Development Co. tested a similar process based on propane hydrate on a 20 000 gpd (76 m³/day) pilot plant at Wrightsville Beach. Projects for a 1 Mgd (3785 m³/day) plant revealed a water cost of 69 cents/kgal (18.2 cents/m³). The estimate for this plant considers the use of multiple units in parallel to attain capacity [41]. A study for the erection of a pilot plant based on propane hydrate was also reported by Pavlov and Medvedev [42].

Martinovskii and Smirnov suggested the use of a turbine and thermocompressor running on a common shaft to power a hydrate conversion plant [43]. The same authors proposed the use of the temperature difference of cold deep water and warm surface water in a hydrate desalination plant, which is expected by this way to be economical [44].

The development and optimization of a hydrate process for desalination of seawater was discussed by Rautenbach and Pennings [45]. The same authors reported on experiments with a counterflow-washing-column wherein the crystals form a concentrated suspension. The crystals rise against gravity and washing water according to density differences between fluid and particles. A cost study, based on the experimental results, led to water costs of 0.04 DM/m³ freshwater at a washing-water consumption-rate of approximately 10% [46].

Literature to 4.3

[1] University of Pittsburgh (Off. Saline Water Res. Develop. Progr. Rept. No. 73 [1963]). — [2] University of Pittsburgh, K. A. Kase (Off. Saline Water Res. Develop. Progr. Rept. No. 121 [1964]). — [3] A. J. Barduhn, H. E. Towlson, Y. C. Hu (Off. Saline Water Res. Develop. Progr. Rept. No. 44 [1960]). — [4] F. A. Briggs, Y. C. Hu, A. J. Barduhn (Off. Saline Water Res. Develop. Progr. Rept. No. 59 [1962]). — [5] Syracuse University Research Institute (Off. Saline Water Res. Develop. Progr. Rept. No. 70 [1963]), A. J. Barduhn, N. Klausutis, R. W. Collette, J. R. Kass (Off. Saline Water Res. Develop. Rept. No. 88 [1964]).

[6] J. R. Kass, R. Fernandez-Martin, H. L. Empie, A. J. Barduhn (Off. Saline Water Res. Develop. Progr. Rept. No. 130 [1965]), R. Fernandez, W. W. Carey, A. T. Bozzo, A. J. Barduhn (Off. Saline Water Res. Develop. Progr. Rept. No. 229 [1967]). — [7] W. W. Carey, N. A. Klausutis, A. J. Barduhn (Desalination 1 [1966] 342/58). — [8] R. G. Latini, O. S. Rouher, P. D. Agrawal, A. J. Barduhn (Off. Saline Water Res. Develop. Progr. Rept. No. 282 [1967]), O. S. Rouher, A. J. Barduhn (Desalination 6 [1969] 57/73). — [9] R. Fernandez, J. B. Pangborn, S. L. Colten, A. J. Barduhn (Off. Saline Water Res. Develop. Progr. Rept. No. 230 [1967], No. 333 [1968]), J. B. Pangborn, A. J. Barduhn (Desalination 8 [1970] 35/68). — [10] A. J. Barduhn, J. M. Poisot, G. M. Roux, T. C. Tsao (Off. Saline Water Res. Develop. Progr. Rept. No. 370 [1968], J. G. Vlahakis, T. C. Tsao, H. A. Richard, A. J. Barduhn (Off. Saline Water Res. Develop. Rept. No. 497 [1969]).

[11] S. L. Colten, F. S. Lin, T. C. Tsao, S. A. Stern, A. J. Barduhn (Off. Saline Water Res. Develop. Progr. Rept. No. 753 [1972]; Desalination 11 [1972] 31/59). — [12] J. G. Vlahakis, H. S. Chen, M. S. Suwandi, A. J. Barduhn (Off. Saline Water Res. Develop. Progr. Rept. No. 830 [1972]). — [13] A. T. Bozzo, H. S. Chen, J. R. Kass, A. J. Barduhn (Proc. 4th Intern. Symp. Fresh Water Sea, Heidelberg 1973, Vol. 3, p. 437./51). — [14] J. Sugi, S. Saito (Desalination 3 [1967] 27/31). — [15] T. Uchida, I. Hayano (Desalination 3 [1967] 373/7).

[16] S. Saito, M. Iijima (Nippon Kaisui Gakkai-Shi 23 No. 2 [1969] 46/53; C. A. 74 [1971] No. 15631). — [17] K. Tabuchi (Ishikawajima-Harima Giho 10 [1970] 481/7; C. A. 74 [1971] No. 89373). — [18] I. Hayano, T. Uchida (Nippon Kaisui Gakkai-Shi 22 [1969] 355/9; C. A. 75 [1971] No. 112762). — [19] I. W. Tester, H. F. Wiegandt (AIChE [Am. Inst. Chem. Engrs.] J. 15 [1969] 239/44). — [20] D. R. Hafeman, S. L. Miller (J. Phys. Chem. 73 [1969] 1392/7).

[21] G. N. Werezak (Chem. Eng. Progr. Symp. Ser. 65 No. 91 [1969] 6/18; C. A. 71 [1969] No. 59819). — [22] J. De Graauw, J. J. Rutten (Proc. 3rd. Intern. Symp. Fresh Water Sea, Dubrovnik 1970, Vol. 3, p. 103/16). — [23] D. A. Wilms, A. Van Haute (Proc. 3rd Intern. Symp. Fresh Water Sea, Dubrovnik 1970, Vol. 3, p. 117/29; Desalination 12 [1973] 379/93). — [24] D. A. Wilms, A. Van Haute (Proc. 4th Intern. Symp. Fresh Water Sea, Heidelberg 1973, Vol. 3, p. 477/84). — [25] A. A. Krasnov, B. V. Klimenok (Zh. Fiz. Khim. 44 [1970] 1342/3; C. A. 73 [1970] No. 92119).

[26] J. R. Backhurst, J. H. Harker (J. Inst. Fuel 43 [1970] 405/6). — [27] D. W. Davidson (Can. J. Chem. 49 [1971] 1224/42). — [28] B. Morris, D. W. Davidson (Can. J. Chem. 49 [1971] 1243/51). — [29] S. M. Iskenderov, R. M. Musaev (Gazov. Delo 1970 No. 12, p. 6/8; C. A. 75 [1971] No. 41238). — [30] D. B. Robinson, B. R. Mehta (J. Can. Petrol. Technol. 10 [1971] 33/5).

[31] S. S. Byk, V. I. Fomina, A. F. Narozhenko (Gazov. Prom. 16 No. 2 [1971] 35/38; C. A. 75 [1971] No. 10901). — [32] V. S. Koshelev, V. J. Fomina, S. S. Byk (Zh. Fiz. Khim. 45 [1971] 2968; Russ. J. Phys. Chem. 45 [1971] 1685). — [33] S. S. Byk, V. J. Fomina (Zh. Fiz. Khim. 46 [1972] 994/6; Russ. J. Phys. Chem. 46 [1972] 576). — [34] W. R. Parrish, J. M. Prausnitz (Ind. Eng. Chem. Process Design Develop. 11 [1972] 26/34). — [35] M. Daee, L. H. Lund, P. L. M. Plummer, J. L. Kassner, B. N. Hale (J. Colloid. Interface Sci. 39 [1972] 65/78).

[36] J. W. Tester, R. L. Bivins, C. C. Herrick (AIChE [Am. Inst. Chem. Engrs.] J. 18 [1972] 1220/30). — [37] N. Gerard, R. Pernolet (Proc. 4th Intern. Symp. Fresh Water Sea, Heidelberg 1973, Vol. 3, p. 453/60). — [38] L. F. Smirnov (Kholodil'n. Tekhn. 1973 No. 2, p. 28/34; C. A. 78 [1973] No. 163861). — [39] C. F. Winans, W. G. Knox, M. Hess, L. Cutter, H. B. Smith (Dechema Monograph. 47 [1963] 839/48), Koppers Company (Off. Saline Water Res. Develop. Progr. Rept. No. 90 [1964]), E. D. Brennan, P. Van der Heem (Off. Saline Water Res. Develop. Progr. Rept. No. 125 [1964]), P. Van der Heem (Proc. 1st Intern. Symp. Water Desalination, Washington, D.C., 1965 [1967], Vol. 2, p. 769/78). — [40] K. Garrison, R. J. Slape, L. L. Snedden (Off. Saline Water Res. Develop. Progr. Rept. No. 368 [1968]).

[41] V. C. Williams, C. L. Roy, R. A. Williams (Proc. 1st Intern. Symp. Water Desalination, Washington, D.C., 1965 [1967], Vol. 3, p. 605/23), V. C. Williams, C. L. Roy, H. Smith, O. B. Battle (Off. Saline Water Res. Develop. Progr. Rept. No. 373 [1968]). — [42] G. D. Pavlov, I. N. Medvedev (Proc. 1st Intern. Symp. Water Desalination, Washington, D.C., 1965 [1967], Vol. 3, p. 123/7). — [43] V. S. Martinovskii, L. F. Smirnov (Progr. Refrig. Sci. Technol. Proc. 12th Intern. Congr. Refrig., Madrid 1967 [1969], Vol. 1, p. 433/40). — [44] V. S. Martinovskii, L. F. Smirnov (Proc. 4th Intern. Symp. Frash Water Sea, Heidelberg 1973, Vol. 3, p. 461/9). — [45] R. Rautenbach, P. Pennings (Chem. Ingr.-Tech. 45 [1973] 259/64).

[46] R. Rautenbach, P. Pennings (Proc. 4th Intern. Symp. Fresh Water Sea, Heidelberg 1973, Vol. 3, p. 471/6).

4.4 Liquid–liquid extraction

This process for desalination of saline waters was mainly developed at the Texas A & M College. Saline water is contacted with an organic solvent to produce a more concentrated raffinate and an extract containing the low salinity water which separates upon heating. The solvent is recycled and the product and raffinate are stripped of the residual solvent content. Secondary and tertiary amines appear to be the best from many types of compounds investigated, except that the resulting high pH precipitates magnesium [1]. Solubility and equilibrium data were reported for a variety of solvents, as well as data for the laboratory model [2].

The effect of temperature, atmosphere, water and time on the degradation of amines was studied by Davison and Smith. Degradation was much higher at 60°C than at room temperature. Amine removal by bacterial action, stripping, ion exchange and ozonization was also investigated [3]. The process was tested in a pilot plant with a capacity of 2000 gpd (7.5 m^3/day) [4].

A potential method of desalination of water with extraction at low temperature was described by Knyaz'kova and Kul'skii. By heating the extract an aqueous layer of lower concentration is formed. Graphs are given of NaCl content in the two layers at various temperatures and concentrations of the original solution [5]. The selectivity of triethylamine and the ratio of chloride concentrations were also reported by the same authors [6].

Potable water and saturated brines were produced by Barton and Fenske from saline waters by extraction with liquid hydrocarbon at 650°F and 2600 psi (343°C and 182 kg/cm²). High selectivity of hydrocarbon for water over salt enables use of a single extraction stage with seawater. Solvent and heat recovery were accomplished in fluidized particle heat exchangers to eliminate scaling problems. Product water contained 120 mg/l of salt and 12 mg/l of C_{11} to C_{12} paraffinic hydrocarbons. A proposed plant producing 10 Mgd of desalted water requires a capital investment of $ 18 million. Operating costs are 99 cents/kgal (26 cents/m^3) of water [7]

Literature to 4.4

[1] D. W. Hood, R. R. Davison (Advan. Chem. Ser. No. 27 [1960] 40/9), R. R. Davison, L. M. Jeffrey, U. G. Whitehouse, D. W. Hood (Off. Saline Water Res. Develop. Progr. Rept. No. 22 [1958]), R. Davison, A. F. Isbell, W. H. Smith, D. W. Hood (Off. Saline Water Res. Develop. Progr. Rept. No. 55 [1961]). — [2] Texas A & M Research Foundation (Off. Saline Water Res. Develop. Progr. Rept. No. 35 [1960]). — [3] R. R. Davison, W. H. Smith (Off. Saline Water Res. Develop. Progr. Rept. No. 371 [1968]). — [4] R. R. Davison, W. B. Harris, W. H. Smith (Desalination 3 [1967] 17/26). — [5] T. V. Knyaz'kova, L. A. Kul'skii (Ukr. Khim. Zh. 35 [1969] 422/5; C. A. 71 [1969] No. 25162).

[6] T. V. Knyaz'kova, L. A. Kul'skii (Ukr. Khim. Zh. 37 [1971] 620/2; C. A. 75 [1971] No. 91214).
— [7] P. Barton, M. R. Fenske (Ind. Eng. Chem. Process Design Develop. 9 [1970] 18/25).

5 Economic considerations

Examining the requirements, supply and production of water by desalination, Othmer reminds that fresh water is most unevenly divided throughout the world. More research and development funds are being spent worldwide to develop desalination technology, than for almost any other non-military purpose except space. The development of many countries will advance rapidly when water will be made available. The less developed countries will be the beneficiaries of desalination research. The highly industrialized countries have an additional problem, namely pollution from the concentration of population and industries [1].

The planning of a seawater desalination plant in a given region involves careful consideration of the local fresh water supply and demand. In many cases it is possible to proceed to a study of the natural resources, that could eventually eliminate the need for desalination. The preliminary study should be further pursued, taking into account that natural resources and desalination systems can be complementary, with the one contributing to the peak demand and the other to the base load. This analysis makes it possible to classify the main consumer regions, according to certain economic criteria, which define their suitability for the use of desalination processes [2]. Any study of the use of desalination techniques to meet water requirements involves comparing the various desalination processes with conventional water supply methods against a background of generally increasing demand. It is essential to have a coherent overall method of assessing the economics of using desalination to meet estimated requirements. A general method of dynamic programming was presented by Thiriet et al. enabling the question of the optimal sizing of desalination facilities to be resolved and bearing in mind economies of scale and possible storage facilities [3].

A decision to postpone development of a desalting plant on the contention that the later development will derive the benefit of technological change could be in error. The experience gained from the building of one plant makes it possible to build subsequent plants with greater confidence in the cost estimates. A basis on which part of the first installation may be shifted to the subsequent projects was developed by El-Ramly and English [4]. A conceptual framework was also provided for taking into account the varying degrees of the confidence professed in alternative schemes. Emphasis was placed upon developing and identifying the broad concept of uncertainty as it pertains to the decision-making process [5].

Estimation of water costs. A standardized procedure was developed by the Office of Saline Water for making initial cost estimates for demineralization of saline waters by various desalting processes [6]. This preliminary procedure was generally used for estimating desalting costs, until engineering data became available from the operation of various plants to develop more accurate cost estimating schemes. A manual to provide recommended guidelines and formats for computing and uniformly presenting estimated costs of desalted water by any process for ease of comparison was then issued. Cost centers taken in consideration refer mainly to the most developed processes, such as distillation, electrodialysis and reverse osmosis. However, they are also applicable to other desalting operations, namely ion exchange, freezing and other processes. The costs associated with desalting plant engineering designs may be grouped into two major cost centers:

      a) Fixed or capital costs   b) Operating and maintenance costs.

Fixed or capital costs may include depreciable costs, such as intake systems, raw water treatment, desalting equipment, replaceable items, steam supply, power supply, structures and improvements, brine disposal and indirect capital costs. Non-depreciable capital costs are land costs and working capital.

Operation and maintenance costs include fuel, steam, electric power, supplies and maintenance materials, chemicals, operating and maintenance labor, administrative expenses, as well as any other current expenses [7].

A cost model of multi-stage flash evaporation relating water costs to parameters such as plant size, energy price, fixed charges rate and load factor was presented by Mawer. All cost estimates were considered in the context of the integration of desalination into overall water supply schemes. The sensitivity of water costs to variation of various operating conditions was also considered [8].

In a more detailed report cost analysis of six water desalting processes was given. These processes are multi-stage flash distillation (MSF), vertical-tube evaporator (VTE) combined with MSF, vapor compression distillation combined with VTE and MSF, vacuum freeze vapor compression, electrodialysis and reverse osmosis. In this report the capital requirements and cost of producing fresh water were determined for the six processes. The major cost elements of each conversion process were identified and the effects of varying important process parameters on water costs were shown for each process. The estimated costs should be viewed with varying degrees of confidence, dependent on size of existing plants, level of knowledge and potential for further improvement. The reliability of cost data is probably highest for the multi-stage flash distillation process. Each process will probably have areas of preferred application, because of capital and operating costs, as well as quality of feed water. An estimate is given of the feed water properties and plant capacities that would provide the most favorable application area for each process [9].

Southwest Research Institute conducted a study of the potential contribution of desalting as a possible solution for municipal water supplies in Texas [10]. Simplified procedures were then developed for calculating the costs of desalting systems and conventional water supply systems. Comparative costs were presented with sufficient accuracy for positive identification of communities, where water from desalting would be less expensive than water from conventional sources [11]. A manual of procedures and methods for calculating brine disposal costs was also presented [12].

In continuing work a manual on desalting cost calculating procedures was published. This manual is one of the most detailed reports of this kind. It enables the user to calculate, for cost comparison purposes, the capital costs, as well as the operating and maintenance costs for five desalting processes, as a preliminary to a possible detailed engineering feasibility analysis. The five processes are multi-stage flash distillation (MSF), vapor compression-vertical tube evaporator—multi-stage flash distillation (VC-VTE-MSF), electrodialysis (ED), reverse osmosis (RO), and vacuum freezing vapor compression (VFVC). This study [13] updated and added to material presented in the previous study [11].

Chapter 7 of the "Desalting Handbook for Planners" contains also desalting cost estimating procedures in detailed form. Cost estimates based on the suggested procedures may be used for comparison studies, preliminary economic analysis and to assist in selection of water supply or augmentation plans for feasibility studies [14]. A part of the cost data, presented in the handbook, is derived from conceptual design studies, laboratory test results, costs associated with construction and operation of either test modules or experimental prototype desalting plants. The handbook is published in leaflet form and periodically updated.

A similar study embodying criteria for application and guidelines for costing with respect to desalination was undertaken in the Resources and Transport Division of the United Nations Department of Economic and Social Affairs [15]. The International Atomic Energy Agency has also prepared a report, which contains a review of the basic principles for costing desalination plants and of the various methods proposed for allocating costs in nuclear dual purpose plants [16].

Factors involved in selecting a desalination process for a particular application and which underlie the many different desalting processes being developed across the world were considered by Kronberger. While water costs from desalting plants still appear high, in many temperate climates, there may be circumstances when desalting plant used in conjunction with or even instead of a natural water scheme may be economic [17].

The costs of items influencing the cost price of potable water were evaluated for distillation, reverse osmosis and electrodialysis installations with production capacities from 10 to 100 000 m³/day. Prices of water leaving the plant range between 0.7 and 15 francs/m³ according to the method used, the size of the unit and the nature of the water to be treated [18].

The second process cost comparison exercise undertaken by the U.K. Atomic Energy Authority Desalination Project has confirmed the cost advantages of the multi-effect vertical tube evaporator process over the conventional multi-stage flash evaporation process and has also confirmed the cost advantages of power consuming processes as a group compared to the above heat consuming processes. Reverse osmosis for seawater conversion, using conservative membrane performance parameters, is now shown to compete on equal terms with the other power consuming processes, vapor compression evaporation and secondary refrigerant freezing. Future escalation of fuel oil prices was considered. It is shown that the presently higher capital cost of power consuming processes is more than outweighed by the savings in fuel which are possible when comparison is made with heat consuming processes [19].

Hybrid systems. The possible application of hybrid desalting plant consisting of vertical-tube distillation units and reverse osmosis units, integrated in large nuclear dual-purpose plants, was analyzed by Glueckstern et al. with respect to improvement in flexibility and economics of non-base load designs. Both brackish and seawater were applied as feedwater to the reverse osmosis units. Potential benefits of combining reverse osmosis units desalting brackish water with one of the existing multi-stage flash evaporator desalting seawater at Eilat on the Red Sea Shore, was also discussed [20].

A study was also performed to determine the technical and economic feasibility of a dual mode desalination device combining a reverse osmosis unit and a multi-stage flash distillation unit. The function of the reverse osmosis unit would be to remove sufficient amounts of scaling salts from intake seawater to allow scale-free operation of the flash distillation unit at temperatures up to 177°C (350°F). A review of membrane technology revealed that at 42 to 56 kg/cm² (600 to 800 psi), specially blended open cellulose acetate membranes or an ion exchange membrane can reject the necessary percentage of scaling salts with sufficient water flux to feed the flash distillation unit. Thus the dual mode desalination appears technically feasible. An economic evaluation of the dual mode device, however, reveals that the cost of its product water would be higher than that from a conventional 121°C (250°F) flash distillation unit [21].

Storage capacity and load factor. The relationship between storage capacity and desalting plant load factor is of importance, when conjunctive use of desalted water and natural water supplies is made. Several typical situations in which a desalter might be involved in a water supply system were explained and illustrated by Golzé. A procedure was programmed for computer solution to show the relationship between storage capacity and the desalting plant load factor and a model to determine optimum operation was presented. Different types of storage facilities were discussed and representative average costs of such storage facilities in California were presented [22]. The United Nations have also published a report on the design of water supply systems based on desalination, with emphasis on the selection of plant sizes and associated storage facilities to meet variations in demand and plant cutages [23].

A simple relationship between load factor of desalination plant augmenting existing natural supplies was developed by Orde and the emergence in practice of comparatively low load factor even in water systems with limited seasonal peaks was demonstrated. The use of dual-purpose plants is questioned under such conditions. The main drawbacks of the multi-stage flash distiller are discussed together with the desirability of developing alternative power consuming processes, which are economical for single-purpose plants [24].

A computation model was developed by Glueckstern et al. for the investigation of water resources development, the testing of new desalting technologies, the economic incentives for water storage capability, the smoothening out seasonal demand variations and required cost to improve water quality standards [25].

Michel has reported on alternatives for storage of desalted water, such as in aquifers, reservoir systems and plastic bags [26] and Adar on storage facilities of water in Israel [27].

Conjunctive use of desalted and natural water. The alternative of a mix between the surface water collection systems and desalination was discussed by English and El-Ramly. The economics of desalination were considered from the total water system of which the desalting plant may be only a component. The optimum mix was determined on the basis of minimizing the actual worth of costs in relation to demand. The effects of the rate of growth in demand on the optimum plant characteristics were shown through the use of continuous discounting and exponential approximations for the demand functions. Capital cost allocation over the plant's life was made in such a way as to be proportional to the demand pattern. Accordingly, the concept of an equivalent marginal pricing was formulated [28].

The operation of a desalination in conjunction with a conventional reservoir can give increased yields at costs as low as 50% of the equivalent base-load desalination cost. Generalized methods were presented by Mawer and Burley for generating operating rules for such systems. Results of two worked examples are included and the technical implications which low load factor operation has on desalination plant design are discussed [29].

The infrequent operation of desalination plant was used to maintain the reliability of a conventional water resource system, whilst effectively increasing the yield of conventional water supply. By the production of very expensive desalted water in drought periods only an additional quantity of almost free water can be taken from an impounding reservoir for the remainder of the time. The over-all effect is to provide the increased supply at costs sometimes as low as 50% of the cost associated with base load desalination. Computer analysis of hydrological data has been used to develop efficient operating rules for the desalination plant. A number of existing reservoirs were considered, as were several types of desalination. Both single- and dual-purpose distillation and electrodialysis were studied for such applications and the particular case of summer-only usage of steam from dual-purpose plant for distillation purposes was considered [30]. Aspects of the flexibility and plant availability of dual-purpose desalting complexes, combined with water storage facilities, in meeting power/water demands subject to short-term variations and long-term growth were discussed. Emphasis was given to the variability of desalted water output arising from plant downtime or from the need to meet peak power demands [31].

Integer linear and nonlinear programming techniques were used in planning a total regional water resources system, where demand for water arises in industrial, municipal and agricultural sectors. Dynamic programming was used to find out how the safe yield from a surface water reservoir can be increased when a dual-purpose desalting plant is operated in conjunction with the surface reservoir [32]. A mathematical model was then constructed to determine the optimum contract levels for firm water and on-peak energy supply from a dual-purpose desalination plant which will be operated in conjunction with an existing multi-purpose surface water reservoir. A computer program was developed to carry out the calculations involved in the exercise of the mathematical model. The model was applied to hypothetical system for illustrative purpose [33]. A number of system configurations have been investigated to demonstrate the feasibility of using dual-purpose desalting plants conjunctively with existing systems of surface water supply. The cost may be as much as 50% lower than that for the case in which the desalting plant is operated as base load plant [34].

An operating rule program for optimum operation of desalting plants as a supplemental source of safe yield was written in Fortran IV computer language and consists of about 1700 statements. It is easily applicable to a wide variety of conditions. Three real water systems were studied: The Cachuma Project near Santa Barbara, California, the Deer Creek Project near Salt Lake City, Utah, and the New York City water supply system. Each system has features different from the others. Sensitivity of the optimum operating rule and the associated costs to changes in various input parameters were described and the influence of intermittent conjunctive operation on the plant design and plant operating features was discussed. The computer program is potentially a practical tool useful to water resources planners to assess the role of desalting plants operating in conjunction with existing water supply systems [35]. Five further objectives were accomplished: The program was applied to a New York City water supply system feasibility study, it was modified to enable assessment of stage construction of desalting units when used in conjunction with a natural water supply system, a separate program was developed to enable analysis of desalting plant operation in conjunction with a natural water supply system having no storage capacity, a training program and a feasibility study of the Norfolk, Virginia, water supply system [36].

Three computational methods for optimizing the design and operation of systems having both conventional water sources and desalting plants, namely an operating rule program, an integer programming approach and optimization of a system with hydroelectric reservoirs and a dual-purpose power-water plant, were summarized by Shiozawa and Spiewak [37].

Investigating the economics of combining distilled seawater and renovated waste water, Porter developed a computerized mathematical model, that uses the network analysis theory as a basis for analyzing complex water and wastewater systems. The model can be used for analyzing existing water and wastewater systems in municipal and in new undeveloped areas. The costs for conveyance of water and wastewater can affect the feasibility of the use of an available water source [38]. A computer program was also developed as an economical tool to evaluate the feasibility of several seawater desalting methods. It provides good relative results for eliminating more costly processes. In addition, the program may be easily modified for future updating of cost-analysis procedures [39].

A detailed systems analysis has been undertaken of the use of desalination for the Barcelona area in conjunction with four regulating reservoirs on the River Llobregat as an interim alternative to the importation of water from the River Ebro. Present worth analysis shows savings ranging up to 8000 million Pesetas (about £ 47 million) for this interim use of conjunctive desalting. A delay of up to 30 years in construction of the Ebro scheme would be achieved, with demand in this interim period being met by construction of up to about 10.5 m³/s (200 Mgd) of desalting plant [40].

Two further case studies were described relating to the integrated use of desalination with conventional water resources. The first study describes work undertaken at the request of the Government of Cyprus to advise on the potential role of desalination in the long-term development of the island's water resources. Special features of the study are the strong influence of irrigation demands and the need to devise plans which are flexible against an only partially defined future. The second study relates to the 1.5 Mgd MSF plant operated by the Jersey New Water Works Company (Channel Islands), where the objective is to maintain overall supply reliability at minimum cost. The plant is operated in conjunction with an unusually complex system of surface water resources which present special difficulties in the optimization of control rules for the desalination plant [41].

California's aqueduct systems supply about half the water used in the state. It was expected that 3800 Mgd (14.4 million m³/day) be delivered to service areas in 1973. This system will meet southern California's water requirements at least through 1990. When additional water is needed in the years to follow on southern California coast, seawater desalting plants may be economically competitive [42].

Previous research which dealt with the desirability of considering water projects in a long range context, the measurement and costs of technological progress and the value of preserving flexibility in future decision making [4, 5] has been combined and applied to the evaluation of the proposed prototype 40 Mgd Diablo Canyon Desalination plant (see section 2.14.4, p. 126). An alternative to the desalination plant is a very large scale aqueduct. The two projects are evaluated together as complementary components of a long range project against the alternative of earlier construction of the aqueduct. The analysis concerns the interest rate, growth rate, demand elasticity, planning horizon, number of planning periods and area growth rate in parametric form [43].

Other recent work on desalting, focussing mainly on the economics, includes papers by Bahari on desalination processes and costs [44], by Bender emphasizing why water desalting will expand [45], by Savage giving a review of desalination processes and product water costs [46], by Tribus and Pezier on the economic value of experimentation in the design of desalting plants [47], by Westbrook et al. on purification techniques including economic brine utilization [48], by Anderson et al. on desalting technology and cost [49], by Beushausen et al. on the economic and technical problems of seawater desalination [50], by Delyannis presenting the state of the art with emphasis on the economics [51], by Fischbeck reviewing the technology and the costs of producing fresh water from the sea [52], by Young discussing desalting in pollution control problems [53], by Laird analyzing the saline water conversion program in California [54], by Dodson et al. presenting a committee report on water desalting with emphasis on applications for public water supplies [55], by Rautenbach and Hoeck on engineering and economic problems [56], by Miller reviewing desalting processes for quality improvement of municipal water supplies with emphasis on the economics [57], by Koelzer [58], by Pugh [59] and by Probstein [60] reporting on the state of the art and associated cost of desalting processes, by Pieper on technical and economic aspects of desalting by means of multi-stage flash evaporation [61], by Appleyard and Shaw on reuse and recycle of water in industry [62], and by Ahlgren on water reclamation from industrial wastes [63].

Literature to 5

[1] D. F. Othmer (Chim. Ind. Genie Chim **103** [1970] 1559/80). — [2] J. Gaussens (Proc. Symp. Nucl. Desalination, Madrid 1968 [1969], p. 605/15). — [3] L. Thiriet, J. J. Libert, J. Lambert (Proc. 4th U.N. Intern. Conf. Peaceful Uses At. Energy, Geneva 1971 [1972], Vol. 6, p. 257/68). — [4] N. El-Ramly, J. M. English (Proc. 3rd Intern. Symp. Fresh Water Sea, Dubrovnik 1970, Vol. 3, p. 433/48). — [5] N. El-Ramly, J. M. English (Univ. Calif. School Eng. Appl. Sci., Los Angeles, Rep. Eng-7071 [1970]).

[6] Office of Saline Water (A Standarized Procedure for Estimating Costs of Saline Water Conversion, PB 161 375 [1956]). — [7] Ralph M. Parsons Company (Off. Saline Water Res. Develop. Progr. Rept. No. 264 [1967]). — [8] P. A. Mawer (Desalination **2** [1967] 99/108). — [9] C. F. Clark (Off. Saline Water Res. Develop. Progr. Rept. No. 495 [1969]). — [10] Southwest Research Institute and Texas Water Development Board (Off. Saline Water Res. Develop. Progr. Rept. No. 250 [1966]).

[11] R. E. Childers, W. L. Prehn (Off. Saline Water Res. Develop. Progr. Rept. No. 257 (Pt. 1) [1966]). — [12] H. D. Holloway, T. R. Weaver (Off. Saline Water Res. Develop. Progr. Rept. No. 257 (Pt. 2) [1966]). — [13] W. L. Prehn, J. L. McGaugh (Off. Saline Water Res. Develop. Progr. Rept.

No. 555 [1970]). — [14] Bureau of Reclamation and Office of Saline Water (Desalting Handbook for Planners, Washington, D.C., 1972). — [15] United Nations (Water Desalination: Proposals for a Costing Procedure and Related Technical and Economic Considerations, New York, N.Y., 1965).

[16] International Atomic Energy Agency (Costing Methods for Nuclear Desalination, Tech. Rept. Ser. Intern. At. Energy Agency No. 69 [1966]). — [17] H. Kronberger (Proc. 3rd Intern. Symp. Fresh Water Sea, Dubrovnik 1970, Vol. 4, p. 253/64). — [18] A. Maurel, P. Vignet (Chim. Ind. Genie Chim. 105 [1972] 1141/57; Proc. 4th Intern. Symp. Fresh Water Sea, Heidelberg 1973, Vol. 2, p. 357/68). — [19] O. Pugh, M. C. Tanner (Proc. 4th Intern. Symp. Fresh Water Sea, Heidelberg 1973, Vol. 2, p. 387/403). — [20] P. Glueckstern, N. Arad, Y. Kantor, M. Greenberger (Proc. 4th Intern. Symp. Fresh Water Sea, Heidelberg 1973, Vol. 2, p. 335/48).

[21] E. E. Cooper (Nav. Civ. Eng. Lab. Port Hueneme, Calif., Rept. TN-1285 [1973]). — [22] A. R. Golzé (Desalination 1 [1967] 267/90). — [23] United Nations (The Design of Water Supply Systems Based on Desalination, Rept. E. 68. II. B. 20 [1968]). — [24] F. K. Orde (Proc. 3rd Intern. Symp. Fresh Water Sea, Dubrovnik 1970, Vol. 3, p. 359/70). — [25] P. Glueckstern, N. Arad, M. Greenberger (Storage and Transport of Water from Nuclear Desalting Plants, Tech. Rept. Ser. Intern. At. Energy Agency No. 141 [1972] 39/48).

[26] J. W. Michel (Storage and Transport of Water from Nuclear Desalting Plants, Tech. Rept. Ser. Intern. At. Energy Agency No. 141 [1972] 125/30). — [27] J. Adar (Storage and Transport of Water from Nuclear Desalting Plants, Tech. Rept. Ser. Intern. At. Energy Agency No. 141 [1972] 161/4). — [28] J. M. English, N. El-Ramly (Desalination 3 [1967] 308/17). — [29] P. A. Mawer, M. J. Burley (Desalination 4 [1968] 141/57). — [30] M. J. Burley, J. C. Clarke (Proc. Symp. Nucl. Desalination, Madrid 1968 [1969], p. 617/30).

[31] P. A. Mawer (Storage and Transport of Water from Nuclear Desalting Plants, Tech. Rept. Ser. Intern. At. Energy Agency No. 141 [1972] 165/82). — [32] F. Mobasheri, J. M. English (Proc. 3rd Intern. Symp. Fresh Water Sea, Dubrovnik 1970, Vol. 3, p. 179/92). — [33] F. Mobasheri, R. Harboe (Desalination 9 [1971] 141/53). — [34] F. Mobasheri, V. S. Budhraja, R. C. Harboe, K. Williams (Off. Saline Water Res. Develop. Progr. Rept. No. 782 [1971]). — [35] C. G. Clyde, W. H. Blood (Off. Saline Water Res. Develop. Progr. Rept. No. 528 [1970]).

[36] C. G. Clyde, W. H. Blood (Off. Saline Water Res. Develop. Progr. Rept. No. 780 [1972]). — [37] S. Shiozawa, I. Spiewak (Storage and Transport of Water from Nuclear Desalting Plants, Tech. Rept. Ser. Intern. At. Energy Agency No. 141 [1972] 183/94). — [38] J. W. Porter (Off. Saline Water Res. Develop. Progr. Rept. No. 617 [1970]). — [39] D. L. Kurtz, R. C. Huntsinger, J. Hatch (J. Am. Water Works Assoc. 64 [1972] 741/5). — [40] P. A. Mawer, J. D. Sherriff (Proc. 3rd Intern. Symp. Fresh Water Sea, Dubrovnik 1970, Vol. 3, p. 163/77).

[41] P. A. Mawer, T. Wyatt (Proc. 4th Intern. Symp. Fresh Water Sea, Heidelberg 1973, Vol. 2, p. 369/79; Desalination 13 [1973] 333/42). — [42] W. E. Thompson (Oak Ridge Natl. Lab. Rept. NDIC-12, UC-80 [1972]). — [43] J. M. English, N. A. Young (Proc. 4th Intern. Symp. Fresh Water Sea, Heidelberg 1973, Vol. 2, p. 323/33; Desalination 13 [1973] 359/71). — [44] E. Bahari (Chem. Process Eng. 50 [1969] 71/5). — [45] R. J. Bender (Power 113 No. 8 [1969] 171/8).

[46] W. F. Savage (Water Resour. Res. 6 [1970] 1449/53). — [47] M. Tribus, J. P. Pezier (Desalination 8 [1970] 311/49). — [48] G. T. Westbrook, H. I. Mahon, W. F. McIlhenny, C. F. MacGowan (Proc. 3rd Intern. Symp. Fresh Water Sea, Dubrovnik 1970, Vol. 4, p. 273/9). — [49] R. T. Anderson, H. L. Sturza, J. J. Strobel (Chem. Eng. Progr. Symp. Ser. 67 No. 107 [1971] 170/7). — [50] J. Beushausen, W. Klose, W. Seifert (Meerestechnik 2 [1971] 73/80, 117/22, 165/71).

[51] A. Delyannis (Dechema Monograph. 64 [1971] 23/38). — [52] K. Fischbeck (Wasser Abwasser 2 [1971] 32/6; Bild Wissenschaft 8 [1971] 579/89). — [53] K. G. Young (J. Am. Water Works Assoc. 63 [1971] 21/4). — [54] A. D. K. Laird (Chem. Eng. Progr. Symp. Ser. 67 No. 107 [1971] 182/3). — [55] R. E. Dodson, J. D. Bakken, W. E. Katz, G. F. Leitner, L. W. Owen, J. Redshaw, R. L. Sanks, E. H. Sieveka, K. S. Spiegler (J. Am. Water Works Assoc. 64 [1972] 690/3).

[56] R. Rautenbach, H. Hoeck (Chem. Ingr.-Tech. 44 [1972] 1145/51). — [57] E. F. Miller (J. Am. Water Works Assoc. 65 [1972] 804/7). — [58] V. A. Koelzer (U.S. Natl. Tech. Inform. Serv. PB 209942 [1972]). — [59] O. Pugh (Process Tech. Intern. 18 No. 1/2 [1973] 57/9; Sci. Tech. Australia 11 No. 2 [1973] 20/6). — [60] R. F. Probstein (Am. Scientist 61 [1973] 280/93).

[61] G. A. Pieper (Ingenieur [The Haag] 85 [1973] 223/9). — [62] C. J. Appleyard, M. G. Shaw (Chem. Ind. [London] 1974 March 16). — [63] R. M. Ahlgren (AIChE [Am. Inst. Chem. Engrs.] Symp. Ser. 70 No. 136 [1974] 539/49).

5.1 Recovery of byproducts

Seawater is an inexhaustible source of supply and also contains many mineral products in common use in industry. It appears, therefore, interesting to couple the production of fresh water with the extraction of minerals. However, consideration of an individual's fresh-water requirements and of the quantity of minerals contained in the corresponding amount of seawater shows that the minerals are present in quantities far in excess of the demand for them. The conclusion is that thorough economic studies are essential for determining to what extent the byproducts of desalination could contribute to the economic development [1].

The economical separation of mineral constituents from seawater and other saline waters was investigated by Salutsky et al. [2]. A feasibility study on recovery of byproducts from saline water conversion plants was presented by Weinberger and DeLapp [3]. Processes involving the removal of potassium from seawater and other saline waters were studied by Salutsky [4] and of potassium and sodium by Salutsky et al. [5]. A more extensive feasibility study was made by Christensen et al. and includes surveys of brine-processing technology, determination of byproduct markets, estimation of the value of brines and the consideration of the economic relationship between byproduct recovery, waste disposal and freshwater recovery [6]. The crystallization from brines was also studied by the same group of authors [7].

The recovery of chemicals from desalting brine effluents was reviewed by Leicerson and Scott, as an economical source for the production of saturated brine, NaCl and other chemicals [8]. Further work is reported by Hanson and Murthy on the components of seawater and their recovery [9], by the same authors on processes for the recovery of minerals from seawater [10], by Wenk on the physical resources of the oceans [11], by Grigorev on the chemical utilization of water desalination residues [12], by Terlizzi on economic considerations of a nuclear energy facility for the production of chemicals from the sea [13], by Shamsul Huq et al. on the extraction of chemicals from seawater [14], by Brice on water desalting and salt production [15], by Sanghavi on the recovery of marine salts from sea bittern [16], by Szymborski on oceanic resources and the technology of extracting them [17], by Kruglikov and Korsunskaya on physicochemical studies of the complex treatment of Caspian Sea water with the purpose of recovery of its mineral components [18], by Timokhina on heterogeneous equilibria in a water-salt system consisting of sodium, potassium and magnesium sulfates and chlorides [19], by Glassett on mineral recovery from concentrated brines [20], by Gaskell on the ocean as an untapped source of raw materials [21], by Hamelin on the ocean as a source of raw materials for the chemical industry [22], by Sharipov on an experimental study in drying of mineral salt solutions in a vibrating fluidized bed [23], by Ridley on minerals recovery from desalination plants [24], by Zdanovskii and Timoshenko on polythermal vaporization of seawater concentrate for the recovery of minerals [25], by Motoyama et al. on the concentration process of seawater and salt deposition [26], by Eriksson on the ocean as raw material source for the chemical industry [27], by Clark on the recovery of chemicals from desalting plant brines [28], and by Schmidt on occurrence and possibilities of obtaining raw materials from the sea [29].

Mixed ionic extractant systems and their possible application to the recovery of byproducts from desalination brines via solvent extraction were studied by Grinstead and Davis. Four possible areas of application were considered: recovery of magnesium from predominantly sodium chloride brines, heavy halide recovery from brines, nitrate removal from waste waters and softening of seawater or other brines for feed to desalination processes [30]. Additional research was presented in a second final report [31]. Byproduct recovery via solvent extraction was also reviewed by Witt and Forbes [32]. The feasibility of the stepwise regeneration of cation exchange resins with chelate solutions in a manner to selectivity separate K, Na, Mg and Ca has been demonstrated by DePree and Weyland [33].

Bruevich has edited a monograph on Chemical Resources of Seas and Oceans [34], the United Nations have published Proceedings of Expert Group Meeting on "Modernization and Mechanization of Salt Industries" [35], and a report on the extraction of chemicals from seawater and brines, as well as on the possibilities of byproduct development using desalting and evaporation [36].

Specific separation processes for various elements are reviewed in the following:

B o r o n. Grinstead has reported on the removal of boron and calcium from magnesium chloride brines by solvent extraction [37], Nikolaev and Ryabinin on the extraction of boron from seawater [38] and Obretenov and Panaiotcv on the extraction of borate from aqueous solutions by ion exchange [39].

Bromine. Fossett has reported on the extraction of bromine from seawater [40] and Foti on the concentration of bromide in seawater by isotopic exchange [41].

Cobalt. Topping reported on the concentration of cobalt from seawater by using 5,7-dibromo-8-hydroxyquinoline supported on an anion exchange resin [42].

Copper. Zeitlin and Kim reported on the separation of copper from seawater by an absorptive colloid flotation process [43].

Lithium. Barett and O'Neill reported on the recovery of lithium from saline brines by using solar evaporation [44].

Magnesium. Mehta et al. reported on the extraction of electrolytic grade magnesium chloride from bittern [45], Orlovskii on the production of magnesium oxide from seawater and brines [46], Berg and Loboda on ceramic and physicochemical characteristics of magnesium oxide obtained from seawater [47], Grinstead and Davis on the extraction of magnesium chloride from seawater concentrates [48], Bakr on the recovery of magnesia from bittern [49], Tsankov et al. [50], Novik and Moshkina [51] and Naidenov on the production of magnesium oxide from brines [52], Tsankov et al. on the effect of the solubility of calcium sulfate dihydrate on the purity of magnesium oxide obtained from seawater [53], and Ali-Kettani and Abdel-Aal on the production of magnesium chloride from brines of desalination plants [54].

Potassium. Chu and Liaw [55] and Kielland [56] reported on the recovery of potassium from seawater and bittern by dipicrylamine, Matsushita and coworkers on the recovery of potassium from seawater and brine [57], Bonazzi on the extraction of potassium chloride from seawater [58], Seshadri et al. on the manufacture of potassium chloride from bittern [59], Vyas on the recovery of potassium chloride and potassium sulfate from seawater and bittern [60], Choudhari on the recovery of kainite as well as of schoenite and of potassium sulfate from marine bittern [61] and on the chemistry of processes for the hot extraction of potassium chloride from marine salts [62], Al-Awadi and Al-Mahdi on potassium sulfate recovery from bittern [63], and Goto et al. on a new approach to potassium recovery from desalination plant effluents [64].

Uranium. Ogata and coworkers reported in several papers on their research for the extraction of uranium from seawater with various adsorbents [65], Ogata on studies for the recovery of uranium [66], Koyanaka on the collection of uranium from seawater by galena [67], Riedel on absorption behavior on synthetic ultramarine [68], Kanno et al. on extraction by titanium hydroxide [69], Yamabe and coworkers on the separation of uranium from seawater by precipitation [70], Ross and George on the recovery of uranium by ion exchange [71], Khan on the extraction from water desalination brines [72], and Ryabinin on sorption by anion exchangers [73].

Zinc. Topping reported on the concentration of zinc by using 5,7-dibromo-8-hydroxyquinoline [42], and Zeitlin and Kim on the separation of zinc by an absorptive colloid flotation process [43].

Literature to 5.1

[1] J. Le Chatelier, P. Charuit (Proc. Symp. Nucl. Energy Costs Econ. Develop., Istanbul 1969 [1970], p. 415/22). — [2] M. L. Salutsky, M. G. Dunseth, O. B. Waters (Off. Saline Water Res. Develop. Progr. Rept. No. 91 [1964]). — [3] A. J. Weinberger, D. F. DeLapp (Off. Saline Water Res. Develop. Progr. Rept. No. 110 [1964]). — [4] M. L. Salutsky (Off. Saline Water Res. Develop. Progr. Rept. No. 137 [1965]). — [5] M. L. Salutsky, J. Block, M. G. Dunseth, O. B. Waters (Off. Saline Water Res. Develop. Progr. Rept. No. 197 [1966]).

[6] J. J. Christensen, W. F. McIlhenny, P. E. Muehlberg, H. G. Smith (Off. Saline Water Res. Develop. Progr. Rept. No. 245 [1967] and No. 246 [1967]). — [7] W. F. McIlhenny, P. E. Muehlberg, H. G. Smith (Chem. Geol. 4 [1969] 9/35). — [8] L. Leicerson, P. C. Scott (Off. Saline Water Res. Develop. Progr. Rept. No. 445 [1969]). — [9] C. Hanson, S. L. N. Murthy (Chem. Ind. [London] 1969 669/81). — [10] C. Hanson, S. L. N. Murthy (Chem. Eng. [London] No. 264 [1972] 295/8).

[11] E. Wenk (Sci. Am. 221 [1969] 166/77). — [12] N. K. Grigorev (Value Agr. High-Qual. Water Nucl. Desalination Rept. Panel, 1967 [1969], p. 65/74; C.A. 71 [1969] No. 103672). — [13] P. M. Terlizzi (Proc. Symp. Nucl. Desalination, Madrid 1968 [1969], p. 645/57). — [14] A. K. M. Shamsul Huq, A. K. Shamsuddin, I. Kamal (Nucleus [Lahore] 6 No. 1 [1969] 17/20). — [15] D. B. Brice (Proc. 3rd Intern. Symp. Fresh Water Sea, Dubrovnik 1970, Vol. 3, p. 461/9).

[16] J. R. Sanghavi (Salt Res. Ind. 7 [1970] 108/9). — [17] S. Szymborski (Podstawowe Probl. Wspolczesnej Tech. 14 [1970] 199/288; C.A. 76 [1972] No. 37195). — [18] A. E. Kruglikov,

E. E. Korsunskaya (Tr. Turkm. Politekhn. Inst. No. 14 [1970] 64/72; C.A. **78** [1973] No. 151488).
— [19] N. I. Timokhina (Nauchn. Tr. Tashkent. Gos. Univ. No. 399 [1970] 194/207; C.A. **77** [1972]
No. 154832). — [20] J. M. Glassett (Off. Saline Water Res. Develop. Progr. Rept. No. 593 [1970]).
[21] T. F. Gaskell (Chem. Ind. [London] **1971** 1149/54). — [22] R. Hamelin (Chem. Ind.
[London] **1971** 1473/81). — [23] Kh. N. Sharipov (Tr. Turkm. Politekhn. Inst. No. 10 [1971]
85/92; C.A. **78** [1973] No. 113206). — [24] R. D. Ridley (AIChE [Am. Inst. Chem. Engrs.] Symp.
Ser. **68** No. 124 [1972] 388/92). — [25] A. B. Zdanovskii, Yu. M. Timoshenko (Zh. Neorgan. Khim.
17 [1972] 259/60; C.A. **76** [1972] No. 87924).

[26] M. Motoyama, M. Kadota, S. Oka (Nippon Kaisui Gakkai-Shi **26** [1972] 241/7; C.A. **77**
[1972] No. 105444). — [27] K. Eriksson (Med. Kemi **1972** No. 10, p. 26/7; C.A. **78** [1973]
No. 62004). — [28] R. L. Clark (Off. Saline Water Res. Develop. Progr. Rept. No. 842 [1973]). —
[29] K. H. Schmidt (Chem. Ind. [Düsseldorf] **26** [1974] 14/7). — [30] R. R. Grinstead, J. C. Davis
(Off. Saline Water Res. Develop. Progr. Rept. No. 320 [1968]).

[31] R. R. Grinstead, J. C. Davis, S. W. Snider (Off. Saline Water Res. Develop. Progr. Rept.
No. 406 [1969]). — [32] P. A. Witt, M. C. Forbes (Chem. Eng. Progr. **67** No. 10 [1971] 90/4). —
[33] D. DePree, H. H. Weyland (Off. Saline Water Res. Develop. Progr. Rept. No. 435 [1969]). —
[34] S. V. Bruevich (Khimicheskie Resursy Morei i Okeanov, Moskva 1970). — [35] United
Nations, UNIDO (Modernization and Mechanization of Salt Industries Based on Seawater in Developing Countries, Publ. E. 70. II. B. 25 [1970]).

[36] United Nations Chemical and Technical Services (Extraction of Chemicals from Seawater,
Inland Brines and Rock Salt Deposits, Publ. E. 71. II. B. 25 [1972]). — [37] R. R. Grinstead (Ind.
Eng. Chem. Prod. Res. Develop. **11** [1972] 454/60). — [38] A. V. Nikolaev, A. I. Ryabinin (Dokl.
Akad. Nauk SSSR **207** [1972] 149/52; Dokl. Chem. Technol. Proc. Acad. Sci. USSR **207** [1973]
220/2). — [39] T. Obretenov, P. Panaiotov (God. Vissh. Khim.-Tekhnol. Inst. Burgas Bulg. **8**
[1972] 119/28; C.A. **79** [1973] No. 70512). — [40] H. Fossett (Chem. Ind. [London] **1971**
1161/71).

[41] S. C. Foti (AD-734383 [1971]). — [42] G. Topping (Limnol. Oceanog. **14** [1969] 798/9).
— [43] H. Zeitlin, Y. S. Kim (Chem. Commun. **1971** 672). — [44] W. T. Barett, B. J. O'Neill (3rd
Symp. Salt, 1969 [1970], Vol. 2, p. 47/50). — [45] M. J. Mehta, B. P. Choudhari, B. K. Shukla
(Salt Res. Ind. **6** [1969] 76/8).

[46] Ya. I. Orlovskii (Probl. Ratsion. Ispol'z. Prir. Bogatstv Sivasha **1969** 96/8; C.A. **72** [1970]
No. 46881). — [47] E. A. T. Berg, F. Loboda (Ceramica [Sao Paulo] **16** [1970] 314/9; C.A. **74**
[1971] No. 115264). — [48] R. R. Grinstead, J. C. Davis (Ind. Eng. Chem. Prod. Res. Develop. **9**
[1970] 66/72). — [49] M. Y. Bakr (Sprechsaal **103** [1970] 1052/6), M. Y. Bakr, H. El-Abd (Sprechsaal **104** [1971] 430, 433/4, 494/8). — [50] I. Tsankov, S. Bagarov, N. Drenska, G. Kalpakchiev,
B. Rakhnev, S. Serbezov, S. Andreev, I. Chakurov (Metalurgiya [Sofia] **1972** No. 4, p. 14/6; C.A.
80 [1974] No. 49795).

[51] V. F. Novik, I. A. Moshkina (Tr. Altai Politekhn. Inst. **1972** No. 17, p. 80/6; C.A. **80** [1974]
No. 72508). — [52] N. Naidenov (Metalurgiya [Sofia] **1972** No. 6, p. 22/5; C.A. **80** [1974]
No. 97951). — [53] I. Tsankov, S. Bagarov, S. Andreev, G. Kalpakchiev, N. Drenska, B. Rakhnev,
S. Serbezov, I. Chakurov (Metalurgiya [Sofia] **1973** No. 4, p. 9/14; C.A. **80** [1974] No. 61591). —
[54] M. Ali-Kettani, H. K. Abdel-Aal (Proc. 4th Intern. Symp. Fresh Water Sea, Heidelberg 1973,
Vol. 2, p. 509/16). — [55] S. K. Chu, C. T. Liaw (Hua Hsueh **1969** No. 4, p. 106/11; C.A. **74** [1971]
No. 5077).

[56] J. Kielland (Chem. Ind. [London] **1971** 1309/13). — [57] H. Matsushita, T. Tagaki
(Nippon Kaisui Gakkai-Shi **22** [1969] 369/79, **23** [1969] 102/6; C.A. **74** [1971] No. 89908, No.
89909), H. Matsushita, T. Takayanagi (Nippon Kaisui Gakkai-Shi **24** [1970] 96/104, **25** [1972]
269/86; C.A. **75** [1971] No. 9767, **77** [1972] No. 131072). — [58] A. Bonazzi (Bol. Acad. Cienc.
Fis. Mat. Nat. Venezuela **30** [1970] 45/105). — [59] K. Seshadri, G. D. Bhat, C. J. Dave, J. R.
Sanghavi (Salt Res. Ind. **7** [1970] 39/44). — [60] S. N. Vyas (Indian Chem. J. **5** No. 10 [1971]
29/34; Chem. Age India **23** [1972] 754/8).

[61] B. P. Choudhari (Appl. Chem. Biotech. **21** [1971] 266/7, 268/9). — [62] B. P. Choudhari
(J. Sci. Ind. Res. [India] **32** [1973] 58/62). — [63] F. Al-Awadi, A. K. Al-Mahdi (Ind. Eng. Chem.
Prod. Res. Develop. **12** [1973] 71/5). — [64] T. Goto, I. Hayano, T. Takeuchi, T. Akiya (Proc.
4th Intern. Symp. Fresh Water Sea, Heidelberg 1973, Vol. 2, p. 517/26). — [65] N. Ogata, H. Kakihana
(Nippon Genshiryoku Gakkaishi **11** [1969] 82/7, 469/76; C.A. **71** [1969] No. 15916, **72** [1970]
No. 23754), N. Ogata, N. Inoue, H. Kakihana (Nippon Kaisui Gakkai-Shi **24** [1970] 68/72; C.A.

74 [1971] No. 102887; Nippon Genshiryoku Gakkaishi **13** [1971] 560/5; C.A. **76** [1972] No. 16851), N. Ogata (Nippon Genshiryoku Gakkaishi **13** [1971] 121/7, 253/9; C.A. **74** [1971] No. 129160, **75** [1971] No. 51740).

[66] N. Ogata (Genshiryoku Kyoyo **16** No. 12 [1970] 19/22; C.A. **74** [1971] No. 89851; Nippon Kaisui Gakkai-Shi **24** [1971] 197/212; C.A. **75** [1971] No. 112744). — [67] Y. Koyanaka (J. Nucl. Sci. Tech. **7** [1970] 426/7). — [68] H. J. Riedel (Kerntechnik **12** [1970] 16/9). — [69] M. Kanno, Y. Ozawa, T. Mukaibo (Nippon Genshiryoku Gakkaishi **12** [1970] 708/14; C.A. **74** [1971] No. 46159). — [70] T. Yamabe, N. Takai (Nippon Kaisui Gakkai-Shi **24** [1970] 16/9; C.A. **74** [1971] No. 91010), N. Takai, T. Yamabe (Mizu Shori Gijutsu **12** No. 4 [1971] 3/8; C.A. **77** [1972] No. 38971), N. Takai, K. Takase, T. Yamabe (Seisan Kenkyu **23** [1971] 76/7; C.A. **74** [1971] No. 144807).

[71] J. R. Ross, D. R. George (U.S. Bur. Mines Rept. Invest. No. 7471 [1971]). — [72] S. Khan (Nucleus [Lahore] **9** [1972] 39/46). — [73] A. I. Ryabinin (Radiokhimiya **15** [1973] 437/40; C.A. **79** [1973] No. 127806).

Reihenfolge (Systemnummern) der im Gesamtwerk behandelten Elemente

Gmelin System of Elements and Compounds

System-Nr.	Symbol	Element		System-Nr.	Symbol	Element
1		Edelgase		37	In	Indium
2	H	Wasserstoff		38	Tl	Thallium
3	O	Sauerstoff		39	Sc	Scandium
4	N	Stickstoff			Y	Yttrium
5	F	Fluor			La	Lanthan
6	**Cl**	**Chlor**			Ce–Lu	Lanthanide
7	Br	Brom		40	Ac	Actinium
8	J	Jod		41	Ti	Titan
	At	Astat		42	Zr	Zirkonium
9	S	Schwefel		43	Hf	Hafnium
10	Se	Selen		44	Th	Thorium
11	Te	Tellur		45	Ge	Germanium
12	Po	Polonium		46	Sn	Zinn
13	B	Bor		47	Pb	Blei
14	C	Kohlenstoff		48	V	Vanadium
15	Si	Silicium		49	Nb	Niob
16	P	Phosphor		50	Ta	Tantal
17	As	Arsen		51	Pa	Protactinium
18	Sb	Antimon		**52**	**Cr**	**Chrom**
19	Bi	Wismut		53	Mo	Molybdän
20	Li	Lithium		54	W	Wolfram
21	Na	Natrium		55	U	Uran
22	K	Kalium		56	Mn	Mangan
23	NH_4	Ammonium		57	Ni	Nickel
24	Rb	Rubidium		58	Co	Kobalt
25	Cs	Caesium		59	Fe	Eisen
	Fr	Francium		60	Cu	Kupfer
26	Be	Beryllium		61	Ag	Silber
27	Mg	Magnesium		62	Au	Gold
28	Ca	Calcium		63	Ru	Ruthenium
29	Sr	Strontium		64	Rh	Rhodium
30	Ba	Barium		65	Pd	Palladium
31	Ra	Radium		66	Os	Osmium
32	**Zn**	**Zink**		67	Ir	Iridium
33	Cd	Cadmium		68	Pt	Platin
34	Hg	Quecksilber		69	Tc	Technetium[1]
35	Al	Aluminium		70	Re	Rhenium
36	Ga	Gallium		71	Np,Pu...	Transurane[2]

HCl

$CrCl_2$

$ZnCrO_4$

$ZnCl_2$

Dem einzelnen Element werden alle Verbindungen mit denjenigen Elementen zugeordnet, die im Gmelin-System vor diesem Element stehen. Bei dem Element Zink mit der System-Nr. 32 stehen z. B. alle Verbindungen mit den Elementen der System-Nr. 1 bis 31.

The material under each element number contains all information on the element itself as well as on all compounds with other elements which preceed this element in the Gmelin System.
For example, zinc (system number 32) as well as all zinc compounds with elements numbered from 1 to 31 are classified under number 32.

[1]) Diese System-Nr. ist im Jahre 1941 unter der Bezeichnung „Masurium" erschienen.
[2]) Bearbeitung erfolgt im Rahmen des Ergänzungswerkes zur 8. Auflage.

Periodensystem der Elemente mit Gmelin Systemnummern siehe Innenseite des vorderen Deckels